Aviation Maintenance Technician Series

Powerplant

Third Edition

DALE CRANE

TERRY MICHMERHUIZEN
Technical Editor

CORNERSTONE AVIATION SERVICES
THE AVIATION DIVISION OF CORNERSTONE COLLEGE
GRAND RAPIDS, MICHIGAN

PAT BENTON
Technical Editor

ASSISTANT PROFESSOR
SCHOOL OF AVIATION SCIENCES
WESTERN MICHIGAN UNIVERSITY

JERRY LEE FOULK
Technical Editor, Second and Third Editions

INSTRUCTOR, SCHOOL OF AERONAUTICAL SCIENCE
LETOURNEAU UNIVERSITY

Aviation Supplies & Academics, Inc.
NEWCASTLE, WASHINGTON

Aviation Maintenance Technician Series: Powerplant
Third Edition

Aviation Supplies & Academics, Inc.
7005 132nd Place SE
Newcastle, Washington 98059-3153
Email: asa@asa2fly.com
Website: www.asa2fly.com

Reader Resources webpage: www.asa2fly.com/amtp
(See this page online for further resources and updates to this book.)

Printed in the United States of America

2016 2015 2014 2013 9 8 7 6 5 4 3 2

Cover photo © Gary Gladstone/The Image Bank

Photo credits and acknowledgments: p. 6—Pratt & Whitney Division, United Technologies Corp.; p. 11—The General Electric Company; p. 44—Jerry Lee Foulk; p. 46—Teledyne-Continental Motors; pp. 207, 211, 216, 219, 229—Bendix Electrical Components Division; p. 494—Champion Aviation Products Division; p. 526—Instrument Technology, Inc.; pp. 527, 528—Machida, Incorporated; p. 536—Milbar Specialty Tools; p. 538—Howell Instruments, Inc.; p. 607—Sundstrand Corporation; p. 709—The General Electric Company; p. 721—TEC Aviation Division

ASA-AMT-P3
ISBN 1-56027-862-5
978-1-56027-862-7

Library of Congress Cataloging-in-Publication Data:
Crane, Dale.
Aviation maintenance technician series. Powerplant / Dale Crane; Terry Michmerhuizen, technical editor.
p. cm.
Includes index.
1. Airplanes—Motors—Maintenance and repair. I. Michmerhuizen, Terry. II. Title.
TL701.5.C725 1995
629.134'35'0288—dc20 95-50012
CIP

14

Dale Crane (1923–2010)

Dale Crane was involved in aviation for more than 50 years. He began his career in the U.S. Navy as a mechanic and flight engineer in PBYs. After college, he worked as an instrument overhaul mechanic, instrument shop manager, and flight test instrumentation engineer. He spent the following 16 years as an instructor, and then became director of an aviation maintenance school. His credentials included Airframe and Powerplant Mechanic, Designated Mechanic Examiner, Commercial Pilot, Flight Instructor (airplanes), and Advanced and Instrument Ground Instructor.

Dale received the FAA's Charles Taylor "Master Mechanic" Award for 50 years of service in and contributions to the aviation maintenance industry, and the recognition of his peers for his excellence in aircraft maintenance as a leader, educator, and aviation safety advocate.

Contents

Preface
to the Third Edition

The training material in the *Aviation Maintenance Technician Series* is chosen to reflect today's required knowledge for the aviation maintenance technician (AMT), and the facts are provided with the backing of personal experience and research. Technicians today need a solid foundation in mechanics, physics, electricity, electronics, and logics, in addition to the information unique to aircraft maintenance and construction. This *Powerplant* textbook, along with the other ASA maintenance volumes, meets these necessary requirements.

ASA is dedicated to providing easy to understand training materials for the AMT certificate applicant. The chapters are carefully chosen to reflect FAA requirements, and chapter structure planned to provide a logical flow that enhances the learning process. The field of aeromechanics is vast covering many complicated subjects; the AMT Series textbooks provide the information necessary to contribute to the building of well-rounded aircraft technicians, who will not only be equipped to understand the workings of aircraft systems, but will have the skills to repair, service, inspect, and troubleshoot them.

Since the printing of the original editions of this textbook series, the FAA has changed the way they refer to subject matter in their certification exams—what was once called Subject Matter Knowledge Codes is now referenced as "Learning Statement Codes" (LSC). Most of these knowledge sources are still the same with some new titles added, such as the FAA's *Aviation Maintenance Technician Handbook—General* (FAA-H-8083-30), which replaced the FAA Advisory Circular 65-9 in 2008. ASA also provides the best collection of AMT-related federal aviation regulation reprints in *FAR for Aviation Maintenance Technicians,* printed yearly and provided with periodic updates on the ASA website (www.asa2fly.com).

Dale Crane was the author of many of the ASA texts and many students over the years came to trust Dale's authorship to not only inform, but to do so in an accurate, concise, and straightforward manner. Those who are currently editing this textbook endeavor to do the same, and are continuing in the effort toward the same quality that Mr. Crane strove for. The current technical editor was educated under Dale Crane's tutelage and pledges to carry on in a manner that would be acceptable to the original author.

Preface to the First and Second Editions

by Dale Crane

Aviation maintenance technology has undergone tremendous changes in the past decades. Modern aircraft, with their advanced engines, complex flight controls and environmental control systems, are some of the most sophisticated devices in use today, and these marvels of engineering must be maintained by knowledgeable technicians. The Federal Aviation Administration, recognizing this new generation of aircraft, has updated the requirements for maintenance technicians and for the schools that provide their training. The FAA has also instituted an Aviation Maintenance Technician Awards Program to encourage technicians to update their training.

New technologies used in modern aircraft increase the importance for maintenance technicians to have a solid foundation in such basic subjects as mathematics, physics, and electricity. The *Aviation Maintenance Technician Series* has been produced by ASA to provide the needed background information for this foundation and to introduce the reader to aircraft structures, powerplants and systems.

These textbooks have been carefully designed to assist a person in preparing for FAA technician certification, and at the same time serve as valuable references for individuals working in the field. The subject matter is organized into categories used by the FAA for the core curriculum in 14 CFR Part 147, Aviation Maintenance Technician Schools, and for the Subject Matter Knowledge Codes used in the written tests for technician certification. In some cases in the ASA series, these categories have been rearranged to provide a more logical progression of learning.

This textbook is part of the ASA series of coordinated maintenance technician training materials. The series consists of the *General, Airframe,* and *Powerplant* textbooks with study questions, the *Fast-Track Test Guides for Aviation Mechanics*, exam software for computerized study for the Aviation Maintenance Technician tests, the *Oral and Practical Exam Guide*, *The Dictionary of Aeronautical Terms*, and the *Aviation Mechanic Handbook.*

To supplement this fundamental training material, ASA reprints the FAA Advisory Circular *Acceptable Methods, Techniques, and Practices—Aircraft Inspection, Repair, and Alteration* (ASA-AC43.13B), and semiannually updated excerpts from the Federal Aviation Regulations that are applicable to the aviation maintenance technician.

ACKNOWLEDGEMENTS

A series of texts such as this *Aviation Maintenance Technician Series* could never be compiled without the assistance of modern industry. Many individuals have been personally helpful, and many companies have been generous with their information. We want to acknowledge this and say thank you to them all.

ACES Systems—TEC Aviation Division, *Knoxville, TN*

Aero Quality International, *Stamford, CT*

Aero-Mach Labs, Inc., *Wichita, KS*

Aeroquip Corporation, *Jackson, MI*

Airborne Division, Parker Hanniflin Corporation, *Elyria, OH*

Allied Signal Aerospace, *Phoenix, AZ*

Allison Engine Company, *Indianapolis, IN*

ASCO Aeronautical, *Columbus, OH*

Aviation Laboratories, Inc., *Houston, TX*

Barfield, Inc., *Atlanta, GA*

Beech Aircraft Corporation, *Wichita, KS*

Bendix Electrical Components Division, *Sidney, NY*

Cessna Aircraft Company, *Wichita, KS*

Chadwick-Helmuth Company, Inc., *El Monte, CA*

Champion Aviation Products Division, *Liberty, SC*

Dowty-Rotol, Inc., *Cheltenham, England*

Dynamic Solutions Systems, Inc., *San Marcos, CA*

Engine Components, Inc., *San Antonio, TX*

General Electric Company, *Cincinnati, OH*

Gulfstream Aerospace, *Savannah, GA*

Hamilton Standard Division of United Technologies, *Windsor Locks, CT*

Howell Instruments, Inc., *Fort Worth, TX*

Machida Incorporated, *Orangeburg, NY*

McCauley Accessory Division Cessna Aircraft Company, *Vandalia, OH*

Milbar Corporation, *Chagrin Falls, OH*

NASA Lewis Research Center, *Cleveland, OH*

Pratt & Whitney Canada, *Longueuil, Quebec, Canada*

Precision Airmotive Corporation, *Everett, WA*

Quan-Tech, *Flanders, NJ*

Ram Aircraft Corporation, *Waco, TX*

Saft America, Inc., *Valdosta, GA*

Slick Aircraft Products, Division of Unison, *Rockford, IL*

Standard Aero, *Winnipeg, Manitoba, Canada*

Stanley-Proto Industrial Tools, *Covington, GA*

Stead Aviation Corporation, *Manchester, NH*

Sundstrand Corp., *Rockford, IL*

Superior Air Parts, Inc., *Addison, TX*

Teledyne-Continental Motors, *Mobile, AL*

Textron Lycoming, *Williamsport, PA*

TRW Hartzell Propeller Products Division, *Piqua, OH*

T.W. Smith Engine Company, Inc., *Cincinnati, OH*

UE Systems, Inc., *Elmsford, NY*

United Technologies, Pratt & Whitney, *East Hartford, CT*

Welch Allyn Imaging Products Division, *Skaneateles Falls, NY*

Development of Aircraft Powerplants

1

Development of Aircraft Powerplants

1

The first man-carrying flights were made in hot air balloons swept along by air currents and without means for the pilot to control the direction of flight. Aircraft had little practical utility until the development of engine-driven propellers. This development of the powerplant has made aviation the vital factor that it is today in the economic world.

powerplant. The complete installation of an aircraft engine, propeller, and all accessories needed for its proper function.

The Principle of Heat Engines

All powered aircraft are driven by some form of heat engine. Chemical energy stored in the fuel is released as heat energy that causes air to expand. The expansion of this air is what performs useful work, driving either a piston or a turbine.

There are two basic types of heat engines: external-combustion and internal-combustion.

heat engine. A mechanical device that converts chemical energy in a fuel into heat energy, and then into mechanical energy.

internal-combustion engine. A form of heat engine in which the fuel and air mixture is burned inside the engine.

External-Combustion Engines

External-combustion engines are most familiar to us as steam engines. Energy released in coal- or gas-fired furnaces or in nuclear reactors is transferred into water, changing it into steam that expands and drives either a piston or a turbine.

Steam engines were used to power experiments in flight made during the late 1800s. Dr. Samuel Langley of the Smithsonian Institution in Washington, D.C. used small steam engines to power a successful series of unmanned machines he called Aerodromes. In 1896, Dr. Langley made a number of powered flights with these models. The most successful had tandem wings with a span of 14 feet, weighed 26 pounds, and was powered by a one-horsepower steam engine. It was launched from a catapult atop a houseboat on the Potomac river, and flew for 90 seconds, traveling more than half a mile.

There was one successful but impractical aircraft steam engine developed in America in 1933 by the Besler brothers, manufacturers of logging locomotives. This 150-horsepower engine, using an oil-fired boiler and having a total installed weight of approximately 500 pounds, was used to power a Travel Air 2000 biplane.

external-combustion engine. A form of heat engine in which the fuel releases its energy outside of the engine.

piston. The movable plug inside the cylinder of a reciprocating engine.

turbine. A wheel fitted with vanes or airfoils radiating out from a central disk. Used to extract energy from a stream of moving fluid.

Aerodrome. The name given by Dr. Samuel Langley to the flying machines built under his supervision between the years of 1891 and 1903.

Otto cycle of energy transformation. The four-stroke, five-event, constant-volume cycle of energy transformation used in a reciprocating engine.

gas turbine engine. An internal combustion engine that burns its fuel in a constant-pressure cycle and uses the expansion of the air to drive a turbine which, in turn, rotates a compressor. Energy beyond that needed to rotate the compressor is used to produce torque or thrust.

turbojet engine. A gas turbine engine that produces thrust by accelerating the mass of air flowing through it.

turbofan engine. A type of gas turbine engine in which lengthened compressor or turbine blades accelerate air around the outside of the core engine.

turboprop engine. A turbine engine in which energy extracted from the accelerated gases is used to drive a propeller.

turboshaft engine. A turbine engine in which energy extracted from the accelerated gases is used to drive helicopter rotors, generators, or pumps.

reciprocating engine. A type of heat engine that changes the reciprocating (back-and-forth) motion of pistons inside the cylinders into rotary motion of a crankshaft.

brake horsepower. The actual horsepower delivered to the propeller shaft of an aircraft engine.

cylinder. The component of a reciprocating engine which houses the piston, valves, and spark plugs and forms the combustion chamber.

Internal-Combustion Engines

The concept of releasing energy from fuel directly inside an engine to heat and expand the air has challenged engineers since the late 1700s. The expanding air can drive reciprocating pistons or spin turbines.

Coal dust, gunpowder, and even turpentine vapors have been exploded inside cylinders, but it was not until 1860 that the French engineer Etienne Lenoir actually built a practical engine that could use illuminating gas as its fuel.

In 1876, Dr. Nikolaus Otto of Germany made practical engines using the four-stroke cycle that bears his name, and it is the principal cycle upon which almost all aircraft reciprocating engines operate. This cycle of energy transformation is discussed in detail in Chapter 2.

Gas turbine engines in the form of turbojet, turbofan, turboprop, and turboshaft engines have revolutionized aviation, and their principle of operation is discussed in Chapter 10.

Aircraft Reciprocating Engines

Throughout the history of aviation, progress has always been dependent upon the development of suitable powerplants.

Aviation as we know it today was born at the beginning of the 1900s with powered flights made by Wilbur and Orville Wright. The Wright brothers approached the problems of flight in a sensible and professional way. They first solved the problem of lift with kites, then the problem of control with gliders, and finally by 1902, they were ready for powered flight. First they painstakingly designed the propellers and then began their search for a suitable engine. Their requirements were for a gasoline engine that would develop 8 or 9 brake horsepower and weigh no more than 180 pounds. No manufacturer had such an engine available, and none were willing to develop one for them. Their only recourse was to design and build it on their own.

The engine, built to their design by Mr. Charles Taylor, had four cylinders in-line and lay on its side. It drove two 8 ½-foot-long wooden propellers through chain drives and developed between 12 and 16 horsepower when it turned at 1,090 RPM. It weighed 179 pounds.

On December 17, 1903, this engine powered the Wright *Flyer* on its historic flight of 59 seconds, covering a distance of 852 feet on the wind-swept sand at Kitty Hawk, North Carolina.

Because of Dr. Langley's success with his Aerodromes, the U.S. government gave him a contract to build a full-scale man-carrying machine. The steam engines used in the models could not be effectively scaled up to power this aircraft, so a better means of propulsion had to be found.

Charles Manly, Dr. Langley's assistant, searched without success, both in the United States and Europe, for a suitable powerplant. The best he found was a three-cylinder rotary radial automobile engine built by Stephen Balzer in New York. This engine was not directly adaptable to the Aerodrome, but Manly, building upon Balzer's work, constructed a suitable engine for it. The Manly-Balzer engine was a five-cylinder, water-cooled static radial engine that produced 52.4 horsepower at 950 RPM and weighed 207.5 pounds complete with water.

On October 7, 1903, the full-scale Aerodrome with Manly as the pilot was launched from atop the houseboat. As the aircraft neared the end of the catapult, it snagged part of the launching mechanism and was dumped into the river. But Manly's engine, which was far ahead of its time, functioned properly and was in no way responsible for the failure of the Aerodrome to achieve powered flight.

Glenn Curtiss was a successful motorcycle builder and racer from western New York state. The use of one of his motorcycle engines in a dirigible in 1907 got Curtiss interested in aviation, and as a result, he became involved in furnishing the powerplants for Dr. Alexander Graham Bell's Aerial Experiment Association. A number of successful aircraft, including the first aircraft to fly in Canada, came from this group.

Curtiss's own company designed and built some of the most important engines in America in the periods before and during World War I and up until 1929, when the Curtiss Aeroplane and Motor Corporation merged with the Wright Aeronautical Corporation to form the giant Curtiss-Wright Corporation.

World War I, between 1914 and 1918, was a time of rapid growth in aviation. The British, French, Germans, and Americans all developed aero engines.

One of the most popular configuration of engines built in this era was the rotary radial engine. With this engine, the crankshaft was attached rigidly to the airframe, and the propeller, crankcase, and cylinders all spun around. Clerget, Gnome, and Rhone in France, Bentley in Britain, Thulin in Sweden, and Oberursel, BMW, Goebel, and Siemens-Halske in Germany all built rotary radial engines. These engines had 5, 7, 9, 11, or 14 cylinders and produced between 80 and 230 horsepower.

The Germans used some very efficient 6-cylinder in-line water-cooled engines built by the Mercedes, Maybach, BMW, Benz, and Austro-Daimler companies. Some of these engines developed up to 300 hp.

Some of the most popular V-8 engines of this time were the French-built 150- to 300-horsepower Hispano-Suizas. These engines were also built under license agreements in Great Britain and the United States.

There were only two aircraft engines designed and built in quantities in the United States during this time, and both were V-engines. Glenn Curtiss's Company built the 90-horsepower, water-cooled V-8 Curtiss OX-5 engine in great numbers, and various automobile manufacturers built the 400-horsepower water-cooled V-12 Liberty engine.

dirigible. A large, cigar-shaped, lighter-than-air flying machine. Dirigibles differ from balloons in that they are powered and can be steered.

rotary radial engine. A form of reciprocating engine in which the crankshaft is rigidly attached to the airframe and the cylinders revolve with propeller.

crankshaft. The central component of a reciprocating engine. This high-strength alloy steel shaft has hardened and polished bearing surfaces that ride in bearings in the crankcase. Offset throws, formed on the crankshaft, have ground and polished surfaces on which the connecting rods ride. The connecting rods change the in-and-out motion of the pistons into rotation of the crankshaft.

crankcase. The housing that encloses the crankshaft, camshaft, and many of the accessory drive gears of a reciprocating engine.

The cylinders are mounted on the crankcase, and the engine attaches to the airframe by the crankcase.

V-engine. A form of reciprocating engine in which the cylinders are arranged in two banks. The banks are separated by an angle of between 45° and 90°.

Pistons in two cylinders, one in each bank, are connected to each throw of the crankshaft.

Curtiss Jenny (Curtiss JN4-D). A World War I training airplane powered by a Curtiss OX-5 engine. It was widely available after the war and helped introduce aviation to the general public.

Standard J-1. A World War I training airplane powered by a Curtiss OX-5 engine.

DeHaviland DH-4. An English designed observation airplane built in large quantities in America during World War I. After the war, surplus DH-4s were used for carrying the U.S. mail.

radial engine (static radial). A form of reciprocating engine in which the cylinders radiate out from a small central crankcase. The pistons in the cylinders drive a central crankshaft which in turn drives the propeller.

Figure 1-1. *The Pratt & Whitney R-4360 Wasp Major, with 28 air-cooled cylinders weighed 3,670 pounds and produced 3,800 horsepower. This engine, with four rows of seven cylinders, was the largest practical aircraft reciprocating engine.*

The years between World Wars I and II are called the golden years of aviation because of the tremendous strides made during this era. Powerplant development was largely responsible for this progress.

At the end of hostilities in 1918, the aviation market was flooded with surplus Curtiss Jennies and Standard J-1s, with their Curtiss OX-5 engines and DeHaviland DH-4 airplanes with Liberty V-12 engines. These airplanes and engines, while limited in utility, were so abundant and cheap that manufacturers were discouraged from developing new engines until these were used up.

Aviation did not become a viable form of transportation until a dependable engine was developed. Beginning in about 1923, Charles Lawrance built a 9-cylinder radial engine that was developed by the Wright Aeronautical Corporation into their famous Whirlwind series of engines, the most famous of which was the 220-horsepower Wright J-5. This is the engine that powered Charles Lindbergh's *Spirit of St. Louis* on its successful 33-hour nonstop flight from New York to Paris in May of 1927. About two weeks later, Clarence Chamberlain, flying a Bellanca, also powered by a Wright J-5 engine, flew nonstop from New York to Germany in 43 hours.

Small 3-, 5-, and 7-cylinder radial engines powered the light airplanes of the 1930s and 1940s, and 7-, 9-, and 14-cylinder radial engines powered the faster private and business airplanes, as well as military and airline aircraft.

During World War II the radial engine was the most popular configuration in the United States. Some fighter airplanes used liquid-cooled V-12 engines, but most aircraft were powered by 9-, 14-, and 18-cylinder radial engines, and by the end of the war, by a popular 28-cylinder engine.

The point of diminishing returns in reciprocating engine development was reached during World War II by the Lycoming XR-7755, a 5,000-horsepower 36-cylinder liquid-cooled radial engine. Fortunately the gas-turbine engine became functional at about this time.

Horizontally opposed engines first became popular as powerplants for very light aircraft in 2- and 4-cylinder models of less than 40 horsepower. This configuration has the advantage of smooth operation, small frontal area, light weight, and dependability. Because of these characteristics, they have been widely produced with 4-, 6-, and even 8-cylinders, with power output of up to 520 horsepower or more.

Horizontally opposed engines have replaced radial engines for almost all reciprocating engine-powered private airplanes and are in the mid 1990s, the only configuration of FAA-certificated reciprocating engines produced in the United States.

Private aviation in the United States has undergone drastic changes since the 1960s. The cost of private aircraft ownership skyrocketed because of the proliferation of product liability lawsuits, and commercial manufacturers virtually stopped producing reciprocating-engine-powered private aircraft in the 1980s. By the mid 1990s, changes in tort reform laws encouraged some manufacturers to re-enter the private aircraft field.

People still want to fly, and therefore the number of amateur-built or homebuilt aircraft has increased dramatically. Much of the emphasis with amateur-built aircraft has changed from the ultra simple machine built without complex tooling and at minimum of cost, to aircraft on the cutting edge of technology. Freedom from some of the FAA constraints under which production aircraft are built and the accompanying reduction of the threat of product liability lawsuits allow private builders to exploit the limitless advantages of composite construction.

Amateur-built aircraft do not require FAA-certificated engines, and as a result, there is a strong movement in the conversion of automobile engines for aircraft use. Some converted automobile engines are truly state-of-the-art powerplants, with electronic ignition and fuel injection. The safety record for these engines is excellent, and it is quite possible that this will continue to be a viable means of developing engines for private aircraft in the future.

amateur-built aircraft. Aircraft built by individuals as a hobby rather than by factories as commercial products. Amateur-built or homebuilt aircraft do not fall under the stringent requirements imposed by the FAA on commercially built aircraft.

As aviation begins its second century, the gasoline reciprocating engine, in spite of its inefficiency, continues to be used, but not without competition. Practically all airline and military aircraft are turbine powered and will continue to be.

Air-cooled, horizontally-opposed gasoline engines will continue to dominate the piston-powered aircraft market for the foreseeable future. There have been, and continue to be, inroads made to develop more fuel-efficient powerplants, but none have risen to the forefront in any significant way to unseat the gasoline-fired mainstay. Some of the ongoing innovations include liquid-cooled gasoline engines, compression-ignition (CI) engines, rotating combustion (RC) engines developed from the Wankel engine, and cam (as opposed to crankshaft) engines.

rotating combustion (RC) engine. A form of internal combustion engine in which a rounded, triangular-shaped rotor with sliding seals at the apexes forms the combustion space inside an hourglass-shaped chamber. Expanding gases from the burning fuel-air mixture push the rotor around and turn a geared drive shaft in its center. The RC engine was conceived in Germany by Felix Wankel in 1955.

The most significant of these improved engines has been the compression-ignition engine, better known as the diesel engine. The diesel or CI engine is about 10% to 15% more fuel efficient than the gasoline engine. This could be a significant savings if that were the only consideration, but the CI engine is considerably heavier than the gasoline-fired engine. This aspect in itself produces considerable inefficiencies when cost per mile is concerned; the search for ideas for more efficient piston engine power therefore continues.

Figure 1-2 highlights the progress made in aircraft reciprocating engines. In only 40 years, engines progressed from almost 15 pounds per horsepower to slightly more than one pound per horsepower.

Manufacturer and Name	Year	Configuration	H.P.	Weight
Wright *Flyer*	1903	4 I L	12-16	179
Manly-Balzer	1903	5 R L	52.4	207
Curtiss OX-5	1910	8V L	90	400
Le Rhone J	1916	9 Ro A	120	323
Liberty V-12	1918	12 V L	400	900
Wright J-5	1925	9 R A	220	510
Pratt & Whitney R-1830	1932	14 R A	1,200	1,467
Wright Turbocompound	1940	18 R A	3,700	2,779
Pratt & Whitney R-4360	1943	28 R A	4,300	3,600
Engines for Private Aircraft				
Continental A-65	1938	4 O A	65	170
Lycoming TIGO-541	1959	6 O A	450	396
I = Inline, R = Radial, V = V, Ro = Rotary, O = Horizontally opposed, L = Liquid cooled, A = Aircooled				

Figure 1-2. *Progress made in aircraft reciprocating engines*

Aircraft Turbine Engines

The principle of using a turbine as a source of power has been known for more than 400 years, since the days of Leonardo da Vinci. Wind-driven turbines in the form of windmills have converted much of the arid wasteland in the western United States into profitable farms and ranches.

Water-driven turbines are used to generate electricity in the huge hydroelectric powerplants, and steam turbines are used to drive electrical generators and propel ocean-going ships.

turbosupercharger. A centrifugal air compressor driven by exhaust gases flowing through a turbine. The compressed air is used to increase the power produced by a reciprocating engine at altitude.

The first practical use of turbines in aviation was the turbosupercharger developed by Dr. Sanford Moss during World War I. A turbine spun by exhaust gases leaving the engine drove a centrifugal compressor that increased the pressure of the air entering the cylinders. Turbosuperchargers allow reciprocating engines to maintain their sea-level power to a high altitude.

centrifugal compressor. An air compressor that uses a scroll-type impeller. Air is taken into the center of the impeller and slung outward by centrifugal force into a diffuser where its velocity is decreased and its pressure is increased.

The gas-turbine engine is a logical progression from a turbosupercharger. A combustion chamber is placed between the turbine wheel and the compressor. Air from the compressor flows through the combustion chamber where fuel is added and burned. The expanding gases drive the turbine, which in turn drives the compressor. Though the compressor requires a tremendous amount of power, the turbine produces enough, with some left over for torque or thrust.

In 1929, Frank Whittle, a brilliant young pilot-officer in the British Royal Air Force, filed a patent for a turbojet airplane engine. Unfortunately, Whittle's genius was not appreciated, and it was not until 1937 that his first jet engine actually ran.

Some scientists in the British Air Ministry were interested in gas-turbine engines, but thought of them only as a source of power to drive propellers.

A propeller produces thrust by delivering a small change in momentum to a large mass of air, but Whittle's concept was that thrust could be produced by a jet engine delivering a far larger change in momentum to a much smaller mass of air. The thrust produced by a turbojet would increase as the aircraft flew faster and would be efficient at high altitude.

Whittle's engine used a turbine-driven centrifugal compressor to move a large mass of air through the engine. Fuel was sprayed into the fast moving air and burned, expanding it and accelerating it enough to produce useful thrust.

torque. A force that produces or tries to produce rotation.

thrust. The aerodynamic force produced by a propeller or turbojet engine as it forces a mass of air to the rear, behind the aircraft.

A propeller produces its thrust by accelerating a large mass of air by a relatively small amount. A turbojet engine produces its thrust by accelerating a smaller mass of air by a much larger amount.

The turbojet engine came about at exactly the correct time. In spite of the lack of interest by the British government, Frank Whittle and his small but devoted crew at Power Jets, Ltd., proved the feasibility of the turbojet engine. In October of 1941, The General Electric Company was licensed to build the Whittle engine in the United States. GE was chosen for two reasons: because of their experience with turbosuperchargers, and because the two primary aircraft engine manufacturers, Pratt & Whitney and Wright Aeronautical, had more than they could handle in the continued development of reciprocating engines that were so desperately needed for the war which, at that time, appeared imminent.

The technology of turbojet engines was so new and the world was so deeply involved in the war, that no great strides in turbine engine development were made until the war was over.

At the end of the war, many reciprocating engines were declared surplus and sold for such low prices that there was little incentive for manufacturers to design and build new reciprocating engines. The gas turbine engine showed so much promise that neither Pratt & Whitney nor Wright Aeronautical felt it wise to continue developing reciprocating engines. Pratt & Whitney transitioned heavily into turbine engines, but Wright Aeronautical did not develop any of their own. They did produce some British engines under license but soon departed entirely from aviation engines.

Turbine engines have a far greater versatility than reciprocating engines because they can be operated either as a thrust or torque producer. Turbojet and turbofan engines produce thrust by accelerating a mass of air. Turboprop and turboshaft engines produce torque to drive propellers or helicopter rotors, or generators and air compressors for auxiliary power units.

power. The time rate of doing work. Power is found by dividing the amount of work done, measured in foot-pounds, by the time in seconds or minutes used to do the work.

Power may be expressed in foot-pounds of work per minute or in horsepower. One horsepower is 33,000 foot-pounds of work done in one minute, or 550 foot-pounds of work done in one second.

thrust horsepower. The horsepower equivalent of the thrust produced by a turbojet engine. Thrust horsepower is found by multiplying the net thrust of the engine, measured in pounds, by the speed of the aircraft, measured in miles per hour, and then dividing this by 375.

There is no direct comparison between turbine engines and reciprocating engines that allows us to visualize the tremendous strides that have been made in aircraft propulsion systems, but we can convert thrust into thrust horsepower and make a power-to-weight comparison.

Power requires movement, so thrust horsepower must take into consideration the speed of the aircraft. Thrust horsepower is found by multiplying the net thrust of the engine measured in pounds, by the speed of the aircraft measured in miles per hour, then dividing this by 375.

$$\text{Thrust horsepower} = \frac{\text{Net thrust (pounds)} \cdot \text{Aircraft speed (miles per hour)}}{\text{375 mile-pound / hour}}$$

The Pratt & Whitney R-1830 engine used in the ubiquitous Douglas DC-3 weighed approximately 1,500 pounds and produced 1,200 brake horsepower for takeoff. This is a power-to-weight ratio of 0.8 horsepower per pound, which is still an acceptable ratio for reciprocating engines.

The Pratt & Whitney JT9D that powers the Boeing 747 weighs approximately 9,000 pounds and produces up to 56,000 pounds of thrust, which at a cruise speed of 550 miles per hour, gives a little over 82,000 thrust horsepower. This is a power-to-weight ratio of a little more than 9 horsepower per pound!

It is easy to see the advantage that turbine engines have over reciprocating engines by comparing two popular torque-producing engines of the same basic power and used in the same types of aircraft. The Pratt & Whitney R-1830 reciprocating engine powers the 21-passenger Douglas DC-3, and the Pratt & Whitney of Canada PT-6 turboprop engine powers the 19-passenger Beech 1900D airliner. The power-to-weight ratio of the turboprop engine is 3.5 times as high as that of the reciprocating engine. *See* Figure 1-3.

	R-1830 Reciprocating	PT-6 Turboprop
Takeoff horsepower	1,200	1,377
Weight	1,500	486
Horsepower/ weight ratio	0.8	2.8

Figure 1-3. *Horsepower to weight ratio comparison between a reciprocating engine and a turboprop engine of comparable power*

Thrust-producing turbine engines have made tremendous progress since their first flight in 1939. Figure 1-4 shows the progress made in a little over fifty years.

Manufacturer Name	Type	Mass Airflow pounds/second	Thrust pounds	Weight pounds	Application
Whittle W1	TJ	22	850	623	E. 28/29
Allison J-33	TJ	90	4,600	1,820	F-80
P&W JT4	TJ	256	17,500	5,100	B-707
P&W JT8D	TF	331	17,400	3,500	B-727
G.E. CF6	TF	1,465	51,000	8,731	DC-10
RR RB.211	TF	1,658	63,000	9,874	B-747

TJ = Turbojet TF = Turbofan

Figure 1-4. *Progress in thrust-producing turbine engines*

Turbofan engines have almost completely replaced turbojet engines, and a new generation of ultra-high-bypass engines shows promise of opening a new niche between the turboprop and the turbofan. UHB engines, such as that in Figure 1-5, drive short, multiblade, contrarotating propellers and have high propulsive efficiency, low noise, low thrust specific fuel consumption, and a high power-to-weight ratio.

TSFC (thrust specific fuel consumption). A measure of the efficiency of a turbojet or turbofan engine. TSFC is the number of pounds of fuel burned per hour for each pound of thrust produced.

Figure 1-5. *The Unducted Fan™ engine is an ultra-high-bypass turbine engine that promises quiet operation with low fuel consumption at a speed higher than that used by turboprop-powered aircraft.*

STUDY QUESTIONS: DEVELOPMENT OF AIRCRAFT POWERPLANTS

Answers are on Page 14. *Page numbers refer to chapter text.*

1. The basic name for an engine that produces mechanical energy by changing chemical energy in the fuel into heat is a/an ______________________ engine. *Page 3*

2. Two types of heat engines are:
   a. ________________________________
   b. ________________________________
   Page 3

3. Two types of internal combustion engines used to power modern aircraft are:
   a. ________________________________
   b. ________________________________
   Page 4

4. A reciprocating engine in which the crankshaft is rigidly attached to the airframe and the cylinders spin with the propeller is called a/an ___________________ radial engine. *Page 5*

5. The most popular configuration of reciprocating engine in the United States from the end of World War I through World War II was the ____________________ engine. *Page 6*

6. The most popular configuration of reciprocating engine for private aircraft built in the United States since World War II is the ________________________ engine. *Page 6*

7. The first practical use of a turbine in aircraft propulsion was the _____________________ . *Page 8*

8. Aircraft turbine engines are used to produce _________________ or ________________ . *Page 9*

9. Two types of thrust-producing aircraft turbine engines are:
   a. _____________________________
   b. _____________________________
   Page 9

10. Two types of torque-producing aircraft turbine engines are:
    a. _____________________________
    b. _____________________________
    Page 9

Answers to Chapter 1 Study Questions

1. heat
2. a. external combustion
   b. internal combustion
3. a. reciprocating engines
   b. turbine engines
4. rotary
5. radial
6. horizontally opposed
7. turbosupercharger
8. thrust, torque
9. a. turbojet
   b. turbofan
10. a. turboprop
    b. turboshaft

Theory & Construction

2

Continued

Theory & Construction

2

Reciprocating Engine Theory

As noted in Chapter 1 the reciprocating engine is the oldest practical type of heat engine used as a powerplant for aircraft. It is an extremely inefficient engine, converting approximately one-third of the heat energy in the fuel that is burned into useful work, but its availability, simplicity, and relatively low cost keep it in use in spite of its shortcomings.

The gas turbine engine has almost completely taken over as the propulsion system for commercial operations, and for military fighters and bombers. Because most military pilots will spend their professional lifetime flying turbine-powered aircraft, the military services use turbine-powered training aircraft. Civilian aviation, on the other hand, is far more cost controlled, and all training and most private flying is done in the less costly reciprocating-engine-powered aircraft.

It is probable that technicians who have gained their initial experience in the military, and who will be working with the airlines, will never work on reciprocating engines. However, the principles of energy transformation are extremely important, and their study should not be slighted, as many of these principles relate to all forms of heat engines.

Energy Transformation

A heat engine does useful work when the chemical energy stored in the fuel is released in the form of heat. The heat enters a parcel of air and causes it to expand, which exerts a mechanical force on either a piston or a turbine.

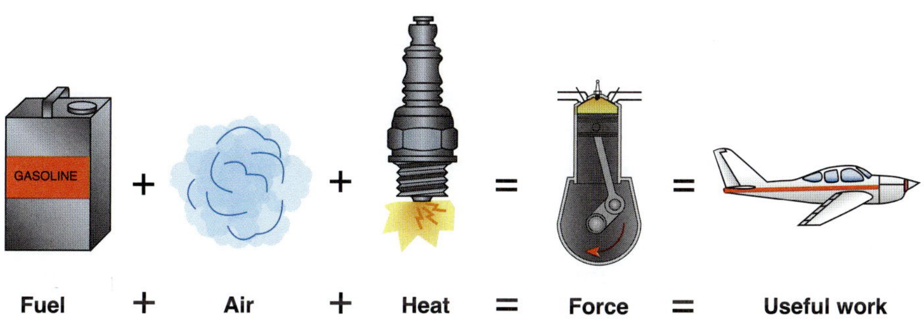

Figure 2-1. *The function of a heat engine is to expand a working fluid, usually air. When the air expands, it produces a force which does useful work.*

stroke. The distance the piston moves inside the cylinder.

The basic difference between the energy-release systems of reciprocating engines and gas turbine engines is in the way in which the heat energy in the fuel is added to the air. In a reciprocating engine, the energy is added in a constant-volume cycle, and in a gas turbine engine, it is added in a constant-pressure cycle. The constant-volume cycle is discussed in this chapter, and the constant-pressure cycle is discussed in Chapter 10, "Turbine Engines, Theory and Construction."

The Constant-Volume Cycle of Energy Release

four-stroke cycle. A constant-volume cycle of energy transformation that has separate strokes for intake, compression, power, and exhaust.

There are two types of reciprocating engines that operate on the constant-volume principle: two-stroke- and four-stroke-cycle engines. The principle of operation of these engines is essentially the same, the difference being in the events that take place as the piston moves in and out and the number of crankshaft revolutions needed to complete the cycle.

The four-stroke-cycle engine is the most widely used to power aircraft, but two-stroke-cycle engines are gaining popularity as the powerplant for ultralight and small kit-produced aircraft.

Four-Stroke, Five-Event Cycle

Otto cycle. The constant-volume cycle of energy transformation used by reciprocating engines. A mixture of fuel and air is drawn into the cylinder as the piston moves to the bottom of its stroke. The mixture is compressed as the piston moves upward in the cylinder, and when the piston is near the top of its stroke, the mixture is electrically ignited and burns. The burning mixture heats and expands the air inside the cylinder and forces the piston down, performing useful work. The piston then moves back up, forcing the burned gases out of the cylinder.

PV diagram. A diagram showing the relationship between the volume and the pressure above the piston during a cycle of engine operation.

The cycle of energy release that describes the way an aircraft reciprocating engine works is the Otto cycle, or four-stroke, five-event cycle. The best way to understand this cycle is to use the simple diagrams of a single-cylinder engine in Figure 2-2 and a Pressure-Volume, or PV, diagram such as the one in Figure 2-3.

The cycle begins as the piston moves downward with the intake valve open. This is called the intake stroke, and you will notice by the trace A-B on the PV diagram that the volume of the cylinder is increasing, but the pressure remains relatively constant.

After passing bottom center, the piston starts upward on the compression stroke, and the intake valve closes. The volume of the cylinder decreases, as seen by trace B-C. Both valves are now closed and the pressure rises. As the piston approaches the end of the compression stroke, an electrical spark jumps the gap in the spark plug and ignites the compressed fuel-air mixture. The mixture burns, and the chemical energy in the fuel is converted into heat energy that causes the air to expand. This mixture does not explode, but rather it burns in a smooth, progressive fashion and is completely consumed shortly after the piston passes top center on the compression stroke and has started down on the power stroke, point C.

The power stroke begins with both valves closed and the pressure inside the cylinder at its maximum value, as seen at point C. The piston is forced down by the pressure, and it is on this power stroke between points C and D that the only useful work is done. The pressure drops along the C-D curve, and as the piston nears the bottom of its stroke, the exhaust valve opens.

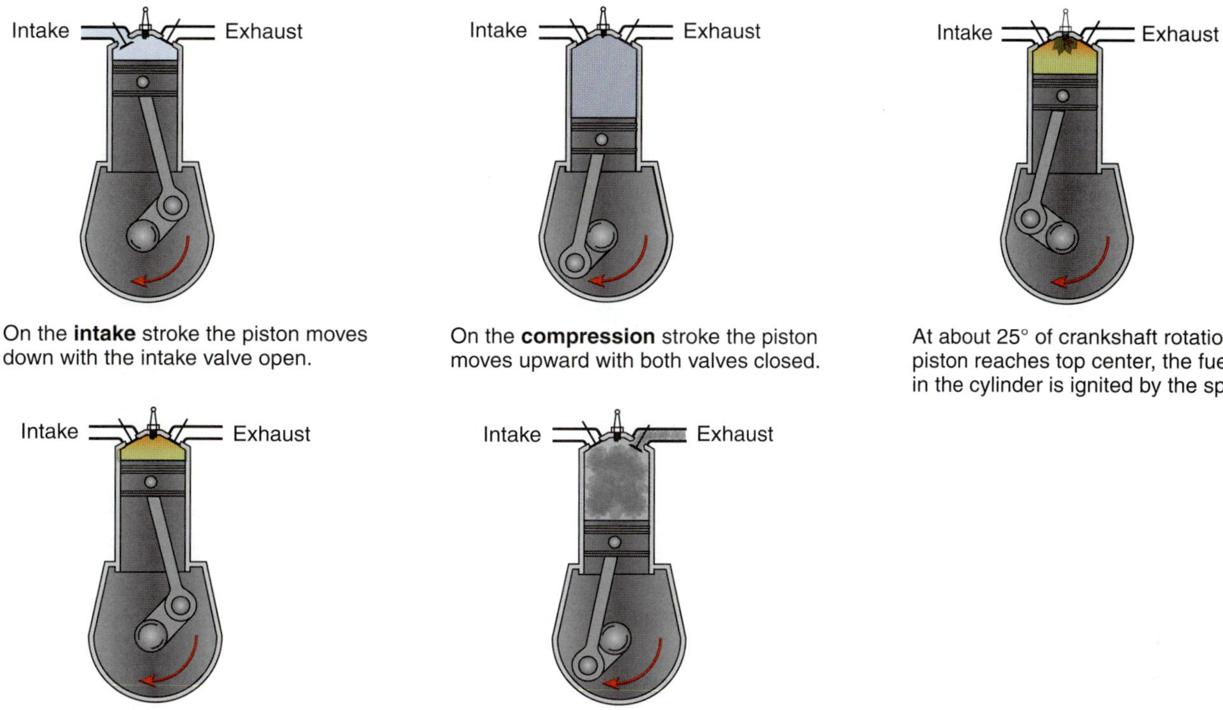

On the **intake** stroke the piston moves down with the intake valve open.

On the **compression** stroke the piston moves upward with both valves closed.

At about 25° of crankshaft rotation before the piston reaches top center, the fuel-air mixture in the cylinder is ignited by the spark plug.

On the **power** stroke, the piston moves down with both valves closed.

On the **exhaust** stroke, the piston moves upward with the exhaust valve open.

Figure 2-2. *The four strokes of an Otto-cycle engine*

As the piston moves upward on the exhaust stroke, along line D-A, the burned exhaust gases are forced out of the cylinder. At point A, the piston begins moving downward, and the cycle is repeated.

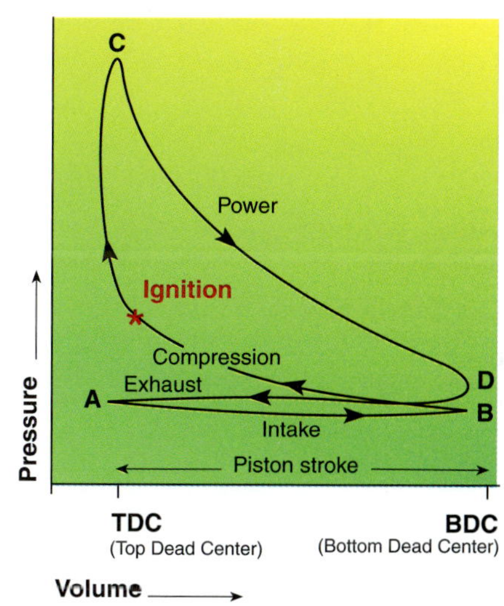

Figure 2-3. *PV diagram of the Otto-cycle of energy release*

Events do not occur instantaneously in a reciprocating engine. The fuel-air mixture flowing into the cylinder has mass and therefore is subjected to the law of inertia, and time is needed to start and stop its flow. The burning mixture requires a definite amount of time for the flame fronts to cross the piston and to heat the air inside the cylinder. The exhaust gases also have inertia, and it takes time for them to be accelerated and leave the cylinder.

Because of the time needed for these events to take place, the valves do not open and close when the piston is at the exact top or bottom of the strokes. Figure 2-4 on the next page illustrates typical timing of the events in a four-stroke-cycle engine.

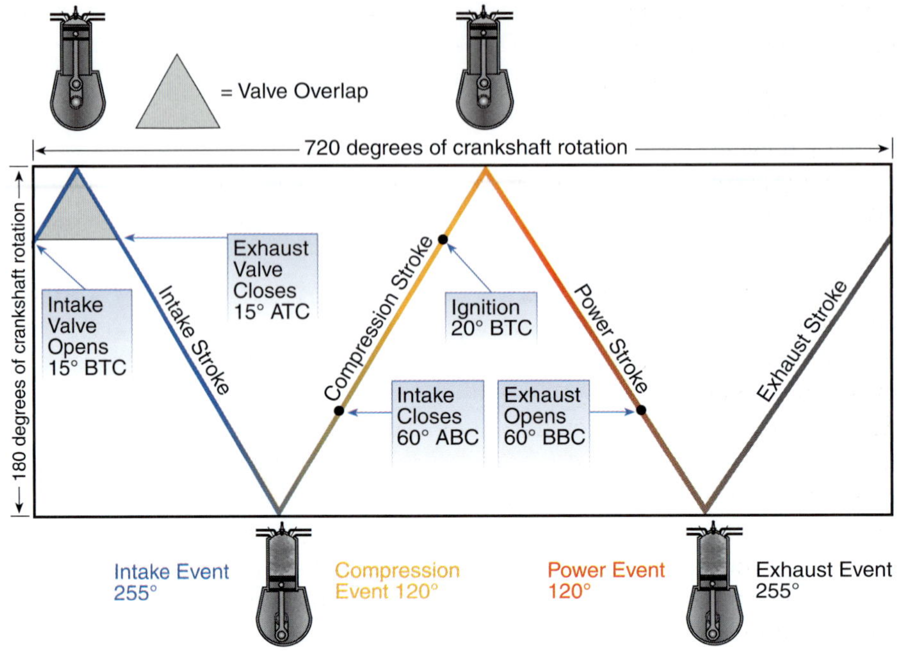

Figure 2-4. *Typical valve opening and closing, and ignition timing in a four-stroke-cycle engine.*

The intake valve opens about 15° of crankshaft rotation before the piston reaches the top of the exhaust stroke. The exhaust valve is also open at this time, and the hot exhaust gases are flowing out of the cylinder. The inertia of the gases leaving the cylinder, and the deceleration of the piston as it approaches top center, creates a slightly low pressure that helps the fuel-air charge begin to move into the cylinder. The exhaust valve closes after the piston has passed over top center, and the crankshaft has rotated about 15° down on the intake stroke. There is very little actual piston movement during this period of valve overlap.

valve overlap. The portion of the operating cycle of a four-stroke-cycle reciprocating engine during which both the intake and exhaust valves are off of their seats at the same time.

The intake valve remains open until after the piston has passed bottom center and the crankshaft has moved up on the compression stroke for about 60°. The valve remains open for this extra length of time to allow the inertia of the flowing gases to get as much fuel-air charge as possible into the cylinder.

On the compression stroke, the cylinder is full of the combustible fuel-air mixture, both valves are closed, and the piston is moving upward, compressing the mixture. When the piston is about 20° of crankshaft rotation before top center, the magneto sends a pulse of high voltage to each spark plug, producing two sparks that ignite the compressed mixture, by two flame fronts as they burn across the face of the piston.

If the mixture ratio and ignition timing are both correct, the fuel-air mixture will be completely burned shortly after the piston passes over top center. The expanding gases caused by absorbing heat from the burning mixture will cause the pressure inside the cylinder to reach its peak.

The high pressure inside the cylinder forces the piston down on the power stroke. This is the only useful work done in this entire energy transformation cycle.

The pressure is still high inside the cylinder when the piston is about 60° of crankshaft rotation before reaching bottom center, and at this time, the exhaust valve opens to allow the high pressure to start the burned gases moving out of the cylinder. At this point in the power stroke, the piston connecting rod assembly has such a poor angle at which to apply energy to the crankshaft, that it is a better use of the pressure left in the cylinder to overcome the inertia of accelerating the exhaust outflow. This acceleration decreases the energy extracted from the crankshaft to push the gases out on the up stroke, and assures more complete scavenging of the spent gases. This early opening reduces engine temperature and the more complete scavenging of the exhaust gases improves the combustion of the next intake mixture. The exhaust valve remains open during all the upward movement of the piston on the exhaust stroke, and until it has passed over top center and has moved downward for about 15° of crankshaft rotation on the intake stroke.

Notice that there is a point when both valves are off of their seats at the same time. The intake valve opens about 15° before the piston reaches top center on the exhaust stroke, and the exhaust valve remains open until the piston has moved downward about 15° on the intake stroke. Having both valves open at the same time allows the inertia of the exiting exhaust gases to aid in drawing the fresh fuel-air charge into the cylinder. This part of the cycle is called valve overlap and is extremely important for volumetric efficiency and proper cooling. Both of these are discussed starting on Page 32.

Two-Stroke Cycle

The two-stroke-cycle engine is the simplest form of reciprocating engine, and it has gained tremendous popularity for use in lawn mowers, snow blowers, chain saws, and many other tools and devices. The advent of ultralight and small kit-built airplanes has brought about aviation applications for these simple engines. Two-stroke engines have two inherent problems: cooling difficulty and spark plug fouling. Since every downward stroke is a power stroke, there is no cooling interval as in a four-stroke engine; and because the crankcase-compression engines are lubricated by mixing oil with the fuel, oil tends to foul the spark plugs.

two-stroke cycle. A constant-volume cycle of energy transformation that completes its operating cycle in two strokes of the piston, one up and one down.

Two-stroke engines are available with one to four cylinders and produce up to 100 horsepower. Every downward stroke of the piston is a power stroke, and these engines normally operate at a high revolutions per minute (RPM).

There are two types of two-stroke engines: one uses crankcase compression to get the fuel-air mixture into the cylinder, and the other uses some form of blower to increase the pressure of the mixture to force it into the cylinder. *See* Figure 2-5.

crankcase. The housing which encloses the crankshaft, camshaft, and many of the accessory drive gears of a reciprocating engine. The cylinders mount on the crankcase, and the engine attaches to the airframe by the crankcase.

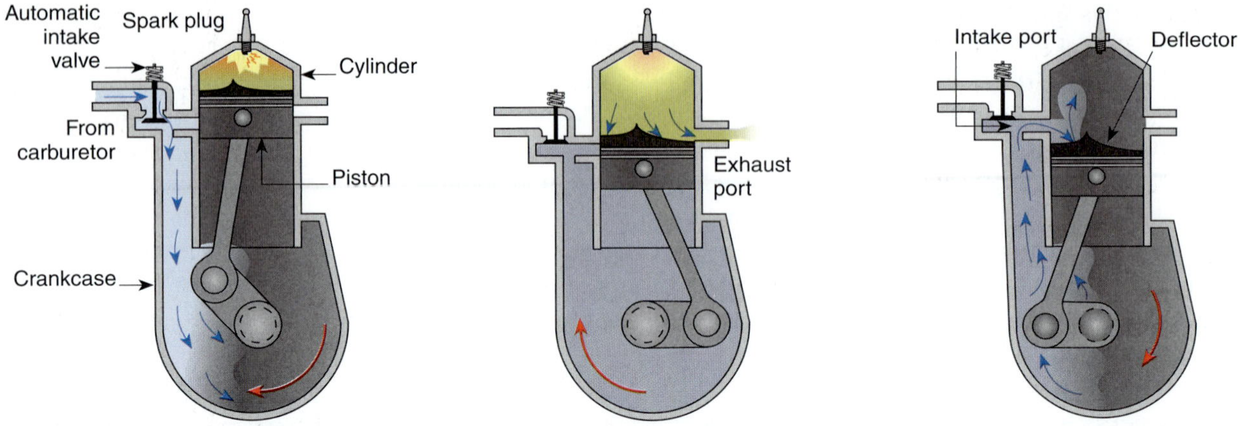

Piston moves upward, compressing mixture in cylinder and lowering pressure in crankcase. This low pressure pulls fresh charge of fuel-air mixture from carburetor through automatic intake valve. When piston is about 30° of crankshaft rotation from top, spark plug ignites mixture.

Gases inside cylinder have expanded and are forcing piston downward. Piston has just passed the exhaust port and the burned gases are flowing out. Pressure on fresh charge in crankcase is increasing.

Piston has passed bottom center and started back up. Compressed mixture in crankcase is flowing into cylinder through intake port. Deflector on top of piston prevents fresh mixture flowing out exhaust port.

Figure 2-5. *Operating principle of a crankcase-compression two-stroke-cycle engine*

automatic intake valve. An intake valve opened by low pressure created inside the cylinder as the piston moves down. There is no mechanical means of opening it.

The two-stroke cycle begins as the fuel-air mixture is drawn into the crankcase through the carburetor and the automatic intake valve, by the low pressure created as the piston moves upward. The mixture inside the cylinder is compressed, and as the piston approaches top center, the spark plug ignites the mixture. The burning is completed by the time the piston passes over top center, and the heat from the burning fuel expands the air and forces the piston down.

As the piston moves down, it compresses the fuel air mixture inside the crankcase, and when the piston passes the exhaust port in the cylinder wall, the burned gases begin to leave the cylinder. The piston continues to move downward, and just after passing the exhaust port, it uncovers the inlet port, and the compressed mixture from the crankcase is forced into the cylinder. The deflector on the head of the piston prevents the fresh charge from flowing out of the cylinder along with the exhaust gases.

After passing bottom center and starting up, the piston covers the inlet port and shuts off the flow of the fuel-air mixture into the cylinder; then it covers the exhaust port and stops the outward flow of burned gases. As the piston continues its upward stroke, the pressure inside the cylinder increases, the pressure in the crankcase decreases, and a fresh charge of fuel-air mixture is pulled in through the automatic intake valve to begin the next cycle.

supercharger. An air compressor used to increase the pressure of the air being taken into the cylinders of a reciprocating engine.

Two-cycle engines using crankcase compression normally have the lubricating oil mixed with the fuel and require no other lubrication system.

Some two-cycle engines have a supercharger, or blower, that increases the pressure of the fuel-air mixture and sends it directly into the cylinder. None of the mixture goes into the crankcase.

When the mixture is ignited and burns, the cylinder pressure is high, and the spring-loaded intake valve is held closed. The piston moves down and uncovers the exhaust ports in the cylinder wall, and as the exhaust gases leave the cylinder, the pressure drops.

As soon as the pressure inside the cylinder is lower than the pressure produced by the blower, fresh fuel-air mixture flows into the cylinder. This flow of fresh mixture continues until the piston passes bottom center and starts back up. After it passes the exhaust port, the pressure builds up higher than the blower pressure, and the intake valve closes.

The spark plug ignites the compressed fuel-air mixture as the piston nears top center in the same way that it does in a crankcase-compression engine.

Blower-equipped two-cycle engines do not have their lubricating oil mixed with the fuel, but have conventional lubrication systems.

There are also two-cycle engines that mechanically open intake, or intake and exhaust valves. These engines tend to be more volumetrically and thermally efficient.

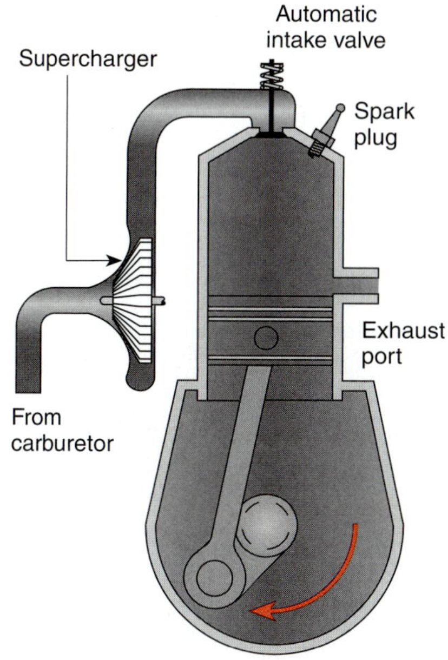

Figure 2-6. *Two-stroke-cycle engine using a supercharger to force the fuel-air mixture into the cylinder*

Work-Power Considerations

The purpose of an aircraft engine is to convert energy in the fuel into useful work. To best understand the way the engine does this, we should review some of the principles regarding work and power. These concepts are covered in more detail in Chapter 3, "Basic Physics," in the *General* textbook of this *Aviation Maintenance Technician Series.*

Work

Work, measured in foot-pounds or inch-pounds, is the product of a force, expressed in pounds, times the distance through which the force acts, expressed in feet or inches.

work. The product of a force times the distance the force is moved.

power. The time rate of doing work.

Work = Force · Distance

When a 20-pound force causes an object to move 10 inches, 200 inch-pounds of work has been done.

Note that in the measurement of work, time is not involved. It takes the same amount of work to move a 100-pound load of sand 10 feet in one trip as it does to move the load in ten trips carrying 10 pounds each trip.

The useful work done with one trip is:

100 feet · 100 pounds = 10,000 foot-pounds

The useful work done with ten trips is the same:

1,000 feet (100 feet · 10 trips) · 10 pounds = 10,000 foot-pounds

The difference between doing the work in one trip or doing it in ten trips is the length of time required, and this determines the amount of power needed.

In the metric system, work is expressed in terms of kilogram-meters, where the force applied is stated in kilograms and the distance is stated in meters. As a conversion, 1 kilogram-meter of work is equal to 7.233 foot-pounds of work.

$$\text{Kilogram-meters} \cdot 7.233 = \text{foot-pounds}$$

Power

Power is the rate of doing work, and power takes time into consideration. In the United States, work is normally measured in foot-pounds, and power is measured in foot-pounds per minute or in horsepower. One horsepower is equal to 33,000 foot-pounds of work done in one minute, or 550 foot-pounds of work done in one second.

In the metric system, power is expressed in watts, with one watt equal to 0.001341 horsepower ($1.341 \cdot 10^{-3}$ horsepower). A more useful unit is the kilowatt (1,000 watts), which is equal to 1.341 horsepower.

horsepower. The most commonly used unit of mechanical power. One horsepower is equal to 33,000 foot-pounds of work done in one minute, or 550 foot-pounds of work done in one second.

watt. The basic unit of power in the metric system. One watt is the amount of power needed to do one joule (0.7376 foot-pound of work) in one second. One watt is 1/746 horsepower.

Horsepower of a Reciprocating Engine

There are a number of types of horsepower involved in the study of reciprocating engines: indicated horsepower (IHP), brake horsepower (BHP), and friction horsepower (FHP). The relationship between these three is:

$$\text{IHP} = \text{FHP} + \text{BHP}$$

IHP (indicated horsepower). The theoretical horsepower a reciprocating engine develops within its cylinders.

BHP (brake horsepower). The actual horsepower delivered to the propeller shaft of a reciprocating engine.

FHP (friction horsepower). The amount of horsepower used to turn the crankshaft, pistons, gears, and accessories in a reciprocating engine and to move the pistons against the pressures of intake, compression, and exhaust.

Indicated Horsepower

Indicated horsepower (IHP) is the horsepower developed in the cylinders of an engine. IHP cannot be measured directly, but can be calculated as follows:

1. Find the area of the piston head that is acted on by the expanding gases. The area, in square inches, is determined by the formula:

   $$A = 0.7854 \cdot D^2$$

   $0.7854 = \text{a constant } (\pi \div 4)$

   $D = \text{cylinder bore measured in inches}$

2. Find the amount of pressure acting on the piston head. This is the mean effective pressure (MEP) and is the average pressure inside the cylinder during the power stroke. Indicated mean effective pressure (IMEP) is actually a measured value. It was originally measured by an instrument called an indicator, thus its name, but today it is measured far more accurately by an electrical pressure transducer. IMEP is expressed in pounds per square inch. *See* Figure 2-7.
3. Find the amount of force acting on the piston by multiplying the area of the piston head in square inches by the amount of pressure in pounds per square inch.
4. Find the number of foot-pounds of work done on each power stroke of a single cylinder. To do this, multiply the stroke of the piston (the distance

bore. The diameter of a reciprocating engine cylinder.

IMEP (indicated mean effective pressure). The average pressure existing inside the cylinder of a reciprocating engine during its power stroke.

the piston travels inside the cylinder) in feet, by the force acting on the piston, in pounds.

5. Find the total number of foot-pounds of work done by all of the power strokes of the engine in one complete cycle by multiplying the foot-pounds of work per cylinder by the number of cylinders in the engine.
6. Find the number of power strokes per minute. In a four-stroke-cycle engine, only every other stroke is a power stroke, so you must divide the engine RPM by 2.
7. Find the total number of foot-pounds of work done in one minute by multiplying the number of foot-pounds of work per cycle by the number of power strokes per minute.
8. Find the indicated horsepower by dividing the foot-pounds of work per minute by 33,000.

This procedure is expressed by the formula:

$$\text{IHP} = \frac{\text{PLANK}}{33{,}000}$$

IHP = Indicated horsepower.
P = Indicated mean effective pressure in psi.
L = Length of the stroke **in feet**. (This is normally expressed in inches and this value will need to be divided by twelve to convert to feet.)
A = Area of the piston head in square inches. This is $0.7854 \cdot \text{bore}^2$.
N = Number of power strokes per minute per cylinder. This is RPM ÷ 2.
K = Number of cylinders.
33,000 = A constant which is the number of foot-pounds per minute of work in one horsepower.

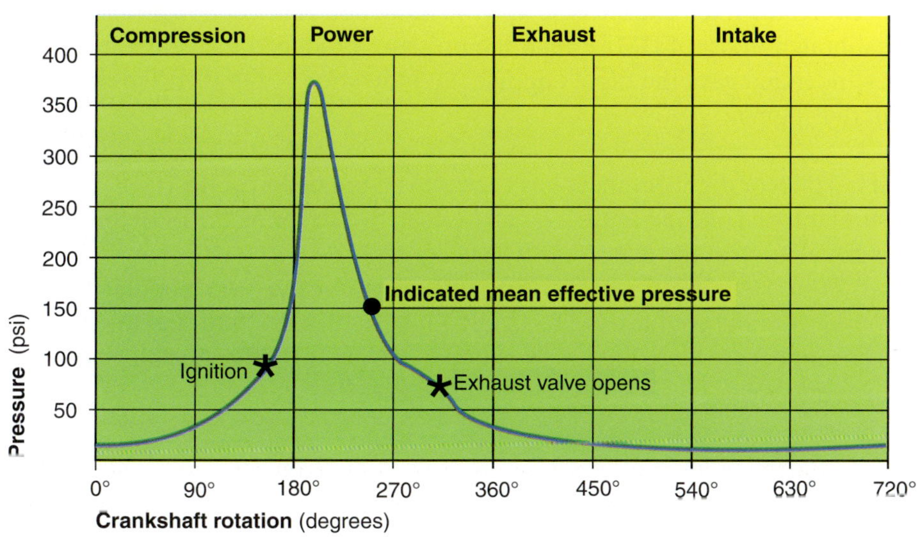

Figure 2-7. *A plot of the indicated pressure inside a cylinder of a reciprocating engine during one complete cycle*

torque. A force that produces or tries to produce rotation.

prony brake. An instrument used to measure the amount of horsepower an engine is delivering to its output shaft. The engine is operated at a specific RPM, and a brake is applied to its output shaft. The amount of torque applied to the brake is measured, and this, with the RPM, is converted into brake horsepower.

BMEP (brake mean effective pressure). The theoretical average pressure inside the cylinder of a reciprocating engine during the power stroke. BMEP relates to the torque produced by the engine, and can be calculated when you know the brake horsepower.

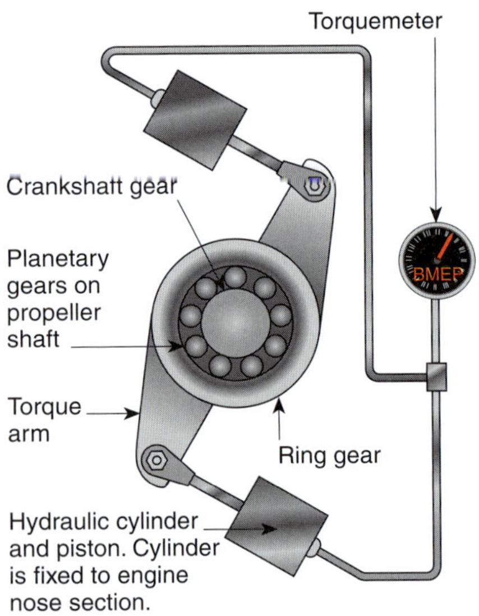

Figure 2-8. *A torquemeter measures the reaction between the fixed engine case and the propeller reduction ring gear. The indicator is calibrated in psi of BMEP.*

Friction Horsepower

Friction horsepower (FHP) is the amount of power used by the engine to sustain its operation. It is the horsepower required to move all the internal parts of the engine, to pull the fuel-air mixture into the cylinders, compress it, and force the burned gases out. It also includes the power expended in opening the valves and operating such devices as the oil pumps and magnetos that are necessary for engine operation. FHP is the difference between IHP and BHP.

Brake Horsepower

Brake horsepower (BHP) is the actual horsepower delivered by the engine to the propeller or helicopter transmission. This horsepower can be measured by running the engine on a dynamometer and measuring the torque the engine produces at a given RPM. The name, brake horsepower, comes from the fact that this power was originally measured with a prony brake that mechanically loaded the engine with friction, and measured the reaction to this friction with a balance scale. Modern dynamometers load the engine with either a hydraulic or electric load.

Brake horsepower is determined by the formula:

$$BHP = \frac{PLANK}{33{,}000}$$

The components of this formula are the same as those for IHP except that P stands for brake mean effective pressure, or BMEP.

BMEP cannot be measured directly, but some of the large reciprocating engines installed on older bombers and airliners had a torque nose and an indicator that gave the flight engineer an indication of the BMEP the engine was developing. Hydraulic cylinders installed between the propeller reduction gear and the nose section of the engine measured the reaction, or torque, between the fixed engine case and the reduction ring gear. A pressure gage calibrated in psi of BMEP read the pressure produced inside the hydraulic cylinder and gave the flight engineer an indication of the BMEP developing in the engine.

If the pilot or flight engineer has access to an indication of the BMEP, a chart similar to the one in Figure 2-9 can be used to determine the BHP the engine is developing at the particular RPM.

Figure 2-9 is a typical BHP/BMEP chart for five very popular Pratt & Whitney radial engines: the R-985, R-1340, R-1830, R-2000, and the R-2800. The numbers in these engine designations relate to their piston displacement in cubic inches.

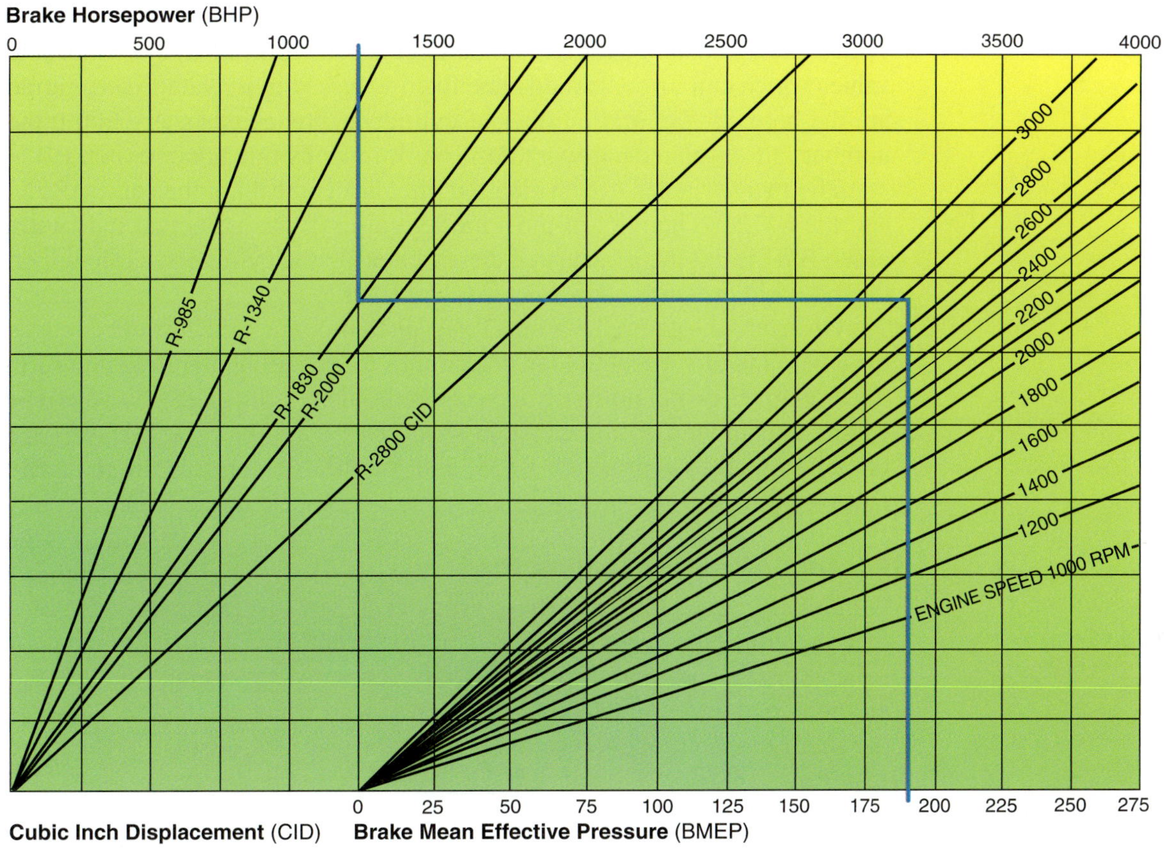

Figure 2-9. *A typical chart relating engine RPM to the BHP and BMEP for five popular Pratt & Whitney radial engines*

Use Figure 2-9 to find the BHP produced by an R-1830 engine that turns at 2,750 RPM for takeoff, if the torquemeter indicates that the BMEP is 190 psi. To use this chart, follow these steps:

1. Draw a line vertically upward from the 190 psi BMEP index until it intersects the imaginary line for 2,750 RPM. This is midway between the diagonal lines for 2,700 and 2,800 RPM.
2. From this point, draw a line horizontally to the left until it intersects the angled line for the 1830 CID engine.
3. From this intersection, draw a line vertically upward to the BHP index. This shows that at this RPM and BMEP, the engine is developing 1,250 BHP.

absolute pressure. Pressure referenced from zero pressure or a vacuum.

Modern aircraft are not equipped with BMEP gages, so the pilot uses a combination of RPM and manifold absolute pressure (MAP) to find the horsepower the engine is developing.

MAP (manifold absolute pressure). The absolute pressure, measured in inches of mercury (in. Hg), of the air inside the induction system of a reciprocating engine.

The horsepower produced by an aircraft engine is determined by three fixed values and a constant over which the pilot has no control, and two variable values which can be controlled. The fixed values which relate to the engine are the bore, stroke, and number of cylinders. The constant is 33,000, the number of foot-pounds of work done per minute by one horsepower.

The two variables over which the pilot has control are the engine RPM and the BMEP. The pilot of most modern aircraft has no actual indication of the BMEP, but the pressure of the air entering the cylinders, as shown on the manifold pressure gage, relates to the BMEP.

On an engine equipped with a fixed-pitch propeller, the throttle controls the RPM, and the propeller load determines the resulting manifold pressure. These engines do not normally have a manifold pressure gage. But when the engine is equipped with a constant-speed propeller, the pilot can vary the propeller pitch to control the RPM, and the throttle controls the manifold pressure, and thus the BMEP.

Factors Affecting Engine Power

thermal efficiency. The ratio of the amount of useful work produced by a heat engine, to the amount of work that could be done by all of the heat energy potential of the fuel.

A number of factors affect the power developed inside the cylinder of a reciprocating engine and delivered by the crankshaft to the propeller or helicopter transmission. These basically relate to the amount of fuel and air taken into the engine, and the efficiency with which the engine converts the energy in the fuel into useful work. Below, we look at some of the more important factors and the way they relate to engine power.

Thermal Efficiency

As mentioned earlier, a reciprocating engine is an exceptionally inefficient device for utilizing the heat energy in the fuel. Normally, it converts approximately one-third of the energy in the fuel into useful work. Figure 2-10 shows a typical distribution of the energy.

Approximately one-half of the energy goes out of the cylinder through the exhaust. Not only is this energy wasted directly, but the exhaust system must be made to accommodate the heat and continued burning and carry it away from the aircraft. There are several different devices, such as turbosuperchargers, or turbochargers, and power recovery turbines, used to recover some of this wasted energy, and these devices are discussed in Chapter 6, "Reciprocating Engine Exhaust Systems."

About 12% of the energy is lost through the cylinder: part is carried away by the air or liquid coolant and part by the lubricating oil. Another 6% is lost in compressing the air in the cylinders and in turning all the moving parts of the engine. This leaves about 31% of the original heat energy to be converted into useful work.

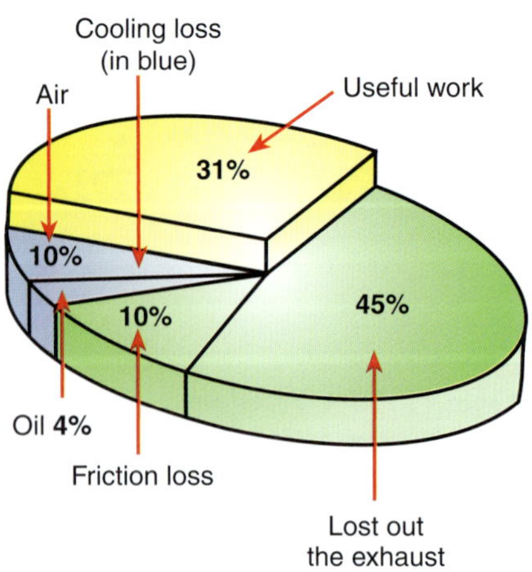

Figure 2-10. *Distribution of energy in the fuel burned in the cylinder of a typical reciprocating engine*

Brake thermal efficiency is the ratio of the heat energy in the fuel converted into useful work, to the amount of heat energy in the fuel that has been burned. To find the brake thermal efficiency of an engine, follow these steps:

1. Multiply the brake horsepower by 33,000 to find the number of foot-pounds of work per minute actually done by the propeller shaft of the engine.
2. Convert the gallons per hour of fuel burned into pounds per minute. Aviation gasoline has a nominal weight of six pounds per gallon, so multiply the gallons per hour by 6 to find the pounds per hour. Divide the pounds per hour by 60 to find the pounds per minute.
3. Find the number of Btu's of heat energy released by the fuel burned in one minute. Aviation gasoline has a nominal heat energy content of 18,880 Btu per pound, so multiply the pounds of fuel burned per minute by 18,880.
4. Find the number of foot-pounds of work produced by fuel burned in one minute. To do this, multiply by 778 the number of Btu released by the fuel burned in one minute, which is the number of foot-pounds of work one Btu will perform.
5. Divide the total number of foot-pounds of work per minute done by the propeller shaft by the number of foot-pounds of work in the fuel burned in one minute. This results in a decimal fraction which must be multiplied by 100 to get the percent BTE.

To illustrate, determine the thermal efficiency of an aircraft engine that burns 13 gallons of aviation gasoline per hour to produce 160 BHP, with the use of the following formula:

$$BTE = BHP \cdot 33{,}000 \text{ ft \# / \# of fuel consumed per minute} \cdot 18{,}880 \cdot 778$$

$$BTE = 160 \cdot 33{,}000 / \{13 \cdot 6 / 60\} \cdot 18{,}880 \cdot 778$$

$$BTE = 5{,}280{,}000 / 1.3 \cdot 18{,}880 \cdot 778$$

$$BTE = 5{,}280{,}000 / 19{,}095{,}232$$

$$BTE = .2765 \cdot 100$$

$$BTE = 27.65\%$$

Therefore, an engine burning 13 gallons of aviation gasoline in one hour and producing 160 BHP has a BTE of 27.65%. If this engine is leaned for cruise, a reduction of approximately 2.6% of the fuel in the mixture ratio, its cruise BTE would be over 30%.

Volumetric Efficiency

volumetric efficiency. The ratio for a reciprocating engine, of the volume of the fuel-air charge inside the cylinder when corrected to existing atmospheric conditions of temperature and pressure, to the single cylinder displacement (volume evacuated by the piston during the intake stroke).

For aviation gasoline to release its stored chemical energy, it must first be vaporized and then mixed with the correct amount of air to obtain the needed oxygen. The ability of an engine to breathe is extremely important, and this ability is expressed as its volumetric efficiency.

The volumetric efficiency of a reciprocating engine is the ratio of the volume of the fuel-air charge taken into the cylinder, to the volume of the charge the cylinder can actually hold under the same atmospheric conditions.

Anything that decreases the amount of air entering the cylinder decreases the volumetric efficiency. Improper valve timing, sharp bends in the induction system, and high carburetor air temperature will decrease the volumetric efficiency.

$$\text{Volumetric efficiency} = \frac{\text{Volume of charge taken into cylinder}}{\text{Single cylinder displacement}}$$

A number of factors affect volumetric efficiency. Some of these are: engine RPM, induction air temperature, cylinder head temperature, valve timing, design of the induction system, and degree of supercharging.

The fuel-air charge has mass, and therefore inertia. Instead of flowing smoothly in the induction system, the charge flows in jerks or pulses. The induction system must be designed in such a way that each cylinder receives the same charge each time the intake valve opens. Long induction pipes of small diameter, sharp bends, less then perfect alignment between sections, and rough internal surfaces restrict the airflow and reduce the volumetric efficiency. In addition, conditions like a dirty air filter, and a restricted exhaust system decreases volumetric efficiency.

The intake valve of a typical engine is open for approximately 255° of crankshaft rotation, and when the engine is turning at 2,300 RPM, the valve is open for approximately 0.0185 second. There is not enough time during each cycle for the mixture to completely fill the cylinder. Improperly timed valves, or valves adjusted with the wrong clearance, do not remain open long enough to thoroughly purge all of the burned exhaust gases, nor to get a full fresh charge into the cylinder.

valve lash. The clearance between the rocker arm face and the tip of the valve stem tip when the piston is on top dead center of the power stroke and the rocker arm is pressed against the pushrod.

Valve overlap, which is the time at the end of the exhaust stroke and the beginning of the intake stroke when both valves are off of their seat, is critical for high volumetric efficiency. If either the intake or exhaust valve clearance is too large, the cam will turn farther before the valve opens, and the valve will close before the cam has rotated to the correct position for it to close. This decreases the number of degrees of crankshaft rotation the valve is off its seat, as well as valve timing, and decreases the valve overlap. Insufficient valve lash increases valve overlap and also changes valve timing.

Adequate valve overlap increases the volumetric efficiency of the engine. It aids in scavenging the burned exhaust gases and gives the engine better cooling characteristics. It does this by assuring that the fuel-air charge in the cylinder is dense enough for proper operation and is not diluted with exhaust gases.

Hot air is less dense than cold air, and either high induction air temperature or high cylinder head temperature will cause the cylinder to be filled with low-density hot air, and there will be less mass of air available to mix with the fuel.

A nonsupercharged, or naturally aspirated, engine can never have a volumetric efficiency of 100%, but a supercharger compresses the air before it enters the engine cylinder, and therefore the volumetric efficiency of a supercharged engine can exceed 100%.

naturally aspirated engine. A nonsupercharged reciprocating engine using atmospheric pressure to force the fuel-air mixture into the cylinders.

Mechanical Efficiency

Mechanical efficiency of a reciprocating engine is the ratio of the amount of power delivered to the propeller shaft or the transmission of a helicopter, to the amount of power developed inside the cylinders. It is the ratio of brake horsepower to indicated horsepower.

$$\text{Mechanical efficiency} = \frac{\text{Brake horsepower}}{\text{Indicated horsepower}}$$

Aircraft reciprocating engines are normally quite efficient mechanically, and a mechanical efficiency of 90% is relatively common.

Piston Displacement

The piston displacement of a reciprocating engine is the total volume swept by the pistons in all cylinders during one revolution of the crankshaft. In the United States, aircraft engine piston displacement is expressed in cubic inches; in many other applications, it is expressed in cubic centimeters (cc) or liters. The greater this volume, the greater the volume of intake energy that can be taken into the engine. The greater the energy volume taken in, the greater the energy out.

piston displacement. The total volume, in cubic inches, cubic centimeters, or liters, swept by all of the pistons of a reciprocating engine as they move in and out in one revolution of the crankshaft.

$$\begin{aligned}\text{Piston displacement} = {} & \text{Piston head area in square inches} \cdot \\ & \text{Stroke in inches} \cdot \\ & \text{Number of cylinders}\end{aligned}$$

To determine piston displacement in cubic inches multiply the area of the piston, in square inches, by the length of stroke, in inches, and multiply this by the number of cylinders in the engine.

Generally, the greater the piston displacement, the more power an engine develops, but there are many variables to consider in determining the way to increase the displacement. Some manufacturers use a few large cylinders, while others use more but smaller cylinders. Some engines have strokes that are longer than the bore and others have shorter strokes, but over the years, "square" engines—those with the same bore and stroke dimensions—have proven to have an excellent dimensional balance between wear rates and efficient torque curves. Currently, most engines are under square with the stroke nearly an inch shorter than the cylinder bore. This gives the engine more efficient torque production in the RPM ranges used by modern engines.

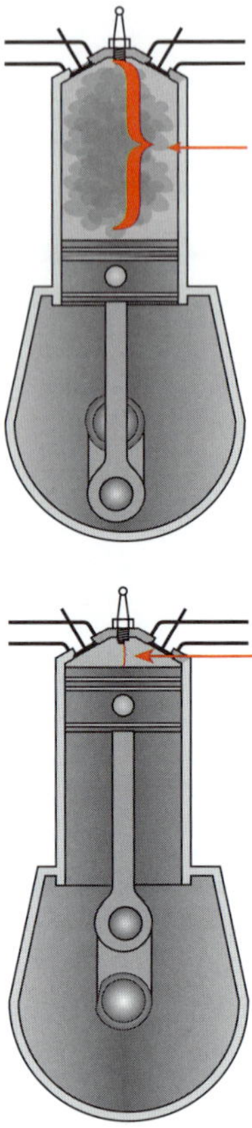

Figure 2-11. *The compression ratio of a reciprocating engine is the ratio of the volume of the cylinder with the piston at the bottom of the stroke, to the volume with the piston at the top of the stroke.*

compression ratio. The ratio of the volume of the cylinder of a reciprocating engine with the piston at the bottom of its stroke, to the volume of the cylinder with the piston at the top of its stroke.

detonation. An uncontrolled explosion inside the cylinder of a reciprocating engine. Detonation occurs when the pressure and temperature of the fuel inside the cylinder exceeds the critical pressure and temperature of the fuel. Detonation may be caused by using fuel that has a lower octane rating or performance number than is specified for the engine.

Compression Ratio

For a heat engine to convert the maximum amount of energy into useful work for the volume taken in, the fuel-air mixture must be compressed into as small a volume as practical before it is ignited. Because of this, the compression ratio of an engine is important in determining the amount of power an engine can produce and the amount it can produce efficiently.

The compression ratio is the ratio of the volume of the cylinder above the piston at the bottom of its stroke, to the volume above the piston at the top of its stroke. FAA materials refer to the space above the piston as the "Combustion Chamber Space." The value is expressed in cubic inches.

$$\text{Compression ratio} = \frac{\text{Displacement of one cylinder} + \text{combustion chamber space}}{\text{Combustion chamber space}}$$

When a highly compressed fuel-air mixture is ignited, it releases its energy rapidly, and produces the maximum cylinder pressure and the maximum amount of push on the piston. By increasing the compression ratio, an engine is able to convert more of its heat energy into useful work, and therefore less energy is lost to the cylinder walls and the combustion chamber surfaces. With less heat energy transferred to the cylinder assembly, the cylinder operates cooler and a greater portion of the heat is spent in expanding the gases.

Manifold absolute pressure is the absolute pressure in the induction system of an engine and is the pressure forcing the air fuel into the cylinder. The pressure on the gases in the cylinder at the beginning of the compression stroke and the amount they are compressed determine the pressure of the mixture at the time of ignition. The higher the pressure, the greater the power the engine can develop.

There is a practical limit to the maximum compression ratio and the maximum manifold pressure any given engine can tolerate, and this is the detonation characteristic of the fuel.

Compression heats a fuel-air charge, and it is possible to compress it until it is hot enough to explode. This explosion is called detonation, and the rapid release of heat causes destructive cylinder pressures and shock waves. Normal combustion flame-front propagation is 25 feet per second to 100 feet per second, idle to full throttle. Detonation flame-fronts travel at approximately 4,000 feet to 9,000 feet per second depending on the severity of the detonation. This rate is comparable to low-order explosives. Waves of this magnitude place sever shock force on the internal cylinder surfaces. If this force is not eliminated quickly, internal engine damage will result.

Detonation is suppressed and the allowable compression ratio increased by increasing the critical pressure and temperature of the fuel. Characteristics of reciprocating-engine fuel are discussed in more detail in Chapter 4, beginning on Page 134.

Ignition Timing

For a reciprocating engine to develop its maximum power, the pressure inside the cylinder caused by the burning gases must peak shortly after the piston passes its top center position and starts down. All FAA-certificated aircraft engines are required to have dual ignition systems which include two spark plugs in each cylinder. While two independent ignition systems enhance safety, the main reason for dual ignition is to provide better burning characteristics in the cylinder, by igniting the mixture in two places.

dual ignition. An ignition system of an aircraft reciprocating engine that has two of every critical unit, including two spark plugs in each cylinder. Dual ignition provides safety in the event of one system malfunctioning, but more important, igniting the fuel-air mixture inside the cylinder at two locations provides more efficient combustion of the air in the cylinder.

The fuel-air charge does not explode, but rather it burns evenly across the combustion chamber, and in a properly-designed cylinder, the mixture is ignited in two locations across the cylinder from each other. The flame fronts started at the two spark plugs progress across the cylinder and meet near the center. The majority of the charge burns in approximately 0.003 seconds at a nominal static RPM of 2,350, and the pressure rises more smoothly than it would if it were ignited by only one spark plug.

Automobile engines have variable ignition timing that changes the point at which ignition occurs, so that regardless of the engine speed, the pressure peaks when the piston is in the correct position. Aircraft engines have fixed timing, and therefore the spark occurs at a time that is a compromise between the optimum for takeoff power and cruise power. This compromise is minimal due to the fact the amount of advance required per RPM rise flattens out as RPM rises. However, one magneto manufacturer is marketing a timing advancing version of its electronic magneto, which produces marginal results when considered against other factors such as cost and liability. There are other options in the works as well.

The spark occurs on most engines about 20° to 25° of crankshaft rotation before the piston reaches top center on the compression stroke. If the ignition occurs too early in the cycle, the engine will lose power, because a greater portion of the pressure is reached while the piston is still moving upward, and the force of the expanding gases will oppose the crankshaft rotation.

If the spark occurs too late, the piston will already be descending and the volume of the cylinder will be increasing as the gases expand, and much of the effect of the push is lost. More seriously, though, when the spark occurs too late, there is not enough time for the mixture to be completely burned before the exhaust valve opens and the burning gases are forced out of the cylinder around the valve. This increases the cylinder head temperature and causes serious damage to the engine.

The heat from the burning gases flowing past the exhaust valve can cause the valve edges, spark plug electrodes, and carbon particles in the combustion chamber to become red hot. These incandescent points will

incandescent. Glowing because of intense heat.

preignition. Ignition of the fuel-air mixture inside the cylinders of a reciprocating engine before the normal spark occurs. Also known as "surface ignition."

ignite the fuel-air mixture before the time of normal ignition. This is called preignition, and it causes the temperature inside the cylinder to rise so high that the fuel-air mixture reaches its critical pressure temperature point before the piston reaches top center, causing detonation.

Special provisions must be made for delaying, or retarding, the spark for starting so the engine will not kick back. These provisions are discussed in detail in Chapter 5, beginning on Page 210.

Power Variations with Altitude

For a pilot to operate an aircraft most effectively and economically, he or she must know the horsepower the engine is developing. The engine manufacturer furnishes a set of curves similar to the one in Figure 2-12 that enables the pilot to find the actual BHP the engine is developing, both at sea level and at altitude, when the RPM, MAP, altitude, and inlet air temperature are known.

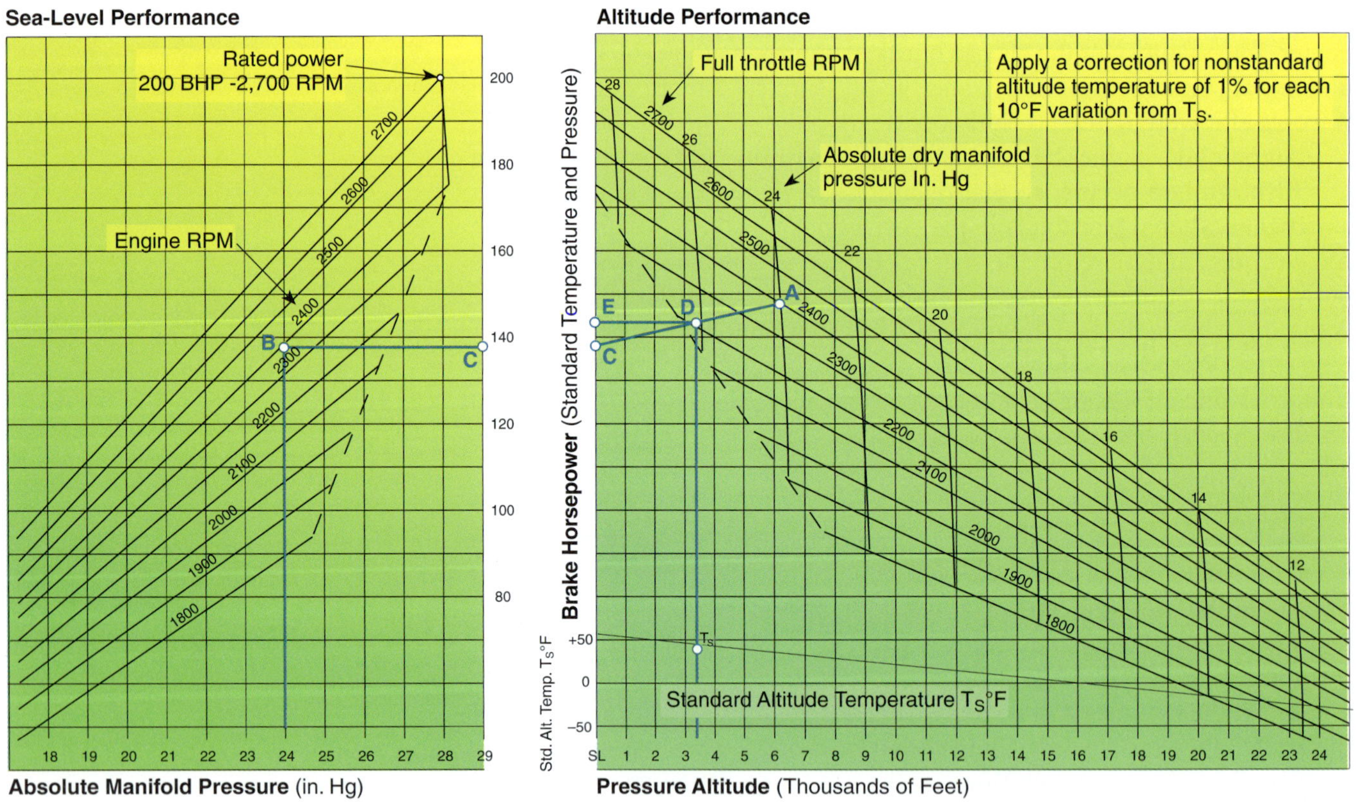

Figure 2-12. *Typical sea-level and altitude horsepower curves for a nonsupercharged fuel-injected engine*

To use this curve, follow these steps:

1. Determine the conditions used to find the horsepower:

   Engine speed = 2,400 RPM

   Manifold pressure = 24 inches Hg

   Pressure altitude = 3,500 feet

   Inlet air temperature = 30°F

pressure altitude. The altitude in standard atmosphere at which the pressure is the same as the existing pressure.

2. Locate point A on the altitude performance curve for 2,400 RPM and 24 in. Hg manifold pressure.
3. Locate point B on the sea-level performance curve for 2,400 RPM and 24 in. Hg manifold pressure.
4. Draw a line horizontally to the right from point B to locate point C on the vertical Brake Horsepower at Standard Temperature and Pressure Index. This engine using these values would produce 138 brake horsepower at sea level.
5. Transfer point C to the Altitude Performance Chart and connect point A with point C.
6. Draw a vertical line upward from 3,500 feet on the Pressure Altitude Scale until it intersects the slanted line between points A and C on the Altitude Performance Curve. Mark point D at the intersection of these two lines.
7. Draw a horizontal line to the left from point D to the vertical horsepower index and note at 3,500 feet with standard temperature, the engine will develop 144 brake horsepower. The horsepower developed at altitude is greater than it is at sea level. This is because there is less exhaust back pressure, and to get the same manifold pressure at altitude, the throttle must be opened more than it is at sea level.
8. Apply a correction of 1% for each 10°F difference between the existing inlet air temperature and the standard air temperature for the altitude flown. If the air is colder than standard, the engine will produce more power; if it is warmer, the power will be less. In this example, the actual inlet air temperature at 3,500 feet is 30°F, and the standard altitude temperature at this altitude, point T_S, is 46°F. Since the air is 16° colder than standard, add 1.6% of 144, or 2.3 horsepower. At 3,500 feet the corrected power is 144 + 2.3 or 146.3 BHP.

Engine Thrust

The purpose of a reciprocating engine in an airplane is to drive a propeller to produce the maximum amount of thrust. As previously stated, the horsepower produced by a reciprocating engine is determined in part by its RPM, with the highest RPM producing the most power. But the propeller's efficiency decreases as the blade tip speed approaches the speed of sound.

Most of the more powerful aircraft engines drive the propeller through a set of reduction gears, which produces the best engine power and propeller efficiency.

Reduction gears allow the engine to turn fast enough to develop the required power while at the same time keeping the propeller speed low. When the propeller tips reach the speed of sound, they become both noisy and aerodynamically inefficient.

Most of the early aircraft engines were heavy and had a large piston displacement. They turned long propellers at a slow speed to produce their thrust. Modern engines produce power at a higher RPM, turn a much smaller propeller, and therefore produce less thrust per horsepower.

An engine with a large piston displacement that developed 90 horsepower at 1,400 RPM produced as much thrust as a modern engine with a smaller displacement that develops 110 horsepower at 2,550 RPM.

Specific Fuel Consumption

One of the more meaningful comparisons of aviation reciprocating engines is the specific fuel consumption. This is the number of pounds of fuel burned per hour to produce one horsepower.

Figure 2-13 is a typical brake horsepower-brake specific fuel consumption curve. To find the number of gallons of fuel burned per hour by this engine under a typical cruise condition of 2,300 RPM, follow these steps:

1. Follow the vertical line for 2,300 RPM, at the bottom of the chart, upward until it intersects the Propeller Load Horsepower Curve.
2. From this intersection, draw a horizontal line to the left until it intersects the Brake Horsepower Index. This intersection occurs at 111 BHP.
3. Follow the vertical line for 2,300 RPM upward until it intersects the Propeller Load Specific Fuel Consumption Curve.
4. From this intersection, draw a horizontal line to the right until it intersects the Specific Fuel Consumption Pound/Brake Horsepower/Hour Index. This intersection occurs at 0.46.
5. When this engine turns at 2,300 RPM, it produces 111 brake horsepower and burns 0.46 pound of fuel per hour for each horsepower.
6. The engine burns 51.06 pounds of fuel per hour ($111 \cdot 0.46 = 51.06$), and since aviation gasoline weighs 6 pounds per gallon, the engine burns 8.51 gallons per hour while producing 111 BHP at 2,300 RPM.

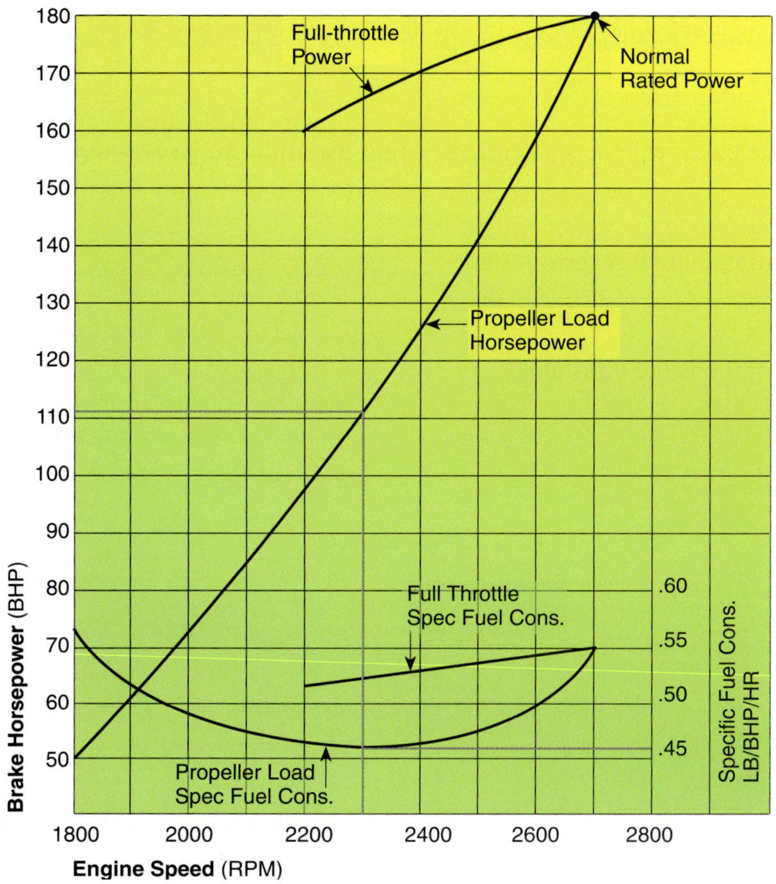

Figure 2-13. *A typical Brake Horsepower – Brake Specific Fuel Consumption curve*

BSFC (brake specific fuel consumption). The number of pounds of fuel burned per hour for each horsepower produced at the propeller shaft.

In Figure 2-13, note that below 2,300 RPM, the engine is not developing as much horsepower as it is capable of, considering the amount of fuel being burned, and above this speed the friction horsepower required to drive the engine has increased and the brake horsepower has decreased. The most efficient speed for this engine is 2,300 RPM.

When the throttle is fully open, the engine will develop 180 BHP, and because of the additional power needed just to turn the engine, the BSFC rises to 0.55. At full throttle, the engine will burn 16.5 gallons of fuel per hour.

STUDY QUESTIONS: RECIPROCATING ENGINE THEORY

Answers are on Page 90. Page numbers refer to chapter text.

1. Useful work is done in a heat engine when the chemical energy is released into the air and causes it to ____________________________ . *Page 19*

2. The energy release cycle used by a reciprocating engine is the constant-____________________________ (pressure or volume) cycle. *Page 20*

3. The energy release cycle used by a turbine engine is the constant-____________________________ (pressure or volume) cycle. *Page 20*

4. The Otto cycle of energy release is a constant-_____________________ (pressure or volume) cycle. *Page 20*

5. The five events in their proper order that take place in an Otto cycle are:
   a. ______________________________
   b. ______________________________
   c. ______________________________
   d. ______________________________
   e. ______________________________
   Page 21

6. The intake valve opens _______________________ (before or after) top center on the ____________________ stroke. *Page 22*

7. The intake valve closes _______________________ (before or after) bottom center on the ____________________ stroke. *Page 22*

8. The exhaust valve opens _______________________ (before or after) bottom center on the ____________________ stroke. *Page 22*

9. The exhaust valve closes _______________________ (before or after) top center on the ____________________ stroke. *Page 22*

10. Ignition occurs ________________________ (before or after) top center on the ____________________ stroke. *Page 22*

11. Valve overlap is the time that both the intake and exhaust valves are off of their seat. This occurs at the end of the ____________________ stroke and the beginning of the ____________________ stroke. *Page 22*

12. It __________________ (is or is not) possible to directly measure the amount of indicated horsepower an engine is producing. *Page 26*

13. It __________________ (is or is not) possible to directly measure the amount of brake horsepower an engine is producing. *Page 28*

14. It __________________ (is or is not) possible to directly measure the amount of indicated mean effective pressure in the cylinder of a reciprocating engine. *Page 26*

15. It ____________________ (is or is not) possible for a naturally aspirated engine to have a volumetric efficiency of 100%. *Page 33*

16. If both the intake and exhaust valve clearances are too great, the volumetric efficiency will be ____________________ (increased or decreased). *Page 32*

17. The bore of a "square" engine is __________________ (longer than, shorter than, or the same as) the stroke. *Page 33*

18. The ratio of the volume of a cylinder above the piston at the bottom of its stroke to the volume with the piston at the top of its stroke is called the ____________________ ratio of the engine. *Page 34*

19. Increasing the compression ratio of an engine ______________________ (increases or decreases) its thermal efficiency. *Page 34*

20. The practical limit to the compression ratio an engine can tolerate is determined by the ________________________ characteristics of the fuel. *Page 34*

21. The two variables over which a pilot has control that determine the horsepower produced by an aircraft engine are:
    a. ______________________________
    b. ______________________________
    Page 30

22. A measure of the number of pounds of fuel burned per hour to produce each brake horsepower is called the ____________________________ of the engine. *Page 38*

Reciprocating Engine Requirements

Engines used in FAA-certificated aircraft must meet the most rigid requirements imposed on any reciprocating engine. Safety is always the paramount consideration, and it takes precedence over all others. But it is also important that the engine meets the requirements established for its particular use. Some of these requirements are discussed below.

Reliability

No engine would be of much value if it could not be depended upon to perform as its manufacturer has promised. Early aircraft engines were not reliable, and pilots had to always be aware of a suitable forced landing spot. It was only with the advent of air-cooled radial engines (such as the 9-cylinder Wright J-5 that allowed Charles Lindbergh to make his famous nonstop 33-hour flight between New York and Paris in 1927), that the public recognized aircraft engines as reliable, and saw aviation as a viable means of transportation.

The early 1930s saw several endurance flights, with the record set by the Key brothers of Mississippi during a flight using air-to-air refueling of 653 hours. Their 5-cylinder Wright J6-5 air-cooled radial engine ran nonstop for more than 27 days. Almost all of the endurance flights in this era were made with air-cooled radial engines.

Durability

TBO (time between overhauls). The recommended maximum number of hours, established by an engine manufacturer, that an engine can run between overhauls without normal wear causing parts of the engine to wear beyond safe limits.

One of the outstanding advancements in aircraft engine development has been durability, which is a measure of engine life.

The comparison in Figure 2-14 between a typical World War I engine and a modern general aviation engine shows the great strides that have been made in durability as well as other engine parameters. The Curtiss OX-5 engine of World War I had a time between overhauls (TBO) of only 100 hours, but

Parameters	Curtiss OX-5	Lycoming O-235
Engine type	V	Horiz. Opposed
Horsepower	90	110
No. cylinders	8	4
Displacement	503 cubic inches	233 cubic inches
Weight	400 pounds	207 pounds
Cooling	Water	Air
RPM	1,400	2,550
Propeller length	96 inches	72 inches
Fuel consumption	9.0 GPH	6.0 GPH
Time Between Overhaul	100 hours	2,000 hours

Figure 2-14. *Comparison between a World War I engine and a modern engine of comparable horsepower*

most did not make that; many modern engines can be expected to run for up to 2,000 hours before they must be overhauled.

While the TBO of an engine is measured in hours of operation, there is more than just hours involved. The type and frequency of operation determine the useful life of an engine. An engine used for student training or agricultural operation has a shorter TBO than an engine operated by experienced pilots for frequent long flights at cruise power. Engines operated for only an hour or so a month will usually have a short TBO because of the rust and corrosion that attack the engine when it is not operating.

Operating Flexibility

Operating flexibility is the ability of an engine to run smoothly and give the desired performance at all speeds from idle to full power.

An engine with operating flexibility is able to function efficiently through all variations of atmospheric conditions the pilot will likely encounter.

Specific Weight

New and stronger materials, and improved design technology have consistently lowered the specific weight, or weight per horsepower, of aircraft engines. The specific weight of an engine — the ratio of its weight to the brake horsepower it produces—has decreased from more than 14 pounds per horsepower for the engine used in the Wright *Flyer*, to less than one pound per horsepower. Most FAA-certificated horizontally opposed engines used in general aviation aircraft have a specific weight of just under two pounds per horsepower.

specific weight. The ratio of the weight of an aircraft engine to the brake horsepower it develops.

Streamlineability

Before airplanes became efficient transportation machines with speed as their important feature, streamlining was not considered essential. Air-cooled engine cylinders were not covered, and protruded into the air stream. Radial engines were especially difficult to streamline because of their large frontal area and the odd shape of the valve mechanism on the cylinder heads.

One of the first successful approaches to reducing the drag of a radial engine was the Townend ring, which was a narrow-chord cowl around the outside of the engine. Townend rings have an airfoil shape that smooths the airflow over the engine. The next major step was the long-chord, full NACA cowling, that has been used almost exclusively over radial engines since the late 1930s. This cowling, developed by the National Advisory Committee for Aeronautics (NACA), the forerunner of NASA, assures a good flow of cooling air over the engine cylinders and actually produces a forward aerodynamic force.

NACA cowling. A long-chord cowling used over a radial engine. The forward portion of this cowling has an aerodynamic shape that produces a forward pull, and the rear portion extends back to fair in with the fuselage. There is a narrow peripheral gap between the rear of the cowling and the fuselage for the cooling air to escape. Some NACA cowlings have controllable flaps over this opening to control the amount of cooling air that flows through the engine.

NACA. National Advisory Committee for Aeronautics. NACA is the forerunner of NASA.

NASA. National Aeronautics and Space Administration.

STUDY QUESTIONS: RECIPROCATING ENGINE REQUIREMENTS

Answers are on Page 90. Page numbers refer to chapter text.

23. The ability of an engine to perform as the manufacturer has promised is known as the ______________________ of the engine. *Page 42*

24. The durability of an aircraft engine may be indicated by the manufacturer's recommended _________________________ . *Page 42*

Reciprocating Engine Configurations

The aircraft industry is different from the automobile industry in that no airframe manufacturers build their own engines. This allows the engine manufacturers a great amount of freedom to develop engines suitable for aircraft built by many different manufacturers.

Cylinder Arrangement

One of the most obvious ways to classify aircraft engines is by the arrangement of the cylinders on the crankcase. Each arrangement has advantages and disadvantages; the most commonly used arrangements are discussed below.

In-Line Engines

One of the simplest configurations is to arrange all the cylinders in a straight line on the crankcase. This configuration has a minimum frontal area and is easy to streamline, but its weight is relatively high because of the long, heavy crankshaft this engine requires.

There are two types of construction used in in-line engines: individual cylinders bolted to the crankcase, or all the cylinders cast into a single aluminum alloy block with steel liners in each of the cylinders.

The cylinders of an in-line engine may be mounted above the crankcase in upright engines or below the crankcase in inverted engines. Upright engines are easier to lubricate, but the pilot will have unrestricted forward visibility and greater ground clearance for propellers with an inverted engine. In-line engines can be either liquid- or air-cooled; air cooled is the more popular.

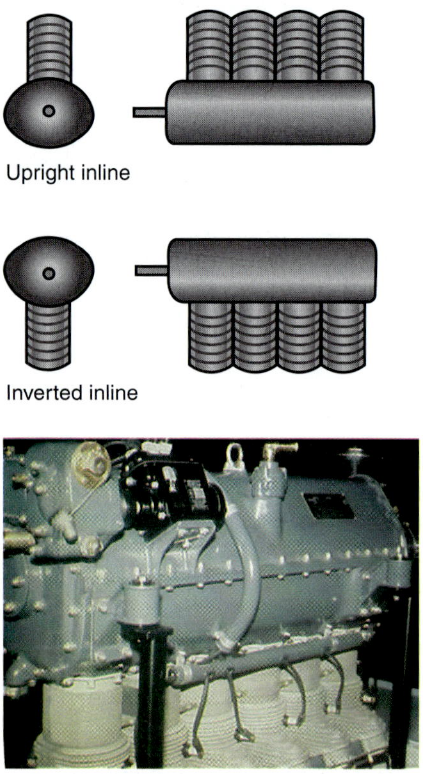

Figure 2-15. *In-line engines have all of the cylinders arranged in a single row, either above or below the crankcase.*

in-line engine. A reciprocating engine with all of the cylinders arranged in a straight line.

V-Engines

A logical step from the in-line engine when more power was needed was the V-engine that had two banks of in-line cylinders mounted on the same crankcase, and separated by an angle of between 45° and 90°. One cylinder in each bank shared a throw of the crankshaft. Some of the earliest successful aircraft engines were liquid-cooled V-engines. This configuration was the most widely used until the air-cooled radial engine took over in the late 1920s. The two most popular engines built in the United States during World War I were the 90-horsepower water-cooled V-8 Curtiss OX-5, and the 400-horsepower, water-cooled Liberty V-12. Almost all V-engines were liquid-cooled, but a few have been air-cooled. V-engines may be upright, with the cylinders above the crankcase or inverted with the cylinders below.

Liquid-cooled V-12 engines made by Allison, Packard, and Rolls-Royce were used as late as World War II to power such fighter airplanes as the Lockheed P-38 Lightning and the North American P-51 Mustang.

Figure 2-16. *V-engines are essentially two banks of in-line cylinders mounted on a single crankcase. The cylinder banks are separated by an angle of between 45° and 90°. One piston in each bank of cylinders connects to each throw of the crankshaft. V-engines may be either upright or inverted. This arrangement improves Weight to Horsepower ratio.*

inverted engine. An in-line or V-engine in which the cylinders are mounted below the crankshaft.

V-engine. A form of reciprocating engine in which the cylinders are arranged in two banks. The banks are separated by an angle between 45° and 90°.

rotary radial engine. A radial engine with the crankshaft attached to the airframe and the propeller mounted rigidly to the crankcase. The propeller, crankcase, and cylinders revolve as a unit.

Radial Engines

The crankshaft in an in-line or V-engine is the heaviest single component, and in an effort to reduce weight, the radial engine was developed. An odd number of cylinders, 3, 5, 7, 9, or 11 are mounted radially, like the spokes of a wheel, around a small central crankcase.

During World War I, one popular configuration of aircraft engine was the rotary radial. The crankshaft was rigidly attached to the airframe, and the propeller and cylinders which were attached to the crankcase, rotated as a unit. The rotation of the cylinders made air cooling practical. These engines turned slowly, and some of them weighed less than two pounds per horsepower, which was exceptionally good for World War I engines.

The crankcase of a static radial engine attaches to the airframe, and the piston in one cylinder connects to the single-throw crankshaft with a master rod; the pistons in all of the other cylinders connect to the master rod by link, or articulating, rods.

Radial engines producing less than 200 horsepower usually have three or five cylinders. Engines producing between 200 and 1,000 horsepower quite often have either seven or nine cylinders. When more horsepower is needed, two rows and sometimes four rows of cylinders are used. Two-row radial engines have two rows of either seven or nine cylinders mounted on the crankcase and drive a two-throw crankshaft. The largest practical radial engine is the Pratt & Whitney R-4360 Wasp Major, which has four rows of seven cylinders driving a four-throw crankshaft. The military version of this engine weighs 3,670 pounds and, with antidetonation injection, produces 3,800 horsepower at 2,800 RPM.

Almost all radial engines are air-cooled, and by the careful use of NACA cowlings, the cooling drag has been reduced to a level compatible with the more streamlined in-line or V-engines.

Figure 2-17. *The cylinders of a radial engine radiate out from the crankcase like the spokes of a wheel.*

Horizontally Opposed Engines

Radial engines have the advantage of being lightweight for the amount of power they produce, but they are difficult to streamline in a narrow fuselage. An in-line engine is easy to streamline in a narrow fuselage, but it has a high weight for its power.

Just before and during World War II, horizontally opposed four-and six-cylinder engines were developed and provided reliable service in many of the liaison aircraft. After the war ended and production of private aircraft resumed, it was seen that a six-cylinder opposed engine could replace a seven-cylinder radial, and be more easily streamlined. The opposed engine allowed much better forward visibility, and its width was no problem since almost all production aircraft had side-by-side seating. Because it is inherently a smooth engine, it uses fewer and larger cylinders than a radial engine of comparable horsepower, and it is similar in weight.

Horizontally opposed engines generally turn faster than radial engines of the same power, and the higher RPMs require a shorter propeller to keep the tip speed below the speed of sound. This works fine for the tricycle landing gears used on most of the airplanes built since World War II.

The horizontally opposed engine came along at the right time and with the right advantages, and today, all FAA-certificated reciprocating engines built in the United States are horizontally opposed. Most of these engines have either four or six cylinders.

liaison aircraft. A type of light military aircraft made popular during World War II because of its ability to land and takeoff from unimproved terrain. Liaison aircraft have been replaced with helicopters.

radial engine. A reciprocating engine in which the cylinders are arranged like the spokes of a wheel radiating out from a small central crankcase.

propeller end. The end of a reciprocating engine to which the propeller is attached.

antipropeller end. The end of a reciprocating engine that does not attach to the propeller.

accessory end. The end of a reciprocating engine on which many of the accessories are mounted.

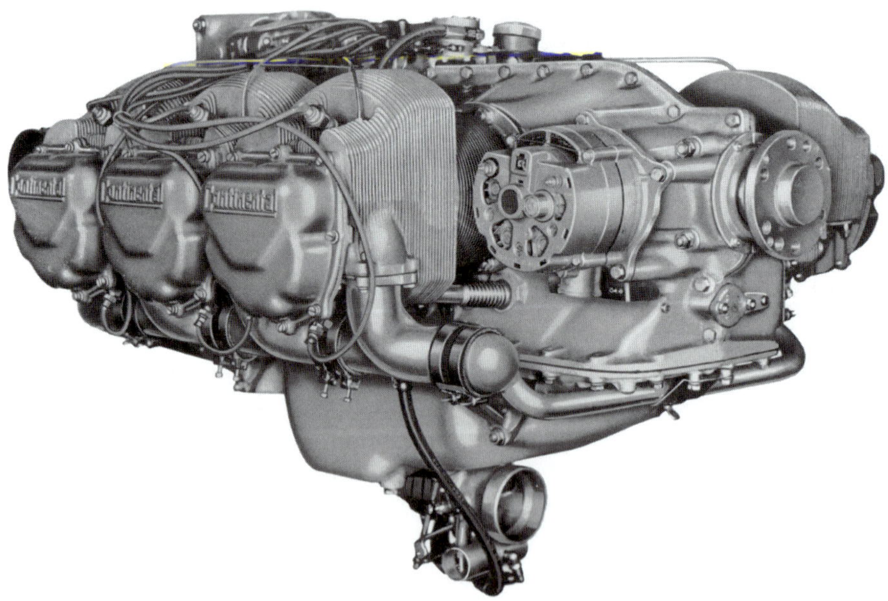

Figure 2-18. *The horizontally opposed engine has the advantage of smoothness of operation, small frontal area, and light weight.*

Cylinder Numbering

It is extremely important when studying the operation of a reciprocating engine that you understand the way the cylinders are numbered. Consider the following ways the cylinders of radial, in-line, V, and horizontally opposed engines are numbered.

Radial Engines

All radial engines number the cylinders with the top cylinder as number 1 and progress consecutively around the engine in the direction of crankshaft rotation. In two-row radial engines, the cylinders with odd numbers are in the rear row, and the even numbers in the front row.

In-Line and V-Engines

The front and rear sides of an engine are not determined by the way the engine is mounted in the aircraft, but are designated as the propeller and antipropeller, or accessory, ends. The left and right side are determined by looking toward the propeller from the accessory end.

Cylinder number one on an in-line engine is the cylinder nearest the accessory end, and the numbers progress toward the propeller.

Cylinders on a V-engine are numbered in the same way, but you identify the left and right banks by looking from the accessory end toward the propeller, regardless of the way the engine is installed in the aircraft.

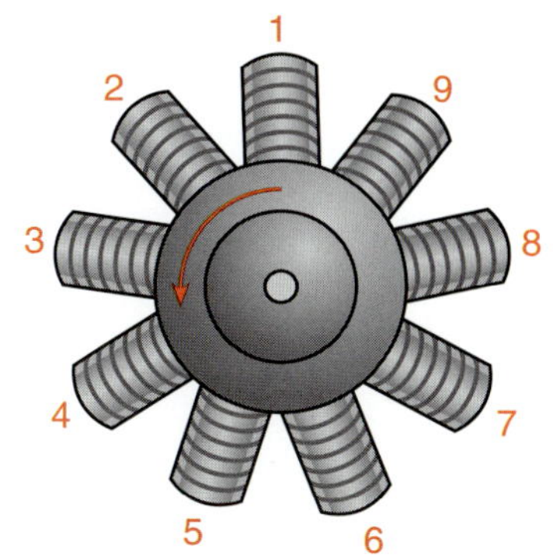

9-Cylinder single row

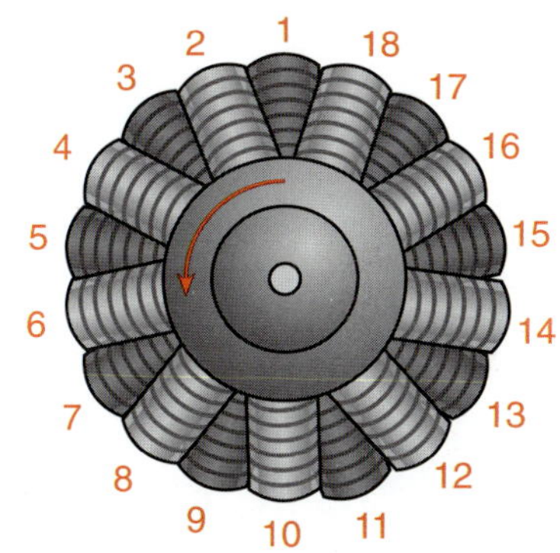

18-Cylinder twin row

Figure 2-19. *The cylinders of a radial engine are numbered consecutively from the top cylinder around the engine in the direction of rotation. In a two-row radial engine all of the odd-numbered cylinders are in the rear row, and all of the even-numbered cylinders are in the front row.*

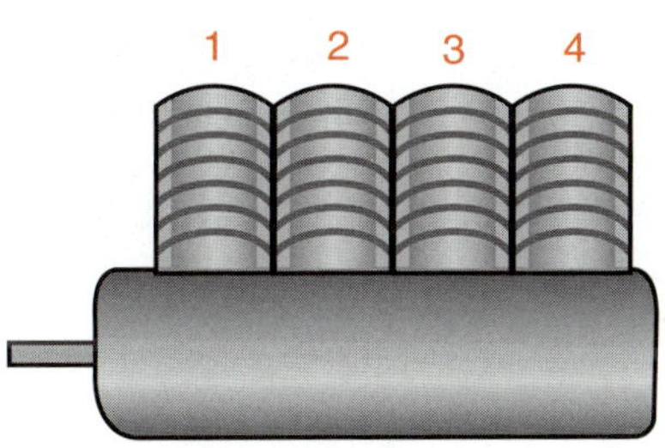

Inline engine

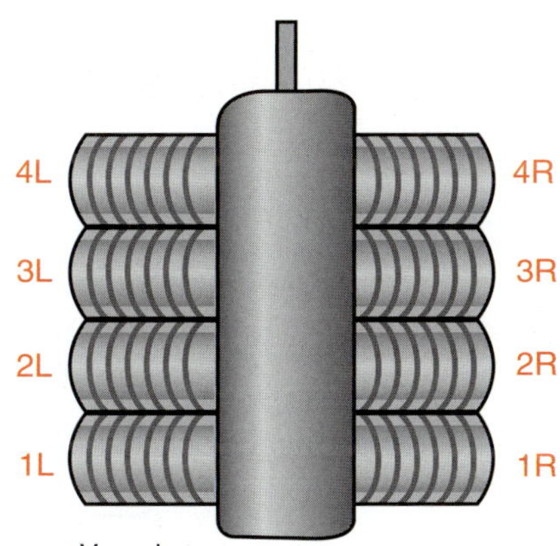

V engine

Figure 2-20. *The cylinders of an in-line or V-engine are numbered beginning at the accessory end of the engine. The left and right banks of a V-engine are determined by looking from the accessory end toward the propeller.*

Horizontally Opposed Engines

The two major manufacturers of horizontally opposed engines in the United States are Teledyne-Continental Motors (TCM) and Textron-Lycoming. The engines produced by both manufacturers have the cylinders on one side slightly offset from those on the opposite side so each connecting rod has its own throw on the crankshaft. The right-hand bank of cylinders on TCM engines is offset to the rear of the cylinders on the left side, and cylinder number 1 is the right rear cylinder. The right-hand bank of cylinders on Textron-Lycoming engines are offset forward, and cylinder number 1 is the right front cylinder.

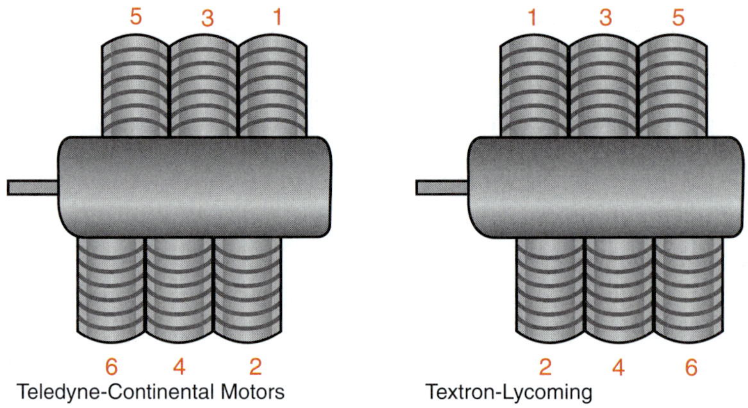

Figure 2-21. *Teledyne-Continental Motors and Textron-Lycoming number the engine cylinders differently, but in both engines, cylinder number 1 is on the right side.*

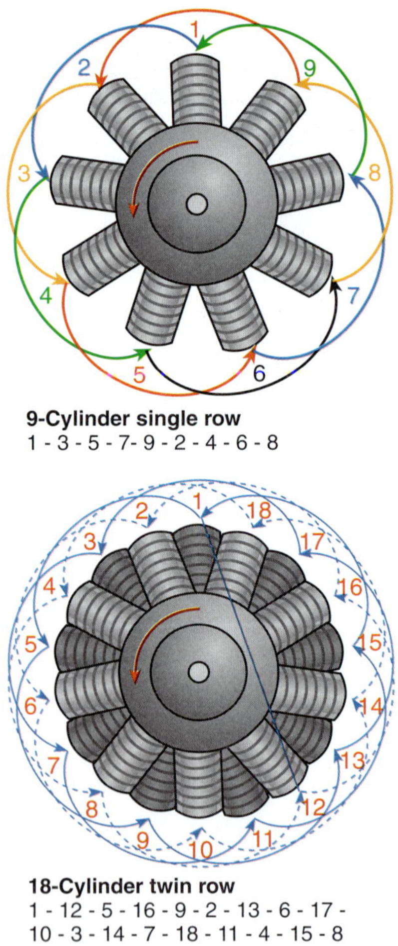

Figure 2-22. *Firing orders and firing sequences for 9- and 18-cylinder radial engines.*

Firing Order

The smoothness with which a reciprocating engine runs is determined by the spacing and the timing of the firing impulses. An engine is inherently smooth when all of the firing impulses are separated by an equal number of degrees of crankshaft rotation; the closer together the firing impulses, the smoother the engine.

Radial Engines

A single-row radial engine fires cylinder number 1, then fires every other cylinder in the direction of rotation. For a nine-cylinder engine, the firing order is 1-3-5-7-9-2-4-6-8. All cylinders fire in two revolutions (720°) of crankshaft rotation.

The firing order for a twin-row radial engine looks complicated, but it is logical when you consider these steps:

1. The top cylinder in the rear row fires first. This is cylinder 1.
2. The next cylinder to fire is in the front row, following the direction of rotation from the cylinder that is directly across the engine from cylinder 1. This is cylinder 12.
3. The next cylinder to fire is in the rear row and is the second cylinder in the direction of rotation from cylinder 1. This is cylinder 5.
4. The next cylinder in the front row to fire is the second cylinder in the direction of rotation from cylinder 12. This is cylinder 16.
5. The sequence continues with every other cylinder in the rear row firing between every other cylinder in the front row. *See* Figure 2-22.

An easy way to determine the firing order of an 18-cylinder twin-row radial engine is to start with cylinder 1 and add 11 or subtract 7, whichever is needed to keep the number between 1 and 18 (1-12-5-16-9-2-13-6-17-10-3-14-7-18-11-4-15-8). The same method can be used to find the firing order of a 14-cylinder twin-row engine. Add 9 or subtract 5 to keep the numbers between 1 and 14 (1-10-5-14-9-4-13-8-3-12-7-2-11-6).

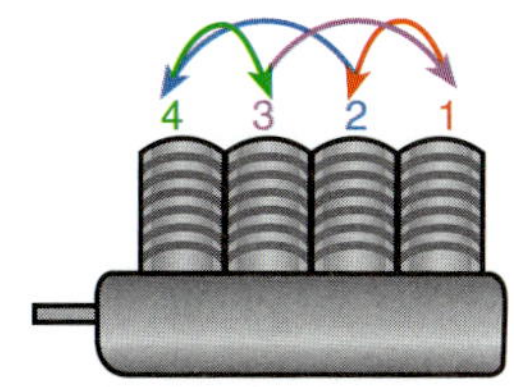

Four-cylinder inline
1 - 2 - 4 - 3

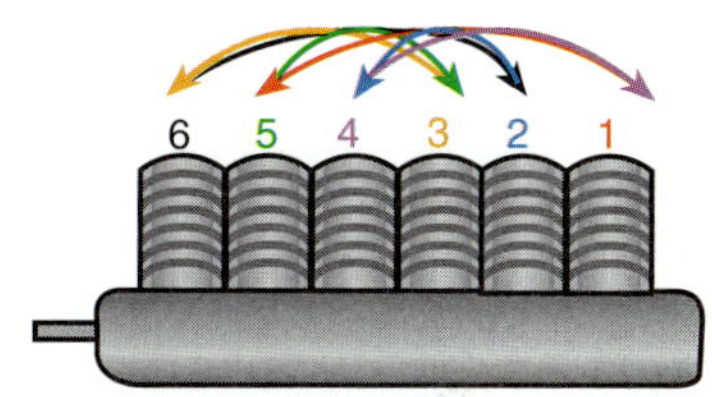

Six-cylinder inline
1 - 5 - 3 - 6 - 2 - 4

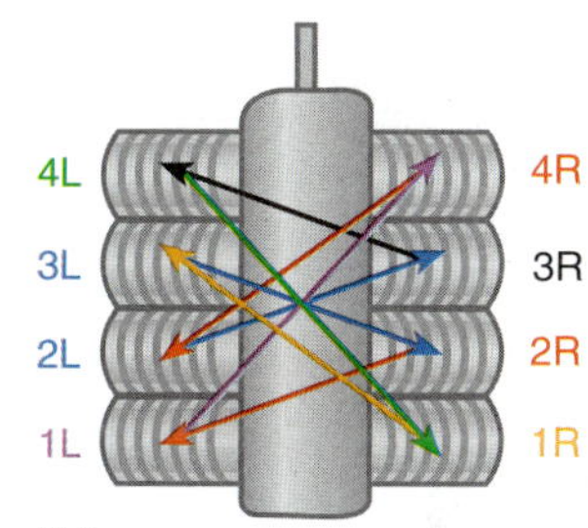

V-8
1L - 4R - 2L - 3R - 4L - 1R - 3L - 2R

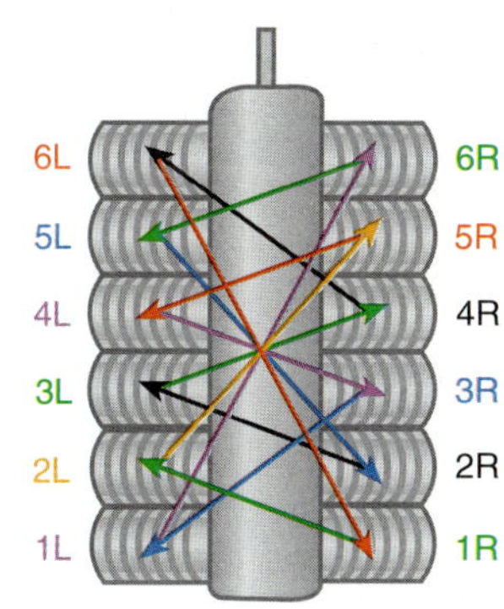

V-12
1L - 6R - 5L - 2R - 3L - 4R - 6L - 1R - 2L - 5R - 4L - 3R

Figure 2-23. *Firing order for in-line and V-engines.*

In-Line and V-Engines

The crankshaft in a four-cylinder in-line engine has two sets of throws, 180° apart. The throws for cylinders 1 and 4 are together, and the throws for cylinders 2 and 3 are together, and are 180° from those for 1 and 4. The pistons in cylinders 1 and 4 are at the top of their stroke at the same time the pistons in cylinders 2 and 3 are at the bottom of their strokes. This movement of the pistons gives the engines a firing order of 1-2-4-3 or 1-3-4-2.

V-8 engines are essentially two four-cylinder banks on a single crankcase, with one cylinder in each bank sharing a crankshaft throw. The left bank fires 1-2-4-3, and the right bank fires 4-3-1-2.

The crankshaft used in a six-cylinder in-line engine has three sets of throws, 120° apart. The throws for cylinders 1 and 6 are together, and 120° from those are the throws for cylinders 2 and 5. 120° from 2 and 5 are the throws for cylinders 3 and 4. With this arrangement, the pistons in cylinders 1 and 6 come to the top of their stroke together; then 120° later, pistons 2 and 5; and then 120° later, pistons 3 and 4. This type of crankshaft gives the engine a firing order of 1-5-3-6-2-4.

A V-12 engine has two banks of six cylinders firing the same sequence, but the right bank starts its firing at the opposite end of the engine.

Horizontally Opposed Engines

Both Teledyne-Continental Motors (TCM) and Textron-Lycoming make four- and six-cylinder horizontally opposed engines. Their four-cylinder engines use a 180° crankshaft and their six-cylinder engines use a 60° crankshaft. Textron-Lycoming also makes an eight-cylinder horizontally opposed engine with a 90° crankshaft.

The right-hand bank of cylinders on TCM engines are offset to the rear of the cylinders on the left side, and cylinder number 1 is the right rear cylinder. The firing order for a four-cylinder TCM engine is 1-3-2-4. The firing order for a six-cylinder TCM engine is 1-6-3-2-5-4.

The right-hand bank of cylinders on Textron-Lycoming engines are offset forward, and cylinder number 1 is the right front cylinder. The firing order for a Textron-Lycoming engine is 1-3-2-4, for a six-cylinder engine it is 1-4-5-2-3-6, and for an eight-cylinder engine it is 1-5-8-3-2-6-7-4.

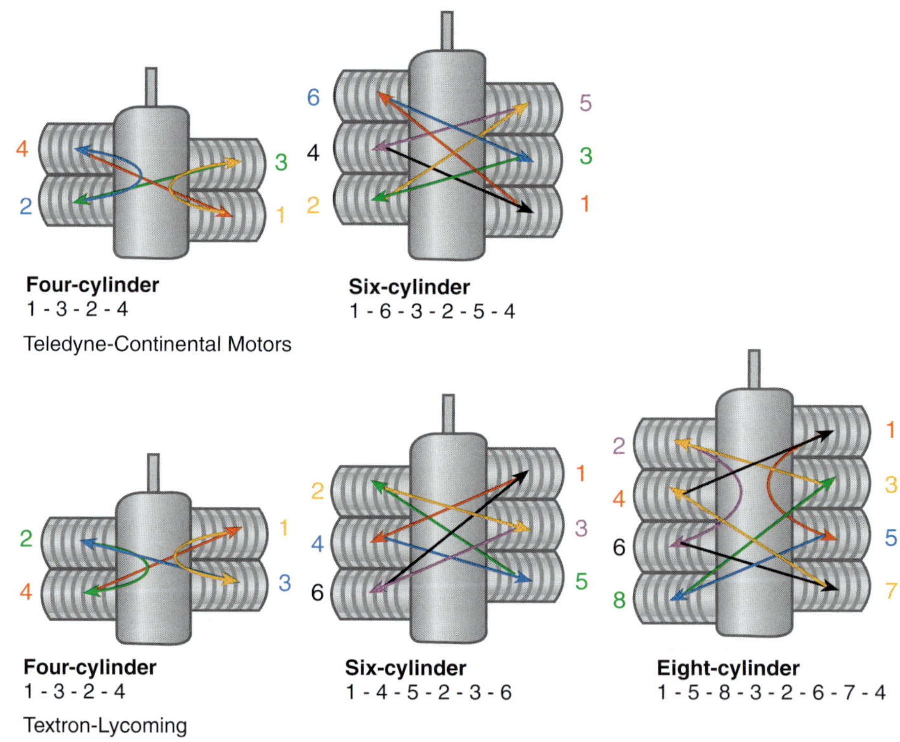

Figure 2-24. *Firing order for horizontally opposed engines*

Cooling Systems

In a reciprocating engine, about 12% of the energy in the burned fuel is absorbed as heat in the engine, and some form of cooling system must be used to remove it.

Most of the heat is removed from the cylinders by either air or liquid cooling. Air cooling transfers the heat directly to the air passing through fins on the cylinders. Liquid cooling first transfers the heat to a coolant flowing through jackets around the cylinders and then carries it outside the aircraft, where it is transferred to the air flowing through a radiator.

The lubrication system also removes some heat from the engine. Lubricating oil inside the engine carries heat from the cylinder walls and the bottom of the pistons to the outside of the engine, where it is transferred into the air flowing through the oil cooler.

Liquid Cooling

Many of the early in-line and V-engines used in aircraft were water-cooled. Water was pumped through jackets around the cylinders to pick up heat, and then pumped out of the engine to a radiator where it gave up this heat to the outside air. Water is an inefficient medium for cooling, as it boils at too low a temperature, especially at altitude, and a large radiator was needed for removing the heat. It was found that by using either pure ethylene glycol, a type of alcohol, or a mixture of ethylene glycol and water, and by holding the cooling system under pressure, that the operating temperature of the engine could be raised, and the radiators could be made much smaller.

thermal shock. The sudden change in engine operating temperature that occurs when engine power is suddenly reduced at the same time the airspeed, thus the cooling, is increased. Thermal shock occurs when an aircraft is required to rapidly descend to a lower altitude.

Almost all radial and horizontally opposed engines have been air-cooled, but Teledyne-Continental has recently introduced a series of liquid-cooled horizontally opposed engines that appears to have a great future. These liquid-cooled engines have a higher compression ratio and a lower specific fuel consumption than comparable air-cooled engines. Liquid-cooled engines are also much less susceptible to thermal shock from rapid descents from altitude.

Air Cooling

Early water-cooled engine installations were plagued with leaks in the plumbing, and were heavy. In the mid-1920s, cylinders were designed that allowed air-cooled engines to develop power comparable to large water-cooled engines. Efficient cowlings and pressure cooling have made air-cooled engines the standard for aircraft powerplants.

pressure cooling. A method of air cooling in which the engine is housed in a tight-fitting cowling that is divided into two compartments. Ram air enters one compartment and is directed by baffles through the fins on the cylinders to the other compartment. The exit cowling from the discharge compartment is flared so air flowing over it produces a low pressure that helps the ram pressure force the maximum amount of air through the fins.

sump (lubricating system component). The reservoir on an aircraft engine that holds the lubricating oil.

Lubrication Systems

The lubrication system of an aircraft engine reduces internal friction, cushions the impacts caused by the power strokes, and removes part of the wasted heat. There are two basic types of lubrication systems used on aircraft reciprocating engines: wet-sump and dry-sump.

Wet-sump engines carry all the lubricating oil in an integral sump that is a part of the crankcase. This is similar to most automobile engines. Many horizontally opposed engines use wet-sump systems.

Dry-sump lubrication systems are normally used by radial engines, with their small crankcases, and inverted in-line and V-engines whose crankcases are above the cylinders. Some horizontally opposed engines mounted in shallow nacelles, or vertically in helicopters, also use dry-sump systems. The oil is stored in an external tank, and pumped through the engine, where it performs its lubrication and cooling functions. It is then returned to the tank and stored until it is pumped through the engine again. The circulation of oil from the tank through the engine and back allows the oil to be continually filtered and cooled.

Engine Identification

Some manufacturers have identified their engine series with names; for example, the Wright Aeronautical Corporation named its engines after meteorological conditions such as Whirlwind and Cyclone. Pratt & Whitney used insect names such as Wasp and Hornet. Presently in the United States, reciprocating engines are identified with a letter to indicate the cylinder arrangement, and a number to indicate the piston displacement in cubic inches. This is followed by other letters to indicate characteristics of the specific engine.

An engine identified as a TSIO-520-C is a turbocharged (TS), fuel injected (I), horizontally opposed (O) engine with 520 cubic inches of piston displacement, and it is the third version (C) of this engine.

Some of the most commonly used identifying letters are:

R – Radial
O – Opposed
A – Aerobatic
G – Geared nose section
H – Horizontal, for helicopter
I – Fuel-injected
L –Left-hand engine rotation
S – Supercharged
T or TS – Turbocharged*
V – Vertical for helicopter

* Currently, Lycoming engines use "T" and Continental uses "TS."

STUDY QUESTIONS: RECIPROCATING ENGINE CONFIGURATIONS

Answers are on Page 90. *Page numbers refer to chapter text.*

25. An engine with all of the cylinders arranged in a single straight row is called a/an ____________________ engine. *Page 44*

26. The odd-numbered cylinders in a twin-row radial engine are in the ____________________ (front or rear) row. *Page 47*

27. The firing order of a six-cylinder Textron-Lycoming engine is ______________________________ . *Page 50*

28. The firing order of a six-cylinder Teledyne-Continental engine is ______________________________ . *Page 50*

29. Liquid-cooled engines use a mixture of water and ____________________ as the coolant. *Page 51*

30. An engine that carries its lubricating oil supply in the engine crankcase is called a ________________ engine. *Page 52*

31. Give the meaning of each of the letters and numbers in the identification of a TIO-360 engine.
    a. T means ______________________________
    b. I means ______________________________
    c. O means ______________________________
    d. 360 means ______________________________
    Page 52

Horizontally Opposed Engine Construction

The horizontally opposed engine is the only configuration of FAA-certificated reciprocating engines currently manufactured in the United States. For this reason we will consider it first. Then we will consider the radial engine, because it is still being used in agricultural applications and for some heavy cargo hauling operations.

Cylinders

The chemical energy in fuel is changed into heat energy, and then into mechanical work in the cylinders of a reciprocating engine. The cylinders receive the greatest amount of stress and are exposed to the highest temperatures.

Most cylinders used on horizontally opposed engines are air-cooled. However, Teledyne-Continental Motors is currently manufacturing liquid-cooled horizontally opposed engines called the Voyager series.

Some of the early horizontally opposed engines had the cylinders cast integrally with the crankcase halves and had removable heads. All the modern engines have separate cylinders with the heads screwed and shrunk onto the barrels.

boss. An enlarged area in a casting or machined part. A boss provides additional strength to the part where holes for mounting or attaching parts are drilled.

A modern air-cooled cylinder assembly consists of a high-strength steel barrel and a cast-aluminum head fitted with valve seats, valve guides, and threaded inserts for the spark plugs. There are also rocker shaft bosses with bushings pressed into them in which the rocker shafts ride. Each of these components will be discussed. *See* Figure 2-25.

Cylinder Barrels

The cylinder barrel is machined from a high-strength chrome-molybdenum or chrome-nickel-molybdenum steel forging. The bottom end of the cylinder barrel has a skirt machined to fit into the crankcase, and the heavy flange just above the skirt is drilled with holes through which the mounting studs pass to attach the cylinder to the crankcase.

The cylinder skirt allows the reduction of the engine width by placing a portion of the cylinder stroke area inside the crankcase. For inverted cylinders, the skirt also reduces the amount of engine oil that would otherwise drain down into the combustion area after engine shutdown.

choke-ground cylinder. A cylinder of a reciprocating engine that is ground so that its diameter at the top of the barrel is slightly smaller than the diameter in the center of the stroke.

The large mass of metal in the cylinder head absorbs enough heat to cause the top end of the barrel to expand more than the rest of the barrel. At normal operating temperature, the diameter of a choke-ground cylinder is uniform throughout.

Cooling fins are machined on the outside of the barrel for most of its length, and the outer surface of the top of the barrel is threaded so it can be screwed into the cylinder head.

Many cylinder barrels are ground so that their diameter at the top of the bore is smaller than the diameter at the center or bottom. The reason for this process, called choke-grinding, is that the highest operating temperature in the cylinder is at the top, where the barrel screws into the head, and when

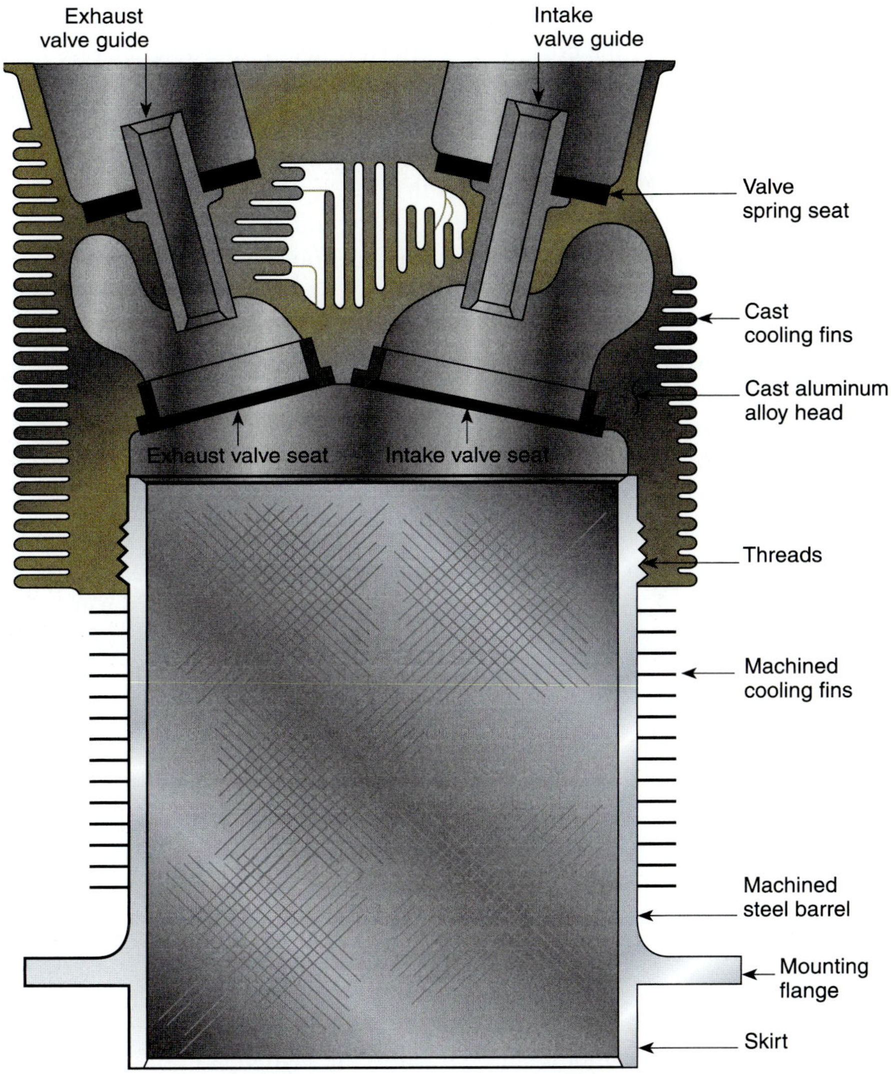

Figure 2-25. *A typical cylinder for an air-cooled horizontally opposed engine*

the engine is hot, the upper end of the cylinder expands more than the rest. The smaller diameter at the top causes the bore to become straight when the cylinder is at its operating temperature. This provision drastically reduces upper piston and ring wear, and is one of the primary changes responsible for extending engine operating time between overhauls (TBO).

The inner surface of the cylinder barrel is ground to a specified dimension, and then the surface is honed with a 45° criss-cross, or cross-hatched, finish that is specified by the engine manufacturer. The degree of surface roughness is extremely important, because it must be smooth enough that it will not cause excessive ring wear, yet rough enough that it will hold oil for

honing (cylinder wall treatment). Scratching the surface of the cylinder wall with an abrasive to produce a series of grooves of microscopic depth and uniform pattern. The honed pattern holds oil to lubricate the cylinder walls.

ring rotation. The small amount of rotation of the piston ring in its groove during each piston stroke. It has been determined that piston rings make one complete revolution approximately every five to seven minutes of engine operation.

microinches rms. A measure used for cylinder wall surface roughness. Twenty microinches rms means that the highest and lowest deviation from the average surface is 20 millionths of an inch.

rms. Root mean square. A dimension that is the square root of the average of an infinite number of varying values. An rms dimension is used to indicate the allowable surface roughness of a reciprocating engine cylinder wall.

profilometer. A precision measuring instrument used to measure the depth of the hone marks in the surface of a cylinder wall.

channel-chromed cylinders. Reciprocating engine cylinders with hard chromium-plated walls. The surface of this chrome plating forms a spider web of tiny stress cracks. Deplating current enlarges the cracks and forms channels that hold lubricating oil on the cylinder wall.

nitriding. A method of case hardening steel. Steel is placed in a retort (a sealed, high-temperature furnace), and heated to a specified temperature while surrounded by ammonia gas (NH_3). The ammonia breaks down into nitrogen and hydrogen, and the nitrogen unites with some of the alloying elements in the steel to form an extremely hard surface.

Nitriding hardens crankshaft bearing surfaces and cylinder walls in reciprocating engines. It takes place at a lower temperature than other forms of case hardening, and does not cause warping.

lubrication and proper ring-face mating to the cylinder wall during engine break-in. The angle of the crosshatch pattern is also important because this will determine the proper rate of ring rotation. If rings do not rotate, severe cylinder wall and ring damage will occur. The exact surface roughness is specified in terms of microinches rms (root mean square). For example, a surface finish of 15-25 μ in. rms means that the highest and lowest deviation from the average surface can range between 15 and 25 millionths of an inch. This is typical for a honed surface. The surface roughness is measured with an instrument called a profilometer.

Some cylinder walls are hardened, either by chrome plating or by a process called nitriding.

Chrome-plated cylinders are ground so their diameter is slightly oversize, and then the inside of the barrel is electrolytically plated with hard chromium to the required dimension. The traditional hard chrome, or channel chrome, used on cylinder walls since World War II has the characteristic of forming a spider web pattern of minute cracks, or channels, on its surface. When the plating is completed, the plating current is reversed, and some of the chrome is removed. The deplating current removes chrome from the cracks at a much faster rate than it does from the crack-free areas, or plateaus. The deplating is continued until the cracks, or channels, are about 0.0045 inch deep and 0.004 inch wide. These channels hold enough oil to assure adequate lubrication.

Chrome-plated cylinders have the advantage of better heat transfer and a surface that is harder and more corrosion- and scuff-resistant than plain steel. They have the disadvantage that the lubricating film on the surface of the chrome is not as strong as the film that forms on plain steel cylinder walls. Chromed cylinders are identified by a band or bands of orange paint around the cylinder base or by orange paint on some of the fins.

Nitriding is a process of case hardening in which the cylinder barrel is heated in an atmosphere of ammonia gas. The alloying elements in the steel combine with the nitrogen from the ammonia to form extremely hard nitrides on the surface.

Nitriding is not a plating, and it causes a dimensional growth of 0.0004 inch or less. It hardens the surface to a depth of about 0.002 inch, with the hardness decreasing gradually from the surface inward. After nitriding a cylinder, it is honed to give it the desired degree of surface roughness. Nitrided cylinders are identified by a band of blue paint around the base or by blue paint on some of the fins.

One drawback to nitriding is the tendency of cylinders to rust or corrode. Nitrided cylinders must be kept covered with a film of engine oil, and if the engine is out of service for any length of time, the cylinder walls should be protected with a coating of a tenacious preservative oil.

In more recent history, CermiChrome and CermiSteel cylinder walls have been used. The CermiChrome process is impregnating a chromed cylinder wall by rolling ceramic particles into the chrome while the cylinder is immersed in a slurry of ceramic. CermiSteel is using the same process on a steel barrel surface. These surfaces have experienced limited success.

The latest and most promising process for barrel wall longevity is CermiNil. This is a nickel-plated barrel inside diameter impregnated with ceramic particles.

The "Cermi" barrel surfaces provide good wear resistance and lubrication. The porous ceramic particles act as the lubrication reservoir in place of the crosshatching filled with carbon residue.

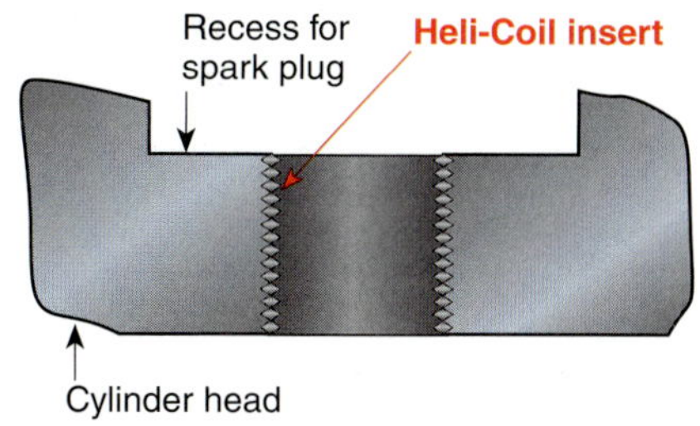

Figure 2-26. *Spark plugs screw into Heli-Coil inserts that are used to protect the threads in the soft cast-aluminum cylinder head.*

Cylinder Heads

The head of an air-cooled cylinder is cast of aluminum alloy, and its open end is internally threaded so it can be screwed onto the steel cylinder barrel.

The aluminum alloy head expands more than the steel barrel, and to prevent the head from loosening when it becomes hot, the threaded joint is designed with an interference fit. The external diameter of the threaded portion of the barrel is larger than the internal diameter of the threads inside the cylinder head. The head is heated to expand it and the barrel is chilled to shrink it. This allows the barrel to be screwed into the head, and when the head and barrel reach the same temperature, the joint will be extremely tight and able to withstand combustion pressures of up to 800 psi without leaking.

All currently produced FAA-certificated reciprocating engines have dual ignition. This requires each cylinder to have two spark plugs located in such a way that they ignite the fuel-air mixture on both sides of the cylinder, so that the flames started by both of the spark plugs will meet in the center of the piston.

Cast-aluminum-alloy cylinder heads would be damaged by the frequent changing if the spark plugs were screwed directly into the heads, so inserts are used to protect the threads. In most of the older engines, the spark plugs were screwed into bronze bushings that were shrunk and pinned into the cylinder head, but most modern engines use Heli-Coil inserts. These are special stainless steel spirals with a diamond-shaped cross section that forms the threads into which the spark plug screws. The spark plug hole is tapped with a special Heli-Coil tap and the insert is screwed into it, using a special insertion tool. Damaged inserts can be easily replaced. *See* Figure 2-26.

Heli-Coil insert. The registered trade name of a special helical insert used to restore threads stripped from a bolt hole, or to reinforce the threads in an aluminum casting.

Most combustion chamber shapes (CCS) are semi-hemispherical. The closer to hemispherical, the more even the pressure is applied to the piston head and the less possibility of detonation. The closer the CCS is to hemispherical the more likely the poppet valve head shape will be tulip or semi-tulip to minimize disruption to the shape.

Almost all modern aircraft engines use poppet valves in the cylinder heads. The cast-aluminum alloy is too soft for the valves to pound against each time they close, and excessive wear would occur as the valve stems continually move up and down when the engine is running. Special valve seats and guides are shrunk into the heads to prevent this wear. Continental is using a venturi-shaped intake valve seat to improve volumetric efficiency in some of their newest engines.

rocker arm. A pivoted arm on the cylinder head of a reciprocating engine. The pushrod forces one end of the rocker arm up, and as the other end moves down, it forces the poppet valve off of its seat.

Most horizontally opposed engines actuate the valves with pushrods and rocker arms. The rocker shaft bosses have bronze bushings pressed into the holes in which the rocker shafts ride.

The most heat in a reciprocating engine is produced in the cylinder head, and fins are cast into the head to carry this heat away. The side of the head containing the exhaust valve has a greater number of fins than the side that contains the intake valve. The fuel-air mixture flowing into the cylinder through the intake valve port helps cool that side of the cylinder. *See* Figure 2-25 on Page 55.

Valve Assemblies

poppet valve. A T-shaped valve with a circular head. Poppet valves are used to cover the intake and exhaust openings in the cylinder head of a reciprocating engine. The valves are held closed by one or more coil springs and are opened by a cam lobe or a rocker arm pushing on the end of the valve stem.

Four-cycle reciprocating engines must have a means of getting the fuel-air mixture into the cylinder and getting the burned gases out. This is done by intake and exhaust valves. There have been a number of different types and arrangements of valves used in aircraft reciprocating engines, sleeve valves and poppet valves being the most popular.

Sleeve valves were used on some of the large British radial engines during World War II and in the immediate postwar era. In a sleeve-valve engine, the piston rides up and down inside a steel sleeve that itself rides inside the cylinder, and is rotated back and forth and up and down by a geared sleeve-drive crank. Holes in the sleeve uncover intake and exhaust ports in the cylinder wall at the proper time in the cycle.

A sleeve-valve engine has wide-open passages through which the gases flow as they move into and out of the cylinder, but a poppet valve obstructs the passage. In spite of this obstruction, the simpler design of the poppet valve makes it the most universally used valve mechanism in modern aircraft engines.

There have been a number of poppet valve configurations used in aircraft reciprocating engines, with the L-head and valve-in-head as the most popular. An L-head cylinder has both the intake and exhaust valve on one side of the cylinder, with the valve stems pointing toward the crankshaft.

As aircraft engines evolved, the valve-in head arrangement has proven superior, and today, almost all engines have both the intake and exhaust valve in the cylinder head. These are actuated by either a pushrod and rocker arm, or in some engines, by overhead cams. In the discussion that follows, we will limit the explanation to cylinders with valves in the head that are actuated by pushrods and rocker arms.

The latest trend is to have the intake port entrance opening on the upward side of the cylinder head rather then on the downward side. This gives the engineers space to design a better-tuned intake system, thus increasing volumetric efficiency, more closely balanced cylinder output and engine BHP. Cylinders with this design are usually referred to as "cross-flow."

Valves

Poppet valves have a hardened and polished stem that rides up and down in the valve guides in the cylinder head. The valve head, at right angles to the stem, has a face which seals against the valve seat that is shrunk into the cylinder head.

Intake valves are usually made from hardened, high-strength alloy steel and may have either a flat, tulip, or semi-tulip head. They are cooled by the fuel and air mixture flowing past them during the intake stroke.

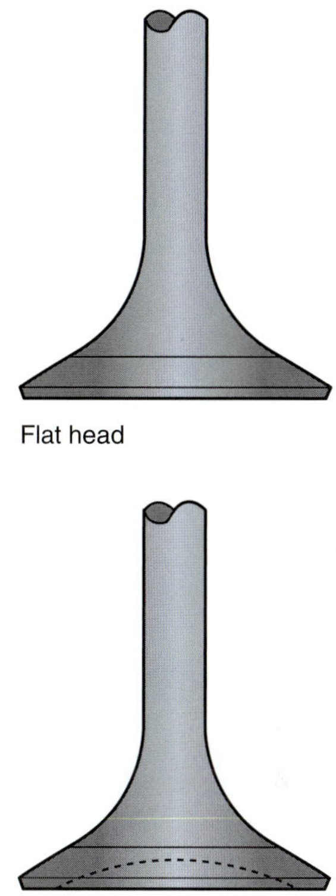

Figure 2-27. *Intake poppet valve head shapes*

The exhaust valves take more punishment than any other component in an engine. They sit directly in the stream of hot gases, and their head is continually pounded against the valve seat. Exhaust valves resemble intake valves, but they are different in their construction and in the material from which they are made. Acceptable exhaust valve head shapes are semi-tulip, flat, and convex (mushroom). The tulip shape has insufficient head-to-stem mass to withstand operating punishment at elevated temperature. Many exhaust valves have stellite inserts welded to their tips and faces. Stellite is a hard, wear-resistant alloy of tungsten, cobalt, and chromium that does not soften at the temperatures encountered on an exhaust valve face.

Many exhaust valves have a hollow stem, and a partially hollowed-out head. The cavity in a hollow valve is partially filled with metallic sodium that aids in cooling the valve. Sodium is a soft metallic chemical element that melts at approximately 98°C (208°F) and is molten when the engine is operating. It sloshes up and down inside the valve stem and absorbs heat from the valve head, then carries it into the stem. From there the heat is transferred into the valve guide, then into the cylinder head and fins, and finally into the air.

stellite. A nonferrous alloy of cobalt, chromium, and tungsten. Stellite is hard, wear resistant, and corrosion resistant, and it does not soften until its temperature is extremely high.

Stellite is welded to the faces of many reciprocating engine exhaust valves that operate at very high temperatures.

The intake and exhaust valves must form an airtight seal with the seats on which they ride. The angle the valve face forms with the valve head is extremely important. Most intake valves are ground with a 30° angle, and the exhaust valves with a 45° angle.

The shallower angle for the intake valve is chosen to provide the maximum amount of airflow into the cylinder when the valve is open, and to provide the longest life for the valve.

The steeper angle used on the exhaust valve increases the seating pressure of the valve and helps prevent carbon build-up on the valve face.

interference angle. The difference between the angles the valve seat and the valve face are ground.

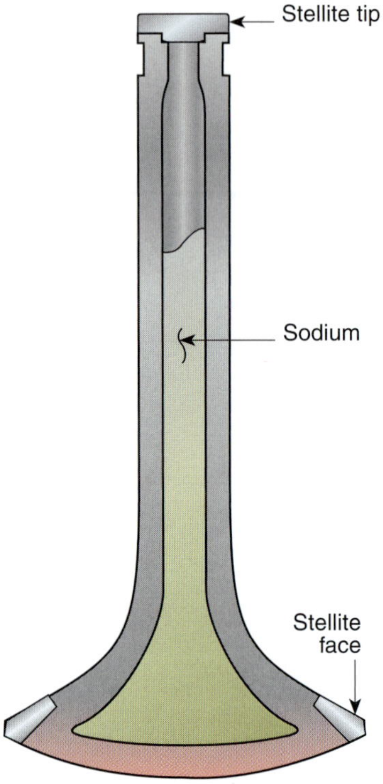

Figure 2-28. *Hollow sodium-filled exhaust valve*

interference fit. An assembly fit in which the hole is smaller than the part that fits into it. The material containing the hole is heated to expand the hole, and the object to fit in the hole is chilled to shrink it.

reamed fit. The fit of a shaft in a hole in which the hole is drilled undersize and cut with a reamer to the correct diameter.

line boring. A method of ensuring that the holes for all bearings for a crankshaft or camshaft are in correct alignment.

The valve faces and valve seats are not always ground to exactly the same angle, but the face of the valve is usually ground 0.5° to 1° flatter than the valve seat. This difference in angle, called an interference angle, allows the valve to form a line contact with the seat that provides a good seal at the operating temperature of the cylinder, decreases the valve lapping time needed to effect a seal, and increasingly improves the sealing integrity with operation time.

Valve Guides

Intake valve guides are made of bronze and exhaust valve guides of a slightly alloyed cast-iron known as Ni-resist, and have an interference fit in the cylinder head. The outside diameter of the guide is larger than the diameter of the hole into which it fits. The head is heated to expand the hole, and the guide is driven into the head with a special drift.

It is extremely important that the valve stem have the proper fit in the guide for it to be able to transfer the heat from the valve into the cylinder head. After the guide has been pressed into place, it is reamed to the fit specified in the engine overhaul manual.

Valve Seats

The head of a poppet valve continually pounds against its seat, and the soft cast-aluminum alloy head would be destroyed in a short time if the valves were seated directly against it. To prevent the valves from damaging the head, rings of steel or bronze alloy are shrunk into the valve opening.

Some engines use forged high-strength alloy steel for the exhaust valve seat, and some have a stellite insert welded into the seat to prevent the hot gases from eroding its surface. Intake valve seats are not subject to such intense heat, and they are usually made of some form of bronze.

The holes for the valve guides and the recesses for the valve seats are line-bored so they have the same center. The outside diameters of the valve seats are several thousandths of an inch larger than the holes into which they are to fit. The head is heated in an oven, and the seat is chilled with dry ice. The seat is then pressed into the seat recess. When the cylinder and seat reach a uniform temperature, the seat is so tight that it will not loosen in normal service.

It is extremely important that the centers of the holes in the valve guide and the valve seat be in perfect alignment so the valve will provide a good seal with the seat. After the guide and seat are installed, the hole in the guide is reamed for the correct size and straightness, and the seat is ground to the required angle, using a tool that is guided by a pilot through the hole in the guide.

When a cylinder is overhauled, both the valve face and seat are ground to provide the best possible seal. The grinding process is discussed in the section on engine overhaul, in Chapter 9 beginning on Page 322.

Valve Springs and Retainers

Poppet valves are opened by pressure from the camshaft acting through the valve lifter, pushrod, and rocker arm. They are closed by the force of two or more coil springs around the valve stem. *See* Figure 2-30.

The valve springs are installed on a lower valve spring seat that fits in the bottom of the rocker box to prevent the spring from wearing the soft aluminum-alloy cylinder head. Two or more springs are installed on the lower seat, and the upper spring seat is installed. The springs are compressed, and two split-cone-type keys, or retainers, are installed between the upper valve seat and a groove machined near the end of the valve stem. When the spring is released, the keys are forced into the groove, and the springs hold the valve tight against its seat.

rocker box. The enclosed part of a reciprocating engine cylinder that houses the rocker arm and valve mechanism.

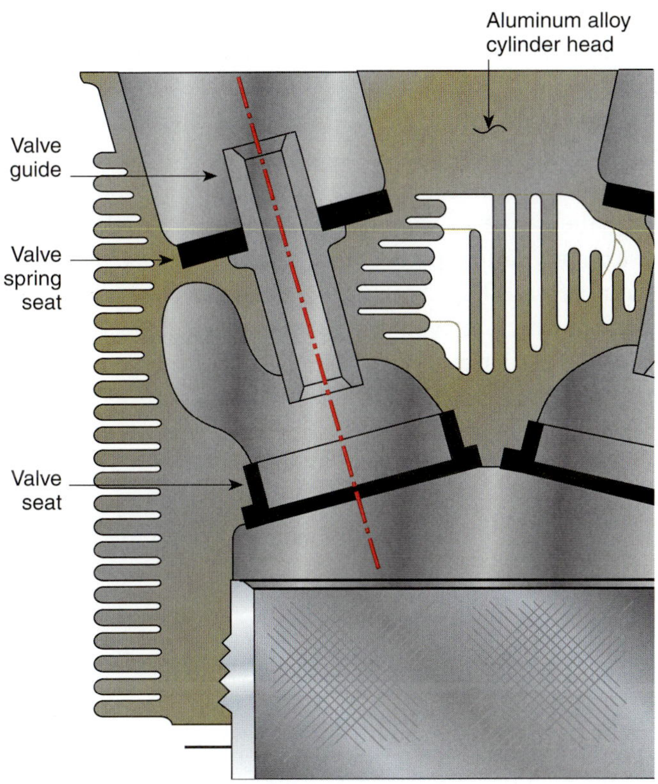

Figure 2-29. *The valve seat and valve guide are both pressed into holes that have been line bored in the aluminum alloy cylinder head, and are reamed so they are perfectly concentric.*

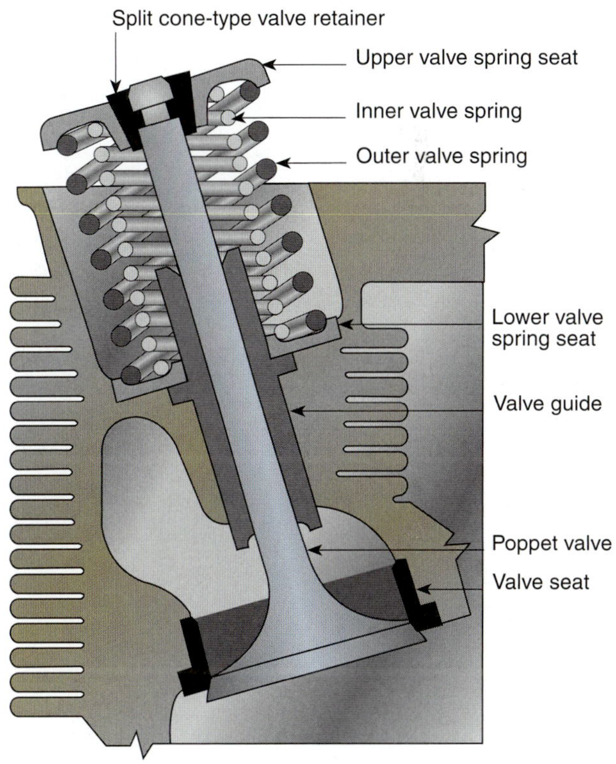

Figure 2-30. *The face of the poppet valve is held tight against its seat by the combined action of the two valve springs. The springs are locked between the upper and lower spring seats by the wedging action of the split-cone valve retainer between the groove in the valve stem and the upper valve spring seat.*

Two or more springs are used on each valve to prevent valve surge, or float. Every coil spring has a natural resonant frequency at which it can be deflected the maximum distance with the minimum amount of force. When impulses are applied to the spring at this particular frequency, the spring loses its ability to oppose them. If a single spring were used around the valve, when the engine RPM was such that the valve opened at the resonant frequency of the spring, the spring would lose its ability to close the valve, and the valve would float. Valve float (within the operation range of the engine) is prevented by using two or three springs, with each made of a different gage wire. The springs have a different diameter and pitch and are wound in opposite directions. The springs have different resonant frequencies, and when the engine operates at a speed at which one spring is resonant, the other spring or springs effectively close the valve.

Some valves have a groove cut in their stem with a safety circlet installed. The circlet is a small snap ring that grips the valve stem so that if the tip should ever break off in operation, the circlet will contact the top of the valve guide and prevent the valve from dropping into the combustion chamber.

Both U.S. engine manufacturers use an exhaust valve rotating system. Lycoming uses an exhaust valve rotator-cap, and Continental incorporates the rotating mechanism into the exhaust spring retainer/outer seat. Both systems take advantage of the torsional effect placed on the valve stem as the spring is compressed and decompressed. The indexing/rotating mechanism provides a momentary disconnect between the valve stem and the spring assembly as the valve starts to open, which is not there as the valve closes. The result is that at each valve closing, it is indexed slightly from its last seating position. This action results in longer sealing integrity. As valves first contact the seat it is not perfectly square with it. If the exhaust valve is not rotated, its face and the seat will wear unevenly, resulting in early leakage. Continental even experimented with indexing the intake valve; but it is suspected that with the cooler operating temperature, the uneven wear was not significant enough to warrant the system.

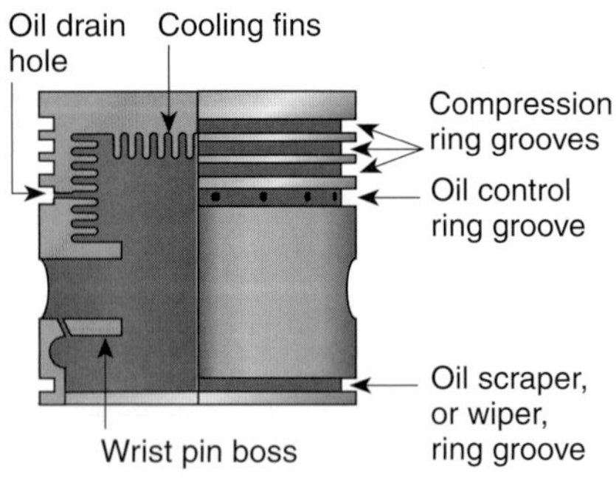

Figure 2-31. *A half-sectioned view of a typical aircraft engine piston*

Pistons

The piston is the device that moves back and forth, or reciprocates, inside the cylinder, and converts the energy in the expanding gases into useful work.

Most aircraft engine pistons are machined of either cast or forged aluminum alloy. Figure 2-31 shows a cross section of an aircraft engine piston. The inside of the head and walls have fins that enable the lubricating oil in the crankcase to pick up some of the heat. Grooves are machined around the outside of the piston and are fitted with rings that seal the piston in the cylinder bore and control the amount of oil allowed to remain on the cylinder walls.

The piston is attached to the connecting rod by the wrist pin which passes through holes in the piston walls in a built-up area called the wrist pin boss. The extra mass of metal in the boss absorbs more heat than the thin wall of the piston, causing the piston to expand nonuniformly. It is important that the piston be as near perfectly round as possible when it is at its operating temperature, so most pistons are cam ground with the dimension perpendicular to the wrist pin a few thousandths of an inch larger than the dimension parallel to the pin. The skirt diameter on most pistons is reduced at the piston-head end, where the greater mass is, for the same reason they are cam-ground. When the piston heats up, the greater mass of metal in the boss area causes it to expand until it is perfectly round.

wrist pin. The hardened steel pin that connects a piston to the small end of a connecting rod.

cam-ground piston. A reciprocating engine piston that is not round, but is ground so that its diameter parallel to the wrist pin is slightly smaller than its diameter perpendicular to the pin.

The mass of metal used in the wrist pin boss, the enlarged area around the wrist pin hole, expands when heated, and when the piston is at its operating temperature, it is perfectly round.

The compression ratio of an engine can easily be changed by changing the pistons. Installing a piston whose head comes closer to the cylinder head decreases the volume of the cylinder when the piston is at the top of its stroke and increases the compression ratio. The heads of many of the pistons are flat, and if the piston comes near enough to the cylinder head that the valves could possibly touch it, recesses may be cut in the piston head to provide the needed clearance. Compression ratio may be increased by using domed pistons or pistons whose heads are in the shape of a truncated (cut off) cone. Figure 2-32 shows some of the typical piston head designs.

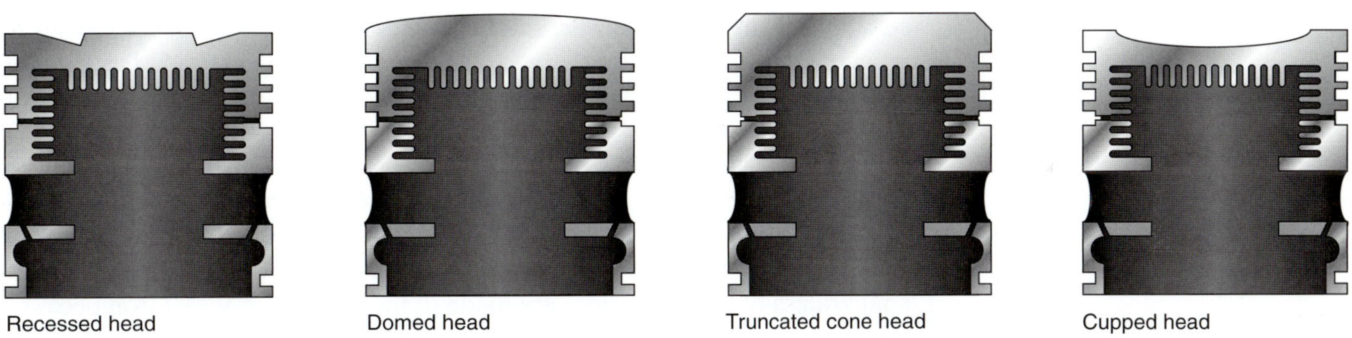

Figure 2-32. *Shapes of aircraft piston heads*

Wrist Pins

The pin that attaches the piston to the small end of the connecting rod is called a wrist pin, or piston pin.

Wrist pins are made of hardened and polished steel, and in most aircraft engines, are free to move within both the connecting rod and the holes in the piston-pin boss. When the pin is free to move in this way, it is called a full-floating pin. Some full-floating pins are prevented from scoring the

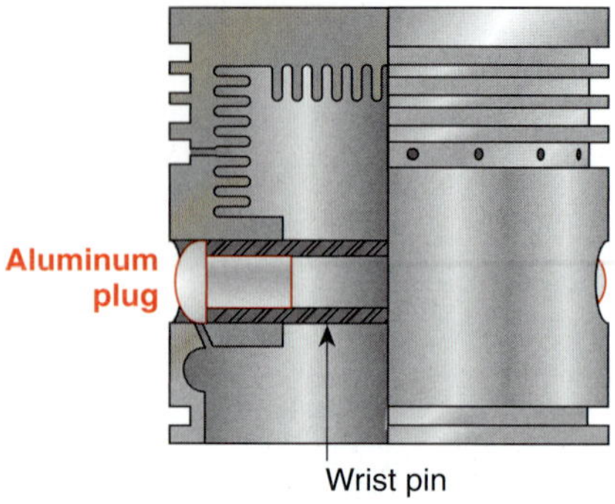

Figure 2-33. *Full-floating wrist pins have aluminum plugs in their ends to prevent scoring the cylinder walls.*

push fit. A fit between pieces in a mechanical assembly that is close enough to require the parts to be pushed together.

A push fit is looser than a press fit, but closer than a free fit.

cylinder walls by spring steel rings that fit into grooves in the piston at each end of the hole in which the pin rides. Most installations, however, protect the cylinder walls by aluminum plugs that fit in the ends of the pin to prevent contact. This style is commonly referred to as "bayonet."

Most wrist pins are installed with a push fit into the holes in the wrist pin boss. They are lubricated through holes in the boss by oil that is sprayed out between the crankpin journal and the big-end bearing, and splashed up by the rotating crankshaft; also, by special oil-spray nozzles on higher output engines.

Piston Rings

The piston must fit into the cylinder loosely enough that the dimensional changes which occur when the engine reaches its operating temperature will not cause it to bind. At the same time it must form a pressure-tight seal with the cylinder wall and the one side of the piston-ring groove to prevent gases from leaking from the combustion chamber into the crankcase. This seal is provided by the piston rings that fit into grooves machined into the outer perimeter of the piston. When the piston travels downward on the intake stroke, the ring seals against the upper-ring groove surface and the cylinder wall. On the power stroke it seals between the lower-ring groove surface and the cylinder wall. This difference is because of the gas pressure against the ring holding it downward. The forces used to seal the ring vary. In short, compression rings seal by the spring force of the ring material, and whatever force is present in the ring grove. For example, on compression the force in the groove is the compression pressure acting outward and downward on the ring. On power, it is the combustion pressure acting in the groove on the ring. The piston rings also help transfer heat from the piston to the cylinder wall, and prevent too much oil from getting into the combustion chamber.

There are three types of rings used on aircraft pistons: compression rings, oil control rings, and oil wiper (scraper) rings. Almost all piston rings, regardless of their function, are made of high-grade gray cast iron because of its ability to remain resilient when extremely hot.

The seal between a piston ring and the cylinder wall requires there to be enough wear between the ring and the wall to smooth the surfaces and allow the rings to "seat." Piston rings used in cylinders with plain steel or nitrided walls can have their wear faces chrome-plated to increase hardness and wear resistance. But piston rings used in chrome-plated cylinders must not be plated, since chrome-plated rings in chrome-plated cylinders will not allow enough wear for the rings to seat and produce the needed seal.

Compression Rings

Compression rings are fitted into the top two or three ring grooves in a piston. In today's engines, these rings come either as wedge or semi-wedge cross-section. This reduces the chance of the ring sticking in the piston ring groove. The older rectangular shape usually became stuck in about 600 hours, thus requiring engine repair. Today, rings also have a decreased face area, which reduces the operation time to seat the ring to the cylinder wall and maintains this seal between overhaul times. This reduction is accomplished by a bevel or stepped shape.

The end gap in piston rings is critical for engine performance. In the past some of the ring-gap joints have been step-cut or angled to enhance the sealing ability, but most modern rings have straight-cut ends. *See* Figure 2-35.

Three things are critical about the fit and condition of a piston ring: its end-gap clearance, its tension, and its side clearance.

The end-gap clearance is measured by placing the ring in the cylinder, and squaring it up at a location specified in the overhaul manual, normally at the cylinder flange. Measure the distance between the ends of the ring with a feeler gage. This measurement is best taken first. Ring gap should be measured before the ring is placed on the piston. Ideally rings only have to be placed on a piston once. Successive times tend to weaken it and increase the chances of ring breakage. When rings are installed, care must be taken not to over-stretch them. Some ring expanding tools have adjustable stops to prevent this, but with others one has to be very careful.

Piston ring tension, in pounds, is measured on the diameter of the ring perpendicular to the gap and is the amount of force needed to close the gap to a specified dimension. An acceptable tool for this is the one used to measure valve spring compression.

The side clearance of a piston ring is the distance, measured with a feeler gage, between the edge of a ring and the side of the ring groove.

When the compression rings are installed on a piston, the joints must be staggered around the perimeter of the piston. If they are not staggered, there will be an excessive amount of blow-by, or flow of combustion gases from the cylinder into the crankcase and oil vapor into the combustion chamber during the intake event.

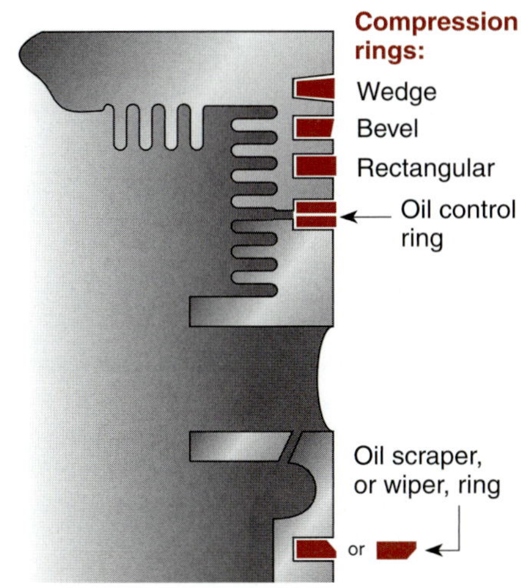

Figure 2-34. *Typical piston ring configuration*

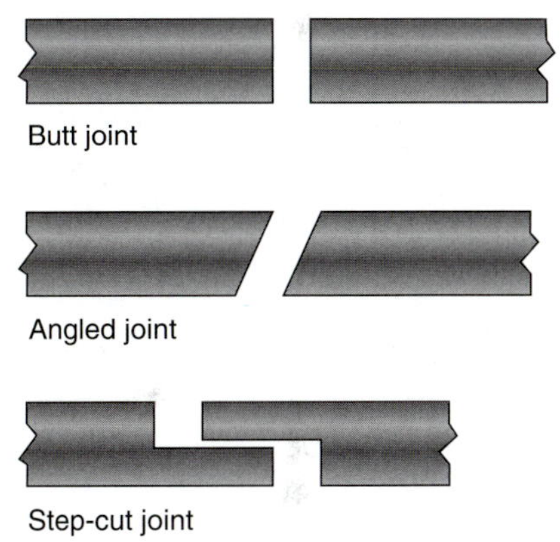

Figure 2-35. *Types of piston ring joints*

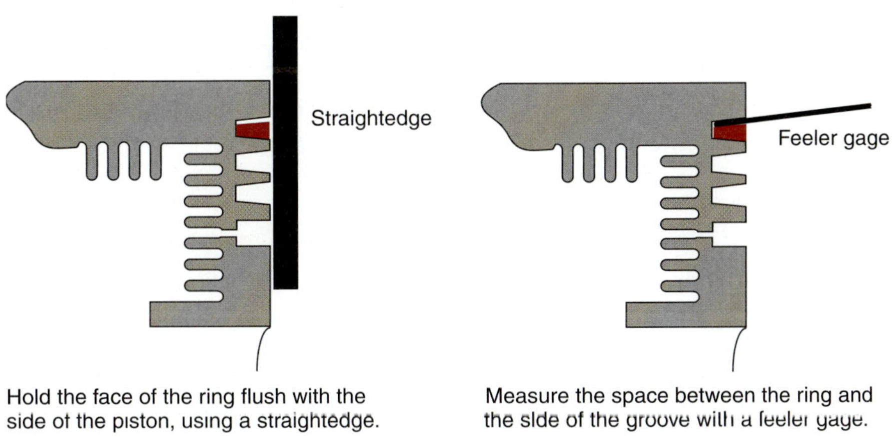

Figure 2-36. *Measurement of piston ring side clearance*

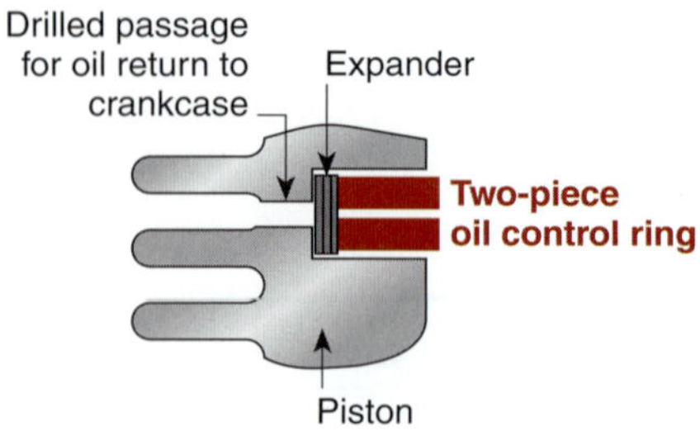

Figure 2-37. *A two-piece oil control ring with an expander behind it controls the amount of oil allowed to remain between the piston and the cylinder wall. Excess oil drains back into the crankcase through the holes in the bottom of the ring groove.*

Oil Control Rings

The oil control ring is installed in the groove immediately below the compression rings; this groove has holes that allow oil to flow from the ring to the inside of the piston and then into the crankcase. Without this provision, oil would have no escape and the ring would be hydraulically lifted from the wall, thus leaving large swatches of oil up around the compression rings and into the combustion chamber. This oil will coke on the rings and inhibit their ability to seal, thus exacerbating the engine's condition. The oil control ring regulates the amount of oil allowed to remain between the piston and the cylinder wall. The excess oil is returned to the crankcase through the holes drilled in the ring groove.

Oil control rings are often made in two parts, and some of them have corrugated expanders installed in the groove behind them. The expander is necessary for additional force to cause the rings to remain seated on the cylinder wall. Even with the oil relief holes, there is significant force attempting to hydraulically lift these rings.

Oil Wiper or Scraper Rings

The ring that is installed at the bottom of the piston is called an oil wiper or scraper. Its function is to ensure that the proper amount of oil is on the cylinder wall. Wiper rings normally have a beveled edge so that they act as effective pumps, and it is extremely important that the bevel be installed in the proper direction relative to the piston head. If the bevel is away from the head, oil will be forced up toward the oil control ring, and if the bevel is toward the piston head, oil will be pumped away from the piston. *See* Figure 2-34 on the previous page.

Connecting Rods

The connecting rod carries the forces produced by the pistons into the crankshaft. These rods must be strong and stiff enough to resist bending during the power stroke, and at the same time, they must be light enough to reduce the inertia as the rod and piston stop, then change direction at the top and bottom of the stroke.

In horizontally opposed engines, the connecting rod is forged of high-strength chrome-nickel-molybdenum steel with an I- or H-shaped cross-section.

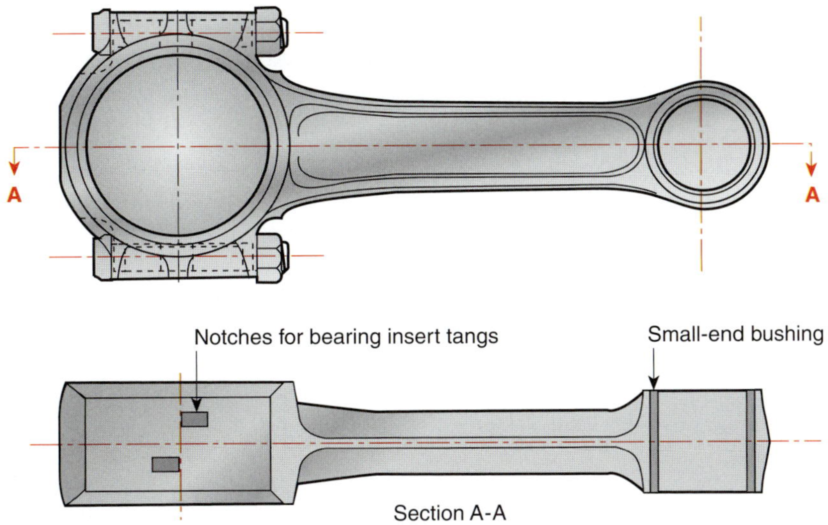

Figure 2-38. *Typical forged connecting rod for a horizontally opposed engine*

The big end of the connecting rod is split, and both the body of the rod and the cap have cylinder identification numbers stamped on them to prevent their being mismatched during overhaul. Two-piece bearing inserts are installed in the big end and are held in place by tangs on the inserts that fit into slots cut into the cap and body. The tangs prevent the insert from spinning inside the rod. Some connecting rods prevent the bearing insert spinning by a short dowel pin pressed into the body of the rod that fits into a hole in the insert. The outside diameter of the bearing insert is approximately a .003" interference fit in the cap and rod bore. This fit helps to prevent the bearing from spinning in the rod, and aids with heat transfer for cooling.

High-strength nuts and bolts hold the cap onto the rod body. The torque applied to the nuts on these bolts is critical, because it must produce a tensile stress in the bolt greater than the stress that will be applied by the hammering action the rod receives in operation.

The small end of the connecting rod has a bronze bushing pressed into its hole, and the bushing is reamed to the correct fit for the wrist pin. The reaming must be precise in order to keep the axis of the bushing parallel to that of the large-end bearing. During overhaul the lack of parallelism is measured as twist and convergence. These limits must be held to approximately .001" off true over 1 inch. If this tolerance is not held, the piston can be pinched in the cylinder, the full surface of the bearings involved will not bear the load, and the rod and piston pin will work back and forth excessively. This condition will cause excessive wear and possible parts failure.

offset throw (crankshaft design). Crank arms on a reciprocating engine crankshaft. The arms, or throws, to which the connecting rods and pistons are attached are offset from the center of the crankshaft to move the pistons in and out of the cylinder. The amount of the offset determines the stroke of the engine.

Crankshaft

The crankshaft is the strongest and heaviest component in an aircraft reciprocating engine. It is the component that transforms the reciprocating motion of the pistons into a rotating motion.

Crankshafts for horizontally opposed engines are forged of chrome-nickel-molybdenum steel with an offset throw and a bearing journal for each connecting rod.

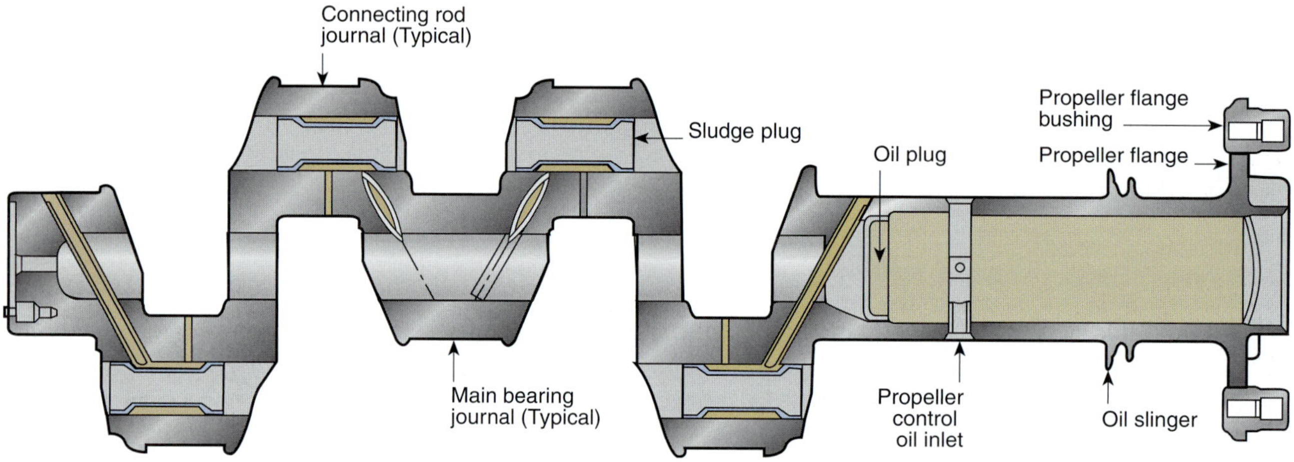

Figure 2-39. *Crankshaft for a typical four-cylinder horizontally opposed engine*

The throws of a four-cylinder crankshaft are paired with the two end throws together and 180° from the two center throws. There is a main bearing journal at the propeller end, the center, and the accessory end of the shaft for Continental engines. Lycoming engines have two main bearings between the propeller attachment and the front cheek. For these engines they are numbered one and two, even though they are most often combined in just two bearing shell halves.

journal (bearing). A hardened and polished surface on a rotating shaft that rides in a plain bearing.

In the crankshaft illustrated in Figure 2-39, the centers of the main bearing journals and crankpins are hollow. Oil picked up at the main bearings flows through drilled passages in the crank throws to lubricate the connecting rod big-end bearings.

Some engines have sludge plugs pressed into each of the hollow crankpins. Any sludge that is carried through the engine in the lubricating oil is compressed to the inside diameter of the crankpin by centrifugal force, and is trapped and held by the sludge plug until the engine is disassembled at overhaul. This adaptation was used by Textron-Lycoming in its mid-generation engines.

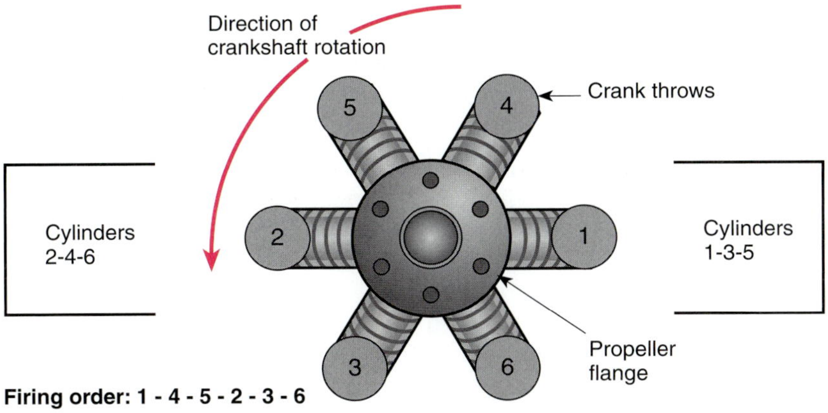

Figure 2-40. *The three pairs of throws in a six-cylinder horizontally opposed crankshaft allows one piston to be in firing position each 120° of crankshaft rotation.*

Otto-cycle engines fire all the cylinders in two revolutions, or 720° of crankshaft rotation. A six-cylinder horizontally opposed Continental crankshaft has six throws with four main bearings, one at the propeller end, one between each pair of throws, and one at the accessory end.

The pairs of throws on the crankshaft in Figure 2-40 are 60° apart, which allows two pistons to come to the top of their stroke every 120° of crankshaft rotation. One piston is at the beginning of the power stroke and the other at the beginning of its intake stroke.

The pistons in cylinders 1 and 2 of a Textron-Lycoming engine come to the top at the same time, with cylinder 1 firing. One hundred and twenty degrees of rotation later, the pistons in cylinders 3 and 4 come to the top, with 4 firing. Then 120° later, 5 and 6 come to the top, with 5 firing. In another 120°, 1 and 2 come to the top again, with 2 firing. Then, 3 and 4, with 3 firing, and finally, 5 and 6, with 6 firing. In another 120°, the cycle starts over.

The same events occur with a six-cylinder Teledyne-Continental engine, the only difference being the way the cylinders and therefore the throws are numbered. On a TCM engine, the pair of throws nearest the propeller is for cylinders 5 and 6, and on a Textron-Lycoming, the throws nearest the propeller are for cylinders 1 and 2.

The Textron-Lycoming IO-720 eight-cylinder engine must fire all cylinders in 720° of crankshaft rotation, or 90° of rotation between firing impulses. An eight-cylinder, horizontally opposed crankshaft has four pairs of throws with main bearings at the propeller end, between each pair of throws, and at the accessory end.

The pairs of throws are 90° apart, which allows two cylinders to come to the top of their stroke every 90° of crankshaft rotation, one at the beginning of the power stroke and the other at the beginning of the intake stroke. *See* Figure 2-41 on the next page.

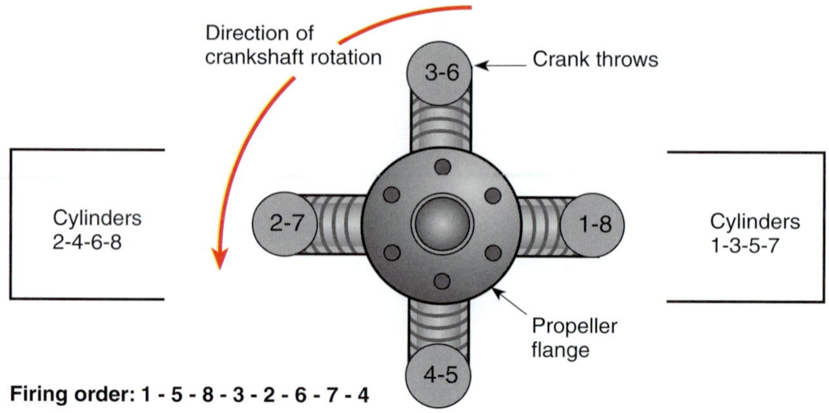

Figure 2-41. *The four pairs of throws in an eight-cylinder crankshaft allow one piston to be in firing position each 90° of crankshaft rotation.*

cheek (crankshaft). The offset portion of a crankshaft that connects the crankpin to the main bearing journals.

The pistons in cylinders 1 and 2 come to the top at the same time, with cylinder 1 firing. Ninety degrees of rotation later, the pistons in cylinders 5 and 6 come to the top, with 5 firing. Then 90° later, 7 and 8 come to the top, with 8 firing. Then 3 and 4 come to top, with 3 firing. Ninety degrees later, 1 and 2 come to the top again, with 2 firing, then 5 and 6 come to the top, with 6 firing. Seven and 8 then come to the top, with 7 firing, and finally 4 and 3 come to the top, with 4 firing. Ninety degrees later the cycle begins again.

The main-bearing and connecting-rod-bearing journals are ground smooth and are hardened by nitriding to form a wear-resistant surface on which the bearings ride. The radius between the journal and the crank cheeks is extremely important, as the torsional stresses in the crankshaft are great enough to break the shaft if its cross-sectional area were to change too rapidly. The large radius prevents this abrupt change.

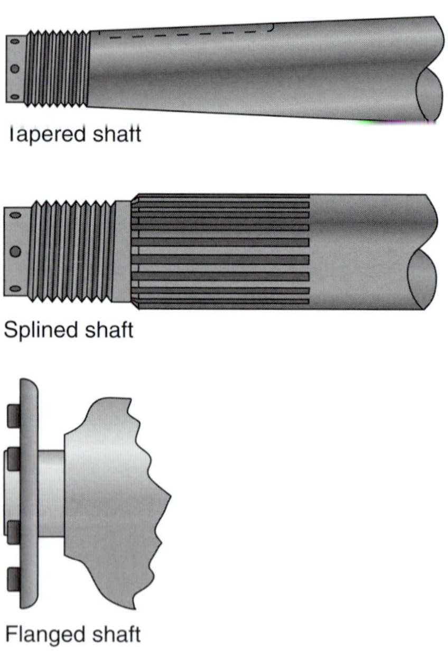

Figure 2-42. *Methods of attaching propellers to reciprocating engine crankshafts.*

Propeller Attachment

There are three main methods for attaching propellers to crankshafts: the shaft may be tapered, flanged, or splined.

Tapered shafts were commonly used by many of the small engines that were fitted with wooden propellers. The propeller was mounted in a steel hub that fit onto the shaft. A steel key fit into a slot in the crankshaft and the propeller hub to prevent the hub from turning on the shaft.

Splined shafts were for many years used on all of the larger engines. These shafts have a series of splines that mate with those in the propeller hub. A master spline, which is much wider than the normal splines, indexes or aligns the propeller so it can be installed on the shaft in only one way. The propeller hub is centered on the shaft between a single-piece bronze rear cone and a two-piece steel front cone.

There are several sizes of splined propeller shafts identified by an SAE (Society of Automotive Engineers) number. SAE 20 splines are used on engines in the 200-horsepower range; SAE 30 splines are used in the 300- and 400-horsepower range; and SAE 40 are used in the 500- and 600-horsepower range. SAE 50 in the 1,000-horsepower range and SAE 60 and 70 are used for larger engines.

SAE (Society of Automotive Engineers). A professional organization that has formulated standards for the automotive and aviation industries.

Almost all current production horizontally opposed engines have flanged propeller shafts. The propeller shaft is forged with a flat flange on its end, and nuts for propeller attachment are pressed into holes in the flange. A large dowel, or guide pin, in the flange indexes the propeller to the shaft. The propeller is attached to the flange with special high-strength bolts.

Dynamic Dampers

Radial engine crankshafts have a single throw for each row of cylinders, and they must have a large heavy counterweight for static balance. Horizontally opposed engines, on the other hand, are built symmetrically and do not need counterweights for static balance, but most of the larger ones use counterweights for dynamic balance, to absorb torsional vibrations.

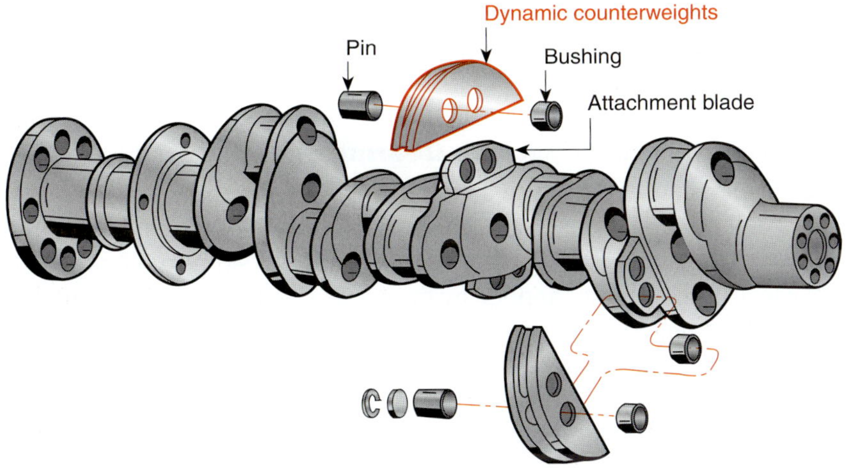

Figure 2-43. *Dynamic counterweights are installed on the crankshafts of some of the higher-powered horizontally opposed engines to absorb torsional vibration.*

Dynamic counterweights are installed on blades that are forged as integral parts of the crankshaft. They are retained by pins with diameters smaller than the holes in the bushings through which they fit. The small pin in the large hole allows the weight to rock back and forth in a pendulum fashion.

The power produced by a reciprocating engine is supplied by the pistons in a series of pushes, or pulses, and when the frequency of these pulses is the same as the resonant frequency of the crankshaft, serious torsional vibration can occur. Dynamic counterweights change the resonant frequency of the crankshaft.

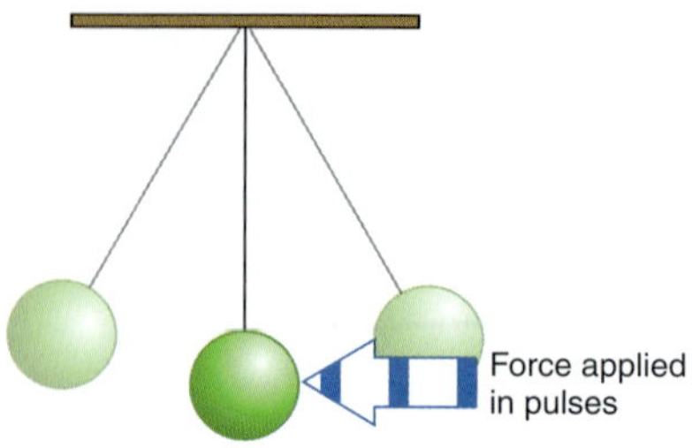

When pulses of energy at the resonant frequency of the pendulum are applied, pendulum swings with a large amplitude.

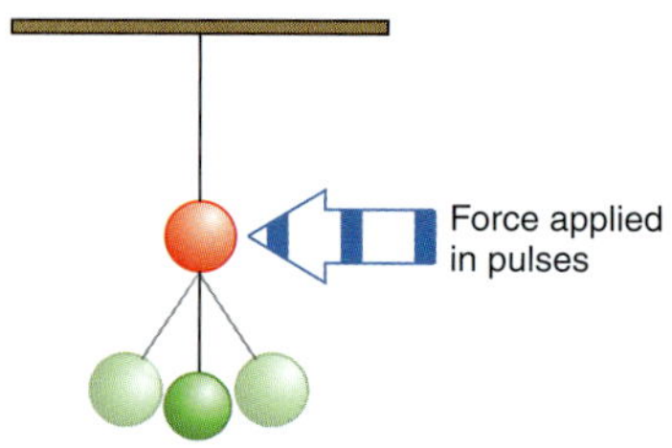

If the weight of the pendulum is divided with the components having different lengths, pulses with the same energy will cause the pendulum to swing with a smaller amplitude.

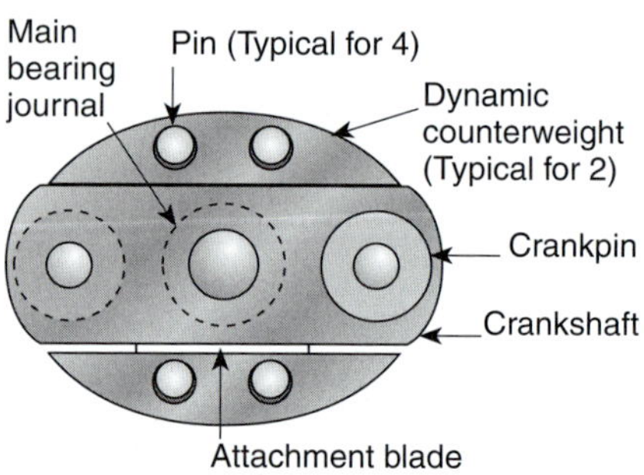

Dynamic counterweights on a crankshaft act in such a way that they change the resonant frequency of the crankshaft.

Figure 2-44. *Dynamic counterweights change the resonant frequency of the crankshaft and decrease the effect of the energy being supplied to the crankshaft in a series of pulses.*

By considering the pendulum in Figure 2-44, you can visualize the way dynamic balances change the resonant frequency of the crankshaft to prevent it from amplifying the pulses from the cylinders.

If the pulses of energy are applied to the freely suspended ball at its resonant frequency, the ball will swing with a large amplitude for the amount of energy in the pulses.

If the weight is divided between different pendulum lengths, and the same pulses of energy are applied, the balls will swing with considerably less amplitude.

Centrifugal force holds the weights away from the crankshaft to the extent allowed by the small pins, and the weights absorb the pulses of energy from the pistons as their large hole and small pin allow them to rock back and forth.

Propeller Reduction Gearing

An aircraft engine must turn at a high speed to develop its maximum power, but propellers become inefficient when their tips approach the speed of sound. The propeller shafts on many of the higher-performance engines are geared so they will turn slower than the crankshaft, which allows the engine to turn fast enough to develop its required power without the propeller operating at an inefficient speed. There are a number of types of gear arrangement used to reduce propeller speed.

External Spur-Type Reduction Gearing

Some of the smaller engines use a simple external spur gear arrangement, with the small drive gear on the crankshaft meshing with a larger driven gear on the propeller shaft. A strong crankcase is needed to support the propeller shaft because of the stresses applied by the precessive forces of the propeller, which acts as a giant gyroscope. Two characteristics of this arrangement are that the propeller is offset from the crankshaft, and the direction of rotation of the propeller is opposite to that of the crankshaft. The amount of reduction is found by the formula:

$$\text{Gear ratio} = \frac{\text{Teeth on drive gear}}{\text{Teeth on driven gear}}$$

Internal Spur-Type Reduction Gearing

The problems inherent in the external spur-type reduction gearing are minimized by using an internal-tooth gear on the propeller shaft. The propeller turns in the same direction as the crankshaft, and the gear ratio is determined using the same formula as for the external spur gear system.

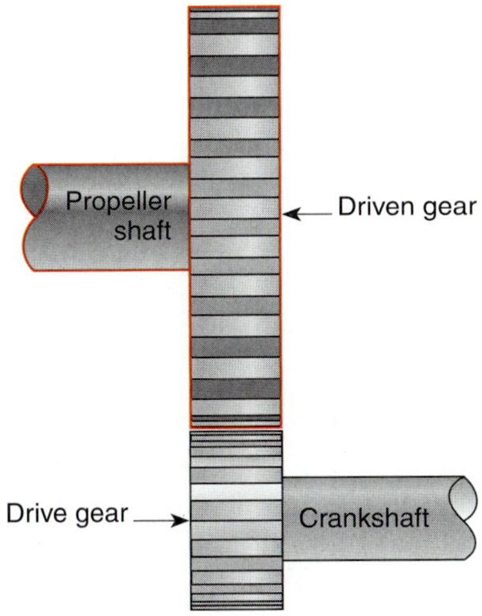

Figure 2-45. *External spur gear-type reduction gearing*

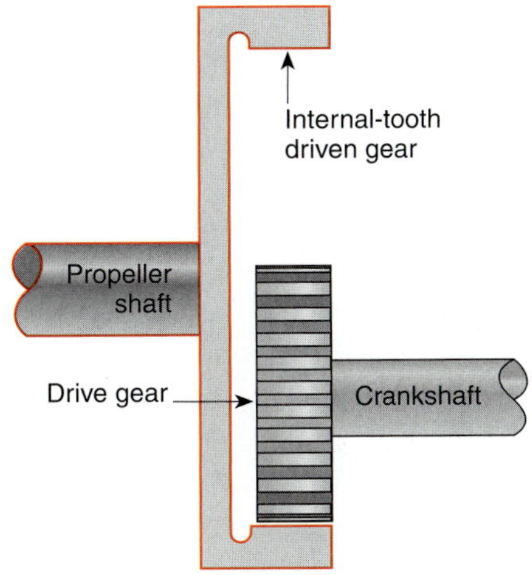

Figure 2-46. *Internal spur gear-type reduction gearing*

Planetary (Epicyclic) Reduction Gearing

When more power must be transferred through the gears than can be handled by simple spur gears, a planetary, or epicyclic, reduction gear system is used. Three types of planetary gear arrangements commonly used in aircraft reciprocating engines are: the bevel-gear system and two types of spur planetary gear arrangements.

epicyclic reduction gears. A gear train in which a series of small planetary gears rotate around a central gear. More commonly called a planetary gear train.

Bevel Planetary Gears

Some engines have used a bevel planetary gear arrangement in which a fixed bevel gear, often called the sun gear, is attached rigidly to the nose section of the engine, and a larger bevel-type drive gear, called a ring gear, is part of the crankshaft. A series of bevel-type pinions are mounted in a cage or carrier that is part of the propeller shaft. The pinions mesh with both the fixed and drive gears, and as they spin around the fixed gear, they rotate the propeller in the same direction as the crankshaft. The number of teeth on the pinions does not enter into the gear ratio. The gear ratio is found by the formula:

$$\text{Gear ratio} = \frac{\text{Teeth on drive gear} + \text{Teeth on fixed gear}}{\text{Teeth on drive gear}}$$

Bevel planetary gears turn the propeller at a speed slower than the crankshaft but in the same direction. The propeller shaft is directly in line with the crankshaft, and this type of gearing requires the smallest nose section of any type of planetary gearing.

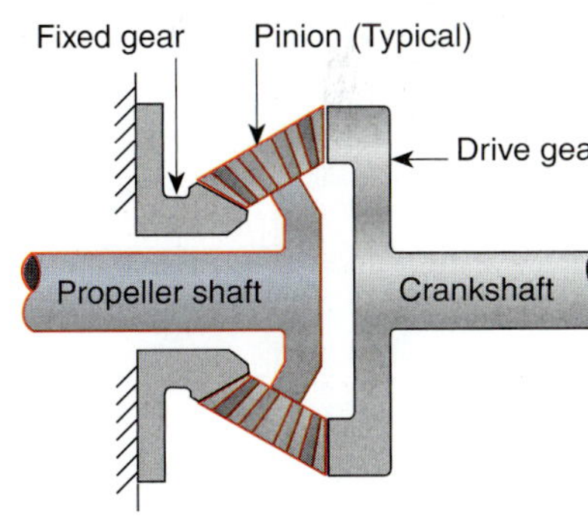

Figure 2-47. *Bevel planetary reduction gears*

quill shaft. A type of shaft used to couple parts of an engine that are subject to torsional loads.

A quill shaft is a long, hardened steel shaft with splines on each end. One end splines into the drive shaft and the other end splines into the device being driven. Torsional vibrations are absorbed by the quill shaft twisting.

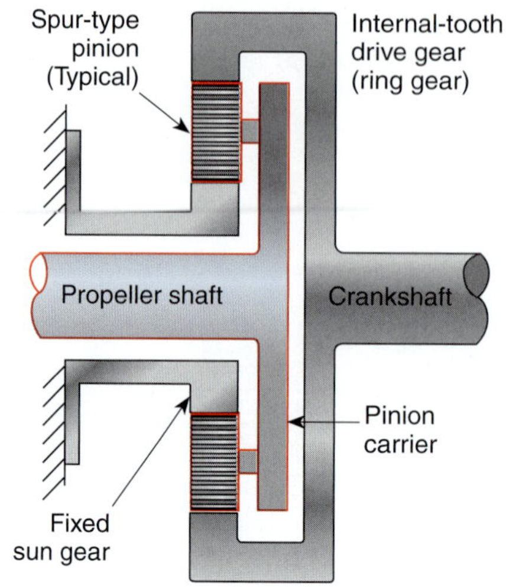

Figure 2-48. *Spur-type planetary reduction gears*

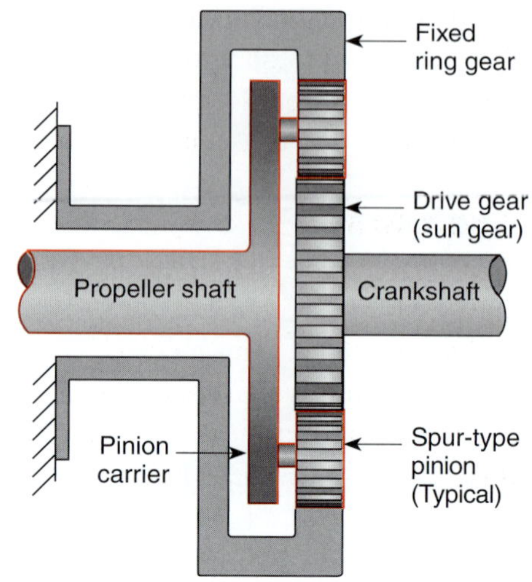

Figure 2-49. *Reverse spur planetary reduction gears*

Spur Planetary Gears

There are two types of spur planetary gear arrangements used to reduce the speed of the propeller relative to that of the crankshaft. One system has the sun gear attached rigidly to the nose section of the engine, and an internal-tooth ring gear is driven by the crankshaft. A strong, forged pinion carrier is part of the propeller shaft, and mounted on it are a series of spur-type pinions that mesh with the teeth on both the sun and ring gears. When the crankshaft turns, the pinions spin around the fixed sun gear and rotate the pinion carrier and the propeller in the same direction, but at a reduced speed. *See* Figure 2-48. The gear ratio of a fixed-sun gear system is determined by the formula:

$$\text{Gear ratio} = \frac{\text{Teeth on ring gear} + \text{Teeth on sun gear}}{\text{Teeth on ring gear}}$$

When a greater speed reduction is needed than can be obtained by the spur reduction gear system, a reverse planetary system similar to the one in Figure 2-49 may be used. In this system, the sun gear is driven by the crankshaft, and the internal-tooth ring gear is fixed in the nose section of the engine. The pinions are mounted on the pinion carrier on the propeller shaft and are driven around inside the ring gear by the sun gear. The gear ratio is found by the formula:

$$\text{Gear ratio} = \frac{\text{Teeth on ring gear} + \text{Teeth on sun gear}}{\text{Teeth on sun gear}}$$

One series of horizontally opposed Continental engines uses a quill shaft to absorb some of the torsional vibrations that intensify when the propeller shaft is geared. *See* Figure 2-50. Splines machined into the inside of the

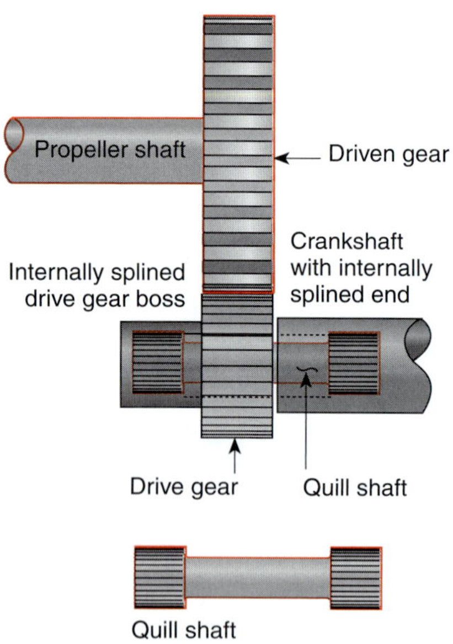

Figure 2-50. *Torsional vibrations between the propeller and the crankshaft are absorbed in this spur-type reduction gearing by the twisting of the quill shaft.*

crankshaft drive a hardened steel quill shaft. Splines on the opposite end of the quill shaft mesh with splines cut into the inside of the hollow drive gear boss. Any torsional vibration between the propeller and the crankshaft is absorbed by the twisting of the quill shaft.

Crankcase

The crankcase is the component that serves as the foundation of the engine and attaches the engine to the airframe. The cylinders mount on the crankcase, and all of the rotating mechanism is housed and protected inside.

Horizontally opposed engine crankcases are cast of aluminum alloy. Some are cast in a sand mold, and others use a permanent mold process that results in thinner and at least equivalent strength walls. Some of these

sand casting. A method of molding metal parts in a mold made of sand.

permanent-mold casting. A casting made in a reusable metal mold. The walls of permanent-mold castings can be made thinner than similar walls made by sand casting.

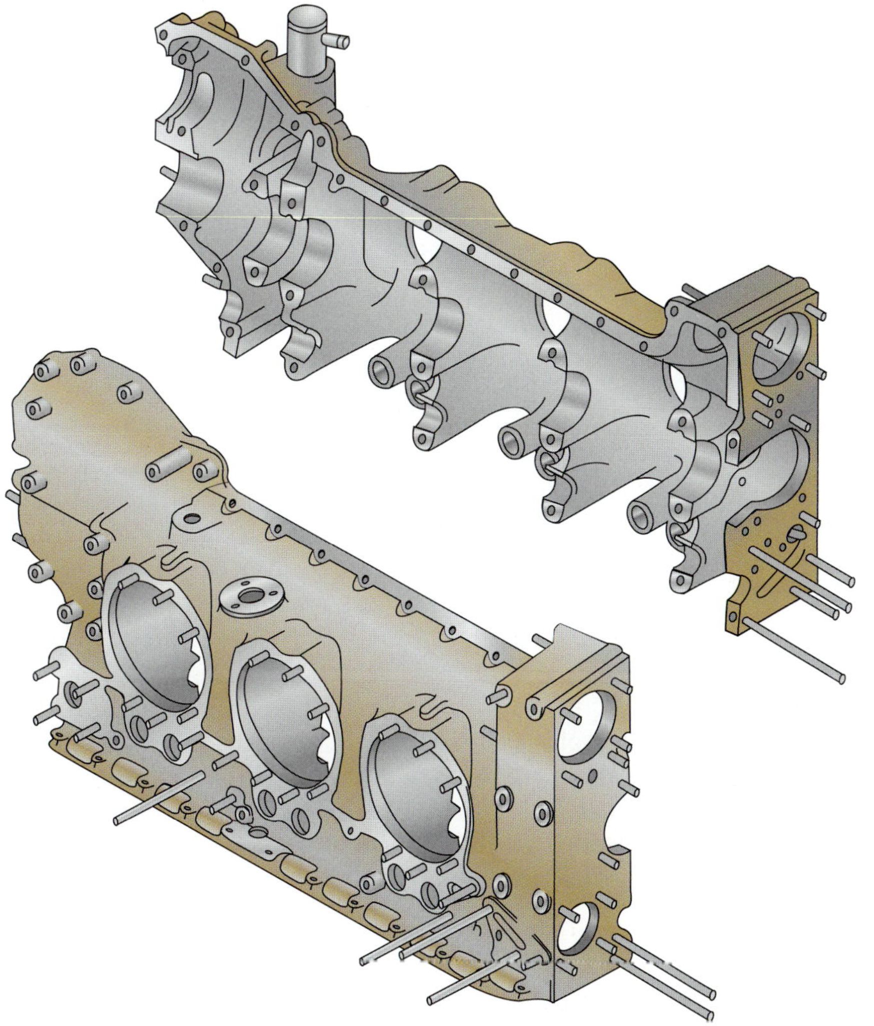

Figure 2-51. *Typical aircraft engine crankcase made of cast-aluminum alloy*

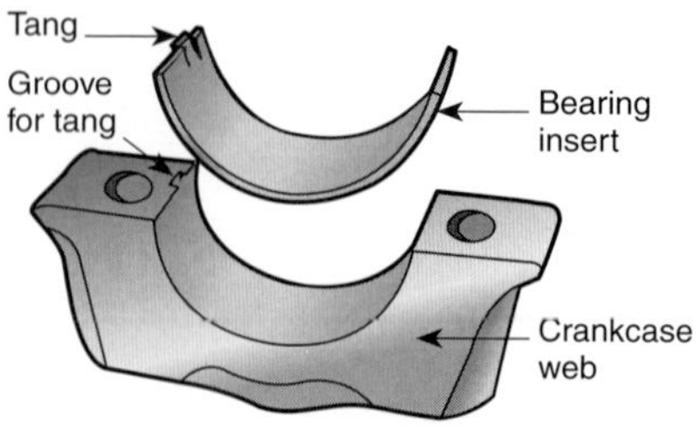

Bearing insert is locked in place with a tang that fits into a groove in crankcase web.

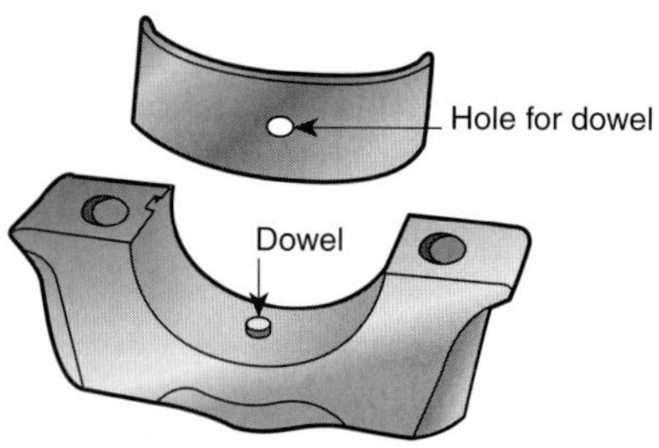

Bearing insert is locked in place by a dowel that is pressed into the crankcase web. The dowel fits into hole in insert.

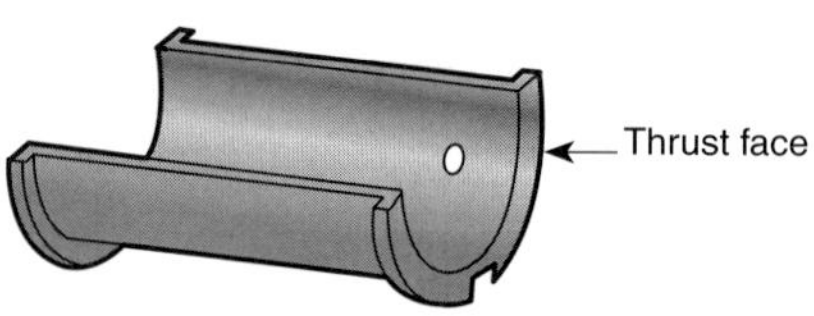

Saddle-type thrust-bearing insert is locked into place by a dowel, and has a thrust face against which flanges on propeller shaft ride.

Figure 2-52. *Bearing inserts of the type normally used to support the crankshaft in a horizontally opposed engine.*

permanent molded cases have proven to be less ductile and have had problems with excessive cracking in service. For a time, this condition was bad enough that one manufacturer issued a bulletin showing where acceptable cracks could be located.

The case is made in two halves, split down the vertical center line. Webs, cast in the two halves, support the bosses for the main and camshaft bearings. The parting surfaces are machined smooth so the two halves will seal without the use of any gaskets. Pads are machined for mounting the cylinders, and holes are drilled for the through studs that mount the cylinders and hold the two halves together. Some arrangements use anchored through studs. These are through studs that are also threaded to one half of the crankcase. With either of these arrangements, smaller through bolts are usually used at the coned section of the case to only hold the case halves together.

Bearings

The crankshaft is supported in the crankcase by plain bearings. The main bearing bosses are line-bored to ensure that the bearing seats are all in perfect alignment, and bearing inserts are fitted into each of the seats. The inserts have a steel backing and use a lead alloy as the bearing material. The bearing inside diameter is then clad with a thin lead or silver coating to provide a very low friction surface. The inserts are prevented from turning in their seats by tangs on one end of the insert that fits into slots in the bearing seat, or by dowel pins pressed into the bearing seats that fit through a hole in the insert. When the case halves are torqued together, the inserts are an interference fit with the bore. This is done for increased heat transfer and bearing retention.

Camshafts of horizontally opposed engines normally ride in line-bored holes through the webs in the crankcase and do not use any type of bearing insert or bushing. This arrangement is satisfactory because camshafts are nearly equally loaded from both sides when in operation. It is not unusual to find the tooling marks still on the case bearing surfaces after one or two TBO periods.

Crankcase Oil Seals

The crankcase of an aircraft engine receives the oil that has been forced through the bearings and splashed onto the cylinder walls. The oil supply is held in the wet sump which is attached to the bottom of the crankcase. The joints in the case are sealed, and all of the rotating shafts protruding through the case ride in spring-loaded flexible seals to prevent the loss of oil. These seals are usually made of neoprene.

The crankcase halves are sealed with a thin coating of a nonhardening gasket compound and a fine silk thread extending the full length of the case. This gasket compound and silk thread make an efficient oil seal, and yet they do not have enough thickness to interfere with the fit of the bearings.

The crankshaft and accessory drive shafts are sealed with spring-loaded seals like the one in Figure 2-53. The seal rides on a hardened and polished shaft, and the spring holds the flexible seal against the shaft to prevent oil flowing past it.

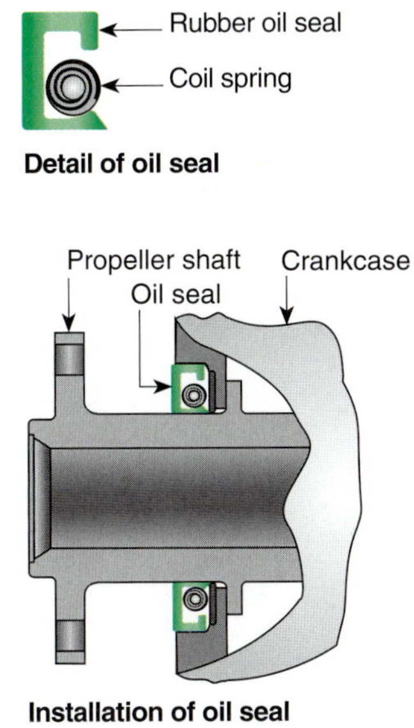

Figure 2-53. *A spring-loaded flexible seal is used around rotating shafts to prevent the loss of oil.*

Valve Operating Mechanism

Almost all of the horizontally opposed engines in modern aircraft use pushrod-operated poppet valves with one intake and one exhaust valve in each cylinder head. Figure 2-54 shows a simplified diagram of the way these valves operate.

The camshaft is driven by gears from the crankshaft, and turns at one-half crankshaft speed for a four-stroke five-event engine. When the valve lifter rides up on the lobe of the cam, it pushes out on the pushrod. The rocker arm is mounted on the rocker arm shaft on the cylinder head, and when the pushrod presses on one side of the rocker arm, the other side pushes down on the valve stem and opens the valve. When the camshaft turns and the valve lifter rides off of the cam lobe, the valve springs close the valve and hold it closed until the cam lobe again opens the valve.

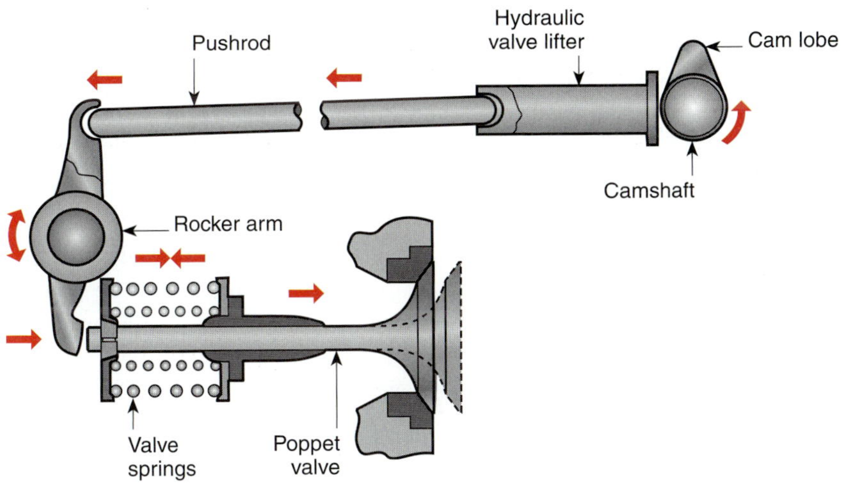

Figure 2-54. *The valve operating mechanism in a typical horizontally opposed engine*

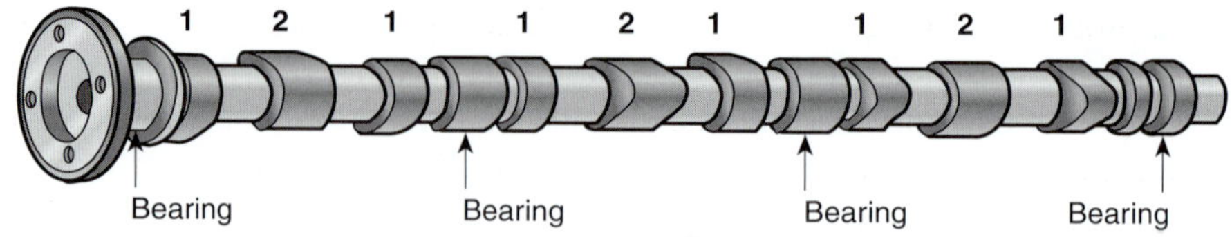

Figure 2-55. *A camshaft for a typical six-cylinder horizontally opposed engine*

Camshaft

The camshaft used on a horizontally opposed engine is a steel shaft with lobes and bearing journals machined on it. The lobes and journals are ground and polished and case hardened to increase their resistance to wear. The body of the camshaft is not case hardened, so it remains tough and resilient. The drive gear that meshes with a gear on the crankshaft is positioned by a dowel pin, or unequally spaced bolt attach holes, that assures that the gear and the lobes on the shaft are in correct alignment. The lobes are tapered from side-to-side across the opening and closing ramps and the apex. This taper causes the lifter to rotate, thus changing the bearing action across the lifter face to a rolling friction rather then a sliding one. This design drastically increases the life of these two surfaces.

The camshaft for a four-cylinder engine has six lobes, with two of the lobes actuating intake valves in cylinders on opposite sides of the engine. The camshaft for a six-cylinder engine has nine lobes, with three of the lobes actuating valves in two cylinders. *See* Figure 2-55.

Valve Lifters

Some of the smaller horizontally opposed engines and all radial engines use solid valve lifters. These lifters require an adjustment on the rocker arm that allows a specific clearance to be maintained in the valve operating train. The majority of horizontally opposed engines use hydraulic valve lifters, and they are discussed here. Solid lifters are discussed with radial engines.

Hydraulic Valve Lifters

A valve lifter body with a hardened and polished face rides on the lobe of the camshaft to change the rotary motion of the shaft into a linear motion. One end of the pushrod fits into a socket that rides in the hollow valve lifter body. A plunger fits in the lifter body under the pushrod socket, and hydraulically maintains zero clearance in the valve operating train. Because of this action, hydraulic valve lifters are called zero-lash valve lifters. Refer to Figure 2-56 to understand the way the hydraulic valve lifter operates. This figure represents a combined cam-follower and hydraulic lifter. Many arrangements use a cam-follower with an independent hydraulic unit placed inside.

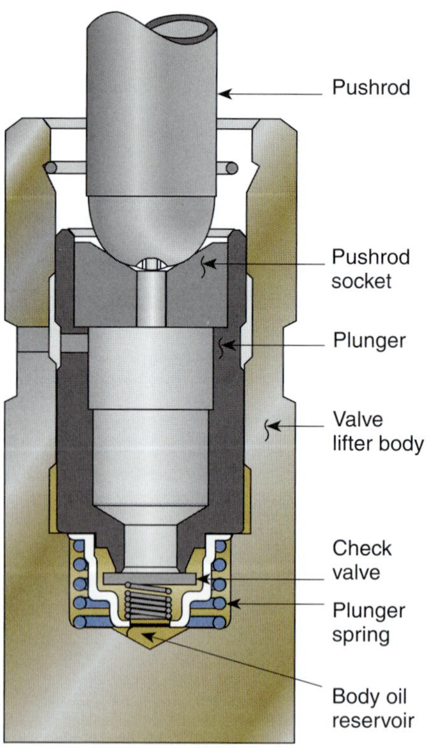

Figure 2-56. *A hydraulic valve lifter maintains zero clearance in the valve operating train.*

The primary advantage of the zero-lash lifter is consistent valve timing over a wide range of cylinder operating temperatures. With solid lifters, as temperature changes, valve lash changes; therefore, valve opening and closing times and valve opening duration and lift change. This is because the cylinder assembly elongates, more than the valve operating components, and extends the rocker arm pivot away from the crankcase.

When the valve lifter is off of the cam lobe, the plunger spring pushes the plunger and pushrod socket up. This maintains zero clearance in the valve train but does not put enough pressure on the pushrod to open the valve. Oil from the engine lubricating system flows into the lifter body and the plunger through drilled passages from the oil gallery, or header, and into the body reservoir through the check valve.

zero-lash valve lifter. A hydraulic valve lifter that maintains zero clearance in the valve actuating mechanism.

When the camshaft rotates and the valve lifter rides up on the cam lobe, the check valve holds the oil trapped in the body reservoir where it acts as a solid member of the valve train. The pushrod is pushed up against the rocker arm, and it rotates to open the valve. The opposition of the valve springs against the rocker arm and pushrod holds a force on the plunger.

The outside of the plunger is ground to fit precisely with the inside of the body forming a matched set, and when the valve lifter is on the lobe of the cam with the valve springs pushing back against it, some of the oil that was trapped in the body reservoir leaks out past the plunger. As the camshaft continues to rotate, the lifter comes off of the lobe and the plunger spring forces the plunger up. Oil from the engine lubricating system flows through the check valve to refill the body oil reservoir.

The action of the oil leakage between the plunger and the body, and the fact that the body reservoir is completely filled each time the lifter is off of the cam lobe, allows the valve train to operate with no clearance between the valve stem and the rocker arm.

Pushrods

Pushrods used in almost all horizontally opposed aircraft engines are made of hollow steel or aluminum alloy tubes with hardened and polished spherical steel inserts pressed into each end. The inserts are drilled so engine lubricating oil can flow through the hollow pushrod to lubricate the valve mechanism in the rocker boxes on the cylinder heads.

The pushrods are housed in thin-wall aluminum shrouds, and in Continental engines the oil from the rocker boxes returns to the engine through these tubes. This is possible because Continental places the camshaft below the crankshaft at the bottom of the crankcase. Lycoming places the camshaft above the crankshaft and has to use rocker box drain tubes to drain oil back to the crankcase.

Engines with hydraulic valve lifters maintain a zero clearance in the valve train, and the pushrod must be the proper length to allow the correct clearance between the valve stem and the rocker arm when the hydraulic unit is completely flat, or empty. This correct clearance ensures the hydraulic lifter will be operating within its limits. Changes such as valve and valve seat grinding may necessitate a change in pushrod length. Continental large-bore engines currently require .060"– .120" of dry lash, and Lycoming .024"– .080".

Engines with solid valve lifters have an adjustment on the rocker arm that allows you to adjust the clearance between the rocker arm and the valve stem.

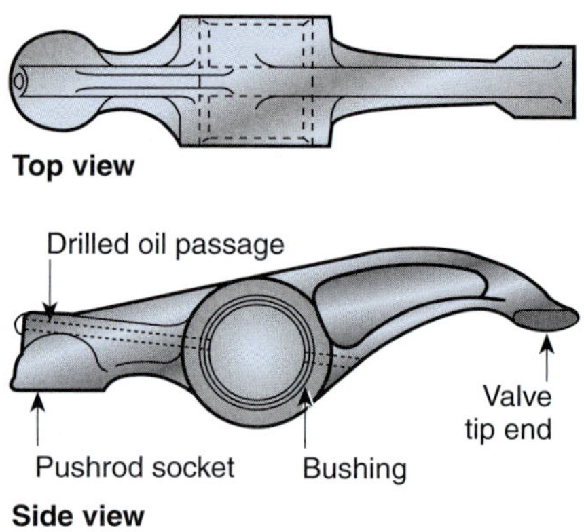

Figure 2-57. *Typical rocker arms for a horizontally opposed engine*

Rocker Arms

The rocker arms used on horizontally opposed engines are made of forged steel with bronze bushings, which ride on the rocker arm shaft, pressed into them. The end in which the pushrod rides has a socket that fits the hemispherical end of the pushrod. A hole is drilled on exhaust arms and Continental intake arms from this socket to the bushing to allow oil that flows through the hollow pushrod to lubricate the bushing.

The end of the rocker arm that rides on the valve stem and pushes the valve open is ground in a smooth curve and polished. This arc decreases the side load placed on the valve guide by the valve stem when the valve is pushed open.

STUDY QUESTIONS: HORIZONTALLY OPPOSED ENGINE CONSTRUCTION

Answers are on Page 90. Page numbers refer to chapter text.

32. The cylinder barrel of a modern reciprocating engine is made of ______________________ or ______________________ steel. *Page 54*

33. The part of a cylinder barrel that fits into the crankcase is called the ______________________ . *Page 54*

34. The diameter at the top of a choke-ground cylinder is ______________________ (greater or smaller) than the diameter in the center. *Page 54*

35. The inside of a cylinder wall ________________ (should or should not) be perfectly smooth. *Page 55*

36. Two ways of hardening the walls of a cylinder barrel are by:
    a. ______________________
    b. ______________________
    Page 56

37. Chrome-plated cylinders are identified by a band of ________________ paint around the cylinder base or on some of the fins. *Page 56*

38. Chrome plating ______________________ (does or does not) provide a perfectly smooth surface on which the piston rings ride. *Page 56*

39. Nitrided cylinders are identified by a band of ________________ paint around the cylinder base or on some of the fins. *Page 56*

40. Nitriding ______________________ (does or does not) appreciably affect the dimensions of a part. *Page 56*

41. Nitriding ______________________ (is or is not) a rustproof finish. *Page 56*

42. The threaded joint between the cylinder head and barrel is made gastight by making the threads a/an ______________________ fit. *Page 57*

43. Spark plug holes in a cast-aluminum alloy cylinder head are protected from wear by the installation of a/an ______________________ insert. *Page 57*

44. The side of a cylinder head that has the most fins is the side that contains the ________________ (exhaust or intake) valve. *Page 58*

Continued

STUDY QUESTIONS: HORIZONTALLY OPPOSED ENGINE CONSTRUCTION *Continued*

45. The tips and faces of some exhaust valves are protected from wear by welding in inserts made of ____________________ . *Page 59*

46. Some exhaust valves are partially filled with metallic ____________________ that is used to help cool them. *Page 59*

47. Most intake valves are ground with a __________ -degree angle between the valve face and the valve head. *Page 59*

48. Most exhaust valves are ground with a __________ -degree angle between the valve face and the valve head. *Page 59*

49. The proper fit between the valve stem and the guide is provided by ____________________ the valve guide after it is installed in the cylinder head. *Page 60*

50. Perfect alignment between the valve seat and the valve guide is ensured by grinding the seat, using a tool that is guided by a pilot riding in the ____________________ . *Page 60*

51. Two or more springs are used on each poppet valve to prevent valve ________________ or ________________ . *Page 61*

52. The diameter of a cam-ground piston perpendicular to the wrist pin is ________________ (largest or smallest). *Page 63*

53. A wrist pin that is not clamped in either the connecting rod nor the piston is called a/an ____________________ wrist pin. *Page 63*

54. The piston rings nearest the piston head are the ________________________ rings. *Page 64*

55. Three functions of a piston ring are:
    a. ______________________________
    b. ______________________________
    c. ______________________________
    Page 64

56. A chrome-plated piston ring ________________ (should or should not) be installed in a chrome-plated cylinder. *Page 64*

57. Three things that are critical about the fit and condition of a piston ring are:

a. ______________________________

b. ______________________________

c. ______________________________

Page 65

58. The piston ring that determines the amount of oil that is allowed to remain between the piston and the cylinder wall is the ______________________ ring. *Page 66*

59. The piston ring that is installed at the bottom of the piston skirt is the __________________ or __________________ ring. *Page 66*

60. If an oil scraper ring is installed in such a way that the bevel is on the side of the ring toward the piston head, the ring will pump oil __________________ (toward or away from) the oil control ring. *Page 66*

61. The bearing inserts used in the big end of a connecting rod are prevented from spinning in the rod by ____________________ or _______________ . *Page 67*

62. The bearing in the small end of a connecting rod is a/an _______________________ . *Page 67*

63. The throws on a crankshaft for a four-cylinder horizontally opposed engine are ___________ degrees apart. *Page 68*

64. The crankshaft for a four-cylinder horizontally opposed engine uses _____________ (how many) main bearings. *Page 68*

65. Sludge in the lubricating oil flowing through a horizontally opposed engine is collected in the sludge plugs that are installed in the hollow ___________________ . *Page 68*

66. The pairs of throws on a crankshaft for a six-cylinder horizontally opposed engine are ___________ degrees apart. *Page 68*

67. The crankshaft for a six-cylinder horizontally opposed engine uses _____________ (how many) main bearings. *Page 69*

68. Three types of propeller shafts used on reciprocating engines are:

a. __________________________

b. __________________________

c. __________________________

Page 70

Continued

STUDY QUESTIONS: HORIZONTALLY OPPOSED ENGINE CONSTRUCTION *Continued*

69. The counterweights on the crankshaft of a horizontally opposed engine are used for ____________________ (static or dynamic) balance. *Page 71*

70. Dynamic counterweights are used to change the ____________________ frequency of the crankshaft assembly. *Page 71*

71. The propeller of a geared aircraft engine turns ____________________ (faster or slower) than the crankshaft. *Page 72*

72. An aircraft engine is geared to prevent the propeller tip exceeding the speed of ____________________ . *Page 72*

73. The component that serves as the foundation of an aircraft engine and attaches the engine to the airframe is the ____________________ . *Page 75*

74. The crankcase for aircraft engines is cast from aluminum alloy and may be cast in either ____________________ or ____________________ molds. *Page 75*

75. The crankshaft in a horizontally opposed engine is supported in ____________________ (plain or ball) bearings. *Page 76*

76. The holes in the crankcase in which the camshaft bearing surfaces ride ____________________ (are or are not) fitted with bronze bushings. *Page 76*

77. The camshaft in a horizontally opposed engine turns at ____________________ the speed of the crankshaft. *Page 77*

78. The purpose of a hydraulic valve lifter is to maintain an operating clearance of ____________________ inch in the valve train. *Page 78*

79. Another name for a hydraulic valve lifter is a/an ____________________ valve lifter. *Page 78*

80. Pushrods for a horizontally opposed engine are made of hollow ____________________ or ____________________ tubes. *Page 79*

81. The bearing used in most rocker arms is a/an ____________________ . *Page 80*

Radial Engine Construction

Radial engines were considered the standard configuration of aircraft engines for several decades until they were replaced by turbine engines for high-power applications, and horizontally opposed engines for low-powered applications. The construction of radial engines is somewhat similar to that of horizontally opposed engines, but there are some significant differences which are discussed below.

Radial engines all have an odd number of cylinders in each row. Three, five, or seven cylinders are common, with nine cylinders being the most that is practical for one row. When more than nine cylinders are needed, two or more rows of cylinders are used. The pistons in each row of cylinders are connected to one throw of the crankshaft through a master rod and link rod arrangement. The largest practical radial engine has 28 cylinders arranged in four rows of seven cylinders each.

Cylinders

Radial engine cylinders are similar to those used on horizontally opposed engines except for the skirts, or the portion that extends into the crankcase. Oil that collects in the crankcase when the engine is not operating tries to fill the lower cylinders. It leaks past the piston rings and into the combustion chamber. If the engine is started with oil filling this chamber, hydraulic lock and serious internal engine damage will result. The skirts on radial engine cylinders are considerably longer than those on horizontally opposed engines, to minimize the amount of oil that flows into the lower cylinders.

Even though longer skirts are used, some oil still leaks into these cylinders, and you should check for a hydraulic lock as a standard prestarting procedure. Pull the propeller through by hand (or, in some installations, with the starter) to rotate the crankshaft through at least two revolutions to be sure there is no oil trapped in the cylinders. If oil is found, remove the most accessible spark plugs from the lowest cylinders and drain the oil out before attempting to start the engine.

hydraulic lock. A condition in which oil drains into the lower cylinders of a reciprocating engine and leaks past the piston rings to fill the combustion chamber. If the oil is not removed before the engine is started, it can cause serious damage.

Cooling was a serious problem for some of the larger radial engines, and the area of the machined fins on the cylinder walls was not great enough to dissipate all of the unwanted heat. These engines had densely finned aluminum alloy muffs, or sleeves, shrunk onto the outside of the cylinder barrel, providing the area for conducting away the extra heat.

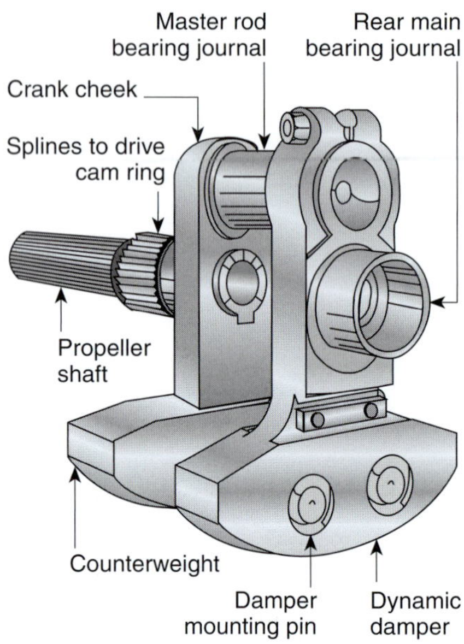

Figure 2-58. *A two-piece crankshaft of a single-row radial engine. The forward counterweight is fixed and the rear counterweight is mounted on two pins so it can rock back and forth and act as a dynamic damper.*

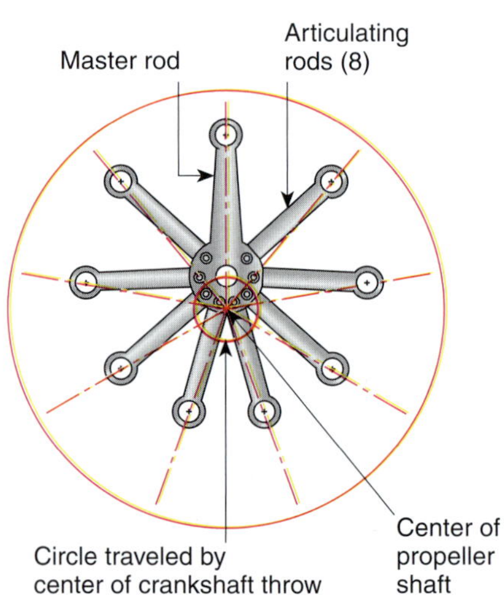

Figure 2-59. *A master rod and articulating rod assembly used on a nine-cylinder radial engine*

Crankcase

The crankcase for a radial engine is very small when compared with the ones used by other types of engines. The cylinders bolt to the power section, which is made of cast or forged aluminum, or in some of the larger engines, forged steel. The propeller reduction gearing is housed in a cast-aluminum alloy nose section bolted to the front of the power section, and most of the accessories are mounted on a cast-aluminum alloy section bolted behind the power section.

Crankshaft

The crankshaft of a single-row radial engine has a single throw, as shown in Figure 2-58. Some crankshafts are made in a single piece of high-strength forged steel, and use a two-piece master rod. Other crankshafts are made in two pieces, and use a single-piece master rod. Heavy counterweights are installed opposite the single throw for static balance; some counterweights have large-diameter holes that ride on small-diameter pins so they can also act as dynamic dampers. Almost all radial engines have splined crankshafts or propeller shaft for attaching the propeller.

Crankshafts used in two-row radial engines have two throws 180° apart. The cylinders in the front row are attached to one throw, and the cylinders in the rear throw are connected to the other.

Connecting Rods

In order to attach nine cylinders to one throw of a crankshaft, a master rod and articulating, or link rods like those in Figure 2-59 are used. A plain bearing in the big end of the master rod rides on the polished bearing journal on the throw of the crankshaft. The small end of the master rod is attached to the piston in one cylinder with a wrist pin.

In a nine-cylinder radial engine, the pistons in eight of the cylinders are connected to the master rod with articulating rods. One end of these rods attaches to the piston with a wrist pin and the other connects to the master rod with a knuckle pin.

Propeller Reduction Gearing

Almost all radial engines use a planetary gear system to reduce the crankshaft rotation RPM to a propeller shaft speed that is efficient for the propeller. Typical reduction gear ratios are 3:2 and 16:9. Two types of planetary reduction gear systems are used: the spur planetary and the bevel planetary, with the spur planetary systems used for the larger reduction ratios. Both systems have a sun gear, ring gear, and a spider with a series of planetary gears attached to it. In both systems the propeller shaft is attached to the spider. Both systems drive the propeller shaft in the same direction as the crankshaft. Planetary gear systems are discussed with horizontally opposed engines, beginning on Page 73.

articulating rod. The rod in a radial engine that connects one of the piston wrist pins to a knuckle pin on the master rod. Also called articulating rods.

Bearings

Radial engines use steel-backed plain bearings in the master rod and bronze bushings pressed into both ends of the link rods. Ball bearings are installed on either end of the crankshaft, and the thrust bearing is a deep-groove ball bearing that can take both radial and thrust loads.

Valve Operating Mechanisms

Radial engines have one intake and one exhaust valve in each cylinder head, and these valves are closed by two or more coil springs and are opened by rocker arms and pushrods, similar to those described in the section on horizontally opposed engines. The basic difference in the valve operating systems for the two types of engines is in the cam and cam followers.

Almost all radial engines have a cam ring or cam plate geared to the crankshaft and turned at a reduced rate from that of the crankshaft. This cam ring, shown in Figure 2-60, has two cam tracks, one for the intake valves and one for the exhaust valves, with lobes ground into them.

Radial engines use solid valve lifters rather than hydraulic lifters like those used in most horizontally opposed engines. Each valve lifter, or tappet, as it is usually called, has a hardened steel roller that rides on the cam ring, and when it rides up on a cam lobe, it pushes up on the pushrod and acts through the rocker arm to open the valve. There must be an accurately maintained clearance in the valve train. If the clearance is too great, the valves will open late and close early, and the valve overlap and the efficiency of the engine will decrease. If the valve clearance is too small, the valves will probably not seat positively when the engine is cold during start and engine warm-up.

There are two clearances specified in the engine maintenance manuals: hot clearance and cold clearance.

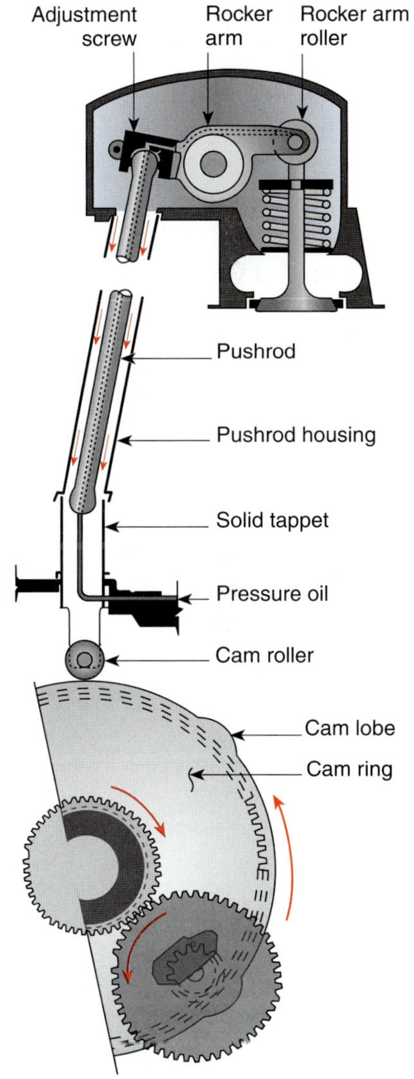

Figure 2-60. *Valve-operating mechanism of a radial engine*

Hot Valve Clearance

The rocker arms and valves are mounted in the cast-aluminum alloy cylinder head, and are opened by a hollow steel pushrod actuated by the cam ring in the crankcase. When the engine is hot, the cylinder head expands a large amount and the rocker arm moves away from the crankcase. Since the pushrod remains essentially the same length, the valve clearance when the engine is hot is greater than when it is cold.

Cold Valve Clearance

The hot clearance determines the time and amount the valves open in normal operation and is therefore the most important clearance, but because it is essentially impossible to adjust the valves to this clearance when the engine is operating, engineers have determined the correct adjustment for a cold engine so that the clearance will be correct when the engine is hot.

The valve timing is checked to determine that the valves open when the crankshaft is in the correct position by adjusting the valves in cylinder number one to their hot, or running, clearance, and setting the cam in the proper relationship to the crankshaft. Then all of the valve clearances are adjusted to the correct cold clearance.

If the valves are adjusted to the hot (running) clearance when the cylinder is cold, the clearance in the valve train will be too great when the engine is at its normal operating temperature. The valves will open late and close early. The cam will have to turn farther to open the valve, and the valve will close before the cam has turned to the normal valve-closing position.

Valve Adjustment: Engines with Floating Cam Rings

Some large radial engines have floating cam rings. These cam rings ride over a shelf-type bearing with a large amount of clearance between the bearing and the ring. When the engine is running, the ring is centered over its bearing by the forces exerted by the valve springs.

When checking the valve clearance on an engine equipped with a floating cam, the bearing clearance must be eliminated by relieving the pressure on the springs of two valves on the opposite side of the engine from the valves being checked.

Depressing the valves removes the valve spring pressure from the cam, allowing the cam ring to move tight against its bearing on the side where the valves are being checked.

Supercharger

Practically all of the larger radial engines have geared centrifugal blowers that compress the fuel-air mixture after it passes through the carburetor to increase the power of the engine. Some of the smaller engines have a centrifugal impeller driven at the crankshaft speed that does not compress the fuel-air mixture, but helps distribute the mixture uniformly to all of the cylinders.

STUDY QUESTIONS: RADIAL ENGINE CONSTRUCTION

Answers are on Page 90. *Page numbers refer to chapter text.*

82. When more cooling fin area is needed than can be provided by machining fins on the cylinder barrel, a muff or sleeve made of ___________________________ can be shrunk on the outside of the barrel. *Page 85*

83. The crankshaft of a single-row radial engine has __________________ (how many) throws. *Page 86*

84. The crankshaft of a 14-cylinder twin-row radial engine has __________________ (how many) throws. *Page 86*

85. The bearing in the big end of a master rod is a _____________________ (ball or plain) bearing. *Page 87*

86. The pin that connects an articulating rod to the master rod of a radial engine is called a/an ___________________ pin. *Page 86*

87. A planetary gear system ____________________ (does or does not) reverse the direction of the propeller shaft relative to the direction of the crankshaft. *Page 87*

88. The thrust bearing used in a radial engine is a ____________________ (ball or plain) bearing. *Page 87*

89. Most radial engines _____________________ (do or do not) use hydraulic valve lifters. *Page 87*

90. The valve clearance when a radial engine is hot is ______________________ (greater or smaller) than the clearance when the engine is cold. *Page 88*

Answers to Chapter 2 Study Questions

1. expand
2. volume
3. pressure
4. volume
5. a. intake
   b. compression
   c. ignition
   d. power
   e. exhaust
6. before, exhaust
7. after, compression
8. before, power
9. after, intake
10. before, compression
11. exhaust, intake
12. is not
13. is
14. is
15. is not
16. decreased
17. the same as
18. compression
19. increases
20. detonation
21. a. RPM
    b. BMEP
22. brake specific fuel consumption
23. reliability
24. time between overhauls (TBO)
25. in-line
26. rear
27. 1-4-5-2-3-6
28. 1-6-3-2-5-4
29. ethylene glycol
30. wet sump
31. a. turbocharged
    b. fuel injected
    c. opposed cylinders
    d. 360 cubic-inch-piston displacement
32. chrome-molybdenum, chrome-nickel-molybdenum
33. skirt
34. smaller
35. should not
36. a. chrome plating
    b. nitriding
37. orange
38. does not
39. blue
40. does not
41. is not
42. interference
43. Heli-Coil
44. exhaust
45. stellite
46. sodium
47. 30
48. 45
49. reaming
50. valve guide
51. surge, float
52. largest
53. full-floating
54. compression
55. a. prevents leakage of gases
    b. transfers heat into the cylinder walls
    c. prevents oil getting into the combustion chamber
56. should not
57. a. end gap clearance
    b. tension
    c. side clearance
58. oil control
59. wiper, scraper
60. away from
61. tangs, dowel pin
62. bronze bushing
63. 180
64. 3
65. crankpins
66. 60
67. 4
68. a. tapered
    b. splined
    c. flanged
69. dynamic
70. resonant
71. slower
72. sound
73. crankcase
74. sand, permanent
75. plain
76. are not
77. one half
78. zero
79. zero-lash
80. steel, aluminum alloy
81. bronze bushing
82. aluminum alloy
83. one
84. two
85. plain
86. knuckle
87. does not
88. ball
89. do not
90. greater

Lubrication Systems

3

Continued

Lubrication Systems

3

Functions of the Lubrication System

The lubrication system of a reciprocating engine is one of its most important and vital systems. An engine operated with insufficient lubrication will, within minutes, drastically overheat and seize. Improper service of the lubricating system can cause excessive wear of the moving parts.

The lubrication system has a number of very important functions. It reduces friction between the moving parts of the engine and provides an effective seal and cushion between moving parts. The oil absorbs heat from the cylinder walls and pistons, and carries it outside the engine where the heat is transferred to the outside air. The oil protects metal parts of the engine against corrosion and picks up contaminants, carrying them into filters where they remain trapped.

Reduces Friction

Even though metal surfaces may appear to be smooth, they are often proven rough when examined under a microscope. Each surface has a series of peaks and valleys, and when two surfaces rub together, the irregularities on one surface lock with those on the other. It requires effort to move one surface over the other. Without adequate lubrication, so much heat is generated that the peaks in one part will weld to the peaks in the other, and continued movement will tear chunks from the metal. These chunks, however small, act as an abrasive causing further wear.

If the surfaces are covered with a lubricant, such as a film of oil, the oil will fill all of the irregularities and hold the surfaces apart so they do not contact each other. The only friction encountered when the parts are moved is the internal friction of the oil.

The friction between the teeth of mating gears could cause a great deal of friction and wear if the gears are not adequately lubricated. Some gears have so much pressure between their teeth that special extreme-pressure (EP) lubricants are used to prevent the film of lubricant from rupturing, which would allow contact between the metals.

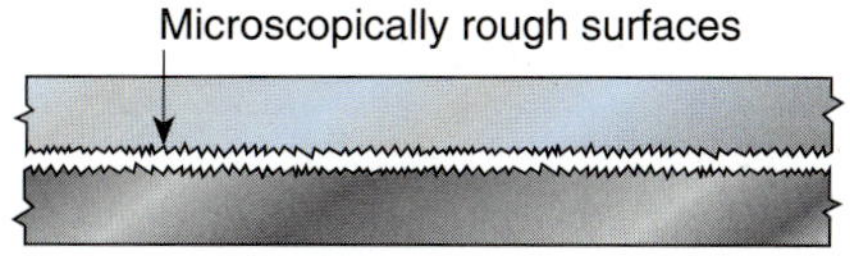

Figure 3-1. *Microscopic roughness on the surface of two pieces of metal rubbing together uses power, produces heat, and wears the metal.*

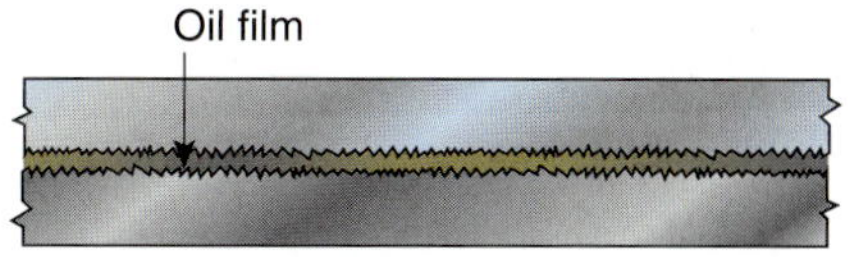

Figure 3-2. *A film of lubricating oil holds the two surfaces apart and reduces friction and wear.*

Seals and Cushions

One of the characteristics of a lubricating oil is its viscosity or stickiness. A viscous oil wets the surfaces where there is relative movement, and provides a seal to prevent air escaping from between them. This type of seal is important between piston rings and the cylinder wall. It is also crucial for forming a seal between the gear teeth and the housing of the lubricating oil pump.

The oil clings to the metal and cushions the impact when surfaces pound together, as the rocker arms pound inside their bushings each time the pushrods ride up on the cam lobes, or as the rocker face pushes into the valve stem.

Removes Heat

Engine lubricating oil absorbs as much heat as possible from all lubricated surfaces, but it absorbs the most heat from the underside of the piston head and from the cylinder walls.

Some larger engine pistons have fins on the underside of their head to increase the surface contacted by the lubricating oil. Some of the oil that is pumped through the hollow crankshaft sprays out between the crank-pin journal and the connecting rod big-end bearing, through a squirt hole in the connecting rod cap, or through special oil spray nozzles placed in the oil galleries. This oil absorbs heat from the piston and cylinder wall, carrying it out of the engine, and into the air that passes through the oil cooler.

Cleans Inside of Engine

Contaminants such as combustion deposits, sludge, dirt, carbon, and particles of metal worn from the moving parts are picked up by the oil as it circulates through the engine. Most oil used in aircraft reciprocating engines is an ashless-dispersant, or AD oil that contains an additive that disperses the contaminants and prevents clumping by giving the particles a like charge. The contaminants are suspended in the oil until trapped in the oil filter, and are removed when the filter is replaced during routine maintenance inspections.

It has been discovered that some of the particles held in suspension are so fine that they pass through the filter and act as polishing compound, thus further reducing wear. With previous coagulating-type detergent oils, these small particles were trapped with the larger ones in the filter, or in small oil supply passages and the polishing effect was lost, not to mention lubrication to some parts.

Protects Against Corrosion

A coating of engine oil on all parts inside the engine prevents oxygen and moisture from reaching the metal and protects it from rust and corrosion. Nitrided crankshafts and cylinder walls are especially susceptible to corrosion and must be protected with a covering of oil.

When an engine is to remain out of service for an extended period of time, the engine must be operated on fresh oil with a percentage of special preservative oil added, and then the cylinder walls should be protected with a 100% concentration of preservative oil that clings to the surface better than ordinary engine oil. If the engine is not operated a short time with clean oil, the acid concentrations that naturally build up as oil is used will corrode critically stressed areas and render the engine unairworthy. The difficulty is that many do not understand this, and operate the engine after storage in a very dangerous condition. Corrosion is the most relentless kind of stress-riser, and it is not detectable until engine component failure occurs, or engine tear-down. This is why manufacturers require the engine to be operated on fresh oil before long-term storage; if it is not, the engine should be considered unairworthy until it is disassembled and inspected.

Performs Hydraulic Action

The vast majority of horizontally opposed engines have hydraulic valve lifters that keep all clearance out of the valve operating mechanism, and almost all of the larger engines have hydraulically actuated constant-speed propellers. The engine lubricating oil acts as the hydraulic fluid for the valve lifters and propeller pitch-change mechanism.

STUDY QUESTIONS: FUNCTIONS OF THE LUBRICATION SYSTEM

Answers are on Page 122. Page numbers refer to chapter text.

1. Six functions of the lubricating system in a reciprocating engine are:
   a. ______________________________
   b. ______________________________
   c. ______________________________
   d. ______________________________
   e. ______________________________
   f. ______________________________
   Pages 93–95

2. When two surfaces are separated by a film of oil, the only friction between the parts is the ______________________ of the oil. *Page 93*

3. Most of the heat picked up by the oil in a reciprocating engine comes from the ____________________ and the ___________________ . *Page 94*

4. The underside of a piston head is ______________________ to increase the surface area from which the oil can absorb heat. *Page 94*

5. The oil that absorbs heat from the cylinder walls is oil that has been sprayed from the crankshaft at the __________________________ . *Page 94*

6. Heat that is picked up from inside the engine is given up to the outside air that passes through the __________________________ . *Page 94*

7. When an engine is to remain out of service for an extended period of time, the cylinder walls should be protected with a special ________________________ . *Page 95*

Reciprocating Engine Lubricating Oils

There are four bases for lubricants used to reduce friction between moving parts: animal, vegetable, mineral, and synthetic.

Animal lubricants are used for special applications. Neat's-foot oil from cattle hooves is used to preserve leather, and oil from the sperm whale is used to lubricate watches and clocks. Animal oil is not suitable for use as an engine lubricant because it becomes chemically unstable at high temperatures, but it can be used in the manufacture of synthetic oils.

Castor oil is a vegetable-base lubricant that was used in rotary radial engines in the World War I era. It has the disadvantage of oxidizing when exposed to the air, and it forms a gummy residue inside the engine.

Mineral oils, which are obtained by the distillation of crude petroleum, are the most widely used lubricants for aircraft engines because they have a much greater chemical stability than either animal- or vegetable-based lubricants.

Synthetic oils are made by synthesizing or changing the molecular structure of animal, vegetable, or mineral oils. Use of synthetic oil is increasing for reciprocating engines, but as yet is not universally approved.

Characteristics of Reciprocating Engine Lubricating Oil

Aviation oils have different characteristics from automotive oils because of the unique requirements of an aircraft engine. Some of these increased requirements for aircraft reciprocating engines are:

- High operating load of the bearings and gears
- Increased rotational speed of the bearings
- Elevated operating temperatures for air-cooled engines

Viscosity

Viscosity is the internal, or fluid, friction of a liquid and the resistance of the material to flow. The viscosity of engine lubricating oil has historically been measured in terms of Saybolt Seconds Universal (SSU) viscosity. A specific volume of oil is placed in a Saybolt Universal viscosimeter and heated to an accurately controlled temperature of 210°F (99°C). The number of seconds required for exactly 60 cubic centimeters of heated oil to flow through a calibrated orifice in the viscosimeter is the SSU viscosity of the oil at that temperature. A second measurement is made at 100°F (37.8°C), and the two points are plotted on a chart. The line between the points is extended to the 0°F line to extrapolate a low-temperature point.

The current method of measuring the viscosity of a lubricating oil is to measure its kinematic viscosity, or the ratio of the absolute viscosity of the oil to its density. Kinematic viscosity is measured in centistokes.

There is another test called cold-cranking simulation which is used to determine the characteristics of an oil at low temperatures. Oils rated by this

viscosity. The resistance of a fluid to flow. Viscosity is the stiffness of the fluid, or its internal friction.

Saybolt Seconds Universal viscosity. A measurement of viscosity (resistance to flow) of a lubricating oil.

The number of seconds needed for 60 milliliters of oil at a specified temperature to flow through a calibrated orifice is measured and is called the Saybolt Seconds Universal (SSU) viscosity of an oil.

The viscosity number used for commercial aviation engine lubricating oil relates closely to the SSU of the oil at 210°F.

viscosimeter. An instrument used to measure the viscosity of a liquid. The time required for a given volume of liquid at a specified temperature to flow through a calibrated orifice is used to indicate the viscosity of the liquid.

kinematic viscosity. The ratio of the absolute viscosity of a fluid to its density. Kinematic viscosity is measured in centistokes.

SAE	Commercial Aviation	Military
SAE 5W		
SAE 10W	Grade 20	
SAE 20W		
SAE 20	Grade 40	
SAE 30	Grade 65	MIL 1065
SAE 15W50	SAE 15W50	
SAE 40	Grade 80	MIL 1080
SAE 20W50	SAE 20W50	
SAE 25W60	SAE 25W60	
SAE 50	Grade 100	MIL 1100
SAE 60	Grade 120	MIL 1120

Figure 3-3. *Comparison of lubricating oil ratings*

cold-cranking simulation. A method used for specifying the characteristics of a lubricating oil at low temperature. Oils rated by this test have the letter W (standing for Winter) in their designation. For example, SAE 15W50.

viscosity index (VI). A measure of change in viscosity of an oil as it changes temperature. The higher the viscosity index, the less the viscosity changes.

test have a viscosity rating such as SAE (Society of Automotive Engineers) 20W50. The letter W stands for Winter.

Temperature has a great effect on the viscosity of a lubricating oil, and the commercial aviation rating is nominally the SSU viscosity as measured at 210°F. Two other rating systems are used for lubricating oil: the SAE rating, and the military rating. Figure 3-3 compares the two ratings.

Oil used in reciprocating engines must have a relatively high viscosity for the following reasons:

- Wide operating clearances due to the relatively large size of moving parts and the large temperature changes associated with air-cooled engines
- Different materials used and different rates of expansion of various materials
- High operating temperatures and high bearing pressures

An ideal engine oil is one with a low viscosity that can easily circulate within the engine; but it must have and maintain a strong enough film that will not break down at high operating pressures and temperatures encountered within the engine.

Viscosity Index

The viscosity index, or VI, of an oil is a measure of change in the oil's viscosity for a given change in temperature of the oil. The higher the viscosity index, the less the viscosity changes as its temperature changes. If an oil recommended for low-ambient-temperature operation is used in an engine when it is operating in high-ambient-temperature conditions, the oil is likely to have too low a viscosity when the engine is at operating temperatures. The oil pressure will be lower than normal and some oil-film breakdown will occur between parts.

Gravity

The specific gravity of a lubricating oil is the ratio of the weight of a definite volume of oil to an equal volume of pure water at its maximum density (at 4°C).

The specific gravity of an engine lubricating oil is normally expressed in terms of API (American Petroleum Institute) gravity. The API gravity number is found by using a chart that compares the specific gravity with the API number.

Ignition Points

The flash point is the temperature of the oil when it gives off enough vapors to ignite and flash, but not continue to burn, when a small flame is passed above it.

The fire point is the temperature to which the oil must be heated for the vapors to ignite and burn continuously when a small flame is passed above it.

Low-Temperature Points

The cloud point is the temperature at which the wax in the oil first becomes visible, and the pour point is the lowest temperature at which the oil will pour without disturbance.

Color

The color of an oil is rated by comparing it with an American Society of Testing Materials (ASTM) standard color chart. The colors range from 1.0, which is pure white, to 8.0, which is darker than claret red.

claret red. A dark purplish pink to a dark gray purplish red color.

Residue

The carbon-residue test measures the weight of carbon left in a test container when a given weight of oil has evaporated at a high temperature under carefully controlled conditions.

The ash test is an extension of the carbon-residue test, and is a measure of the weight of residue left after all carbon has evaporated and burned.

Types of Reciprocating Engine Lubricating Oil

There are six types of oil that are or have been used in aircraft engines: straight mineral, detergent, ashless-dispersant, multiviscosity, synthetic, and semisynthetic.

Straight Mineral Oil

Straight mineral oil is obtained by fractional distillation of crude petroleum. Two bases of crude oil produced in the United States are asphaltic and paraffinic. Asphaltic- or naphthenic-base crude oil comes from California or the Gulf of Mexico, and paraffinic-base crude oil comes from oil fields in and around Pennsylvania.

Straight mineral oil was the standard oil for reciprocating engines for many years, but it is not an ideal oil for modern engines because it oxidizes when exposed to high temperatures. It also combines with partially burned fuel, water, and lead compounds to form sludge that clogs the oil strainers and scores, or scratches, the engine bearings.

When a turbocharged engine is improperly shut down by not allowing an adequate cooling-down time, the turbocharger housing is so hot that the oil, which is no longer circulating, forms carbon, or coke, in the bearings.

coke. The solid carbon residue left when all volatile parts of a mineral oil have been evaporated by heat.

Straight mineral oil, meeting MIL-L-6082 and SAE J 1966 specifications, is no longer used as the principal lubricating oil for aircraft reciprocating engines, but most engine manufacturers recommend its use in new and freshly overhauled engines for about the first 10 to 50 hours, or until oil consumption stabilizes. After this break-in period, an ashless-dispersant oil is used. There are some notable exceptions to this procedure for some Lycoming engines and engines with some of the newer cylinder wall treatments.

detergent oil. A type of mineral oil with metallic-ash-forming additives that protects the inside of an engine from sludge and varnish buildup. Used in automotive engines, it has proven unsuitable for use in aircraft engines.

Metallic-Ash Detergent Oil

Detergent oils have been used successfully in automotive engines for years, but their use in aircraft engines has proven to be less than satisfactory. These lubricants contain ash-forming additives to improve their antioxidation characteristics, but leftover ash deposits can build up in the cylinders and absorb enough heat to cause preignition.

These additives have a strong detergent action that loosens sludge and carbon deposits which could then flow through the lubrication system and clog oil passages and filters.

Detergent oils have, in the past, been approved for some aircraft engines but they are no longer used.

AD oil. Ashless-dispersant oil. A mineral oil with nonmetallic dispersant additives that hold the contaminants in suspension, preventing them from clumping together and forming sludge deposits inside the engine.

Ashless-Dispersant (AD) Oil

The main lubricant used in aircraft reciprocating engines is an ashless-dispersant or AD oil that meets MIL-L-22851 and SAE J 1899 specifications. The additives in AD oil do not prevent the formation of carbon, and they do not break loose any sludge or carbon deposits that have formed in the engine. The dispersant additives cause the contaminants that the oil picks up to repel each other so they do not form a screen-clogging sludge, but rather, remain suspended in the oil until collected in the filters.

AD oils have such good lubricating properties that they are not recommended for the break-in period in new engines with steel or nitrided cylinder walls, and should be used only after the rings have seated and the oil consumption stabilizes. However, AD oil is recommended for breaking in some cylinders with modern cylinder-wall treatments. When installing any new or reconditioned cylinder, you must follow in detail the procedures recommended by the engine manufacturer or cylinder overhauler.

Multiviscosity Oil

The viscosity of a liquid is its resistance to flow. Water has a low viscosity, and it flows readily, but a liquid such as honey has a high viscosity, because it flows very slowly, especially when cold.

The viscosity index of an oil is a measure of the change in viscosity with a change in temperature. An oil with a high VI changes viscosity very little with changes in temperature, but one with a low VI changes viscosity appreciably as its temperature changes.

viscosity index improver. An additive used to produce a multiviscosity lubricating oil. The polymer additive expands as temperature increases and contracts as temperature decreases. VI improvers have an overall effect of minimizing viscosity change with temperature change.

An additive called a viscosity index improver (VI improver) can be mixed with a lubricating oil that will decrease its viscosity when cold and increase it when hot.

Figure 3-4 shows how the viscosity of three different oils changes with temperature. The lower curve shows that the viscosity of a typical SAE 20 oil decreases as its temperature increases. The upper curve shows the same characteristics of an SAE 50 oil. Although it is more viscous than the SAE 20 oil, its viscosity also decreases as it gets hot. The middle curve shows that the viscosity of an SAE 20W50 oil remains relatively constant over the same temperature range.

When SAE 20W50 oil is cold, it acts much like a cold SAE 20 oil, but when it gets hot, it acts like a hot SAE 50 oil.

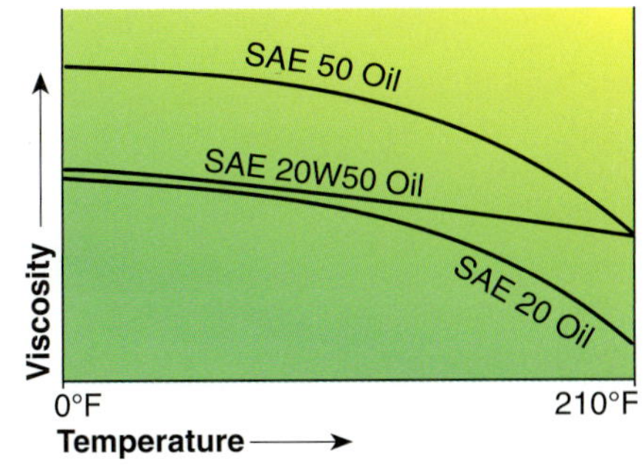

Figure 3-4. *Viscosity changes with temperature changes for an SAE 20 oil, an SAE 50 oil, and an SAE 20W50 multiviscosity oil*

Synthetic Oil

Synthetic oil is made by synthesizing or changing the molecular structure of certain animal, vegetable, or mineral bases to form a new type of oil base. Synthetic oils have superior characteristics for high temperatures and are used almost exclusively for turbine engines. They have two characteristics that make them desirable for use in reciprocating engines: They have a superior resistance to oxidation, which allows a longer period between changes, and they have low internal friction. One problem with synthetic oils is their tendency toward sludge buildup, especially in engines that are not used frequently.

Synthetic oils are not universally approved for use in reciprocating engines, but this is subject to change as further study and developments are made.

Semisynthetic Oil

A combination of mineral oil and synthetic oil with the proper additives has proven to have the characteristics of a multiviscosity oil, and does not absorb the lead salts that cause the sludge buildup in synthetic oils.

Extreme Pressure (EP) Lubricants

There are applications in aircraft engines and helicopter transmissions in which the film strength of an ordinary lubricating oil is not strong enough to withstand the high tooth pressures and high rubbing velocities encountered. For these applications, an EP lubricant is required.

EP lubricants contain additives that form iron chlorides, sulfides, or phosphides on the surface of a steel part. These surfaces give the lubricant an extremely high-strength bond with the metal.

extreme pressure (EP) lubricant. A lubricant that reacts with iron to form iron chlorides, sulfides, or phosphides on the surface of a steel part. These compounds reduce wear and damage to surfaces in heavy rubbing contact. EP lubricants are specially suited for lubricating gear trains.

Compatibility of Lubricating Oils

All mineral-base lubricating oils approved for use in aircraft reciprocating engines are compatible with each other. When a straight mineral-base oil is mixed with an AD oil, the AD characteristics decrease, but there are no problems caused by the mixing.

The additives in oils designed for automotive engines are different from those in aviation oils. Automotive oil should not be used in aircraft engines, nor should automotive oil be mixed with aviation oil. For example: If aviation oil is used in an automotive engine, the first noticeable damage is usually the cam-lobes and cam-follower faces. Automotive lobes and faces are usually softer and require additional additives than the more durable surfaces of those used in aviation. The largest difference is the concentration of a zinc derivative (usually ZnDTP) needed in the auto oil to protect these areas.

Turbine engines use synthetic-base oil that is formulated for the specific requirements of turbine engines and approved under MIL-L-7808 specifications. These requirements are different from those in reciprocating engines, and turbine engine oil should not be used in a reciprocating engine.

STUDY QUESTIONS: RECIPROCATING ENGINE LUBRICATING OILS

Answers are on Page 122. Page numbers refer to chapter text.

8. Four bases of lubricants used to reduce friction between moving parts are:
   a. ____________________
   b. ____________________
   c. ____________________
   d. ____________________
   Page 97

9. One of the problems with using animal-base lubricants in aircraft engines is that they become ____________________ at high temperatures. *Page 97*

10. Two bases for mineral oils used in aircraft engines are:
    a. ____________________
    b. ____________________
    Page 99

11. Six types of oil that are or have been used in aircraft engines are:
    a. ____________________
    b. ____________________
    c. ____________________
    d. ____________________
    e. ____________________
    f. ____________________
    Page 99

12. The type of oil recommended by most engine manufacturers for use during the initial break-in period is ____________________ oil. *Page 99*

13. Detergent oils ___________________ (are or are not) one of the more popular lubricants for use in modern aircraft engines. *Page 100*

14. The most popular lubricating oil for use in modern aircraft engines is a/an ______________________ oil. *Page 100*

15. The measure of the opposition of a liquid to flow is called its _____________________ . *Page 97*

16. The measure of the relationship between the change in the viscosity of a liquid to a change in its temperature is the _________________________ of the liquid. *Page 100*

17. An oil with a high viscosity index changes its viscosity very __________________ (little or much) with changes in temperature. *Page 100*

18. A multiviscosity oil is one that has a/an ________________________ additive mixed with it. *Page 100*

19. Two advantages of a synthetic oil for use in reciprocating engines are:
    a. ______________________________
    b. ______________________________
    Page 101

20. The type of lubricant needed when there is a high tooth pressure or a high rubbing velocity is a/an __________________ lubricant. *Page 101*

21. It ______________________ (is or is not) a recommended procedure to mix automotive oil with aviation oil if they have the same SAE number. *Page 102*

22. It ______________________ (is or is not) a recommended procedure to use turbine engine oil in a reciprocating engine. *Page 102*

23. The number of seconds required for exactly 60 cubic centimeters of a heated oil to flow through a calibrated orifice in a viscosimeter is the _______________________ of the oil at that temperature. *Page 97*

24. The SSU viscosity of a lubricating oil as measured at 210°F is used as the basis for the ____________________ (commercial aviation or SAE) rating of the oil. *Page 98*

Continued

STUDY QUESTIONS: RECIPROCATING ENGINE LUBRICATING OILS *Continued*

25. SAE 40 engine oil has the same viscosity of commercial aviation grade ________________ oil. *Page 98*

26. An ideal engine lubricating oil is one with as ________________ (high or low) a viscosity as possible consistent with a high film strength. *Page 98*

27. The temperature at which an oil gives off enough vapors to ignite and flash, but not continue to burn when a small flame is passed above it, is called the ________________ point of the oil. *Page 98*

28. The temperature at which the wax in an oil first becomes visible is called the ________________ point of the oil. *Page 99*

Reciprocating Engine Lubrication Systems

The lubrication system is one of the most important systems in an aircraft engine. If for any reason the proper amount of lubricating oil is unable to reach the engine parts to lubricate and cool them, the engine life will be very short. It is important to understand the lubricating systems so you can properly service them and diagnose any malfunctions.

Types of Lubrication Systems

There are two ways the lubrication systems of reciprocating engines can be classified: the location in which the oil supply is carried, and the method of lubrication within the engine itself.

Oil Supply Storage

sump. A low point in an aircraft engine in which the oil collects and is stored or from which it is pumped from the engine into an external tank.

dry-sump engine. An engine that carries its lubricating oil supply in a tank external to the engine.

wet-sump engine. An engine that carries its lubricating oil supply in a reservoir that is part of the engine itself.

The oil supply can be carried inside the engine itself in a wet-sump system, or it can be carried in a separate tank outside the engine in a dry-sump system. Both types of systems are used in modern aircraft. A comprehensive description can be found in the section on oil reservoirs beginning on Page 116.

Wet-Sump

The oil is carried in the crankcase of a wet-sump engine in the same way it is carried in almost all automobile engines. The oil is picked up from the sump by the oil pump, and is forced through the engine, as discussed in the internal lubricating system. When it has served its lubricating functions, it drains back into the sump and is picked up and recirculated through the engine.

Dry-Sump

The crankcase of a radial engine is too small to carry the oil supply, and some horizontally opposed engines are mounted in shallow nacelles where there is not enough room for a deep oil sump. Some airplanes designed for acrobatic flight cannot use wet-sump engines.

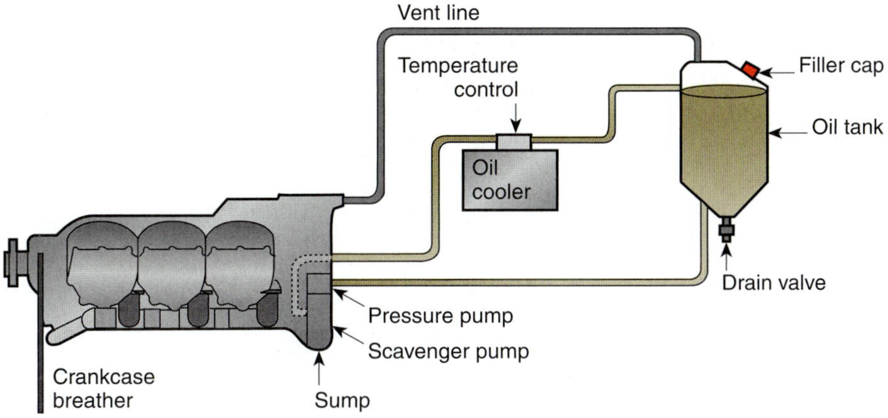

Figure 3-5. *A horizontally opposed engine using a dry-sump lubrication system*

Oil is carried in an external oil tank that is higher than the oil inlet to the engine, and it flows to the inlet of the oil pressure pump by gravity. The pump forces oil through the engine where it lubricates and cools, and then drains down into a small collection sump where it is picked up by the scavenger pump and returned to the tank.

The scavenged oil is hot, and it normally contains some air so its volume is greater than the cooler oil that is forced through the engine by the pressure pump. The scavenger pump, therefore, must have a considerably larger volume than the pressure pump.

An oil cooler with a temperature control is mounted in the line between the scavenger pump and the oil tank. If the oil does not need cooling, it passes around the core of the cooler, but if it is too hot, it is forced to flow through the cooler core where it transfers its heat to the air flowing through the cooler.

The normal procedure is to vent the oil tank to the engine crankcase, which is in turn, vented to the outside air through the crankcase breather line. This method of venting provides adequate ventilation of the tank, and prevents oil loss that could occur if the tank were vented directly.

One very popular series of Continental horizontally opposed engines has a semi-dry sump system. The oil drains into a kidney-shaped steel sump attached to the bottom of the engine, which does not require a scavenger pump. The oil is picked up by an oil pickup tube that extends down into the sump.

Most engines installed in airplanes that are to be flown inverted use dry-sump lubrication systems. The oil pickup tube in the reservoir is flexible and weighted so it can pick up oil even when the airplane is inverted.

Internal Lubrication

All moving parts of an aircraft engine must receive the proper amount of oil under the proper pressure to ensure lubrication to cushion any impact loads that are present, and carry away heat released by the burning fuel-air mixture.

The crankshaft, camshaft, and propeller shaft bearings, as well as the rocker arm bushings, are lubricated by pressure oil through drilled passages. Oil sprayed out between the crankshaft and the big-end bearing of the connecting rod lubricates the wrist pin in both the piston and the small end of the connecting rod. Cylinder walls are lubricated by oil that is sprayed from the bleed-off of the main and connecting rod bearings, and for larger opposed engines, also sprayed with oil from cylinder lubricating nozzles. Gear teeth in the accessory section are lubricated by oil that is sprayed from the accessory shaft bearings by centrifugal force and splashed around inside the accessory case.

In the wet-sump lubrication system of a typical horizontally opposed engine, such as the one in Figure 3-6, a spur-gear-type pump picks up the oil from the reservoir, or sump (1).

thermostatic valve. A temperature-sensitive valve that controls the temperature of oil in an aircraft engine. When the oil is cold, the valve directs it through the engine, but when it is hot, the valve shifts and directs the oil through the oil cooler.

From the discharge of the oil pressure pump (2), the oil flows through a thermostatic oil cooler bypass valve (3). If the oil temperature is low enough, the valve opens and the oil passes on to the pressure screen (4). But if the temperature is too high, the bypass valve closes, and the oil is forced to flow through the oil cooler (5) before going to the pressure screen.

From the pressure screen the oil goes to the oil pressure relief valve (6). This is a spring-loaded valve that is forced off its seat when the oil pressure exceeds a preset value and the oil that caused the excess pressure drains down into the sump. Oil at the regulated pressure lubricates the drive for the vane-type vacuum pump (7). Drilled passages are provided for lubricating the vanes of a wet-type vacuum pump (8) if one is used. Oil flows from the relief valve to the crankcase headers, or oil galleries (9).

The oil flows from the right header through drilled passages to lubricate the three main crankshaft bearings (10), and from these bearings through the hollow crankshaft to the crankpin bearings (11), and then sprays out to lubricate the pistons, wrist pins, cam lobes, and cylinder walls. It then drains back into the sump. The oil also flows under pressure to the camshaft bearings (12).

Oil under pressure flows into the valve lifters, or tappets (13), on the right bank of cylinders, and from the tappets it flows through the hollow pushrods (14) into the rocker boxes, then through drilled holes in the rocker arms to lubricate the rocker arm bushings. The oil sprays out from the rocker arm bushings and lubricates the rocker arms, valve stems and springs, and then drains back into the sump through oil drain tubes. Engines with the pushrods below the cylinders use the pushrod shroud tubes for the oil return.

Oil from the right header also goes into the propeller governor (15) to control the pitch of the propeller. From the governor, the oil is either directed to the propeller or it drains back into the engine sump.

Oil from the left header flows through the tappets and pushrods into the rocker boxes on the left bank of cylinders, and drains into the sump through oil drain tubes.

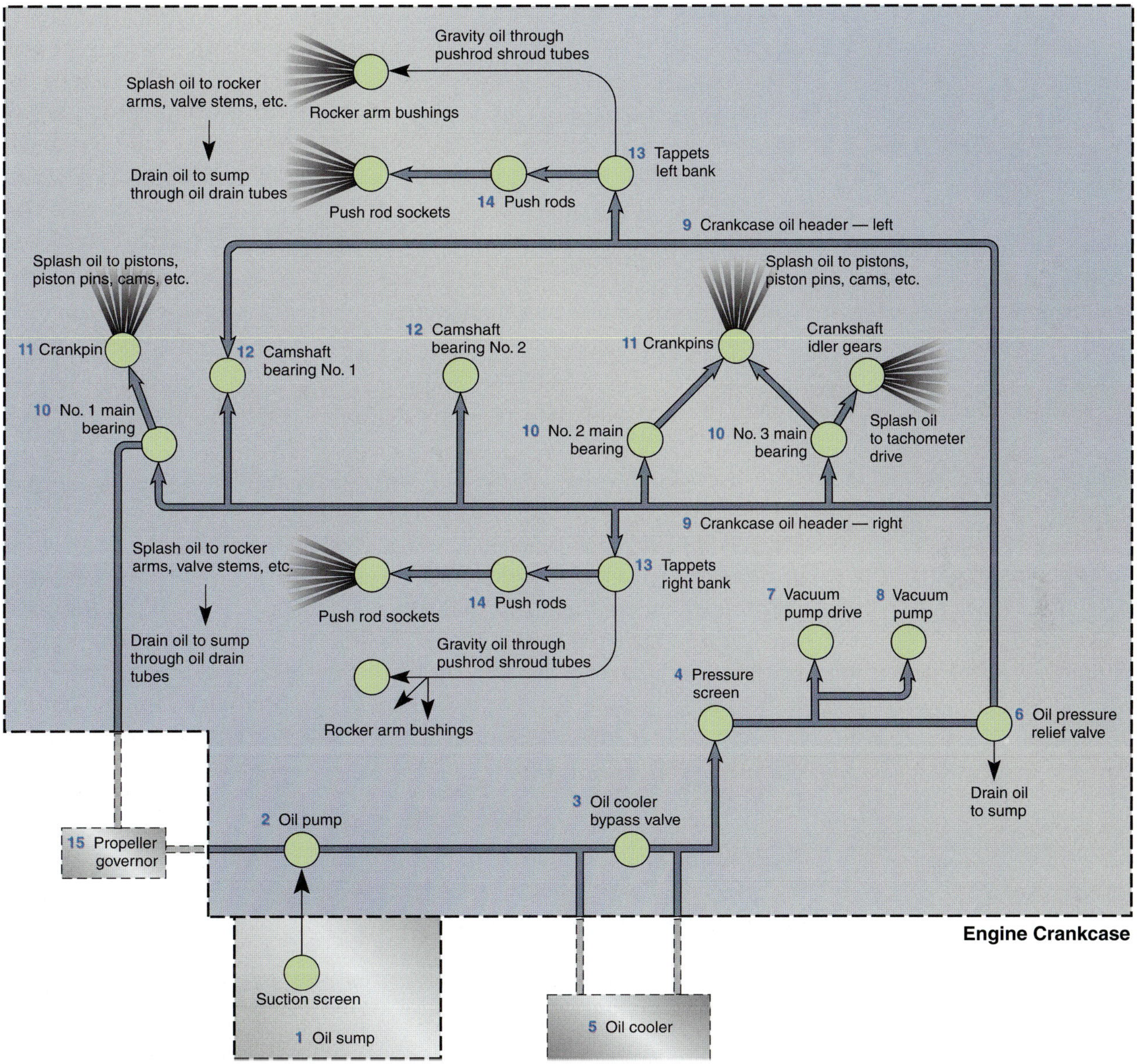

Figure 3-6. *A typical internal lubrication system for a wet-sump four-cylinder horizontally opposed engine*

STUDY QUESTIONS: RECIPROCATING ENGINE LUBRICATION SYSTEMS

Answers are on Page 122. Page numbers refer to chapter text.

29. Two types of lubricating systems based on the oil supply storage are:
    a. ______________________________
    b. ______________________________
    Page 104

30. The oil supply is carried inside the engine in a ____________________ (wet or dry) -sump engine. *Page 104*

31. The oil pump in a dry-sump engine that has the greater capacity is the ____________________ (pressure or scavenger) pump. *Page 105*

32. The oil cooler in a dry-sump engine installation is installed between the ____________________________ (scavenger pump and the oil tank or oil tank and the pressure pump). *Page 105*

33. The oil tank in a dry-sump engine is vented to the outside air through the ________________________ . *Page 105*

34. The crankcase of a reciprocating engine is vented to the outside air through a/an ______________________ line. *Page 105*

35. The wrist pins are lubricated by ________________ (pressure or spray) lubrication. *Page 106*

36. The desired oil pressure in an engine lubricating system is maintained by a pressure ______________________ (regulator or relief valve). *Page 106*

37. When the oil temperature is low, the thermostatic bypass valve is __________________________ (open or closed). *Page 106*

38. Oil used to lubricate rocker arms and valve stems is carried to the rocker boxes through hollow __________________________ . *Page 106*

Lubrication System Components

We have seen the basic lubrication system for a typical aircraft reciprocating engine, and in this section, we want to consider the various system components.

Pumps

An oil pump is not a pressure producer; it is a fluid mover. A constant-displacement pump moves a given volume of oil each time it rotates, and if there were no opposition to this flow, there would be no pressure. But the oil does encounter opposition in the engine as it is forced between the bearings and the crankshaft journals and other parts of the engine it lubricates. Oil pumps have a high capacity, and when turned at a high speed, they can force so much oil through the engine that the oil consumption can be excessive and hydraulic lifters inflate to the point that they hold valves unseated. Because of this, some form of relief valve is needed to maintain the pressure at the value specified by the engine manufacturer.

Reciprocating engines use constant-displacement oil pumps designed to move a medium amount of fluid under medium pressure, and a form of gear pump is best suited for this type of operation.

The two most commonly used types of pumps are the spur-gear pump and the gerotor pump.

constant-displacement pump. A fluid pump that moves a specific volume of fluid each time it rotates.

spur-gear pump. A form of constant-displacement fluid pump that uses two meshing spur-gears mounted in a close-fitting housing.

Fluid is moved by carrying it around the housing between the teeth of the gears as they rotate.

Spur-Gear Pump

A spur-gear oil pump, like the one in Figure 3-7, uses two external-tooth gears of the same size that mesh and rotate inside a close-fitting housing. One, driven by an accessory drive from the engine, drives the other gear. The inlet side of the pump is connected by a drilled passage or a pickup tube to the oil sump.

When the engine is turning, the two gears rotate in the direction shown in Figure 3-7. As they rotate, the teeth on the inlet side of the pump unmesh, and the volume of the inlet cavity increases, which decreases the pressure at that point.

Atmospheric pressure forces oil from the sump to fill the cavity in the pump. As the gears rotate, the oil is trapped between the teeth and the housing and carried around to the outlet side of the pump, where the teeth mesh and the volume of the cavity becomes smaller. The oil in the outlet cavity is forced out of the pump.

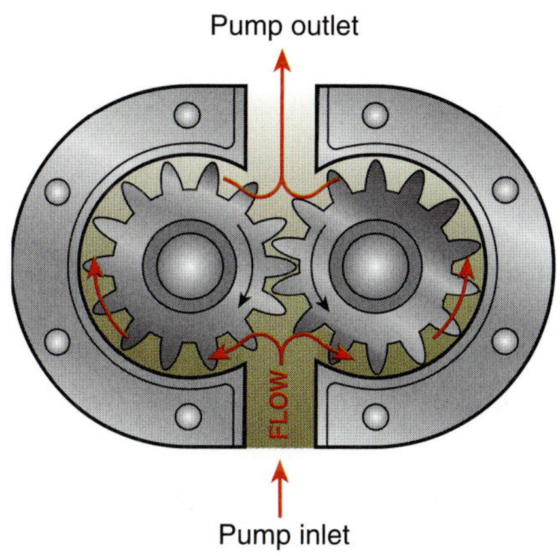

Figure 3-7. *A spur-gear oil pump*

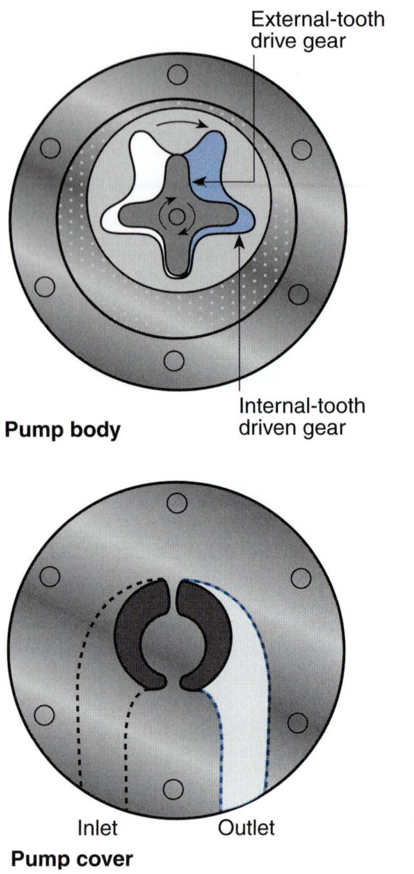

Figure 3-8. *A gerotor oil pump*

gerotor pump. A form of constant-displacement pump that uses an external-tooth drive gear that meshes with and drives an internal-tooth gear that has one more space for a tooth than there are teeth on the drive gear. Both gears turn inside a close-tolerance housing.

Gerotor Pump

A gerotor pump, like that in Figure 3-8, is a type of gear pump. A specially shaped spur gear driven by an accessory drive shaft turns inside of and drives a rotor. The rotor gear has one more internal tooth than the spur gear has external teeth. The rotor and drive gear ride inside a close-fitting housing that has two kidney-shaped openings. One opening is connected to the oil inlet from the sump, and the other is connected to the oil passages in the engine.

As the pump turns, the space between the tooth of the drive gear and the cavity in the rotor becomes larger under the inlet opening and smaller under the outlet opening. Oil is pulled into the pump as the space enlarges and is forced out as the space decreases.

Pressure Relief Valve

It is important that the oil pressure be maintained within the limits established by the engine manufacturer. When the pressure is too low, not enough oil is forced through the engine to properly lubricate and cool it. The oil temperature rises and the wear becomes excessive. Pressure higher than that recommended by the engine manufacturer should be avoided because excessive pressure can damage the oil cooler, burst the oil lines, cause excessive leakage at the oil seals, cause excessive oil consumption by supplying excessive amounts of oil to the cylinder-wall area, and cause the hydraulic lifters to keep the valves from seating.

A pressure relief valve, as shown in Figure 3-9, maintains a constant pressure in the lubricating system as the engine speed changes. At all but idling speeds, the pump moves more oil than is needed by the engine, and restriction to the flow is great enough to raise the pressure to a value higher than that specified by the engine manufacturer.

As soon as the pressure at the inlet to the relief valve creates a force greater than the force of the spring, the valve moves off its seat, and enough oil returns to the sump to maintain pressure in the oil system at the value specified by the engine manufacturer.

One advantage of using a relief valve to maintain the oil pressure in an engine is that the flow of oil through the engine increases as the clearances between moving parts of the engine increase through normal wear and as the viscosity of the oil changes with temperature. The output volume of the pressure pump remains constant for any constant RPM. As the engine parts wear, more oil flows out through the bearings and other moving parts, and less oil is returned to the sump through the relief valve.

Figure 3-9 shows two additional valves that are important to the operation of the lubrication system: the bypass valve and the check valve inside the strainer. If the strainer element should ever become clogged, the output of the pump will produce enough pressure differential across the strainer to offseat the

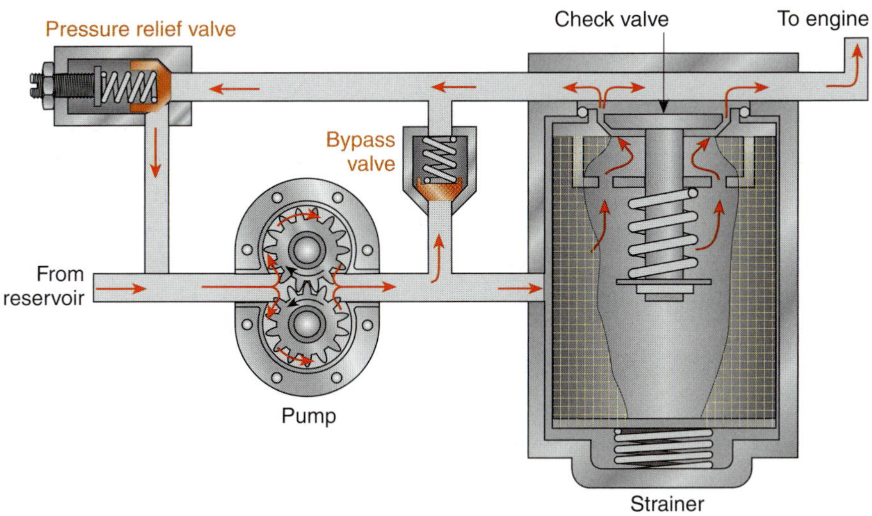

Figure 3-9. *A typical oil pressure system for an aircraft reciprocating engine*

bypass valve, and unfiltered oil will flow through the system rather than depriving the system of all oil.

The oil reservoir used in dry-sump lubrication systems is required to be higher than the inlet of the pressure pump, so the oil is forced into the pump by gravity. If the engine is not operated for a long period of time, such as overnight, oil can seep past the gears and flood the engine. To prevent this, strainers used with this type of system have a check valve in them that prevents the oil flowing into the engine until the engine starts. The gravitational force of the oil in the tank is insufficient to unseat the check valve, but the oil pump forces enough oil into the strainer to create a pressure that opens the check valve and allows oil to flow to the engine.

Some large reciprocating engines require a very high oil pressure to supply oil to the bearings when the oil is cold, but this pressure must be reduced when the oil warms up. These engines use compensated pressure relief valves that have two springs holding the valve on its seat when the oil is cold.

A bimetallic, thermostatic valve senses the oil temperature, and when the oil is sufficiently warm, it opens a passage that allows the oil to act on a piston in the relief valve that removes the force of one of the springs. For normal operation, only one spring holds the pressure relief valve on its seat, and the pressure drops to the normal operating range.

The initial setting of an oil-pressure relief valve for a newly overhauled engine is made in the overhaul shop. If the pressure is not correct when the engine is started, it can then be adjusted with the engine running and the oil at operating temperature.

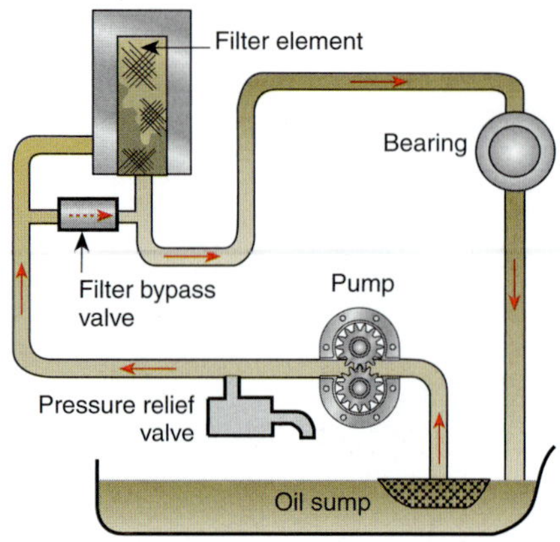

Figure 3-10. *A full-flow oil filter system*

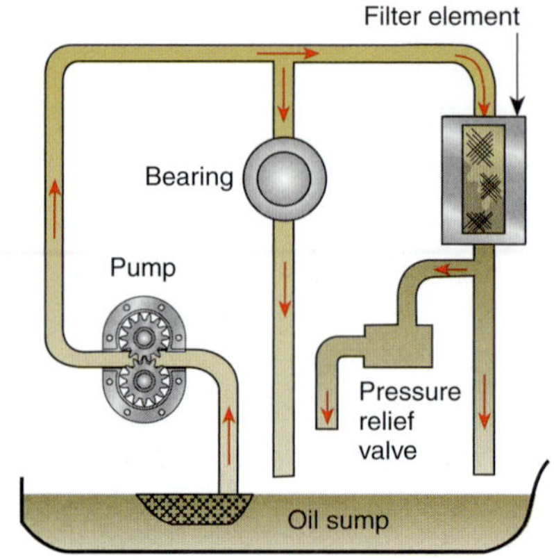

Figure 3-11. *A bypass oil filter system*

Oil Filter Systems

One function of a lubricating system is to clean the inside of the engine. As the oil flows through the engine, it picks up contaminants and carries them to the filters where they are trapped until the filter is cleaned or replaced during routine maintenance.

Oil filters trap solid contaminants, but there are acids and chemical contaminants that cannot be trapped by the types of filtering systems used, and it is because of the accumulation of these harmful substances that the oil should be periodically drained and replaced with fresh oil.

There are two basic types of oil filter arrangements: full-flow and bypass.

The full-flow filter shown in Figure 3-10, is the more common type. It is installed between the oil pump and the engine bearings. In this location, the filter can remove any contaminants from the oil before it enters the bearings.

Federal Aviation Regulation Part 33 *Airworthiness Standards for Aircraft Engines* requires that an aircraft engine oil strainer must be constructed and installed so that oil will flow through the engine at its normal rate, even when the strainer or filter element is completely blocked. Some filters have a spring-loaded bypass valve built into them, while others are designed to collapse, allowing the oil to circulate through the engine without being filtered. It is better to lubricate the engine with dirty oil than with no oil at all.

Bypass oil filters are installed in parallel with the bearings and filter only a small percentage of oil each time it passes through the engine. The oil flowing through the bearings is not filtered, but eventually, all of it is filtered with a bypass system.

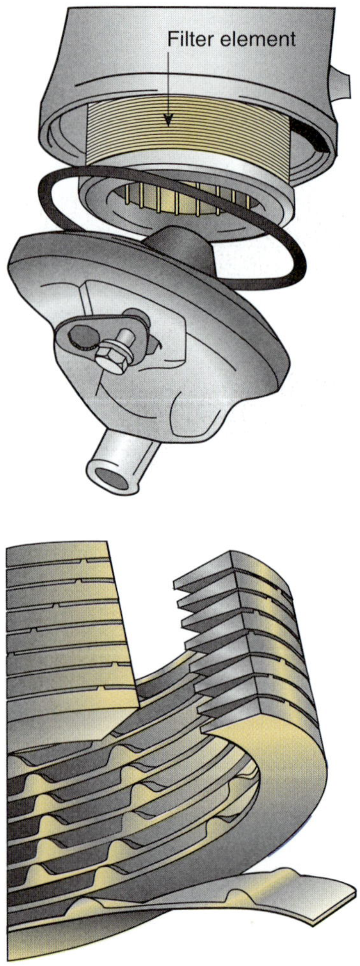

Detail of filter element

Figure 3-12. *A spiral-wound edge-type oil filter element*

Oil Filter Elements

A number of types of filter elements are used in aircraft reciprocating engines. Some elements can be cleaned and reused; others are used once and then thrown away.

Edge filters may be of either the spiral-wound or the Cuno type. A spiral-wound filter is a long strip of metal with a wedge-shaped cross section wound into a tight spiral. The thick edge of the strip forms the outside of the filter, and the turns are separated by ridges formed across the strip. These ridges hold the edges apart a uniform amount, enough to allow the oil to flow through freely, but close enough to trap solid contaminants. Contaminants collect on the outside of the filter element and are removed when the filter is cleaned during routine maintenance inspections.

Cuno filters were popular on many of the large, high-powered reciprocating engines. The filter element consists of a stack of disks made of thin sheet metal. Each disk has a center hole into which a shaft fits, and spokes join the area around the center hole with the rim. A large number of the thin disks are stacked on the center shaft, and between each disk is an even thinner spacer that is fixed rigidly to the filter housing. The thickness of the spacers determines the size of contaminants that are allowed to pass between the disks.

Engine oil flows into the filter housing on the outside of the stack of disks and flows between the disks through openings caused by the spacers. Any contaminants in the oil are trapped on the outside of and between the disks. Periodically the stack of disks is rotated by a handle attached to the shaft, and the contaminants are scraped away by the spacers. Contaminants collect in the bottom of the filter housing and are removed during periodic maintenance inspections.

Some engines that use Cuno oil filters have a small hydraulic motor built into the filter housing that slowly and continuously rotates the stack of disks to keep the contaminants scraped away.

Many engines use a screen like the one shown in Figures 3-9 and 3-13 to remove contaminants from the oil system. These screens are made with a fine wire mesh supported on a heavier wire screen. This type of filtering is referred to as surface filtration.

Most modern engines use disposable semi-depth-type filter cartridges that either fit into a filter can mounted on the outside of the engine, or are of the spin-on type similar to those used on automobile engines that screw into an adapter on the engine crankcase.

The disposable filtering element is a pleated sheet of resin-impregnated fibers that collects any contaminants from the oil that flows through it. The pleating provides a large surface area for the filtering element, which extends its service life without clogging and going into the bypass mode. This also makes it possible for the needed surface area to be housed in a reasonably sized container. The thinness of the material makes it possible for uniformity of the small filtering openings to be held consistently across the element. This improves the efficiency of the filtering element to keep larger-sized particles from finding their way through and into the engine.

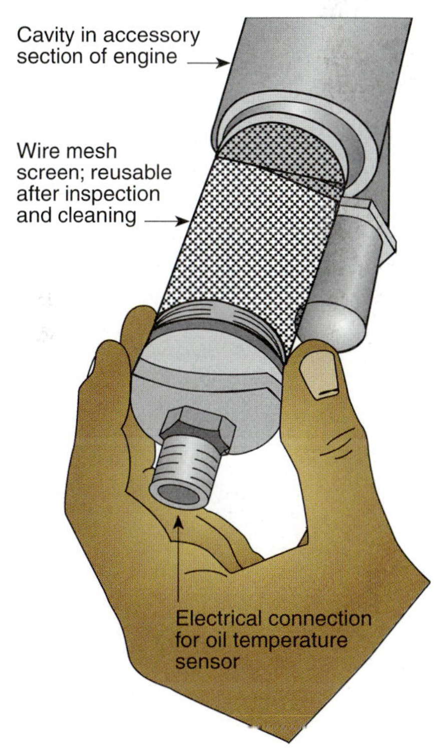

Figure 3-13. *A typical oil pressure screen installed in the accessory section of a reciprocating engine*

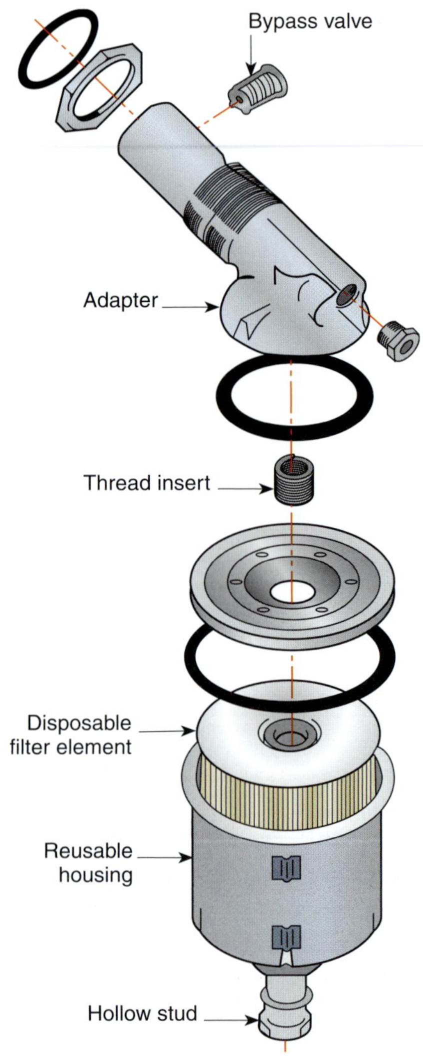

Disposable element in a reusable housing

Spin-on filter
Element and housing are disposable

Figure 3-14. *Typical disposable oil filters*

After removing a disposable filter during a routine maintenance inspection, cut the element open with a knife to eliminate the possibility of creating filter element fragments in the sample, wash the outside of the pleats into a clean container with a clean solvent, and strain the container contents through a standard white coffee filter. Under good light, look at the contaminants with a magnifying glass. Run a magnet under the coffee filter to check for any ferrous metal particles. Filter the contents of the housing, in the same manner, for particles not trapped by the filter. It is normal for there to be some ferrous and/or nonferrous metallic shavings in the filter on a new or freshly overhauled engine, but if there are any chunks of metal you must find their source. After engine break-in, only a very few extremely thin metal shavings, very miniscule in size, are acceptable in a filter sample.

Don't confuse carbon particles for metal, this is where a magnet will help to differentiate. The only nonferrous metal that looks like carbon is fine particles of aluminum oxide. This is formed in an engine when aluminum is fretted away, such as when a cylinder valve spring seat is left out at overhaul and is fretting into the base of the rocker-box. Other areas that have caused significant aluminum oxide in the filter sample are: valve guides oscillating in and out with a poppet valve, crankcase halves fretting at the parting surface, and spinning main bearing inserts.

The metal can of the spin-on filters must be cut open with a special wheel-type cutter, which does not leave any metal shavings that could be mistaken for metal particles from the engine.

Oil Coolers

Oil coolers are used in a lubricating system because it is important that the oil entering the engine be cool enough to remove the excessive heat, and to maintain the oil temperature in the correct range.

An oil cooler is an oil-to-air heat exchanger. Figure 3-6 on Page 107 shows that the oil, after leaving the pressure pump, flows through an oil-cooler bypass valve. If the oil is cool enough to pick up the heat as it passes through the engine, it bypasses the cooler and goes directly to the oil pressure screen and into the oil passages. If, however, it is too hot, the bypass valve closes, and the oil must leave the engine and flow through the oil cooler before it goes to the screen and the internal oil passages.

Some of the more powerful reciprocating engines with dry-sump lubrication systems used a rather large oil cooler in the return line between the engine and the oil tank. Figure 3-15 shows such a cooler.

This oil cooler has three noteworthy features:

- There is a built-in thermostatic flow control valve. When the oil is cool, it flows around the cooler between the core and the outer shell in what is known as the warming-jacket.

- When the oil is hot, the closed bypass valve forces the oil to flow through the core and transfers its heat to the outside air.
- If the weather is very cold and the engine is started with the cooler full of congealed oil, the surge-protection valve offseats and allows the oil to bypass the entire cooler.

In Figure 3-15A, the oil is cool, and the thermostatic flow control valve has contracted enough for the oil to flow around the core of the cooler and not be cooled. After flowing past the thermostatic valve, the oil lifts the check valve off of its seat and flows to the oil tank.

In Figure 3-15B, the oil is hot, and the thermostatic flow control valve has expanded and shut off the return passage from the outer jacket. The oil must now flow through the core of the cooler where it gives up its heat to the outside air. After passing through the core, the oil lifts the check valve off of its seat and flows to the tank.

In Figure 3-15C, the oil in the cooler has congealed, and the surge-protection valve protects the cooler from rupturing due to pressure from the scavenger pump. The high scavenger oil pressure pushes up on the surge valve and opens a passage between the inlet and outlet of the cooler. The oil flows directly to the tank until it is warm enough to liquefy the congealed oil in the cooler, allowing normal function.

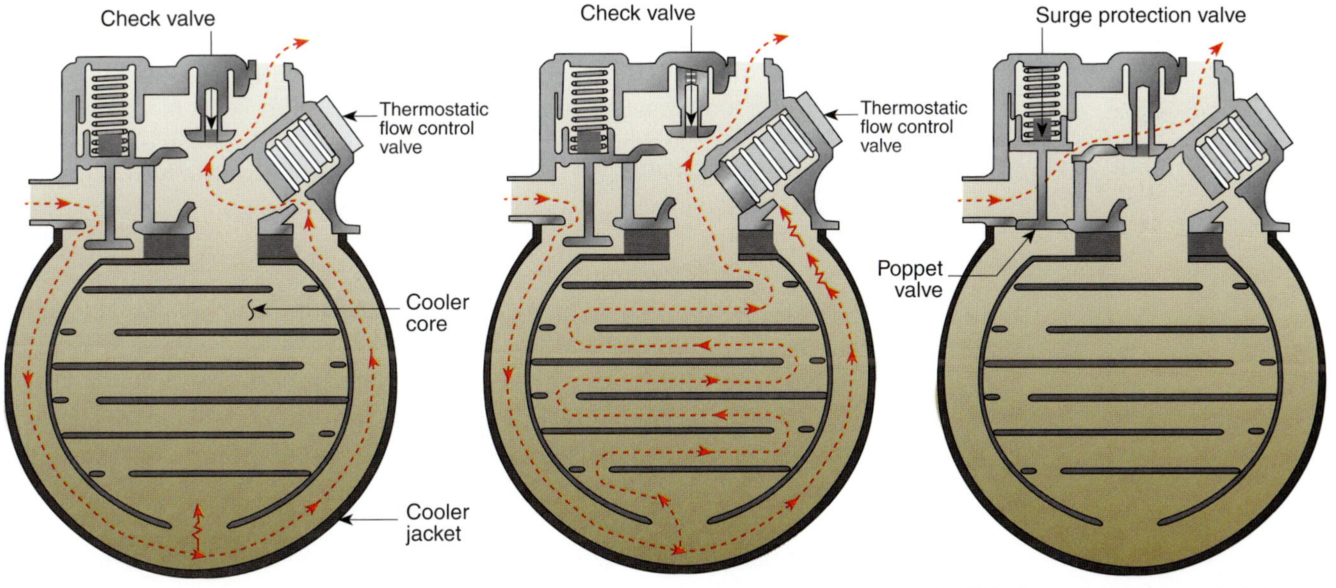

A Oil is cool and the thermostatic flow control valve is contracted, allowing oil to flow around core without further cooling.

B Oil is hot and the thermostatic valve has expanded, shutting off flow from jacket. Oil is forced to flow through core where it is cooled.

C Oil in the cooler has congealed. Pressure of oil entering the cooler forces surge protection valve up, opening the passage between cooler inlet and outlet. Poppet valve closes, preventing oil flow into cooler.

Figure 3-15. *An oil cooler for a large reciprocating engine with a dry-sump lubrication system.*

Oil Reservoirs

14 CFR Part 33, *Airworthiness Standards: Aircraft Engines* requires that the lubrication system must function properly in all flight attitudes and atmospheric conditions in which the aircraft is expected to operate. There must be enough oil carried to provide proper lubrication under critical operating conditions for the endurance of the airplane.

The two types of reservoirs used for the engine oil are wet-sumps and dry-sumps.

Wet-Sump

dipstick. A gage, made in the form of a thin metal rod, used to measure the level of liquid in a reservoir.

Wet-sump engines use the crankcase or a separate sump attached to the crankcase as a reservoir for the engine lubricating oil. Engines that carry oil in this way must meet the requirements for proper lubrication with only one-half of the maximum supply of oil in the engine.

The filler cap must be marked with the word "OIL." The oil quantity is normally measured on a wet-sump engine by a dipstick, which is often attached to the oil filler cap.

Dry-Sump

Radial engines with their small crankcase and horizontally opposed engines that fit into shallow nacelles use dry-sump lubrication systems. The oil is carried in an external oil tank that is not physically attached to the engine.

14 CFR §23.1013 requires that oil tanks must have an expansion space of not less than 10% of the tank capacity or one-half gallon, whichever is greater. The tank must be vented to the engine crankcase from the top of the expansion space so that the vent cannot be covered during any normal flight conditions. Venting the oil tank to the engine prevents heating and expansion of the oil, causing a buildup of air pressure inside the tank.

Oil flows from the oil tank to the inlet of the engine pressure pump by gravity, and some engines can be flooded by oil seeping into the engine when it is not operating. To prevent this, oil tanks used with these engines may be equipped with spring-loaded check valves that prevent the oil flowing from the tank when the engine is not running. As soon as the engine oil pressure pump begins to move the oil, the valve opens and oil flows into the engine. *See* Figure 3-9 on Page 111.

Some airplanes equipped for aerobatic flight, have engines with an oil system that prevents loss of oil when the aircraft is inverted. These oil tanks have special oil pickups that provide oil during any type of maneuver. The crankcase and oil tank are vented to the breather lines through an oil separator.

Most radial engines carry a large supply of engine oil, and in cold weather, it is inefficient to have to warm up all the oil in the tank before takeoff. Oil tanks used to hasten engine warm-up are equipped with hoppers. A hopper, seen in Figure 3-16, is a sheet metal cylinder, open at both ends and with holes in its side. It is installed inside the tank between the oil return line and the oil outlet. Oil from the scavenger pump returns to the tank at the top of the hopper, and oil leaves the tank from the bottom of the hopper, going to the pressure pump.

When the engine is started, the oil in the hopper circulates through the engine and back to the tank. As this oil becomes warm, it warms the oil surrounding the hopper, which then flows to the engine. The hopper decreases the warm-up time appreciably by allowing a small amount of oil to flow through the engine several times until the oil in the tank is uniformly heated.

For some aircraft that operate in extremely cold conditions, provisions are made to dilute the engine oil with gasoline from the aircraft fuel tanks before the engine is shut down at night. The diluted oil remains in the hopper and is the first oil used when the engine is started. As the engine runs, the warm oil returns to the hopper and warms the surrounding oil, which becomes thin enough to flow to the inlet of the pressure pump. As the engine warms up, the gasoline evaporates, and the oil regains its normal viscosity. Engine TBO is usually reduced for engines that use the dilution system.

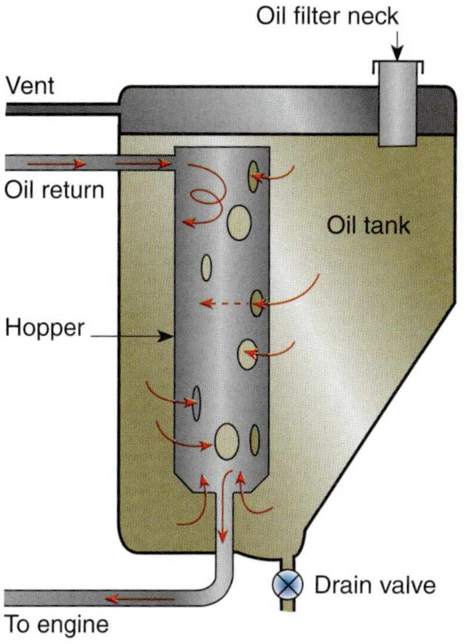

Figure 3-16. *Hoppers are used in large oil tanks to hasten the warm-up of the engine oil.*

Oil tanks used with engines equipped with Hamilton Standard Hydromatic feathering propellers have two outlets. These propellers require a supply of engine oil to feather the blades. The outlet that feeds the engine oil pump is located on the side of the tank, or the tank is equipped with a standpipe so the engine does not have access to all the oil in the tank. The outlet going to the propeller feathering pump is at the bottom of the tank that has access to all of the oil. Should the engine have a serious oil leak and lose all of the oil avail-able to the engine oil pump, there will still be enough oil in the tank to feather the propeller and thus prevent the engine from seizing and possibly shearing the propeller shaft.

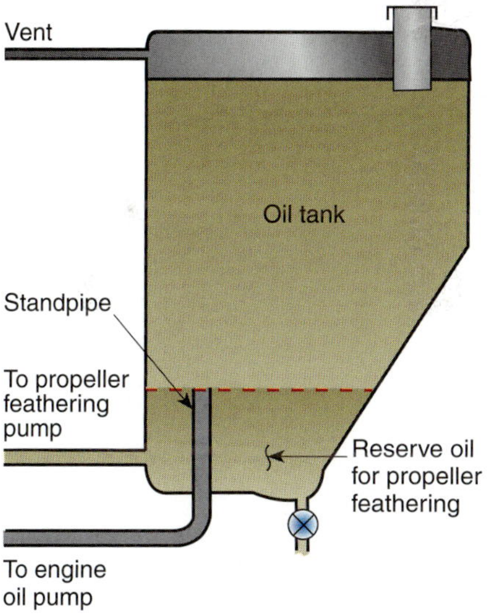

Figure 3-17. *The oil tank for an engine equipped with a Hydromatic feathering propeller has a standpipe that prevents the engine-driven oil pump from pumping all of the oil overboard in the event of a major oil leak.*

Lubrication System Instrumentation

oil dilution. A method of temporarily lowering the viscosity of the lubricating oil by diluting it with gasoline from the aircraft fuel tanks, to make it possible to start a reciprocating engine when the temperature is very low.

The lubrication system of an engine is so important that the FAA requires two instruments to show the condition of this system: an oil pressure gage for each engine that uses a pressure lubrication system, and an oil temperature gage for each air-cooled engine.

Oil Pressure Measurement

Almost all oil pressure gages are of the Bourdon tube type described in Chapter 16, beginning on Page 552.

There are two locations in an engine where the oil pressure may be measured. Some engines measure it immediately downstream of the oil pump, and others measure it at the end of the oil galleries, some distance from pump.

standpipe. A pipe which protrudes upward from the base of an oil tank and through which oil used for normal engine lubrication is drawn. In the event of a catastrophic leak when all oil available to the engine-driven pump is lost overboard, enough oil is available from an outlet below the standpipe to feather the propeller.

The oil pressure indication is taken from a restricted fitting (usually referred to as a snubber) in the engine, and the line to the instrument is normally filled with a lightweight oil or kerosine. The restricted fitting helps dampen any pulses in the oil pressure from the pump, reduces the loss of oil if the oil line cracks or comes loose, and retains the light oil in the oil gauge line for extended periods of time. The lightweight oil prevents the instrument reading from lagging behind the actual pressure changes that would occur if the line were filled with the high-viscosity oil used in the engine.

Oil Temperature Measurement

The oil temperature is measured on almost all smaller single-engine aircraft with a Bourdon tube pressure gage. The Bourdon tube is sealed as a unit with a capillary tube and a temperature-sensing bulb filled with a volatile liquid. The vapor pressure above the liquid varies with the oil temperature, and it is measured with the Bourdon tube.

Larger aircraft and almost all multiengine aircraft use electrical temperature measuring instruments, usually of the resistance-change type, and they are described in Chapter 16, beginning on Page 556.

The oil temperature indication is normally picked up before the oil enters the actual engine itself. In many horizontally opposed engines, the oil temperature bulb is in the oil pressure screen. Here it measures the temperature of the oil after it has passed through the oil cooler, but before it goes to the oil pressure relief valve and bearings. On some engines with dry-sump systems, the oil temperature is measured in the line between the oil tank and the engine. Continental large-bore, sand-cast crankcase engines measure the oil temperature at the end of the right-hand oil gallery, just after the oil cooler. This is a variation from 14 CFR Part 33's requirement to sample the temperature before or as it enters the engine. The engine was type-certificated this way.

STUDY QUESTIONS: LUBRICATION SYSTEM COMPONENTS

Answers are on Page 122. Page numbers refer to chapter text.

39. The oil pump used in an aircraft reciprocating engine is a ____________________ (constant or variable) -displacement pump. *Page 109*

40. Two types of oil pumps used in reciprocating engine lubrication systems are:
    a. ______________________________
    b. ______________________________
    Page 109

41. In a spur-gear oil pump, the oil moves through the pump __________________________ (between or around) the gears. *Page 109*

42. A gerotor pump __________________ (is or is not) a type of gear pump. *Page 109*

43. A constant-displacement oil pump __________________ (does or does not) require a pressure relief valve. *Page 109*

44. When the pressure produced by an oil pump is greater than that held by the pressure relief valve, the excess oil is returned to the _____________ . *Page 110*

45. When the clearance between the bearings in a reciprocating engine increases due to wear, the oil flow through the bearings increases. The output volume of the pump ______________________ (does or does not) increase to provide this greater flow. *Page 110*

46. A compensated oil pressure relief valve uses a thermostatic valve to ________________ (increase or decrease) the oil pressure when the oil is cold. *Page 111*

47. The initial setting of an oil-pressure relief valve for a newly overhauled engine is made in the _______________ (overhaul shop or test cell). *Page 111*

48. A surge-protection valve in a large-engine oil cooler protects the oil cooler from damage when the oil is too __________________ (cold or hot). *Page 115*

49. The oil tank used with a dry-sump lubrication system is vented to the outside air through the engine ___________________ . *Page 116*

50. The oil warm-up time for an engine having a large-volume oil tank is decreased by having a/an _________________ installed inside the tank. *Page 117*

Continued

STUDY QUESTIONS: LUBRICATION SYSTEM COMPONENTS *Continued*

51. When an aircraft is operated in extremely cold weather conditions, the oil may be diluted with ____________________ to make the engine easy to start. *Page 117*

52. When engine oil that has been diluted with gasoline warms up, the gasoline is removed by ______________________. *Page 117*

53. Oil tanks for engines equipped with a Hydromatic feathering propeller have two outlets for the oil. The outlet at the bottom of the tank goes to the ____________________ (engine oil pump or propeller feathering pump). *Page 117*

54. Two basic types of oil filter arrangement are:
    a. ______________________________
    b. ______________________________
    Page 112

55. The more common type of oil filter arrangement used in aircraft reciprocating engines is the ______________________ filter. *Page 112*

56. The size of contaminants that can be trapped by a Cuno filter is determined by the thickness of the ____________________ (disks or spacers). *Page 113*

57. Most modern aircraft reciprocating engines use ____________________ (cleanable or disposable) oil filters. *Page 113*

58. Two instruments relating to the engine lubrication system required for an air-cooled aircraft engine are:
    a. ____________________________
    b. ____________________________
    Page 118

59. Pulsations in the oil pressure gage indication are minimized by the use of a/an _________________ in the oil line to the engine. *Page 118*

60. The oil pressure line between the engine and the oil pressure gage is filled with _________________ or _________________. *Page 118*

61. Most of the electrical oil temperature measuring instruments are of the ______________________ (resistance-change or thermocouple) type. *Page 118*

62. The oil temperature shown on the instrument in the cockpit is the temperature of the oil as it ____________________ (enters or leaves) the engine. *Page 118*

Spectrometric Oil Analysis

The condition of lubricating oil after it has served its function in an engine is a good indicator of the internal condition of the engine. Although the filters are cut open and checked for particles of metal they may have collected, by the time the engine is actually producing particles large enough to be readily visible, serious damage has probably already occurred.

A spectrometric oil analysis program, or SOAP, as it is generally called, consists of taking a sample of oil at each oil change and sending it to a lab where it is diluted with an appropriate solvent and burned in an electric arc. The wavelength of light produced by the burning oil is analyzed by a computer, and the number of parts per million of certain elements in the oil is printed out. Normally the elements detected are iron, copper, chromium, aluminum, nickel, tin, and silicon.

The oil sample should be taken in the manner specified by the lab, and placed in a container furnished by the lab. The sample is normally obtained when the oil is hot, and is taken from the oil stream during the drain, not from the first, or last oil that comes from the sump. Samples should be taken at regular oil-change intervals and the results of all tests plotted in a graph.

The real value of an oil analysis is not in the amount of trace elements themselves, but rather in the change in the amount of any particular element over time.

An increase of iron could indicate wear of the piston rings or cylinder walls. An increase of aluminum could indicate problems with the pistons, and an increase in copper and tin points to bearing wear. Some conditions that are more difficult to analyze have been: a steep rise in silicon which turned out to be a ¼-inch thin slit in the flexible duct between the filter and the carburetor air-box, and an excessive copper rise when a rocker bearing wore through the steel jacket of the Continental rocker shaft. Many do not know the shaft's core is copper.

A SOAP does not by itself tell the entire story about engine condition; but when it is used properly and consistently, and analyzed along with careful inspection of the filter, consideration of the total time in service, and results of compression checks, oil analysis is a valuable maintenance tool.

Answers to Chapter 3 Study Questions

1. a. reduces friction
   b. seals and cushions
   c. removes heat
   d. cleans inside of engine
   e. protects against corrosion
   f. performs hydraulic action
2. internal friction
3. pistons, cylinder walls
4. finned
5. connecting rod bearings
6. oil cooler
7. preservative oil
8. a. animal
   b. vegetable
   c. mineral
   d. synthetic
9. chemically unstable
10. a. paraffinic
    b. naphthenic
11. a. straight mineral oil
    b. metallic-ash detergent oil
    c. ashless-dispersant oil
    d. multiviscosity oil
    e. synthetic oil
    f. semisynthetic
12. straight mineral
13. are not
14. ashless dispersant
15. viscosity
16. viscosity index
17. little
18. viscosity index improver
19. a. superior resistance to oxidation
    b. low internal friction
20. extreme pressure
21. is not
22. is not
23. SSU viscosity
24. commercial aviation
25. 80
26. low
27. flash
28. cloud
29. a. wet-sump system
    b. dry-sump system
30. wet
31. scavenger
32. scavenger pump and the oil tank
33. engine crankcase
34. breather
35. spray
36. relief valve
37. open
38. pushrods
39. constant
40. a. spur-gear pump
    b. gerotor pump
41. around
42. is
43. does
44. sump
45. does not
46. increase
47. overhaul shop
48. cold
49. crankcase
50. hopper
51. gasoline
52. evaporation
53. propeller feathering pump
54. a. full-flow
    b. bypass
55. full-flow
56. spacers
57. disposable
58. a. oil pressure gage
    b. oil temperature gage
59. restricted fitting
60. lightweight oil, kerosine
61. resistance-change
62. enters

Fuel Metering & Induction Systems

4

Continued

Reciprocating Engine Fuel Metering Systems *(Continued)*

FUEL METERING & INDUCTION SYSTEMS

4

Transformation of Energy

Aircraft reciprocating engines are a type of heat engine that changes the chemical energy in gasoline and air into heat energy, and then changes this heat energy into useful work to drive the pistons.

Gasoline is a liquid hydrocarbon compound obtained by fractional distillation of crude oil. Its chemical formula is C_8H_{18}, it nominally weighs six pounds per gallon, and contains approximately 20,000 Btu of heat energy per pound.

Heat energy in the gasoline is released by a chemical reaction between the hydrogen and carbon in the gasoline and the oxygen in the air. When the proper amounts of gasoline and air are mixed and the temperature of the mixture is raised to its kindling point, the carbon and some of the oxygen combine to form carbon dioxide, and the hydrogen and the rest of the oxygen combine to form water. This reaction takes place so rapidly that a great deal of heat is released, and it is this heat that performs useful work.

In the perfect combination of gasoline and air, two molecules of gasoline (C_8H_{18}) and 25 molecules of oxygen (O_2) combine to form 16 molecules of carbon dioxide (CO_2) and 18 molecules of water (H_2O), plus a large amount of heat.

fractional distillation. Procedure used for separating various components from a physical mixture of liquids.

Crude oil is a mixture of many different types of hydrocarbon fuels which can be separated by carefully raising its temperature. The first products to be released, those having the lowest boiling points, are some of the gaseous fuels; next are gasoline, kerosine, diesel fuel, heavy fuel oils, lubricating oils, and finally, tar and asphalt.

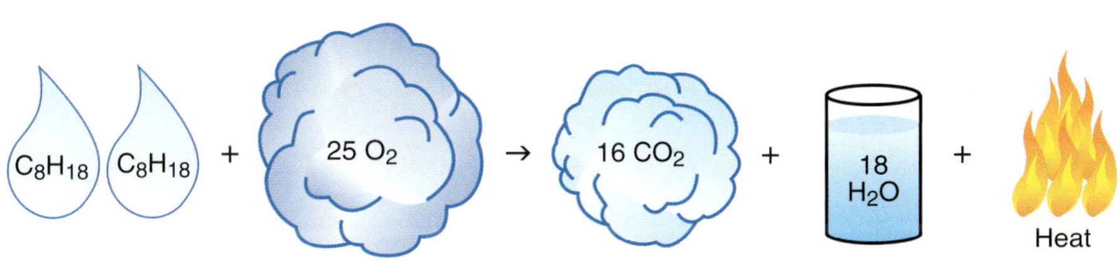

$$2\ C_8H_{18} + 25\ O_2 \rightarrow 16\ CO_2 + 18\ H_2O + \text{Heat}$$

Figure 4-1. *Gasoline and oxygen from the air combine to form carbon dioxide and water, and in the process, release a large amount of heat.*

Approximately fifteen pounds of air is needed to provide enough oxygen for complete combination with one pound of gasoline. This mixture ratio is expressed as an air-fuel ratio of 15:1, or a fuel-air ratio of 0.067 (1:15). This ratio, which provides the exact and correct number of oxygen molecules to unite with all of hydrocarbon molecules in the gasoline, is called a stoichiometric mixture.

air-fuel mixture ratio. The ratio of the weight of the air to that of the fuel in the mixture fed into the cylinders of an engine.

stoichiometric mixture. The fuel-air mixture ratio that, when burned, leaves no uncombined oxygen nor any free carbon. It releases the maximum amount of heat, and therefore produces the highest exhaust gas temperature. A stoichiometric mixture of gasoline and air contains 15 pounds of air for 1 pound of gasoline.

Gasoline will burn with a mixture as rich as 8:1 (0.125) to one as lean as 18:1 (0.056) when atomized fuel is injected into engine induction air. This ratio band is known as the "limits of flammability" for reciprocating aircraft engines. When burned in a mixture richer than 15:1, there is not enough oxygen in the mixture for the fuel to release all of its energy, and an excess of carbon will appear as black smoke and soot. When the mixture is leaner than 15:1, there is more oxygen than needed for the fuel, so the fuel cannot release all of its energy.

Air is a compressible fluid which is a physical mixture of gases. It is made up of approximately 21% oxygen and 78% nitrogen, by weight, with traces of other gases. The percentages of these gases remain relatively constant throughout the atmosphere, but the pressure produced by each gas decreases with altitude. This fact is very important in the operation of aircraft engines. The power produced by an engine at sea level will decrease as the aircraft goes up in altitude because the air density lessens, resulting in fewer pounds of oxygen to combine with the gasoline.

Since the purpose of a reciprocating engine is to convert the maximum amount of heat energy in the fuel into useful work, it would appear that the engine should always be operated with an air-fuel mixture ratio of 15:1. This is true in theory, but many variables prevent this from actually happening. These variables are discussed next.

Thermal Efficiency

thermal efficiency. The ratio of the amount of useful work produced by a heat engine to the amount of work that could be done by the fuel that was burned.

British thermal unit (Btu). The amount of heat energy needed to raise the temperature of one pound of pure water from 60°F to 61°F.

brake horsepower (BHP). The horsepower actually delivered by a reciprocating engine to its output shaft. BHP is actually the useful horsepower.

A reciprocating engine is an extremely inefficient machine, converting less than one-third of the heat energy available in the fuel burned into useful work. The thermal efficiency of an engine is the ratio of the amount of work actually done, to the amount of work that could be done by the fuel that is burned.

Aviation gasoline has a nominal weight of six pounds per gallon, and each pound of fuel contains approximately 20,000 Btu of heat energy. Each Btu can do 778 foot-pounds of work.

If an aircraft engine burns 9.25 gallons of gasoline per hour to produce 118 brake horsepower, it has a thermal efficiency of 27%.

Power in the fuel:

118 HP · 33,000 ft. pounds per HP ÷ 778 ft. pounds work per Btu =
5,005 Btu per minute required to produce 118 HP

This quantity divided by:

9.25 Gallons per hour · 6 pounds per gallon for gasoline =
55.5 pounds of gasoline per hour

55.5 pounds per hour ÷ 60 = 0.925 pounds fuel consumed per minute.

0.925 · 20,000 Btu per pound of gasoline = 18,500 Btu in the fuel
burned per minute by the engine

5005 Btu divided by 18,500 Btu yields 0.27 or
27% thermal efficiency.

The burned fuel contained enough energy to produce 436.15 horsepower, but the engine only produced 118 brake horsepower. The thermal efficiency is therefore 118 ÷ 436.15 = 0.27 or 27%.

Specific Fuel Consumption

The thermal efficiency of an engine is an important consideration, but the specific fuel consumption, which is the number of pounds of fuel burned per hour to produce each brake horsepower, is a clearer measure of engine performance.

Figure 2-13 on Page 39 is a specific fuel consumption-horsepower curve for an aircraft engine that shows the relationship between fuel consumption, engine RPM, and the power the engine is producing.

Notice in Figure 2-13 that the BSFC is lowest at 2,300 RPM, which produces 110 brake horsepower, or 61% of the engine's rated power. For the most economical cruise on the basis of BSFC, the pilot would choose an engine speed of 2,300 RPM.

brake specific fuel consumption (BSFC). A measure of the amount of fuel used for a given amount of power developed by a heat engine. BSFC is expressed in pounds of fuel burned per hour for each brake horsepower the engine develops.

At speeds below 2,300 RPM, the engine is not turning fast enough to produce its power efficiently. Above this speed, much of the power is used in overcoming increased friction, increased air pumping losses, and the necessary increased richness of the fuel mixture to cool the internal surfaces of the combustion chamber. All of these lower efficiency.

Mixture Ratio and Engine Power

A mixture of 15 pounds of air for every pound of gasoline burns with no excess oxygen, hydrogen, or carbon, and it would appear that the fuel metering system should be adjusted to produce this mixture. But the design of the induction system, the valve timing, and the amount of heat the engine can tolerate, all enter into choosing the correct mixture ratio for the existing conditions.

A carburetor is a very inefficient device for getting a uniform combustible mixture into the cylinders. The correct amount of fuel is metered and sprayed into the throat of the carburetor as droplets of liquid gasoline which evaporate to become fuel vapor. There is approximately 9,000 times the volume of air flowing into the engine as there is fuel, and the fuel vapor and the air do not mix perfectly.

The carburetor is connected to the intake valve ports of various cylinders by induction pipes, which vary in length and have several bends in them. The air and fuel vapor flowing into the cylinders does not move in a steady stream, but in a series of pulses caused by the low pressure each time an intake valve opens. These pulses cause the velocity, and therefore the pressure, to vary along the intake pipes. Ideally, the fuel vapor and air would be perfectly mixed, and the pressure of the mixture would be maximum at the intake valve in order to get the greatest amount of mixture into the cylinder. But the pressure and amount of mixing change all along the induction system,

detonation. Uncontrolled burning, or explosion, of the fuel-air mixture inside the cylinder of a reciprocating engine. Detonation occurs when the pressure and temperature inside the cylinder become higher than the critical pressure and temperature of the fuel.

The pressure rise inside the cylinder caused by the fast-moving flame front can heat and compress the unburned fuel-air mixture enough for it to explode, or release its energy almost immediately.

lean mixture. A fuel-air mixture that contains more than 15 parts of air to 1 part of fuel, by weight.

fuel-air mixture ratio. Ratio of the number of pounds of fuel to the number of pounds of air in the mixture burned in cylinders of a reciprocating engine.

and they also change with engine speed. As a result, some cylinders receive a rich mixture and others receive a much leaner mixture. This uneven fuel-air mixture distribution has been a serious problem with aircraft reciprocating engines since the earliest days, but is minimized to some extent by fuel injection systems.

An air-fuel mixture leaner than 15:1 has more air than is needed for the fuel, so the mixture does not burn completely during the time the piston is at the top of the compression/power stroke. It is still burning as the piston moves down on the power stroke, and even after the exhaust valve opens. This long period of burning raises the temperature of the cylinder enough to cause severe damage and detonation. Due to the danger of operating any cylinder with an overly lean mixture, engines are normally operated with a mixture that is inefficiently rich.

One instrument that has made powerplant management much more precise than was previously possible is the EGT, or Exhaust Gas Temperature, indicator. A thermocouple probe in the exhaust pipe near the exhaust valve measures the temperature of the gases as they leave the cylinder. The actual temperature of these gases is not as important as the trend in the temperature changes, with changes in the fuel-air mixture ratio at specific engine power levels. EGT indicators with probes for all of the cylinders detect inconsistency in the fuel-air mixture between the cylinders.

Figure 4-2 shows how the EGT increases as the mixture is leaned until the stoichiometric mixture ratio of 15:1 (0.067) is reached. When the mixture is richer than stoichiometric, there is not enough oxygen for the fuel, and the EGT is low. As the mixture is leaned, the EGT rises to a peak, and as the leaning continues, there is no longer enough fuel for the air and the temperature begins to drop off. The trend in the changes in EGT with the changes in mixture ratio is always the same, but the actual temperature depends upon the volume of fuel being burned.

Theoretically, the engine should produce the maximum amount of power when the mixture ratio releases the maximum amount of heat, but in an actual engine, this is not the case. The design of the induction system and the valve timing require the fuel-air mixture to be slightly richer than stoichiometric to produce maximum power. The fuel-air ratio that normally produces maximum power is about 0.083, or an air-fuel ratio of 12:1.

The amount of power an engine produces is determined by the number of pounds of fuel and air the engine burns. The air has its greatest density at sea level, and the volume of the air taken into the engine is at its maximum weight. The volume of air flowing through the carburetor determines the amount of gasoline it discharges. When an airplane goes up in altitude, the air becomes less dense. For all practical purposes, for a given throttle setting the volume of air flowing through the carburetor remains constant, and

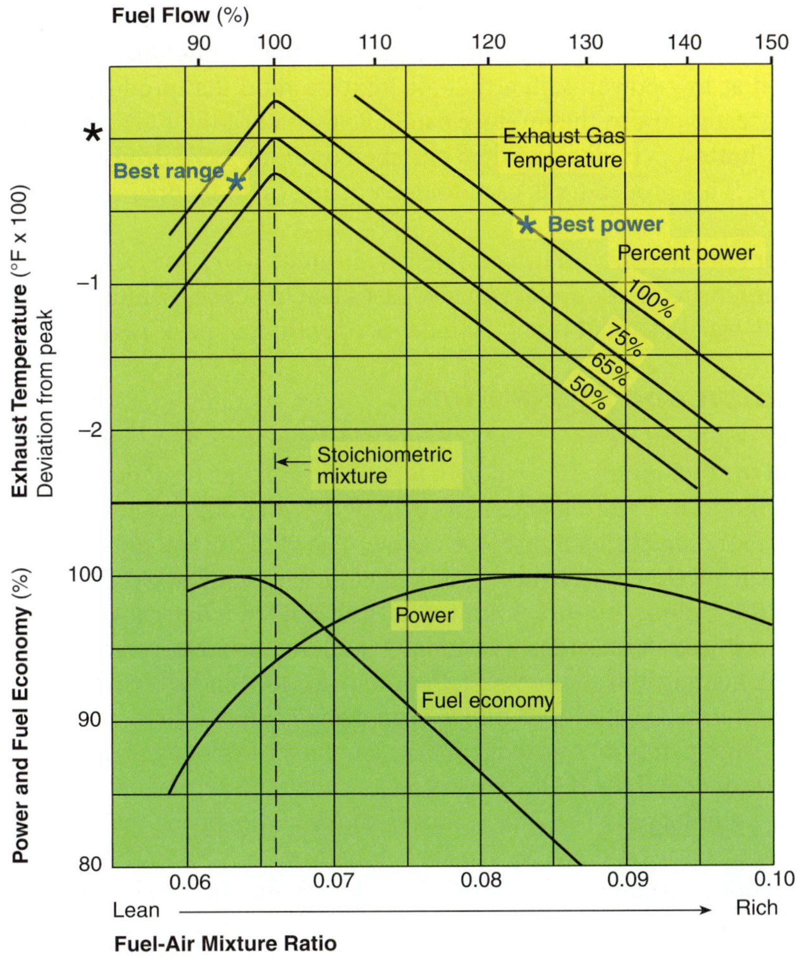

Figure 4-2. *Relationship between fuel-air mixture ratio, exhaust gas temperature, engine economy, and engine power*

Figure 4-3. *The dial of a typical exhaust gas temperature indicator. The actual temperature is not as important as the trend in the temperature changes as the mixture is changed.*

therefore the amount of gasoline discharged into the engine remains the same. However, there are fewer pounds of air per cubic foot and more importantly, less oxygen molecules per cubic foot to react with the fuel, and the mixture therefore becomes richer. It is for this reason that aircraft carburetors are equipped with a mixture control. As the mixture grows richer, the pilot can decrease the amount of gasoline the carburetor discharges into the induction air, and the mixture ratio can be kept relatively constant.

Carburetor ice is an ever-present danger with a carbureted engine, and to prevent it, warm air from around the exhaust system is directed into the carburetor. This warm air is less dense than cold, and when carburetor heat is used, the mixture becomes richer.

The maximum fuel economy is obtained by using a mixture that is approximately stoichiometric. Engine manufacturers normally allow the engine to be operated at low power with a fuel-air mixture ratio that produces peak EGT. For greater power, the mixture can be leaned until the EGT peaks on the leanest/hottest cylinder, and then enriched until the EGT drops a specified amount. This procedure is used to prevent any cylinder operating at its peak EGT.

Modern engines are often equipped with multi-point EGT indicators that measure the temperature of exhaust gases leaving each cylinder. These instruments warn the pilot if any cylinder is operating at peak EGT.

Detonation and Preignition

preignition. Ignition of the fuel-air mixture inside the cylinder of an engine before the time for normal ignition. Preignition is often caused by incandescent objects inside the cylinder.

compression ratio. The ratio of the volume of a reciprocating engine cylinder with the piston at the bottom of its stroke, to the volume of the cylinder with the piston at the top of its stroke.

Detonation is audible because flame-front propagation is 100 feet per second or lower during normal combustion, but during detonation the flame-front speed can reach 9,000 fps. When this front comes in contact with the internal surfaces at this velocity, its collision generates an audible sound; this rapid application of pressure waves destroys engine parts.

The power produced by an aircraft engine can be increased by raising the pressure of the fuel-air mixture before it is ignited. This can be done by either compressing the air before it enters the cylinders with a supercharger, or by increasing the compression ratio of the engine. The ultimate amount of power an engine can develop is determined by detonation characteristics of the fuel.

When the fuel-air mixture is ignited in the cylinder, it does not explode, but rather it burns in an orderly fashion. The flame progresses across the piston head, heating and compressing the unburned fuel and air in front of it. When the unburned fuel-air mixture is heated and compressed to its critical pressure and temperature, it no longer burns, but rather explodes, releasing its energy immediately. This explosion causes an instantaneous and tremendous increase in cylinder pressure, and produces shock waves inside the cylinder which can cause severe damage. This is detonation.

Detonation in an automobile engine is easily heard as the familiar pinging or knocking. In an aircraft engine, it is not normally heard over the exhaust and propeller noise, but can be detected by a loss of power, vibration, and a decrease in EGT. This is followed by a slow increase in cylinder head temperature.

Preignition and detonation are often thought of as being the same; and although definitely related, they are quite different. Preignition is the ignition of fuel-air mixture inside the cylinder before the time of normal ignition.

Preignition is caused by local hot spots inside the cylinder. These can be incandescent particles of carbon or red-hot edges of a valve. If the mixture is ignited and begins to burn while the piston is moving upward too early on the compression stroke, the air inside the cylinder heats and begins to expand, opposing the movement of the piston and decreasing the engine's power. The long burning period of the fuel-air mixture releases more heat into the engine than it can dissipate, and the cylinder head temperature increases. The unburned fuel and air in the cylinder is heated and compressed to its critical temperature and pressure. This causes it to explode or detonate. Detonation acts like a hammer blow at a high temperature and it can actually burn a hole in the piston head. Severe detonation can destroy an engine within a few seconds.

STUDY QUESTIONS: TRANSFORMATION OF ENERGY

Answers begin on Page 197. Page numbers refer to chapter text.

1. The air-fuel ratio that provides the correct amount of oxygen to burn all of the fuel is _______:1. *Page 127*

2. A 15:1 air-fuel mixture ratio is the same as a fuel-air ratio of _______________ . *Page 127*

3. The two chemical compounds that are produced when aviation gasoline is burned are __________________ and ________________ . *Page 127*

4. The air-fuel ratios that are considered to be the richest and leanest that will burn are _________________ for rich mixture and _________________ for lean mixture. *Page 128*

5. Air is a _______________________ (chemical compound or physical mixture). *Page 128*

6. Oxygen accounts for approximately ___________ % of the weight of the air we breathe. *Page 128*

7. If an engine burns 10 gallons of gasoline per hour to produce 120 brake horsepower, its thermal efficiency is ________________ percent. *Page 128*

8. If an engine has a brake specific fuel consumption of 0.50 pound per horsepower per hour and produces 120 brake horsepower, it will burn ____________ gallons of gasoline per hour. *Page 129*

9. A fuel-air mixture ratio is a ratio of the ________________ (weight or volume) of the fuel to the air. *Page 129*

10. As an aircraft goes up in altitude, the fuel-air mixture has a tendency to become ______________ (leaner or richer). *Page 130*

11. When carburetor heat is used on an aircraft engine, the mixture becomes ______________________ (leaner or richer). *Page 131*

12. The explosion of fuel-air mixture inside the cylinder of an aircraft engine is called ____________________ . *Page 130*

13. The ignition of fuel-air mixture inside the cylinder before the normal time of ignition is called ________________________ . *Page 132*

catalyst. A substance used to change the speed, or rate, of a chemical action without being chemically changed itself.

Reciprocating Engine Fuels

Almost all spark-ignition reciprocating engines used in aircraft are fueled with gasoline. Gasoline is a hydrocarbon fuel obtained from crude oil by the process of fractional distillation. The crude oil, with suitable catalysts added, is heated, and at specific temperatures, fractions of the oil boil off. These fractions are condensed into such products as gasoline, kerosine, turbine engine fuel, diesel fuel, and heating oil. Gasoline produced by fractional distillation is called straight-run gasoline, and is the base of practically all gasoline used in aviation engines.

Aviation Gasoline

Gasoline straight from the fractional distillation process is not suitable for either automotive or aviation applications. It must be combined with certain additives to make it suitable for its intended use.

The early airplane engines ran on just about any gasoline that was available because optimum performance was not an issue. During World War I, it was discovered that aircraft engines could function normally on some gasoline, but would seriously overheat, lose power, and fail structurally when another gasoline was used. These problems were caused by detonation, which at that time was a little understood phenomenon. Then someone discovered that by adding about 20% benzene to the fuel, the problems disappeared.

By 1919, research had proven that when a fuel-air mixture burned inside a cylinder, some heat was absorbed by the unburned fuel. When the fuel absorbed more heat than it could dissipate, the entire remaining mass of fuel ignited spontaneously and exploded, creating shock waves and extremely high pressure and temperature. Researchers had discovered the reason for the engine failures. The solution was to find something that would prevent the fuel-air mixture from exploding under intense heat and pressure.

In 1921, it was discovered that a very small amount of a poisonous organic compound, tetraethyl lead ($Pb(C_2H_5)_4$) added to the fuel, would dramatically improve the antidetonation characteristics. For the half century following the introduction of tetraethyl lead (TEL) to gasoline, it was a standard and important additive used in both aviation and automotive engines. Only in the last few decades has there been a concerted effort to eliminate leaded fuels because of environmental considerations.

Gasoline devised specifically for aviation use has been of major importance to the petroleum industry, but this is now changing. The use of aviation gasoline peaked during World War II when more than 30,000,000 gallons per day were used. After the war, airlines expanded, and their high-power reciprocating engines used large amounts of aviation gasoline.

Beginning in the 1960s, turbine engines started to replace reciprocating engines in military and airline aircraft, and the need for aviation gasoline decreased tremendously. General aviation does not create enough demand to make aviation gasoline an economical product for retailers.

general aviation. A term used to describe the total field of aviation operation except the military and airlines.

Gasoline Specifications

There are a number of specifications important to understanding reciprocating engine fuels. The following are some of the more important specifications and how they apply to aviation gasoline.

Heat Energy Content

Aviation gasoline has a nominal heat energy content of 20,000 Btu per pound, and a nominal weight of 6 pounds per gallon. Kerosine, which is a major component of turbine engine fuel, has a nominal heat energy content of 18,500 Btu per pound, and a nominal weight of 6.7 pounds per gallon.

Vapor Pressure

The vapor pressure of a liquid is the amount of pressure needed above the liquid to prevent the fuel from evaporating. Vapor pressure is determined by the temperature of the fuel and is rated according to its Reid vapor pressure (RVP), which is measured at 100°F.

If a fuel has an RVP of 15 psi at 100°F, any time the atmospheric pressure is 15 psi (approximately normal sea level pressure) or less, and the temperature of the fuel is 100°F, the fuel will vaporize. If the RVP is 5 psi, the atmospheric pressure must decrease to 5 psi, which is standard for an altitude of approximately 26,000 feet, before the fuel will vaporize.

Aviation gasoline is required to have an RVP of 5.5 to 7 psi. If the RVP is too high, the fuel vaporizes too easily; when the temperature of the fuel is high or the atmospheric pressure in the tank is low, vapors are released from the fuel, and are likely to cause a vapor lock in the lines to the engine.

If the vapor pressure of the fuel is too low, the fuel will not readily vaporize when discharged from the carburetor. The engine will be hard to start, the fuel-air distribution to the cylinders will not be uniform, the engine may not respond rapidly to additional power demands, the lubrication film will be weakened in the upper cylinder range, and fuel efficiency will be decreased.

vapor pressure. The amount of pressure needed above a liquid to prevent it from evaporating.

Reid vapor pressure. The amount of pressure that must be exerted on a liquid to keep it from vaporizing. Reid vapor pressure is measured at 100°F.

vapor lock. A condition of fuel starvation that can occur in a reciprocating engine fuel system. If the fuel in the line between the tank and carburetor is heated enough for the fuel to vaporize, a bubble will form in the line. If the vapor pressure of the bubble is high enough, it will block the fuel and keep it from flowing to the engine.

Critical Pressure and Temperature

It is important to rate a gasoline according to its ability to resist detonation, and this resistance is determined by its critical pressure and temperature.

Antidetonation characteristics of a fuel are measured by the octane rating or the performance number of the fuel. These ratings are discussed in more detail in the *General* textbook of this *Aviation Maintenance Technician Series*. The octane rating of a fuel is measured by comparing its performance in a special laboratory test engine to that of a mixture of two hydrocarbon products, iso-octane and heptane.

octane rating. A system used to rate the antidetonation characteristics of a reciprocating engine fuel.

Fuel with an octane rating of 80 performs in a laboratory test engine the same as fuel made of a mixture of 80% iso-octane and 20% heptane.

iso-octane. An organic compound, $(CH_3)_2CHCH_2C(CH_3)_3$, used as the high reference fuel for rating the antidetonation characteristics of aviation gasoline.

heptane. An organic compound, $CH_3(CH_2)_5CH_3$, that is used as the low reference fuel for rating the antidetonation characteristics of aviation gasoline.

performance number. The rating of antidetonation characteristics of a reciprocating engine fuel that is better than the high rating reference fuel, iso-octane. Performance numbers are greater than 100.

ethylene dibromide. A colorless poisonous liquid $BrCH_2CH_2Br$ that is blended with leaded gasoline to help scavenge lead oxides.

tricresyl phosphate (TCP). A colorless, combustible compound, $(CH_3C_6H_4O)_3PO$, that is used as a plasticizer in aircraft dope and an additive in gasoline and lubricating oil. TCP aids in scavenging lead deposits left in the cylinders when leaded fuel is burned.

aromatic compound. A chemical compound such as toluene, xylene, and benzene that is blended with gasoline to improve its antidetonation characteristics.

benzene. A colorless, volatile, flammable, aromatic hydrocarbon liquid which has the chemical formula C_6H_6.

Benzene, which is sometimes called benzoil, is used as a solvent, a cleaning fluid, and a fuel for some special types of reciprocating engines.

Iso-octane is a hydrocarbon compound with an exceptionally high critical pressure and temperature that is used as a high-end reference for antidetonation rating, and it has an octane rating of 100. Heptane, as the low reference, has an octane rating of zero. If a fuel performs in the same way as a mixture of 80% iso-octane and 20% heptane, it is given an octane rating of 80.

Fuels that perform better than 100% iso-octane are compared with iso-octane that contains varying amounts of tetraethyl lead as the reference, and are rated in performance numbers rather than octane ratings. Performance numbers are greater than 100.

Gasoline Additives

The antidetonation characteristics of aviation gasoline are improved when tetraethyl lead is added. But when fuel containing tetraethyl lead is burned, lead oxides are left in the cylinder, and these collect in the firing end of the spark plugs, and provide a conductive bridge for high voltage to leak across. A lead-fouled spark plug will not ignite the fuel-air mixture.

Ethylene dibromide is often mixed with the gasoline. When it burns, its residue combines with the majority of the lead oxides and converts them into volatile lead bromides, which leave the cylinder with the exhaust gases rather than fouling the spark plugs. Tricresyl phosphate (TCP) may also be added to aviation gasoline to minimize lead fouling of spark plugs. It converts the lead deposits into a nonconductive lead phosphate, which is easier to eliminate from the cylinder than is lead bromide.

Aromatic compounds such as benzene, toluene, and xylene have been added to aviation gasoline to increase its antidetonation characteristics. However, these additives attack rubber products in the fuel system and must only be used in systems specifically approved for them. These would be systems where the sealing materials are not susceptible to aromatic-compound deterioration.

Gasoline Ratings

Aviation gasoline is rated by grade according to the amount of tetraethyl lead it contains and is dyed to identify its grade. Shortly after the end of World War II before turbine engines became so important, five grades of aviation gasoline were available: red 80/87-octane, blue 91/98-octane, green 100/130-performance, brown 108/135-performance, and purple 115/145-performance.

With the tremendous decrease in aviation gasoline requirements, the 108/135 and 115/145 gasoline for large engines and the 91/98 for smaller engines were discontinued. The TEL content for the 100/130 gasoline has been decreased from 4.0 to 2.0 milliliters per gallon, and this low-lead fuel is now called grade 100LL. It is dyed blue to distinguish it from the higher-lead green grade 100/130. The extremely small demand for grade 80 gasoline makes it difficult to find, and engines that were designed to run on this fuel are being switched to 100LL or automobile gasoline.

Old designations for aviation gasoline grades were dual number ratings. The first number was the octane or performance rating of the fuel with a lean mixture ratio, and the second number was the rating with a rich mixture as would be used for takeoff.

rich mixture. A fuel-air mixture that contains less than 15 parts of air to 1 part of fuel, by weight.

Figure 4-4 shows three grades of aviation gasoline currently available with their color and lead content. Grades 80 and 100 are being phased out.

Grade	Old Rating	Color	Maximum TEL (Milliliters/gallon)
80	80/87	Red	0.5
100	100/130	Green	4.0
100 LL	100 Low lead	Blue	2.0

Figure 4-4. *Grade identification of aviation gasoline*

Automobile Gasoline

For many years automobile gasoline was considered to be totally unsuited for use in aircraft engines, but it was used illegally in a number of instances. The FAA definitely disapproved the use of automobile gasoline, and the engine manufacturers made it very clear that its use would void any engine warranty.

Most small engines were designed for grade 80 gasoline with its 0.5 milliliter per gallon of TEL. When grade 80 became unavailable, operators of these engines found that the 2.0 ml/gal. of lead in 100LL gasoline was too much, and there was a rash of serious problems with lead-fouled spark plugs and sticking valves. Even the addition of extra TCP to help scavenge the lead deposits did not completely solve the problem.

The shortage of available grade 80 fuel with its attendant high price forced many airplane owners to either quit flying or switch to a less costly fuel, and automobile gasoline was the logical choice. The Experimental Aircraft Association (EAA) made an in-depth study of the use of automobile gasoline starting in the late 1970s. Then in the mid-1990s the FAA first approved its use, under certain controlled conditions, by the issuance of a Supplemental Type Certificate (STC) for engines designed to operate on grade 80 gasoline.

Automobile gasoline is blended both for geographical regions and for the time of year, and this creates major inconsistencies between batches. There are five different vapor pressure ratings approved for gasolines used in automobiles, all higher than the maximum vapor pressure approved for aviation gasoline. Additives used in automobile gasoline are different from those in aviation gasoline, and many are incompatible with aircraft fuel system components. Gasoline containing methanol (wood alcohol) can damage seals, gaskets, hoses, and bladder-type fuel tanks. Research is being done with ethanol, or grain alcohol, which may prove to be an important fuel in the future.

methanol. Alcohol made from wood.

ethanol. Alcohol made from cereal grains such as corn.

The sophistication of automobile electronic fuel injection systems has caused gasoline producers to initiate tighter quality control over production than they have in the past. More than 12 years of research into the use of automobile gasoline in airplane engines designed for operation on grade 80 aviation gasoline has shown that when used in compliance with an STC, the use of automobile gasoline can be as safe as, and more economical than operating on 100LL aviation gasoline.

Currently, the FAA, SAE, fuel companies, and engine manufacturers are working together to produce an acceptable nonleaded fuel for aircraft usage. Due to the small demand, and increasing environmental concerns regarding lead, it is only a matter of time before all leaded fuel is removed from the market.

Importance of Proper Fuel Grade

When the FAA issues an Approved Type Certificate (ATC) for an aircraft engine, the minimum allowable fuel grade is specified, and only this grade of fuel may be used.

If fuel having an octane or performance rating lower than that specified is used, there is a real probability that when the high power required for takeoff is applied, the fuel will detonate and the engine and aircraft will be destroyed.

If the correct grade of fuel is not available, the rule of thumb has been that a higher grade of fuel could be substituted. It is true that fuel with a higher rating will not cause detonation, but the additional TEL used to give the fuel its higher rating will foul the spark plugs sooner and can cause valve sticking. Using fuel with too low TEL content, even though the octane rating is high enough, can also cause valve problems because there is not enough lead to adequately lubricate the valve stems.

There have been a number of aircraft accidents attributed to the inadvertent fueling of a reciprocating engine airplane with turbine engine fuel. This fuel will cause severe detonation and can destroy an engine in a very short time.

If an aircraft has been serviced with turbine fuel and the mistake is discovered before the engine is run, the fuel should be drained, all lines flushed with gasoline, and the tanks filled with the proper grade of fuel.

If the engine has already been run with turbine fuel, there is danger of damage from detonation, and the following should be done:

- Drain the affected tanks, lines, and carburetors and flush the system with gasoline.
- Drain the oil and examine all strainers and filters for evidence of engine damage.
- Perform a compression check on all cylinders.
- Inspect the inside of all cylinders with a borescope for indication of damage.
- Fill tanks with the proper grade of fuel.
- Conduct a complete engine run-up check.

STUDY QUESTIONS: RECIPROCATING ENGINE FUELS

Answers begin on Page 197. Page numbers refer to chapter text.

14. The major additive used to suppress detonation in aviation gasoline has been ______________________ . *Page 134*

15. Aviation gasoline has a nominal heat energy content of ______________________ Btu per pound. *Page 135*

16. Aviation gasoline has a nominal weight of ____________ pounds per gallon. *Page 135*

17. The amount of air pressure above the fuel in a tank that is needed to prevent the fuel from vaporizing is called the ________________________ of the fuel. *Page 135*

18. Vapor pressure specified for aviation gasoline is measured at a temperature of ___________°F. *Page 135*

19. The maximum Reid vapor pressure allowed for aviation gasoline is ___________ psi at 100°F. *Page 135*

20. An engine will be hard to start if the fuel has a vapor pressure that is too ____________________ (high or low). *Page 135*

21. Fuel will tend to vaporize in the fuel lines and cause a vapor lock if the vapor pressure is too ____________________ (high or low). *Page 135*

22. Two ratings that are used to indicate the antidetonation characteristics of aviation gasoline are ___________________ and _______________________ . *Page 135*

23. The hydrocarbon that is used as the high reference in octane rating is _______________ . *Page 136*

24. The hydrocarbon that is used as the low reference in octane rating is _______________ . *Page 136*

25. Fuel with antidetonation characteristics that are better than those of iso-octane is rated with a/an ________________________ . *Page 136*

26. Two additives that may be used with aviation gasoline to minimize the problems caused by the lead deposits from the tetraethyl lead are _________________ and _________________ . *Page 136*

27. Aromatic additives such as benzene are damaging to components in the fuel system that are made of _____________________ . *Page 136*

Continued

STUDY QUESTIONS: RECIPROCATING ENGINE FUELS *Continued*

28. Different grades of aviation gasoline are identified by their ________________ . *Page 136*

29. Grade 100LL gasoline is dyed ____________ and it has a maximum lead content of __________ milliliters per gallon. *Page 136*

30. In the old dual number system for identifying grades of gasoline, the first number was the octane or performance rating of the fuel when it was burned with a ______________ (lean or rich) mixture. *Page 137*

31. Automobile gasoline can be legally used in some aircraft engines when the requirements for a/an ______________________ are followed. *Page 137*

32. If a reciprocating engine has been run on turbine engine fuel, the inside of the cylinders should be inspected with a/an ____________________ . *Page 138*

Reciprocating Engine Fuel Metering Systems

The basic purpose of a fuel metering system is to measure the amount of air entering the engine and to meter into this air the correct amount of fuel for the type of operation being conducted. The liquid gasoline must then be vaporized, mixed with the air, and uniformly distributed to the various cylinders.

Reciprocating engine fuel metering systems have gone through a number of steps in their evolution. In the original Wright engine of 1903, fuel was dripped onto a hot portion of the water jacket surrounding the cylinders. The fuel vaporized and was pulled into the cylinders through automatic intake valves when the pistons moved downward on the intake strokes.

Float carburetors were devised that served as the standard fuel metering system until the pressure injection carburetor of World War II was developed to overcome some of their basic limitations.

Direct fuel injection systems similar to those used on a diesel engine were used on some large engines during World War II, but their complexity prevented widespread usage. The need for further development in direct fuel injection systems disappeared with the advent of the turbine engine.

Uneven fuel-air mixture distribution has been a real problem with carburetor-equipped engines, and the continuous-flow fuel injection system has become popular for horizontally opposed aircraft engines.

At the time of this text, float carburetors and continuous-flow fuel injection systems are the two most commonly used fuel metering systems installed on FAA-certificated aircraft; but electronic fuel injection systems such as those used in automobiles are likely to become the fuel metering system of choice for the next generation of aircraft engines.

Float Carburetors

Float carburetors are essentially fuel metering devices that utilize a float-actuated needle valve to maintain a fuel supply level slightly below the edge of a fuel discharge nozzle. An aircraft reciprocating engine fuel metering system must perform a number of functions vital to the operation of the engine. Some of these functions are:

- Measure the amount of air entering the engine.
- Meter into this air the correct amount of atomized liquid gasoline.
- Convert the liquid gasoline into gasoline vapors and distribute them uniformly to all cylinders.
- Provide a constant fuel-air mixture ratio with changes in air density and volume.
- Provide an overly rich mixture when the engine is operating at peak power to remove some of the excessive heat.
- Provide a temporarily rich mixture when the engine is rapidly accelerated.
- Provide for effective fuel metering when the engine is idling and the airflow through the carburetor is not sufficient for normal metering.

A float carburetor accomplishes these functions with five systems: main metering, idling, acceleration, mixture control, and power-enrichment, or economizer. The examples used are the Marvel-Schebler MA4-5 carburetor used on engines in the 200 horsepower range, and the Bendix NAS-3 carburetor used on engines up to about 100 horsepower. The Marvel-Schebler carburetor has both a main and a boost venturi, and the smaller Bendix carburetor has only a single main venturi.

Main Metering System

The main metering system measures the amount of air entering the engine and meters into this air the correct amount of liquid gasoline in the cruise to full power range. The liquid sprays out into the air stream and is atomized or broken into tiny droplets that can evaporate more readily.

The venturi is the heart of the main metering system. Air flowing into the engine must pass through the venturi, and as the passage in the throat of the venturi becomes smaller, the air must speed up to get through. As it speeds up, its pressure decreases. This is in accordance with Bernoulli's principle, which states that if energy is neither added to nor taken from a fluid in motion, any increase in velocity will result in a corresponding decrease in pressure. The carburetor in Figure 4-5 has a boost venturi whose exit is in the main venturi at the point of lowest pressure. This increases the velocity of air flowing through the boost venturi and produces a pressure in its center that is lower than that in the main venturi.

float carburetor. A fuel metering device that uses a float-actuated needle valve to maintain fuel level slightly below the edge of the discharge nozzle.

venturi. A specially shaped restrictor in a tube designed to speed up the flow of fluid passing through it. According to Bernoulli's principle, any time the flow of fluid speeds up without losing or gaining any energy from the outside, the pressure of the fluid decreases.

Bernoulli's principle. A physical principle that explains the relationship between kinetic and potential energy in a stream of moving fluid. When energy is neither added to nor taken from the fluid, any increase in its velocity (kinetic energy) will result in a corresponding decrease in its pressure (potential energy).

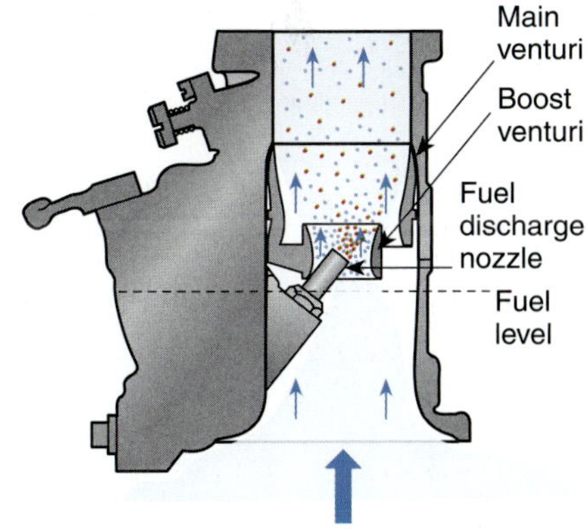

Figure 4-5. *The low pressure at the discharge nozzle caused by air flowing through the restriction of the boost venturi allows atmospheric pressure in the float bowl to force fuel out into the air stream.*

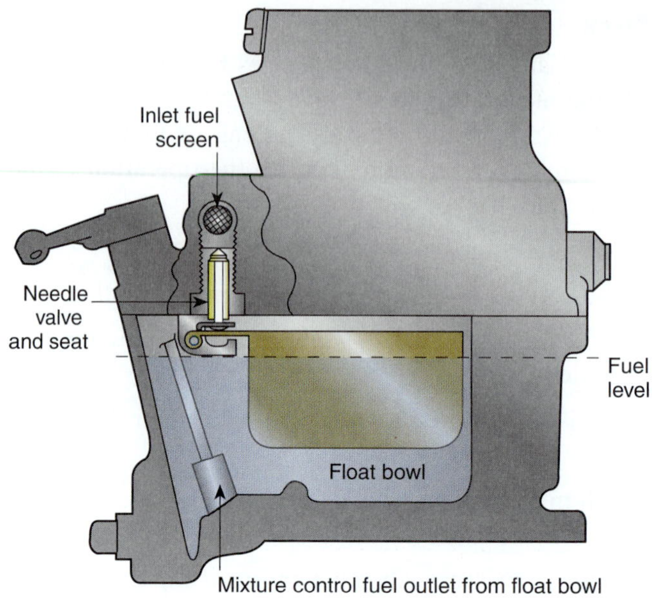

Figure 4-6. *A float actuated needle valve maintains the fuel level in the float bowl at a constant level.*

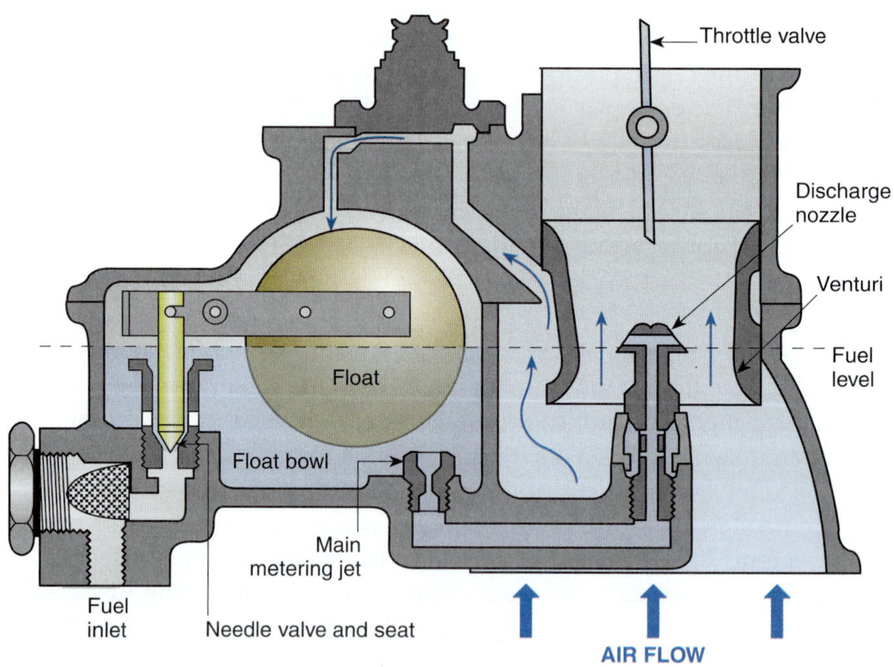

Figure 4-7. *The amount of fuel discharged into the air is determined by the difference between air pressures in the float bowl and at the discharge nozzle. The fuel-air mixture becomes richer as the air flowing into the engine increases.*

Liquid gasoline is fed into the carburetor float chamber from the aircraft fuel tank either by gravity in a high-wing airplane, or by an electric or engine-driven pump in a low-wing airplane. As the engine uses fuel, a float-operated needle valve in the carburetor maintains the fuel at a constant level in the float bowl. *See* Figure 4-6.

The carburetor in Figure 4-7 uses a single venturi and a discharge nozzle that sprays fuel out at right angles to the airflow. The discharge nozzle is located in the throat of the venturi at the point the air pressure is the lowest, and the float bowl is vented to the inlet ram air. There is therefore a difference in pressure between that in the float bowl and at the discharge nozzle. This pressure difference meters fuel from the float bowl to the discharge nozzle proportional to the volume of air flowing into the engine.

ram air. Air whose pressure has been increased by the forward motion of the aircraft. Ram air pressure is the same as pitot pressure.

Main Air Bleed

The simple metering system just described has a serious limitation. The fuel mixed with air at the discharge nozzle increases as the air flowing into the engine increases. But the fuel increases disproportionally more than the air, so the mixture becomes progressively richer as the airflow increases.

If the discharge nozzle were vented to the atmosphere with an unrestricted air bleed, as in Figure 4-8, the low pressure would pull a large amount of air into the discharge nozzle. As the airflow into the engine increases, the mixture becomes progressively leaner.

air bleed (carburetor component). A small hole in the fuel passage between the float bowl and the discharge nozzle of a float carburetor. Air drawn into the liquid fuel through the air bleed breaks the fuel up into an emulsion, making it easy to atomize and vaporize.

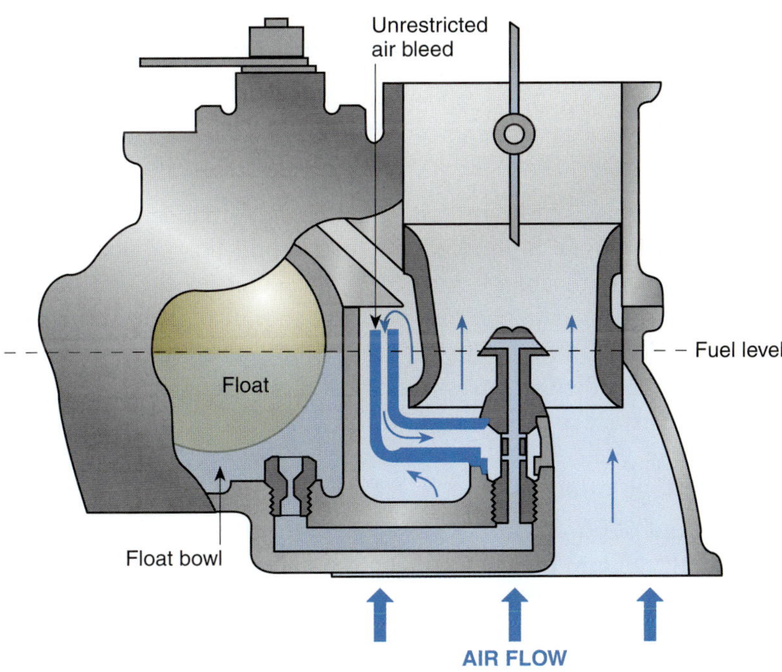

Figure 4-8. *It is easier for the low pressure at the discharge nozzle to pull air through an unrestricted air bleed than it is to pull fuel from the float bowl. The fuel-air mixture becomes leaner as air flowing into the engine increases.*

By installing a bleed restrictor with the correct size orifice, just enough air will be metered into the fuel to keep the fuel-air mixture ratio constant as the amount of air flowing into the engine changes.

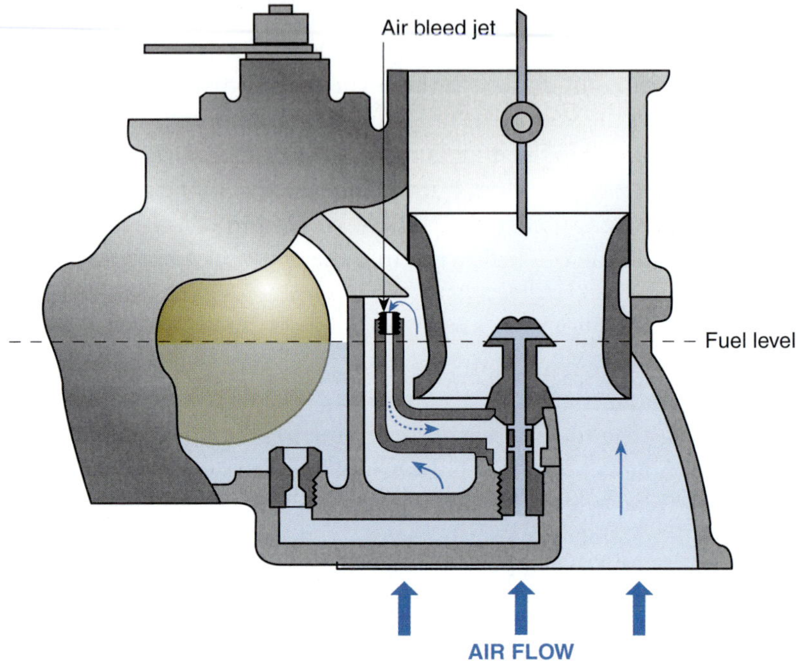

Figure 4-9. *An air bleed jet, or orifice, of the correct size will meter just enough air into the fuel at the discharge nozzle to keep the fuel-air mixture ratio constant as the amount of air flowing into the engine changes.*

emulsion. A suspension of small globules of one fluid in a second fluid with which it does not mix. Tiny bubbles of air in a column of liquid gasoline form an emulsion.

An air bleed serves more functions than just maintaining a constant fuel-air mixture ratio above high idle. The air introduced as bubbles into the fuel upstream of the discharge nozzle outlet decreases the density of the fuel and makes it easier for the low pressure to pull fuel from the nozzle. This decrease allows a more realistic venturi pressure drop to initiate fuel flow by the time the idle system is phasing out. The fuel-air emulsion provides a large fuel surface area for rapid vaporization.

butterfly valve. A flat, disk-shaped valve used to control the flow of fluid in a round pipe or tube. When the butterfly valve is across the tube, the flow is shut off, and when it is parallel with the tube, the obstruction caused by the valve is minimum, and the flow is at its greatest.

Butterfly-type throttle valves are used to control the airflow through the fuel metering system.

Airflow Regulation

The amount of power produced by a reciprocating engine is determined by the quantity of air and fuel entering the cylinders.

Air flowing into the engine at conditions other than full throttle is controlled by a circular butterfly-type valve actuated by the throttle control in the cockpit. When the throttle is wide open, the butterfly valve is parallel with the airflow, and it offers minimum restriction. When the throttle is closed, the valve nearly shuts off the flow of air into the engine, but an adjustable stop screw prevents it from completely blocking the airflow. The amount of air flowing past the nearly closed valve determines the idle speed of the engine.

The size of the venturi limits the maximum amount of air that can flow into the engine under wide-open throttle conditions. The same model of carburetor can be used on engines of various sizes by changing the venturi, the main metering jet, the main air bleed restrictor, and idle metering tube.

Idling System

When the throttle is closed, there is not enough air flowing past the main discharge nozzle to produce a pressure low enough to pull fuel from the float bowl through the main metering jet, so a separate metering system must be used for idling.

When the engine is idling, all air that flows into the cylinders must pass around the edge of the butterfly valve.

Airflow around the throttle butterfly valve is restricted, causing it to travel at a high velocity past the edge of the valve. This high velocity results in a low pressure.

There is a series of idle discharge holes in the throttle body located where pressure is the lowest when the throttle valve is in the idle range. In the largest of these holes is an idle mixture adjustment needle valve and it is located were the throttle valve pressure is the lowest at low idle. When this valve is screwed in, flow from the largest hole is shut off, and as it is screwed out, an increasing amount of fuel discharges into the air stream. The secondary, tertiary, and at times a quad opening, act as additional air bleeds at low idle and as additional fuel discharge ports when the throttle transitions from low to high idle. As the throttle opens and its edge passes the secondary, and progressively the other openings, it/they become exposed to the low pressure and transition to discharging fuel. This additional fuel is needed because the primary idle port fuel flow is decreasing as the throttle opens and additional air passes through the induction system. The throttle edge is moving away from the bore wall causing less squeeze on the air rushing by and consequently a decrease in pressure drop. As all the idle ports are just ceasing fuel flow, due to the increasing pressure rise, the main metering system begins to flow out the discharge nozzle because of the dropping venturi pressure.

A drilled passage, containing an idle emulsion tube, connects the idle discharge holes to an annulus, or ring, just above the main metering jet that is filled with fuel from the float bowl. In this passage there is a perforated idle metering tube and its entrance contains the idle metering jet. This tube screws into an enlarged hole in the carburetor body. In the side of the tube is a small hole that serves as the idle air-bleed. Air from the upper annulus formed between the venturi and the air bore supplies air at approximately ambient pressure, to a cavity on the outside of the idle tube bleed opening.

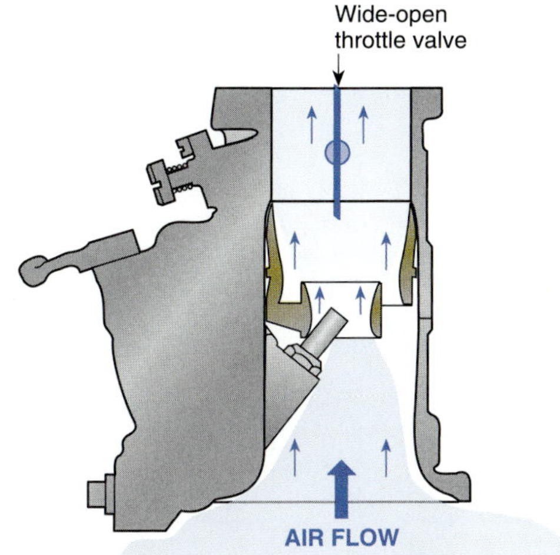

When throttle is wide open, the maximum amount of air that can flow into the engine is determined by the size of the venturi.

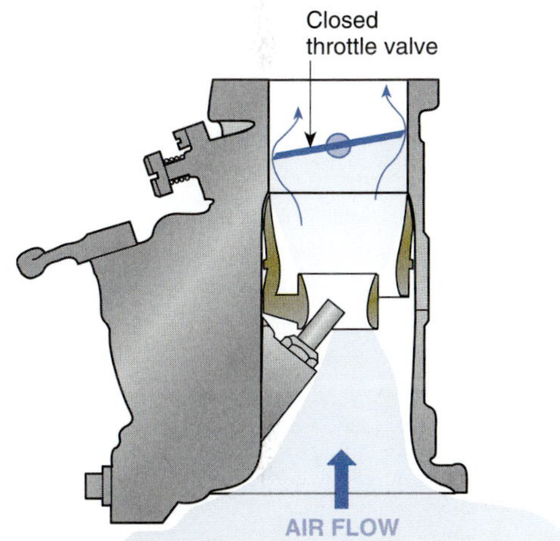

When throttle is closed, the space between butterfly valve and wall of throttle body determines the amount of air that can flow into the engine.

Figure 4-10. *Airflow regulation*

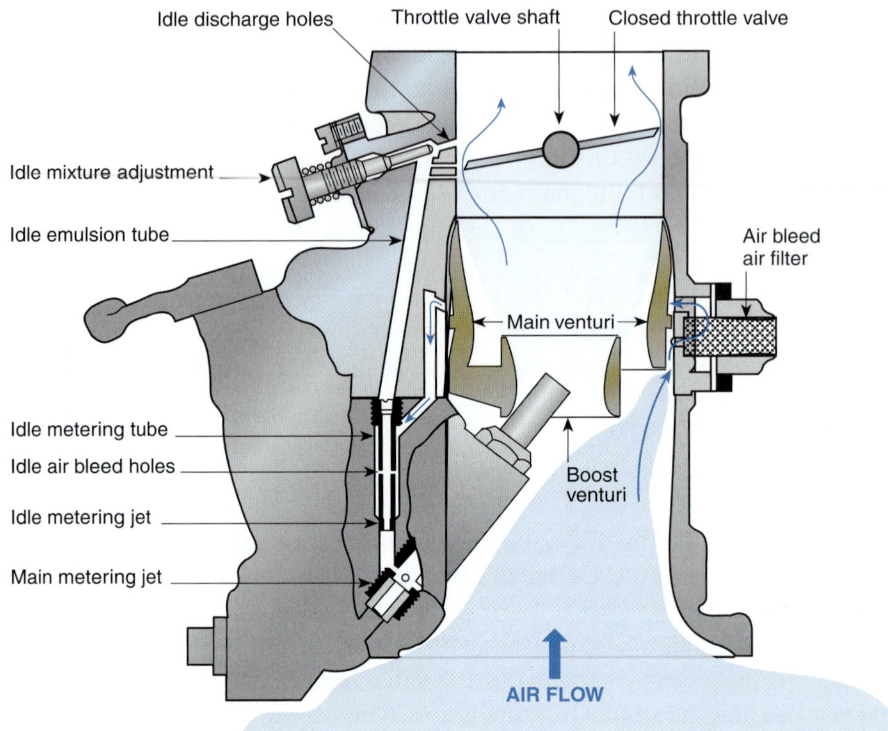

Figure 4-11. *Idling system of a typical float carburetor*

The bleed air enters the lower annulus behind the venturi, passes through a bleed air-filtering screen, and then enters the upper venturi annulus. This air supply not only feeds the idle bleed in the carburetor, depicted in Figure 4-11, but also acts as the vent for the fuel bowl.

When the throttle is closed, low pressure at the edge of the butterfly valve pulls fuel up through the idle metering jet. At the same time, it pulls air from behind the venturi through the air bleed holes in the idle metering tube. The air and fuel form an emulsion that is pulled up to the idle discharge holes and discharged into the air going into the cylinders.

Setting the idling conditions of an engine calls for two adjustments: the closed position of the throttle valve adjusts the idle speed, or RPM, and the amount of fuel-air mixture discharged determines how smoothly the engine idles. Figure 4-12 shows the controls for a typical float carburetor.

With the engine warmed up and all systems operating properly, hold the throttle in a position that produces the desired idle RPM, and adjust the idle mixture needle valve until the engine RPM peaks. From this mixture position, enrich the mixture to provide a 50 RPM decrease. With the throttle reset at the desired idle speed, open it slowly to cruise. If the transition is smooth and without hesitation, the mixture is correct but if there is hesitation in the transition, richen or lean the idle mixture slightly to correct the situation. Next, idle at the desired RPM, apply full carburetor heat, and make sure the engine continues to idle without the tendency to stall (stop operating). When the idle mixture is properly set, screw the idle RPM adjustment screw in until it contacts the idle stop on the throttle arm. Advance the throttle until the engine runs in its cruise RPM range to clear the spark plugs of any fouling caused by the idling, and then pull the throttle back. If the controls have been properly adjusted, the engine should return to a smooth idle at the speed for which it was adjusted. If it does not idle properly, repeat the process.

An overly-rich idle mixture actually aids in the transition from idle to main metering operation, aids with additional cooling at idle, smooths engine idle by making the leanest cylinders rich enough to fire consistently, and compensates for cylinder misfire due to poor exhaust scavenging at low RPM.

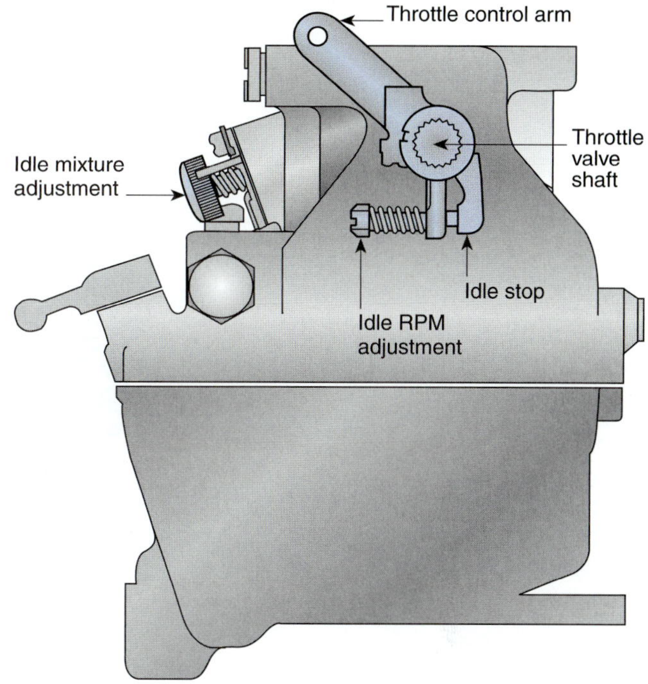

Figure 4-12. *Controls for adjusting the idle RPM and mixture on a typical float carburetor.*

Acceleration System

Both the idle system and the main metering system require a specific amount of airflow for their proper operation. When advancing the throttle from its idling position to the cruise position, there is a range in which the throttle valve is open too much for the idling system to function adequately, and not enough for the main metering system to be in full operation. This so-called flat spot will cause the engine to hesitate when the throttle is opened, but this hesitation can be eliminated by using an acceleration system.

FAA literature states that "the acceleration system compensates for the lag in fuel flow *acceleration* upon rapid opening of the throttle." The fuel flow in the main metering system is denser then the air in the induction system and has less of a differential applied across it. Therefore, the fuel acceleration rate lags that of the induction air.

annulus. A ring or groove around the outside of a circular body or shaft, or around the inside of a cylindrical hole.

The simplest acceleration system is an enlarged annulus, or groove, around the main discharge nozzle as is seen in Figure 4-13 on the next page.

When the engine is idling, all fuel entering the cylinders is metered through the idle system and the acceleration well fills with fuel. When the throttle is suddenly opened, all fuel in the well is pulled out through the main discharge nozzle and the engine receives a momentarily rich mixture that causes acceleration. As soon as the RPM builds up, air flows in through the main air bleed and the fuel metering returns to normal.

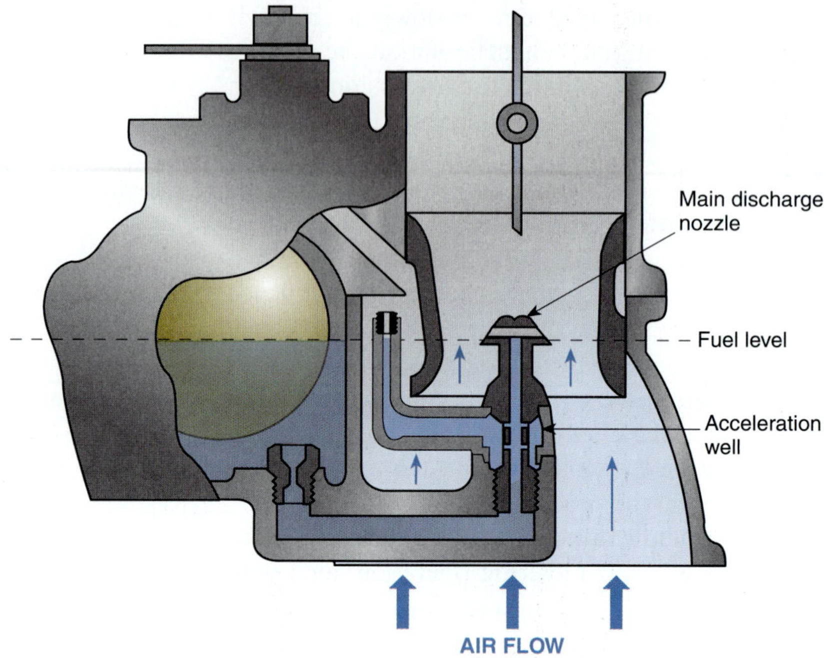

Figure 4-13. *The acceleration well fills with fuel when the engine is idling and discharges fuel through the main discharge nozzle when the throttle is first opened. This provides the momentarily rich mixture needed to transition between the idling system and the main metering system.*

Acceleration wells are used in carburetors generally mounted on engines of less than about 200 cubic inch displacement. Larger engines require a positive discharge of fuel to enrich the mixture during the transition from the idling system to the main metering system. This is because the larger venturi openings required are less responsive in the lower RPMs.

Figure 4-14 shows a typical piston-type acceleration pump. When the engine is idling, the throttle is closed, the pump plunger is at the top of its stroke, and the pump chamber is full of fuel that was pulled in from the float bowl through the pump inlet check valve. When the throttle is opened, the piston is forced down. The pump inlet check valve closes, and the fuel is pumped out through the pump discharge check valve and is sprayed into the air flowing into the engine through the venturi. The outlet check is needed to prevent induction air from being drawn into the pump during the pump's intake stroke, and the inlet check is necessary to prevent a significant portion of the fuel from returning to the fuel bowl when the pump is discharging.

The pump piston is mounted on a spring-loaded telescoping shaft. The restriction caused by the pump discharge nozzle prevents all fuel from discharging immediately when the throttle is opened. But the spring compresses, and then produces a sustained discharge as it extends to provide the engine with a rich mixture until its speed builds up enough for the main metering

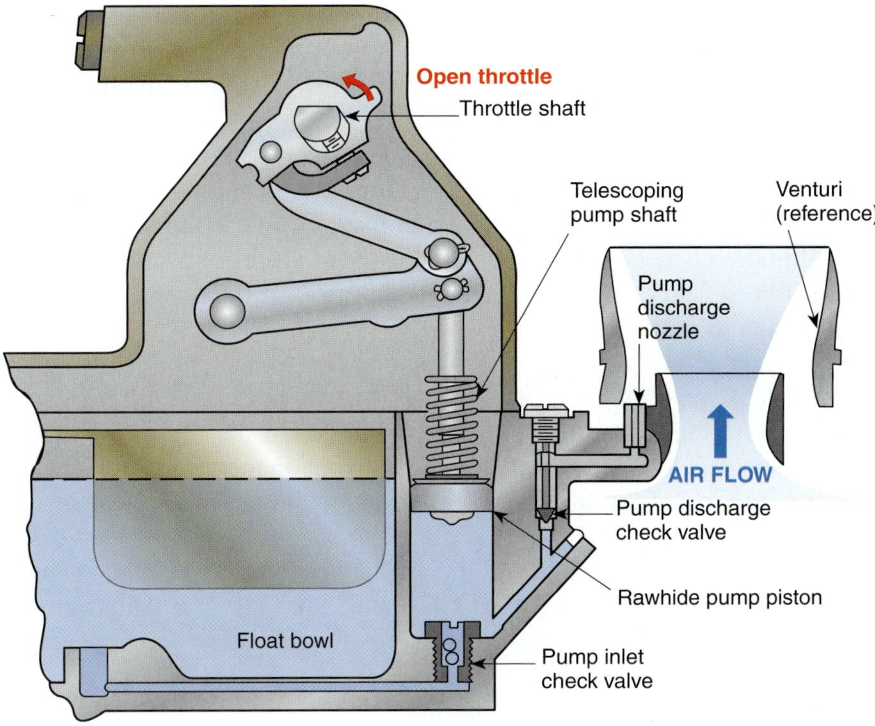

Figure 4-14. *The piston-type acceleration pump provides a sustained discharge of fuel that provides a rich mixture until the engine speed builds up enough for the main metering system to function.*

system to function. This arrangement also allows the pilot to open the throttle rapidly without damage to the acceleration linkage. The discharge valve is either weighted or spring-loaded to prevent the syphoning of fuel out of the pump discharge during main metering operation. If this valve leaks it will cause an enriching of the mixture above idle until it is compensated for with the manual mixture control system.

Mixture Control System

One of the primary limitations of a float carburetor is that it meters fuel on the basis of the volume of air entering the engine, but the power produced by the engine is a function of the mass, or weight, of the air to the appropriate amount of fuel.

An engine takes in the same volume of air for a given RPM at sea level as it does at altitude, but at altitude, the air is less dense; it has less mass. If the mixture ratio is correct for sea level air density and the airplane goes up in altitude, the same number of cubic feet of air is drawn into the cylinders, but there are fewer pounds of air. If the same amount of fuel is metered into this less dense air, the fuel-air mixture will become richer, and the engine will lose power. Because combustion is a reaction involving the oxygen in the air and the fuel hydrocarbons, as the aircraft gains altitude and the air becomes lighter it contains significantly fewer oxygen molecules per volume; for all practical purposes, the fuel contains the same quantity of hydrocarbons to react with the oxygen. Thus, if the imbalance is not corrected, the fuel mixture is too rich and engine performance is decreased.

back-suction mixture control. A type of mixture control used in some float carburetors that regulates the fuel-air mixture by varying the air pressure above the fuel in the float bowl.

Aircraft carburetors are equipped with mixture controls that allow the pilot to reduce the amount of fuel, or lean the mixture, as the airplane goes up in altitude. There are two principal types of mixture controls in common use on aircraft float carburetors: the variable-orifice mixture control and the back-suction mixture control.

The variable-orifice mixture control consists of a needle and seat in series with the main metering jet or a rotary valve in the float bowl in series with the main metering jet, as shown in Figure 4-15. When the engine is running at a high speed at low altitude, the orifice in the main metering jet is smaller than the opening of the mixture control valve, and the main metering jet does the metering.

As the aircraft goes to altitude, the mixture becomes richer and the pilot closes the mixture control valve until its opening has less area than the main metering jet orifice. Under these conditions the mixture control does the metering.

To shut the engine down, the mixture control is rotated to the idle cutoff (ICO) position, which stops all fuel flow to the engine, and it stops.

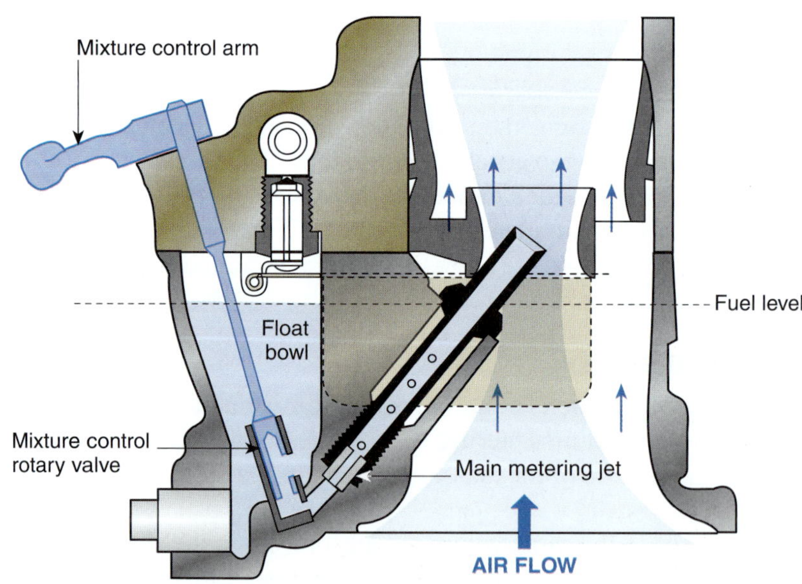

Figure 4-15. *A variable-orifice mixture control is actually a variable-area valve in series with the fixed-area main metering jet. The smaller orifice does the metering.*

A back-suction mixture control operates on the principle of varying the pressure drop across the fixed main metering jet to control the amount of fuel that flows to the main discharge nozzle.

Figure 4-16 shows the back-suction mixture control on one of the older carburetors for small aircraft engines. For rich operation, the float bowl is vented with air from the carburetor inlet that flows behind the venturi and through the open disk-type mixture control valve.

At altitude, the pilot moves the mixture control toward the lean position. This restricts the disk valve and subjects the float bowl to a slightly low pressure from the edge of the venturi. The lower pressure in the float bowl decreases the pressure drop across the main metering jet, decreasing the amount of fuel that flows through it, and leans the mixture.

Some carburetors with back-suction mixture controls have a cutoff valve that vents the float bowl of low pressure taken from the center of the venturi. This pressure is the same as that at the main discharge nozzle, and no fuel will flow through the main metering jet when these pressures are the same.

Some float-carbureted engines, particularly helicopter engines, incorporate an automatic mixture control (AMC). This is an aneroid-controlled needle valve in series with the main metering jet. As the barometric pressure decreases with altitude the aneroid expands placing the needle valve closer to its seat thus restricting the fuel flow more then the main metering jet.

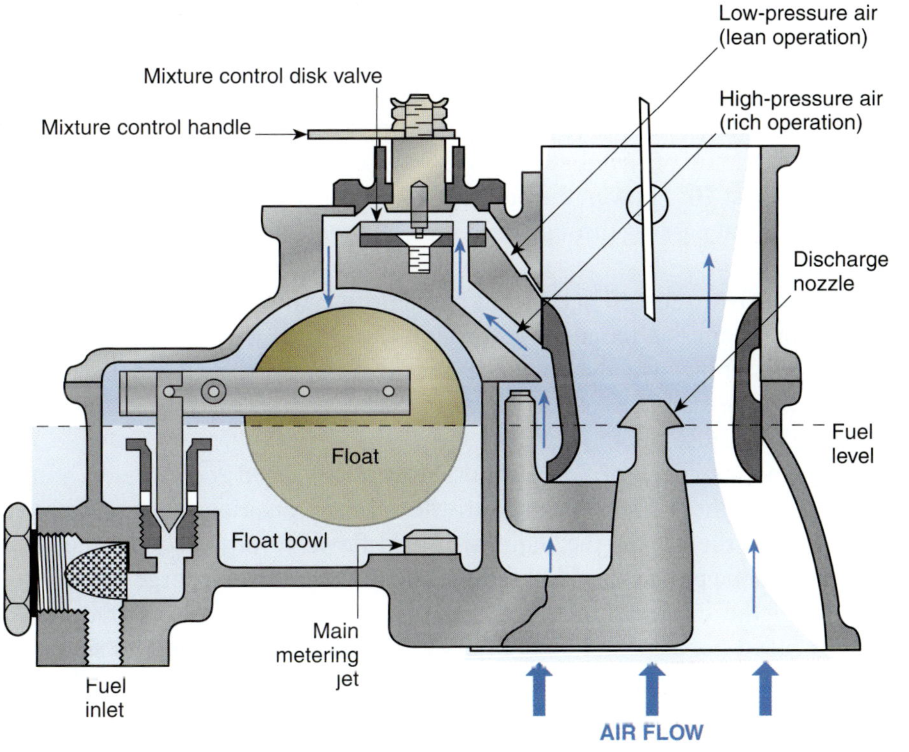

Figure 4-16. *A back-suction mixture control varies the mixture ratio by controlling the pressure of air above the fuel in the float bowl.*

Power Enrichment or Economizer System

Aircraft carburetors are designed so that they meter the correct amount of fuel into the air for economical cruise flight. But provisions must be made to remove some of the excess heat that develops in the engine when run above 70% power. The internal surfaces of the combustion chamber can only conduct heat at a certain rate. As the BTU release increases with engine power the surfaces reach a temperature where abnormal combustion will occur if additional cooling is not provided. If this additional cooling is not provided, an engine would be limited to approximately 70% of its rated power. The exact point at which additional cooling is needed depends on a number of factors, two of which are cylinder head cooling-fin density, and fuel octane.

The simplest and most effective way of removing this heat is to meter additional fuel into the air. When this fuel changes from liquid into vapor, it absorbs heat, and the exhaust gas temperature drops. It should be pointed out that approximately 30% of the fuel is not vaporized at the point of ignition with the best of fuel metering devices. Therefore, with additional fuel added there is significantly greater internal cooling during the combustion process.

The systems that provide this additional fuel are called either power enrichment systems because they enrich the mixture under full power conditions, or economizer systems because they allow the engine to operate with a lean, economical mixture when the engine is not developing full power.

There are three types of power enrichment systems commonly used: the needle-type, the air bleed-restricting type, and the back-suction type.

The needle-type power-enrichment system has an economizer, or power enrichment, valve and jet in parallel with the main metering jet. When the throttle is in a position other than fully open, the main metering jet alone limits the amount of fuel that can flow to the cylinders. But when the throttle is opened past 70% power, the economizer valve is open, and the fuel flow is the sum of that passing through both the main metering and economizer jets.

As previously discussed, a large air bleed orifice allows the fuel-air mixture to become leaner as the airflow through the engine increases. A small air bleed or no air bleed at all causes the mixture to become richer as the airflow increases. The size of the main air bleed is chosen to provide an economical mixture for cruise operation.

One popular series of carburetors uses an air bleed economizer system as seen in Figure 4-17. When the throttle is in any position other than above 70% power, the air for the main air bleed flows from the top of the float chamber to the passage around the outside of the main discharge nozzle and in through the air bleed holes. When the throttle is opened above 70% power, the air bleed needle drastically restricts the metering orifice and shuts off the air to the main air bleed. The mixture is enriched as long as the throttle is opened to this extent.

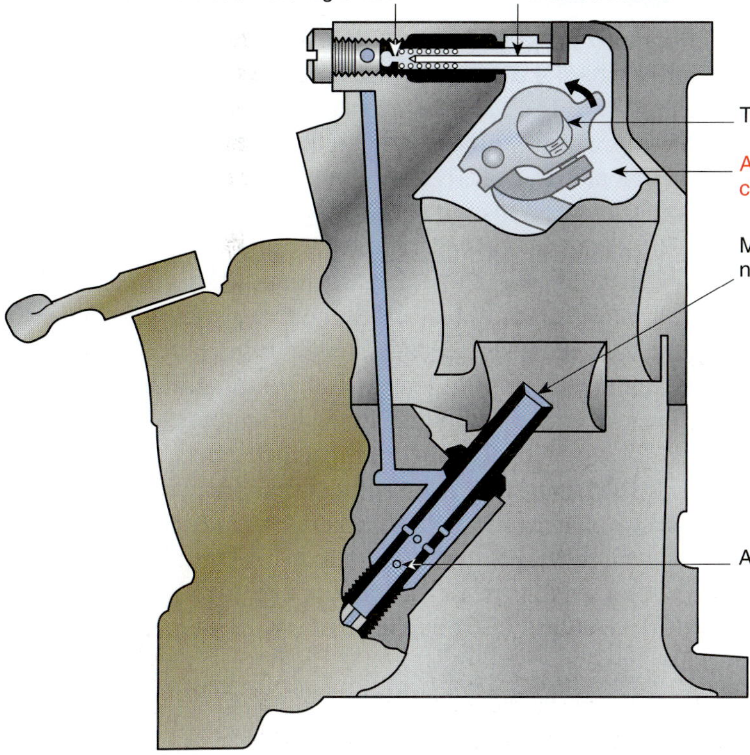

Figure 4-17. *The air bleed enrichment system restricts the main air bleed when the throttle is opened 70% and above. Restricting the air bleed causes the mixture to become richer.*

As discussed with the idle system, when the throttle plate edge forces air to accelerate past an opening the pressure decreases. In the back-suction style power-enrichment system, as used in the smaller Precision Airmotive Corporation float carburetors, the float-chamber vent is influenced by the low-pressure passage, the end of which originates in the vicinity of the lower throttle plate edge in the cruise range. With this low pressure in the back-suction passage decreasing the fuel bowl pressure, the MMJ is sized to provide the 12:1 cruise mixture ratio. As the plate edge moves away from the back-suction opening toward open throttle, the pressure rises in the bowl and the mixture is enriched to the power-enrichment ratio of 10:1.

Service and Maintenance of Float Carburetors

Float carburetors, like other accessories on an aircraft engine, do not have a specified time for overhaul, but good operating practice requires that they be overhauled at the same time as the engine. The premier aircraft float carburetor manufacturer specifies their units be overhauled with the engine.

The overhaul of a float carburetor is not a complex operation, but, like any other aircraft maintenance, the instructions of the carburetor manufacturer must be followed in detail, and all special tools called for in the overhaul procedure must be used. All applicable Airworthiness Directives must be complied with at the time of overhaul. For liability purposes, the manufacturer's overhaul manual and applicable service bulletins should be closely followed.

The float-operated needle valve is the critical component in a float carburetor. The position of the float when the needle valve shuts off the flow of fuel determines the fuel level relative to the discharge nozzle; this is a major factor in determining the amount of fuel that is metered into the induction air.

There are two ways of determining the correct setting of the float needle valve: one is by measuring the level of the fuel in the float bowl, and the other is by measuring the position of the float when the needle valve is closed.

The fuel level is measured in carburetors in which the needle valve is mounted in the float bowl.

Mount the carburetor bowl on a level surface and connect the fuel line. Supply fuel to the bowl under the pressure specified in the carburetor overhaul manual. Measure the distance between the parting surface of the float bowl and the level of fuel. Measure this distance at some point away from the float or float bowl wall because capillary attraction causes fuel to ride up these surfaces. If the level is not correct, adjust it by adding or removing shims from beneath the needle valve seat.

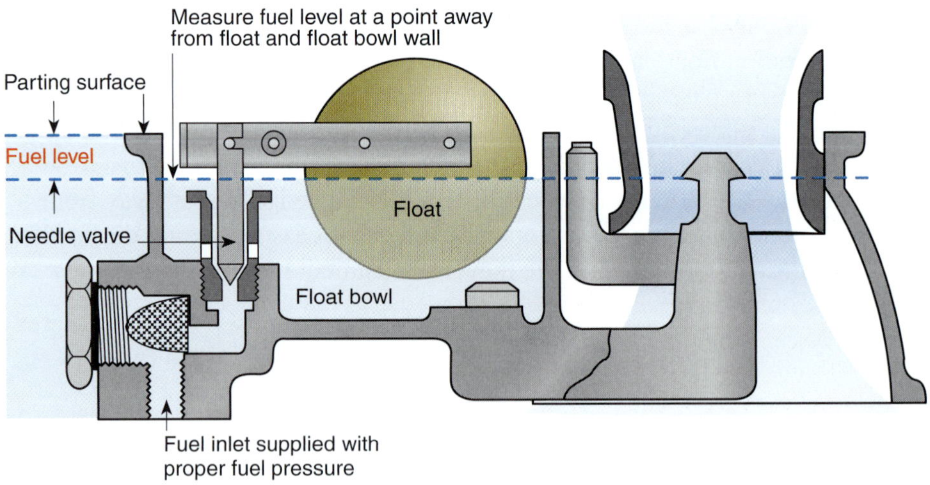

Figure 4-18. *Measurement of the fuel level in a float carburetor in which the needle valve is mounted in the float bowl.*

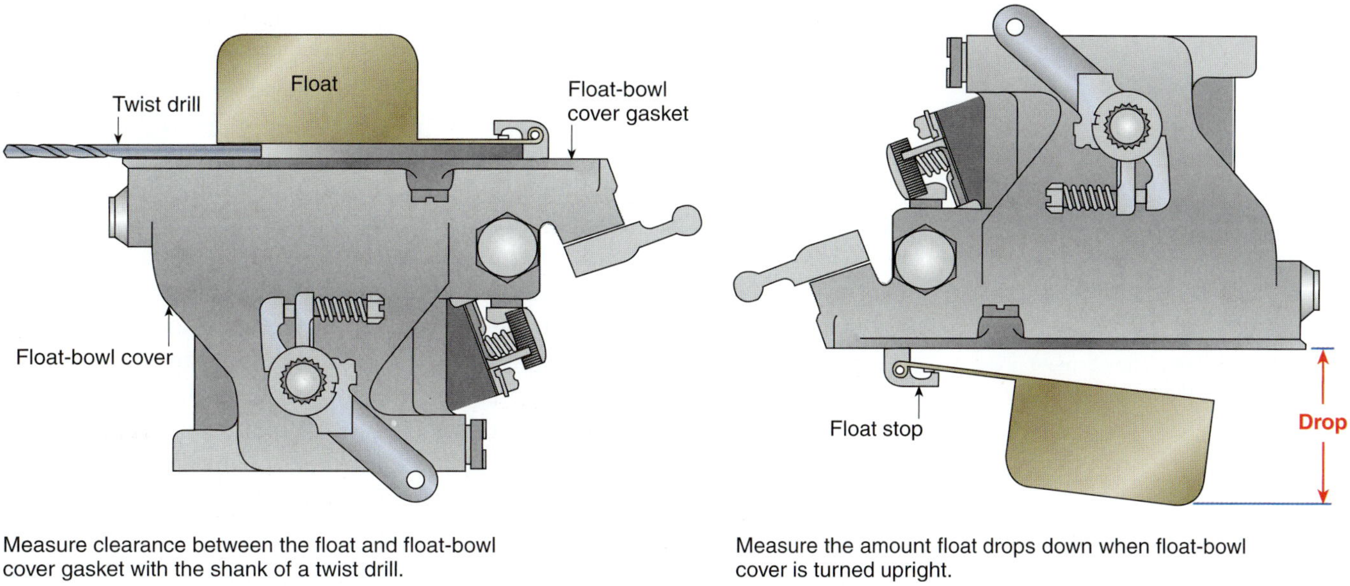

Figure 4-19. *The float level is measured in a carburetor in which the float is mounted on a bracket on the float-bowl cover.*

The float level is checked in carburetors that have the float mounted on a bracket on the float bowl cover, as shown in Figure 4-19.

Place a new gasket on the float bowl cover and install the needle in the seat and the float in its bracket. Hold the cover inverted and level, and measure the distance between the top of the float (the bottom when it is inverted) and the surface of the gasket. Use the shank of a twist drill, the size of which is specified in the carburetor overhaul manual. If this distance is significantly incorrect, adjust it by adding or removing shims under the needle seat. For small changes, very slightly adjust the tab located on the float bracket where the needle head rests.

Turn the float-bowl cover upright and check the distance the float drops down. If the drop distance is not that specified in the carburetor overhaul manual, adjust the float stop to get the correct amount of drop.

Run-up and Idle Adjustment

After the carburetor has been overhauled according to manufacturer's specifications, install it on the engine and adjust the idling mixture and RPM. Before proceeding, make sure the idle mixture is positioned as specified in the overhaul manual and the throttle stop is backed out to allow idle RPM to be adjusted by throttle position during the procedure.

An engine equipped with a float carburetor is started with the airplane headed into the wind and the mixture control in the FULL RICH position. Prime the engine by spraying gasoline into the intake valve chamber, using the number of strokes of the primer pump recommended for the engine. Advance the throttle slightly, turn the ignition switch to BOTH, and engage the starter. When the engine starts, adjust the throttle until it runs smoothest; this is normally around 1,000 RPM. Allow the engine to warm up until the oil temperature is in the normal operating range, and then advance the throttle to full power and check to be sure that the engine is producing the correct static RPM. Throttle back to the correct speed for a magneto check; check them to be sure both are operating properly and there are no fouled spark plugs. You can now adjust the idling speed and mixture.

static RPM. The number of revolutions per minute an aircraft engine can produce when the aircraft is not moving.

Hold the throttle so the engine idles at the desired RPM, and adjust the idling mixture needle valve until the engine runs at the highest smooth RPM for the fixed throttle position and then richen the mixture to produce a 50 RPM drop. If the engine is equipped with a manifold pressure gage, adjust the mixture until the manifold pressure is at its lowest for the RPM held by the throttle, and then richen it to provide a ¼-inch rise in MAP. Run the engine up to near full throttle to clear the spark plugs of any fouling that may have occurred during idling, and slowly bring the throttle back to the desired RPM. Adjust the throttle stop screw until it contacts the throttle stop. Run the engine up again and pull the throttle back until it contacts the stop. The RPM should be the same each time after the throttle is pulled back and the engine will continue to idle with the carburetor heat on. The throttle linkage should be such that the stop on the carburetor is contacted just before the stop in the cockpit. The cockpit control should move a small amount after the carburetor stop is reached, and then spring back.

manifold pressure. The absolute pressure of the air inside the induction system of a reciprocating engine.

springback. A condition in the rigging of an aircraft engine control in which the stop at the engine is reached before the stop in the cockpit. The cockpit control moves slightly after the stop in the engine is reached, and when it is released, it springs back slightly.

Make a final check of the idle mixture by watching the RPM as you slowly pull the mixture control back to the IDLE CUTOFF position. The RPM should increase approximately 50 RPM, depending upon the engine, before the engine stops running. This rise in RPM shows that the idle mixture is slightly richer than ideal. As the mixture is leaned, it passes through the ideal ratio and the RPM increases, then drops off as the mixture becomes excessively lean. This additional richness of the idling mixture ensures that all cylinders are receiving a mixture that is rich enough for them to fire consistently, aid in the transition from idle to main metering operation, aid with additional cooling at idle, and compensate for cylinder misfire due to poor exhaust scavenging at low RPM.

After the carburetor has been installed and adjusted, check carefully for any fuel leaks or improper rigging, and be sure that all required safetying is done according to the airframe manufacturer's specifications.

STUDY QUESTIONS: FLOAT CARBURETORS

Answers begin on Page 197. Page numbers refer to chapter text.

33. The device in the main metering system that senses the amount of air entering the engine is the ______________________ . *Page 141*

34. The level of fluid in the float bowl is maintained constant by a float-actuated ______________________ . *Page 143*

35. The discharge nozzle is located near the narrowest portion of the venturi where the pressure is ______________________ (highest or lowest). *Page 143*

36. A float carburetor meters the fuel on the basis of the ___________________ (volume or mass) of the air entering the engine. *Page 149*

37. The device that controls the amount of air flowing into the engine at other than full power is the butterfly-type ____________________ valve. *Page 144*

38. The device that limits the amount of air that can flow into an engine under full throttle conditions is the _____________________ . *Page 145*

39. When the same model carburetor is used on different size engines, the components that are different are the _____________ , _________________ , ________________ , and the ________________ . *Page 145*

40. The two adjustments that must be made when setting the idling conditions of a reciprocating engine are:
    a. ________________________________
    b. ________________________________
    Page 146

41. An acceleration system is needed to provide a momentarily rich mixture to sustain the engine during the time the carburetor is transitioning between the ______________ metering system and the ______________ metering system. *Page 147*

42. The variable-orifice mixture control valve is in ___________________ (series or parallel) with the main metering jet. *Page 150*

43. Another name for a power enrichment system is a/an __________________________ system. *Page 152*

44. The economizer jet in a needle-type power enrichment system is in __________________ (series or parallel) with the main metering jet. *Page 152*

Continued

STUDY QUESTIONS: FLOAT CARBURETORS *Continued*

45. The fuel level in a float carburetor is adjusted by adding or removing shims from beneath the ____________________ . *Page 154*

46. An engine with a float carburetor is started with the mixture control in the ____________________ (FULL RICH, LEAN, or CUTOFF) position. *Page 155*

47. The correct idling mixture ratio is slightly ____________________ (richer or leaner) than the perfect mixture ratio. *Page 156*

Pressure Carburetors

pressure carburetor. A carburetor installed on some aircraft reciprocating engines that uses the pressure difference between air inside the venturi and ram air entering the carburetor to produce a fuel-metering force. Pressure carburetors have been generally replaced with continuous-flow fuel injection systems.

automatic mixture control (AMC). The device in a fuel metering system such as a carburetor or fuel injection system that keeps the fuel-air mixture ratio constant as the density of air changes with altitude.

pressure-injection carburetor. A multibarrel pressure carburetor used on large radial and V-engines. Fuel is metered on the basis of air mass flowing into the engine and is sprayed under pressure into the eye, or center, of the internal supercharger impeller.

Float carburetors have limitations that have been overcome by pressure carburetors:

- Float carburetors are susceptible to carburetor icing because fuel is discharged into the venturi where it evaporates rapidly there or at the throttle plate. Pressure carburetors do not discharge their fuel in the venturi but rather downstream of the throttle.
- Float carburetors are sensitive to certain maneuvers. Fuel can surge to the top of the float chamber and shut off the fuel flow to the engine. Pressure carburetors meter the fuel by measuring the amount of air flowing into the engine and spraying an appropriate amount of fuel under pressure into this air. These carburetors can even function properly, inverted, for limited periods of time.
- Float carburetors produce an increasingly rich mixture as the aircraft goes up in altitude. Some pressure carburetors are equipped with automatic mixture controls that hold the mixture ratio constant as altitude changes. Carburetors so-equipped become truly mass metering devices.

Large, high-powered reciprocating engines used during World War II required a fuel metering system that was superior to the float carburetor, in that it did not ice up and did not shut off fuel flow to the engine during negative-G maneuvers. Direct fuel injection was used on some engines, but the most successful and popular fuel metering system was the Stromberg pressure-injection carburetor.

This carburetor measured the mass of air entering the engine and metered the correct amount of fuel. The fuel was injected under pressure into the center, or eye, of the internal supercharger impeller. Some of these carburetors were fitted with antidetonation injection systems that sprayed an alcohol-water mixture into the supercharger along with the fuel. The water evaporated and increased the density of air flowing into the engine.

Modern pressure carburetors are adapted from the pressure injection carburetor but are simplified to meet the needs of smaller engines.

Pressure carburetors have the same basic systems as float carburetors: main metering, idling, acceleration, mixture control, and power enrichment.

Bendix PS Pressure Carburetors

Single-barrel Bendix PS pressure carburetors like that in Figure 4-20 are used on some horizontally opposed engines.

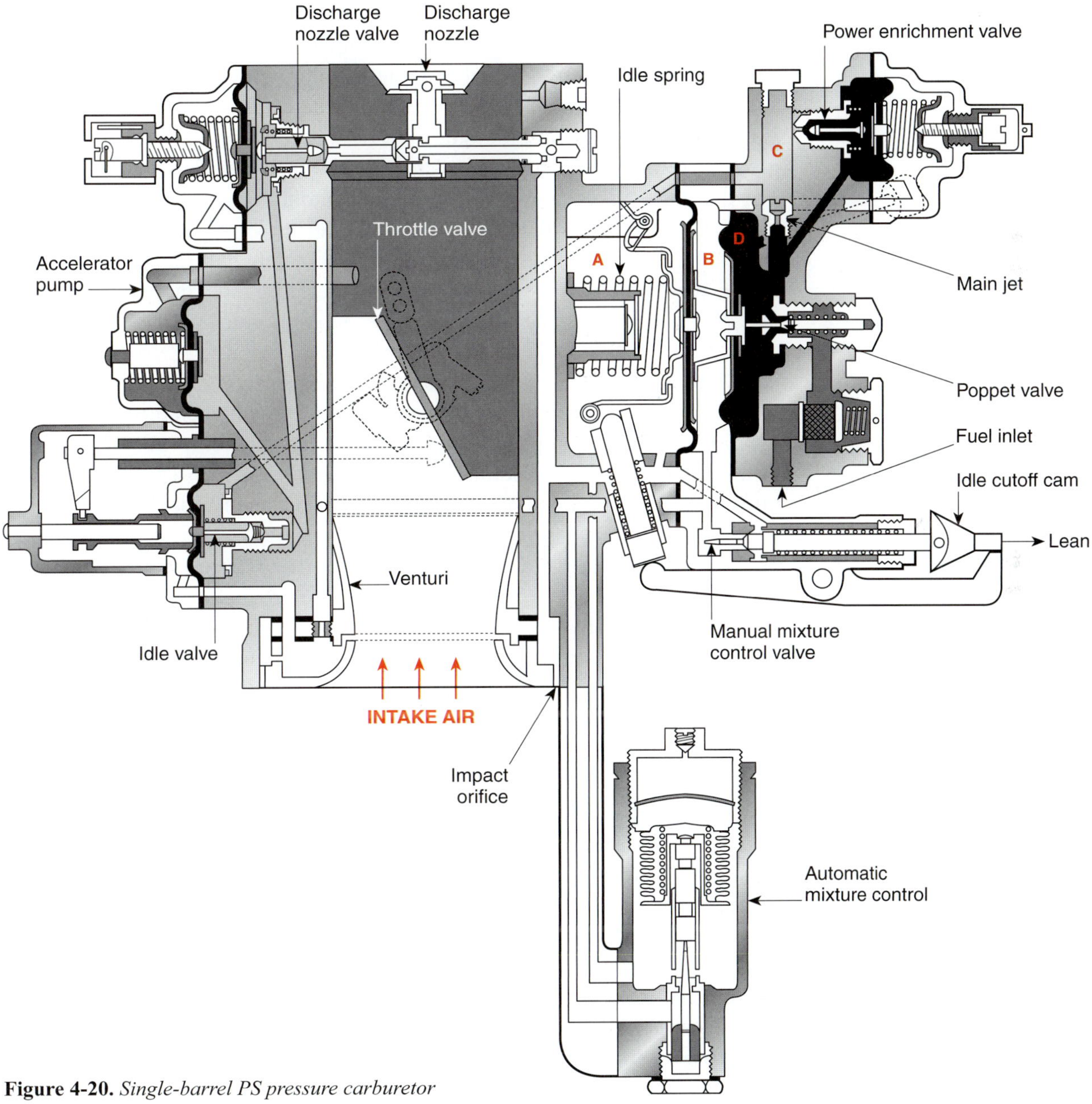

Figure 4-20. *Single-barrel PS pressure carburetor*

Main Metering System

Low pressure from the venturi in chamber B and high pressure from the impact orifice in chamber A create an air metering force on the air diaphragm that moves the diaphragm to the right to open the poppet valve. (*See* Figure 4-20 on Page 159.)

A fuel metering force is made up of regulated and metered fuel pressures. The regulated fuel in chamber D has passed through the poppet valve, and acting on the B-D diaphragm, it tries to close the poppet valve. Metered fuel pressure in chamber C is held constant by the discharge nozzle valve and places a constant pressure on the downstream side of the main metering jet.

As the throttle is opened, the air-metering force increases and further opens the regulating poppet valve. D chamber pressure rises until its pressure acting on the B-D diaphragm is sufficient to arrest further opening. The end result is a greater differential and fuel flow across the main metering jet, a greater opening at the discharge valve, and increased fuel flow to the engine. The discharge valve opens due to the tendency for metered fuel pressure to rise with greater flow through the main metering jet.

Due to the B-D diaphragm being approximately half the cross-sectional area as that of the A-B diaphragm, the fuel metering force increases are twice that of each air metering force increase. This provides good fuel metering response to throttle increases.

More advanced models also sized the discharge nozzle diaphragm so that the metered fuel pressure decreased slightly with each throttle opening increment, rather than just compensating for spring compression increases to hold the pressure constant under the same situation. This change also improved metering response and metering consistency above idle.

Fuel vapors that enter the carburetor or are formed there are sent back from D chamber to one of the fuel tanks (usually the main tank for that engine), along with a small amount of fuel that continuously flows through a fixed restrictor and a vapor return line. On aircraft with multiple tanks per engine, the tank receiving the vapor return must be drawn from for a minimum of usually one-half hour before using from another tank. This prevents overfilling this tank.

Idling System

The idling fuel pressure is produced by a coil-type idle spring in chamber A that holds the poppet valve slightly open when the air flowing through the carburetor is insufficient to produce a stable metering force.

The idle valve is a needle valve set in the fuel line between the main jet and the discharge valve. This valve is actuated by the throttle control. When the throttle is closed, the needle is almost in its seat; the amount it remains off its seat is adjustable from the outside of the carburetor, and it controls the idling mixture to 10% power.

Acceleration System

At idle, low manifold pressure from the engine side of the throttle butterfly valve pulls against the diaphragm in the accelerator pump and compresses the spring. Fuel from downstream of the idle valve flows into the chamber on the fuel side of the diaphragm.

When the throttle is opened suddenly, the manifold pressure rises, and the spring moves the diaphragm over, forcing the fuel out of the chamber and into the line to the discharge nozzle. The discharge nozzle responds by opening to keep the metered fuel pressure from rising and, therefore, an immediate increase in fuel flow is provided until the main metering flow is accelerated.

Mixture Control System

The PS carburetor has both a manual and an automatic mixture control that vary the pressure drop across the air diaphragm to change the fuel-air mixture ratio.

The manual mixture control is a needle valve controlled from the cockpit. When the mixture is full rich, the needle is seated in the valve orifice, and there is no airflow between chambers A and B. When the mixture control is moved toward the LEAN position, the needle is pulled out of the orifice and air bleeds from chamber A into B. This decreases the air metering force and allows the poppet valve to close slightly, leaning the mixture.

When the mixture control is moved to the IDLE CUTOFF position, the needle valve is pulled completely out of the orifice, and the idle spring is compressed so the poppet valve can close completely and shut off all fuel to the cylinders.

The automatic mixture control contains a bellows that moves a reverse-taper needle valve in a passage between chambers A and B that is parallel to the passage provided by the manual mixture control. When the atmospheric pressure is high, as it is at sea level, the bellows contracts, and the thick part of the needle is in the seat. Very little air flows between chambers A and B. The air metering force is maximum and the mixture is rich. As the airplane climbs to altitude, the air pressure decreases and the bellows expands, moving the smaller part of the needle into the seat. This allows more air to bleed between the chambers, and the poppet valve to close slightly. Less fuel flows into the engine, and the mixture is leaned.

Power-Enrichment System

The power-enrichment valve is in parallel with the main metering jet. The valve is held closed with a spring and is opened by regulated but unmetered fuel pressure from chamber D, acting on a diaphragm. When the throttle is nearly wide open, the regulated fuel pressure is high enough to open the power-enrichment valve, allowing additional fuel to flow to the discharge valve.

Pressure Carburetor Installation and Maintenance

14 CFR Part 43, Appendix A, classifies the overhaul of a pressure carburetor as an appliance major repair. A proper overhaul and calibration requires special tools and equipment, and is normally done by the manufacturer or an FAA-approved repair station that specializes in and is approved for this type of overhaul.

After a carburetor is overhauled, the fuel chambers are filled with a preservative oil which must be removed before the engine can be run. Install the carburetor on the engine and connect the fuel line. Remove the drain plug from the bottom of the regulator and turn on the boost pump. Move the mixture control to the RICH position and allow fuel to flow through the regulator until there is no longer any trace of the preservative oil. Place the mixture control in the CUTOFF position and reinstall the drain plug. Now, with the mixture rich and the boost pump on, allow flow to purge the rest of the fuel chambers and passages. With the mixture in IDLE CUTOFF, allow the carburetor to sit with fuel in the regulator for as long as practical before the engine is started. Eight hours is recommended as the minimum to restore the same degree of pliability to the diaphragms that they had when the carburetor was calibrated.

Run-up and Idle Adjustment

Engines with pressure carburetors are started on fuel from the primer system with the mixture control in the IDLE CUTOFF position. When the engine starts, move the mixture control to the RICH position for the ground run-up and takeoff.

When the engine has warmed up and has been checked to determine that it produces the correct static RPM, and when a check of both magnetos shows that the ignition systems are working properly, the idling RPM and mixture can then be adjusted.

Hold the throttle to produce the desired idle speed, usually around 600 RPM. Watch the tachometer as you move the mixture control toward the IDLE CUTOFF position. The engine speed should increase slightly before it begins to die. Be sure to return the mixture control to the RICH position before the engine stops completely.

If the RPM increase is more than is specified for the particular engine, the idling mixture is too rich. If there is no increase in RPM before the engine dies, the idling mixture is too lean.

If the engine is equipped with a manifold pressure gage, it may be observed when adjusting the idling mixture. The correct mixture is indicated by the manifold pressure remaining steady as the mixture control is pulled toward IDLE CUTOFF; it then rises when the engine stops. If the mixture is too rich, the manifold pressure will decrease and then rise. If it is too lean, the manifold pressure will rise as soon as the mixture control is moved toward IDLE CUTOFF.

When adjusting the idle mixture, move the mixture screw one detent at a time in the required direction and check the results. Between each adjustment, run the engine up to approximately 2,000 RPM to clean the spark plugs of fouling and close the throttle slowly to prevent reloading of the plugs. Continue this procedure until the proper engine RPM and mixture adjustment are obtained.

STUDY QUESTIONS: PRESSURE CARBURETORS

Answers begin on Page 197. Page numbers refer to chapter text.

48. A pressure-injection carburetor discharges its fuel into the center, or eye, of the internal ______________________ . *Page 158*

49. Two components of the air metering force are _________________ pressure and _____________________ pressure. *Page 158*

50. Metered fuel is held at a constant pressure by the spring-loaded __________________ valve. *Page 160*

51. The device that automatically adjusts the fuel-air mixture ratio to keep it constant as the altitude changes is the _________________________________ . *Page 158*

52. Fuel vapors that collect in a pressure carburetor are returned with a small flow of fuel to one of the aircraft ________________________ . *Page 160*

53. Idling fuel pressure is produced by a ______________ in chamber A which holds the poppet valve open when there is only a small amount of air flowing through the carburetor. *Page 160*

54. The acceleration pump diaphragm is held against spring pressure by __________________ (low or high) manifold pressure. *Page 161*

55. The manual mixture control in a pressure carburetor is an adjustable ______________________ between chambers A and B. *Page 161*

56. The automatic mixture control _________________ (enriches or leans) the fuel-air mixture as the aircraft goes up in altitude. *Page 161*

57. The power-enrichment system adds additional fuel when the engine is operating at _________________ (high or low) power. *Page 161*

58. The amount the idle fuel valve remains off its seat can be adjusted to change the idle ______________________ (mixture or speed). *Page 160*

Continued

STUDY QUESTIONS: PRESSURE CARBURETORS *Continued*

59. When the throttle is suddenly opened, the fuel from the acceleration pump is forced out into the air stream by __________________ (a spring or air pressure). *Page 161*

60. The automatic mixture control is _____________________ (parallel or series) with the manual mixture control. *Page 161*

61. The power-enrichment valve on a PS carburetor is _______________________ (parallel or series) with the main metering jet. *Page 161*

62. Overhaul of a pressure carburetor is a __________________ (major or minor) repair. *Page 162*

63. Before returning an engine to service after the installation of a freshly overhauled pressure carburetor, the fuel supply should be connected and the carburetor allowed to sit with fuel in it for at least ______________ hours. *Page 162*

64. An engine equipped with a pressure carburetor is started with the mixture control in the _________________ (AUTO LEAN, AUTO RICH or IDLE CUTOFF) position. *Page 162*

65. When adjusting the idle mixture control, pull the mixture control toward IDLE CUTOFF and watch the tachometer. Before the engine dies, the RPM should momentarily _________________ (increase or decrease). *Page 162*

Fuel Injection Systems

Two major limitations of a carburetor are that the fuel-air mixture is not uniform among all the cylinders, and carburetors are subject to the formation of ice. With large engines having long induction pipes of varying lengths and shapes, fuel vapors and air do not mix uniformly, and some cylinders receive a rich mixture while others receive a lean one. The pressure-injection carburetor of World War II did a lot to solve this problem by spraying its fuel into the center of the internal supercharger impeller where fuel and air were thoroughly mixed and then forced into the cylinders.

Diesel engines use a direct injection system in which the fuel is sprayed under extremely high pressure directly into the combustion chamber. Some gasoline engines have used direct injection to provide a uniform mixture, but the high precision required for the manufacture of these systems has made them too expensive to be practical.

The fuel injection systems used on modern aircraft engines are not direct injection systems, but are continuous-flow systems that deliver a low-pressure continuous flow of fuel to nozzles installed in the intake valve chambers of the cylinders, just outside the valve. Fuel collects inside the nozzle when the intake valve is closed, and when it opens, this fuel, along with air from air bleed holes in the nozzle, is pulled into the cylinder. All of the fuel vaporizes and is mixed with air from the induction system to provide a uniform fuel-air mixture to each cylinder. Vaporization of the fuel takes place in the hot intake valve chamber where there is no possibility of ice forming. These systems are uncomplicated and efficient.

Two types of continuous-flow fuel injection systems used on modern engines are the Precision Airmotive, formerly Bendix, RSA system, and the Teledyne-Continental system. Both are discussed below.

Precision Airmotive RSA Fuel Injection System

The Precision Airmotive RSA fuel injection system, originally the Bendix system, operates on somewhat the same principle as the Bendix PS carburetor just discussed. This system consists of four basic subsystems: the servo regulator, the fuel control, the flow divider, and the nozzles. The same functions performed by float and pressure carburetors are also accomplished in this system.

flow divider. The valve in an RSA fuel injection system that divides the fuel from the fuel control unit and distributes it to all cylinders. It is comparable to the manifold valve in a Teledyne-Continental fuel injection system.

Main Metering System

The amount of fuel delivered to the flow divider is determined by the pressure drop across the main metering jet and the mixture control orifice, which are in series. The pressure drop is determined by the amount of air flowing into the engine.

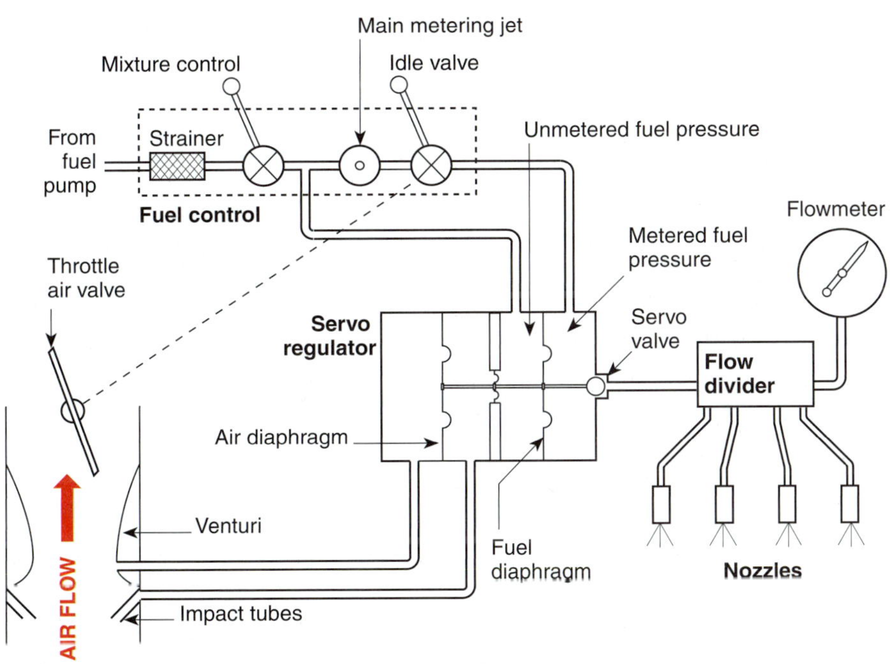

Figure 4-21. *Simplified diagram of the RSA fuel injection system*

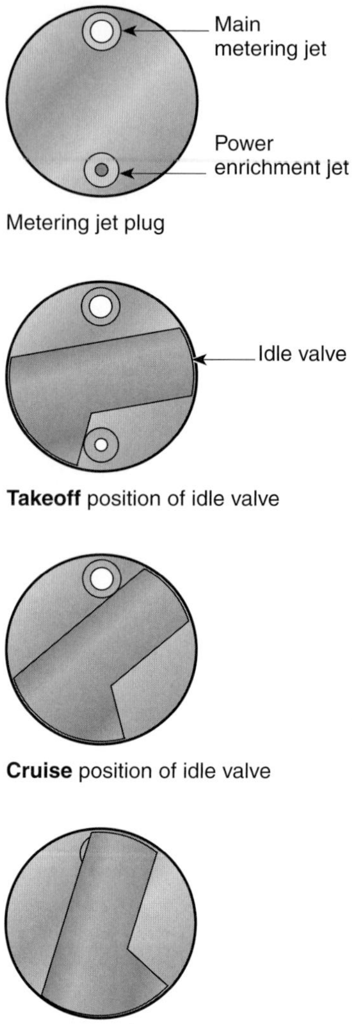

Figure 4-22. *The idle valve restricts the exit from the main metering jet to vary the amount of fuel that reaches the cylinders. When the throttle is in the idling position, the valve produces an extremely small orifice through which the idling fuel flows.*

The regulator consists of an air diaphragm and a fuel diaphragm on a shaft with a ball-type servo valve in the metered fuel passage to the flow divider.

An air metering force is made up of low pressure from the throat of the venturi and high pressure from impact tubes in the inlet to the venturi. These two pressures act on opposite sides of the air diaphragm to move it in the direction that opens the servo valve as the amount of air flowing into the engine cylinders increases.

Fuel flows from the engine-driven fuel pump into the fuel control unit through a wire mesh strainer and orifice of the mixture control valve. From there, a sample of unmetered fuel goes to the regulator unit on the side of the fuel diaphragm that closes the servo valve.

The majority of fuel that flows through the mixture control orifice continues on and passes through the main metering jet and throttle valve into the regulator on the side of the diaphragm that opens the servo valve.

A balance of forces is key to the operation of this system. This balance consists of venturi air pressure and impact air pressure, assisted by metered fuel pressure trying to open the servo valve, and unmetered fuel pressure trying to close it.

For any given amount of air flowing into the engine, the correct amount of fuel is sent through the servo valve to the flow divider and nozzles, and the pressures are balanced.

When the throttle is advanced, allowing more air to flow into the cylinders, the impact air pressure increases and the venturi pressure decreases. The air diaphragm moves over and pulls the servo valve farther away from its seat, causing the metered fuel pressure to drop. The unmetered fuel pressure remains constant, so the pressure drop across the metering jet increases, and more fuel is sent through the jet and the servo valve to the nozzles.

The increased pressure drop across the metering jet and the fuel diaphragm now balances the increased air metering force, and the servo valve stops moving and remains in its new position. The servo valve is in the passage to the flow divider, but it does not do any metering; its function is to produce a pressure differential across the metering jet that is proportional to the amount of air flowing into the engine.

The throttle-controlled "idle valve" in Figure 4-22 begins to uncover the power enrichment passage as the engine power level demands increased cooling with a richer mixture.

The more prevalent design is one that incorporates only one metering passage. With this single-passage metering plug the flat cam-plate incorporates a varying cross-sectional slot that uncovers the metering jet outlet. This variable slot provides the mixture ratio change for the power enrichment function.

Idling System

The idling system has two requirements: there must be an orifice of the proper size to meter the fuel, and there must be the proper fuel pressure across the orifice.

The main metering jet in the fuel control unit is an orifice of fixed size that is covered by the flat cam-plate. As the throttle is closed, the power enrichment orifice is covered, and the main metering orifice is progressively covered, until the throttle is fully back in the idle position. Here, the throttle valve covers almost all of the orifice, leaving only a small opening whose area is varied by the idle mixture adjustment.

The idle fuel pressure is maintained by the constant head spring that holds the servo valve slightly off its seat when the air through the engine is insufficient to produce a consistent air metering force (AMF).

Figure 4-23 shows how the idling fuel pressure is produced. There is a weak, large diameter constant-effort spring between the air diaphragm and regulator body. A smaller diameter and slightly stronger constant-head spring acts between the other side of the air diaphragm and the servo valve stem. When only a small amount of air flows into the engine, the constant-head spring holds the servo valve slightly off of its seat and maintains a constant head of pressure across the fuel control.

head of pressure. Pressure exerted by a column of fluid and created by the height of the column.

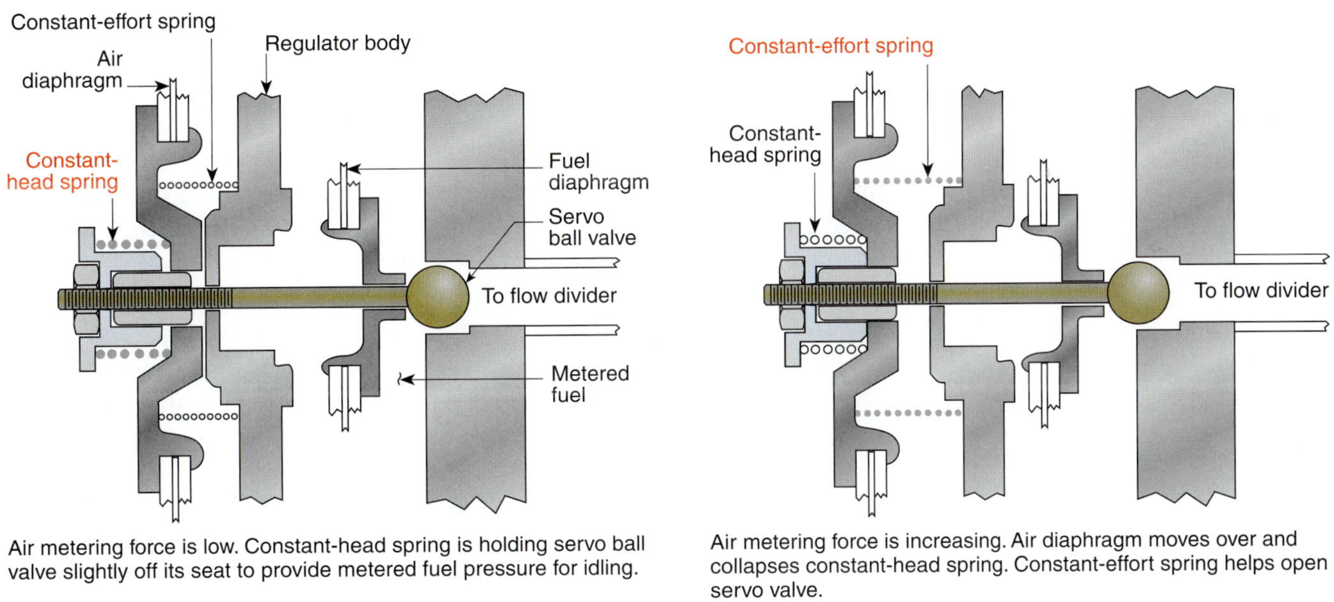

Figure 4-23. *Diagram of a servo regulator for the RSA fuel injection system showing the air diaphragm between the constant-head and constant-effort springs*

As the airflow increases, the air diaphragm moves over and compresses the constant-head spring until the diaphragm bushing makes solid contact with the servo valve shaft. The servo valve then acts as though it were fixed directly to the air diaphragm.

The constant-effort spring preloads the air diaphragm and aids it in moving smoothly between the idle range and the range of higher air flows.

Mixture Control System

The manual mixture control of this fuel injection system is a spring-loaded flat plate valve that rides against the opposite end of the main metering jet plug from the idle valve. When the mixture control is placed in the IDLE CUTOFF position, passage to the main metering jet is completely closed and no fuel can flow to the jet. When in the FULL RICH position, the mixture control orifice is completely uncovered, and is larger than the main metering jet, so the main jet limits the flow. When the mixture control is in an intermediate position, the opening is smaller than the main jet and it becomes the flow-limiting device.

The automatic mixture control works in the same way as the one used on the PS pressure carburetor. Refer to Figure 4-20 on Page 159. An evacuated bellows controls a reverse-tapered needle valve in a line between the two sides of the air diaphragm. When the aircraft is flying at low altitude, the atmospheric pressure is high and the bellows is collapsed, causing the needle valve to close the orifice. As the aircraft goes to altitude, the air pressure decreases because of lower atmospheric pressure and the bellows expands, moving the smaller part of the needle into the orifice, which decreases the pressure difference across the air diaphragm and allows the servo valve to move toward its seat. This decreases the pressure drop across the metering jet and leans the mixture.

Power-Enrichment System

The power-enrichment system consists of a drilled passage in the metering plug that is parallel with the main metering jet. *See* Figure 4-22. When the engine is operating at conditions other than full throttle, only the main metering jet orifice is uncovered. But when the throttle is opened wide, the power-enrichment passage is also uncovered. This effectively increases the size of the main metering jet and allows additional fuel to flow to the engine. This extra fuel removes heat that would otherwise damage the engine.

One very popular design incorporates only one metering passage. With this single-passage metering plug the flat cam-plate uses a varying cross-sectional slot that uncovers the metering jet outlet. This variable slot provides the mixture ratio change for the power enrichment function.

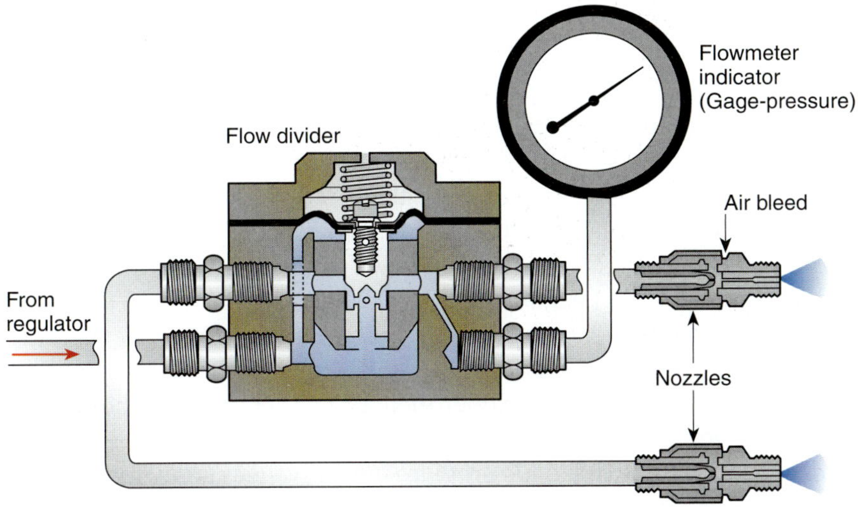

Figure 4-24. *Flow divider for an RSA fuel injection system*

Flow Divider

Metered fuel flows from the regulator to the flow divider and then to the injector nozzles, one in each cylinder head. *See* Figure 4-24.

The flow divider serves three basic functions: it aids in consistent idling, it distributes fuel evenly to all cylinders, and it shuts off the fuel sharply and completely when the engine is shut down with the use of idle cutoff.

The flow divider is mounted on the top center of the crankcase and is connected to the servo regulator with a fuel line usually enclosed in a fire sleeve. All nozzles are connected to the flow divider with small diameter stainless steel lines. A pressure gage is connected to one port of the flow divider so that it measures the pressure drop across the nozzles. This instrument is calibrated in terms of flow, operating on the basis that the flow through a fixed nozzle is directly related to the pressure drop across it.

Fuel flows from the servo regulator to the flow divider, where it acts on the diaphragm and forces the valve up, opening the passage to the nozzles.

For flow conditions other than idling, the nozzles provide sufficient restriction to produce a back pressure that ensures consistent metering. When the engine is idling, there is not enough fuel flow to produce consistent pressure across the nozzles, and metering would be erratic. But the spring-loaded valve in the flow divider produces the correct amount of opposition for consistent metering during idling.

Fuel from the servo regulator is distributed to all cylinders uniformly by the flow divider, and when the mixture control is placed in the IDLE CUTOFF position, pressure drops and the valve immediately shuts off all flow to the nozzles and traps fuel in the lines to the regulator.

fire sleeve. A covering of fire-resistant fabric used to protect flexible fluid lines that are routed through areas subject to high temperature.

vibration loop. A loop in a rigid fluid line used to prevent vibration from concentrating stresses that could cause the line to break.

gage pressure. Pressure referenced from existing atmospheric pressure.

ambient air pressure. The pressure of the air that surrounds an object.

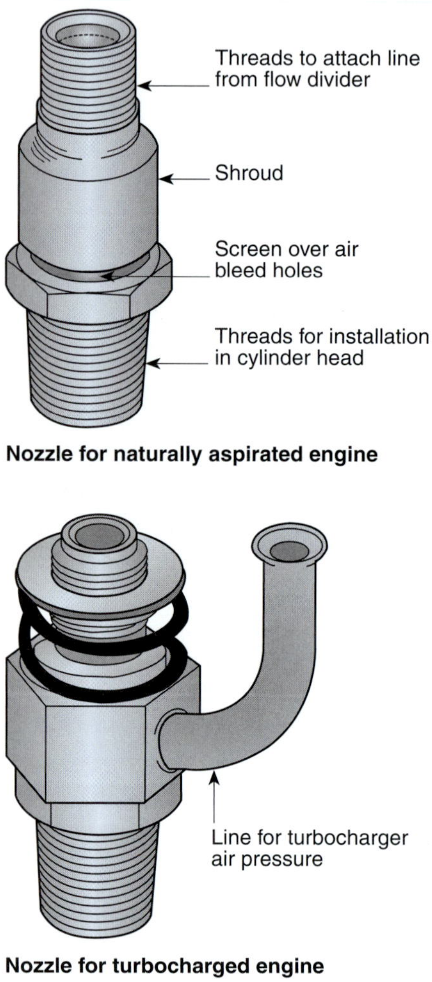

Figure 4-25. *Air bleed-type nozzles used with an RSA fuel injection system*

Nozzles

Nozzles used with the RSA fuel injection system are of the air bleed type as shown in Figure 4-25. Fuel from the regulator flows continuously into the nozzle through the calibrated orifice, or restrictor. Fuel collects in the chamber at the center of the nozzle when the intake valve is closed, and when the valve opens, low pressure pulls the fuel along with air from the air bleed holes into the cylinder in the form of an air-fuel emulsion. Heat from the cylinder head immediately vaporizes the fuel, and the cylinder receives a uniform amount of fuel vapor.

With turbocharged induction systems, insufficient ambient air pressure exists above idle which would result in fuel flowing out the nozzle air vents when the intake valve is closed rather then air venting in, so turbocharged engines route some upper-deck air from the compressor discharge into the shrouds around the nozzles. This provides a positive air supply to mix with fuel each time the intake valve opens.

Installation and Maintenance of RSA Fuel Injection System

The RSA servo regulator mounts on the engine in the location normally used by the carburetor and the flow divider mounts above the center of the crankcase.

Fuel is supplied to the regulator from the aircraft fuel system through the main strainer, a boost, or auxiliary, fuel pump, and an engine-driven pump of either the sliding vane or diaphragm type.

Normal maintenance consists of cleaning the fuel strainer after the first 25 hours of operation and every 50 hours after that. The strainer must be removed from the fuel-supply-line end rather than the end with the block-off plug. Removing the strainer from the plug end allows contaminants to flow from the center of the filter into the system as the strainer is withdrawn. This incorrect procedure has been known to cause malfunctioning of the system. Precision Airmotive has issued special instructions concerning this procedure because it is easer to remove the block-off plug than the fuel line and fitting from the other end. Other primary service items are the cleaning and flow-checking of the injection nozzles. Check the entire installation for any indications of fuel leakage, and check the throttle and mixture control linkage for freedom of movement.

Run-up and Idle Adjustment

Starting an engine equipped with an RSA fuel injection system uses a different procedure from that used with either a pressure or float carburetor. Start the engine using the following procedures:

1. Place the mixture control in the IDLE CUTOFF position.
2. Open the throttle about $\frac{1}{8}$ of the way.
3. Turn on the master switch and auxiliary fuel pump.

4. Prime the engine by moving the mixture control to FULL RICH and watching the fuel flowmeter. At the first indication of flow, return the mixture control to IDLE CUTOFF.
5. Turn the ignition switch to BOTH and engage the starter.
6. As soon as the engine starts, move the mixture control to FULL RICH. For a hot, recently shut down engine, the priming step is usually eliminated. Priming under these conditions usually floods the engine. Depending on the airframe manufacturer, this procedure is sometimes modified to fit the specific installation. The manufacturer's procedures should always be followed.

The idle RPM and mixture are adjusted by basically following the same procedures used with the float carburetor.

Start the engine and check to be sure that all systems are operating properly and it develops its proper static RPM. Move the throttle to the correct idling speed, and adjust the idling mixture by changing the length of the connecting rod between the throttle air valve and the throttle fuel valve. This is done by turning the wheel in the center of the rod. *See* Figure 4-26. The threads are coarse on one end of the wheel shaft and fine on the other end. Lengthening the rod by screwing the coarse threads out enriches the mixture.

Check the idle mixture for proper adjustment by pulling the mixture control to the IDLE CUTOFF position. The RPM should increase approximately 50 RPM before the engine stops. After the mixture is adjusted satisfactorily, set the idle speed adjustment so the engine will always idle at the same RPM when the throttle is pulled all the way back.

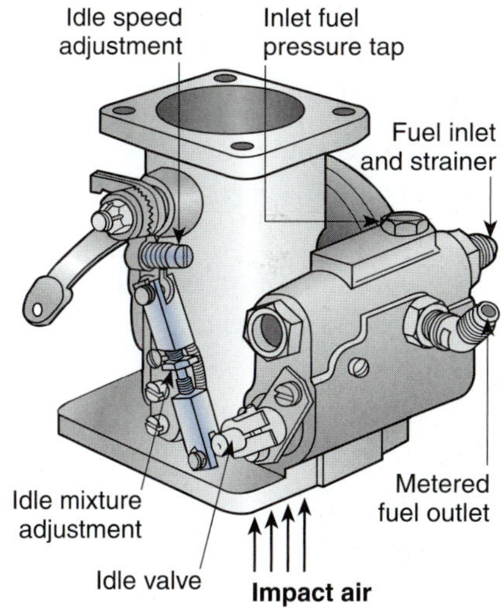

Figure 4-26. *Idle mixture adjustments on an RSA fuel injection system*

Many problems with a fuel injection system are caused by dirty or plugged nozzles. To clean the nozzles, remove them from the engine and soak them in Hoppes No. 9 gun cleaning solvent. Rinse them with Stoddard solvent, and dry them with clean compressed air. Be sure to direct compressed air into the outlet only. First let the air blow straight through and out the inlet. Next, plug off the inlet with your finger to force compressed air out the screen. Examine the orifices with a 10-power magnifying glass to be sure that there are absolutely no restrictions. Never probe into the orifice with a piece of safety wire or drill, or any stiff object. This will change the metering characteristics of the unit!

Check the nozzles' uniformity of flow by screwing them onto their lines and placing each nozzle in a small bottle or jar. Turn on the fuel and auxiliary pump, place the mixture control in the RICH position, and open the throttle. Allow the fuel to flow until there is enough in each bottle to determine whether or not all of the nozzles are flowing the same. If the flow is different, there is a good probability that a fuel line is kinked or a nozzle is clogged or damaged.

When troubleshooting a fuel injection system, remember that the fuel flowmeter is a pressure gage that measures the pressure drop across the injector nozzles. If a nozzle is clogged, less fuel will flow than in a clear nozzle, but the flowmeter will indicate an *increased* flow overall.

STUDY QUESTIONS: RSA FUEL INJECTION SYSTEM

Answers begin on Page 197. Page numbers refer to chapter text.

66. The four basic subsystems for an RSA fuel injection system are:
    a. ______________________________
    b. ______________________________
    c. ______________________________
    d. ______________________________
    Page 165

67. The mixture control orifice and main metering jet of an RSA fuel injection system are in ______________________ (series or parallel). *Page 165*

68. The air metering force in an RSA fuel injection system tends to ______________________ (open or close) the servo valve. *Page 166*

69. Unmetered fuel pressure acting on one side of the fuel diaphragm tends to ______________________ (open or close) the servo valve. *Page 166*

70. When airflow into the engine increases, the pressure drop across the main metering jet ______________________ (increases or decreases). *Page 166*

71. The servo valve in the RSA fuel injection system ______________________ (does or does not) meter fuel. *Page 166*

72. The RSA fuel injection system ______________ (does or does not) have a separate idle metering jet. *Page 167*

73. The servo ball valve is held off its seat during idle conditions by the constant-___________________ (effort or head) spring. *Page 167*

74. The movement of the air diaphragm is assisted when transitioning from idle to cruise power by the constant- ______________________ (effort or head) spring. *Page 167*

75. The automatic mixture control ________________ (increases or decreases) the pressure difference across the air diaphragm when the aircraft goes up in altitude. *Page 168*

76. The power-enrichment jet is in ________________ (series or parallel) with the main metering jet. *Page 168*

77. Three functions of the flow divider are:
   a. ______________________________
   b. ______________________________
   c. ______________________________
   Page 169

78. An engine equipped with an RSA fuel injection system is started with the mixture control in the ______________________ (LEAN, RICH, or IDLE CUTOFF) position. *Page 170*

79. Fuel for starting the engine is sprayed into the intake valve chamber by momentarily moving the mixture control to the ________________ position with the boost pump turned on. *Page 171*

80. Increasing the length of the rod connecting the throttle air valve with the throttle fuel valve causes the idle mixture to become ________________ (leaner or richer). *Page 171*

81. A plugged fuel injector nozzle will indicate on the flowmeter as a/an ______________________ (increased or decreased) flow. *Page 171*

Teledyne-Continental Fuel Injection System

The Teledyne-Continental, or TCM, fuel injection system and the RSA system are both continuous-flow injection systems that do the same job, but they accomplish it in different ways. The RSA system measures the volume of air flowing into the engine to determine the amount of fuel to mix with it. The TCM system uses the engine RPM to determine the amount of fuel to send to the injector nozzles.

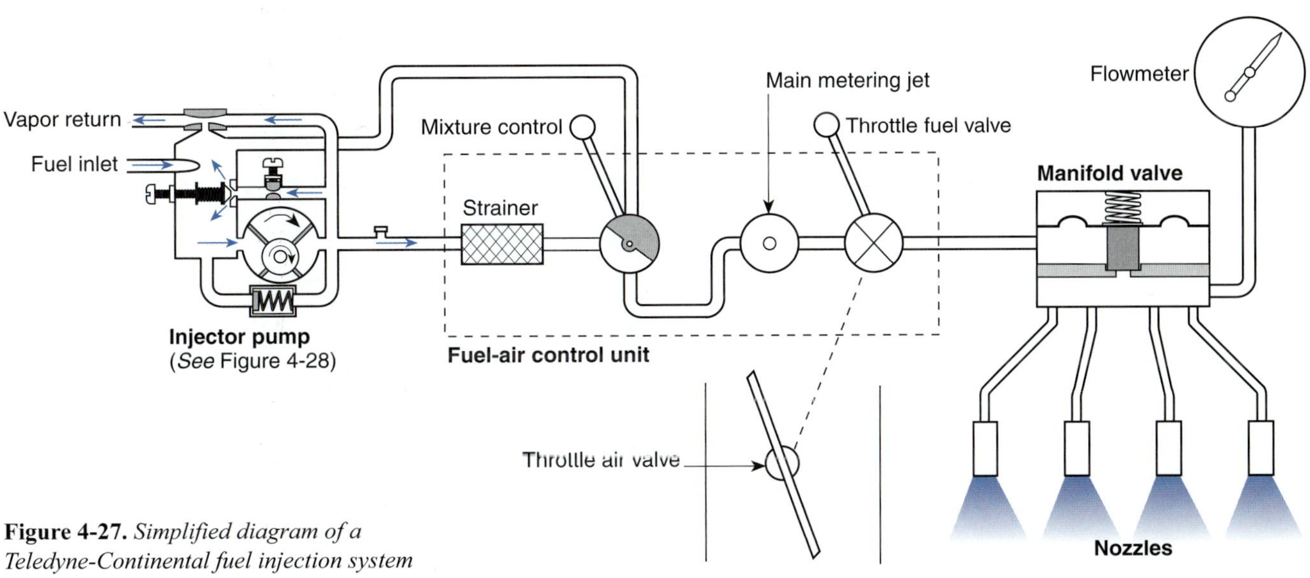

Figure 4-27. *Simplified diagram of a Teledyne-Continental fuel injection system*

The TCM system has four basic components: the engine-driven injector pump, the fuel-air control unit, the fuel manifold valve, and the nozzles. We will consider each of these components and how they fit into the complete system. Figure 4-27 is a line diagram of the TCM system.

Injector Pump

The engine-driven injector pump is the heart of the TCM system. This is a vane-type constant-displacement pump with some unique features that allow it to accomplish these functions:

- Provides fuel flow that increases with engine speed. This increase in output is modified to furnish enough fuel when the pump turns at idling speed, yet not move too much fuel when it turns at takeoff speed.
- Removes vapors from the fuel and sends them to one of the fuel tanks.
- Incorporates a bypass valve that allows fuel from the auxiliary pump to flow to the engine for starting and for operation should the engine-driven pump malfunction.
- Provides a regulated low pressure when the engine is idling.
- Provides a regulated high pressure when the engine is operating at high speed.

Figure 4-28 is a simplified diagram of the TCM fuel injector pump for a naturally aspirated engine. Fuel from the auxiliary pump enters the vapor-separator tower in a swirl that spins vapors out of the liquid fuel. Vapors rise to the top of the chamber, and then return to one of the fuel tanks.

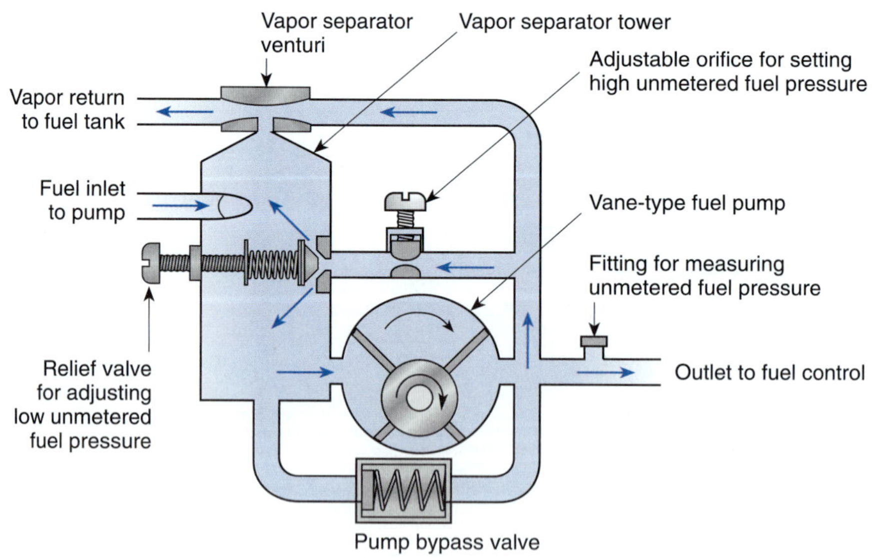

Figure 4-28. *Simplified diagram of a TCM engine-driven fuel injector pump for a naturally aspirated engine*

When the engine is running, vanes in the pump move fuel to the outlet. Since this is a constant-displacement pump, the faster the pump turns, the greater the fuel volume moved. This pump always moves more fuel than the engine requires, but the excess is routed back to the inlet side of the pump. Some of the excess fuel returns to a fuel tank through the vapor-separator venturi. This return fuel produces low pressure in the top of the vapor-separator chamber, which picks up the vapors released from the fuel and carries them back to the fuel tank. When the engine is operating at low speed, some of the fuel from the discharge side of the pump forces the spring-loaded relief valve off its seat and returns to the inlet side of the pump. Pressure held on the output fuel by the relief valve is called the "low unmetered" fuel pressure, and it is one of the pressures that must be adjusted when setting up this system.

low unmetered fuel pressure. Pressure in a Teledyne-Continental fuel injector pump that is adjusted by the relief valve.

An adjustable orifice in series with the relief valve determines the output fuel pressure when the engine is operating at high speed. Return fuel flows through the orifice on the way to the relief valve, and the restriction caused by the orifice at higher pump speed determines the output fuel pressure. This pressure is called the "high unmetered" fuel pressure, and it must be adjusted when setting up this system.

high unmetered fuel pressure. Pressure in a Teledyne-Continental fuel injector pump that is adjusted by the variable orifice.

When the engine is being started, fuel from the auxiliary pump forces the pump bypass valve off its seat, and fuel flows around the pump mechanism to the fuel control.

Many TCM fuel injection systems are used on engines equipped with turbochargers. Turbocharged engines present a unique problem to their fuel metering systems. When the throttle is opened, fuel flow to the engine increases immediately. The turbocharger is driven by exhaust gases, and its

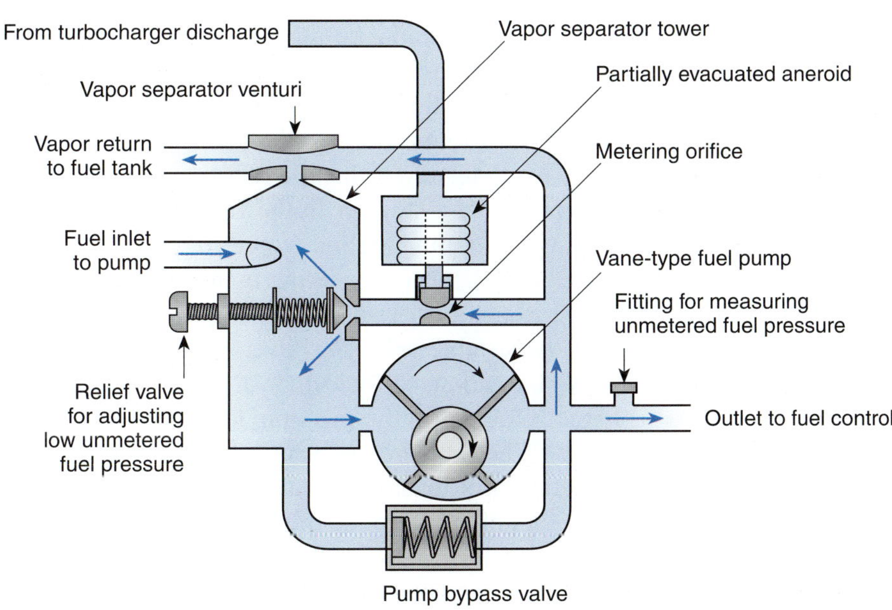

Figure 4-29. *Simplified diagram of a TCM injector pump used on a turbocharged engine*

speed increase lags considerably behind that of the engine. Since the engine receives its increased fuel before the airflow increases, in a turbocharged engine, the mixture becomes overly rich until the turbocharger comes up to speed.

To prevent this momentary over-enrichment, the fuel pumps installed on turbocharged engines use a bellows, or aneroid, evacuated to approximately 28 inches Hg., to control the size of the metering orifice. This bellows is mounted in a compartment inside the pump where it senses the upper-deck, or turbocharger discharge, pressure. *See* Figure 4-29. When the throttle is opened, the fuel pump turns faster, and discharges only a slight increase in fuel flow. Before the turbocharger increases in speed, the upper-deck pressure is low and the aneroid is expanded, which leaves the size of the return orifice unrestricted so that the pump output does not increase significantly. As the turbocharger speed increases, the upper-deck air pressure increases, and the aneroid compresses, decreasing the size of the orifice in the return line and increasing the pump output to match the rising induction airflow.

Fuel-Air Control Unit

Fuel flows from the fuel injector pump into the fuel portion of the fuel-air control unit and the manifold valve. *See* Figure 4-27. Air flows from the inlet air filter through the air portion of the fuel-air control unit into the intake manifold.

Float carburetors, pressure carburetors, and the RSA fuel injection systems all have venturis in the air passage to the cylinders to measure the amount of air flowing into the engine, but the TCM system does not use a venturi. The air passage is a smooth tube whose diameter allows it to supply an adequate amount of air into the cylinders during maximum power conditions. The throttle air valve is a circular butterfly valve across this tube that controls the amount of air entering the engine. An adjustable linkage connects the throttle air valve with the throttle fuel valve. An adjustable stop determines the amount the air valve remains open when the throttle is pulled back to the idling position. This determines the idle RPM.

The fuel control unit consists of a strainer in the fuel inlet, a mixture control valve, a metering plug containing the main metering orifice (jet), a throttle valve, a metered fuel outlet connected to the manifold valve, and a return fuel outlet that returns fuel to the inlet side of the pump. *See* Figure 4-30.

The mixture control is a variable selector valve. When the mixture control is in the FULL RICH position, the selector valve sends all fuel to the throttle. When it is in the IDLE CUTOFF position, it sends all fuel back to the inlet side of the pump. When it is in any position between these two extremes, some fuel goes to the throttle valve and some returns to the pump thus modulating unmetered fuel pressure to the metering jet.

The throttle full valve is a metering valve that covers the metering orifice outlet. When the throttle is wide open, the outlet is uncovered, and the main metering orifice does the metering. When the throttle is closed, the

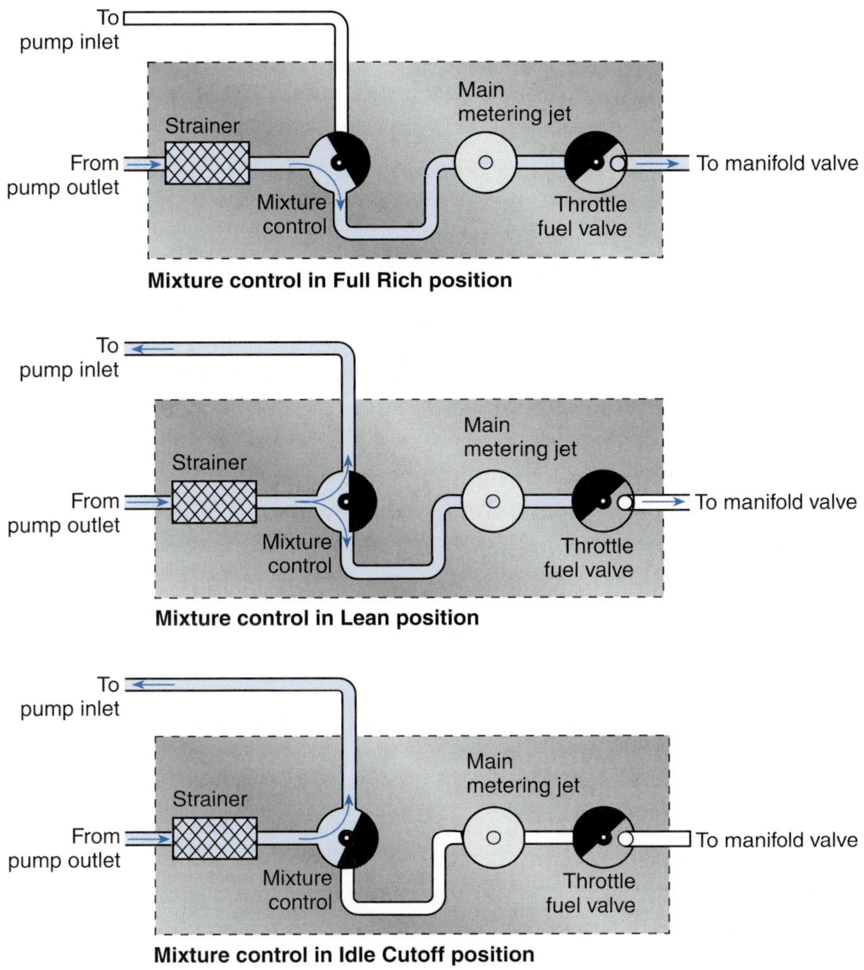

Figure 4-30. *Simplified diagram of the fuel-air control unit of a TCM fuel injector system*

valve allows only enough of the outlet to be uncovered to meter the correct amount of fuel for idling. The linkage between the throttle air valve and the fuel valve is adjusted to control the idling fuel-air mixture.

Fuel Manifold Valve

The fuel manifold valve on the TCM system is the same type of device as the flow divider on the RSA system. Metered fuel flows from the fuel-control unit to the fuel manifold valve, and in this valve, it passes through a filter screen and applies an upward force against a diaphragm. The diaphragm raises the cutoff valve off its seat, but the spring-loaded poppet valve remains seated, preventing fuel from flowing to the nozzles until the cutoff valve is fully open. This prevents the cutoff valve from metering. When the poppet valve is unseated, fuel flows to the nozzles. Under idling fuel flow conditions,

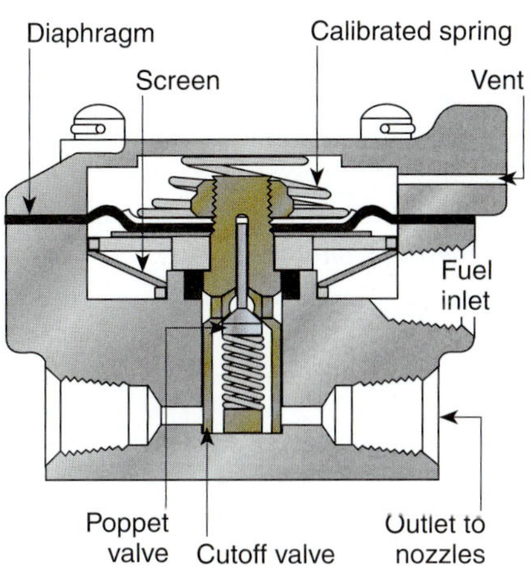

Figure 4-31. *Fuel manifold valve for a TCM fuel injection system*

the poppet valve provides a stable pressure downstream of the metering jet to allow stable fuel metering. At fuel flows above idling, the valve is fully open, and it presents no opposition to the fuel flow. Opposition under these conditions is provided by the nozzles. The manifolds come with either plus or minus spring pressure. The different spring pressure is used to modify flow when the system is calibrated to improve flow characteristics through the operating range for a specific engine. One should take care when servicing this system on twin engine aircraft. The engines may not have the same spring pressure on the manifolds and if they were switched would cause metering irregularities.

Lines connecting the fuel manifold valve to the nozzles are made of small-diameter stainless steel. Early systems are sensitive to injector line length for identical metering to all cylinders, therefore all injector lines are of equal length. Usually, the two center lines utilize a vibration loop because the distance to these cylinders is closer than the others. The latest systems are using lines of differing length. (The reason for this change have not been shared with the public.) In all cases, the lines either have a vibration loop or a double bend, over their length, to compensate for vibration and expansion.

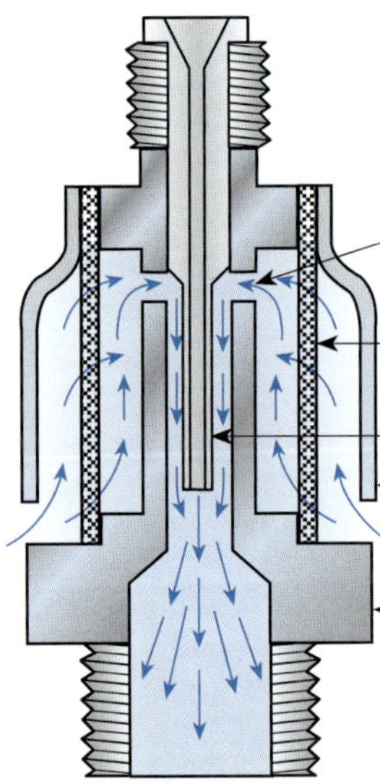

Figure 4-32. *TCM fuel injector nozzle for a naturally aspirated engine*

Injector Nozzles

TCM injector nozzles are similar to those used with an RSA system. Each nozzle contains a calibrated orifice and air bleed holes. The nozzle has a filter screen around it and is enclosed in a steel shield, or shroud. *See* Figure 4-32.

Fuel flows continuously from the fuel manifold valve and is sprayed from the calibrated orifice into the intake port along with air that flows into the nozzle through the air bleed holes. This forms an emulsion and is drawn into the cylinder when the intake valve opens, where it immediately vaporizes and mixes with air taken into the cylinder through the intake manifold.

Except at idle, turbocharged engines operate with a manifold pressure above that of ambient. This condition would cause a fuel outflow through the nozzle vent rather then a bleed air inflow. Therefore, nozzles used in these engines have an air line connected to the turbocharger side of the throttle valve that supplies pressurized air to the nozzle air bleed.

Installation and Maintenance of TCM Fuel Injection System

Starting an engine equipped with a TCM fuel injection system is similar to that with an RSA system. Follow the normal procedure to prepare the engine for starting. Turn the fuel ON and open the throttle about $\frac{1}{8}$ of the way. Then place the mixture control in the FULL RICH position and turn the boost pump on HIGH. As soon as there is an indication of flow on the flowmeter, turn the ignition switch to BOTH and engage the starter. This procedure varies somewhat depending on the airframe manufacturer's preference and their specifics should always be followed.

Fuel-injected engines have a reputation for being difficult to start when they are hot. This is caused by engine heat evaporating the fuel in the injector lines. If the normal start procedure is used, the engine will not get enough fuel to the nozzles to start it. Continued cranking will cause the engine to pass from a starved to a flooded condition, and it will not start until all excess fuel is cleared from the cylinders.

A hot engine can be started by first removing fuel vapors from the system's fuel lines. This is done by placing the mixture control in the IDLE CUTOFF position with the throttle wide open and the boost pump on HIGH. Allow fuel to circulate through the pump for about 15 or 20 seconds, then turn the boost pump off. Place the mixture control in FULL RICH, close the throttle to the correct position for starting, and turn the ignition switch to BOTH. Then engage the starter. The fuel lines will be full of liquid fuel, and the engine should start.

Adjusting High and Low Unmetered Fuel Pressure

Two of the most important things to check when installing a TCM system, or when there is any fuel metering problem, are adjustments of the high and low unmetered fuel pressure, while remaining within the specified metered fuel pressure limits for idle and full throttle.

Some TCM systems are equipped with T-fittings installed where the unmetered fuel pressure is to be measured. Connect an accurate pressure gage in the line between the pump and fuel control, and between the FCU and the fuel manifold. The first indicates unmetered pressure and the latter the metered pressure. Measure the pressure with the engine started, warmed up, and with the idle RPM and mixture properly adjusted. When the engine is idling at the correct RPM, the pressure must be within the limits specified in the most recent TCM service manual. If it is not, adjust the relief valve to bring it within the allowable limits.

When the low unmetered pressure is adjusted, adjust the idle RPM and mixture using procedures discussed for previous systems. Slowly move the mixture control toward the IDLE CUTOFF position and watch the tachometer. The speed should increase approximately 50 RPM before the engine dies. If the rise is less than this, the mixture must be enriched; if it is more, the mixture must be leaned.

With the low unmetered fuel pressure and the idle RPM and mixture adjusted, check the high unmetered fuel pressure. Operate the engine at full throttle, maximum RPM, and check the test fuel pressure gages. These pressures must be within the range of TCM specifications. If it is not, adjust the orifice. Continental says that best results will be obtained if the low unmetered pressure is adjusted as low as the low metered pressure range will allow and the high unmetered pressure as high as the metered high pressure range will allow. Always follow the specific instructions in the latest revision of the Continental Service Bulletin when performing this procedure.

When the low and high unmetered fuel pressures have been adjusted and are correct, fly the aircraft and check the high metered fuel pressure as is indicated on the installed fuel flowmeter. This pressure is checked with the engine operating at full throttle, maximum RPM, after the engine conditions have been allowed to stabilize, by operating at cruise power for about 10 or 15 minutes.

STUDY QUESTIONS: TELEDYNE-CONTINENTAL FUEL INJECTION SYSTEMS

Answers begin on Page 197. Page numbers refer to chapter text.

82. The four basic components in a TCM fuel injection system are:
   a. ______________________________
   b. ______________________________
   c. ______________________________
   d. ______________________________
   Page 174

83. The device in a TCM fuel injection system that varies the pressure of fuel going to the metering jet is the ______________________ . *Page 174*

84. Vapors are separated from fuel in a TCM fuel injection system in the ______________________________ . *Page 174*

85. Fuel from the auxiliary pump for starting the engine flows around the injector pump vanes through the _________________________ . *Page 174*

86. The low unmetered fuel pressure is set by adjusting the ________________________ in the injector pump. *Page 175*

87. The high unmetered fuel pressure is set by adjusting the ________________________ in the injector pump. *Page 175*

88. The bellows in the injection pump on a turbocharged engine controls the ____________________ (relief valve or orifice). *Page 176*

89. The TCM fuel-air control unit _________________ (does or does not) use a venturi to sense the amount of air flowing into the engine. *Page 176*

90. The adjustable stop that controls the amount the throttle air valve remains open when the throttle is all the way back determines the idle ______________________ (RPM or mixture). *Page 176*

91. The opening of the orifice in series with the relief valve in a TCM fuel pump used on a turbocharged engine is controlled by a bellows that senses the ______________________________ pressure. *Page 176*

92. A stable fuel pressure downstream of the metering jets under idling conditions is provided by the ______________________ valve in the fuel manifold valve. *Page 177*

93. When checking the low unmetered fuel pressure, the engine is operated at its ________________ (idle, cruise, or takeoff) RPM. *Page 179*

Antidetonation Injection Systems

Detonation is the uncontrolled burning of the fuel-air mixture inside the cylinder of a reciprocating engine. It is actually an explosion that occurs when the fuel absorbs more heat than it can dissipate and reaches a critical temperature/pressure point on the detonation curve. Power-enrichment systems prevent detonation by providing an excessively rich mixture. Some of the excessive heat produced when the engine operates at full power is used to vaporize the extra fuel. Use of an overly rich mixture cools the engine, but causes it to operate inefficiently, and consequently does not produce the maximum power.

Some large reciprocating engines that were required to produce the maximum power without detonation, used water injection systems, more accurately called antidetonation injection, or ADI, systems.

ADI (antidetonation injection) system. A system used with some large reciprocating engines in which a mixture of water and alcohol is sprayed into the engine with the fuel when operating at extremely high power. The fuel-air mixture is leaned to allow the engine to develop its maximum power, and the ADI fluid absorbs excessive heat when it vaporizes.

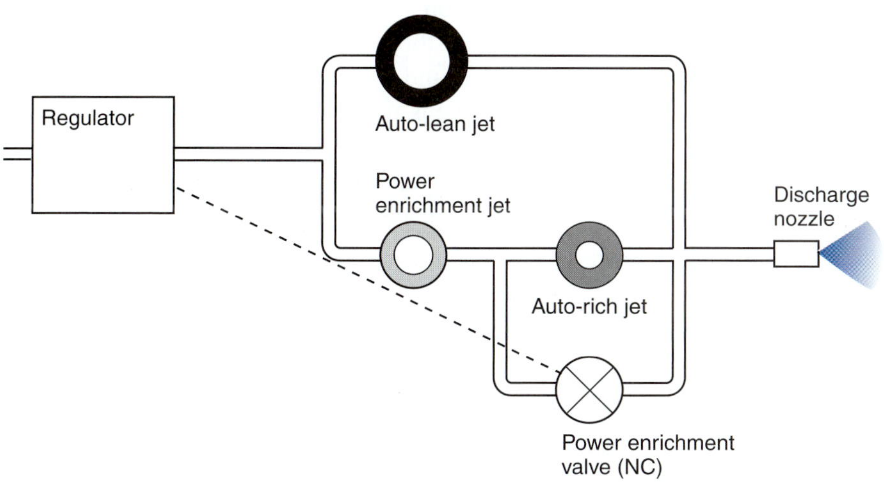

Figure 4-33. *Fuel flow route for a large reciprocating engine using a pressure-injection carburetor without antidetonation injection*

ADI (antidetonation injection) fluid. A mixture of water and methanol with a small amount of water-soluble oil added for corrosion prevention. Methanol is used primarily to prevent water from freezing at high altitude.

An ADI system leans the fuel-air mixture to a ratio that produces the best power and sprays ADI fluid into the engine through the main discharge nozzle. Heat that would normally cause detonation is absorbed by converting the water in the ADI fluid into water vapor.

ADI fluid is a water-methanol mixture with some water-soluble oil added to help minimize corrosion. Methanol is used primarily as an anti-freeze additive.

Fuel metering with a pressure-injection carburetor is done by the regulator section varying the pressure drop across three fixed jets. *See* Figure 4-33.

- Any time the engine is running, fuel flows through the largest jet, which is the auto-lean jet.
- For takeoff and applications requiring high power, the mixture control is placed in the AUTO RICH position. Fuel continues to flow through the auto-lean jet and through the auto-rich jet which is in series with the power-enrichment jet. Under normal AUTO RICH operation, the auto-lean and auto-rich jets do the metering.
- Additional fuel is metered for full power operation when the normally closed (NC) power-enrichment valve opens. The power-enrichment jet is in series with, and is larger than, the auto-rich jet, and when the power-enrichment valve is closed, it serves only as a fuel passage. When the fuel regulator calls for the maximum amount of fuel, the power-enrichment valve, which is in parallel with the auto-rich jet, opens, and additional fuel flows to the discharge nozzle. Metering is now done by the auto-lean and power-enrichment jets.

Pressure-injection carburetors used on large engines equipped with ADI systems are somewhat different from those used on engines without ADI. The auto-lean jet is slightly smaller, and a derichment jet in series with a normally open (NO) derichment valve is in parallel with the auto lean jet.

When the need for maximum power is anticipated, the pilot arms the ADI system and the following events take place:

- The pilot advances the throttle, and when the manifold pressure reaches the preset value, the ADI pump turns on and directs a flow of ADI fluid to the discharge nozzle where it is sprayed out with the fuel.
- At the same time the ADI fluid is directed to the discharge nozzle, the ADI pressure closes the normally open derichment valve and leans the mixture to the ratio that gives maximum power. Heat that would normally damage the engine is absorbed when it changes the water and methanol in the ADI fluid into vapor.
- When the water supply is exhausted, or when the pilot turns the ADI system off, the derichment valve opens and the mixture returns to its normally rich ratio.

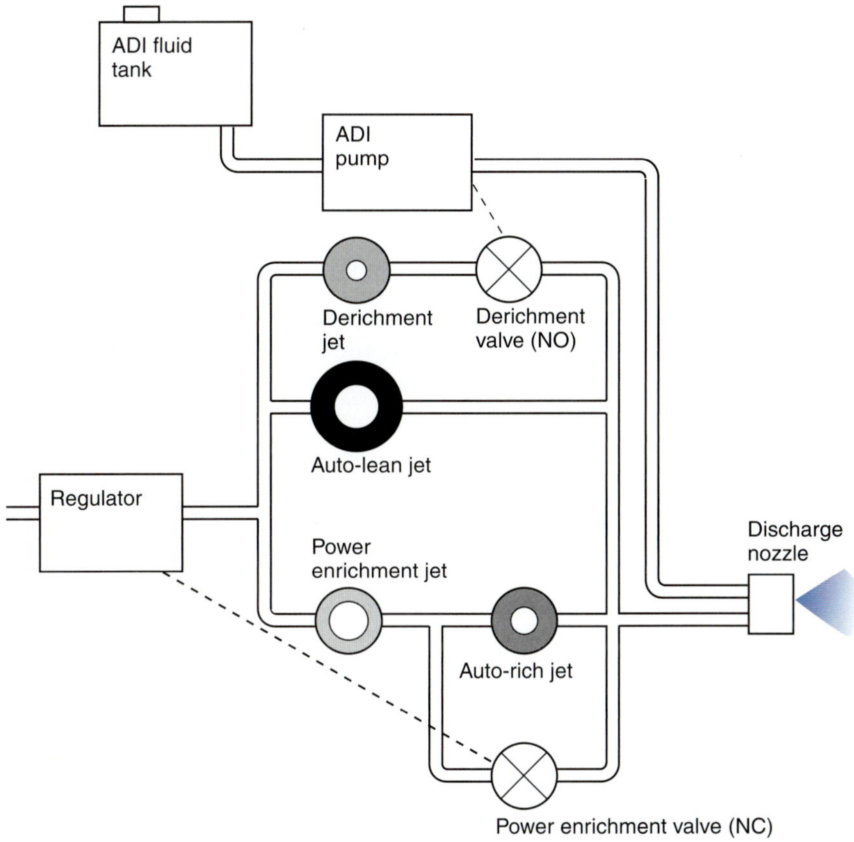

Figure 4-34. *Fuel flow route for a large reciprocating engine using a pressure-injection carburetor equipped with an antidetonation injection system.*

STUDY QUESTIONS: ANTIDETONATION INJECTION SYSTEMS

Answers begin on Page 197. Page numbers refer to chapter text.

94. Antidetonation fluid is a mixture of water and ________________ with some water-soluble oil. *Page 182*

95. For full-power operation without water injection, the metering is done by the ________________ and the ________________ jets. *Page 182*

96. The auto-lean jet on an engine equipped with an ADI system is ________________ (larger or smaller) than the auto-lean jet on an engine without an ADI. *Page 182*

97. When using the ADI system, the derichment valve causes the fuel-air ratio to become ________________ (richer or leaner). *Page 182*

98. When the ADI fluid is exhausted, the derichment valve ________________ (opens or closes). *Page 182*

Reciprocating Engine Induction Systems

supercharged engine. A reciprocating engine that uses a mechanically driven air compressor to increase the air pressure before it enters the engine cylinders.

For an engine to develop its designed power, it must receive an adequate amount of fuel and air. We have just seen how the fuel is supplied, and now we will consider the air supply.

Reciprocating engines are classified according to their induction systems in two ways: naturally aspirated and supercharged.

Naturally Aspirated Engine Induction Systems

naturally aspirated engine. A reciprocating engine that uses atmospheric pressure to force the charge of fuel-air mixture into the cylinders.

The induction system of a naturally aspirated, or nonsupercharged, engine consists of an air scoop, induction air filter, alternate air valve with a source of heat, fuel metering device, and induction pipes to the cylinders.

The air scoop is used to convert some ram air pressure from the forward motion of the aircraft into a pressure that forces the fuel-air mixture into the cylinders. A well-designed air scoop can increase the power of an engine by as much as 5%.

flock. Pulverized wood or cotton fibers mixed with an adhesive. Flock, attached to a wire screen, acts as an effective induction air filter for small reciprocating engines.

It is very important that the air entering an aircraft engine be filtered to remove dust and dirt particles that can drastically shorten the engine life. Filters for some smaller engines are made of a metal screen wire covered with tiny thread-like fibers called flock. These filters are coated with clean engine oil to attract and hold the dust and dirt particles. They are washed with reverse flow during normal maintenance inspections, reoiled, and used again.

Some engines use pleated paper filters similar to those used on automobile engines and they are normally replaced on a time interval basis. They may be cleaned with 35 psi compressed air in the opposite direction of induction air flow about 30 times, or cleaned with a mild detergent and warm water about 20 times. A General Airworthiness Directive requires replacement of all paper filters after 500 hours in service.

Some paper filter elements may be replaced with an element made of polyurethane foam that is impregnated with a glycol solution. The glycol aids these filters in attracting and holding dust. It is not recommended that foam filters be cleaned, but should be replaced on an operating time interval basis recommended by the airframe manufacturer. There are Supplemental Type Certificates for converting an aircraft engine system to foam filtering. This type of conversion is a major alteration and requires a Major Repair and Alteration form.

Alternate Air System

One main disadvantage of a float carburetor is its tendency to form carburetor ice. The venturi decreases both the air pressure and temperature, and when liquid fuel is sprayed into the throat of the carburetor, it vaporizes and the air temperature drops as much as 70°F. Carburetor ice can form when the outside air temperature is as high as 90°F, and if the humidity is high, even with no visible moisture in the air.

FAA materials specify three forms of induction system ice. One is "throttle ice" that usually forms when the throttle is between idle and one third open. The fuel of a float carburetor evaporates rapidly as it passes into the turbulence, and the pressure drop that exists when the throttle is in this area causes the air's moisture to condense out and freeze. The second form of ice is that which forms at the fuel discharge in the venturi, usually when the throttle is above one third open. This is due to the rapid evaporation of the fuel here under this power level. This is referred to as "refrigerant ice." The third type of ice is that which collects across the front of the air filter from super-cooled water and ice crystals in the atmosphere. The most common terms used for ice forming in this area is "rime ice" or "impact ice" depending on the atmospheric conditions causing it.

14 CFR Part 23 *Airworthiness Standards: Normal, Utility, Acrobatic, and Commuter Category Airplanes* requires all sea level engines using a conventional venturi carburetor to have a preheater that can provide a temperature rise of 90°F when the engine is operating at 75% power. The preheater of an altitude (supercharged) engine that uses a venturi carburetor must produce a temperature rise of 120°F at 75% power. This heat is normally provided by passing the induction air through a muff around some part of the exhaust system as shown in Figure 4-35.

Air entering the engine when the carburetor heat control is in the HOT position is not filtered. For this reason, carburetor heat should not be used on the ground, except as required in the pre-takeoff checklist.

Pressure carburetors and fuel injection systems are not subject to carburetor ice, but there is a possibility that ice can form on the inlet air filter and shut off air to the engine. These engines normally have a manually operated or automatic alternate air valve that allows the engine to take in warm air from inside the lower engine cowling.

Automatic alternate air valves have a spring-loaded, or magnetic-catch door. If the inlet air filter ices over, the low pressure in the intake manifold will pull the door open and take in warm air, slightly heated from cooling the cylinders.

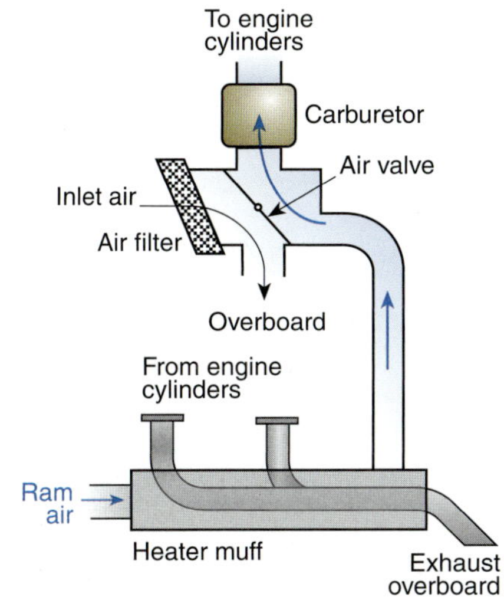

When air valve is in **hot** position, heated but unfiltered air flows into carburetor and filtered air is dumped overboard.

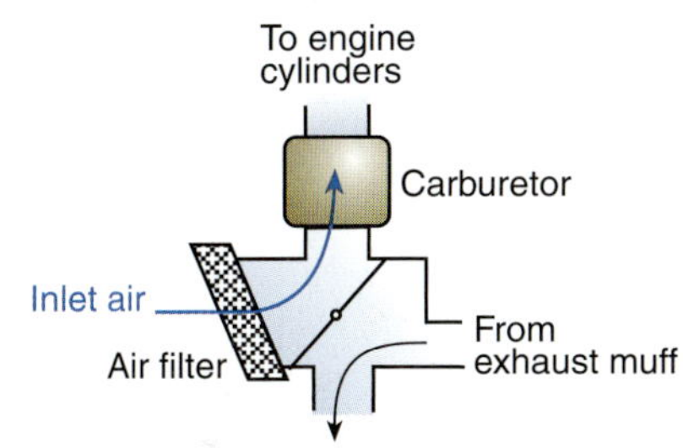

When air valve is in **cold** position, filtered air flows into carburetor and heated air is dumped overboard.

Figure 4-35. *Heated air for preventing carburetor ice is obtained by passing induction air around part of the exhaust system.*

Superchargers

volumetric efficiency. The ratio of the volume of the charge of the fuel and air inside the cylinder of a reciprocating engine to the total physical volume of the cylinder.

The volumetric efficiency of an engine is a measure of its ability to breathe. A naturally aspirated engine cannot have a volumetric efficiency of 100%. However, it is possible by using an air pump to compress the air before it enters the cylinders to increase the volumetric efficiency to more than 100%, thereby increasing the engine horsepower. This is called supercharging. Two types of superchargers used to increase the power of an engine are internal gear-driven superchargers and external exhaust gas-driven turbochargers.

Internal Superchargers

ground-boosted engine. An engine whose sea level horsepower is increased by the use of a supercharger.

boost. A term for manifold pressure that has been increased above the ambient atmospheric pressure by a supercharger.

The most commonly used internal supercharger is a gear-driven centrifugal impeller that turns at approximately ten times the crankshaft speed. Air is taken in through the carburetor and mixed with fuel. The fuel-air mixture is compressed in the supercharger and directed into the cylinders through the thin metal intake pipes.

The sea level power of the engine is increased by a single-stage internal supercharger, but the power of such a ground-boosted engine drops with altitude as seen in Figure 4-36.

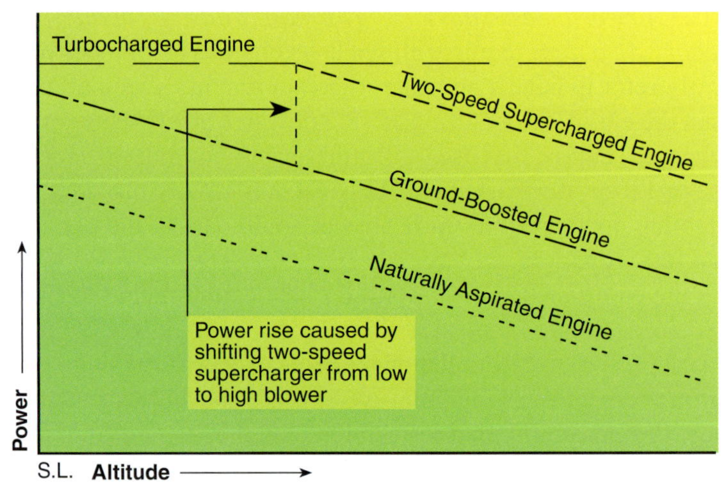

Figure 4-36. *The relationship between engine power and altitude for engines having different types of superchargers*

Some large reciprocating engines have two-speed superchargers. Takeoff is made with the supercharger control in the LOW BLOWER position. The power drops with altitude as it does with a single-speed system. But at a specified altitude, the pilot can shift the control to HIGH BLOWER, and the supercharger speeds up and increases the induction system pressure enough to produce sea-level power. From there, the power drops off with altitude.

Gear-driven superchargers use a great deal of engine power, but they enable the engine to produce more power than they use.

turbosupercharger. The original name for a turbocharger.

turbocharger. An exhaust-driven air compressor used to increase the power of a reciprocating engine. A turbocharger uses a small radial inflow turbine in the exhaust system to drive a centrifugal-type air compressor on the turbine shaft. The compressed air is directed into the engine cylinders to increase power.

intercooler. An air-to-air heat exchanger installed between a turbosupercharger and the carburetor. Intercoolers decrease the temperature of compressed air to prevent detonation.

Turbosuperchargers and Turbochargers

Turbosuperchargers were experimented with as early as 1918, but it was only with the need for high altitude bombers and fighters in World War II during the 1940s, that they became practical.

Exhaust gases from the large radial or V-engines spun a turbine that drove a centrifugal air compressor. The compressed air was passed through an intercooler which lowered its temperature enough to prevent the fuel-air mixture from detonating. The compressed air was then sent through a pressure-injection carburetor and mixed with fuel, and then further compressed by a gear-driven internal supercharger before it was directed into the cylinders.

The much smaller turbocharger used with modern horizontally opposed aircraft engines is an outgrowth of the turbochargers used on diesel-engine trucks.

Figure 4-37 shows the basic operating principle of the aircraft installation of a turbocharger. Exhaust gases flow from the cylinder through the exhaust system and through a radial-inflow turbine, then out the exhaust tail pipe. The turbine drives a centrifugal compressor that takes in induction system air and compresses it, then sends it into the cylinders. The turbine in the exhaust system decreases the power slightly because of the exhaust back pressure it causes, but the power increase is so high that turbochargers are efficient.

Turbochargers are simple in their concept, but complex in their construction. They operate in extremely hot exhaust gases, and the rotor with the turbine and compressor spin at speeds in excess of 100,000 RPM.

A typical turbocharger is illustrated in Figure 4-38. It consists of the compressor housing, center housing, and turbine housing. The turbine wheel assembly with the compressor impeller mounted on it ride in aluminum alloy bearings in the center housing.

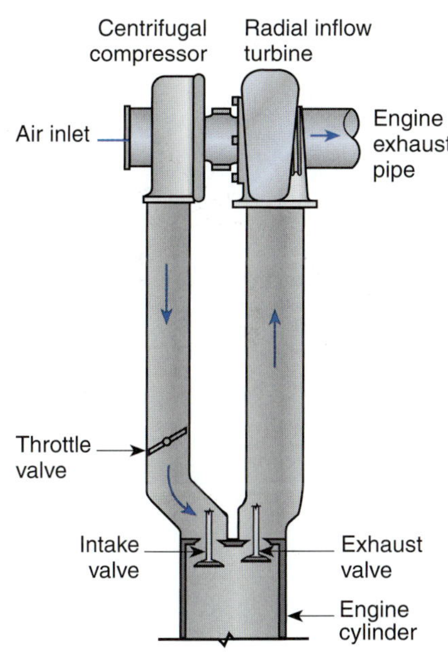

Figure 4-37. *Basic operation of a turbocharger on a reciprocating engine*

radial-inflow turbine. A turbine driven by a flow of gases that enters from the rim and flows radially inward, discharging at the center.

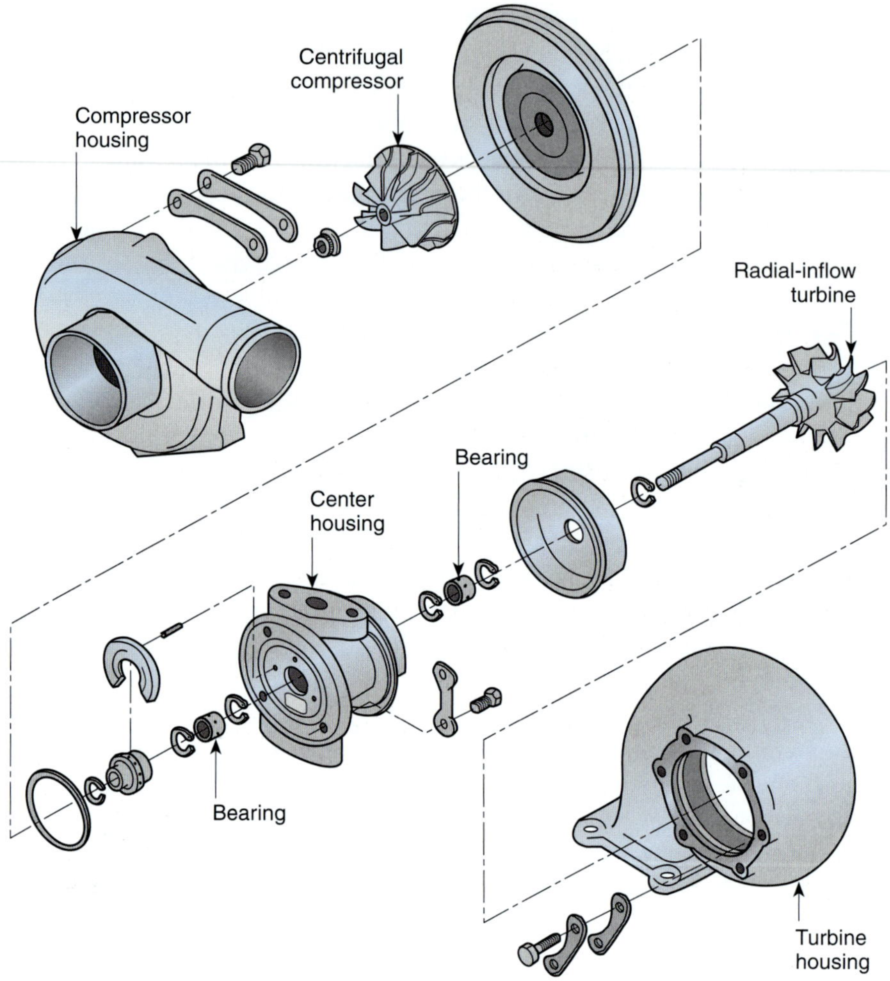

Figure 4-38. *Exploded view of a typical aircraft turbocharger*

Lubricating oil from the engine flows into the center housing and around two aluminum alloy bearings that are free to rotate in the housings. Oil flows between the housing, the bearings, and through holes drilled in the bearings themselves. This provides a continuous flow of oil between the shaft and bearings, and supports the shaft on a film of oil. The oil, used to lubricate the bearings and remove heat from the housing, drains down into a sump, and from there is drawn back into the engine by a scavenger pump. Allowing the bearings to rotate in the housing reduces the velocity between the bearing and the shaft. This helps reduce the temperature on the shaft-bearing surface and thus reduces oil coking. Coking freezes the shaft in the bearing. Floating the bearing in oil helps eliminate pressure points by providing self-alignment between the bearing and shaft, thus eliminating hot-spot coking and bearing failure.

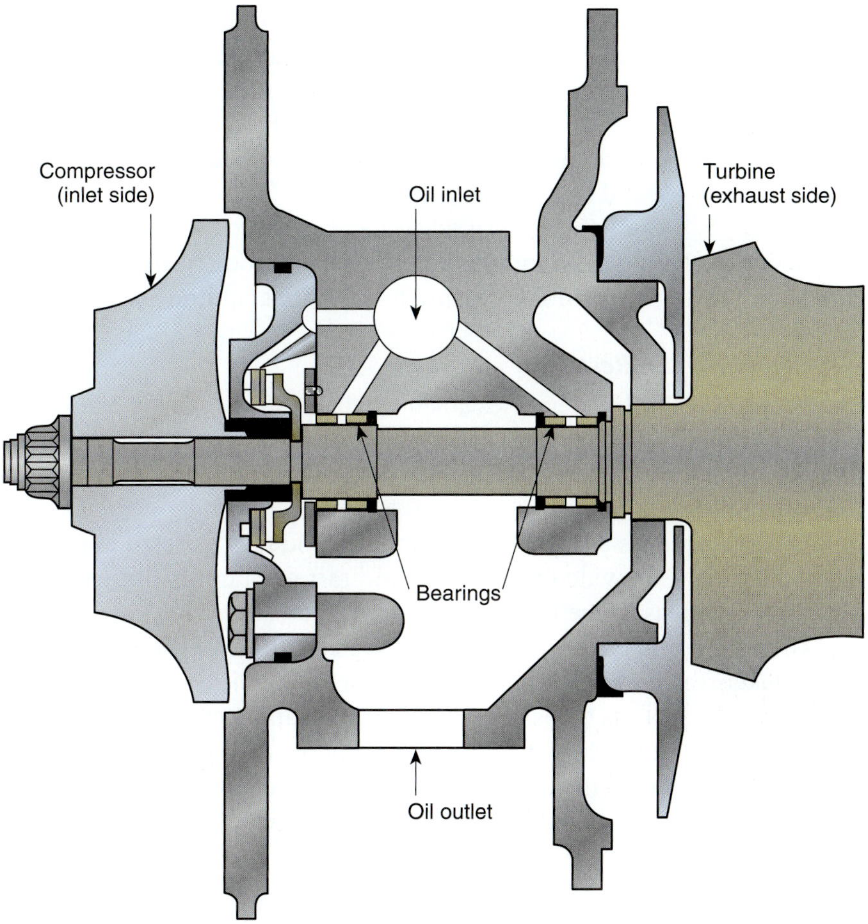

Figure 4-39. *Typical turbocharger center housing showing the oil inlet, bearings, and oil outlet.*

Turbocharger Controls

Turbochargers increase the mass of air flowing into an engine, and if there were no control of this air, the manifold pressure would increase excessively as the speed of the turbocharger increased, and detonation would likely cause serious damage to the engine.

Controlling the amount of exhaust gases allowed to flow through the turbine controls the speed (and therefore the manifold pressure) of turbochargers. Turbosuperchargers used on the large engines of the World War II era were controlled with a waste-gate valve located downstream of the back side of the turbine disk. When the waste gate valve was open, all gases flowed through it and none went through the turbine blades. When the valve was closed, the gases had to flow through the turbine blades and then overboard. Electronically controlled waste-gate motors sensed the engine manifold pressure to control the opening of the waste gate to maintain the desired manifold pressure.

waste gate. A controllable butterfly valve in the exhaust pipe of a reciprocating engine equipped with an exhaust-driven turbocharger.

When the waste gate is open, exhaust gases leave the engine through the exhaust pipe, and when it is closed, they leave through the turbine.

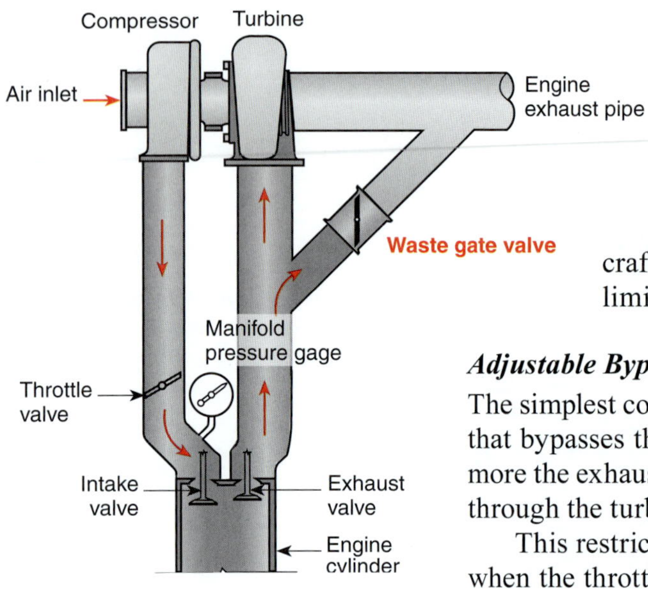

Figure 4-40. *The speed of a turbocharger is controlled by a waste-gate valve in the exhaust system, parallel with the turbine. When the waste-gate valve is open, the exhaust gases bypass the turbocharger, but when it is closed, the gases are forced to flow through the turbine.*

critical altitude. The altitude above which a reciprocating engine will no longer produce its rated horsepower with its throttle wide open.

overboost. A condition of excessive manifold pressure in a reciprocating engine. Overboosting occurs when the supercharger is operated at too high a speed.

Modern turbochargers are controlled in the same basic way. The waste-gate valve is parallel with the turbine as seen in Figure 4-40. Turbocharged engines are always started with the waste gate valve wide open to prevent excessive manifold pressure that could cause detonation and destroy the engine.

Manually controlled systems are used on some smaller aircraft, and more sophisticated systems automatically maintain the limiting manifold pressure to the critical altitude of the aircraft.

Adjustable Bypass Restrictor

The simplest control system uses an adjustable restrictor in the exhaust line that bypasses the turbocharger. The more this restrictor is screwed in, the more the exhaust is restricted, and the greater the amount of gases that flow through the turbocharger.

This restrictor is adjusted so the engine produces its rated horsepower when the throttle is wide open. As the aircraft goes up in altitude and the air density decreases, the manifold pressure decreases.

There is a problem with this simple system. If the outside air temperature at takeoff is lower than standard, and/or the ambient pressure is higher than standard, there is a possibility the engine can be overboosted, and the manifold pressure can be too high. To prevent overboosting, an upper-deck pressure relief valve is installed that relieves the excessive pressure, but with the inefficiency of more back pressure than necessary.

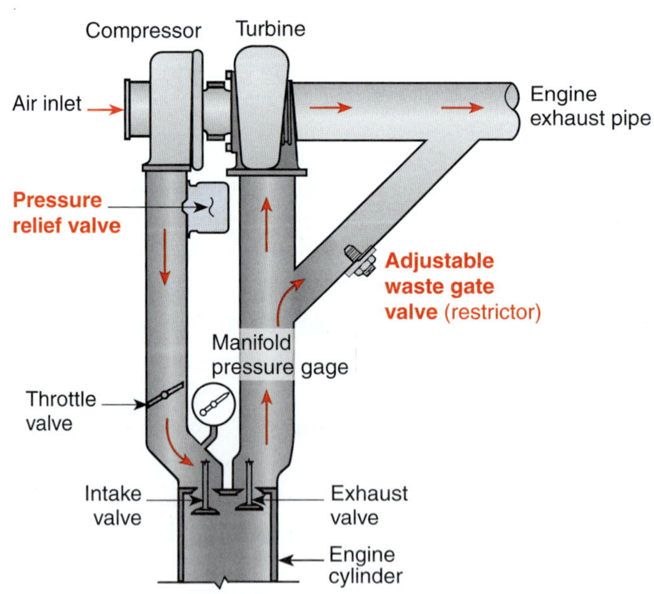

Figure 4-41. *Turbocharger control system using an adjustable restrictor and a pressure relief valve.*

Manually-Controlled Waste Gate

One type of manually-controlled waste gate has a mechanical slide-type linkage between the throttle valve and the waste gate valve. For takeoff, the throttle is advanced to the takeoff position, which is less than wide open, and the engine develops takeoff power with the waste gate fully open. As the airplane goes up in altitude, the engine power is maintained by further opening the throttle and closing the waste gate. An altitude will finally be reached when a wide-open throttle will not produce the engine's rated power. When the throttle is advanced beyond takeoff position, the additional movement begins to close the waste gate, and the power increases to the rated horsepower. The waste gate opens when the throttle is retarded. Some systems using a manually-controlled waste gate have two controls: one for the throttle and the other for the waste gate.

Automatic Turbocharger Control Systems

The two major manufacturers of modern aircraft reciprocating engines, Teledyne-Continental Motors (TCM) and Textron-Lycoming, use slightly different types of control systems on their turbocharged engines. We will consider both systems.

Teledyne-Continental Motors (TCM) System

The TCM system uses four basic types of controllers: an absolute pressure controller (APC), a rate-of-change controller, a pressure-ratio controller, and a variable absolute pressure controller (VAPC).

APC. Absolute pressure controller.

The absolute pressure controller, or APC, senses the upper-deck pressure and regulates the position of the waste gate valve to maintain the desired upper-deck pressure up to the critical altitude of the engine.

upper-deck pressure. The absolute pressure of air at the inlet to the fuel metering system of a turbocharged engine. Upper-deck pressure is the same as the turbocharger discharge pressure.

The waste gate valve in Figure 4-42 is held open by a spring and is closed when oil pressure acts on the piston in the waste gate actuator. Oil from the engine oil pump flows into the actuator cylinder through a capillary-tube restrictor and out through the APC valve, and back into the engine oil sump.

When the engine is idling, the throttle is closed, the manifold pressure is low, and the upper-deck pressure is near that of the atmosphere. The bellows in the APC senses the upper-deck pressure, and when this pressure is low, it holds the valve closed, allowing oil pressure to build up in the actuator and close the waste gate. The closed waste gate forces gases to flow through the turbine and spin the compressor, increasing the upper-deck and the manifold pressure.

When the throttle is opened for takeoff, the turbine speeds up, and both the upper-deck pressure and manifold pressure increase. When the upper-deck pressure reaches a value approximately one inch of mercury above the desired maximum manifold pressure, the bellows begins to collapse and open the valve. Oil can flow from the actuator faster than the capillary tube restrictor allows it to flow in, and the spring partially opens the waste gate.

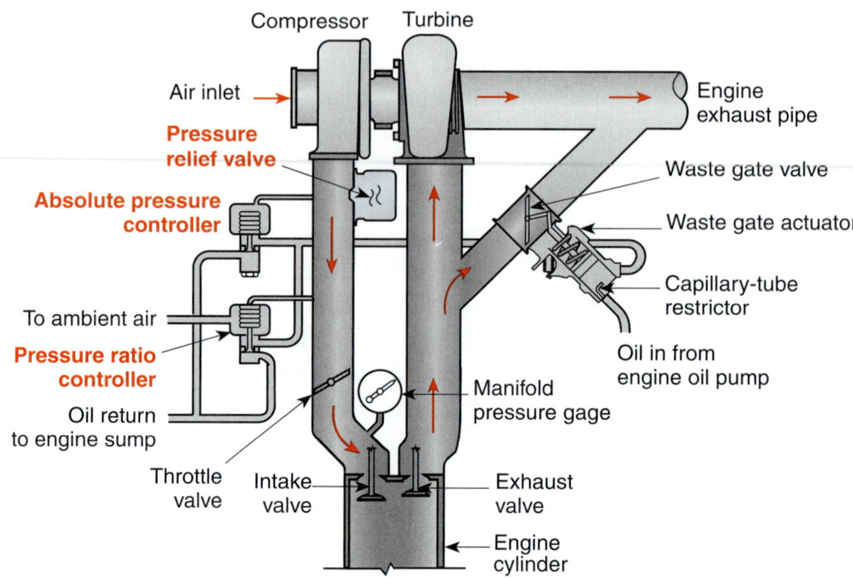

Figure 4-42. *Turbocharger control system incorporating an absolute pressure controller, pressure ratio controller, and pressure relief valve*

As the waste gate opens, some of the exhaust gases bypass the turbine, allowing the compressor to slow down enough to hold the upper-deck pressure constant at the maximum allowable value. In this way, the manifold pressure is held approximately one inch below upper-deck pressure with the throttle wide open.

As the altitude increases, the air becomes less dense and the upper-deck pressure decreases. The APC senses this decrease and closes the valve enough to restrict the oil and close the waste gate enough to speed up the turbine and compressor and bring the manifold pressure back to its maximum allowable value.

As the altitude continues to increase, the upper-deck pressure decreases, and the waste gate continues to close until critical altitude is reached. With today's turbos and only an APC controlling the system, the APC/turbocharger combination is capable of exceeding the safe critical altitude of the aircraft. The pilot must be careful to limit further increased turbo output which could result in abnormal combustion and possible engine failure.

The maximum induction air temperature is the limiting factor that the engine can tolerate at maximum manifold pressure and still have normal combustion. This requires setting the aircraft critical altitude below the capability of the turbocharging system. The pilot must limit the output by decreasing the throttle sufficiently to cause a one-inch drop in manifold pressure for each 1,000-foot increase above the critical altitude.

Some engines are restricted to the maximum altitude at which they are allowed to maintain the maximum rated manifold pressure by having a stop installed on the waste gate that limits the amount it can close. This prevents the engine reaching its maximum rated manifold pressure above its critical altitude.

Leakage in either the induction or exhaust system will prevent the engine reaching its critical altitude.

To eliminate the pilot's need to monitor and manually maintain the safe critical altitude for the engine, a pressure-ratio controller may be installed in parallel with the APC.

The pressure-ratio controller is installed in parallel with the APC in the oil line between the waste gate actuator and the engine sump. Therefore, the opening of either controller's valve will cause the waste gate to begin to open.

Assume that the turbocharger control system is designed to limit the maximum manifold pressure to 36 inches of mercury at 16,000 feet. The waste gate is partially closed, maintaining the upper-deck pressure at 37 in. Hg (one inch above the desired manifold pressure). As the altitude is increased, the waste gate progressively closes under the regulation of the APC, and would continue to do so beyond 16,000 feet. At approximately 16,000 feet the pressure-ratio controller senses that upper-deck pressure is 2.2 times that of the atmosphere at 16,000 feet and it begins to bleed off oil pressure in the waste-gate actuator, even though the APC continues to close, trying to further increase actuator pressure.

The pressure-ratio controller remains closed as long as the upper-deck pressure is no more than 2.2 times the ambient air pressure. At 16,000 feet, the standard ambient pressure is 16.22 in. Hg, and the ratio controller remains closed until the upper-deck pressure reaches 35.7 in. Hg ($2.2 \cdot 16.22$). As the airplane continues to climb, the ratio controller will continue to open the waste gate to maintain the upper-deck pressure that is no more than 2.2 times the ambient pressure. At 18,000 feet, the upper-deck pressure will be 32.9 in. Hg, and at 20,000 feet it will be down to 30.25 in. Hg.

A rate-of-change controller may be installed in parallel with the APC and the pressure-ratio controller. It remains closed until the upper-deck pressure increases at a rate faster than the manufacturer allows. If the throttle is opened too rapidly, the upper-deck pressure will rise at its maximum rate, and at this point, the rate-of-change controller will open the oil line and allow the spring in the waste gate actuator to open the waste gate and slow the compressor.

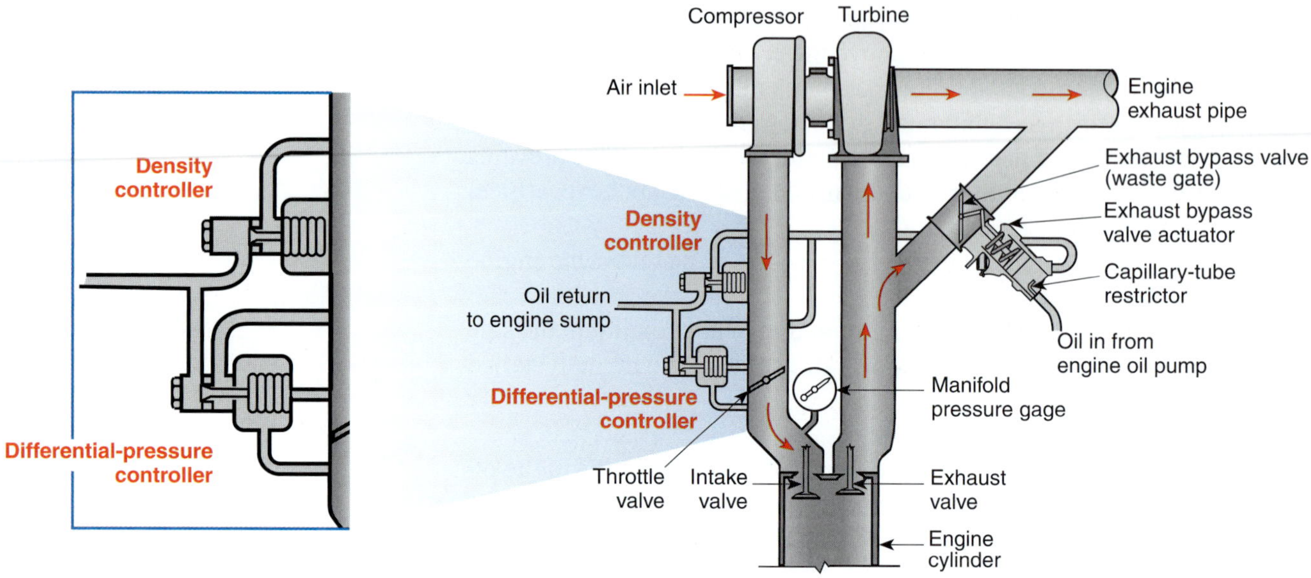

Figure 4-43. *Turbocharger control system using a density controller and a differential-pressure controller.*

Some installations replace the APC with a variable absolute pressure controller. Its function is similar to that of the APC except that the throttle controls the position of the seat of the oil drain valve. The VAPC maintains a different constant upper-deck pressure for every position of the throttle.

Textron-Lycoming Engine Turbocharger Controls

The turbochargers installed on some Textron-Lycoming sea-level boosted engines are controlled in the same way as the system described for TCM engines, by controlling the oil allowed to flow from the cylinder of the exhaust bypass valve (the waste gate valve). The valves used to control this flow are slightly different; they are the density controller and the differential-pressure controller. *See* Figure 4-43.

The density controller senses the pressure and temperature, thus the density, of air in the upper-deck zone. If the density is low, it restricts the oil returning to the engine crankcase. This causes pressure to build up in the exhaust bypass valve actuator cylinder and close the exhaust bypass valve, forcing more exhaust to flow through the turbine. The turbine speeds up and the compressor increases the density of air going into the engine. The density controller limits the manifold pressure below the turbocharger's critical altitude and regulates the bleed oil only during full-throttle operation.

When the throttle is less than wide open, the turbocharger speed is regulated by the differential-pressure controller which measures the pressure drop across the throttle valve. This pressure drop is the difference between the turbocharger discharge pressure and the manifold pressure.

If the turbocharger discharge pressure is more than 2 to 4 inches Hg greater than the manifold pressure, the differential-pressure controller oil valve will open and allow the spring to open the exhaust bypass valve, slowing the turbine and reducing the turbocharger discharge pressure.

The differential-pressure controller reduces the unstable condition known as bootstrapping. When the exhaust bypass valve is fully closed, the differential-pressure controller no longer has any effect on its position, and any slight change in power, such as that caused by a change of temperature or an RPM fluctuation, will be magnified and will cause the turbine to speed up and the manifold pressure to increase. This increase in manifold pressure will cause the engine to produce more exhaust gas, and the turbine speed to increase further. Under conditions other than full throttle, when the waste gate is fully closed, the differential-pressure controller regulates the position of the waste gate and prevents bootstrapping.

bootstrapping. An action that is self-initiating or self-sustaining. In a turbocharger system, bootstrapping describes a transient increase in engine power that causes the turbocharger to speed up, which in turn causes the engine to produce more power.

STUDY QUESTIONS: RECIPROCATING ENGINE INDUCTION SYSTEMS

Answers begin on Page 197. Page numbers refer to chapter text.

99. Three types of air filters for normally aspirated engines are:
 a. ____________________
 b. ____________________
 c. ____________________
 Page 184

100. The heat used to prevent carburetor ice is obtained by routing induction air around parts of the ____________________. *Page 185*

101. The carburetor heat system installed on a naturally aspirated engine with a float-type carburetor must provide a heat rise of __________°F when the engine is operated at 75% power. *Page 185*

102. Air that flows into the engine when the carburetor heat control is in the HOT position __________ (is or is not) filtered. *Page 185*

103. An internal supercharger compresses the air ______________ (before or after) it passes through the carburetor. *Page 186*

104. A turbocharger compresses the air ______________ (before or after) it passes through the fuel metering system. *Page 187*

Continued

STUDY QUESTIONS: RECIPROCATING ENGINE INDUCTION SYSTEMS *Continued*

105. Two functions of the lubricating system in a turbocharger are:
   a. ______________________________
   b. ______________________________
   Page 188

106. The waste gate on a turbocharger installation on a modern horizontally opposed reciprocating engine is in ____________________ (series or parallel) with the turbine. *Page 190*

107. When a turbocharger waste gate is wide open, the manifold pressure is _________________ (maximum or minimum). *Page 190*

108. When a turbocharger is equipped with a manually controlled waste gate, the engine will develop its takeoff power with the waste gate fully ____________________ (open or closed). *Page 190*

109. An engine equipped with a turbocharger should be started with the waste gate ________________ (open or closed). *Page 190*

110. Turbochargers equipped with an adjustable waste gate restrictor have a manifold pressure _____________ __________ to prevent the engine overboosting when the outside air temperature is low. *Page 190*

111. An APC regulates the position of the waste gate to maintain desired manifold pressure up to the ______________________ altitude of the engine. *Page 191*

112. The density controller limits the maximum manifold pressure produced by the turbocharger at _____________________ (full throttle or other than full throttle). *Page 194*

113. Upper-deck pressure is normally ____________________ (higher or lower) than the corresponding manifold pressure. *Page 193*

114. The pressure-ratio controller is installed in ________________ (series or parallel) with the APC in the oil line between the waste gate actuator and engine oil sump. *Page 193*

115. The ratio controller functions only at ________________ (low or high) altitude. *Page 193*

116. The variable absolute pressure controller maintains a constant upper-deck pressure for any position of the _____________________ . *Page 194*

Answers to Chapter 4 Study Questions

1. 15
2. 0.067 or 1:15
3. carbon dioxide, water
4. 8:1, 18:1
5. physical mixture
6. 21
7. 25.4
8. 10
9. weight
10. richer
11. richer
12. detonation
13. preignition
14. tetraethyl lead
15. 20,000
16. 6
17. vapor pressure
18. 100
19. 7
20. low
21. high
22. octane rating, performance number
23. iso-octane
24. heptane
25. performance number
26. ethylene dibromide, tricresyl phosphate (TCP)
27. rubber
28. color
29. blue, 2.0
30. lean
31. Supplemental Type Certificate (STC)
32. borescope
33. venturi
34. needle valve
35. lowest
36. volume
37. throttle
38. venturi
39. venturi, main metering jet, main air bleed restrictor, idle metering tube
40. a. Idle RPM (speed)
    b. Idle mixture (smoothness)
41. idle, main
42. series
43. economizer
44. parallel
45. needle valve seat
46. FULL RICH
47. richer
48. supercharger impeller
49. venturi, impact air
50. discharge nozzle
51. automatic mixture control
52. fuel tanks
53. spring
54. low
55. needle valve
56. leans
57. high
58. mixture
59. a spring
60. parallel
61. parallel
62. major
63. 8
64. IDLE CUTOFF
65. increase
66. a. servo regulator
    b. fuel control
    c. flow divider
    d. nozzles
67. series
68. open
69. close
70. increases
71. does not
72. does not
73. head
74. effort
75. decreases
76. parallel
77. a. aids in consistent idling
    b. distributes fuel evenly to all cylinders
    c. provides a positive shutoff when the engine is shut down
78. IDLE CUTOFF
79. FULL RICH
80. richer
81. increased
82. a. engine-driven injection pump
    b. fuel-air control unit
    c. fuel manifold valve
    d. nozzles
83. injector pump
84. injector pump
85. bypass valve

Continued

Answers to Chapter 4 Study Questions *(Continued)*

86. relief valve
87. orifice
88. orifice
89. does not
90. RPM
91. upper deck or turbocharger discharge
92. poppet
93. idle
94. methanol
95. auto-lean, power-enrichment
96. smaller
97. leaner
98. opens
99. a. metal screen covered with flock
    b. pleated paper
    c. polyurethane foam
100. exhaust system
101. 90
102. is not
103. after
104. before
105. a. to lubricate the bearings
     b. to remove heat from the housing
106. parallel
107. minimum
108. open
109. open
110. relief valve
111. critical
112. full throttle
113. higher
114. parallel
115. high
116. throttle

IGNITION SYSTEMS

5

Continued

Ignition Systems

5

Introduction

The *electro-mechanical* magneto ignition system is one of the most important systems in a reciprocating engine. All FAA-certificated aircraft reciprocating engines must have dual ignition with two spark plugs in each cylinder. Dual ignition not only provides safety in the event one system fails, but also, by igniting the fuel-air mixture in two places, usually on opposite sides of the cylinder, the mixture burns quickly and completely. Combustion is complete and there is maximum pressure inside the cylinder by the time the piston begins to move downward on the power stroke. This gives the engine maximum power and allows it to run at its coolest.

An ignition system must be completely dependable, as any malfunction can cause serious safety-of-flight problems. If one spark plug fails to ignite the mixture in a cylinder, the combustion process will be slow. There will be a loss of power because the mixture will still be burning as the piston moves down on the power stroke. This longer burning time will increase cylinder temperature and possibly cause detonation which can destroy the engine.

A magneto ignition system has fixed timing, and it always fires when the piston is at the same angular travel of the crankshaft from top center. This is inefficient, because as the engine speed changes, the ignition timing should also change. While magneto systems are inefficient, they are dependable, and are still the most widely used ignition system for FAA-certificated reciprocating engines. Electronic ignition systems, with their high-voltage spark and timing automatically varied to match the operating conditions of the engine, promise to be the ignition system of the future.

Battery Ignition Systems

The purpose of an ignition system is to produce a hot spark inside the cylinder, which ignites the fuel air mixture at the proper time. This can be done using a battery as the source of electrical energy in a system such as the one in Figure 5-1.

Aircraft batteries have not been traditionally regarded as a highly dependable source of electrical energy, so for years, the self-contained magneto has been the most popular ignition system. A magneto does have a serious drawback in that it does not produce a hot enough spark at low RPM to start the engine. Because of this limitation, some engines have used a dual ignition system that uses both a battery system and a magneto. The engine is started on the battery system alone, but both systems are used at the same time for normal operation.

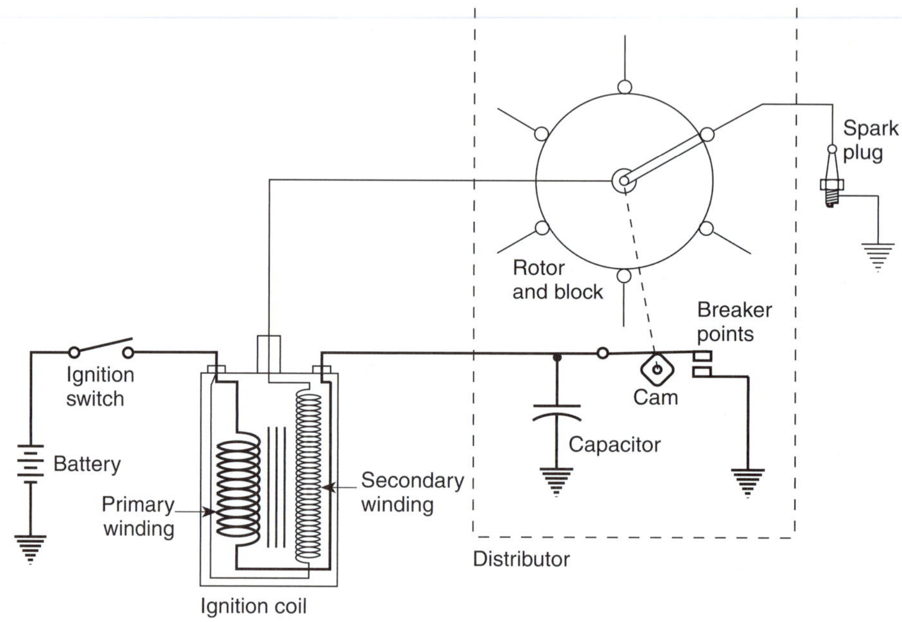

Figure 5-1. *Typical battery ignition system*

cam. An eccentric, or lobe, on a rotating shaft that changes rotary motion into linear motion.

capacitor. An electrical component used to store electrical energy in electrostatic fields.

distributor. A high-voltage selector switch, driven from the rotating magnet shaft in a magneto. The distributor directs the high voltage to the proper spark plug.

electrical steel. A low-carbon iron alloy that contains some silicon. It has high permeability and low retentivity.

permeability. A measure of the ease with which lines of magnetic flux can pass through a material.

retentivity. The ability of a magnetizable material to retain the alignment of the magnetic domains after the magnetizing force has been removed.

magnetic flux. Lines of magnetic force that are assumed to leave a magnet at its north end and return to its south end.

To understand the operation of a battery ignition system, follow the circuit in Figure 5-1. When the ignition switch is closed (ON), current from the battery flows to ground through the switch contacts, the primary winding of the ignition coil, and the closed breaker points inside the distributor. A strong magnetic field surrounds the primary winding and extends beyond the many turns of wire in the secondary winding.

At the correct time for the spark to occur inside the cylinder, the distributor cam opens the breaker points. Current can no longer flow through the primary winding, and its magnetic field collapses. As it collapses, it cuts across all turns in the secondary winding and induces a very high voltage. This high voltage is directed through the rotor in the distributor to the correct spark plug, where the high voltage ionizes the air between electrodes, and current flows to ground across the gap, producing the hot spark that ignites the fuel-air mixture.

A capacitor installed in parallel with the breaker points helps stop the flow of current when the points begin to open. This hastens the collapse of the primary magnetic field and prevents arcing across the breaker points.

Magneto Ignition Systems

Early batteries were heavy and not dependable enough for anything so vital as an aircraft's ignition system, so in most cases two magnetos were used as the source of electrical energy.

There are two types of magneto ignition systems: high-tension and low-tension. The high-tension system is discussed first as it is by far the more popular. The low-tension system was used for aircraft that flew at high altitude, but now, most aircraft that need this type of system are powered by turbine engines.

High-Tension Magneto Systems

A high-tension magneto ignition system consists of the following components:

- A permanent-magnet AC generator
- A high-voltage step-up transformer
- A set of cam-actuated breaker points that interrupts the AC flowing in the primary winding of the transformer at a specific time
- A capacitor to assist the collapse of the magnetic field when the primary current is interrupted
- A distributor, which is a high-voltage selector switch
- Ignition leads that carry high-voltage current to the spark plugs
- Spark plugs that conduct the high voltage into the combustion chambers and provide a gap across which the current flows to ground to produce a hot spark
- An ignition switch that grounds the magneto primary circuit to prevent the production of high voltage

Magnetos contain a magnetic circuit and two electrical circuits, a primary and a secondary.

The magnetic circuit consists of a high-strength permanent magnet rotor driven by gears from the engine accessory section. This magnet rotates in a frame made of laminations of special electrical steel that carries the magnetic flux to the core of the coil. This steel has a high permeability which concentrates the lines of magnetic flux and allows them to pass with the minimum of opposition and low retentivity that allows the flux lines to reverse their direction as the magnet turns. Part of the frame acts as the core of the transformer coils.

When the rotating magnet is in its full-register position as shown in Figure 5-2A, the maximum amount of flux passes through the core in the direction shown by the arrows. As the magnet rotates, the number of flux lines passing through the core decreases until at 90° (Figure 5-2B), the magnet is in its neutral position, and there is no flux in the core. Another 90° of rotation

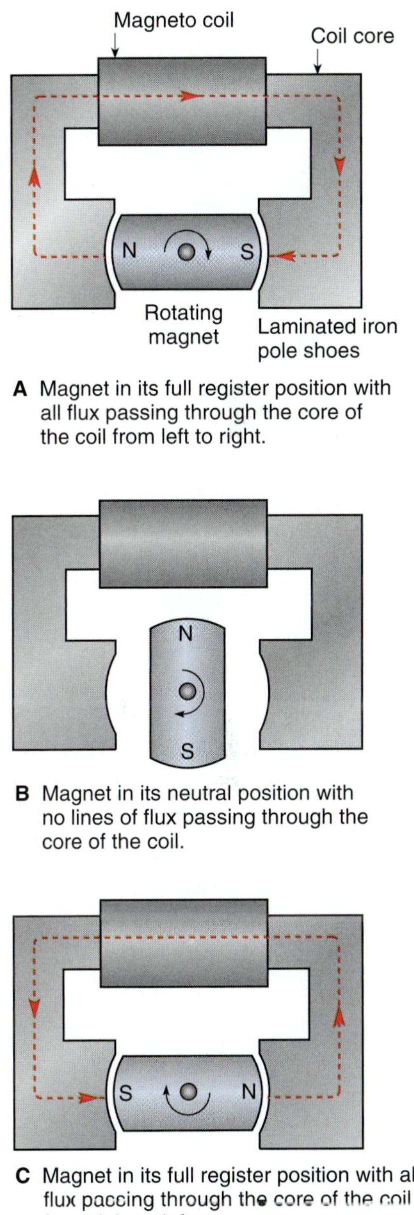

A Magnet in its full register position with all flux passing through the core of the coil from left to right.

B Magnet in its neutral position with no lines of flux passing through the core of the coil.

C Magnet in its full register position with all flux passing through the core of the coil from right to left.

Figure 5-2. *The magnetic circuit of a magneto*

brings the magnet back into alignment, as seen in Figure 5-2C, and there is again a maximum number of flux lines in the core, only this time they are passing in the opposite direction.

The primary electrical circuit of a magneto consists of the primary winding of the coil, cam-actuated breaker points, a capacitor, and the ignition switch. Figure 5-3 shows the electrical relationship of these components.

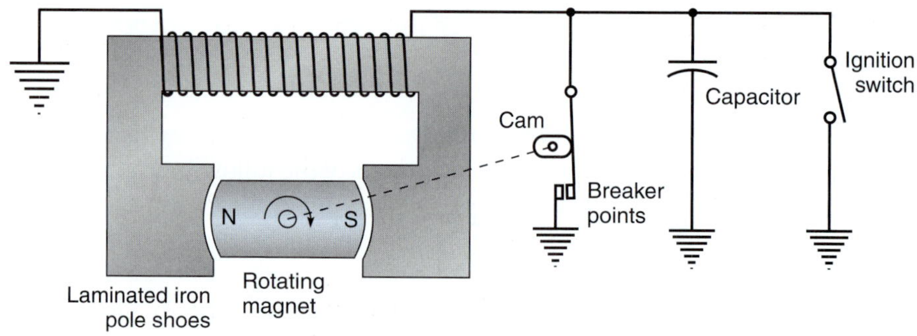

Figure 5-3. *The primary electrical circuit of a magneto*

Windings of the primary coil are wound around the laminated steel core so the changing flux cuts across the turns of coil and induces an AC voltage in them. The amount of voltage induced into the primary coil is determined by the rate at which the flux through the core changes and by the number of turns of wire in the coil.

When the breaker points are closed, primary current flows to ground. Current builds up as the magnet rotates, and when the magnet is in its full-register position, there is no change in the number of flux lines, and the primary current is zero. When the magnet rotates through its neutral position, the flux lines stop flowing in one direction and start flowing in the opposite direction. The *amount* of flux is minimum in the neutral position, but the *change* in flux is maximum. It is the rate of change that determines the amount of current, so the primary current is the greatest near the neutral position of the magnet. *See* Figure 5-4.

When current flows in the primary winding of the coil, a magnetic field surrounds each turn. The flux from this magnetic field is out of phase with the static flux from the magnet, and the two fields interact to produce the resultant flux as seen in Figure 5-4. This resultant flux causes the peak of primary current to occur, not exactly at the neutral position of the magnet, but a few degrees of rotation later.

The breaker points are opened by a cam on the same shaft as the magnet, so the magnet is in a definite position when the breaker points open. This position is called the E-gap (efficiency gap) angle or, more generally, the E-gap position.

full-register position. The position of a magnet in a magneto when its poles are aligned with the pole shoes and the maximum amount of magnetic flux flows through the magnetic circuit.

neutral position. The position of the magnet in a magneto when its poles are between the pole shoes and no lines of flux are flowing through the magnetic circuit.

primary winding. The winding in a magneto or ignition coil between the source of voltage and the breaker points.

E-gap angle. The position of the rotating magnet in a magneto when the breaker points are timed to open.

When the magnet is in its E-gap position, it has rotated just a few degrees beyond its neutral position, and as is seen in Figure 5-4, the combination of static flux and flux from the primary current have produced a resultant flux that lags behind the static flux. The breaker points open and stop the flow of primary current at the point of maximum field stress. Since there is no flow of primary current, the resultant flux drops immediately to the static flux curve.

When the magnetic field collapses, its flux cuts across all turns in the secondary winding and induces a voltage. The amount of this voltage is determined by the rate at which the field collapses and by the ratio between the number of turns in the secondary winding and the number of turns in the primary.

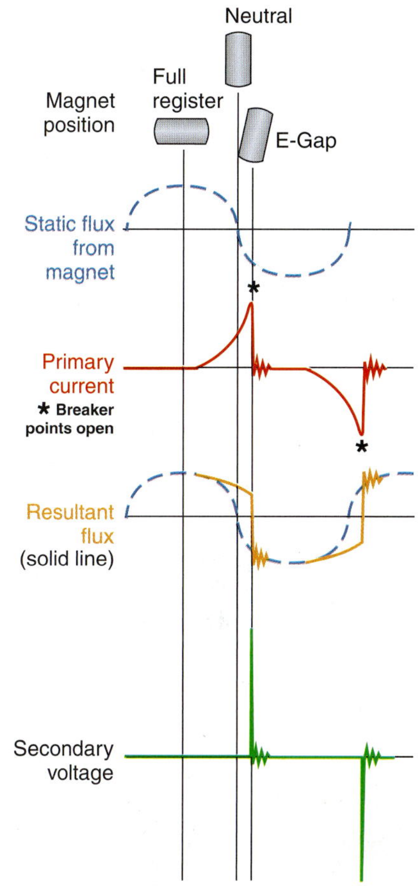

Figure 5-4. *Relationship between magnet position, static flux, primary current, resultant flux, and secondary voltage in a magneto*

The capacitor seen in Figure 5-3 is installed in parallel with the breaker points to aid the collapse of the magnetic field and prevent arcing across the breaker points as they begin to open.

It is alternating current that flows to ground through the primary winding and the breaker points. At the peak of the current, the cam begins to open the points. If there were no capacitor in the circuit, the current would continue to flow, and an arc would form as the points begin to separate. The collapsing magnetic field would sustain this flow, which would seriously burn the points. A proper size capacitor installed in parallel with the points will prevent arcing. As the points begin to separate and their resistance increases, the current finds what appears to be a low-resistance path to ground through the capacitor. By the time the capacitor is fully charged, the points are open and no current can flow to ground. Since it is AC flowing in the primary winding, the capacitor discharge aids the buildup of current in the next alternation.

The ignition switch is in parallel with the breaker points and capacitor. When the switch is open (ON), the magneto operates as just described, but when the switch is closed (OFF), all primary current flows to ground through it, and no voltage is induced into the secondary winding.

A magneto acts much like a radio transmitter. The collapse of the magnetic field when the breaker points open induces a high voltage in all components in the primary circuit as well as those in the secondary circuit. Current caused by this high voltage creates electromagnetic radiation from all the wires, which if not suppressed, would cause serious radio interference. Radiation from the ignition leads in the secondary circuit is captured in the shielding described in the discussion of ignition harnesses.

The wire from the breaker points to the ignition switch is called the P-lead (primary lead), and since it has an AC voltage in it when the engine is running, it acts as an antenna and radiates electromagnetic energy. To prevent this radiated energy causing radio interference, the P-lead must be shielded. It is encased in a metal wire braid that intercepts the radiated energy and passes it to ground. Some modern magnetos also use a feed-through capacitor which serves a dual purpose. It acts as the normal capacitor to prevent arcing at the breaker points, and since one of its plates is in series with the ignition

secondary winding. The winding in a magneto or ignition coil that connects to the distributor rotor.

electromagnetic radiation. A method of transmitting energy from one location to another by electromagnetic waves.

shielding. The electrically conductive covering placed around an electrical component to intercept and conduct to ground any electromagnetic energy radiated from the device.

P-lead. Primary lead. The wire that connects the primary winding of a magneto to the ignition switch.

lead, it shunts any electromagnetic energy in the primary lead to ground, preventing it from causing radio interference. The feed-through capacitor in Figure 5-5B is installed inside the magneto housing with its pigtail connected to the insulated breaker point. The threaded terminal on the outside housing connects to the magneto switch, and the capacitor case grounds to the magneto housing.

When current flowing through the primary winding is interrupted, its magnetic field collapses instantly and cuts across the secondary winding. It induces a voltage determined by the rate of change in the primary flux, and by the ratio of the number of turns in the secondary winding to the number of turns in the primary winding.

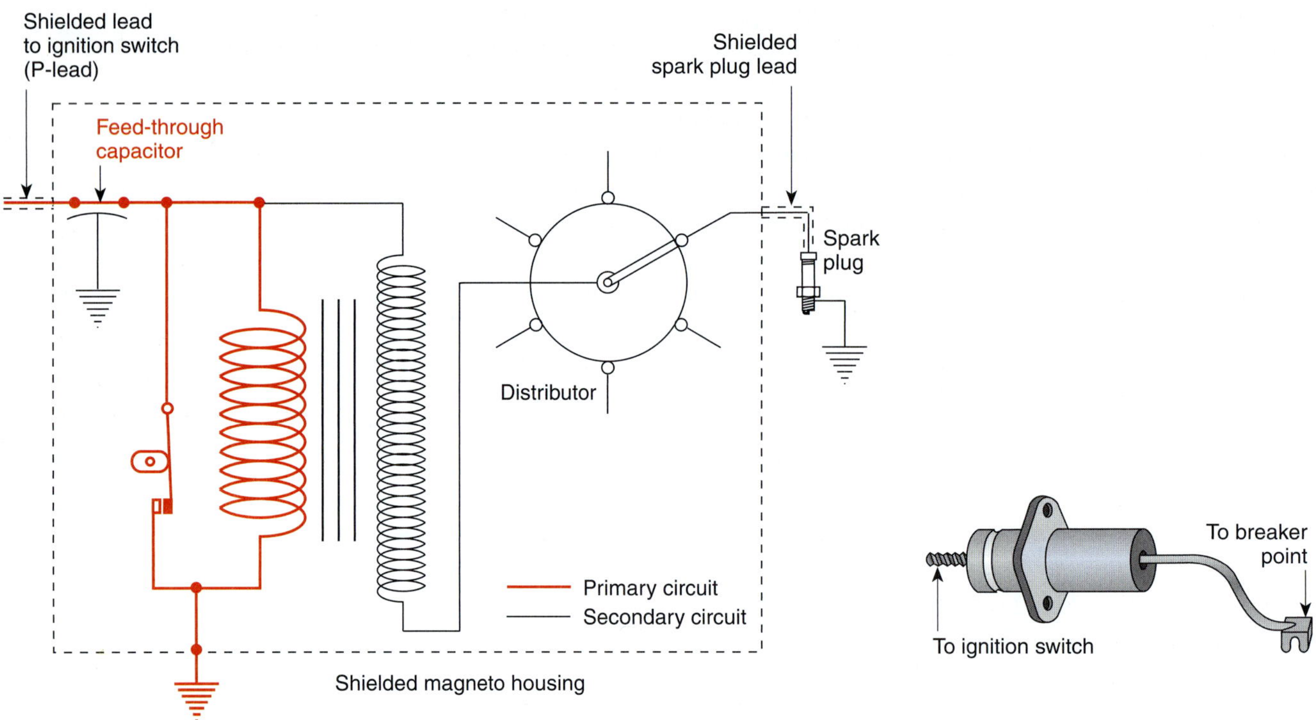

A Feed-through capacitor prevents arcing at breaker points and also passes electromagnetic energy in the primary lead to ground, preventing radio interference

B Physical appearance of feed-through capacitor

Figure 5-5. *Feed-through capacitor in the primary circuit of a magneto*

The secondary winding is made up of thousands of turns of very fine insulated copper wire wound on top of the primary winding. One end of the secondary winding is connected to the ungrounded end of the primary winding, and the other end goes to an external high-voltage terminal on the side of the coil. *See* Figure 5-6.

The high-voltage terminal on the coil is contacted by a spring-loaded carbon brush in the center of the distributor rotor. High voltage is conducted from the coil through the distributor brush to a finger on the distributor rotor. The rotor is gear driven from the magneto shaft at such a speed that it aligns with the high-voltage terminals for all cylinders when the engine crankshaft makes two rotations. This allows all spark plugs to fire in 720° of crankshaft rotation.

The high-voltage terminals on the distributor are numbered, with the numbers representing the sparking order of the magneto. When connecting the spark plug leads to these terminals, the firing order, not the cylinder number, must be matched to the distributor number. When connecting the ignition leads from the distributor to an engine with a firing order of 1-4-5-2-3-6, connect the lead for the spark plug in cylinder number 1 to distributor terminal number 1; the lead to the spark plug in cylinder number 4 to distributor terminal number 2; and so on in the order shown in Figure 5-7.

The secondary voltage rises until it ionizes the air in the gap between the spark plug electrodes. At this point it stops rising, and current flows across the ionized gap and becomes the spark that ignites the fuel-air mixture. If a spark plug lead should become disconnected, the voltage will continue to rise until it is high enough to damage the secondary insulation and destroy the coil. To prevent this type of damage, there is usually a location within the magneto where secondary current can jump to ground before it builds up enough to cause damage. This location is called the safety gap.

Figure 5-6. *Typical coil installed in a magneto. The wire on the left of the coil goes to the breaker points, and the tab in the center of the coil is the high-voltage terminal that is contacted by the spring-loaded distributor brush.*

Firing order	Distributor number
1	1
4	2
5	3
2	4
3	5
6	6

Figure 5-7. *Numbers on a distributor are the sparking order of the magneto, and they must be matched with the ignition lead according to the firing order of the engine.*

safety gap. A location in a magneto that allows a spark to jump to ground from the secondary circuit before the voltage rises high enough to damage the secondary insulation.

Low-Tension Magneto Systems

Air at high altitude is not an effective electrical insulator, and high voltage in the magneto distributor can jump to the wrong electrode and cause the engine to misfire. This condition, known as flashover, results in loss of power, rough running, and excessive heat. Several different approaches have been taken to cure this problem. Some engines have distributors that are physically large, and others pressurize the distributor with air from the engine supercharger to make it more difficult for the spark to jump to the wrong electrode.

One cure for high altitude flashover is the low-tension magneto system. A low-tension magneto has a magnetic circuit and a primary electrical circuit similar to those described for the high-tension magneto, but it has no actual secondary circuit as such.

flashover. An ignition system malfunction in which the high voltage in the magneto distributor jumps to the wrong distributor terminal.

The primary current is routed to the rotor of a brush-type distributor and from there to the primary winding inside a high-tension transformer for each spark plug. These transformers, which have a primary and a secondary winding, are mounted as near to the spark plugs as is practical to keep the wires carrying high voltage as short as possible.

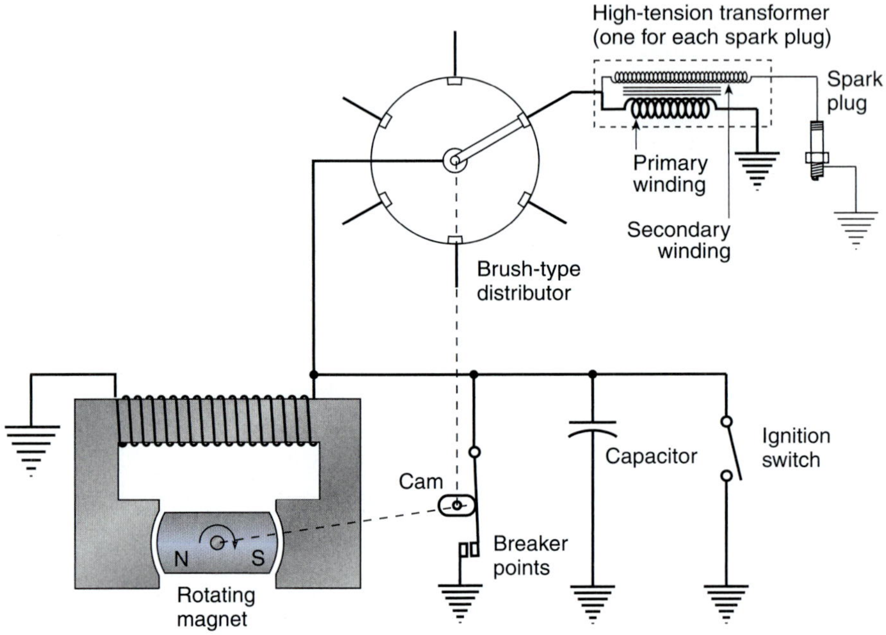

Figure 5-8. *Low-tension magneto system*

STUDY QUESTIONS: INTRODUCTION TO RECIPROCATING ENGINE IGNITION SYSTEMS

Answers are on Page 244. *Page numbers refer to chapter text.*

1. The spark produced by a battery ignition system occurs when the breaker points ________________ (open or close). *Page 202*

2. The capacitor in a battery ignition system is installed in ________________ (series or parallel) with the breaker points. *Page 202*

3. The ignition switch of a battery ignition system is installed in ________________ (series or parallel) with the breaker points. *Page 202*

4. When an engine is equipped with both a magneto and battery ignition system, the engine is started on the ________________ (battery or magneto) system. *Page 202*

5. Two types of magneto ignition systems are:
   a. ______________________________
   b. ______________________________
   Page 203

6. The three circuits in a magneto are the:
   a. ______________________________
   b. ______________________________
   c. ______________________________
   Page 203

7. When a magneto magnet is in its neutral position, there is a _________________________ (maximum or minimum) of magnetic flux flowing through the core of the coil. *Page 203*

8. The amount of voltage induced into the secondary winding of a magneto is determined by two things. These are:
   a. ________________________________
   b. ________________________________
   Page 205

9. Flux caused by current flowing in the primary winding of the magneto coil interacts with flux from the rotating magnet to cause the maximum amount of primary current flow to occur slightly ____________________ (before or after) the magneto has passed its neutral position. *Page 204*

10. Magneto breaker points are timed to open when the magnet has rotated a few degrees beyond its neutral position. This position is called the ________________ position. *Page 204*

11. Arcing across the breaker points as they open is prevented by installing the proper size ________________________ across the points. *Page 205*

12. The capacitor in the primary circuit of a magneto is installed in ____________________ (series or parallel) with the breaker points. *Page 205*

13. A magneto ignition switch is installed in __________________ (series or parallel) with the breaker points. *Page 205*

14. When a magneto ignition switch is open, it is ________________ (ON or OFF). *Page 205*

15. Electromagnetic energy is prevented from radiating from the P-lead by ____________________ the lead. *Page 205*

Continued

STUDY QUESTIONS: INTRODUCTION TO RECIPROCATING ENGINE IGNITION SYSTEMS *Continued*

16. The current in the primary winding of a magneto is ________________ (AC or DC). *Page 205*

17. The finger of the magneto distributor makes one revolution while the engine crankshaft is making ________________ (how many) revolutions. *Page 207*

18. Numbers on the magneto distributor are the __________________________________ (firing order of the engine or sparking order of the magneto). *Page 207*

19. When the breaker points open, the secondary voltage rises until a spark jumps the gap in the ______________________ . *Page 207*

20. If a spark plug lead is not installed, the secondary voltage will rise until a spark jumps the _____________________ in the magneto. *Page 207*

21. A low-tension magneto has one high-tension _________________ for each spark plug. *Page 208*

22. A six-cylinder engine with a low-tension magneto ignition system has ________________ (how many) high-tension transformers. *Page 208*

Auxiliary Starting Systems

Two major problems with a magneto ignition system have to do with starting the engine. Voltage produced by the slow-turning magneto is too low to successfully start the engine, and the spark occurs while the piston is so far before top center that the engine will kick back rather than start.

Voltage produced by a magneto is determined in part by the rotational speed of the magnet. When the starter is cranking the engine, the magnet is rotating slowly, and the voltage produced is not high enough to successfully start the engine.

Magnetos have fixed timing. The spark occurs when the piston is a specified number of degrees of crankshaft rotation before the piston reaches top center on the compression stroke. When the engine is being started, this early ignition allows the mixture to burn and produce a high pressure before the slow-moving piston reaches the top of its stroke, and it forces the piston back down and makes the engine try to run backward.

retarded sparks. Ignition sparks for starting an engine, timed to occur later than the normal sparks.

These two problems are solved by auxiliary systems that allow the magneto to produce a hot and retarded spark for starting.

Impulse Coupling

The most popular system for providing a hot and retarded spark for starting smaller aircraft engines is an impulse coupling, located on the shaft of the magneto.

impulse coupling. A spring-loaded coupling between a magneto shaft and the drive gear inside the engine, used to produce a hot, late spark for starting the engine.

Usually only one of the magnetos, normally the left one, has a spring-loaded impulse coupling between the engine drive gear and the magnet shaft. The magneto drive gear in the engine turns the impulse coupling body, and the cam turns with the magnet. The spring forms a connection between the body and the cam.

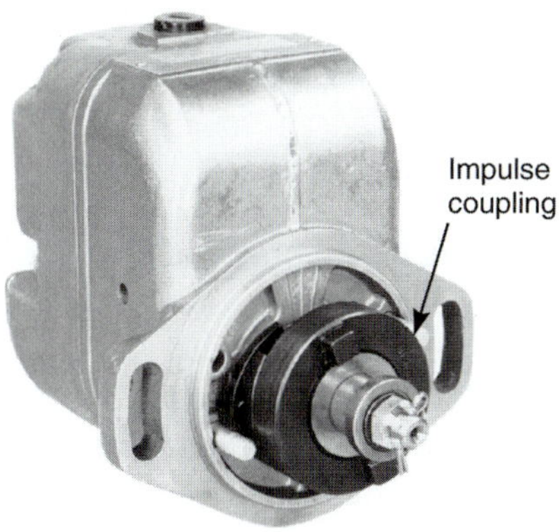

Figure 5-9. *Magneto with an impulse coupling on its magnet shaft*

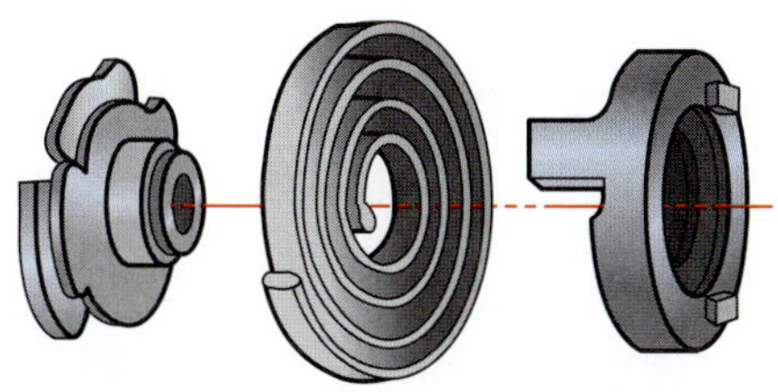

Figure 5-10. *The cam, spring, and body of an impulse coupling*

The body of the impulse coupling in Figure 5-9 has two ears that fit between a pair of hard rubber drive cushions held inside an adapter in the magneto drive gear.

Figure 5-12 shows the end of the magnet shaft and the cam that turns with it. When the engine is turned with the starter, the upper flyweight drops down so its toe section extends beyond the rim of the body. As the body

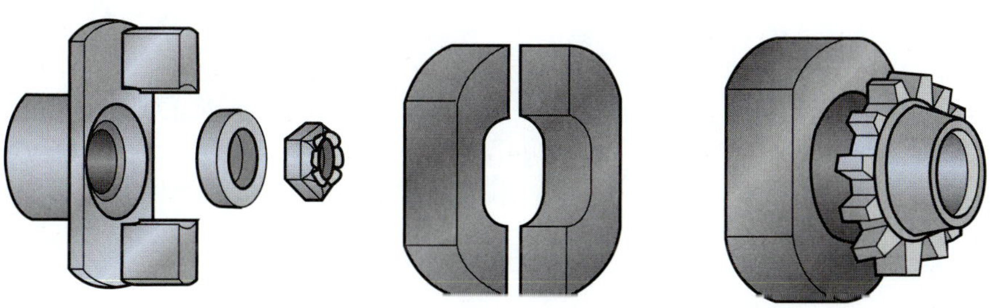

Figure 5-11. *The two ears on the magneto coupling fit between two rubber cushions held in an adapter inside the magneto drive gear.*

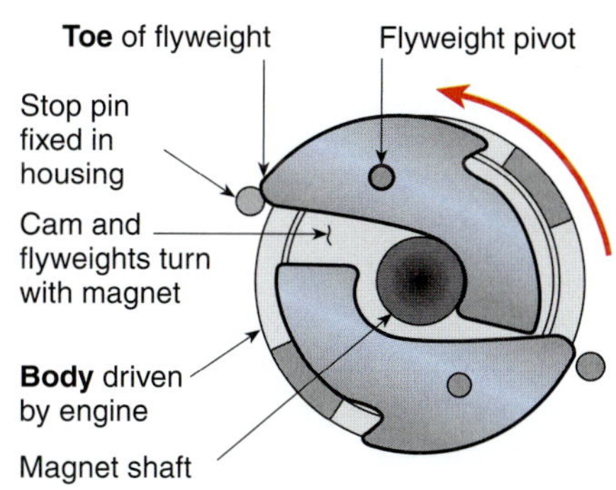

When starter begins to rotate engine, toe of upper flyweight contacts the stop pin and holds magnet. The body turns and winds up spring between cam and body.

Figure 5-12. *The impulse coupling as the starter begins to rotate the crankshaft*

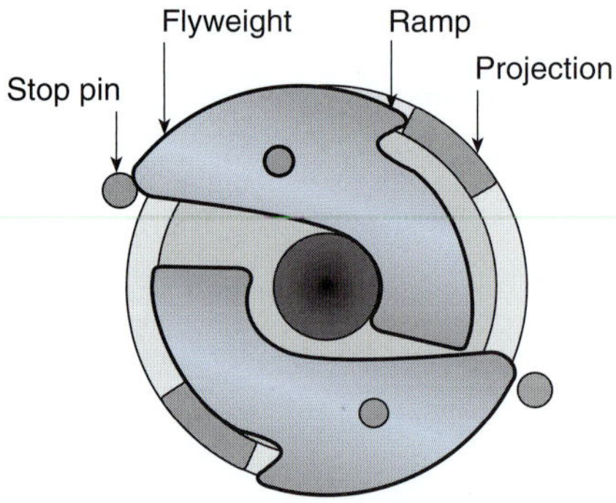

When piston is near top center, the projection on body presses against ramp on flyweight and forces it away from stop pin. The spring spins magnet and produces a hot, retarded spark to start engine.

Figure 5-13. *The impulse coupling at the point the spring is released*

and cam turn, the flyweight contacts the stop pin that is part of the magneto housing. The stop pin holds the cam and prevents the magnet from turning as the drive gear continues to turn the body and wind up the spiral spring (shown in Figure 5-10) between the body and the cam. *See* Figure 5-12.

The starter continues to turn the crankshaft until the piston is near its top-center position. At this time, the projection on the body contacts the ramp on the back of the flyweight and forces the toe of the flyweight away from the stop pin. The spring spins the magnet and produces a hot, retarded spark to start the engine. *See* Figure 5-13.

As soon as the engine begins to run, centrifugal force acting on the tails of the flyweights moves them outward so the noses are held away from the stop pins, and the engine drives the magneto so that its spark occurs with its normal advanced timing. *See* Figure 5-14.

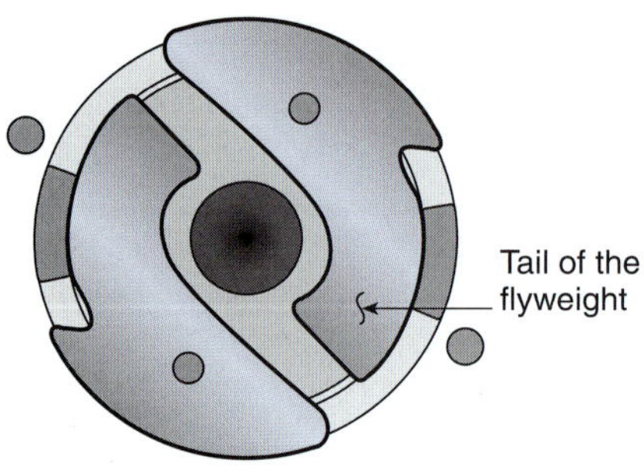

When engine is running, centrifugal force acting on the tail of flyweights holds them out so toe does not contact stop pin. The engine drives magneto so it produces a normal advanced spark.

Figure 5-14. *The impulse coupling when the engine is running*

Induction Vibrator System

Many modern reciprocating engines use an induction vibrator and a set of retard breaker points to provide a sustained flow of late, high-voltage current across the electrodes of the spark plugs when starting the engine.

A popular induction vibrator system introduced by Bendix is called the Shower of Sparks system. The ignition switch for this system has five positions: OFF, LEFT, RIGHT, BOTH, and a spring-loaded START position.

When the switch is in the OFF position, the primary windings of both magnetos are grounded, and the switch between the battery bus and the starting vibrator is open.

Shower of Sparks ignition system. The registered trade name of a popular type of induction vibrator starting system for a reciprocating engine.

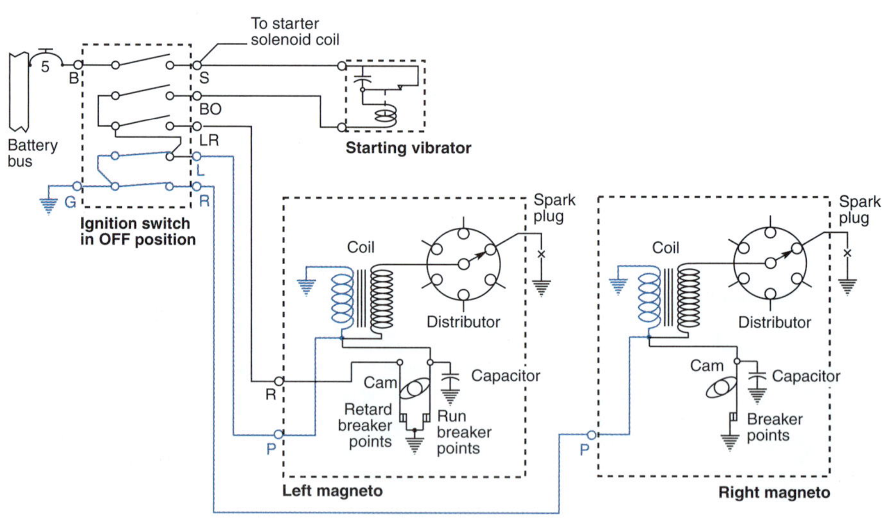

Figure 5-15. *Induction vibrator ignition system with the ignition switch in the OFF position*

To start an engine equipped with this system, prepare the engine for starting and hold the ignition switch in the spring-loaded START position. In this position the following will happen:

- The P-lead of the right magneto is grounded. This prevents it from igniting the mixture inside the cylinder at the normal advanced timing position.
- The P-lead of the left magneto is connected to the retard breaker points and to the coil of the starting vibrator.
- The battery switch is closed, allowing current to flow to the starter solenoid coil to close the starter solenoid and allow the starter to crank the engine.
- Current flows through the contacts of the starting vibrator and through its coil. When current energizes the coil, its magnetic field pulls the contacts open and current stops flowing. When current stops, the spring-loaded points close and current flows again. This action changes straight DC into pulsating DC.
- The pulsating DC flows to ground through both the run breaker points and retard breaker points. It does not flow to ground through the left magneto primary winding because resistance through the breaker points is so much less.
- When the starter rotates the engine to the normal advanced timing position, the run breaker points open. But the retard points are still closed, and current continues to flow to ground through them.

Continued

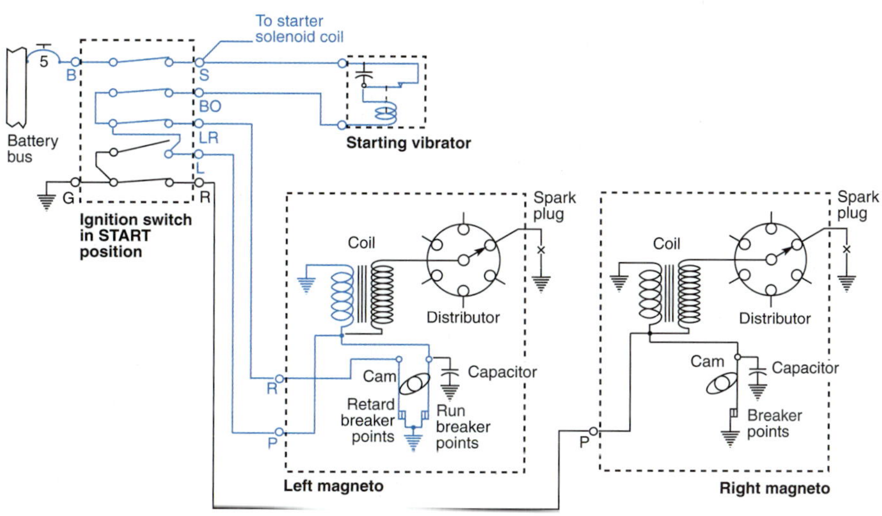

Figure 5-16. *Induction vibrator ignition system with the ignition switch in the START position*

- When the crankshaft has turned enough for the piston to be near its top center position, the cam opens the retard breaker points. The current now must flow to ground through the primary winding of the left magneto.
- As the starter continues to crank the engine, the pulsating DC flowing in the primary winding induces a high voltage in the secondary winding, and it is sent to the correct spark plug by the distributor each time the retard breaker points open. *See* Figure 5-16.

After the engine starts, the switch is released and the spring returns it to the BOTH position. Now all contacts are open. *See* Figure 5-17.

When the magneto check is made before takeoff, the engine is run up to the correct RPM, and the switch is moved to the LEFT position (the switch position farthest from BOTH). The right magneto is grounded, and the engine is running on the left magneto alone. The RPM should drop by the amount specified in the Pilot's Operating Handbook. The switch is returned to BOTH until the RPM comes back to its original value. Then the switch is moved to RIGHT. The left magneto is grounded and the engine is running on the right magneto alone. The RPM drop is noted and the switch is returned to the BOTH position, and the engine returns to the run-up RPM. The switch remains in the BOTH position for the flight.

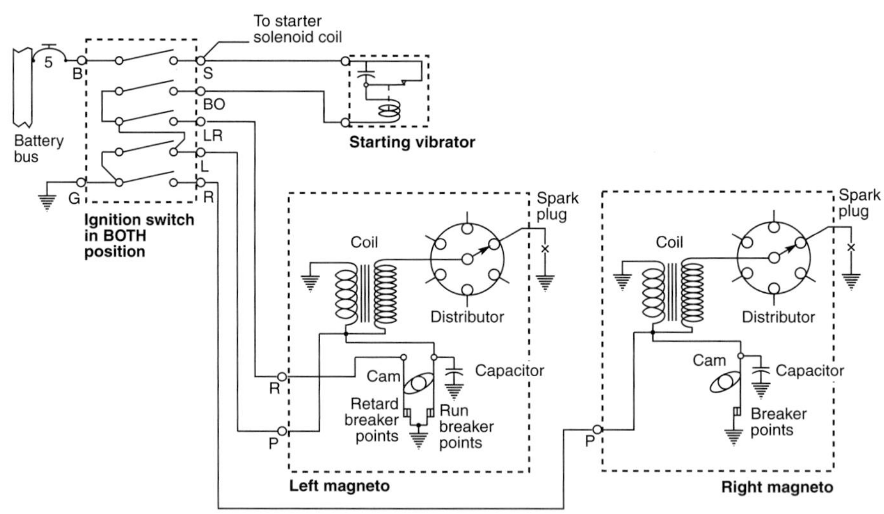

Figure 5-17. *Induction vibrator ignition system with the ignition switch in the BOTH position*

STUDY QUESTIONS: AUXILIARY STARTING SYSTEMS

Answers are on Page 244. Page numbers refer to chapter text.

23. The spark used for starting an aircraft engine must occur ________________ (earlier or later) than the normal spark. *Page 210*

24. The mechanical system that uses a spring to spin the magnet to produce a hot retarded spark for starting the engine is called a/an ______________________ coupling. *Page 211*

25. The positions for the ignition switch used with an induction vibrator ignition system are:
    a. __________________________
    b. __________________________
    c. __________________________
    d. __________________________
    e. __________________________
    Page 212

26. An engine equipped with an induction vibrator ignition system is started on the __________________ (right or left) magneto. *Page 213*

27. The induction vibrator (starting vibrator) changes DC into ___________________ (AC or pulsating DC). *Page 213*

28. When the ignition switch for an engine equipped with an induction vibrator ignition system is held in the START position, sparks occur in the spark plug gap when the ___________________ (retard or run) breaker points open. *Page 214*

Special Types of Magnetos

The magneto described to this point has been the type most widely used, a typical fixed-timing, two-pole, rotating-magnet, high-tension magneto with a built-in distributor. Relatively few different types of magnetos have been used over the years.

14 CFR Part 33. Federal Aviation Regulation Part 33 *Airworthiness Standards: Aircraft Engines.*

Engines up into World War I used a simple wound-rotor magneto in which the primary and secondary coils and the condenser (now called the capacitor) were rotated in the magnetic field produced by two or more heavy horseshoe magnets.

Between the World Wars and through World War II, two types of magnetos were popular: the four-pole rotating-magnet magneto and the polar-inductor magneto. The polar-inductor magneto had two strong fixed permanent magnets, and the engine turned a soft-iron rotor between the magnets to reverse the direction of flux flowing through the coil. *See* Figures 5-18 and 5-19.

Double Magnetos

14 CFR §33.37 requires that "Each spark ignition engine must have a dual ignition system with at least two spark plugs for each cylinder and two separate electrical circuits with separate sources of electrical energy, or have an ignition system of equivalent inflight reliability." This requirement is normally met by using two separate magnetos, but double magnetos have been used successfully on both large and small engines.

Double magnetos have two separate and independent electrical circuits and distributors, but share a common rotating magnet and magneto housing.

Regardless of the way a double magneto is mounted on the engine, the side of the magneto on which the nameplate is mounted is always considered to be the "top." The left and right magneto are determined by looking at the magneto from the distributor end with the nameplate on top. *See* Figure 5-20.

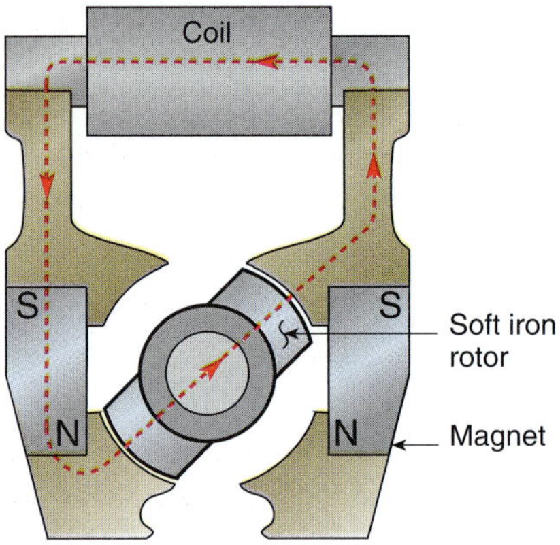

In this rotor position, lines of flux pass through the coil core from right to left.

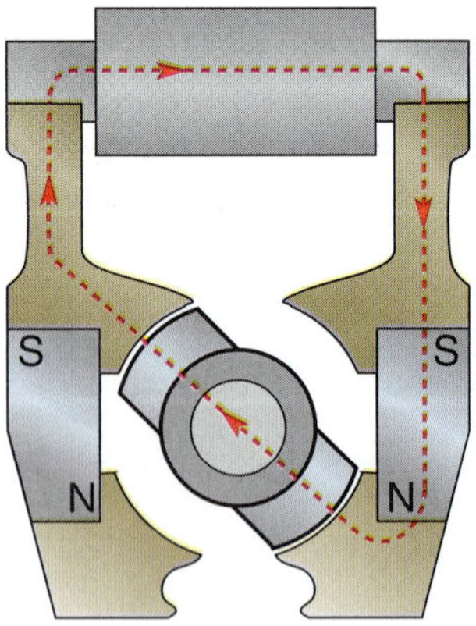

In this rotor position, lines of flux pass through the coil core from left to right.

Figure 5-18. *Polar-inductor magneto*

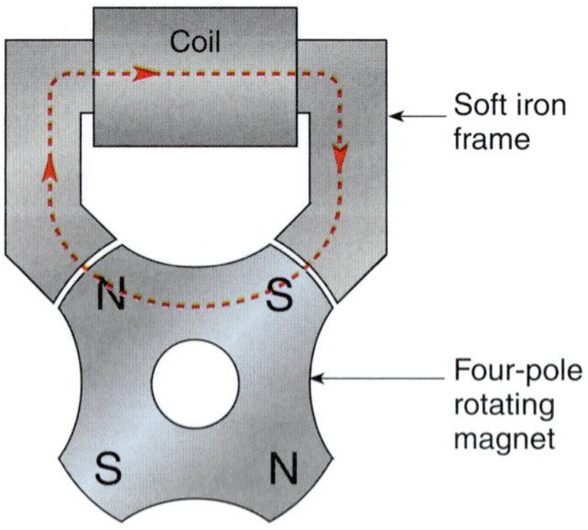

Figure 5-19. *Four-pole magnet magneto*

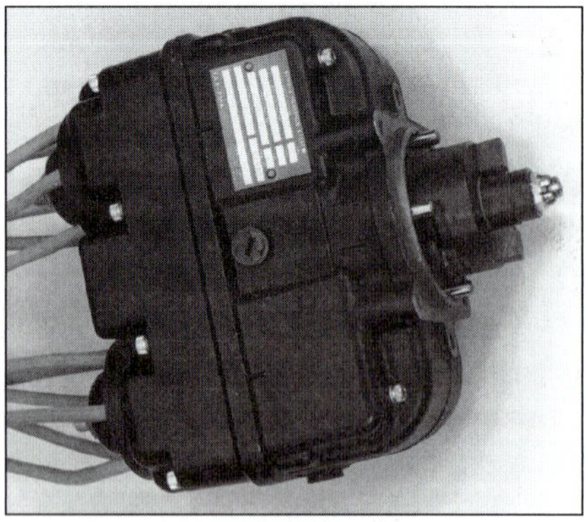

Figure 5-20. *Double magneto. The side of the magneto with the nameplate, or dataplate, is considered to be the "top" side.*

Magnetos with Compensated Cams

Large, high-performance radial engines have a special ignition requirement. The geometry of the master rod and link rods causes the pistons to be at slightly different linear distances from the top of their strokes for a given number of degrees of crankshaft rotation. For engines of about 300 horsepower or less, the difference in piston location is of little consequence. But larger engines can be made appreciably more efficient if the spark occurs in each cylinder when the piston is at the same linear distance from the top of its stroke. The spark must occur when the crankshaft is a slightly different number of degrees from top center for each cylinder. This timing difference is accomplished by using a breaker point cam that has a lobe for each individual cylinder, with each lobe ground slightly different.

Figure 5-21 shows a typical compensated cam for a 9-cylinder radial engine. Note that the cam is ground for a particular engine, a Wright R-1820. The master rod for this engine is in cylinder number 1, the E-gap is 12° after the neutral position of the magnet, and the arrow shows that the cam rotates counterclockwise. The dot identifies the lobe for cylinder number 1.

Figure 5-21. *Compensated cam for a magneto installed on a Wright R-1820, a 1,000-horsepower, 9-cylinder radial engine*

STUDY QUESTIONS: SPECIAL TYPES OF MAGNETOS

Answers are on Page 244. Page numbers refer to chapter text.

29. The north poles in a four-pole rotating magnet are ________________ (90 or 180) degrees apart. *Page 216*

30. The rotating element in a polar inductor magneto is a ________________ (magnet or soft-iron rotor). *Page 216*

31. A double magneto has ____________ (one or two) rotating magnet/s. *Page 216*

32. A double magneto has _____________ (one or two) capacitor/s. *Page 216*

33. The "left" magneto in a dual magneto is the one on the left side when viewing the magneto with the nameplate on _________________ (top or bottom). *Page 216*

34. A compensated magneto cam for a 9-cylinder radial engine has _________________ (how many) lobes. *Page 217*

Magneto Installation

Magnetos are different from any other engine accessory in that their moving parts must be internally timed so the magneto will produce the highest induced voltage. The complete magneto and distributor must then be timed to the engine so high voltage will be directed to the proper spark plug when the piston is in the correct position for ignition to produce maximum power.

Internal Timing

internal timing. The adjustment of the breaker points of a magneto so they will begin to open at the time the magnet is in its E-gap position.

When preparing a magneto for installation on an engine, the internal timing must be checked and adjusted. Internal timing requires:

- The rotating magnet must be in its proper E-gap position when the breaker points begin to open.
- A marked tooth on the distributor gear must be meshed with a corresponding marked tooth on the drive gear on the magnet shaft. This indicates that the distributor is in the correct position to fire the spark plug in cylinder No. 1.

The rotating magnet must be turned to its E-gap position. This is the correct number of degrees of rotation beyond its neutral position.

Some magnetos have a hole in the bottom of the housing and one in the rotating magnet shaft. To time these magnetos, they must first be "sparked out" to place them in the correct position for firing cylinder No. 1.

Hold the ignition lead for cylinder No. 1 about $\frac{1}{8}$ inch from the magneto frame and rotate the drive until the impulse coupling snaps and a hot spark jumps from the lead to the case. This shows that the magneto is in the position that has just fired cylinder No. 1. Rotate the magnet backward until you can insert a timing pin through the hole in the housing and the magnet shaft. This locks the magnet in its E-gap position. With the pin in place, adjust the breaker points so they just begin to open, as is indicated by a magneto timing light. Other magnetos by this same manufacturer require special tools to locate the E-gap position and hold the magnet while adjusting the breaker points. When timing any magneto, be sure to follow the manufacturer's recommendations exactly, and use the special tools specified.

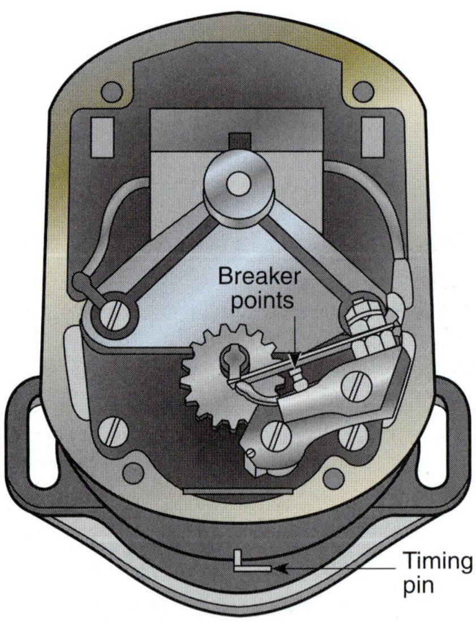

Figure 5-22. *The magnet in some magnetos is held in its E-gap position by inserting a timing pin through a hole in the bottom of the housing and into a hole in the magnet shaft.*

Many Bendix magnetos are put into their E-gap position by following these steps:

1. Install a rotor-holding tool on the magnet shaft, but do not lock the rotor in place yet. *See* Figure 5-23.
2. Install a timing plate on the magneto housing, and clip the pointer onto the cam screw as shown in Figure 5-24.
3. Remove the timing inspection plug from the top of the magneto and turn the rotor in the direction of normal rotation until the painted chamfered tooth on the distributor gear is centered in the inspection window. *See* Figure 5-25.
4. Put the magnet in its neutral position by moving it a few degrees back from the position that centered the chamfered tooth in the window, and releasing it. The magnet has a natural tendency to pull into its neutral position. Lock the rotor-holding tool so the rotor cannot move from its neutral position.
5. Move the pointer until it aligns with the zero mark on the timing plate.
6. Slightly loosen the knob on the rotor-holding tool and turn the magnet in the direction of normal rotation until the pointer indicates the magnet has moved the correct number of degrees specified in the magneto service manual for the E-gap position. At this position, tighten the knob of the rotor-holding tool enough to prevent the rotor from turning.
7. Adjust the breaker points until they are just beginning to open as is indicated by a magneto timing light.

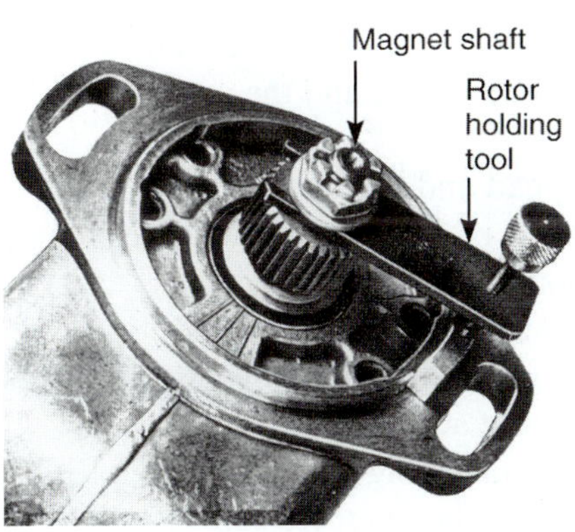

Figure 5-23. *Rotor-holding tool installed on the magnet shaft*

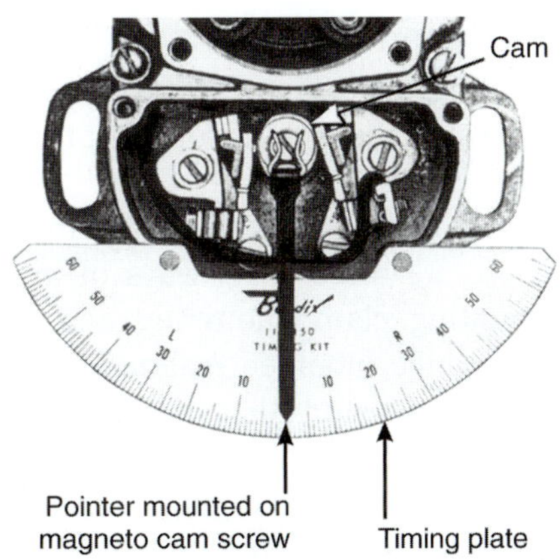

Figure 5-24. *Timing plate and pointer installed on the breaker compartment*

Figure 5-25. *The inspection plug has been removed from the top of this magneto, and the painted, chamfered tooth is visible in the hole. This tooth shows that the distributor is in the correct position to fire the spark plug in cylinder number 1.*

Timing the Magneto to the Engine

When the magneto has been correctly timed internally, it will produce the hottest possible spark. Now, the magneto must be timed to the engine so these hot sparks will occur at the proper time relative to the position of the pistons.

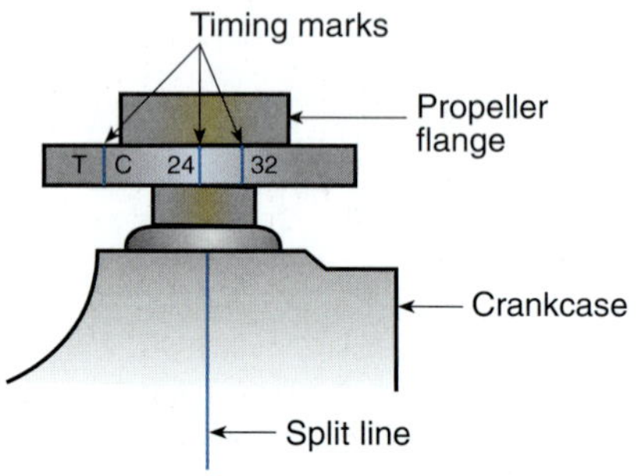

Figure 5-26. *Timing marks on the propeller flange are aligned with the split line of the crankcase to locate the proper crankshaft position for magneto timing.*

Preparing the Engine

Prepare the engine by putting the piston in cylinder No. 1 on its compression stroke. Remove the more accessible spark plug from this cylinder, and place your thumb over the hole. Rotate the propeller in the direction of normal rotation until air pressure tries to push your thumb away from the hole. This indicates that both valves are closed and the piston is moving outward. Rotate the propeller until the piston is in the correct position for the magneto to fire the spark plug. You can locate this position on some engines by aligning marks on the propeller flange with the split line of the crankcase. *See* Figure 5-26.

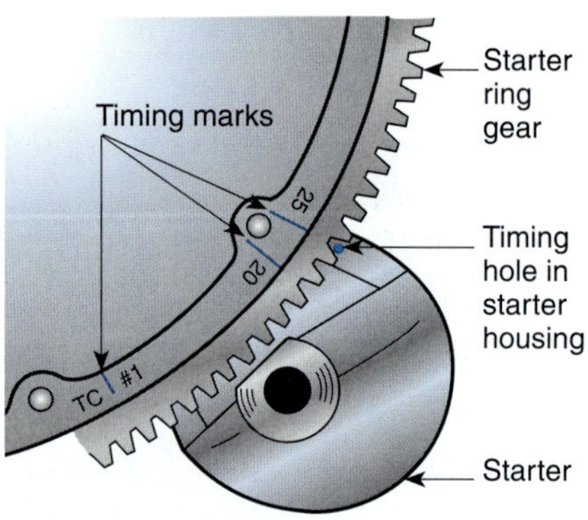

Figure 5-27. *Timing marks on the starter ring gear are aligned with a small hole in the front face of the starter housing.*

Engines with a large starter ring gear on the front of the crankshaft are timed by rotating the propeller until the proper advance timing mark on the front face of the ring gear is aligned with a small hole located at the two o'clock position of the front face of the starter housing. The ring gear is marked at 20° and 25°, so the engine manual will have to be consulted to find the correct timing for the particular engine. *See* Figure 5-27.

Some engines do not have timing marks, and the correct position of the piston will have to be determined by using a top-dead center indicator and a timing protractor, or a piston-position indicator such as a Time-Rite indicator.

Time-Rite indicator. A patented piston-position indicator used to find the position of the piston in the cylinder of a reciprocating engine.

Installing Magnetos without Impulse Couplings

Place the piston in cylinder No. 1 in its firing position, and check the magneto to be sure the distributor is set to fire cylinder No. 1 and the magnet is held in its E-gap position.

Slide the magneto onto its mounting pad and secure it loosely in place. There are three basic ways of mounting a magneto on an engine: flange mounting, clamp mounting, and base mounting.

Flange-mounted magnetos are attached to the accessory section with nuts on two studs that stick through slots in the magneto housing. The slots allow about 15° of movement for final timing adjustment.

Clamp-mounted magnetos have a circular flange that fits into a recess in the magneto mounting pad on the engine. The magneto can be rotated in its mounting pad for final timing adjustment and then clamped in place with magneto-holding washers and nuts.

Base-mounted magnetos were used on many World War II vintage engines. The magneto is secured to the engine and cannot be rotated. Timing adjustment is made with a vernier coupling. Figure 5-28 shows a vernier coupling between a magneto drive gear and the magnet shaft. The coupling has a set of serrations on either side. One side has, for example, 35 serrations and the other side has 36. Advancing the 35-serration side by one notch advances the timing 10.28°. Moving the 36-serration side back one notch retards the timing by 10°. This type of coupling allows the timing to be varied by increments of 0.28°.

With the magneto snug, but not yet tightened in position, clip the leads of a magneto timing light across the breaker points, and bump the magneto with a soft-faced mallet until it moves in its mounting enough for the points to open. Tighten the clamps or the mounting nuts. Move the propeller back about 90° in the direction opposite its normal rotation. Then, move it slowly in the direction of normal rotation until the timing light indicates the points have just opened. The timing marks should indicate that the piston is in the correct position for ignition to occur.

Note: *Some timing lights turn on when the points open, and others turn off. When using a timing light, be sure you understand which condition indicates the opening of the points.* See Figure 5-29.

Make a final check of magneto timing by connecting the leads of the timing light to the insulated breaker points of both magnetos. Place the ignition switch in the BOTH position so neither magneto is grounded. Move the propeller back about a quarter turn, and then slowly move it forward until the lights indicate that the breaker points in both magnetos open at the same time. If the breaker points of either magneto do not open at the correct time, the magneto should be slightly loosened in its mounting and tapped to move it enough to cause the points to open at the correct position of the crankshaft. When the timing procedure is completed, tighten the mounting hardware as specified in the engine maintenance manual.

Some engines have poor scavenging of exhaust gases from the cylinders, and these engines use staggered ignition timing. The magneto that supplies the spark plug nearest the exhaust valves fires a few degrees of crankshaft rotation before the magneto that fires the spark plug nearest the intake valve. Improper scavenging causes the fuel-air mixture on the side of the cylinder with the exhaust valve to be leaner than the mixture on the other side. A lean mixture burns slower than a correct mixture, and in order for the flame front started by both spark plugs to meet near the center of the piston, the flame in the lean mixture must begin first.

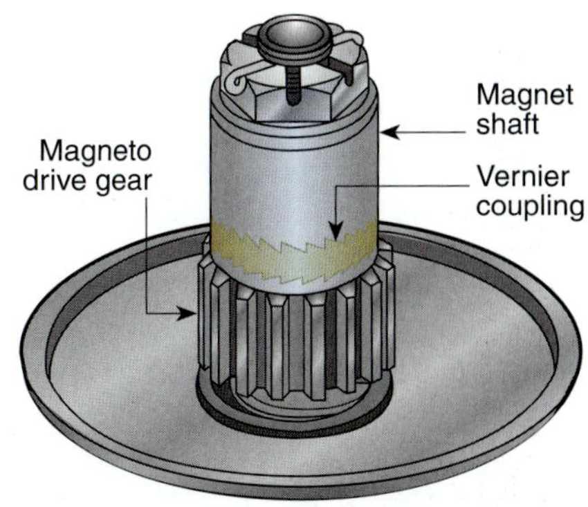

Figure 5-28. *A vernier coupling allows the timing of base-mounted magnetos to be varied in increments of much less than one degree.*

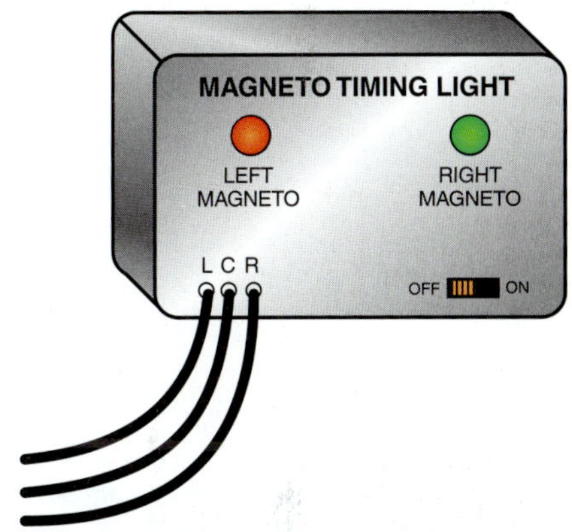

Figure 5-29. *A timing light used to indicate the opening of the magneto breaker points. Some timing lights incorporate an oscillator or buzzer that changes its tone when the breaker points open. This allows the technician to concentrate on the magneto without having to watch the timing light.*

vernier coupling. A timing coupling used with base-mounted magnetos that allows the timing to be adjusted in increments of less than one degree.

timing light. An indicator light used when timing magnetos to an engine to indicate when the breaker points open.

staggered timing. Ignition timing that causes the spark plug nearest the exhaust valve to fire a few degrees of crankshaft rotation before the spark plug nearest the intake valve.

Installing Magnetos with Impulse Couplings

An impulse coupling holds the rotating magnet until the piston reaches its approximate top center position. When timing a magneto equipped with an impulse coupling, follow these steps:

1. Turn the propeller slowly in the direction of normal rotation until the piston in cylinder number 1 is moving outward on its compression stroke and the impulse coupling snaps.
2. Connect the timing light across the breaker points.
3. Move the propeller backward beyond the position that the breaker points close, but not far enough to allow the impulse coupling to engage.
4. Slowly turn the propeller in the direction of normal rotation until timing marks show that the crankshaft is in the proper advanced position for the magneto to fire.
5. Tap the magneto with a soft-faced mallet to rotate it enough for the breaker points to open, indicated by the timing lights.
6. Tighten the clamps or mounting nuts to secure the magneto to the engine.
7. Make a final check of the timing after the magneto has been tightened down.

Note: *An impulse coupling spins a magneto so fast that it produces enough primary current to damage some timing lights. For this reason, the timing light should not be left connected across the points of an impulse coupled magneto when the coupling is snapped.*

STUDY QUESTIONS: MAGNETO INSTALLATION

Answers are on Page 244. Page numbers refer to chapter text.

35. Two types of timing required for a magneto are:
    a. ________________________________
    b. ________________________________
    Page 218

36. The breaker points must just begin to ____________________ (open or close) when the magnet is in its E-gap position. *Page 218*

37. A marked tooth on the distributor gear is aligned with a specified mark to indicate that the distributor is in the correct position to fire the spark plug in cylinder number ____________________ . *Page 218*

38. The spark plug should ignite the mixture inside a cylinder when the piston is a few degrees of crankshaft rotation ________________________ (before or after) top center on the compression stroke. *Page 220*

39. Three ways of mounting a magneto on an engine are:
   a. ______________________________
   b. ______________________________
   c. ______________________________
   Page 220

40. Base-mounted magnetos are timed by using a ________________ coupling. *Page 221*

41. When making the final magneto timing check on the engine, the ignition switch must be in the ____________________ (BOTH or OFF) position. *Page 221*

42. If an engine has staggered timing, the spark plug on the exhaust side of the cylinder fires _______________________ (first or last). *Page 221*

Magneto Inspection and Servicing

Magnetos must be carefully checked on each annual or 100-hour inspection. Remove the breaker point cover and check the condition of the points. There should be no oil in the magneto housing and no evidence of burning or pitting of the points. Normal operating points will generally be smooth and flat, and the surface will have a dull gray sandblasted, or frosted, appearance.

Almost all modern magnetos use pivotless breaker points. As the cam rotates, its lobes press upward on the plastic (nylon or phenolic) cam follower. This lifts the upper insulated contact. When the cam rotates enough for the follower to be off the lobe, the spring forces the points together. The fixed point is mounted on a slotted bracket that can be moved to adjust the maximum amount the points can open.

Felt pads on the cam follower should be lubricated with just a trace of the oil specified in the magneto overhaul manual. Too little oil will cause the cam follower to wear, and too much will get on the points and burn.

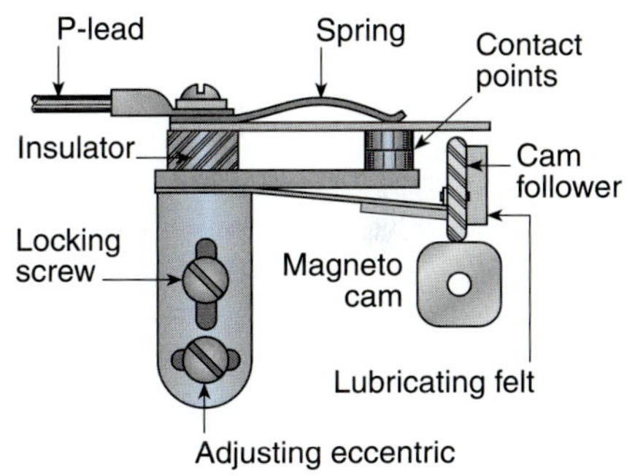

Figure 5-30. *A pivotless breaker point assembly*

The timing of the magnetos to the engine should be checked every 100 hours of operation or sooner if the engine is consistently hard to start or if it appears to be losing power.

Two things that cause the timing of a magneto to drift are erosion of the breaker points and wear of the cam follower. Erosion of the breaker points causes the timing to drift early, and wear of the cam follower causes it to drift late. Ideally, these two wear conditions should cancel each other.

Most wear and timing drift occurs during the first few hours after new points assemblies are installed in a magneto. For this reason, timing should be checked at the first 25 and 50 hours after installation of new points. If the timing is outside the tolerance allowed by the engine manufacturer, remove the magneto and check the internal timing.

Some technicians have been known to loosen a magneto and bump its timing in the advance direction to cure an excessive RPM drop caused by late timing drift. This sometimes cures the problem, but the correct procedure is to remove the magneto and adjust its internal timing, then set the timing of the magneto to the engine according to the latest timing recommendation in service bulletins issued by the engine manufacturer.

Magneto Overhaul

Magnetos do not normally have a specified time between overhauls, but the standard operating practice is to overhaul all accessories each time the engine is overhauled.

The number of special tools and elaborate test equipment required, and the liability incurred if any of the manufacturer's service bulletins should be missed, make it a normal procedure for most magneto overhaul to be done by shops that are fully equipped and experienced in this work. To emphasize the importance of magneto overhaul done only in well-equipped shops with highly experienced personnel, one manufacturer introduced a series of magnetos that could not be overhauled in the field, but instead were designed to be replaced when they needed major servicing. The magnetos described in this text are all designed so that they can be overhauled rather than replaced.

When component replacement is done in the field, it must be done in strict accordance with instructions in the magneto manufacturer's overhaul manual, and all replacement parts must be those specified by the manufacturer.

If a magnet is removed from the magneto, it should have soft iron keepers placed across the poles to prevent loss of strength. The magnet's strength can be measured by placing it on a test stand and operating it at a specified speed. Place a piece of insulating paper between the breaker points and measure the primary current with an AC ammeter.

carbon track. A trail of carbon deposited by an arc across a high-voltage component such as a distributor block.

The distributor in a magneto consists of the rotor and the block. Since these components are in the high-tension circuit of the magneto, they are susceptible to the formation of carbon tracks that result when moisture has allowed the current to burn its way across the surface. The distributor components can be cleaned with a solvent and protected with a special shellac that has a high dielectric strength to prevent absorption of moisture and high voltage leakage.

Magneto Check on Engine Run-up

Aircraft engines are required to have two ignition systems. The safety factor of two systems is very important, but the engine operating efficiency is also improved by igniting the mixture inside the cylinder in two locations, usually across from each other. Igniting the fuel-air mixture in this way assures the most complete combustion and allows the engine to develop maximum power.

A vital part of preflight inspection is the determination that all cylinders are firing on both magnetos. When the engine is operated on only one magneto, the combustion will not be as complete and the RPM will decrease. The magneto check is therefore a comparative check to assure that the engine operates the same with either magneto alone. The actual amount of RPM drop is not necessarily significant, as it is affected by such variables as humidity, air temperature, and airport altitude. The drop, however, should be within the limits specified in the Pilot's Operating Handbook for the particular aircraft, and it should be nearly the same on both magnetos.

If the magneto check shows no drop on either magneto, there is a possibility that the ignition switch is not grounding the magneto as it should. If any of the cylinders are not firing, the engine will run extremely rough, and the cause must be determined.

A typical magneto check should be done in this manner:

1. Start and run the engine until the oil and cylinder head temperatures are in their correct operating range.
2. Place the propeller pitch control in its low-pitch, high-RPM setting.
3. Advance the engine speed to 1,700 RPM.
4. Move the ignition switch to the RIGHT position and note the amount of RPM drop.
5. Return the switch to the BOTH position and allow the RPM to build back up. This will clear the spark plugs of any fouling.
6. Move the ignition switch to the LEFT position and note the RPM drop and then return it to the BOTH position.
7. The RPM drop should not exceed 150 RPM on either magneto, nor should there be more than a 50 RPM difference between the two magnetos.

Not only is the amount of RPM drop and the difference between the magnetos significant, but so is the way the drop occurs. A smooth RPM drop beyond normal is usually an indication that the mixture is either too rich or too lean. A sharp RPM drop beyond normal is likely caused by a fouled spark plug, a defective harness, or a magneto out of time. An excessive drop on one magneto and normal drop on the other may indicate the magneto is improperly timed, either internally or to the engine.

If a magneto check leaves any doubt as to the condition of the engine, a check at a higher RPM or with a leaned mixture will normally verify whether or not a problem actually exists.

Magneto Safety Check

If there is no drop in RPM when switching from BOTH magnetos to either of the magnetos alone, there is a possibility that the P-lead for a magneto is disconnected, or that the timing on a magneto is set too far advanced.

To check for a disconnected P-lead or malfunctioning ignition switch, pull the throttle back to idle and move the ignition switch to OFF. The engine should stop firing. Return the switch to the BOTH position before the engine stops turning. If the engine does not stop firing, either one of the P-leads is disconnected, or the switch is faulty. Be sure to check for this at the lowest speed possible; otherwise the engine can backfire and cause serious damage.

STUDY QUESTIONS: MAGNETO INSPECTION AND SERVICING

Answers are on Page 244. Page numbers refer to chapter text.

43. The breaker points used in most modern magnetos are of the ____________________ (pivot or pivotless) type. *Page 223*

44. The normal interval for checking the magneto-to-engine timing is every _________________ hours. *Page 223*

45. Erosion of the breaker points will cause the magneto to engine timing to drift _______________ (early or late). *Page 223*

46. Wear of the cam follower will cause the magneto to engine timing to drift _______________ (early or late). *Page 223*

47. The magneto to engine timing should be checked at the first ___________ hours of operation after a new breaker point assembly has been installed. *Page 223*

48. An engine is likely to have an excessive magneto drop if the timing has drifted _________________ (early or late). *Page 224*

49. Magnetos ________________ (do or do not) normally have a specified time between overhaul. *Page 224*

50. When a magnet is removed from a magneto, a keeper made of ________________ should be placed across its poles. *Page 224*

51. Magnet strength in a magneto can be measured by running the magneto on a test stand at a specified speed, with a piece of paper between the breaker points, and by measuring the ______________ (primary or secondary) current. *Page 224*

52. The two components in a magneto distributor are the ________________ and the ______________ . *Page 224*

53. A distributor block can be prevented from absorbing moisture by coating the block with a special high dielectric-strength __________________ . *Page 224*

54. The actual number of RPMs dropped in a magneto check ________________ (is or is not) the most significant factor in determining if the magneto is functioning properly. *Page 225*

55. A fouled spark plug will cause the RPM to drop __________________ (smoothly or sharply) beyond the allowable limit when performing a magneto check. *Page 225*

56. If there is little or no RPM drop when switching from BOTH to either magneto by itself, there is a possibility that the timing of that magneto is too far ______________________ (advanced or retarded). *Page 225*

57. If a P-lead is disconnected from a magneto, when that magneto is selected, the RPM will __________________ (drop to zero or not change). *Page 226*

58. When checking the magneto switch to be sure that it stops the engine firing, the check should be made at ________________ (idle or cruise) RPM. *Page 226*

Ignition Harness

The ignition harness is a vital part of an aircraft ignition system. It must carry high voltage from the distributor to the spark plug, and it must be shielded so that the electromagnetic radiation does not cause interference in the radio equipment. All electromagnetic energy radiated from the central conductor is intercepted by the shielding and carried to ground.

Ignition leads are easily damaged and they are subject to severe operating conditions. The leads installed must be those specified in the aircraft parts list or ones specifically approved as a replacement. They must be routed and supported exactly as the aircraft manufacturer installed them.

Two types of high-tension leads used in the ignition harnesses for modern aircraft engines are stranded wire and coiled conductor. Both leads have high-voltage insulation around the conductor and a braided wire shield around the insulation. The shield is protected by a plastic abrasion jacket.

Ignition leads were at one time 7-mm in diameter, and could not be bent with a small radius, so the spark plug terminals were made with several different angles. All modern leads have a 5-mm diameter, and there are basically only two types of terminals available: straight, as shown in Figure 5-31, and angled as shown in Figure 5-32. It is possible, even with the smaller diameter lead, to bend it so sharply that the insulation will be stressed enough to leak high-voltage current through to ground, when the lead gets hot. When a lead attached to a spark plug requires a relatively sharp bend, it is bent with a radius large enough that the insulation is not overstressed and is clamped as shown in Figure 5-33.

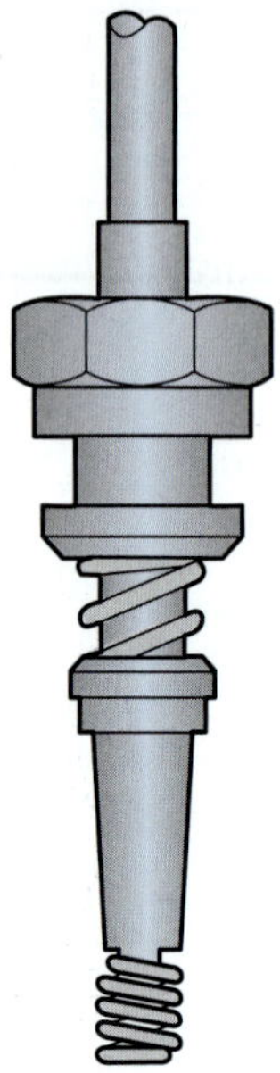

Figure 5-31. *Spark plug end of an ignition lead with a straight terminal. This terminal is for a ¾-20 all-weather spark plug.*

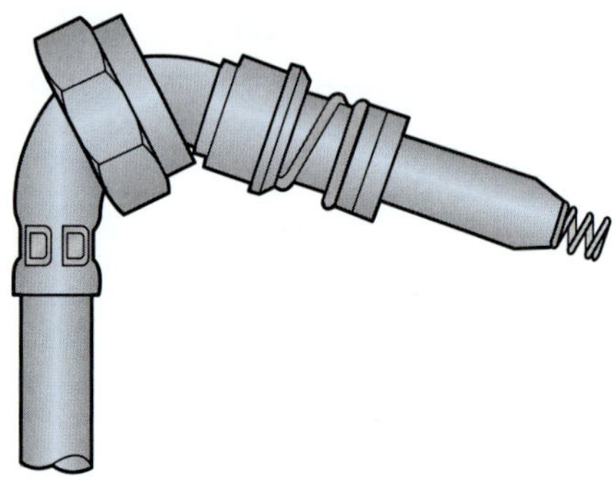

Figure 5-32. *Spark plug end of an ignition lead with an angled terminal.*

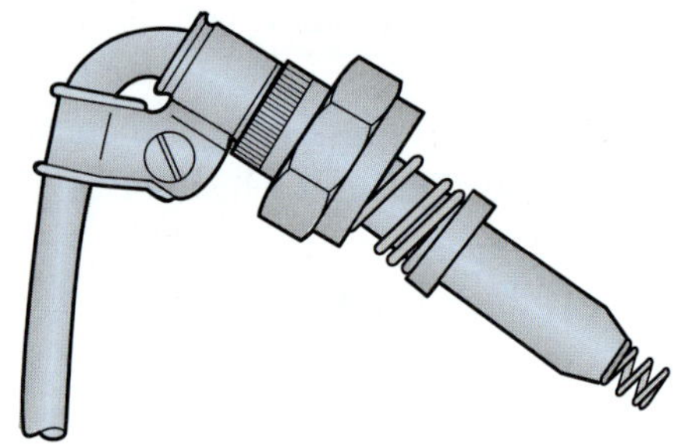

Figure 5-33. *Spark plug end of an ignition lead with a straight terminal bent to allow straight entry into the spark plug. The bend in the lead is supported with a clamp.*

all-weather spark plug. A shielded spark plug designed for high-altitude operation. A resilient grommet on the ignition harness forms a water-tight seal in the recess in the shell.

Ignition harnesses available fit two basic types of spark plugs: the standard ⅝-24 spark plug and the ¾-20 all-weather, or high-altitude, spark plug. The lead for the ¾-20 spark plug has a resilient seal that prevents water from getting into the terminal cavity.

If a single lead becomes damaged, it can be replaced without having to replace the entire harness. Manufacturers sell replacement leads that have the spark plug end installed, the magneto end ready to be pushed through the distributor plate and grommet, and the terminals attached.

Ignition troubles are often blamed on the magnetos or spark plugs when the trouble is actually in the harness. It is possible for high voltage to leak to ground through the lead and prevent the spark plug from firing. Leakage may be detected by applying high voltage across the lead insulation and measuring the leakage current with a microammeter. One handy tester that works on a different principle is shown in Figure 5-34. One lead from the tester is attached to the center conductor and the other lead is connected to the outer shielding. Then the Press-to-Test button is depressed. If the lead is good, the indicator light will illuminate and a spark will jump across the gap in the tester. If there is excessive leakage, the voltage will not build up high enough to jump the gap.

When an ignition harness check indicates several bad leads, it is possible that the problem is in the distributor block. Moisture on the block surface can give the same indication as a faulty lead. Treatment of the distributor block must be done in strict accordance with the magneto manufacturer's recommendations.

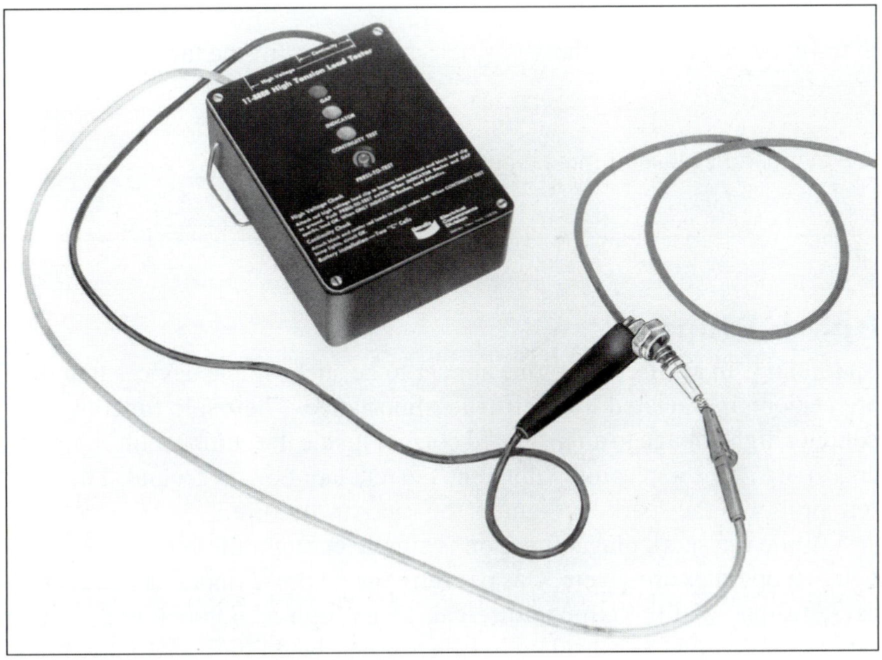

Figure 5-34. *This ignition harness tester shows when an ignition lead has an excessive leakage of high-voltage current.*

Leads from the distributor carry high voltage to the spark plugs in proper sequence. The lead connected to the No. 1 distributor terminal goes to the first cylinder in the engine firing order. The lead from the next terminal in the direction of distributor rotation goes to the next cylinder to fire. Figure 5-35 shows how the leads from two distributors are connected to the spark plugs of a typical six-cylinder horizontally opposed engine with a firing order of 1-4-5-2-3-6. Notice that the leads from each distributor go to three spark plugs on top (T) and three on the bottom (B). The lower spark plugs are most susceptible to fouling, and this arrangement prevents either magneto from firing just the lower spark plugs.

Firing order 1 - 4 - 5 - 2 - 3 - 6			
Left Distributor	**Spark Plug**	**Right Distributor**	**Spark Plug**
1	1-T	1	1-B
2	4-B	2	4-T
3	5-T	3	5-B
4	2-B	4	2-T
5	3-T	5	3-B
6	6-B	6	6-T

Figure 5-35. *Ignition lead connection between magneto distributors and spark plugs on a typical six-cylinder horizontally opposed engine*

STUDY QUESTIONS: IGNITION HARNESS

Answers are on Page 244. Page numbers refer to chapter text.

59. Radio interference from the ignition system is prevented by enclosing the leads between the distributor and the spark plug in a metal braid. This braid is called ____________________ . *Page 227*

60. An ignition lead tester is used to determine whether or not the ____________________ of the high-voltage current is excessive. *Page 228*

61. Sharp bends in ignition leads must be avoided because of the danger of weakening the ________________ (conductor or insulator). *Page 228*

Spark Plugs

Spark plugs in an aircraft engine appear to be such simple devices that they are seldom appreciated for their true importance. Their sole function is to conduct high voltage to an insulated contact inside the combustion chamber, and to provide a gap across which the current can flow to ground. This current produces the spark that ignites the fuel-air mixture.

While the spark plug's function is simple enough, the conditions under which it operates are severe. Gas pressure inside the cylinder can be as high as 2,000 psi, and the temperature can be as high as 3,000°F and change rapidly. Voltage at the insulated electrode can be as high as 18,000 volts. The arc across the electrodes which occurs each time the spark jumps, acts in the same way as an electric arc welder and removes metal from one of the electrodes. These sparks occur in each spark plug more than six million times during the normal 100-hour interval between spark plug servicing.

Spark Plug Design

Aviation spark plugs are more complex than the typical automotive spark plug, and their design requirements are more stringent. Here, we will look at some of their design features and then will consider their actual construction and servicing.

Spark Plug Size

Aviation spark plug sizes are rated according to the metric diameter of the shell threads that screws into the cylinder head. There are two sizes of spark plugs used in aircraft engines: plugs with 14-mm shells and those with 18-mm shells. The 18-mm spark plug is by far the more popular size. 14-mm spark plugs are almost obsolete, as the only popular FAA-certificated engine using them is no longer manufactured in the United States.

Electrodes

The electrodes of a spark plug are the reason for its existence. They are its most important component and are designed to meet the following requirements:

- They must allow the spark to jump at the lowest possible voltage.
- The spark must be intense enough (have enough energy) to ignite the fuel-air mixture.
- The electrodes must have the maximum operating life without changing the size of the gap.
- There must be a minimum opposition to the flow of combustion gases around the center electrode. This allows the lead-forming gases to be scavenged before they solidify into lead deposits.

ice bridging. A spark plug failure that occurs when starting a reciprocating engine in extremely cold weather. Water vapor from the combustion process condenses on the spark plug electrodes and freezes, bridging the electrode gap and prevents the plug firing.

Massive Electrodes

Some early spark plugs had a single ground electrode similar to those on automobile spark plugs. Most massive electrode aircraft spark plugs have two, three, or four nickel alloy ground electrodes, arranged around the center electrode to form gaps for the spark to jump. *See* Figure 5-36.

As was mentioned earlier, the arc across the spark plug gap acts in the same way as the arc produced by an electric arc welder. During the spark produced when the center electrode is positive, the ground electrode is worn more than the center electrode, but during the spark when the center is negative, the center electrode is the most severely eroded.

Horizontally opposed engines have an even number of cylinders, and since the spark occurs in a spark plug every other time the crankshaft rotates, the polarity of the spark is the same each time the plug fires. The center electrode in half of the spark plugs will be most severely worn, while it is the ground electrodes in the other plugs that will show the most wear. To even this wear, the spark plugs are returned to a different cylinder each time after servicing. There is more information on this in the section on spark plug servicing later in this chapter. Radial engines with an odd number of cylinders do not have this kind of electrode wear, as the polarity of the sparks is opposite each time the spark plug fires. The electrode wear is the same on all of the spark plugs.

Massive-electrode spark plugs are relatively low in cost and are efficient enough for many operations, but excessive amounts of lead in the fuel can cause them to foul.

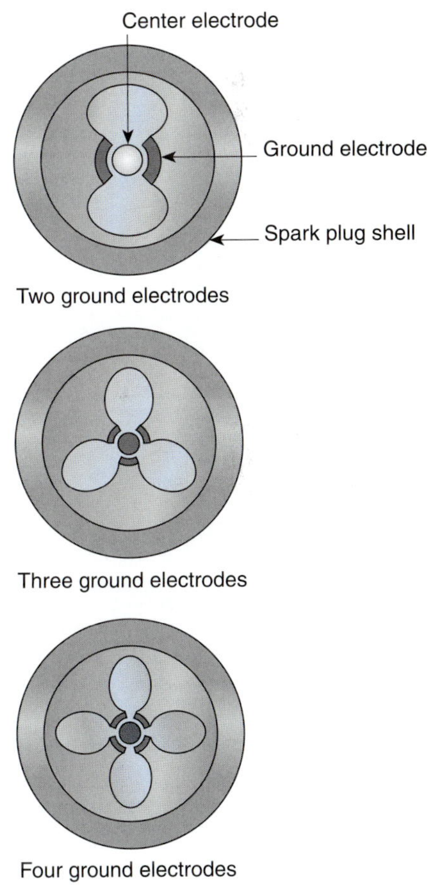

Figure 5-36. *Most aircraft spark plugs have more than one massive ground electrode to provide a large amount of metal to wear before the gap becomes excessive.*

Fine-Wire Electrodes

During World War II, severe operating conditions encountered by aircraft engines in Arctic and desert conditions led to many spark plug failures caused by lead fouling and ice bridging. Spark plug manufacturers developed fine-wire spark plugs to solve these problems. A small, round platinum wire is used as the insulated center electrode. Two platinum electrodes having

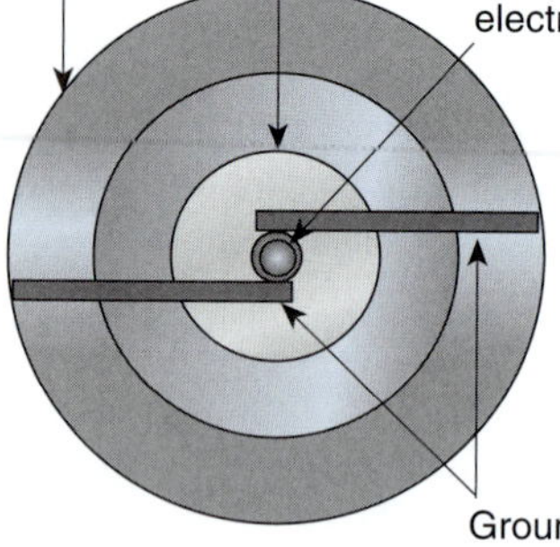

Figure 5-37. *A fine-wire spark plug has a wide open end that allows the lead-forming gases to be purged before they solidify and form conductive lead deposits.*

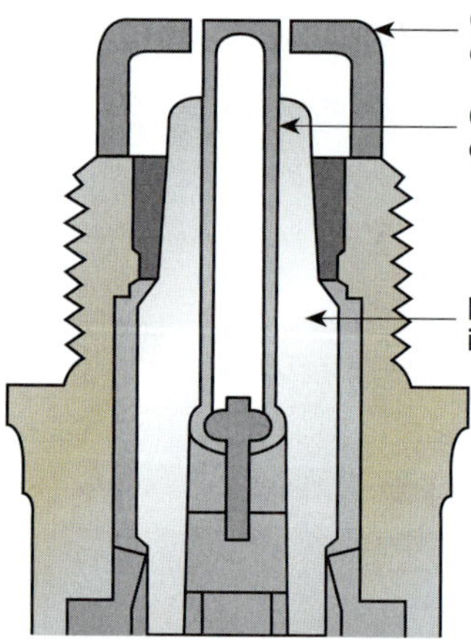

Figure 5-38. *The Champion REM37BY spark plug is designed to minimize lead fouling when used in engines that are forced to use fuel with more lead than the engine was designed to accommodate.*

a square cross section about the same size as the center electrode are spot welded to the rim of the firing end cavity. The ground electrodes are moved near the center electrode to form two accurately spaced gaps across which the spark can jump.

Fine-wire electrodes allow combustion gases containing lead to be scavenged from the spark plug before they solidify and form a low-resistance path for high voltage to leak to ground. *See* Figure 5-37.

Fine-wire platinum electrodes did much to minimize the lead fouling and ice bridging problem, but platinum is eroded by lead contamination at high temperatures. This erosion is prevented by making the electrodes from iridium wire. Some fine-wire spark plugs have a platinum center electrode and iridium ground electrode. These are called hybrid spark plugs. Fine-wire spark plugs are significantly more expensive than massive electrode spark plugs.

Projecting Electrodes

The Champion Spark Plug Company and Teledyne-Lycoming Engines have worked on the problem of lead fouling of spark plugs in engines required to use fuel with more lead than initially designed to accommodate. The result of this collaboration is the projected electrode REM37BY spark plug shown in Figure 5-38. Notice that the electrodes extend well beyond the firing-end cavity, and there are no places for combustion gases to be trapped and form lead deposits.

Insulation

Spark plugs must withstand the high voltage that causes the spark to jump across the electrode gap. The first spark plugs used porcelain for the insulators. It was discovered in the late 1930s that laminated mica was superior to porcelain as an insulator, and mica spark plugs became the standard up to the beginning of World War II.

Mica had two drawbacks for spark plug use. It absorbed oil, and the physical inconsistencies of the natural material made volume production difficult. During the early years of World War II, a ceramic material made of aluminum oxide was developed that was superior to either porcelain or mica. This is presently the standard spark plug insulator.

Shielding

Ignition systems for aircraft without radio equipment may remain unshielded; but when radio equipment is installed, the ignition system must be shielded to prevent the electromagnetic radiation from the spark plug leads and the magneto P-lead from causing radio interference. Almost all modern aircraft have some form of radio equipment, and for this reason unshielded ignition systems are seldom seen. In this text, only shielded spark plugs are discussed.

Some shielded spark plugs have a barrel with $\frac{5}{8}$-24 threads for the ignition lead connection. These spark plugs have an operational altitude limit of approximately 30,000 feet. Military service by the end of World War II needed a spark plug that would allow high-altitude operation, so spark plugs with the larger ¾-20 threaded lead connection were developed. This larger spark plug, called an all-weather or high-altitude spark plug, has become the standard for almost all operations.

iridium. A very hard, brittle, highly corrosion-resistant, whitish-yellow, metallic chemical element.

hybrid spark plug. A fine-wire spark plug that has a platinum center electrode and iridium ground electrodes.

porcelain. A hard, white, translucent ceramic material used as the insulator in some early aircraft spark plugs.

ceramic. Any of several hard, brittle, heat-resistant, noncorrosive materials made by shaping and then firing a mineral, such as clay, at a high temperature.

resistor spark plug. A shielded spark plug with a resistor between the ignition lead terminal and the center electrode.

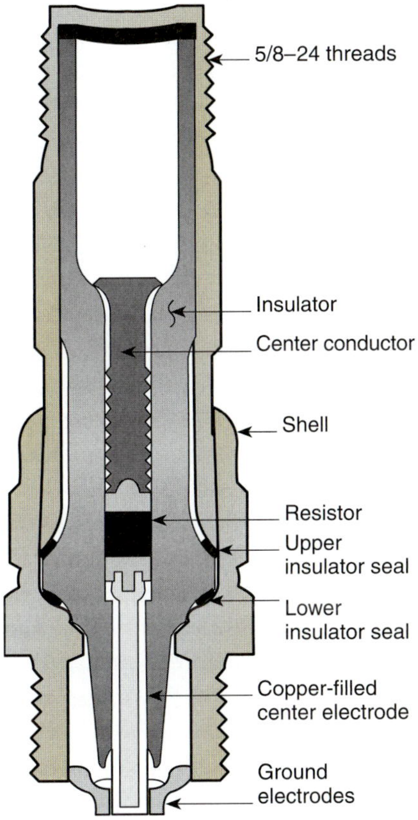

Figure 5-39. *A massive-electrode spark plug with a ⅝-24 shielding*

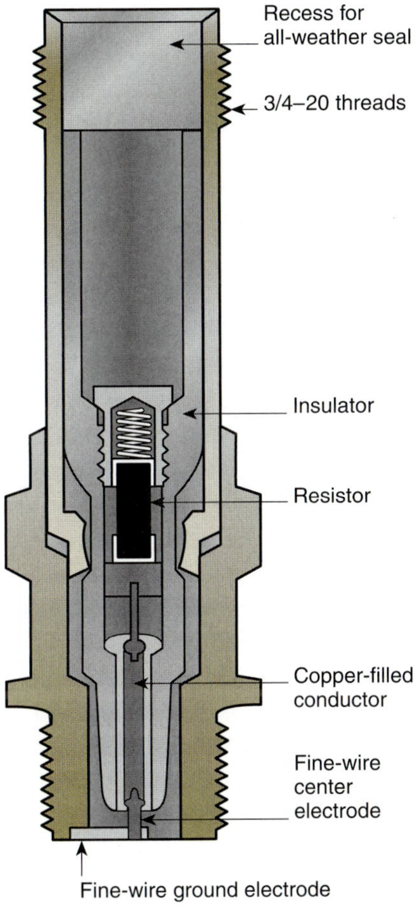

Figure 5-40. *A fine-wire electrode spark plug with ¾-20 shielding*

Resistors

A shielded ignition lead consists of a central conductor, an insulator, and an outer conductive shielding. These three components make an effective capacitor, which stores electrical energy in electrostatic fields.

When a high voltage is induced into the secondary winding of the magneto coil and is carried through the ignition lead to the spark plug, it ionizes the air between the electrodes and allows the current to flow to ground. The high voltage creates an electrostatic field in the shielded lead which stores electrical energy.

capacitance afterfiring. The continuation of the spark across the gap in a shielded spark plug after the fuel-air mixture in the cylinder is ignited.

As soon as the current across the spark plug gap begins to decrease, the capacitance returns the stored energy to try to maintain the flow. The sustained spark, called capacitance afterfiring, is high in heat energy, and erodes the electrodes. The initial spark ignites the fuel-air mixture, but the current that continues after ignition serves no useful purpose, and shortens the spark plug life.

A resistor of between 1,500 and 5,000 ohms is built into most shielded spark plugs between the ignition lead and the center electrode. *See* Figures 5-39 and 5-40. The resistor causes the spark to stop when the secondary voltage drops to a specified low value, thus preventing capacitive afterfiring.

Reach

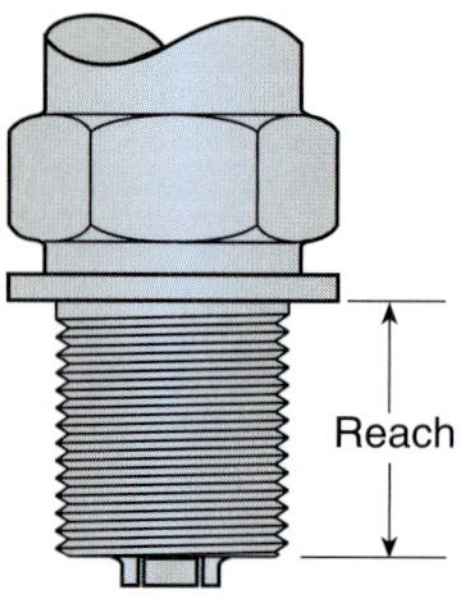

Figure 5-41. *The reach of a spark plug is the length of the threaded portion of the shell that screws into the cylinder head.*

reach. The length of the threads on the shell of a spark plug.

The reach of a spark plug is the length of the threaded portion of the shell, and there are two standard reaches for 18-mm spark plugs. A long-reach spark plug has $^{13}/_{16}$ inch of threads, and a short-reach spark plug has only $^{1}/_{2}$ inch of threads. It is important that the reach of a spark plug be chosen to fit the thickness of the cylinder head into which it fits.

If a long-reach spark plug is installed in a cylinder requiring a short-reach plug, the threads will protrude into the cylinder. Heat will damage them, and carbon will form in them. When the spark plug is removed from the cylinder, the carbon will likely damage the bushing or the Heli-Coil insert in the cylinder head. This same type of damage can occur if the spark plug gasket is left off when the plug is installed in the cylinder.

If a short-reach spark plug is installed in a cylinder requiring a long-reach plug, the bottom of the threads in the spark plug hole will be exposed to the combustion gases, and carbon will form in them. Later, when a spark plug with the correct reach is installed, it will bottom on the carbon, and the specified installation torque may be improperly reached before the spark plug seats correctly on its gasket. If this occurs heat cannot be properly transferred

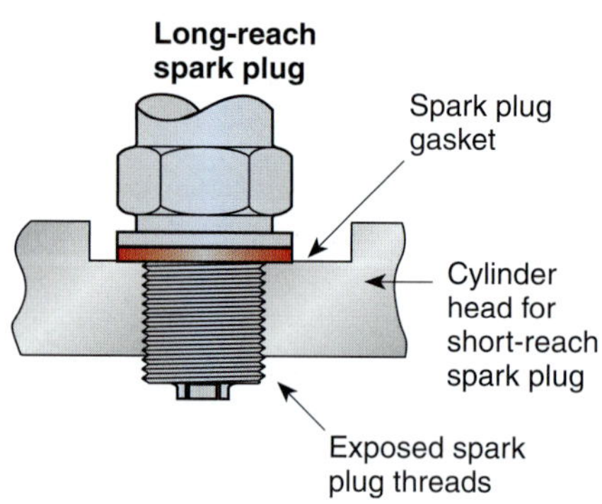

Figure 5-42. *If a long-reach spark plug is installed in an engine that requires a short-reach plug, the lower threads will be damaged and can damage the bushing or Heli-Coil insert when the plug is removed.*

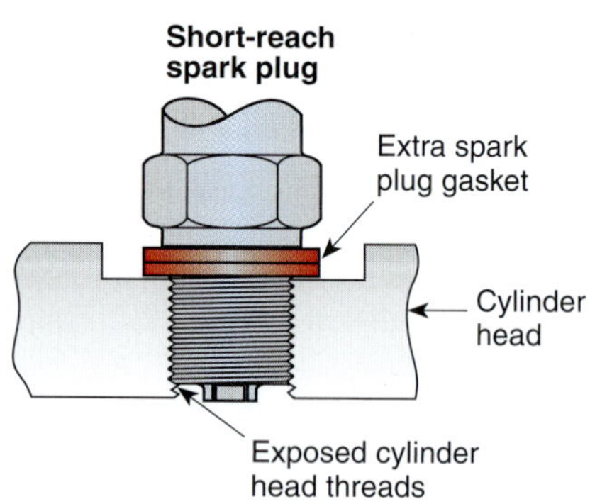

Figure 5-43. *If a short-reach spark plug is installed in an engine requiring a long-reach plug, or if two gaskets are used, the lower threads in the cylinder head will fill with carbon. The correct spark plug cannot be installed until the carbon is removed.*

from the spark plug into the cylinder head, and it is possible for hot combustion gases to escape around the threads and damage the cylinder head.

The same problems when using a short-reach spark plug in a cylinder requiring a long-reach plug can occur if two spark plug gaskets are used. These problems also occur when a regular spark plug gasket is installed on top of a gasket-type cylinder head temperature thermocouple pickup.

Heat Range

Research has shown that lead bromide and carbon deposits form on the insulator nose of a spark plug when its temperature is below 900°F. Lead fouling from the tetraethyl lead in the fuel causes most problems in the spark plug when the temperature is above 1,300°F. Deposits that form at temperatures between 1,000°F and 1,300°F are low in volume, low in electrical conductivity, and least likely to cause spark plug fouling.

In order to keep fouling to a minimum, spark plugs are designed so that the insulator nose maintains a temperature within the range that causes fewest deposits.

The heat range of a spark plug is a measure of its ability to dissipate heat received from the combustion chamber into the cylinder head. This heat is then transferred into the cooling air that passes through the cylinder head fins.

There are two classifications of heat ranges for aircraft spark plugs: hot plugs and cold plugs. Within these two classifications are degrees of heat range. When selecting a spark plug for a particular engine, use only plugs that are listed in the Type Certificate Data Sheet for the engine and are recommended in the most recent engine manufacturer's applicable service bulletin.

A hot spark plug has a long insulator tip that provides a long path for the heat to travel to get into the cylinder head; therefore, it is slow to dissipate the heat. Hot spark plugs dissipate heat slowly from their long insulator tip, and are used in engines whose cylinder heads dissipate heat rapidly and thus operate with low cylinder head temperature.

A cold spark plug has a short insulator tip, and quickly dissipates the heat it picks up into the cylinder head. Cold spark plugs dissipate heat rapidly from their short insulator tip, and are used in engines whose cylinder heads dissipate heat slowly and thus operate with high cylinder head temperature.

In short, a spark plug should conduct heat at a rate that keeps its operating tip temperature above 900°F when the engine is idling, and below 1,300°F degrees when the engine is producing maximum power. This minimizes the formation of deposits that tend to foul the plug, and prevents the edges of the firing end from glowing and causing pre-ignition. To accomplish this, plugs used in different engines need to conduct heat at varying rates.

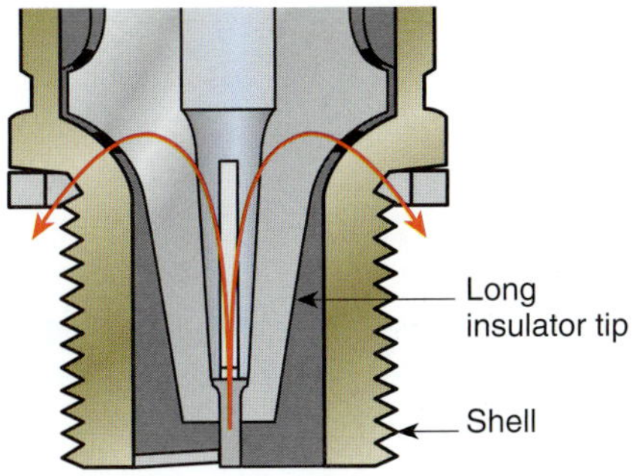

Figure 5-44. *A hot spark plug has a long insulator tip which is slow to dissipate the heat. Hot spark plugs are used in engines that operate with low cylinder temperatures.*

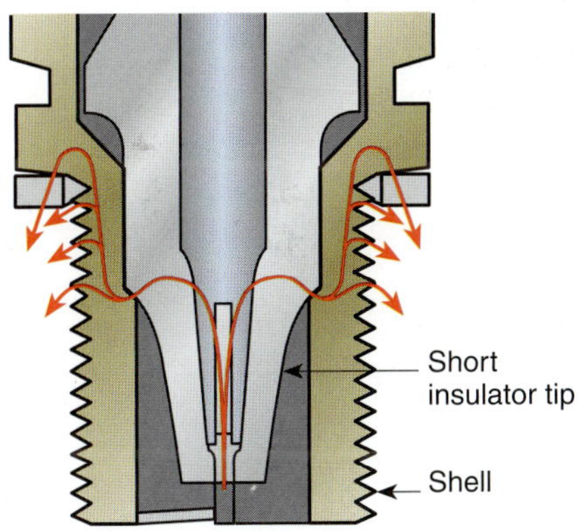

Figure 5-45. *A cold spark plug has a short insulator tip which rapidly dissipates the heat. Cold spark plugs are used in engines that operate with high cylinder temperatures to minimize accumulation of lead deposits.*

Spark Plug Construction

The insulator is made of a special ceramic material with a hardness close to that of a diamond. It is especially designed to have excellent heat dissipating characteristics that allow the heat to be conducted away from the insulator tip.

The center electrode is locked into the insulator and makes contact with the resistor. Some massive electrode spark plugs have solid nickel alloy center electrodes, and others use a nickel alloy sheath filled with copper to hasten the transfer of heat from the tip.

Fine-wire spark plugs have a solid platinum or iridium electrode locked into the insulator tip.

The shell and shield of a spark plug are machined of steel. The shell of a 14-mm spark plug has a $^{13}/_{16}$ hex, and an 18-mm spark plug has a $^{7}/_{8}$-inch hex. Ground electrodes are bonded into the open end of the shell.

The insulator is assembled into the shell with a lower insulator seal. An upper insulator seal is put in place and the shield is inserted; then the shell is crimped over the shield to form a one-piece leakproof assembly. *See* Figure 5-39 on Page 233.

STUDY QUESTIONS: SPARK PLUGS

Answers are on Page 244. Page numbers refer to chapter text.

62. The two sizes of spark plug shells used in aircraft engines are ____________ mm and ____________ mm. *Page 230*

63. Two basic types of electrodes used on aircraft spark plugs are:
    a. ____________
    b. ____________
    Page 231

64. Massive ground electrodes are made of ____________ alloy. *Page 231*

65. Two metals used for fine-wire electrodes are:
    a. ____________
    b. ____________
    Page 232

66. The material used as insulator in most modern spark plugs is ____________ (porcelain, mica, or aluminum oxide). *Page 232*

67. Two sizes of spark plug shielding threads are:
    a. ____________
    b. ____________
    Page 232

68. A spark plug with ¾-20 shielding is called a/an ____________ spark plug. *Page 233*

69. A resistor is installed in a shielded spark plug to prevent electrode erosion caused by _______________________________ . *Page 234*

70. The length of the threaded portion of a spark plug shell is called the ___________________ of the spark plug. *Page 234*

71. The most prevalent types of fouling that form on a spark plug at low temperature are ________________ and ____________________ fouling. *Page 235*

72. The most prevalent type of fouling that forms on a spark plug at high temperature is ____________________ fouling. *Page 235*

73. The two classifications of spark plugs according to their heat range are:
   a. __________________________
   b. __________________________
   Page 235

74. A hot spark plug has a _____________________ (long or short) path for the heat to follow between the insulator tip and cylinder head. *Page 235*

75. The correct heat range spark plug to use on an engine with low cylinder temperature is a ___________________ (cold or hot) spark plug. *Page 235*

76. If an engine must be operated with fuel having a higher TEL content than it is designed to use, the recommended spark plug is _________________ (colder or hotter) than that used with the proper fuel. *Page 235*

Spark Plug Servicing

Spark plug servicing is one of the most time-consuming portions of a routine maintenance inspection, especially in engines having many cylinders. The seemingly simple task of removing, cleaning, gapping, and replacing the spark plugs often causes this function to get less than its deserved care.

Here, we will consider the procedure of servicing spark plugs as recommended by some manufacturers.

In normal engine operation, carbon, lead, and silicon form deposits inside the firing end of the spark plug that allow current to leak to ground. This leakage prevents the voltage from building up enough to jump the electrode gap. Metal is removed from an electrode each time a spark occurs, and the electrode gap increases. A higher voltage is required before the spark can jump the enlarged gap.

Two basic elements of spark plug servicing are removing the contaminants from the firing end and restoring the electrode gap to the dimension specified by the spark plug manufacturer.

Removal

Spark plugs are designed and built to withstand a tremendous amount of abuse in the engine, but they are actually quite delicate and can be damaged beyond repair by careless handling during routine servicing.

The lead terminal must be removed by using the proper type of open-end or crowfoot wrench. Hold the lead with one hand and loosen the terminal nut with the other. When the terminal nut is loose, unscrew the nut completely and pull the terminal straight out of the cavity. Some leads have ceramic terminals, and it is possible to crack them if they are allowed to cock to one side as they are pulled out.

Use a special six-point, deep-socket spark plug wrench to loosen and remove the spark plug. Be sure the socket is square on the hex of the spark plug and that the wrench handle provides a good straight pull while supporting the socket with the free hand. If the wrench cocks on the spark plug, there is a good probability that the insulator will be cracked during the removal process.

Remove the spark plugs and place them in numbered holes in a spark plug tray. This is very important because the spark plugs indicate the operating conditions in the cylinders and it is essential to be able to identify the cylinder and position from which each spark plug was removed. Also, because of the polarity of the sparks, the plugs should be replaced in a different cylinder after they have been serviced.

Inspection

When all spark plugs have been removed from the engine and are in their numbered tray, examine them carefully before cleaning.

The insulator of a normally operating spark plug will be covered with a dull brown deposit. There may be some lead deposits in the cavity, but it will not be filled. The electrode gap will be slightly wide, but the electrodes will not show excessive wear. Spark plugs should be replaced when the electrodes have worn to one half of their original dimension.

Electrodes that show severe wear could indicate abnormal engine operation. Examine the cylinder from which the spark plug was removed for signs of detonation, and if the engine is fuel injected, check the injector nozzle to be sure that it allows the correct amount of fuel to flow.

Spark plugs that show an excessive amount of black fluffy carbon indicate that the cylinder has been operating with an overly rich mixture. This could be caused by excessive idling or by a carburetor or fuel injection system that is metering too much fuel.

Black, wet, oily deposits on the insulator tip are signs of the cylinder pumping too much oil. This could be caused by worn or faulty piston rings or worn intake valve guides. Oily deposits are conductive at low temperatures and will prevent the spark plug from firing.

If the firing-end cavity is filled with hard, brittle deposits, the spark plug is lead fouled. This is caused by using fuel that contains more tetraethyl lead than the engine was designed to accommodate, or the engine may be operating too cold for the amount of lead in the fuel.

Excessive lead fouling may also be caused by improper fuel distribution in a carbureted engine.

A shiny, dark brown deposit on the insulator tip can be silicon contamination. Lead deposits have a strong affinity for silicon, and if the aircraft has a leaking carburetor air filter that allows dust or sand to be ingested, the silicon will unite with the lead deposits and act as a flux, lowering the melting point of oxides and converting them into lead silicates on the insulator tip. Lead silicates are insulators at low temperature, but when hot, they become conductive and leak the secondary voltage off to ground before it builds up enough to jump the electrode gap.

Cleaning

Spark plugs must be cleaned in two ways: they must be thoroughly degreased, and all contaminants must be removed from the firing-end cavity.

Plugs can be degreased by putting them in a metal rack with their electrode end down and placing the rack in a container of Stoddard solvent, varsol, or naphtha for 20 to 30 minutes. Be careful not to allow any solvent to get into the terminal cavity. After soaking the plugs for the recommended time, remove them and drain for a minute or two, then blow them completely dry with compressed air.

Loosen the cinder-like lead deposits from the firing-end cavity using a vibrator-type cleaner with the correct blades for the spark plug being cleaned. Insert the blades into the cavity and gently work the spark plug back and forth over the vibrating blades with a semi-rotating motion until all deposits are loosened.

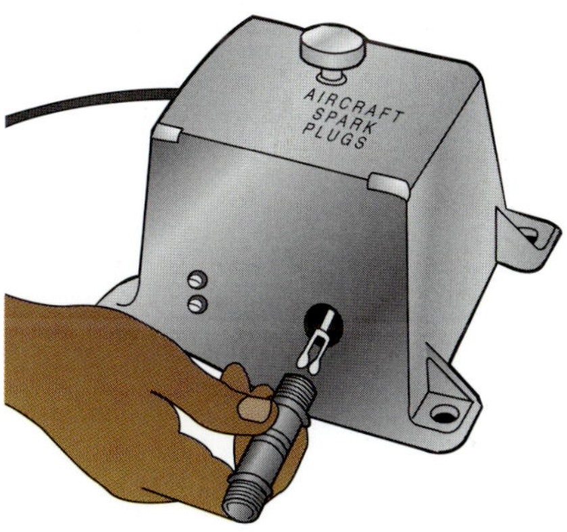

Figure 5-46. *Lead deposits may be loosened from the firing-end cavity of a spark plug with a vibrator-type cleaner.*

When all lead deposits have been broken out of the firing end, remove loosened residue and all other contaminants with an abrasive blast.

Use only glass beads or aluminum oxide abrasive for cleaning the spark plugs, never silica sand because of the danger of silicon contamination. Do not blast the electrodes for more than three to five seconds, as excessive blasting can erode them equivalent to hundreds of hours of operation. It is important to follow the instructions as designated by the cleaning equipment manufacturer, as the machines and abrasives differ.

Clean the terminal cavity with a cotton or felt swab saturated with acetone. If there are carbon stains inside the barrel that are not removed by the solvent, it is permissible to use a fine abrasive such as Bon Ami on the swab. Scrub the insulator with a twisting motion until all of the stain has been removed. Remove all residue with a clean swab, damp with solvent, and blow out the barrel with compressed air.

Clean the threads with a hand wire brush or a fine-wire power-driven brush. Do not brush the insulator or electrodes.

Gapping

One of the most important parts of spark plug servicing is setting the gap. A few special tools make the job easy and prevent damage to the spark plugs.

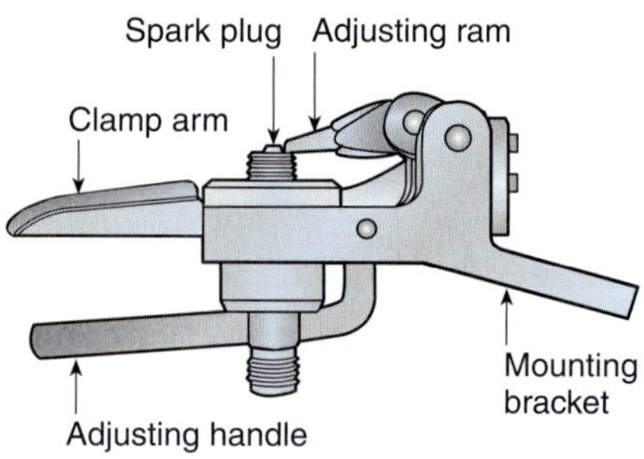

Figure 5-47. *A gapping tool for massive electrode spark plugs moves the ground electrode over so it remains parallel with the center electrode.*

Massive Electrode Spark Plugs

The gap in a massive electrode spark plug is adjusted by moving the ground electrode over so that it remains parallel with the center electrode. Screw the spark plug into a gapping tool like that in Figure 5-47, and very carefully move the ground electrode over until the gap is correct, measured with a round wire gage.

The wire gage must not be between the electrodes when the ground electrode is moved over. This would place a side load on the center electrode and would likely crack the insulator. Move the ground electrode over slightly and check the gap. If it is not close enough, move it a bit more and check it again.

It is not recommended that a gap be widened if it has been inadvertently closed too much. There is too much danger of damaging the insulator.

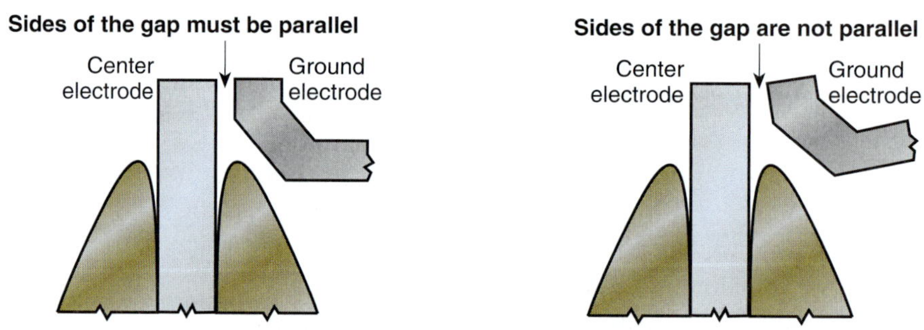

Figure 5-48. *If the ground electrode is not parallel with the center electrode, only a little erosion is needed to widen the gap beyond acceptable limits.*

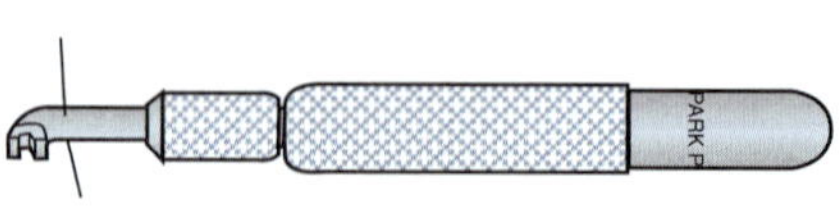

Figure 5-49. *Gapping tool for fine-wire spark plugs*

Fine-Wire Spark Plugs

The fine wire of which these electrodes are made makes adjustment of the gap appear simple, but the wires are brittle and special care must be taken to prevent breakage. A gapping and measuring tool such as the one in Figure 5-49 is recommended for servicing fine-wire spark plugs. Very carefully bend the ground electrodes toward the center electrode until the gap between the two is the same in diameter as the wire gage that is part of the tool. Do not bend these electrodes too much, as they should not be moved back. It could weaken them enough that they could fail in operation.

Testing

Proper servicing of a spark plug includes testing the plug under pressure. The most accurate tester, and the type that is used in most large spark plug overhaul facilities, is a bomb tester. It places the electrodes under nitrogen pressure of approximately 200 psi, and places a high voltage across the electrodes. If the spark plug fires consistently under this pressure, it will operate satisfactorily in the engine.

Most maintenance shops use a tester that is part of the spark plug cleaning machine. The plug is screwed into the tester only finger tight, and high-voltage is applied to a contactor inserted into the terminal cavity. The gap is viewed in a mirror, and the air pressure is adjusted to the value specified for the particular spark plug gap.

The loose fit of the spark plug in the tester allows air to leak out around the threads. This leakage removes the ionized air from the test chamber and aids in better control of the air pressure. Plugs that fire consistently in the tester will operate satisfactorily in the engine.

If cleaned and tested spark plugs are not to be immediately installed in an engine, the threads on both the shell and shielding connection should be protected by a light coat of rust-inhibiting oil and the plugs placed in cartons similar to those used for new spark plugs. If they are to be stored for some time in a humid area, they should be kept in a closed cabinet with a low-wattage light bulb to inhibit moisture.

Installation

As previously mentioned, spark plugs can be easily damaged by careless handling. If a spark plug is dropped, it may not appear to be damaged, but the insulation may be cracked enough to cause failure when subjected to heat and vibration of normal use.

Spark plugs removed from an engine on a routine maintenance inspection should be kept in numbered holes in a spark plug rack so they can be installed in the proper cylinder and position when they are reinstalled.

Magnetos on horizontally opposed engines produce sparks of alternating voltage, and since each spark plug fires on every other revolution of the crankshaft, all sparks produced in any given spark plug have the same polarity, and the electrode erosion will originate from either the center or ground electrode. Wear can be evened out over the lifetime of the plug by ensuring that the polarity of the sparks is reversed every time they are changed. To reverse this polarity, replace the plugs in the next cylinder in the firing order of the engine, and swap them from bottom to top. In a six-cylinder engine with a firing order of 1-6-3-2-5-4, the spark plug that is removed from cylinder 1 top would be replaced in cylinder 6 bottom. *See* Figure 5-50.

Firing Order 1 - 6 - 3 - 2 - 5 - 4												
Remove plug from	1T	6B	3T	2B	5T	4B	1B	6T	3B	2T	5B	4T
Replace plug in	6B	3T	2B	5T	4B	1T	6T	3B	2T	5B	4T	1B

Figure 5-50. *When replacing spark plugs in an engine during a routine maintenance inspection, install them in the next cylinder in firing order, and swap them from top to bottom.*

Because spark plugs operate in extremely harsh environmental conditions, they can seize inside the cylinder and be extremely difficult to remove. To assure easy removal on the next inspection, apply a small amount of the antiseize compound recommended by the spark plug manufacturer to the second thread from the end of the shell. Do not allow any compound to get into the end threads, as it can run down into the electrodes and foul them.

When installing spark plugs in the engine, put a new solid copper gasket on the shell and screw the plug into the cylinder. Threads in the cylinder should be clean so the spark plug can be screwed all the way to the gasket without using any tool. If the plug will not screw in completely, carbon is probably in the threads, and a thread-cleaning tool should be screwed into the hole to remove it. If a cylinder head temperature thermocouple with a spark plug gasket-type pickup is used, do not put a regular gasket on the spark plug with the thermocouple.

When all spark plugs are in the cylinders, use an accurately calibrated torque wrench to apply the torque specified in the engine service manual. Never attempt to torque a spark plug into a cylinder by feel. If the plug is installed too loosely, there is danger of gas leaking past the threads and damaging the cylinder. If installed too tight, the shell can be distorted, the gap changed, or the insulator cracked.

With all spark plugs installed and torqued, wipe the terminals on the ignition leads with a cloth moistened with acetone. Inspect them, and if the terminals are clean and undamaged, slip them into the terminal end of the spark plug. Screw the terminal nut down finger tight and use the proper lead wrench to tighten it about an additional $\frac{1}{8}$ of a turn.

STUDY QUESTIONS: SPARK PLUG SERVICING

Answers are on Page 244. Page numbers refer to chapter text.

77. The two basic elements of spark plug servicing are:
    a. ______________________________
    b. ______________________________
    Page 237

78. Spark plugs should be removed from the engine using a _____ (six or twelve)-point deep socket. *Page 238*

79. A spark plug is considered to be worn out when its electrodes have worn to _______________ of their original dimension. *Page 238*

80. The cylinder from which spark plugs show an excessive amount of electrode wear should be inspected for signs of ______________________ . *Page 238*

81. A black, wet, oily deposit in a spark plug can be caused by worn piston rings or worn __________________ . *Page 238*

82. Lead deposits are loosened from the firing-end cavity of a spark plug with a ________________-type cleaner. *Page 239*

83. The terminal cavity is cleaned with a cotton or felt swab saturated with ____________ . *Page 239*

84. The electrode gap in a spark plug is measured with a ________________ (flat or round) gage. *Page 240*

85. Both platinum and iridium used as the electrodes in fine-wire spark plugs are __________________ (brittle or ductile). *Page 240*

86. The amount of air pressure used on a spark plug tester is determined by the spark plug ______ . *Page 241*

87. The spark plug removed from cylinder 1 top on a six-cylinder horizontally opposed engine with a firing order of 1-6-3-2-5-4 should be replaced in cylinder No. ___________ , ___________ (top or bottom). *Page 242*

88. When installing a spark plug in a cylinder, antiseize compound should be applied to the ___________ (first or second) thread from the end of the shell. *Page 242*

89. Terminals on an ignition lead should be wiped with a cloth dampened with ______________ before installing them in the spark plug. *Page 242*

Answers to Chapter 5 Study Questions

1. open
2. parallel
3. series
4. battery
5. a. high-tension system
   b. low-tension system
6. a. magnetic
   b. primary
   c. secondary
7. minimum
8. a. the rate of change in the flux
   b. ratio of turns on secondary winding to turns on primary
9. after
10. E-gap
11. capacitor
12. parallel
13. parallel
14. ON
15. shielding
16. AC
17. two
18. sparking order of the magneto
19. spark plug
20. safety gap
21. transformer
22. 12
23. later
24. impulse
25. a. OFF
    b. LEFT
    c. RIGHT
    d. BOTH
    e. START
26. left
27. pulsating DC
28. retard
29. 180
30. soft-iron rotor
31. one
32. two
33. top
34. 9
35. a. internal timing
    b. timing to the engine
36. open
37. 1
38. before
39. a. flange mount
    b. clamp mount
    c. base mount
40. vernier
41. BOTH
42. first
43. pivotless
44. 100
45. early
46. late
47. 25
48. late
49. do not
50. soft iron
51. primary
52. block, rotor
53. shellac
54. is not
55. sharply
56. advanced
57. not change
58. idle
59. shielding
60. leakage
61. insulator
62. 14, 18
63. a. massive electrodes
    b. fine-wire electrodes
64. nickel
65. a. platinum
    b. iridium
66. aluminum oxide
67. a. $\frac{5}{8}$-24
    b. $\frac{3}{4}$-20
68. all-weather
69. capacitive afterfiring
70. reach
71. carbon, lead bromide
72. lead
73. a. hot
    b. cold
74. long
75. hot
76. colder
77. a. removing contaminants
    b. adjusting electrode gap
78. six
79. one half
80. detonation
81. valve guides
82. vibrator
83. acetone
84. round
85. brittle
86. gap
87. 6, bottom
88. second
89. acetone

Exhaust Systems

6

Exhaust Systems

6

Evolution of Reciprocating Engine Exhaust Systems

Approximately one half of the heat energy in fuel burned in an aircraft engine is wasted as it is carried out of the engine through the exhaust system. Not only is this energy lost, but this heat must also be disposed of without damaging the aircraft structure. This is achieved with the exhaust system.

The exhaust system must be open enough that it does not produce enough back pressure to impede the flow of exhaust gases from the cylinders. At the same time, it must protect the hot exhaust valve from a direct blast of cold air which could warp it. The system must also decrease the engine exhaust noise.

The very earliest stationary aircraft engines and the rotary radial engines had no exhaust system as such. The exhaust gases were forced from the cylinders directly into the air surrounding the engine.

Rotary engines were designed so that the exhaust valve was open when the cylinder passed below the fuselage. Exhaust from these engines contained so much unburned oil that a cowling was needed to prevent oil from covering the entire aircraft as well as the pilot.

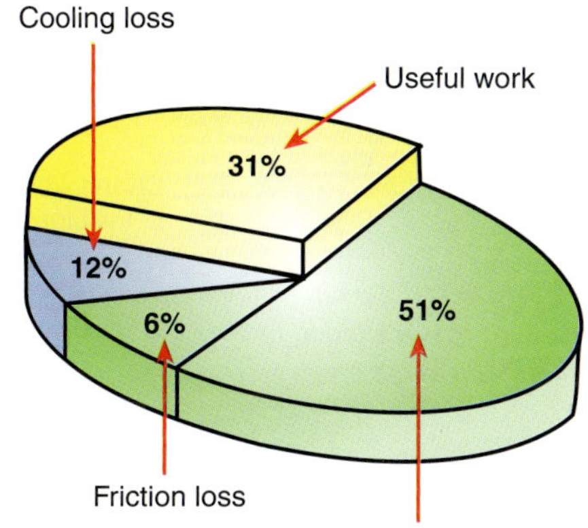

Figure 6-1. *Approximately one half of the heat energy in fuel is lost through the exhaust system.*

In an effort to keep exhaust back pressure to a minimum, many engines built during the 1930s had short exhaust stacks—short lengths of steel tubing that directed exhaust gases away from the engine. This system got rid of the gases with a minimum of back pressure, but it had two distinct disadvantages: the noise was excessive, and it caused the exhaust valves to warp. The hot exhaust valves were exposed and when the aircraft was put into a side slip, cold air was forced into the exhaust port where it shock-cooled the valves and caused them to warp.

Both of these problems were minimized with bayonet, or pinched-slot, stacks. Gases exit the stack through a long, narrow slot at right angles to the stack. Air flowing over the bayonet stack creates a slightly low pressure that aids gases in leaving the cylinder and thus decreases back pressure. Cold air can no longer reach the valves to warp them, and the bayonet decreases the exhaust noise.

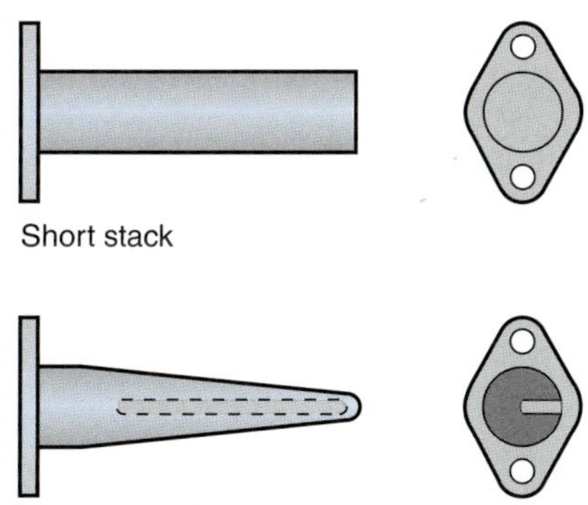

Figure 6-2. *Bayonet stacks were used on some early engines to decrease exhaust back pressure, minimize valve warpage, and decrease exhaust noise.*

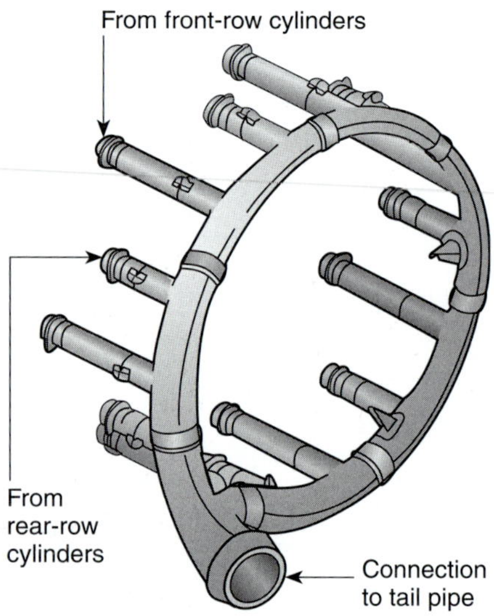

Figure 6-3. *A typical exhaust collector ring for a 14-cylinder, twin-row radial engine*

As engine size increased and airplanes were flown at night as well as during the day, the exhaust system underwent an evolutionary change. Exhaust gases from all cylinders were collected and discharged overboard from one location where the exhaust flame would not interfere with the pilot's night vision.

Exhaust gases from individual cylinders in a radial engine flow into a collector ring that extends around the engine. The collector ring terminates with a large-diameter tail pipe. Figure 6-3 shows a typical exhaust collector ring for a 14-cylinder twin-row, radial engine. This collector ring is mounted immediately behind the cylinders, ahead of the firewall in a stainless steel channel. The tail pipe on a multi-engine airplane discharges on the side of the engine that is away from the fuselage. On airplanes with turbosuperchargers, the tail pipe carries exhaust gases back to the supercharger unit.

The exhaust system for a horizontally opposed engine collects exhaust gases from cylinders on each side of the engine. In some smaller aircraft, the exhaust is carried away from the aircraft through two exhaust tail pipes, one for each side. Exhaust gases from the two sides of the engine in most larger aircraft are collected and sent through a single muffler or are used to drive a single turbocharger.

Figure 6-4 shows a typical exhaust system for a turbocharged airplane powered by a six-cylinder horizontally opposed engine. Gases flow out of the left-side cylinders through a crossover tube which passes in front of the engine and joins the exhaust collector on the right side of the engine. Gases

collector ring. A ring made of thin corrosion-resistant steel tubing that encircles a radial engine and collects exhaust gases from each cylinder. The ring ends with a connection to the exhaust tail pipe.

turbocharger. A turbosupercharger that consists of a centrifugal air compressor on the same shaft with an exhaust-driven radial-inflow turbine. Compressed air is directed into the engine cylinders to increase power produced by the engine.

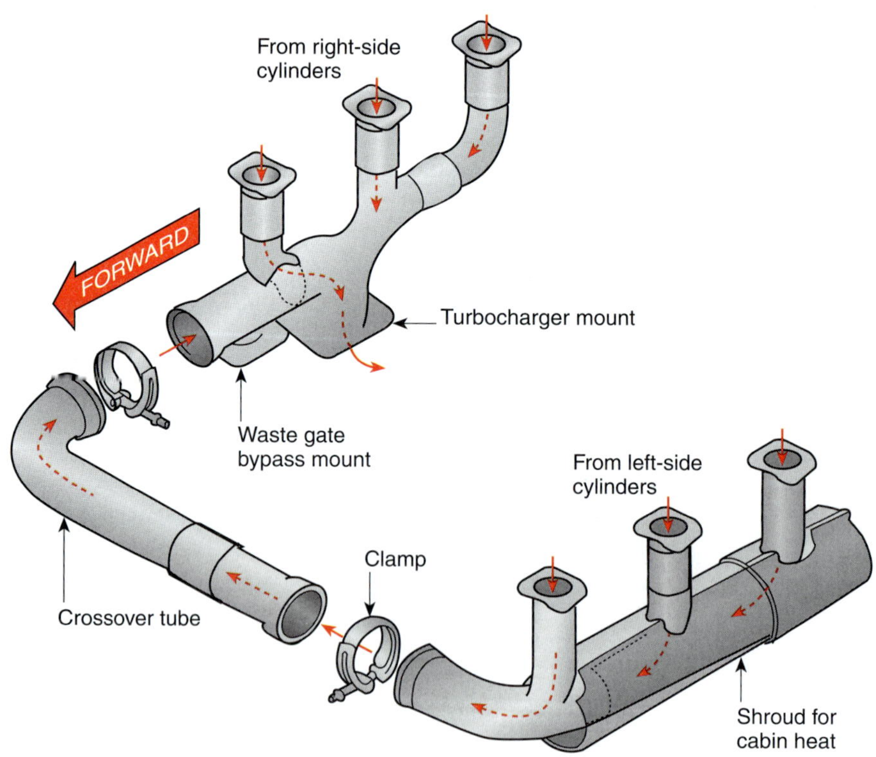

Figure 6-4. *A typical exhaust system for a six-cylinder horizontally opposed, turbocharged engine.*

from all six cylinders exit the engine compartment through the tail pipe, which is connected to the turbocharger and the waste gate.

Cabin and Carburetor Heat Provisions

Some heat energy in the exhaust gases is used to heat the aircraft cabin and some is used to prevent the formation of carburetor ice, or melt it if it has formed. A shroud made of thin sheet stainless steel is fastened around the muffler or some part of the exhaust pipe so that air can flow through it and through a flexible duct to the cabin or carburetor heat box.

Heat boxes incorporate pilot-actuated valves that direct heated air overboard when it is not required to prevent the overheating of the shrouds.

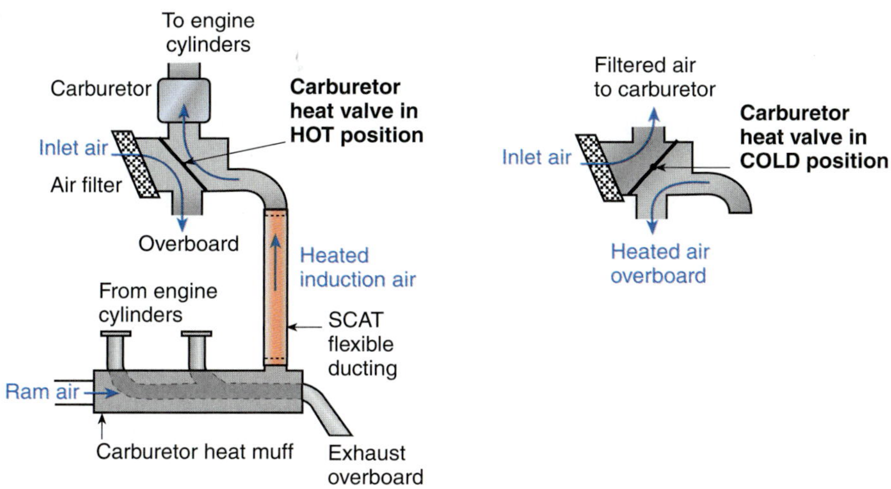

Figure 6-5. *Heat taken from part of the exhaust system warms the induction air to prevent or remove carburetor ice.*

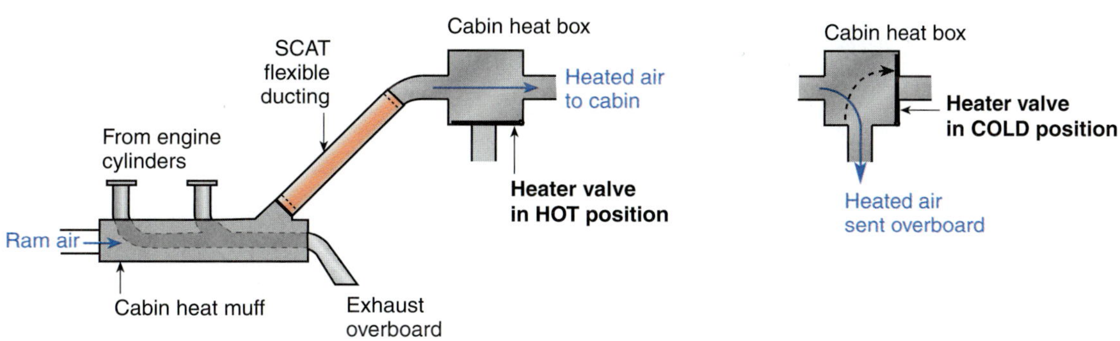

Figure 6-6. *Heat taken from another part of the exhaust system is used to heat the cabin.*

The flexible ducting that carries the cold air to the muffs is made of a single layer of special fiberglass cloth tubing impregnated with black neoprene rubber. The ducting that carries the hot air from the muffs to the heat boxes is impregnated with red silicone rubber. This ducting is supported with a spiral-wound, copper-coated, spring steel wire. The black ducting is called "CAT," and is designed to withstand temperatures from -65°F to 300°F. The red ducting is called "SCAT," and is used for temperatures up to 450°F. There is also a special two-ply ducting called "SCEET" that is used for temperatures of around 500°F.

Mufflers

exhaust back pressure. The pressure in the exhaust system of a reciprocating engine that opposes the flow of exhaust gases as they are forced out of the cylinders when the exhaust valve is open.

Mufflers are used on almost all general aviation aircraft that are not turbocharged. The muffler fits in the exhaust system in such a way that it receives exhaust gases and absorbs much of the sound energy, then directs the gases out through the tail pipe. The muffler must be matched to the engine so it produces the minimum amount of back pressure to the flow of exhaust gases. Some mufflers have heat transfer knobs or rings welded to their outer surface to increase their surface area. This allows air flowing over the muffler on its way to the carburetor or cabin to absorb the maximum amount of heat.

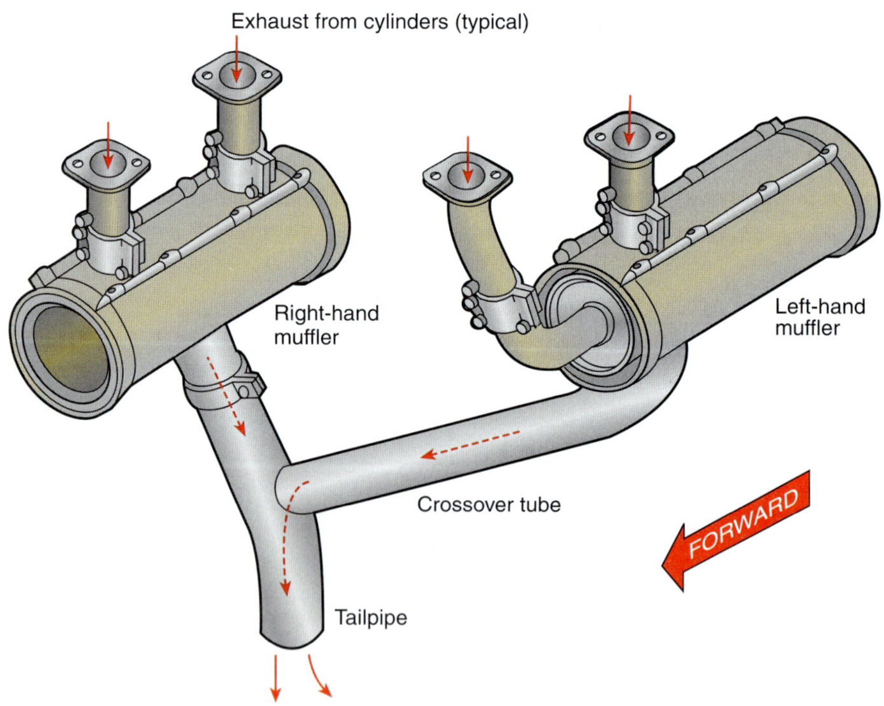

Figure 6-7. *Some smaller aircraft use two mufflers whose outputs are joined together by a crossover tube so that all exhaust leaves the aircraft from a single tail pipe.*

Augmentor Tubes

An augmentor tube uses some of the velocity energy in exhaust gases to increase the cooling airflow through the engine and decrease the exhaust back pressure. It is a stainless steel tube with a slightly flared, or bell mouth, opening inside the engine cowling.

Exhaust from the cylinders is collected and discharged into the bell-mouth of the tube. The shape of the tube causes the gases to accelerate, and according to Bernoulli's principle, as the velocity of the gases increases, their pressure decreases. This low pressure below the engine draws cooling air into the cowling and through the cylinder fins to remove heat. Low pressure at the end of the exhaust pipe decreases the exhaust back pressure and allows the engine to breathe more freely.

augmentor tube. A long, specially shaped stainless steel tube mounted around the exhaust tail pipe of a reciprocating engine. As exhaust gases flow through the augmentor tube, they produce a low pressure in the engine compartment that draws in cooling air through the cylinder fins.

bell mouth. The shape of the inlet of an augmentor tube that forms a smooth converging duct. The bell mouth shape allows the maximum amount of air to be drawn into the tube.

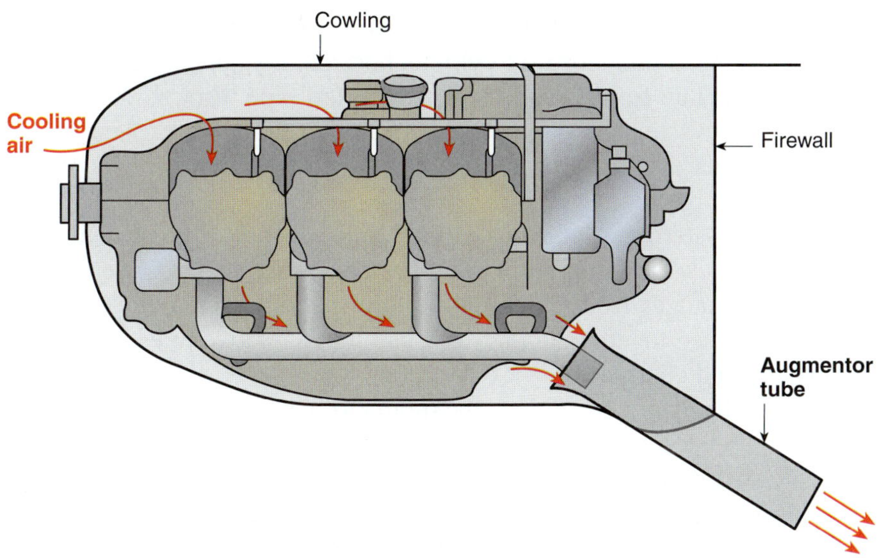

Figure 6-8. *An augmentor tube uses the velocity of exhaust gases leaving the engine to increase the flow of cooling air through the cylinder fins and to decrease the exhaust back pressure.*

Power Recovery Devices

We cannot forget that approximately half of the energy in fuel being burned is wasted out the exhaust. For decades engineers have pondered ways of recovering some of this energy; two systems that have emerged are turbochargers and power recovery turbines (PRTs) the latter being only somewhat successful.

Turbochargers

Fighter and bomber aircraft in World War II were able to reach the high altitude demanded for their operations because they were equipped with exhaust-driven turbosuperchargers. Modern horizontally opposed engines can maintain their sea level horsepower to a specified altitude by the use of small turbochargers.

A turbocharger consists of a centrifugal compressor driven by the exhaust gases flowing through a radial-inflow turbine. Figure 6-9 shows a typical turbocharger installation for a six-cylinder horizontally opposed engine. The exhaust gases flow from the cylinders to the tail pipe through the turbine or through the waste gate to bypass the turbine. A complete explanation of the turbocharger system is in the text on induction systems in Chapter 4, beginning on Page 184.

Exhaust system components expand and contract a great deal with extreme changes in temperature. The system, designed to prevent these dimensional changes from cracking the thin metal, is made of a series of small units joined together so that they can move relative to each other as they expand and contract.

In naturally aspirated engines the tubing, of which various sections are made, usually telescopes together and is held in place with clamps that are loose when the system is cold. This is a simple method of manufacture, but it does allow some slight leakage of exhaust gases. A turbocharger produces some back pressure that increases the leakage between the joints of a simple slip-together system, so these systems have ball joints and slip joints that allow the components to move, but do not allow any leakage of gases. Some turbocharged engines use corrugated metal bellows rather than slip joints.

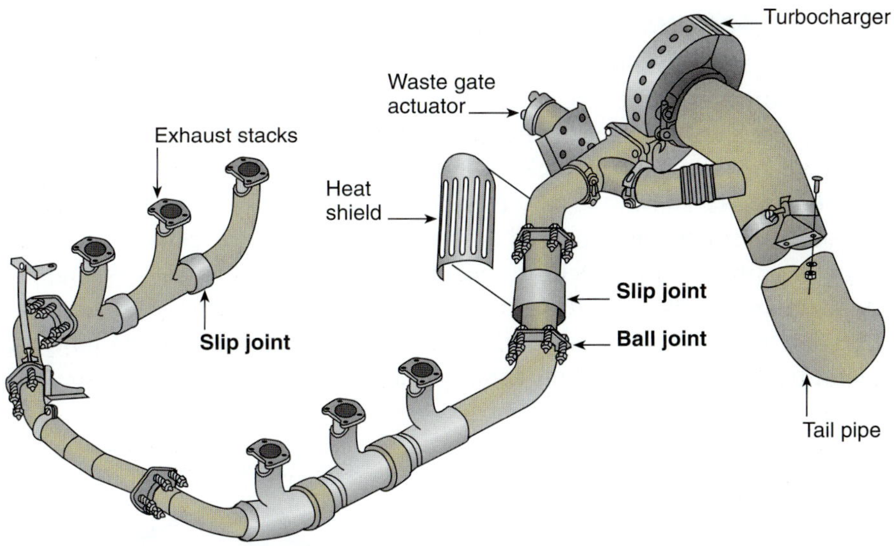

Figure 6-9. *The exhaust system of a turbocharged six-cylinder horizontally opposed engine*

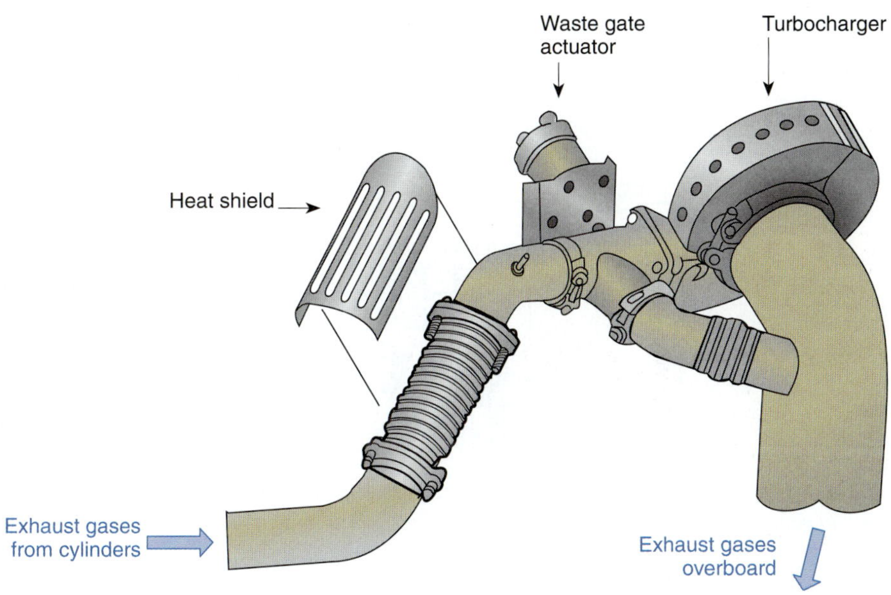

Figure 6-10. *Some turbocharged engines use corrugated metal bellows rather than slip joints in the exhaust system.*

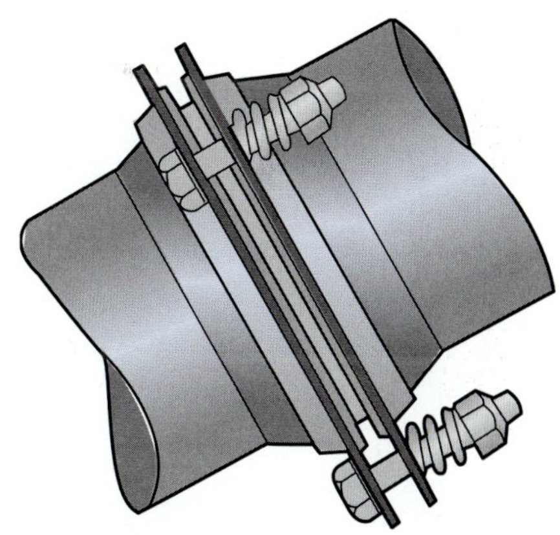

Figure 6-11. *A typical ball joint in the exhaust system of a turbocharged engine*

PRT (power recovery turbine). A turbine driven by exhaust gases from several cylinders of a reciprocating engine. Energy extracted from exhaust gases by the turbine is coupled, through a fluid clutch, to the engine crankshaft.

velocity turbine. A turbine driven by forces produced by the velocity, rather than the pressure, of gases flowing through the vanes.

Power Recovery Turbines

The 18-cylinder Wright R-3350 turbocompound engine is an example of the practical limit of power an aircraft reciprocating engine can develop. This engine, with the assistance of three power recovery turbines (PRTs), develops 3,700 horsepower with a weight of less than 2,800 pounds.

Figure 6-12 shows the way this system operates. Three velocity-type, two-stage turbines are mounted around the engine, with exhaust from six cylinders directed into each of the turbines. The turbine shaft is coupled to the engine crankshaft through a fluid coupling which prevents torsional vibration from being transmitted into the crankshaft, and under certain engine operating conditions, prevents the crankshaft from driving the turbine.

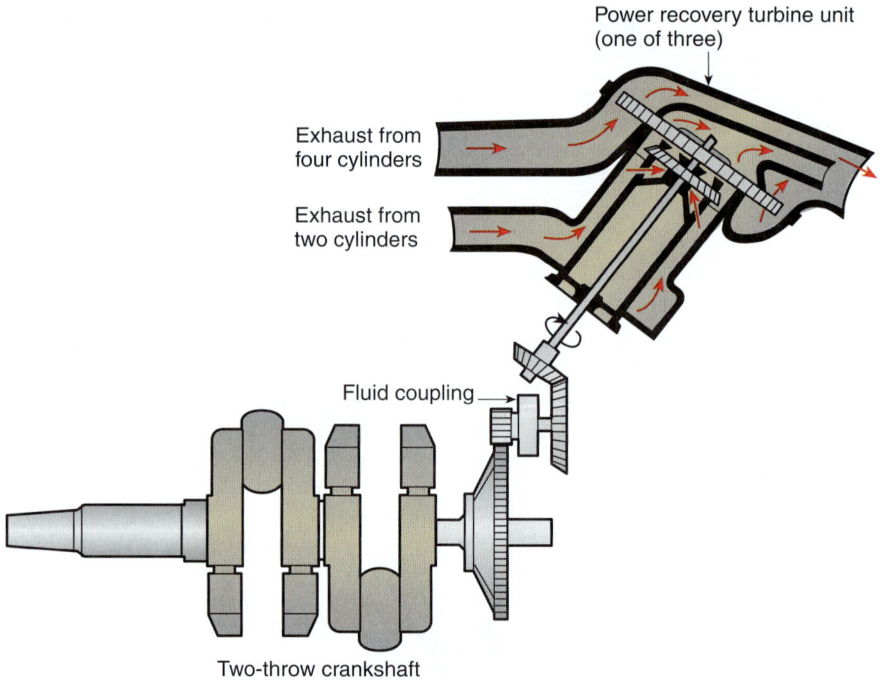

Figure 6-12. *One of the three power recovery turbines on a Wright turbocompound engine used to couple energy recovered from the exhaust gases to the engine crankshaft*

Exhaust System Inspection and Repair

The exhaust system of an aircraft engine is a critical system. Most of its parts are made of thin corrosion-resistant steel and are subjected to extremely high temperatures that change often and rapidly. Part of the system is connected to the shock-mounted engine, and in some installations part is connected to the stationary airframe. Vibrations inherent in this arrangement and sudden temperature changes create cracks that leak hot exhaust gases, causing a serious fire hazard. If cracks occur in the part of the exhaust system that provides heat for the aircraft cabin, the leaking gases will contain enough carbon monoxide to endanger the occupants.

Carbon Monoxide Detection

Aircraft whose cabin heat is supplied by the engine exhaust system should be equipped with carbon monoxide, CO, detectors. These are simply small containers of colored chemical crystals that change color in the presence of CO. By noting the color of the crystals, aircraft occupants can be warned of the presence of CO long before they could detect it by other means.

CO is a colorless, odorless, tasteless, unstable gas always present in exhaust gases of an aircraft engine. Hemoglobin is a part of our blood that carries oxygen from our lungs to our brain and to all of our organs that must have oxygen to function properly. But hemoglobin has a greater affinity for CO than it has for oxygen. If CO is present in the air, the hemoglobin will absorb it rather than the oxygen.

The effect of CO poisoning is cumulative; that is, breathing air that is even slightly contaminated with CO over a prolonged period of time is as bad as breathing a heavy concentration for a short period of time. Either will affect the ability to safely operate an aircraft.

Exhaust System Inspection

A critical part of every 100-hour and annual inspection is a careful and extensive examination of the entire exhaust system. This inspection requires knowledge and skill, as there are special precautions that must be taken. Before beginning the inspection, remove all of the heater muffs, or shrouds, so you can have unrestricted access to the components themselves.

Many exhaust system components on large engines operate at such high temperatures that they normally glow red hot. Any marks made on an exhaust system component with a lead pencil, or any zinc deposits left by contact with a piece of galvanized metal will cause the component to crack. The carbon or zinc will infuse into the steel at high temperature and weaken the metal enough that vibration will cause it to crack.

A leak in an exhaust system component shows up as feather-shaped streaks of dull light gray or sooty black. Leaks normally occur at a crack or where two mating components are not in perfect alignment.

Clamps used in an exhaust system should not be overtightened. When you discover a leak, loosen the clamps and realign the components, and then tighten the clamps only enough to remove the slack. If the clamp is too tight, the bolts may break when the components expand as they are heated. Many of these installations have specific torques that need to be adhered to.

Inspect the exhaust components on turbocharged engines with special care, paying close attention to the ball joints, the slip joints or bellows, and the heat shields.

As a general inspection, the discharge side of a vacuum cleaner can be used to pressurize an exhaust system to check for cracks or leaks. With pressure applied to the inside of the system, cover all areas of suspected leaks with a soap and water solution and watch for the appearance of bubbles. Carefully examine the areas around welds and around the heat-transfer rings or pins on the mufflers for possible cracks. Some installations require a much higher, and specific pressure for proper inspection. For this reason always consult the manufacturer's information for proper test pressures.

Exhaust System Repairs

Exhaust system components are made of special metals that retain their strength under extreme temperature and vibration. Cracks and other damage can render these components unairworthy. Generally, most maintenance performed is the replacement of damaged components with new units from the aircraft manufacturer. But because these components are normally very expensive, some may be repaired by an FAA-certificated repair station that specializes in exhaust systems. These shops have the welding equipment and skilled technicians that can repair components and approve them for return to service.

The installation of new or repaired components must be done in strict accordance with the aircraft service manual. New gaskets and the correct type of nuts should be used, and all nuts should be torqued to the values specified in the aircraft service manual.

STUDY QUESTIONS: RECIPROCATING ENGINE EXHAUST SYSTEMS

Answers are on Page 258. *Page numbers refer to chapter text.*

1. Approximately ________________ (what percent) of the heat energy in fuel burned is lost out the exhaust system. *Page 247*

2. Two problems caused by short exhaust stacks are:
   a. ____________________________
   b. ____________________________
   Page 247

3. The large pipe into which all exhaust gases from a radial engine flow is called a/an ______________________ . *Page 248*

4. The name of the exhaust system component through which exhaust gases leave the aircraft is the ____________________ . *Page 248*

5. Two uses for some of the heat taken from the exhaust gases are:
   a. ____________________________
   b. ____________________________
   Page 249

6. The flexible ducting used for carrying the heated air from the shroud around the muffler to the carburetor heat box is ________________ (black or red). *Page 250*

7. A stainless steel tube through which some of the exhaust gases pass after leaving the tail pipe is called a/an ___________________________ tube. *Page 251*

8. Two devices used to recover some of the energy from the exhaust gases are:
   a. ____________________________
   b. ____________________________
   Page 252

9. Two methods of allowing exhaust system components to move relative to each other without leaking exhaust gases are:
   a. ____________________________
   b. ____________________________
   Page 252

Continued

STUDY QUESTIONS: RECIPROCATING ENGINE EXHAUST SYSTEMS *Continued*

10. Energy extracted from exhaust gases by a power recovery turbine is coupled to the crankshaft through a ________________________ . *Page 254*

11. An exhaust system component must be protected from marks made by galvanized metal because the ____________________ deposits will diffuse into the metal and cause cracking. *Page 255*

12. Exhaust system components must never be marked with a lead pencil because the __________________ in the mark will diffuse into the metal and cause cracking. *Page 255*

13. The air pressure used to pressure-check an exhaust system for leaks may be produced by the discharge of a/an ________________________ . *Page 256*

14. Carbon monoxide detectors are packets of chemical crystals that change __________________ in the presence of CO. *Page 255*

Answers to Chapter 6 Study Questions

1. 50
2. a. excessive noise
   b. exhaust valves warpage
3. collector ring
4. tail pipe
5. a. cabin heat
   b. carburetor heat
6. red
7. augmentor
8. a. turbochargers
   b. power recovery turbines
9. a. slip joints
   b. metal bellows
10. fluid coupling
11. zinc
12. carbon
13. vacuum cleaner
14. color

COOLING SYSTEMS

7

Cooling Systems

7

Evolution of Reciprocating Engine Cooling Systems

As mentioned earlier in this text, a reciprocating engine is an inefficient machine. Only about one-third of the energy from fuel burned is converted into useful work. Half of the total energy is lost out the exhaust, and about 12% is lost through the cooling system.

Part of this heat is carried away by the lubrication system, as explained in Chapter 3. In this chapter, we discuss the cooling systems, which are designed solely to remove unwanted heat. The two types of cooling systems used are air cooling and liquid cooling.

Many early liquid-cooled engines had thin sheet metal jackets around the cylinders through which water flowed. This water absorbed heat from the cylinders and carried it into a radiator where it was transferred into the air. Water cooling was popular up through the 1920s on engines that had been produced several years earlier during the first World War. These engines required large radiators and lots of tubing and hoses, and the cooling system was one of the more troublesome systems on the airplanes. It leaked, and the large radiators created a tremendous amount of drag. But at the low speeds attainable by these airplanes, the high drag could be tolerated.

In an effort to make water cooling practical for high-speed airplanes, engineers experimented with skin radiators. These radiators used double sheets of thin brass over much of the wing surface, and the cooling water passed between the sheets to dissipate its heat.

They also experimented with steam cooling, or evaporative cooling. Cooling water was allowed to boil, and then the steam was condensed into a liquid. Condensing the steam released much more heat than passing hot water through a radiator.

Air cooling promised many advantages over water cooling, as no radiator was needed and there was no complicated plumbing. Fins were cast or machined on the cylinder barrels and heads, but at the slow speed flown by these aircraft, not enough air flowed over the engine to carry away the heat.

The potential advantages of air cooling were instrumental in the development of the rotary radial engine. In these engines, the cylinders themselves spin around the crankshaft at about 1,200 RPM, and this rotation produces sufficient airflow to remove the unwanted heat.

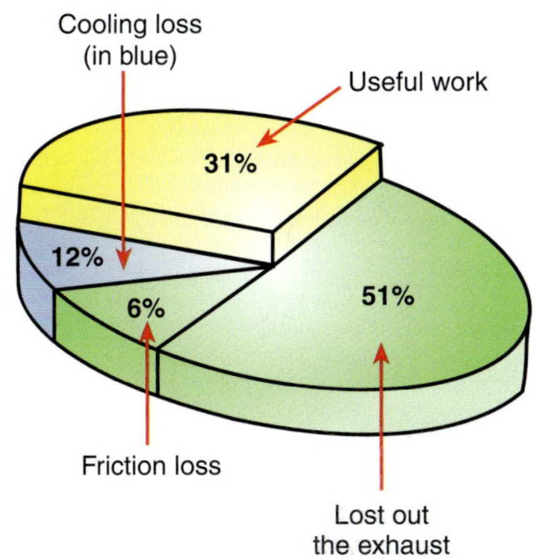

Figure 7-1. *Approximately 12% of the heat energy in fuel burned in an aircraft engine is carried away by the cooling system.*

skin radiator. A type of radiator used on some early liquid-cooled racing airplanes. The radiator was made of two thin sheets of brass, slightly separated so the heated coolant could flow between them. Skin radiators were mounted on the surface of the wing, on the sides of the fuselage, or on the floats of seaplanes. Air flowing over the smooth surface of the radiator removed heat from the coolant.

steam cooling. A method of liquid cooling in which the coolant, normally water, is allowed to absorb enough heat that it boils. The steam gives up its heat when it condenses back into a liquid.

air cooling. The removal of unwanted heat from an aircraft engine by transferring the heat directly into the air flowing over the engine components.

It was only when the successful design of an air-cooled cylinder made the famous Wright J-5 Whirlwind engines possible that air-cooled engines became truly reliable. In 1927, these engines made flights across both the Atlantic and Pacific oceans possible, and caused aviation to become accepted as a dependable means of transportation.

baffle. A thin sheet metal shroud or bulkhead used to direct the flow of cooling air between and around the cylinder fins of an air-cooled reciprocating engine.

turbocompound engine. A reciprocating engine that has power recovery turbines in its exhaust system. The power extracted from the exhaust by these turbines is directed into the engine crankshaft through a fluid coupling.

Air-Cooled Engines

The vast majority of reciprocating engines in the modern general aviation fleet are air cooled. The cylinders have evolved with efficient fin designs, but the cowling and baffles are also important components that make air cooling efficient and successful.

Fins

Most of the heat in an engine cylinder is in the area of the head, and the amount of heat that can be removed by air cooling is affected by the amount of exposed surface area.

Almost all air-cooled engines have alloy steel cylinder barrels with fins machined onto their surface. This barrel is screwed into a cast aluminum alloy head that has fins cast into it. Fins on the early engines were coarse and had a relatively small amount of surface area, but as casting and machining techniques have improved, fins have been made thinner and deeper with a marked increase in surface area.

The 18-cylinder Wright R-3350 engine represents the peak in power per cylinder, with the turbocompound version of this engine producing more than 200 horsepower per cylinder for takeoff. The cylinder heads on these engines have extremely deep, thin fins. Finned muffs made of thin aluminum alloy have been shrunk around the steel cylinder barrel to dissipate the heat. *See* Figure 7-2.

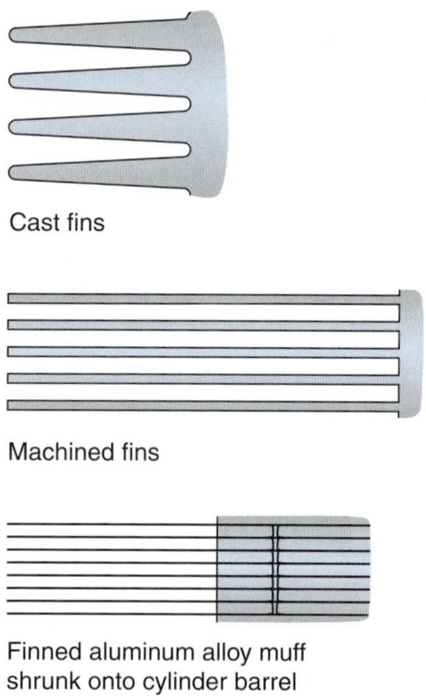

Figure 7-2. *Relative size of the cooling fins for air-cooled cylinders*

Cowling

The cylinders of early air-cooled engines were not cowled and were cooled by the air flowing over them. The cooling was generally adequate, but as airspeeds increased beyond about 120 miles per hour, the drag penalty proved excessive.

Radial Engine Cowling

By 1929 the Townend ring, or speed ring, was accepted as a drag-reducing expedient. This narrow-chord ring around the outside of the cylinders has an airfoil-shaped cross section that produces a forward aerodynamic force, as well as improves the uniformity of air flowing through the cylinder fins. A properly designed and installed Townend ring decreased the drag by about 11%.

cowling. The removable cover that encloses an aircraft engine.

Townend ring. A type of ring cowling used over a single-row radial engine. The cross section of the ring is in the form of an airfoil that produces enough forward thrust to compensate for the cooling drag of the engine. In the United States, Townend rings are often called speed rings.

The NACA (National Advisory Committee for Aeronautics) developed a cowling that, like the Townend ring, has an airfoil shape, but has a long chord and extends back behind the cylinders and fairs into the fuselage or the engine nacelle. A narrow gap between the cowling and the nacelle or fuselage allows the cooling air to exit uniformly around the engine. The aerodynamic shape of an NACA cowling produces enough forward thrust that it over-comes the cooling drag. NACA cowlings became the standard, and have been used on almost all radial engines installed in fighters, bombers, and transport aircraft.

The amount of cooling air that flows through the cylinder fins of a radial engine enclosed in an NACA cowling is controlled by adjustable cowl flaps that cover the exit gap between the rear of the cowling and the nacelle or fuselage. For all ground operation, the flaps should be wide open, and in flight, they should be closed. Some larger installations control the cowl flaps automatically to maintain the cylinder head temperature within the desired range.

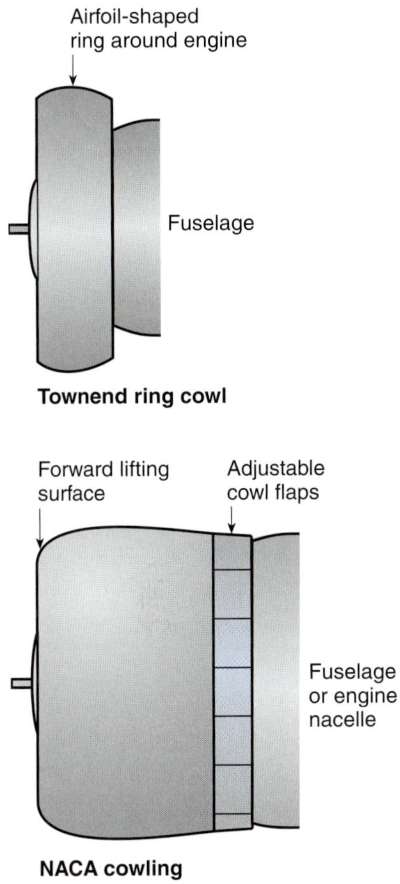

Figure 7-3. *Typical cowlings for radial engines*

Horizontally Opposed Engine Cowling

Like the radial engine, the first air-cooled horizontally opposed engines had the cylinders sticking out into the air stream. Most smaller opposed engines had a hood, or scoop, over the top of the cylinders on each side of the engine that collected the air and forced it downward through the cylinder fins.

As opposed engines became more powerful and airplanes became faster, more efficient cooling was required, which resulted in the development of a pressure cooling system. *See* Figure 7-4.

The cowling around a modern horizontally opposed engine is divided into two compartments. The compartment above the engine is formed by a vertical baffle, or bulkhead, at the rear of the engine and a horizontal baffle approximately along the center line of the cylinders. Both the vertical and horizontal baffles are sealed against the cowling with a strip of rubber or rubberized fabric. Air enters the upper compartment through large, forward-facing air inlets at a pressure above ambient, because of the ram effect from the forward speed of the aircraft and the acceleration of the air by the propeller. A low pressure below the engine caused by air flowing over the flared cooling air exit or over the adjustable cowl flaps pulls the air downward into the lower compartment through the cylinder fins. Close-fitting intercylinder baffles and shrouds around the cylinders force all the air to flow through the fins so it can remove the maximum amount of heat.

NACA. National Advisory Committee for Aeronautics. This organization, dedicated to the technical development of aviation, has been superseded by NASA, the National Aeronautics and Space Administration.

nacelle. An enclosed compartment, normally in the leading edge of the wing, into which an aircraft engine is mounted.

NACA cowling. A long-chord ring cowling whose trailing edge fairs smoothly into the fuselage or engine nacelle. NACA cowlings are used on the vast majority of radial engine installations.

Airflow over the cowling below the engine is deflected by the adjustable cowl flap. This produces the low pressure inside the cowling determined by the amount the flap is open. The more the flap is deflected, the lower the pressure and the more air that flows through the fins. Cowl flaps are opened wide while the aircraft is on the ground and closed in flight to maintain the cylinder head temperature within the desired limits.

pressure cooling. A method of air cooling an engine in which the cylinders are enclosed in tight-fitting shrouds. Ram air is directed into one compartment and the pressure in the other is decreased by air flowing over a flared exit or adjustable cowl flaps. The pressure difference across the cylinders causes cooling air to be drawn through the fins to remove the unwanted heat.

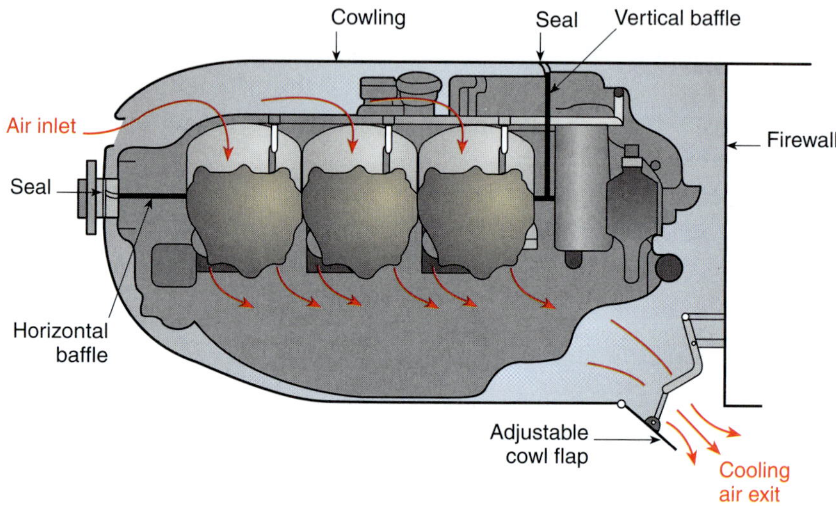

Figure 7-4. *Pressure cooling system for a horizontally opposed engine*

Augmentor Tubes

Augmentor tubes, such as shown in Figure 6-8 on Page 251, are used to increase the flow of cooling air through the cylinder fins. Engines using this form of cooling have two exhaust tail pipes, one for the cylinders on each side of the engine. Around the end of each tail pipe is a long stainless steel augmentor tube. These tubes have a slightly flared opening on the end inside the cowling, and the other end extends outside the aircraft. They act as a venturi converting the high velocity of the exhaust gases into a low pressure inside the cowling below the engine. This low pressure improves the cooling by increasing the airflow through the cylinder fins.

Liquid-Cooled Engines

Water-cooled engines have two drawbacks: they require a large radiator and they limit the temperature the engine can produce. These drawbacks have been overcome by using ethylene glycol or a mixture of ethylene glycol and water as the coolant. Pressurizing the system increases the temperature and allows a smaller radiator to be used.

liquid cooling. The removal of unwanted heat from an aircraft engine by transferring the heat into a liquid and then passing the heated liquid through a liquid-to-air heat exchanger (radiator) to transfer the heat into the ambient air.

ethylene glycol. A form of alcohol used as a coolant for liquid-cooled aircraft engines. It is also used in automobile engines as a permanent antifreeze.

During World War II, the United States favored air-cooled radial engines for most of the bombers and transport aircraft, and many of the fighters, but two of the most effective fighters, the Lockheed Lightning and the North American Mustang, used liquid-cooled V-12 engines. The British and Germans made extensive use of liquid-cooled V-type engines that were cooled with a mixture of ethylene glycol and water under pressure. Heat from the cylinders was transferred into the air through radiators that were enclosed in streamlined ducts. The temperature was controlled by a thermostatic valve in the water line between the coolant pump and the radiator. When the temperature was too high, the coolant was routed through the radiator, but when it was low enough, the thermostatic valve routed the coolant back through the engine.

The vast majority of modern general aviation engines are air cooled, but the Teledyne-Continental Voyager series is reviving interest in liquid cooled engines.

The Voyager engines use alloy steel cylinder barrels screwed into cast aluminum alloy heads. A mixture of 60% glycol and 40% water flows at a high velocity through coolant passages in the cylinder head. The cylinder walls and pistons are cooled by a high-volume spray of oil directed against the undersides of the pistons. A Voyager engine can be installed in a closed cowling with a small duct for the induction air to the turbocharger, and with flush inlets for the oil cooler and a turbocharger intercooler. The small radiator required for this engine can be installed in the aft portion of the engine nacelle with cooling air ducted through it.

Voyager engines exploit some of the more important advantages of liquid cooling. These are:

- *Improved cooling.* The even distribution of the coolant between all cylinder heads reduces the temperature differences that are characteristic of air-cooled engines. Prolonged fast idling on the ground does not cause the engine to overheat.
- *Reduced cooling drag.* A well designed, small water/glycol radiator can be placed at any location on the aircraft, with cooling air ducted through it. This allows the basic cowling to be sealed.
- *Lower fuel consumption.* Even and efficient cooling allows the engine to be operated at all times with an economical mixture ratio rather than having to use an overly rich mixture to remove heat for high-power operations.

Continued

- *Better wear characteristics.* Even cooling allows component parts to be built with smaller clearances that improve their life. Cylinder heads are cooled by a high-velocity flow of coolant while pistons and cylinder barrels are cooled by a high-volume spray of oil.
- *High altitude capability.* The aircraft can be held at high altitude for a longer period of time and then let down with a steep descent without subjecting the cylinders to the thermal shock that causes so much damage to air-cooled cylinders.

Cooling System Inspection and Maintenance

Modern air-cooled engines are designed to extract the maximum amount of energy from fuel burned. To do this, the cooling system must function as designed. Any broken fins or missing baffles or seals can cause a cooling loss that creates hot spots in a cylinder and causes detonation, which can destroy the engine.

When inspecting the cooling system, remove the cowling and carefully examine all cylinders for any sign of burned paint. This could indicate a local hot spot which could cause or be the result of detonation.

Check the rear vertical baffle and the horizontal baffles for any missing sections, cracks, or sealing strips that are not folded upward or forward.

The cowl flaps should operate freely and through their complete range of travel. There should be no looseness caused by worn hinge pins or hinge brackets.

The condition of the fins on any air-cooled cylinder is extremely important. Small cracks may be found, and if these do not extend into the head itself, they may be stop-drilled at their end to prevent further cracking. Rough edges on a fin may be removed with a fine file to prevent any cracks from forming. Before removing any fin area by filing, or profiling, as this procedure is called, consult the engine service manual to determine the maximum amount of fin area the manufacturer allows to be removed.

Bent cast aluminum fins that do not prevent the airflow reaching the cylinder should not be straightened, as there is too much danger of breaking the brittle fin.

STUDY QUESTIONS: RECIPROCATING ENGINE COOLING SYSTEMS

Answers are on Page 268. Page numbers refer to chapter text.

1. The two types of systems used to remove excess heat from an aircraft engine are:
   a. ______________________________
   b. ____________________________
   Page 261

2. The area from which heat can be dissipated from a cylinder is increased by _________________ that are cast or machined onto the surface. *Page 261*

3. Three components of an air-cooled engine installation must work together for most efficient cooling. These are:
   a. ______________________________
   b. ______________________________
   c. ______________________________
   Page 262

4. The cylinder head of most air-cooled engines is made of ______________________________ . *Page 262*

5. The cylinder barrels of the Wright Turbocompound engine dissipated heat from the cylinder barrel through finned muffs made of __________________________ that were shrunk around the steel cylinder barrel. *Page 262*

6. The most popular type of cowling for radial engines is the ____________________ cowling. *Page 263*

7. The amount of air flowing through an NACA cowling is controlled by adjustable ______________________ over the air exit gap. *Page 263*

8. For all ground operations, the cowl flaps should be fully _________________ (open or closed). *Page 263*

9. The type of cooling used by most modern air-cooled horizontally opposed engines is ______________________ cooling. *Page 263*

10. The cowling for a horizontally opposed engine is divided into two compartments by_________ . *Page 263*

11. Air is forced to flow through the cylinder fins by close-fitting shrouds and ______________________ baffles. *Page 263*

12. Air is prevented from leaking past the baffles by a flexible seal made of __________________ . *Page 263*

Continued

STUDY QUESTIONS: RECIPROCATING ENGINE COOLING SYSTEMS *Continued*

13. The pressure of the air below the engine in a pressure cowling is controlled by the ________________ . *Page 264*

14. Opening the cowl flaps in flight ________________ (increases or decreases) the air pressure below the engine. *Page 264*

15. An augmentor tube increases the flow of cooling air through the engine cylinder fins because of low pressure caused by the high-velocity ________________ gases flowing through the tube. *Page 264*

16. The coolant used in modern liquid-cooled engines is ________________ . *Page 265*

17. The pistons and cylinder barrels of a TCM Voyager engine are cooled with ________________ (ethylene glycol or oil). *Page 265*

18. Burned paint on a cylinder can be an indication of ________________ . *Page 266*

19. The rubber sealing strip on the vertical baffle on a horizontally opposed engine installation should fold ________________ (forward or backward). *Page 266*

20. Small cracks in a cylinder head fin can be prevented from continuing by ________________ the end of the crack. *Page 266*

21. The maximum amount of fin area that can be removed from a cylinder is specified in the engine ________________ . *Page 266*

22. Bent cast aluminum cylinder head fins ________________ (should or should not) be straightened. *Page 266*

Answers to Chapter 7 Study Questions

1. a. air cooling
   b. liquid cooling
2. fins
3. a. fins
   b. cowling
   c. baffles
4. cast aluminum alloy
5. aluminum alloy
6. NACA
7. cowl flaps
8. open
9. pressure
10. baffles
11. intercylinder
12. rubber
13. cowl flaps
14. decreases
15. exhaust
16. ethylene glycol
17. oil
18. detonation
19. forward
20. stop-drilling
21. service manual
22. should not

Starting Systems

8

STARTING SYSTEMS

8

The Evolution of Reciprocating Engine Starting Systems

Early aircraft reciprocating engines were started by pulling the propeller through by hand until the cylinders began to fire. The engine was prepared by pulling the propeller through two or three revolutions with the ignition switch OFF to be sure that no cylinders had filled with oil leaking past the rings. Then the hand primer pump was worked the specified number of strokes to spray raw fuel into the induction pipes. The aircraft wheels were chocked, and when everything was ready and the controls were placed in the correct position, the person at the propeller called "Contact." The person in the cockpit answered "Contact," and then turned the ignition switch to BOTH. The propeller was pulled sharply to rotate the engine through a compression stroke, and hopefully the cylinder fired and the engine started.

Hand-propping is dangerous if not done correctly, and as engines grew in size and compression ratios increased, two or three persons were required to cooperate in pulling the propeller through so that no one fell into it as the engine started.

"Contact." The term used between a person hand-propping an aircraft engine and the person in the cockpit. When the person is ready to spin the propeller, he calls "contact." The person in the cockpit turns on the fuel, slightly opens the throttle, applies the brakes, and replies "contact," and then turns the ignition switch to BOTH. The propeller is then pulled through to start the engine.

Hand-starting large engines was sometimes accomplished with bungee starters, a length of bungee cord (from 10 to 20 feet, depending upon the aircraft) attached to a leather or canvas sock. The engine was prepared for starting and the propeller was stopped horizontally. The sock was placed over the tip of one propeller blade and the bungee cord was stretched, parallel to the propeller blades, over the propeller hub and pulled by several persons, or was attached to a ground vehicle and pulled. When the cord was stretched tight, the propeller was flipped just enough to allow the bungee cord to pull on the propeller blade and rotate the engine, causing several cylinders to fire.

bungee cord. An elastic cord made of small strips of rubber encased in a loosely braided cloth tube that holds and protects the rubber, yet allows it to stretch. The energy in a stretched bungee cord may be used to crank a large aircraft engine.

Some engines were started by priming a cylinder with gasoline or ether, placing the piston just past top center on the power stroke, and cranking a booster magneto that directed high-voltage alternating current to the spark plug. This caused a hot and retarded spark that ignited the fuel-air charge and started the crankshaft rotating.

ether. A volatile, highly flammable liquid that may be used to prime the cylinders of an aircraft engine when starting under extremely cold conditions.

Compressed air was used to start some engines. An engine-driven air compressor charged a storage bottle with compressed air. When the engine was ready to be started, the starter handle in the cockpit was pulled and the compressed air was discharged. This air flowed through a distributor valve and check valves into the cylinders just as the pistons passed top center on the power stroke. The flow of air picked up some liquid gasoline and sprayed it into the cylinder to provide a rich mixture for starting.

cartridge starter. A self-contained starter used on some military aircraft. A cartridge similar in size to a shotgun shell is ignited in the starter breech. The expanding gases drive a piston attached to a helical spline that converts the linear movement of the piston into rotary motion to rotate the crankshaft.

inertia starter. A starter for a large reciprocating engine that uses energy stored in a rapidly spinning flywheel to turn the crankshaft.

series-wound motor. An electric motor with field coils connected in series with the armature.

Many carrier-borne Naval fighters of World War II were equipped with cartridge starters. These lightweight, self-contained starters used a cartridge similar in size and appearance to a large shotgun cartridge. To start the engine, the cartridge was put into the breech of the starter and the trigger was pulled. Powder was ignited, and as it burned, it generated enough pressure to force down a piston that drove a helical spline to convert the piston's linear motion into rotary motion with enough torque to turn the engine through several revolutions.

The first starters to gain popularity were inertia starters. A series-wound DC electric motor spun a heavy flywheel inside the starter. These starters were controlled by a two-position switch. For normal operation the switch was placed in the ENERGIZE position to bring the flywheel up to speed. When it reached its peak speed, the switch was moved to the ENGAGE position, and a solenoid moved a ratcheting jaw forward to engage a mating jaw on the rear of the crankshaft. A spring-loaded clutch slipped enough to absorb the shock when the two sets of jaws engaged. As soon as the engine started, the starter jaw no longer drove the engine, and the ratcheting action removed the pressure between the jaws. With this pressure gone, a spring pulled the starter jaw back and disconnected it from the engine.

If the battery was too low to spin the starter motor, a hand crank could be inserted into a receptacle and the flywheel brought up to speed by hand. When the speed was high enough, the crank was removed and a hand toggle was pulled to engage the starter jaw with the engine. Starters energized by electricity and by hand are called hand/electric inertia starters.

By far the most popular starters since World War II have been direct-cranking electric starters.

Electric Starters for Large Engines

planetary gears. A type of large-ratio reduction gearing. A series of small planetary gears are mounted on a spider attached to the output shaft. The planetary gears rotate between a fixed sun gear and a driven ring gear.

helical spline. A spline that twists, or winds, around the periphery of a shaft. Helical splines are used to change linear motion into rotary motion of the shaft on which the splines are cut.

By the end of World War II, series-wound electric motors had been developed that were strong enough to give a lightweight motor sufficient torque to crank a large engine directly through a set of planetary reduction gears.

When the starter switch was moved to the START position, the armature began to spin, and a ratcheting jaw moved forward along a set of helical splines to engage the jaw on the crankshaft. A heavy-duty clutch between the jaws and the starter housing absorbed the shock caused by the sudden application of torque.

As soon as the engine started, the starter switch was released and the motor stopped turning. Pressure between the jaws ceased, and a spring pulled the starter jaw back, disengaging it from the engine.

Electric Starters for Small Engines

The horizontally opposed engines used in the general aviation fleet are usually started with direct-cranking electric starters. Three types of starters are commonly used: starters with overrunning clutches, starters with Bendix drives, and starters with right-angle adapter drives.

Starters with an Overrunning Clutch

One popular starter for some smaller Continental engines is seen in Figure 8-1. This starter uses a series-wound motor with a long-splined armature shaft that allows the overrunning clutch assembly to remain engaged with the motor as it is moved outward. A shift lever, operated by a toggle and cable or by a long-throw solenoid, moves the overrunning clutch assembly outward until the pinion slides into engagement with a starter gear in the accessory case of the engine. Continued movement of the shift lever compresses the meshing spring (*see* Figure 8-2) and allows the stud on the shift lever to depress the starter switch and energize the motor. The armature speed is geared down

overrunning clutch. A type of clutch that couples an input shaft with an output shaft. When the input shaft is driven, the output shaft rotates with it. But when the output shaft is driven, the input shaft does not turn.

toggle. A T-shaped handle fitted onto the end of a cable used to engage a simple starter with an overrunning clutch.

pinion. A small gear that meshes with and drives a larger gear.

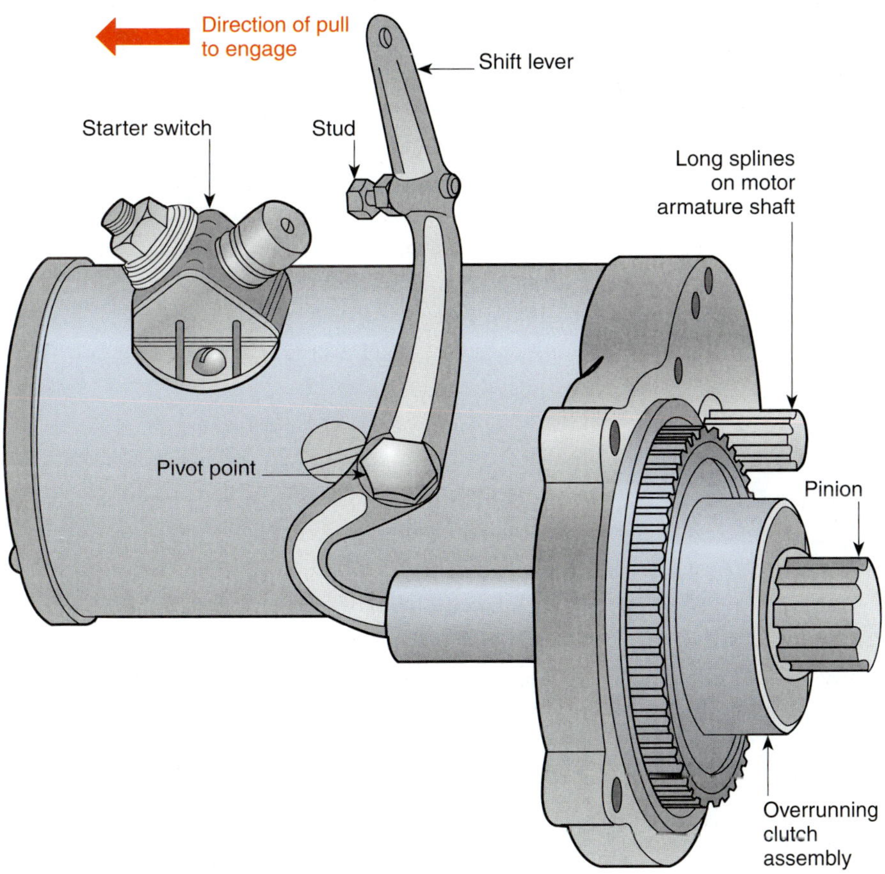

Figure 8-1. *An overrunning clutch-type starter used on some small horizontally opposed Continental engines*

by the large gear on the overrunning clutch and by the gear ratio between the pinion and the starter gear. This speed reduction provides enough torque to allow the small motor to rotate the crankshaft and start the engine.

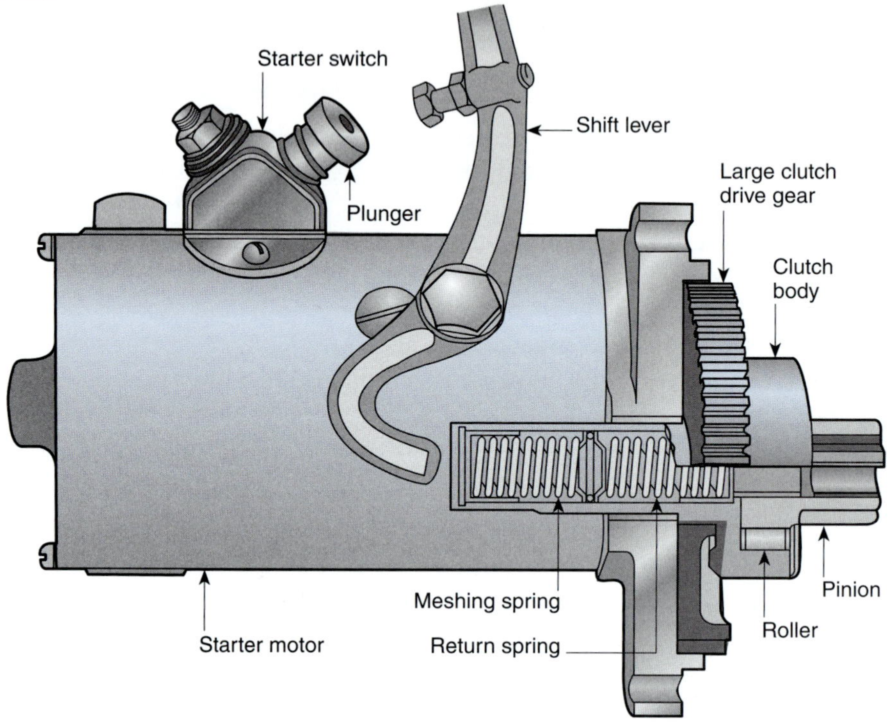

Figure 8-2. *The meshing spring forces the pinion into the starter gear inside the engine and then allows the shift lever to move and depress the plunger on the starter switch. The return spring pulls the pinion away from the starter gear when the shift lever is released.*

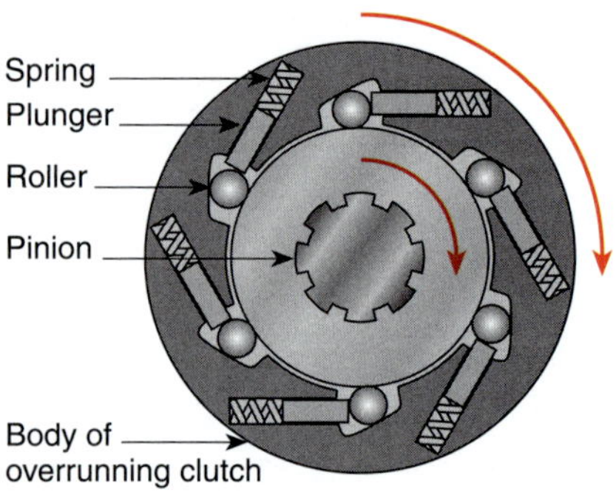

Starter motor is driving the body of overrunning clutch through a large gear that is integral with body. Rollers wedge between body and pinion to force the pinion to turn starter gear inside engine.

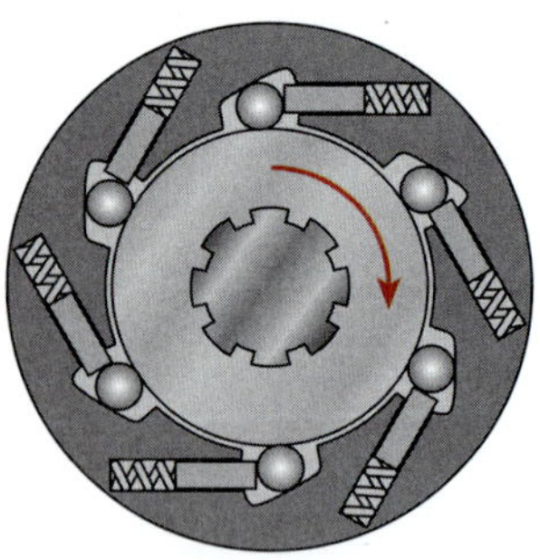

Engine has started and is driving pinion. Rollers are forced back against plungers and springs. Body of clutch and motor armature are not turning.

Figure 8-3. *Principle operation of an overrunning clutch*

As soon as the engine starts, the overrunning clutch, similar in principle to the one in Figure 8-3, disconnects the pinion from the drive gear. When the toggle is released or the solenoid is de-energized, the return spring pulls the pinion out of mesh with the starter gear.

Starters with Bendix Drive

Many Lycoming engines have a large starter gear around the propeller end of the crankshaft. The starter uses a Bendix drive to engage and disengage. *See* Figure 8-4.

The starter motor is series-wound, or compound-wound, and the Bendix drive unit is mounted on its shaft. The Bendix drive pinion fits loosely around the drive sleeve, which is turned by the armature shaft or spur-gear reduction and has external helical splines that mate with the internal splines in the drive pinion.

When the starter switch is closed, the armature begins to rotate. The drive pinion does not accelerate as fast as the armature, so the helical splines force the pinion to move endwise on the shaft and to mesh with the teeth on the starter gear. As soon as the pinion contacts the pinion stop, it begins to rotate with the armature and crank the engine.

When the engine starts, the starter gear drives the pinion and spins it faster than the armature. The Bendix drive spring and the helical splines pull the pinion back out of mesh with the starter gear. The antidrift spring prevents vibration from causing the pinion to move forward on the sleeve and contacting the starter gear when the engine is running.

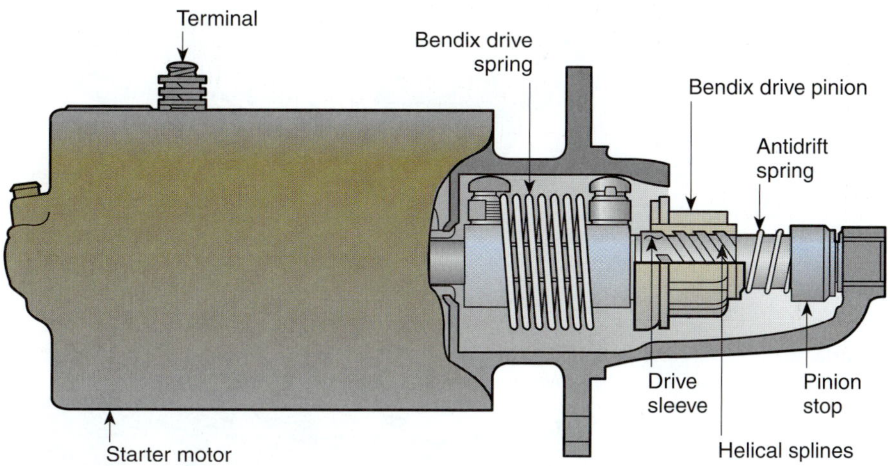

Figure 8-4. *An aircraft engine starter using a Bendix drive*

Starters with Right-Angle Drive Adapter

Some larger Teledyne-Continental horizontally opposed engines use a starter with a right-angle drive adapter. Figure 8-5 shows how this starter works, and Figure 8-6 is an exploded view of the starter motor and drive adapter.

The series-wound starter motor drives a worm gear that meshes with a worm-gear wheel, which rides over, but is not connected to, the starter shaft.

One end of the heavy helical spring is attached to the worm-gear wheel, while the other end is unconnected, but rides inside a close-fitting steel sleeve inside the starter adapter housing. *See* Figure 8-6. A knurled starter clutch drum that is part of the starter shaft rotates freely inside the spring.

When the starter switch is closed, the worm gear drives the worm-gear wheel and the helical spring. Friction between the outside of the spring and the inside of the sleeve in the adapter housing causes the spring to wind up, decreasing its diameter enough that the outside pulls away from the sleeve and the inside grips the starter shaft drum. This locks the starter shaft to the worm-gear wheel, and turns the crankshaft.

helical spring. A spring wound in the form of a helix, or coil.

As soon as the engine starts, the starter shaft turns faster than the worm-gear wheel, and the spring unwinds, increasing its diameter so that it no longer contacts the clutch drum.

The pulley for the generator is keyed to the end of the starter shaft, which is geared to the crankshaft and rotates when the engine is running.

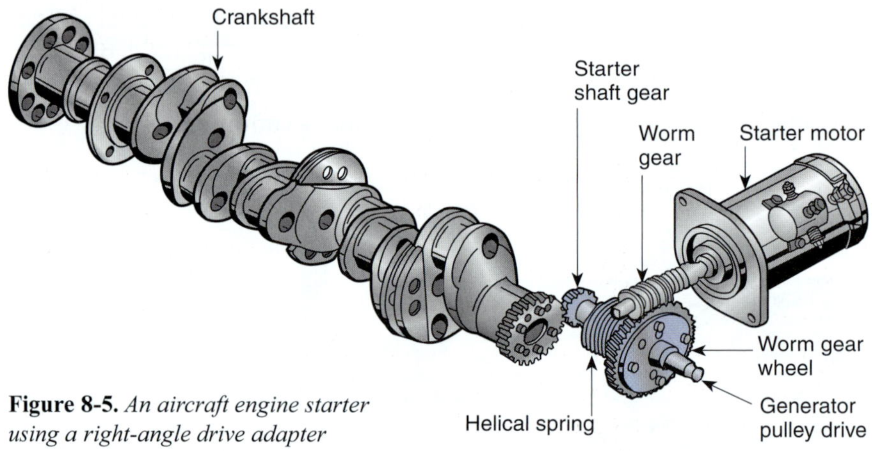

Figure 8-5. *An aircraft engine starter using a right-angle drive adapter*

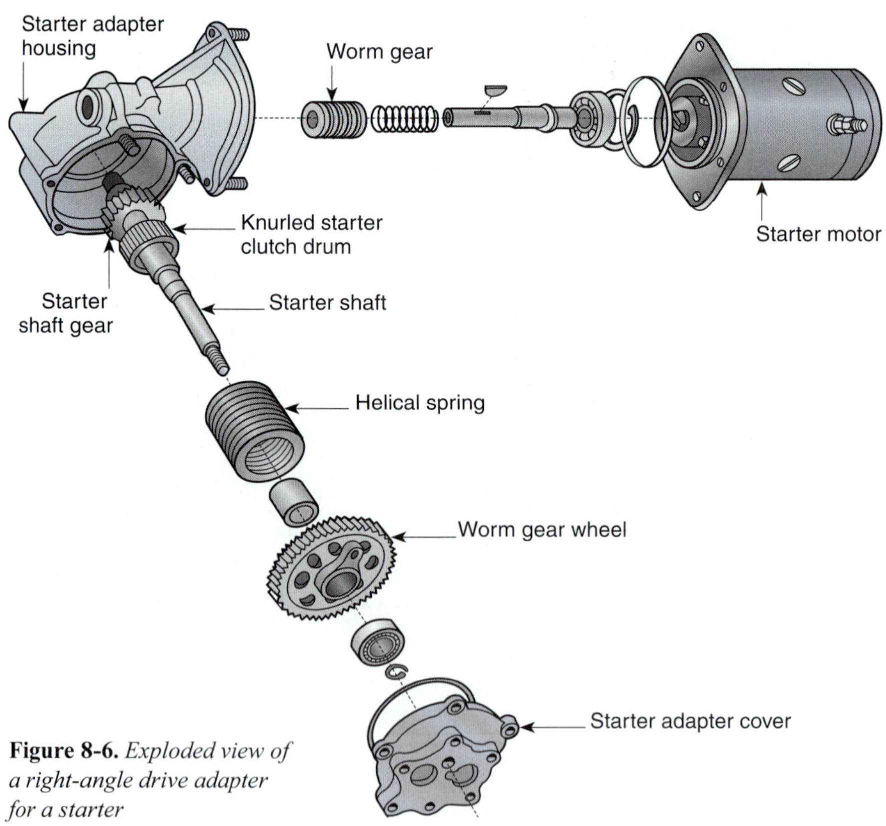

Figure 8-6. *Exploded view of a right-angle drive adapter for a starter*

Electric Starter Troubleshooting and Maintenance

Starters are relatively simple devices and when properly maintained, they create few problems. Starters, like all other engine accessories, are normally overhauled at the time the engine is overhauled.

The most common operational problems with aircraft starters are associated with the power circuit. Starters require so much current that the battery must be fully charged, and all connections between the battery and starter terminal must be clean, tight, and free of corrosion. The battery ground connection must also be clean and tight.

Starter solenoids normally resemble battery contactors, but when replacing a solenoid, be sure to use only one having the correct part number. Starter solenoids are intermittent duty devices, whereas battery contactors are designed for continuous operation.

A low battery is indicated by a clicking sound or chattering noise coming from the starter solenoid. The battery has enough power to close the solenoid, but as soon as the starter is connected and takes its current, the voltage drops too much to supply enough current to hold the solenoid energized, and it opens, disconnecting the starter. The voltage rises and the solenoid again closes, repeating the cycle and producing the chattering.

The starter switch is sometimes included in the ignition switch. The popular "Shower of Sparks" ignition system has a switch with five positions: OFF, LEFT, RIGHT, BOTH, and a spring-loaded START position. The main starter circuit does not incorporate any fuse or circuit breaker, but the starter solenoid coil receives its power through a circuit breaker, often from the one used for instrument lights or cabin lights. Check the aircraft's wiring diagrams to determine which circuit breaker is used.

Shower of Sparks ignition system. A patented ignition system for reciprocating engines. An induction vibrator sends pulsating direct current into a set of retard breaker points in one of the magnetos. This provides a hot and retarded spark for starting the engine.

Starter motors are all series-wound to provide the highest possible starting torque. The commutator and brushes of these high-current motors can become worn and pitted. Examine the brushes. They should be replaced when they have worn to approximately one-half of their original length. The commutator should be smooth and free of oil. Some armatures for smaller starters are supported in bronze bushings that are subject to wear. Worn bushings can cause the armature to drag against the pole shoes and prevent the motor from running as it should. Other armature shafts are supported in ball and/or roller bearings.

commutator. The component on the armature of an electric motor that changes the direct current input to the armature into alternating current in the armature coils. The brushes that bring the current into the armature ride on the commutator.

As with any aircraft maintenance, overhaul of a starter should be done in accordance with the manufacturer's service manual, and all replacement parts should have the part number specified for the particular starter. Parts for automobile starters may look like the ones in an aircraft starter, but they are not approved for installation on an FAA-certificated engine.

STUDY QUESTIONS: RECIPROCATING ENGINE STARTING SYSTEMS

Answers are on Page 280. *Page numbers refer to chapter text.*

1. The first popular aircraft engine starter was the ________________ starter. *Page 272*

2. An inertia starter cranks an engine by energy stored in a rapidly spinning ______________ . *Page 272*

3. The shock that occurs when an inertia starter engages the crankshaft is minimized by the use of a spring-loaded ____________________ . *Page 272*

4. An inertia starter is coupled to the crankshaft through a set of ratcheting _____________ . *Page 272*

5. Electric motors used in an aircraft engine starter are _________________ (series, shunt, or compound) -wound. *Page 272*

6. Three types of electric starters for general aviation aircraft are starters with:
   a. __________________________
   b. __________________________
   c. __________________________
   Page 273

7. A starter with an overrunning clutch is engaged by a long-throw solenoid or a ________________ and ________________ . *Page 273*

8. In a starter with an overrunning clutch, the starter motor drives the ____________________ (clutch body or pinion). *Page 274*

9. When the toggle for a starter with an overrunning clutch is released, the ____________________ spring pulls the pinion out of mesh with the starter gear. *Page 274*

10. The starter gear for a Lycoming engine using a starter with a Bendix drive is _______________________ (inside the engine or external). *Page 274*

11. The Bendix drive pinion is moved forward when the motor starts by helical splines on the _____________________ . *Page 274*

12. The Bendix drive pinion rotates with the armature as soon as the pinion contacts the ____________________ . *Page 275*

13. The Bendix drive pinion is prevented from moving into contact with the starter gear when the engine is running by the _____________________. *Page 275*

14. High torque is produced in a starter with a right-angle drive by the starter motor driving the starter gear through a _________________________ on the starter motor shaft. *Page 275*

15. The torque from the starter motor with a right-angle drive adapter is applied to the crankshaft by the spring winding up and gripping the knurled starter clutch _______________ on the starter shaft. *Page 275*

16. When an engine equipped with a starter having a right-angle drive adapter starts, the spring ____________________ (winds or unwinds), and its inside diameter increases enough that it releases its grip on the drum on the starter shaft. *Page 276*

17. A pulley on the end of the starter shaft of a right-angle drive adapter drives the ___________________ . *Page 276*

18. The main electrical circuit for a starter motor ________________ (does or does not) have a fuse or circuit breaker. *Page 277*

19. A starter solenoid is a _____________________ (continuous or intermittent) -duty device. *Page 277*

20. A starter solenoid coil normally ________________ (does or does not) have its own circuit breaker. *Page 277*

21. The starting torque of a series motor is _________________ (high or low). *Page 277*

22. The brushes in a starter motor are normally replaced when they have worn to _______________ (what portion) of their original length. *Page 277*

Answers to Chapter 8 Study Questions

1. inertia
2. flywheel
3. clutch
4. jaws
5. series
6. a. an overrunning clutch
   b. a Bendix drive
   c. a right-angle drive adapter
7. toggle and cable
8. clutch body
9. return
10. external
11. drive sleeve
12. pinion stop
13. antidrift spring
14. worm gear
15. drum
16. unwinds
17. generator
18. does not
19. intermittent
20. does not
21. high
22. one-half

Operation & Maintenance

9

Continued

Operation & Maintenance

9

Engine Service Life

Reliability and durability are both important qualities of an aircraft engine that relate to its effectiveness and service life. Reliability is the ability of an aircraft engine to continue to meet the manufacturer's specifications under widely varying flight attitudes and extreme weather conditions. Durability is closely related to reliability and is a measure of engine service life. Durability is normally measured in hours between required overhauls.

reliability. The ability of an aircraft engine to perform its designed functions under widely varying operating conditions.

durability. A measure of engine life. Durability is normally measured in TBO hours.

Aircraft reciprocating engines are precision devices, and their service life has increased dramatically as new metals have been developed and manufacturing procedures have been perfected. Most of the engines built before World War II were routinely overhauled when they had accumulated between 300 and 600 hours of time in service.

Almost all recently manufactured engines can be expected to operate for 1,400 to 1,600 hours before they need an overhaul, and many of them can safely exceed 2,000 hours.

The costs of engine overhaul and replacement parts have risen astronomically in the past few decades, and engine durability has increased in importance.

Engine manufacturers have a measure of durability called the TBO (Time Between Overhauls). This is an indication of the number of hours a particular engine should be able to operate normally without excessive wear of the parts.

TBO (time between overhauls). A time period specified by the manufacturer of an aircraft engine as the maximum length of time an engine should be run between overhauls without normal wear causing parts of the engine to be worn beyond safe limits.

The TBO is a *recommended* number of hours, and the manufacturer does not guarantee that an engine will reach TBO before some parts fail, nor does TBO mean that the engine must be given a major overhaul when this specific number of hours is reached.

Requirements are written into the operations manuals for some commercial operators that do require engines to be overhauled when the manufacturer's TBO hours have been reached, but there is no such requirement for private owners.

major overhaul. The disassembly, cleaning, and inspection of an engine and the repair and replacement of all parts that do not meet the manufacturer's specifications.

TBO, as determined by the manufacturer, is based on engine operation under average conditions and all maintenance done in accordance with the manufacturer's recommendations. If the engine is operated continually at high power settings, but not exceeding the allowable RPM or temperatures, and if only the correct fuel and lubricants are used, and all filters are changed as recommended, the engine should reach TBO with none of the reusable parts worn beyond serviceable limits.

TBO can be shortened drastically by using an improper grade of fuel, exceeding the recommended oil change interval, or by allowing cooling baffles to break or misalign. Operation under extremely dusty or sandy conditions, and operating in a training environment with a large percentage of the time at a high power setting can shorten the TBO.

Another way of shortening the TBO, is to fly the aircraft infrequently. When the engine is not operated for long periods of time without being properly preserved, moisture from the surrounding air will cause cylinder walls and other exposed parts to rust or corrode, and acid will build up in the engine lubricating oil.

Reciprocating Engine Operation

Aircraft reciprocating engines have evolved into high-powered devices of minimum weight. This makes them sensitive to proper operating procedures. It is easy to damage an engine if it is not started and shut down properly, or if it is operated improperly on the ground or in the air.

Starting Reciprocating Engines

POH (Pilot's Operating Handbook). A document published by the airframe manufacturer and approved by the FAA that lists the operating conditions for a particular model of aircraft. Engine operating parameters are included in the POH.

All aircraft engines have certain starting procedures that must be followed for best results. This text covers the basic procedures. But detailed procedures differ from one installation to another, so instructions in the Pilot's Operating Handbook (POH) for each particular aircraft should be followed when starting an engine.

Starting Engines with Float Carburetors

Float carburetors are used on older small trainer aircraft and on many older aircraft with engines of up to 200 horsepower or slightly more. These are general steps that when followed, should result in a successful start.

Prepare the aircraft:

1. Position the aircraft where there is no loose gravel.
2. Place chocks in front of the main wheels and set the brakes.
3. Locate the nearest fire extinguisher.
4. Face the aircraft into the wind.
5. Make sure that nothing aft of the aircraft can be damaged by the propeller blast.

Prepare the engine:

1. Turn the battery master switch to ON.
2. Turn the fuel tank selector to a full tank.
3. Set the carburetor heat control to COLD.
4. If the aircraft is equipped with a constant-speed propeller, set the propeller pitch control to the LOW PITCH (HIGH RPM) position.

5. Set the mixture control to FULL RICH.
6. Open the throttle slightly (approximately ⅛ of its travel).
7. For low-wing airplanes, turn the auxiliary fuel pump ON.

Start the engine:

1. Check to be sure no one is standing near the propeller. Call out the word "Clear" to warn all personnel to stand clear of the propeller.
2. Turn the ignition switch to the START position. This is a spring-loaded position that removes the ground from both magnetos and engages the starter.
3. As soon as the engine starts, release the ignition switch, and it will return to the BOTH position. Watch for an indication of oil pressure, and adjust the throttle to the speed at which the engine runs the smoothest. This is normally around 1,000 RPM.

If there is no indication of oil pressure within 30 seconds (60 seconds in extremely cold weather), shut the engine down and determine the reason for the lack of pressure.

Starting Engines with the TCM Fuel Injection System

Fuel-injected engines are started by using slightly different procedures from those with a float carburetor. The airplane is prepared in the same way, and the engine is prepared by following these steps:

1. Turn the battery master switch to ON.
2. Turn the fuel tank selector to the tank specified in the Pilot's Operating Handbook.
3. Set the alternate air control to DIRECT (or off).
4. Set the propeller pitch control to the LOW PITCH (HIGH RPM) position.
5. Open the throttle slightly (approximately ⅛ of its travel).
6. Set the mixture control to FULL RICH.
7. Check to be sure no one is standing near the propeller. Call out the word "Clear" to warn all personnel to stand clear of the propeller.
8. Turn the boost pump to HIGH, and watch the fuel flowmeter. When flow is indicated, secure the boost pump and move the ignition switch to the START position.
9. As soon as the engine starts, release the ignition switch, and it will return to the BOTH position. Watch for an indication of oil pressure, and adjust the throttle to the position in which the engine runs the smoothest. This is normally around 1,000 RPM.

If there is no indication of oil pressure within 30 seconds, shut the engine down and determine the reason for the lack of pressure.

Starting Engines with the RSA Fuel Injection System

The different principle of operation between the TCM and RSA fuel injection systems is the reason they require different starting procedures. Prepare the aircraft as previously described and follow these steps to start the engine:

1. Turn the battery master switch to ON.
2. Turn the fuel tank selector to the tank specified in the Pilot's Operating Handbook.
3. Set the alternate air control to DIRECT.
4. Set the propeller pitch control to the LOW PITCH (HIGH RPM) position.
5. Open the throttle slightly (approximately ⅛ of its travel).
6. Set the mixture control to IDLE CUTOFF.
7. Check to be sure that no one is standing near the propeller. Call out the word "Clear" to warn all personnel to stand clear of the propeller.
8. Turn the boost pump ON, and move the mixture control to the FULL RICH position until there is an indication of flow shown on the flow-meter. Then, move the mixture control back to IDLE CUTOFF and the ignition switch to the START position.
9. As soon as the engine starts, move the mixture control to the FULL RICH position and release the ignition switch. It will return to the BOTH position. Watch for an indication of oil pressure, and adjust the throttle to the position in which the engine runs the smoothest. This is normally around 1,000 RPM.

If there is no indication of oil pressure within 30 seconds, shut the engine down and determine the reason for the lack of pressure.

Engine Ground Fires

When starting aircraft engines, there is always the possibility of an engine fire. If the engine is difficult to start, priming fuel can run into the induction system and a backfire will ignite it. When starting large radial engines, it is customary to have a fire guard with a fire extinguisher stationed by the engine.

Most operations for general aviation aircraft do not use a fire guard, but any time you are starting an engine, you should know the location of the nearest fire extinguisher.

When an induction system fire does occur, keep cranking the engine to try to get it started so it will suck the fire into the engine and put it out. If this does not extinguish the fire and you can no longer crank the engine, shut off the fuel, get a CO_2 fire extinguisher, and discharge it into the induction filter to extinguish the fire.

Reciprocating Engine Ground Check

The ground check of a reciprocating engine tells much about its condition, and it should be conducted in an area that is free of rocks or pebbles that could be drawn into the propeller. Position the aircraft so the slipstream will not blow on other aircraft or into any building, and head the aircraft into the wind to assist in cooling.

Start the engine and allow it to warm up until the oil temperature gage shows an indication. A complete ground check consists of these items:

- *Alternator or generator*—Check the ammeter, loadmeter, or voltmeter to ensure that the alternator or generator is functioning properly.
- *Ignition switch ground*—With the engine operating at its lowest speed, place the ignition switch in the OFF position. Both magnetos should stop firing and the engine should quit. Before it stops turning, return the switch to the BOTH position.
- *Magneto check*—Bring the engine RPM up to the speed recommended in the POH. Move the ignition switch from BOTH to LEFT (the switch position furthest from BOTH). Note the amount of RPM drop. Move the switch back to BOTH and allow the engine to regain its normal speed. Move the switch from BOTH to RIGHT and note the amount of drop.

  The RPM drops should be within the limits specified for the engine. They should be nearly the same for both magnetos, and the drop should be relatively smooth.
- *Fuel flow*—With the engine operating at the speed for the magneto check, note the fuel flow indication, and compare this with the flow specified for this RPM.
- *Propeller governor*—Advance the RPM to the 2,200 to 2,300 RPM range and slowly move the propeller pitch control back in the HIGH PITCH (LOW RPM) direction. Allow the RPM to drop about 50 RPM and then increase the throttle. The RPM should remain steady and the manifold pressure should increase.
- *Static RPM*—Having placed the propeller control back in the max RPM position, advance the throttle to full forward and check the RPM and manifold pressure. They should be the values specified in the POH.

STUDY QUESTIONS: ENGINE SERVICE LIFE AND RECIPROCATING ENGINE OPERATION

Answers begin on Page 341. Page numbers refer to chapter text.

1. All FAA-certificated aircraft engines must be overhauled when they reach the number of hours specified by the manufacturer as their TBO (Time Between Overhauls). This statement is ________________ (true or false). *Page 285*

2. An aircraft that is flown infrequently has a ______________________ (better or poorer) chance of reaching the recommended TBO than one flown consistently. *Page 286*

3. Engines equipped with float carburetors are started with the mixture control in the ___________________ position. *Page 287*

4. When starting an engine equipped with float carburetors, the carburetor heat control should be in the ____________________ (HOT or COLD) position. *Page 286*

5. When starting an engine equipped with a constant-speed propeller, the propeller pitch control is placed in the __________________ (LOW or HIGH) -PITCH position. *Page 286*

6. An engine equipped with a TCM fuel injection system is started with the mixture control in the __________________ position. *Page 287*

7. An engine equipped with an RSA fuel injection system is started with the mixture control in the __________________ position. *Page 288*

8. When an induction system fire occurs, the best procedure is to _________________________ (keep cranking the engine or shut the engine down). *Page 288*

9. The best fire extinguishing agent for an induction system fire that occurs when starting the engine is _________________ . *Page 288*

10. When checking to be sure that the ignition switch grounds both magnetos in the OFF position, the RPM should be _______________ (low or high). *Page 289*

11. When a constant-speed propeller is operating in its governing range, forward movement of the throttle will result in the RPM ______________________________ (increasing, decreasing, or remaining the same), and the manifold pressure __________________ (increasing, decreasing, or remaining the same). *Page 289*

Reciprocating Engine Maintenance Inspections

One of the more important aspects of an aviation maintenance technician's work is inspections. Their purpose is to determine that the aircraft and its engines are in a condition to safely accomplish the flight the pilot intends. One-hundred-hour and annual inspections are performed to determine that the aircraft and its engines meet all requirements for their certification.

one-hundred-hour inspection. An inspection required by 14 CFR §91.409 for FAA-certificated aircraft operated for hire or used for flight instruction for hire. A 100-hour inspection is identical in content to an annual inspection, but can be conducted by an aviation maintenance technician who holds an Airframe and Powerplant rating, but does not have an Inspection Authorization.

One-Hundred-Hour and Annual Inspections

The FAA requires that all certificated aircraft operated under the provisions of Federal Aviation Regulations Part 91 have a complete inspection every 12 calendar months, and if the aircraft is operated for hire, or is used to give flight instruction for hire, it must have an inspection of the same depth each 100 hours of operation. The annual and 100-hour inspection are identical in scope, the difference being in the person authorized to conduct them. An AMT that holds an Inspection Authorization is authorized to conduct either an annual or a 100-hour inspection and approve the aircraft for return to service. An AMT that holds both an Airframe and Powerplant rating, but no Inspection Authorization, can conduct a 100-hour inspection and approve the aircraft for return to service.

annual inspection. A complete inspection of the airframe and powerplant required for FAA-certificated aircraft operating under 14 CFR Part 91 *General Operating and Flight Rules*, and not on one of the authorized special inspection programs. An annual inspection must be conducted every 12 calendar months, and it must be conducted by an aviation maintenance technician who holds an Airframe and Powerplant rating and an Inspection Authorization.

The scope of an annual inspection is the same as that of a 100-hour inspection.

An aircraft inspection includes both the airframe and the powerplant, and requires the technician conducting them to have both the Airframe and Powerplant rating. A technician who holds only a Powerplant rating is authorized to conduct the powerplant portion of a 100-hour inspection and approve the powerplant for return to service. The powerplant, for this purpose, is described as the complete installation of an aircraft engine, propeller, and all accessories needed for its proper operation.

calendar month. The measurement of time used by the FAA for inspection and certification purposes. One calendar month from a given date extends from that date until midnight of the last day of that month.

The regulatory factors regarding annual and 100-hour inspections are covered in the *General* textbook of this *Aviation Maintenance Technician Series*, and the procedures for the airframe portion of the inspections are detailed in the *Airframe* textbook of this series. Here we consider only the aspects of the inspections that are applicable to the powerplant.

Title 14 of the Code of Federal Regulations (14 CFR) Part 43 *Maintenance, Preventive Maintenance, Rebuilding, and Alteration* requires that the technician conducting an annual or 100-hour inspection use a checklist to ensure that none of the items listed in 14 CFR Part 43, Appendix D are missed. These items are included on the typical powerplant inspection checklist in Figure 9-1.

The FAA does not require routine maintenance inspections more often than every 100 hours, but manufacturers normally specify certain items be inspected more often. The checklist in Figure 9-1 is adapted from the service manual for a fuel-injected turbocharged, single-engine airplane and is typical for this type of aircraft. Note that some items require inspection at not only each 100-hour inspection, but also at each 50 hours of operation.

Powerplant Maintenance Inspections

Items to be checked each 50 hours of operation	
Item	**AMT Initial**
Engine oil screen, filler cap, dipstick, drain plug, and external filter element	
Induction air filter	
Engine baffles	
Hoses, metal lines, and fittings	
Intake and exhaust systems	
Fuel Injection system	
Fuel/air control unit	
Discharge nozzles	
Fuel flow gage	
Engine cowl flaps and controls	
Turbocharger mounting brackets and linkages	
Alternator support bracket	
Propeller spinner	
Propeller blades	
Anti-icing system wiring, brushes, slip rings and boots	
Items to be checked each 100 hours of operation	
Item	**AMT Initial**
Oil cooler	
Induction air box, air valves, doors and controls	
Hot and cold air hoses	
Cylinders, rocker box covers and push rod housings	
Crankcase, oil sump, accessory section and front crankshaft seal	
Ignition harness	
Spark plugs	
Cylinder compression check	
Crankcase and vacuum system breather lines	
Electrical wiring	
Vacuum pump	
Vacuum relief valve filter	
Engine controls and linkage	
Engine shock mounts, mount structure, and ground straps	
Cabin heat valves, doors, and controls	
Starter, solenoid, and electrical connections	
Starter brushes, brush leads, and commutator	
Alternator and electrical connections	
Alternator belt tension	
Alternator brushes, brush leads and slip rings	
Voltage regulator mounting and electrical leads	
Magnetos (externally) and electrical connections	
Magneto-to-engine timing	
Firewall	
Engine cowling	
Turbocharger	
Waste gate, actuator, controller, and oil lines	
Propeller spinner bulkhead	
Propeller attachment bolts and nuts	
Propeller hub	
Propeller governor and control	

Figure 9-1. *A typical checklist for the powerplant portion of an annual or 100-hour inspection for a turbocharged single-engine airplane*

Preparation for Inspection

Paperwork is one of the more important aspects of a 100-hour or annual inspection. All Airworthiness Directives, General Aviation Airworthiness Alerts, manufacturer's service bulletins, and service letters must be checked to be sure nothing is overlooked.

In the past, standard procedure was for each maintenance shop to have all of this information on file. However, this data has become so complex and its compilation so important for avoiding possible lawsuits, that most shops subscribe to one of the services that furnish copies of all applicable documents. This information is available over the telephone lines via modems connected to personal computers.

The shop work order is initiated, and the aircraft maintenance records are checked to determine that all accessories on the engine are approved, and that all items on the equipment list are actually installed.

The propeller hub and blades must be approved for the aircraft, and must be the ones listed in the aircraft records. The total time on any life-limited components must be checked to determine if any should be retired.

If there are any FAA Form 337s (*Major Repair and Alteration*) on the engine or propeller, check that the work has actually been done as described on the form.

Determine if there are any recurring Airworthiness Directives on the engine or propeller that must be complied with on this inspection.

Airworthiness Directive. A regulatory notice sent out by the FAA to the registered owner of an aircraft informing him or her of the discovery of a condition that keeps the aircraft from continuing to meet its conditions for airworthiness. Also called AD notes, they are covered by Federal Aviation Regulations Part 39 *Airworthiness Directives*. They must be complied with within the specified time, and the fact, date, and method of compliance must be recorded in the aircraft maintenance records.

General Aviation Airworthiness Alerts. Documents published by the FAA as Advisory Circulars 43.16. These are used to alert technicians of problems that have been found in specific models of aircraft, and reported on Malfunction and Defect Reports. Airworthiness Alerts suggest corrective action, but compliance with the suggestion is not mandatory.

FAA Form 337. The *Major Repair and Alteration* form that must be completed when an FAA-certificated aircraft or engine has been given a major repair or major alteration.

Give the engine a thorough preinspection run-up to determine its condition and get the oil warm and the cylinder walls covered with oil so the compression check will be meaningful.

Check and record all pressures and temperatures and determine that the engine is developing its proper static RPM. Determine that the magneto drop is within the range specified by the engine manufacturer, and that the drops are nearly the same for each magneto.

Check for unusual noises and vibrations, and observe how the engine responds to throttle changes. Check the operation of the propeller. It should change pitch smoothly and should recover to the low-pitch RPM within the specified time.

Check the engine idling. It should be smooth, and when the mixture control is moved to the IDLE CUTOFF position, the RPM should rise by the amount specified in the aircraft service manual before the engine stops firing.

The alternator should put out the correct voltage, and the engine instruments should indicate that all systems are functioning properly.

The fuel flow indicator on fuel-injected engines should display the correct fuel flow. This indication is actually the pressure drop across the injector nozzles.

When everything is checked and discrepancies are noted for attention, shut down the engine and begin the inspection.

The Actual Inspection

To begin the actual inspection, bring the aircraft into the hangar and remove the cowling.

Compression Check

One of the most important and meaningful portions of a powerplant inspection is the compression check. This test gives a good indication of the sealing properties of the valves and piston rings.

There are two types of compression checks: the direct check as used for automobile engines, and the differential check. The differential check is more significant, as it gives much more information about the condition of the engine. This is the type of compression check described in this text.

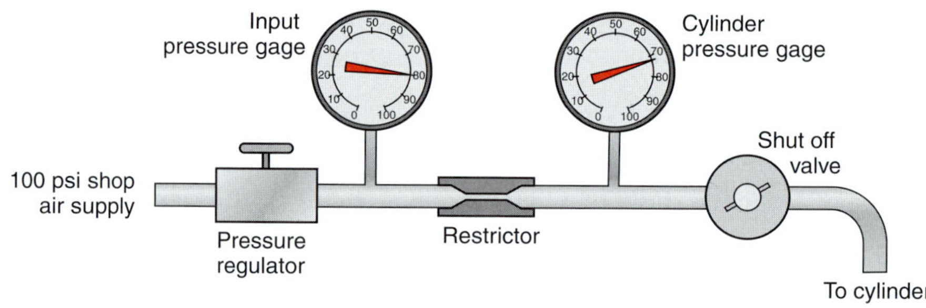

Differential compression tester showing gage readings for a good cylinder.

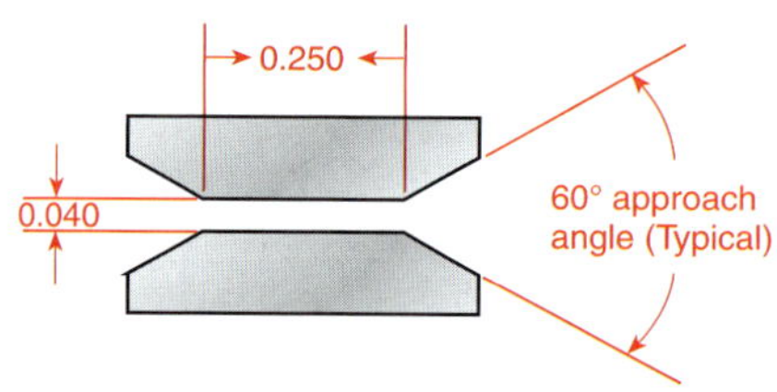

Dimensions of the restrictor orifice for engines up to 1,000 cubic inches displacement.

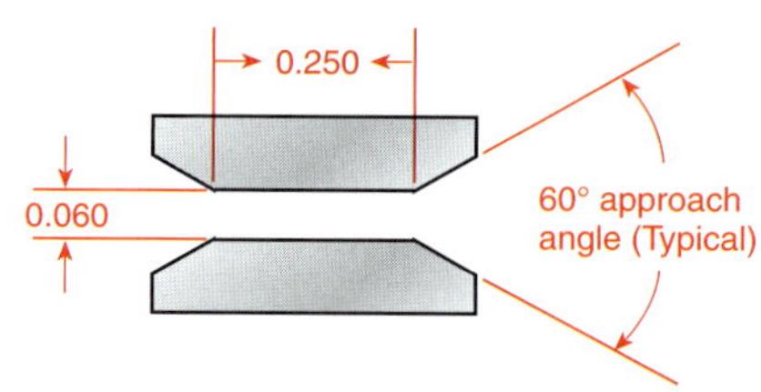

Dimensions of the restrictor orifice for engines of more than 1,000 cubic inches displacement.

Figure 9-2. *A differential compression tester*

With the engine still warm from the preinspection run-up, remove the more accessible spark plug from each cylinder, and screw the compression tester adapter into the spark plug hole of the first cylinder to be checked. Be sure, when removing the spark plugs from the engine, to place them in a numbered rack to identify the cylinder and position from which they were removed. The sequence of testing the cylinders is not important. Some technicians check them according to the engine firing order; others check all on one side, and then the other.

With the tester connected to one cylinder, the shop air turned off, and the shut-off valve open, rotate the propeller until the needle on the cylinder-pressure gage moves up. This shows the cylinder is on the compression stroke.

Rotate the propeller backward until the piston is near the bottom of its stroke. Turn the air on and adjust the regulator to get about 20 psi inside the cylinder. This pressure will push back on the piston, but not enough to prevent you from turning the propeller. Slowly rotate the propeller until the air in the cylinder ceases to force the piston down. The piston is now in its top dead center position with the piston rings pressing against the lands, or edges of the piston ring grooves. Do not rock the propeller in this position, as it will disturb the rings from their seating in the ring grooves. Increase the pressure on the tester to 80 psi, and read the cylinder-pressure gage.

land. The portion of a piston between the ring grooves.

Caution: *You must be very careful when conducting a differential compression check that the propeller does not move even slightly off of the top dead center position when there is air pressure inside the cylinder. All personnel must remain clear of the propeller, and there should be no stands or tools too close to the propeller. Eighty psi of air pressure can swing the propeller hard enough to cause injury.*

The standard allowable difference between the two gages is 25%. This means that the cylinder must hold 75% of the air put into it. Any pressure reading on the cylinder gage lower than 60 psi shows that the cylinder is questionable, and the reason for the low reading should be determined. Some engines are not allowed this much leakage, so use the latest service information by the manufacturer to determine the lowest reading acceptable for the engine on which you are working.

With 80-psi pressure held on the cylinder, listen for sounds of leaking air. If the intake valve is leaking, a hissing will be heard at the carburetor or fuel injector air inlet. If the exhaust valve is leaking, it will be heard at the exhaust stack. If air is leaking past the piston rings, hissing will be heard at the crankcase breather or the oil filler neck.

Record the compression readings, and keep them with previous inspections records so trends can be studied.

Lubrication System

After the compression check is completed, and while the oil is still warm, remove the drain plug from the oil sump and drain the oil. After the oil begins to flow, catch a sample to send to the laboratory for an oil analysis. Be sure to keep a record of all oil analysis reports, and look for trends in the growth or increase in the amount of any metal present in the oil. It is not the amount of metal that matters, but how the amount changes as the engine accumulates hours of operation. Review the section on oil analysis, in Chapter 3, beginning on Page 121.

oil analysis. A method of measuring the contents in parts per million of various chemical elements in oil.

A sample of the oil is burned in an electric arc, and the resulting light is analyzed with a spectroscope which identifies the chemical elements in the oil and gives an indication of the amount of each element. This type of oil analysis is called a spectrometric oil analysis program, or SOAP.

Remove the oil filter. If it is the can type, use a special cutter to open the can. If it is a disposable element in a reusable housing, cut the element open with a knife. Spread the pleated element out on a piece of clean paper and examine it carefully for signs of metal particles. Run a clean magnet through each fold to pick up any ferrous particles that are too small to be identified by sight alone.

Dilute the oil that remains in the filter can or housing with clean aviation gasoline and pour it through a pad of cheesecloth. This will help you find any particles that were not trapped in the filter element.

If there are a large number of metal particles in the filter, remove the engine sump and examine the oil suction screen for large chunks of metal. The type of metal particles found in the oil filter indicates the type of failure the engine is encountering. Figure 9-3 shows the source of some metals that may be found in a filter.

babbitt. A soft, silvery metal used for main bearing inserts in aircraft reciprocating engines. Babbitt is made of tin with small amounts of copper and antimony.

Metal	Possible Source
Aluminum	Pistons, piston pin plugs, cylinder heads
Steel or iron	Cam or cam followers, piston rings, rocker shafts, gears
Bronze	Connecting rod bushings, tachometer bushings
Babbitt	Bearing inserts
Chromium	Piston rings, cylinder walls

Figure 9-3. *Typical sources of metal likely to be found in an oil filter*

When installing a new filter, check the mounting surface to be sure there are no nicks, gouges, or parts of old gasket. Use only new gaskets having the correct part number for the installation, and lubricate them with the kind of lubricant specified in the aircraft service manual. Some filter gaskets can be lubricated with clean engine oil or general-purpose grease, and others require Dow Corning DC-4 silicone grease. Be sure to use only the lubricant specified for the particular filter. When the filter is screwed into its mount, torque it in place using an accurate torque wrench and the torque specified in the service manual, and safety-wire the filter in place according to the method specified in the aircraft service manual.

Check the crankcase, oil cooler, and all the oil lines for any indication of leaks. Carefully examine the nose section of the crankcase for indications of leakage around the crankshaft oil seal, and check the rocker box covers and pushrod housings for leakage. Replace and safety the oil drain plug and put the specified amount of the recommended type of oil in the sump.

After the engine is run, carefully examine the entire filter area for indications of oil leakage. Do not recheck the filter torque. Repeated applications of the same torque over-compresses the thick rubber gasket, which can make removal difficult and possibly result in the gasket leaking.

Ignition System

Remove the spark plugs that were left in the engine for the compression check. Place them in the numbered tray and examine carefully for signs of cylinder distress. Examine their condition, and clean, gap, and test them as explained in the section on Spark Plug Servicing in Chapter 5, beginning on Page 237.

Reinstall the spark plugs using new gaskets. Return them to the next cylinder in the firing order, after the cylinder from which they were removed, and swap them from top to bottom. Prepare the spark plugs by putting a small amount of an approved spark plug antiseize compound on the second thread from the firing end, and screw them down with your fingers until they contact the gasket. Then torque them to the value specified in the aircraft service manual with an accurately calibrated torque wrench. Wipe the lead terminal cigarettes with a rag lightly dampened with acetone to remove any oil left from your fingers. Slide the cigarette in straight, and screw the terminal nut down with your fingers until it is snug, and then turn it with a properly-fitting wrench about one hex (60°) more.

cigarette. A commonly used name for a spark plug terminal connector used with a shielded spark plug.

Carefully inspect the leads between the spark plugs and magnetos. There should be no indication of rubbing or chafing, and the leads should all be supported in the clamps specified in the aircraft service manual.

Check the exterior of the magnetos and the connection for the P-lead. Remove the cover from the magneto breaker housing and examine the inside for oil that could have leaked in through a faulty oil seal. Check the condition of the breaker points and check the magneto-to-engine timing. If the timing is off, remove the magneto and check its internal timing. Adjust if necessary, and retime the magneto to the engine as described in Chapter 5, beginning on Page 220.

Fuel System

Turn the fuel off at the selector valve and remove and clean the main fuel strainer and the strainer in the carburetor or fuel injection system inlet. Reinstall the strainers using new gaskets. Turn on the fuel and the boost pump and check the entire system for leaks. Check the condition of the flexible hoses and fire shields installed around these hoses.

springback. A method used to ensure that the control arm of a device actually reaches its full-travel stop. Springback ensures that the stop on the carburetor is contacted before the stop in the cockpit.

Check the carburetor or fuel injection system for fuel dye stains that would indicate a fuel leak. Check the movement of all the controls to be sure that the arm on the engine component reaches its stops in both directions, and that the cockpit control has a slight amount of springback at either end.

On fuel-injected engines, check the distributor valve (or manifold valve) for security and cleanliness. Be sure that the vent on a manifold valve is open and has no indication of fuel stain. Check this with a piece of thread, not with a piece of wire.

Check all lines between the distributor valve and the nozzles to be sure that none are loose or leaking. Check the injector nozzles for fuel dye stain. A plugged nozzle would cause fuel to flow out the air bleed hole and evaporate, leaving the dye stain.

If there is an indication that any nozzles are plugged, remove and soak them in a suitable solvent such as a good grade of gun-cleaning solvent. Do not probe into the orifice with metal wire, but soak the contaminants and blow the nozzles clean with compressed air.

Induction System

Remove and clean or replace the induction air filter. Check the entire system for sign of damage. Be very careful that there are no loose screws or rivets that could fall into the system and be sucked into the cylinders.

Check the carburetor heat or alternate air valves to be sure they open and close fully, and do not allow unfiltered air to flow into the engine when the valve is in the COLD position.

Check all sections of flexible hose that connect the carburetor heat or alternate air boxes to the fuel metering device and to the source of alternate air. These hoses should show no indication of chafing or wear. No fluid lines or controls should rub against the hoses. If hoses show any evidence of collapse or wear, they should be replaced.

Exhaust System

Carefully examine the exhaust stacks as they leave the cylinder for indication of loose nuts, leakage, blown gaskets, or cracks. Check the exhaust gas temperature probe for indication of leakage around it. Check the thermocouple wiring to be sure it is properly supported and is not chafing on anything.

Remove the shroud from around the muffler. Examine the muffler very carefully for cracks, as it takes only a small crack to release deadly carbon monoxide into the cabin. If there is a service bulletin or AD that specifies a particular inspection for the muffler, be sure to comply with it.

Turbocharger

The turbocharger operates at such high temperature and high speed that it must be inspected with extreme care. Remove the cover and examine the turbocharger for signs of cracking or leakage. Examine the mounting brackets for cracks or other damage. Check the waste gate actuator, controller, relief valve, and all oil lines and linkages. Any special inspection procedures specified by the airframe manufacturer must be followed in detail.

Cooling System

Check all engine baffles for indication of cracks or missing pieces. Check the condition of the sealing strips, and be sure they fold over in the correct direction to provide a seal.

Check the entire cowling for cracks or missing screws. Check the cowl flaps for proper actuation and for indication of wear or looseness in the hinges. Lubricate the cowl flap system as specified in the aircraft service manual.

Electrical System

Electrical systems have grown in complexity and importance, and they must be properly maintained so they will continue to provide the dependability and efficiency designed into them.

Check the generator or alternator mount for security. Alternator mounts are especially prone to cracking because of jolts to the alternator when a large electrical load is suddenly connected to it. Check the voltage regulator and any solenoids or relays for security of mounting and integrity of the wiring connected to them. Check all electrical wire bundles and connectors. The bundles must be secured in the manner specified by the airframe manufacturer. Check the condition and tension of the generator or alternator belt. If the tension is not that specified in the service manual, adjust it. If the proper tension cannot be obtained, replace the belt.

Check the condition of the generator or alternator brushes and the commutator or slip ring. Replace the brushes when they have worn to the length specified by the manufacturer.

Check the starter for its security of mounting, and check the commutator and brushes for wear or indication of excessive sparking.

Check the switches in the cockpit that control the starter relay, the ignition, and any device that applies directly to the powerplant.

Check for the integrity of the ground strap around the engine shock-absorbing mount. The ground strap must be in good condition and there must be no corrosion at either end.

Instrument Systems

Check the engine-driven air pump for security of mounting and for indication of damage. Clean or replace the air filters for the pump and the vacuum relief valve.

Check the thermocouples for the EGT and cylinder head temperature for their condition and security of mounting. Check the probe and wiring for the oil temperature and carburetor air temperature. Check the integrity of the fluid lines for the oil pressure, fuel pressure, and fuel flow indicators.

Check the powerplant instruments in the cockpit for condition, security of mounting, and proper marking.

General Condition

Carefully examine all hoses, metal lines, and fittings ahead of the firewall. Check the crankcase breather for security of mounting to be sure that its open end is free of any obstruction and that its vacuum relief port has not inadvertently been covered.

Check all controls to be sure the stops on the engine component are reached before the stop on the cockpit end. The cockpit control should have the specified amount of springback.

Check the condition of the engine mount shock absorbers. They should show no sign of deterioration, and all connecting bolts should be properly safetied.

The heat duct and control box for the cabin heater should be in good condition, and the heater controls should work smoothly through their full travel.

Check the condition of the firewall. It should have no cracks or holes, and all lines and controls that pass through it should be sealed in the manner specified by the manufacturer.

Propeller

The entire propeller installation should be carefully examined. If the manufacturer recommends lubrication, it should be done in the manner and with the lubricants specified in the service manual.

burnish. To smooth the surface of a metal part that has been damaged by a deep scratch or gouge. Metal piled at the edge of the damage should be removed and not pushed back into the damaged area. The burnishing should be the smoothing of the intact base metal, with a smooth-shanked tool, using a motion that causes the surface of the metal to flow from the sharp pointed base outward until a smooth rounded depression results.

Check the blades for indication of wear or binding, and the hub for chafing or wear. Check the condition of the mounting bolts and nuts. Carefully burnish out any pits along the leading edge of the blades. Work out any large nicks or pits only to the extent allowed by the propeller manufacturer. Because of the extreme stresses imposed on a propeller, any questionable condition should be referred to an FAA-certificated propeller repair station.

Check the condition of the propeller spinner bulkhead and the spinner itself. Cracks cannot be tolerated in these critical areas, and if any are found, they must be repaired in the manner specified by the airframe manufacturer, or the component must be replaced. Repairs that might alter the balance of the spinner must be avoided.

Check the propeller governor for condition, security of mounting, and full and unrestricted travel of its control.

Check the anti-icing system for condition of the boots and for the integrity of mounting of the brushes and slip rings. Brushes should be replaced as specified in the service manual.

Post-Inspection Run-Up and Records

When the actual inspection is completed, carefully wash the engine down, reinstall the cowling, and give the engine a careful post-inspection run-up. Check all systems and compare engine instrument readings with those obtained on the preinspection run-up.

If everything checks out on the run-up, record the inspection in the aircraft maintenance records. The record for this inspection should include the following items:

- Date the inspection was completed
- TOTAL TIME on the engine
- Description of the work done. This usually includes an identifying reference to the shop work order for the inspection.

- List of any ADs complied with and the method of compliance. Note the date or operational hours for compliance with any recurring AD.
- A statement in the powerplant records similar to this:

> Date ______________________
>
> I certify that this powerplant has been inspected in accordance with a 100-hour inspection and is determined to be in an airworthy condition.
>
> Signed ____
>
> Certificate type and number ______________________________

If the powerplant was determined to be in an unairworthy condition, a signed and dated list of all reasons the powerplant failed the inspection must be given to the registered owner or operator of the aircraft. A statement similar to this should be placed in the records.

> Date ______________________
>
> I certify that this powerplant has been inspected in accordance with a 100-hour inspection, and a list of the discrepancies and unworthy items dated ______________________ has been provided for the aircraft owner or operator.
>
> Signed ____
>
> Certificate type and number ______________________________

STUDY QUESTIONS: RECIPROCATING ENGINE MAINTENANCE INSPECTIONS

Answers begin on Page 341. Page numbers refer to chapter text.

12. Two maintenance inspections required for all aircraft operating for hire under 14 CFR Part 91 are:
    a. ______________________________
    b. ______________________________
    Page 291

13. All FAA-certificated aircraft operating under 14 CFR Part 91 are required to have a complete inspection every ________________ calendar months. *Page 291*

14. All FAA-certificated aircraft operating for hire under 14 CFR Part 91 are required to have a complete inspection every ____________________ hours of operation. *Page 291*

15. An AMT with both Airframe and Powerplant ratings __________________ (is or is not) required to have an Inspection Authorization to conduct a 100-hour inspection and approve the aircraft for return to service. *Page 291*

16. An AMT with only a Powerplant rating _____________ (is or is not) authorized to perform the powerplant inspection portion of a 100-hour inspection. *Page 291*

17. The propeller __________________ (is or is not) a portion of an aircraft powerplant. *Page 291*

18. The use of a checklist __________________ (is or is not) required when conducting an annual or 100-hour inspection. *Page 291*

19. Two types of compression checks are the direct check and the differential check. The one that gives the most information is the _________________________ compression check. *Page 294*

20. In order to perform a compression check, the cylinder must be on the __________________ (power or compression) stroke. *Page 295*

21. A compression check is made with _____________ psi of air pressure applied to the cylinder. *Page 295*

22. A cylinder is normally considered to fail the compression test if it does not hold at least ______________ psi of air pressure. *Page 295*

23. When a compression check is performed, a hissing sound heard at the carburetor or fuel injector air inlet is an indication that air is leaking past the _____________________ . *Page 295*

24. When a compression check is performed, a hissing sound heard at the exhaust outlet is an indication that air is leaking past the _____________________ . *Page 295*

25. When a compression check is performed, a hissing sound heard at the crankcase breather is an indication air is leaking past the ____________________ . *Page 295*

26. The oil sample to be used for oil analysis should be taken ________________________________ (at the beginning of the flow or after the flow has started). *Page 295*

27. The important information furnished by an oil analysis is the ______________________ (growth trend or actual amount) of the various metals in the engine oil. *Page 295*

28. Particles of babbitt found in an oil filter would likely have come from the ________________________ . *Page 296*

29. When preparing a spark plug for installation, a small amount of antiseize compound should be placed on the ______________________ (end or next to the end) thread. *Page 297*

30. If the magneto-to-engine timing is off, the ___________________ timing of the magneto should be checked. *Page 297*

31. An old fuel leak is normally indicated by stains from the ____________________ where the fuel leaked and evaporated. *Page 297*

32. When adjusting the throttle control, the stops on the carburetor or fuel injection device should be contacted, and the control in the cockpit should have a slight bit of ______________________ . *Page 297*

33. A plugged fuel injector nozzle would cause fuel dye stains around the nozzle because of the fuel leaving the nozzle through the _______________ holes. *Page 297*

34. When the carburetor heat valve is in the COLD position, the air that enters the engine is ____________________ (filtered or unfiltered). *Page 298*

35. Alternator mounts should be carefully inspected for _____________________ (what kind of damage). *Page 299*

36. Check both ends of electrical ground straps around the shock absorbing engine mounts for indication of ______________________ . *Page 299*

37. If a powerplant fails its 100-hour inspection, a signed and dated list of discrepancies that caused the engine to be unairworthy must be given to the ________________________ or ________________________ of the aircraft. *Page 301*

Reciprocating Engine Overhaul

In the years when aircraft engines had to be overhauled every 300 to 600 hours, they were simple in design, and most maintenance shops were equipped to do the overhaul. Modern engines are so highly developed and produce so much power from such a light structure that overhaul has become a highly sophisticated procedure. Modern engine overhaul is most generally done by facilities that specialize in this type of work.

Three types of engine reconditioning widely used today are top overhaul, major overhaul, and rebuilding.

Top Overhaul

A top overhaul is just what the name implies—an overhaul of the top end of the engine. The crankcase is not opened.

top overhaul. An overhaul of the cylinders of an aircraft engine. The valves, pistons, and cylinders are overhauled, but the crankcase is not opened.

There is no recommended time for a top overhaul. A top overhaul cannot be used to extend the TBO of the engine, but is used when there has been some localized damage to the cylinders, or if the engine has had some unusual operating conditions. It is seldom economically wise to top overhaul an engine when it is near its TBO, as it will have to be done again when the engine receives a major overhaul.

run in. A time of controlled operation of a new or freshly overhauled engine that allows the moving parts to wear together.

At the minimum, a top overhaul consists of removing the cylinders, replacing the valve guides, grinding and reseating the valves, honing the cylinder walls, and replacing the piston rings. After a top overhaul, the engine must be properly run in to seat the piston rings to the cylinder walls.

Major Overhaul and Rebuilding

A major overhaul of an aircraft reciprocating engine includes complete disassembly and cleaning, and visual, dimensional, and other types of inspections. Parts are repaired as necessary or replaced, and the engine is reassembled, tested, and then approved for return to service.

An engine overhaul manual includes two types of limits in its table of limits: new-parts minimum and maximum, and serviceable limits. The excerpt from a typical table in Figure 9-4 on Page 306 shows that the cylinder bore of a new cylinder can be anywhere between 5.2510 and 5.2530 inches. The cylinder can be considered serviceable until it has worn to a diameter of 5.2560.

A major overhaul to new-parts limits, as done in many of the major overhaul facilities, measures all components, and any that fall outside of new-parts limits are not reused. When these limits are recognized, the engine has a good possibility of running to TBO.

Some overhaulers measure all parts, and reinstall those that measure within the serviceable limit. For example, if the cylinder bore measures 5.2550 inches, it is outside of the new-parts limit, but it is within the serviceable limit dimension of 5.2560 inches. It is legal to use this cylinder, but not economical, because an engine assembled with these parts has little possibility of reaching TBO.

When an engine receives a major overhaul, its time in service continues from its original time. For example, if an engine has 1,600 hours total time in service and is given a major overhaul with all parts having new-parts dimensions, its total time is still 1,600 hours, but its time since major overhaul is now zero hours SMOH (since major overhaul).

According to 14 CFR §91.421, the engine manufacturer or an agency approved by the manufacturer can rebuild an engine and issue a new maintenance record with zero hours of operation.

An engine to be rebuilt is disassembled, cleaned, and all components inspected according to the manufacturer's recommendations. All components must conform to new-parts tolerances and limits, or to approved oversize or undersize dimensions; all AD notes, service bulletins, and service letters are complied with. The engine is reassembled, tested, and run in the same way as a new engine.

The crankshaft may be reconditioned by grinding the crankpins and journals undersize, and fitting the crankcase with undersize bearings so the allowable new-parts fit is maintained. *See* Figure 9-4 on Page 306. Some rebuilders will not put undersize crankshafts in a rebuilt engine even though the new-parts fit is maintained.

When an engine is overhauled, all the parts stay with the particular engine, but not so when it is rebuilt. Instead, the engine is assembled from a stock of parts in the same way as a new engine.

rebuilt engine. A used engine that has been completely disassembled, inspected, repaired as necessary, and reassembled, tested, and approved in the same manner and to the same tolerances and limits as a new engine, using either new or used parts. However, all parts used must conform to all production drawings, tolerances, and limits for new parts, or be of approved oversize or undersize dimensions for a new engine.

According to 14 CFR §91.421, a rebuilt engine is considered to have no previous operating history and may be issued a zero-time logbook. Only the engine manufacturer or an agent approved by the manufacturer can rebuild an engine and issue a zero-time record.

Major Overhaul Procedures

In this discussion, we follow a typical horizontally opposed engine as it travels through an overhaul facility. When an engine is overhauled, all manufacturer's recommendations must be followed in detail, and all latest service information must be consulted and recommendations complied with.

Here, we will consider that the engine is to be maintained as a unit and returned to the owner with all serviceable original components installed.

When the engine is received, a shop work order begins with recording the name and telephone number of the owner and the serial numbers of all components to be kept with the engine.

Engines sent to be overhauled normally have the carburetor or fuel injection system complete with all induction pipes installed, as well as the magnetos, spark plugs, and ignition harness. The intercylinder baffles are included with the engine.

The propeller, propeller governor, vacuum pump, hydraulic pump, starter, generator or alternator, and exhaust system are not considered as part of the actual engine and are not usually on the engine when it is received in the shop. Most of these components are overhauled at the time the engine is overhauled, and they travel through the overhaul facility with paperwork separate from that of the engine.

Table of Limits

Ref. No.	Chart No.	Model	Description	Serviceable Limit	New Parts Minimum	New Parts Maximum
			Cylinder and Head Assembly			
1	1	All	Cylinder bore (lower 4-1/4" of barrel) ... Diameter:	5.2560	5.2510	5.2530
2	1	All	Cylinder bore choke (at 5.75" from open end of barrel ... Taper:	0.0010	0.0030	0.0050
3	1	All	Cylinder bore out-of-round ... :	0.0030	0.0000	0.0020
4	1	All	* Cylinder bore ... Allowable Oversize:	5.2700	5.2660	5.2680
5	1	All	Cylinder bore surface roughness ... RMS:	—	15	25
6	1	All	Cylinder barrel in crankcase ... Diameter:	—	0.0040L	0.0100L
7	1	All	Intake valve seat insert in cylinder head ... Diameter:	—	0.0090T	0.0120T
8	1	All	Intake valve guide in cylinder head ... Diameter:	—	0.0010T	0.0025T
9	1	All	Exhaust valve guide in cylinder head ... Diameter:	—	0.0010T	0.0025T
10	1	All	Exhaust valve seat insert in cylinder head ... Diameter:	—	0.0070T	0.0100T
11	1	All	Intake valve seat ... Width:	—	0.0630	0.1400
12	1	All	Exhaust valve seat ... Width:	—	0.0630	0.1400
	1	All	Exhaust valve seat-to-valve guide axis ... Angle:	—	44° 30'	45° 00'
	1	All	Intake valve seat-to-valve guide axis ... Angle:	—	59° 30'	60° 00'
			Rocker Arms and Shafts			
13	1	All	Rocker shaft in cylinder head bosses ... Diameter:	0.0030L	0.0000	0.0015L
14	1	All	Rocker shaft in rocker arm bearing ... Diameter:	0.0060L	0.0010L	0.0025L
15	1	All	Rocker arm bearing in rocker arm ... Diameter:	—	0.0020T	0.0040T
16	1	All	Rocker arm ... Side Clearance:	0.0350L	0.0020L	0.0015L
17	1	All	Intake valve in guide ... Diameter:	0.0050L	0.0012L	0.0027L
18	1	All	Exhaust valve in guide ... Diameter:	0.0070L	0.0030L	0.0050L
19	1	All	Intake valve face (to stem axis) ... Angle:	—	59° 45'	60° 15'
20	1	All	Exhaust valve face (to stem axis) ... Angle:	—	45° 00'	45° 30'
21	1	All	Intake valve (max. tip regrind .015) ... Length:	4.7890	4.8040	4.8240
22	1	All	Exhaust valve (max. tip regrind .015) ... Length:	4.7910	4.8060	4.8260
23	1	All	Intake and exhaust valve (full indicator reading) concentricity ... :	0.0040	0.0000	0.0020
	1	All	Hydraulic lifter-to-valve stem ... :	—	0.0600	0.2000
			Pistons, Rings and Pins			
24	1	All	Piston (2nd & 3rd lands) in cylinder ... Diameter:	—	0.0310L	0.0350L
25	1	All	Piston (bottom of skirt) in cylinder ... Diameter:	0.0160L	0.0100L	0.0130L
26	1	All	Top piston ring in groove ... Side Clearance:	0.0080L	0.0040L	0.0060L
27	1	All	Second piston ring in groove ... Side Clearance:	0.0080L	0.0040L	0.0060L
28	1	All	Third piston ring in groove ... Side Clearance:	0.0075L	0.0035L	0.0055L
29	1	All	Fourth piston ring in groove ... Side Clearance:	0.0100L	0.0060L	0.0080L
30	1	All	Top ring gap (in cylinder barrel) ... Gap:	0.0570	0.0330	0.0490
31	1	All	Second ring gap (in cylinder barrel) ... Gap:	0.0480	0.0240	0.0400
32	1	All	Third ring gap (in cylinder barrel) ... Gap:	0.0480	0.0240	0.0400
33	1	All	Fourth ring gap (in cylinder barrel ... Gap:	0.0570	0.0330	0.0490
	1	All	** Top & 2nd ring (standard gap) ... Tension:	12 lbs.	13 lbs.	17 lbs.
	1	All	** Third ring assembly (standard gap) ... Tension:	11 lbs.	12 lbs.	16 lbs.
	1	All	** Fourth ring (standard gap) ... Tension:	8 lbs.	9 lbs.	13 lbs.
34	1	All	Piston pin in piston (standard or 0.005" oversize) ... Diameter:	0.0013L	0.0001L	0.0007L
35	1	All	Piston pin ... Diameter:	—	1.1243	1.1245
			Piston pin (0.005" oversize) ... Diameter:	—	1.1293	1.1295
36	1	All	Piston pin in cylinder ... End Clearance:	0.0300L	0.0310L	0.0480L
37	1	All	Piston pin in connecting rod bushing ... Diameter:	0.0040L	0.0022L	0.0026L
38	1	All	Bushing in connecting rod ... Diameter:	—	0.0025T	0.0050T
39	1	All	Bolt in connecting rod ... Diameter:	—	0.0000	0.0018L
40	1	All	Connecting rod bearing on crankpin ... Diameter:	0.0060L	0.0009L	0.0034L
41	1	All	Connecting rod on crankpin ... End Clearance:	0.0160	0.0060	0.0100
42	1	All	Connecting bearing and bushing twist or convergence per inch of length ... :	0.0010	0.0000	0.0005
			Crankshaft			
43	2/3	All	Crankshaft in main bearings ... Diameter:	0.0061L	0.0018L	0.0047L
44	2/3	All	*** Crankpins ... Out-of-Round:	0.0015	0.0000	0.0005
45	2/3	All	*** Main journals ... Out-of-Round:	0.0015	0.0000	0.0005
46	2/3	All	Crankshaft main and thrust journals ... Diameter:	2.3720	2.3740	2.3750
47	2/3	All	Crankpins ... Diameter:	2.2470	2.2490	2.2500
48	2/3	All	Crankshaft run-out at center main journals (shaft supported at thrust and rear journals) full indicator reading ... :	0.0150	0.0000	0.0150

Figure 9-4. *An excerpt from an engine overhaul manual showing new-parts and serviceable limits*

Table of Limits *(continued)*

Ref. No.	Chart No.	Model	Description	Serviceable Limit	New Parts	
					Minimum	Maximum
			Crankshaft *(continued)*			
49	**2/3**	**All**	**Crankshaft run-out at propeller flange (when supported at front and rear main journals) full indicator reading .... :**	**0.0050**	**0.0000**	**0.0050**
50	2/3	All	Damper pin bushing in crankcheek ext. .... Diameter:	—	0.0015T	0.0030T
51	2/3	All	Damper pin bushing in counterweight .... Diameter:	—	0.0015T	0.0030T
52	2/3	All	Damper pin in counterweight .... End Clearance:	0.0400	0.0010	0.0250
53	2/3	All	Crankcheek in counterweight .... Side Clearance:	0.0170	0.0070	0.0120
54	3	P	Alternator gear on crankshaft .... Diameter:	—	0.0010T	0.0040T
55	2/3	All	Crankshaft gear on crankshaft .... Diameter:	—	0.0000	0.0020T
56	3	P	Crankshaft in thrust bearing .... End Clearance:	0.0200	0.0060	0.0140
57	2	S	Crankshaft in thrust bearing .... End Clearance:	0.0250	0.0080	0.0180
			Camshaft			
58	2/3	All	Camshaft journals in crankcase .... Diameter:	0.0050L	0.0010L	0.0030L
59	2/3	All	Camshaft in crankcase .... End Clearance:	0.0140	0.0050	0.0090
60	2/3	All	Camshaft run-out at center journals (shaft supported at end journals) full indicator reading .... :	0.0030	0.0000	0.0010
61	2/3	All	Camshaft gear on camshaft flange .... Diameter:	—	0.0005T	0.0015L
62	3	P	Governor drive gear on camshaft .... Diameter:	0.0060L	0.0010L	0.0030L
63	2	S	Governor drive gear on camshaft .... Diameter:	0.0060L	0.0002L	0.0020L
			Crankcase and Attached Parts			
64	2/3	All	Through bolt in crankcase .... Diameter:	—	0.0005T	0.0013L
65	1	All	Hydraulic lifter in crankcase .... Diameter:	0.0035L	0.0010L	0.0025L
66	2/3	All	Governor drive shaft in crankcase .... Diameter:	0.0050L	0.0014L	0.0034L
			Spring Test Data			
	3	All	Oil temperature control valve 0.090 inches minimum travel at .... Temperature:	—	120°	170°
	3	All	Oil temperature control valve must close between .... Oil Temperature:	—	168°	172°
	2	All	Relief valve spring (634150) compressed to 1.25 inch length .... Load:	30 lbs.	32 lbs.	37 lbs.
	2		Oil filter by-pass valve spring (631478) in pump compressed to 1.09 inch length .... Load:	5.0 lbs.	5.6 lbs.	—
	1	All	Inner valve spring (631521) compressed to 1.230 inch length .... Load:	82 lbs.	87 lbs.	97 lbs.
	1	All	Inner valve spring (631521) compressed to 1.746 inch length .... Load:	29 lbs.	32 lbs.	38 lbs.
	1	**All**	**Outer valve spring (631520) compressed to 1.275 inch length .... Load:**	**109 lbs.**	**117 lbs.**	**133 lbs.**
	1	**All**	**Outer valve spring (631520) compressed to 1.791 inch length .... Load:**	**46 lbs.**	**49 lbs.**	**55 lbs.**

* *Use 0.005" oversize piston and 0.005" oversize rings.*

** ***Measure piston ring tension on diameter perpendicular to gap when ring is compressed to specified inch gap.***

*** *If crankshafts are worn beyond these limits they may be repaired by grinding crankpins and journals to 0.010 under new shaft limits and renitriding the crankshafts.*

Figure 9-4. *An excerpt from an engine overhaul manual showing new-parts and serviceable limits (continued)*

Disassembly

The engine is taken to the disassembly area and mounted vertically on a stand. The ignition and fuel metering components are removed and tagged and sent to the proper shops for their overhaul. The engine is completely disassembled, and all parts are placed in a rack so they can be kept together during their trip through the shop.

All parts that are routinely replaced during an overhaul are discarded. These parts usually include all gaskets, oil seals, bearing inserts, stressed nuts and bolts, exhaust valves, pistons rings, and ignition cables.

A note is made of any obvious damage or loose studs that should be further investigated.

Cleaning

When the engine is completely disassembled, all parts are degreased by soaking or spraying them with varsol or Stoddard solvent. Water-based degreasers are not used, because the alkali which is a normal part of these degreasers is difficult to remove from the pores of the metal. When the engine is returned to service, the alkali reacts with the hot lubricating oil to form a soap that causes the oil to foam and the lubricating system to fail.

Degreasing removes dirt, grease, and some of the softer carbon deposits, but many components have hard carbon on them that degreasing solvents cannot remove.

Hard carbon is removed by soaking the part in a decarbonizing solution, which is often heated. Decarbonizer solutions are highly reactive to lightweight metals, so instructions furnished with the solution must be followed in detail. Magnesium parts must never be soaked in a decarbonizer unless it is proven to be safe for magnesium.

The solution softens the carbon so it can be removed from the part by blasting it with wet steam. If this does not remove all traces of carbon, the part may be dry-blasted with plastic pellets, glass beads, or an organic material such as crushed walnut shells. All the machined surfaces must be masked off, and all internal passages tightly plugged. Blasting should be done with the lowest effective air pressure and discontinued as soon as the carbon is removed.

Pistons are highly stressed parts of an engine, and they must be cleaned with special care. Carbon should never be removed by scraping, so it is softened by soaking the piston in a chemical decarbonizer and removed with a soft abrasive blast. The skirt is a critical part of a piston, and it must not be scratched or abraded when it is being cleaned.

Piston ring grooves must never be cleaned with an automotive-type ring groove cleaner, as it will destroy the critical radius at the bottom of the groove and damage the bottom surface. Remove the softened carbon deposits by drawing a hard-twist cotton string through the groove.

The sludge plugs, if installed, must be removed from the crankshaft and all passages thoroughly cleaned by flushing them with wet steam and then spraying them with varsol.

When all the parts have been cleaned, protect all steel parts from rust by coating them with a film of clean oil.

sludge plugs. Spool-shaped sheet metal plugs installed in the hollow throws of some engine crankshafts.

Inspection

The inspection phase of an overhaul is one of the more critical phases. It is here that the decision is made to use a part as it is, repair it, or discard it.

By the time the engine reaches this step, it should be accompanied with an overhaul form with space for recording the results of the inspection. All manufacturer's overhaul manuals, service bulletins, and service letters must be available as well as the proper inspection equipment.

All measuring instruments must be accurately calibrated within the time period specified for the particular instrument, and the equipment used for calibration must be traceable to the National Institute of Standards and Technology (NIST), formerly known as the National Bureau of Standards.

Visual Inspection

A careful visual inspection allows classification of parts into three categories: parts that must be discarded, parts that must be repaired, and parts that appear to be serviceable.

The parts should be inspected under a good light and with the aid of magnifying glasses for indication of deep scratches, warping or distortion, or burned or pitted areas. Studs should be inspected for looseness, damaged threads, or evidence of bending. Gear teeth and bearing surfaces should be carefully examined, as these highly stressed parts are subject to pits, brinelling, and chipping. Figure 9-5 describes types of damage that may be found on a visual inspection.

Nondestructive Inspections

Nondestructive inspection allows us to detect flaws and defects in engine parts without affecting their structural integrity. The principles of NDI are discussed in the *General* textbook of the *Aviation Maintenance Technician Series*, and here we consider the two procedures that are most generally used in reciprocating engine overhaul: magnetic particle inspection and fluorescent penetrant inspection.

magnetic particle inspection. A method of nondestructive inspection for ferrous metal components.

The part being inspected is magnetized and then flooded with a solution of iron oxide suspended in a light oil much like kerosine. Any flaw, either on the surface or just below the surface, forms a north and south pole, and the iron oxide attracted to these poles helps locate the flaw.

The iron oxide is normally treated with a fluorescent dye, and the inspection is conducted in a darkened booth. When an ultraviolet light (black light) is shone on the part, the treated iron oxide shows up as a brilliant line.

Magnetic Particle Inspection

Magnetic particle inspection is one of the most important inspections for engine components made of steel. In this procedure, the steel part is magnetized, and a liquid similar to kerosine, that contains iron oxides, is flowed over the part being inspected. A fault, either subsurface or one that extends to the surface, will disrupt the magnetic lines of force within the metal and form north and south magnetic poles which attract the iron oxide. The oxides are

Defects or Damage Found by Visual Inspection

abrasion — An area of rough scratches or marks that are usually caused by foreign material between moving parts or surfaces.

brinelling — Indentations on a bearing race that are caused by high static loads or excessive force being applied to the bearing. The indentations are smooth, and rounded or spherical.

burning — Surface damage caused by excessive heat. Burning is usually caused by an improper fit, inadequate lubrication, or overtemperature operation.

burnishing — Polishing one surface by sliding contact with a smooth, harder surface. Burnishing does not remove any of the metal.

burr — A sharp or rough projection on the edge of a piece of metal. Burrs are usually the result of some cutting operation such as shearing, drilling, or sawing.

chafing — A condition of wear caused by two parts rubbing together under a light pressure.

chipping — Breaking away of pieces of material caused by excessive stress concentration.

corrosion — An electrolytic action that changes a metal into a chemical salt.

crack — A partial separation of material that is caused by internal stresses or vibration. Once a crack has started, it will continue to grow.

cut — A separation of the material caused by mechanical means. A blow from a sharp-edged object produces a cut.

dent — A small, rounded depression that is caused by a surface being struck by a rounded object. Dents may destroy the rigidity of a part, but it does not necessarily destroy its strength.

erosion — Removal of surface metal by mechanical action of foreign objects such as grit or fine sand. The leading edge of propellers installed on seaplanes are eroded by water.

flaking — The breaking loose of small pieces of the hardened surface of a bearing race, journal, or chrome-plated cylinder wall. Flaking is usually caused by improper lubrication or excessive pressure.

fretting — A condition of surface erosion that is caused by minute movement between two parts that are riveted or clamped together with considerable pressure.

galling — A severe condition of fretting or chafing in which metal from one surface is transferred to the other. Galling is caused by limited movement between highly loaded parts.

gouging — A condition in which a rough-edged channel has been dug in a surface. The material removed from the channel is deposited along its edges.

grooving — A condition in which smooth-edged channels are formed in a surface. Grooving is often caused by the misalignment of parts.

inclusion — Foreign materials in a piece of metal that were left when the metal was poured, rolled, or forged.

nick — A sharp-edged gouge or depression in a metal component that is caused by careless handling.

oxidation — The reaction of a metal with the oxygen in the air. Rust is the porous product of the oxidation of a ferrous metal. Oxidation of aluminum and magnesium produces an airtight film that prevents further reaction between the oxygen and the metal.

peening — A series of blunt depressions in a surface that results from continued pounding by rounded objects. Some highly stressed surfaces are peened with steel shot to produce a uniform compressive stress on the surface to help overcome the tensile stresses to which the part will be exposed.

pitting — Small, irregular shaped hollows in the surface of a piece of metal. Pitting is often caused when corrosion replaces some of the metal with the salts of corrosion.

rust —The rough, red, porous surface that results when a ferrous metal is converted into iron oxide.

scoring — A series of deep scratches caused by foreign particles between moving parts.

scratches — Thin grooves on a part caused by abrasive foreign matter.

scuffing — A buildup of metal from one area onto another. Scuffing is normally caused by insufficient lubrication or clearance between the parts.

spalling — A form of damage in which chips are broken from the surface of a case-hardened material such as a cam lobe or a tappet face. Spalling occurs when the hardened surface is placed under a load great enough to distort the softer inner part of the metal, and cause the hard, brittle surface to crack. Once the crack starts in the surface, chips break out.

stain — A localized change in the color of a material. Stains do not normally change the characteristics of the material.

Figure 9-5. *Types of defects or damage that may be detected by careful visual inspection*

treated with a fluorescent dye that causes them to glow when viewed under an ultraviolet (black) light. The inspection is conducted in a darkened area, and an ultraviolet light is shone on the part being inspected. True faults show up as green lines that, when wiped off, will immediately re-form.

Two methods are used for magnetic particle inspection: continuous and residual. In the continuous method, the part is subjected to the magnetizing force the whole time it is being flooded with the iron oxide solution. In the residual method, the part is magnetized and then removed from the magnetizing force, flooded, and inspected. When either method is used, the parts must be thoroughly demagnetized before reinstalled in the engine.

continuous magnetic particle inspection. A method of magnetic particle inspection in which the part is inspected by flowing a fluid containing particles of iron oxide over the part while the magnetizing current is flowing.

residual magnetic particle inspection. A form of magnetic particle inspection for small steel parts that have a high degree of retentivity. The part is magnetized, removed from, flooded with the iron oxide solution and inspected away from the magnetizing machine.

Two ways a part can be magnetized are circularly and longitudinally, and two inspections are usually conducted using each type of magnetization. This allows faults with any orientation to be detected.

Circular Magnetization

A part is circularly magnetized by placing it between the heads of the magnetizing machine and passing direct current through it. *See* Figure 9-6. Lines of magnetic flux caused by and surrounding the current will magnetize the part as shown. Faults extending lengthwise in the part show the best, and those that extend across the part show the least.

circular magnetism. A method of magnetizing a part for magnetic particle inspection. Current is passed through the part, and the lines of magnetic flux surround it. Circular magnetism makes it possible to detect faults that extend lengthwise through the part.

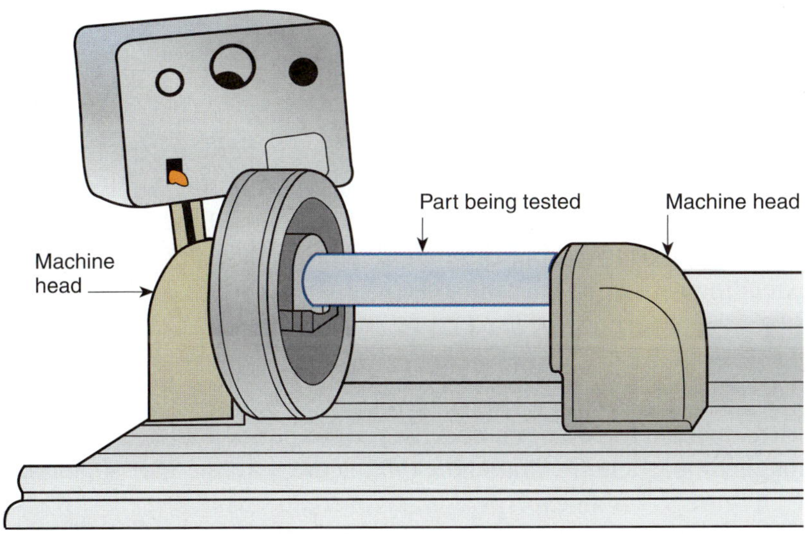

Part is circularly magnetized by placing it between the heads of the magnetizing machine and passing direct current through it.

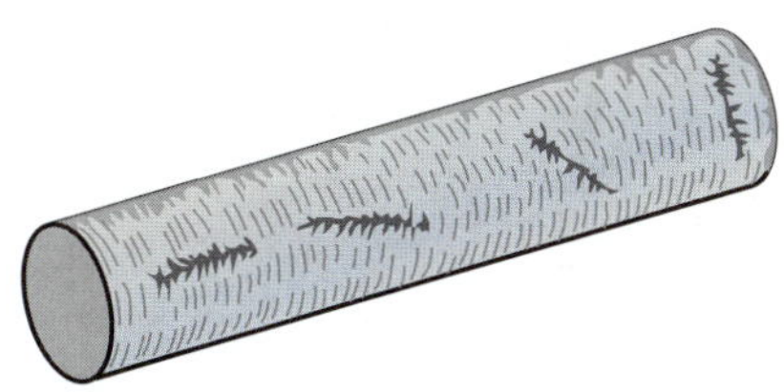
Lines of flux encircle the magnetized part.

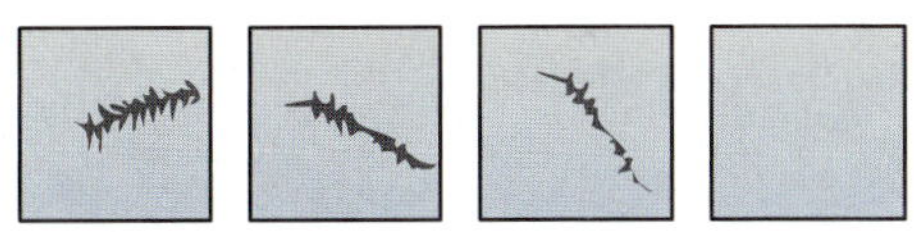
Circular magnetism causes faults that are parallel to the length of the part to show up best.

Figure 9-6. *Circular magnetization*

wrist pin. The hardened steel pin that attaches a piston to the small end of a connecting rod.

Small parts such as wrist pins may be circularly magnetized by placing them over a brass or copper rod clamped between the heads of the machine. Current flowing through the rod produces a magnetic field that magnetizes the part. *See* Figure 9-7.

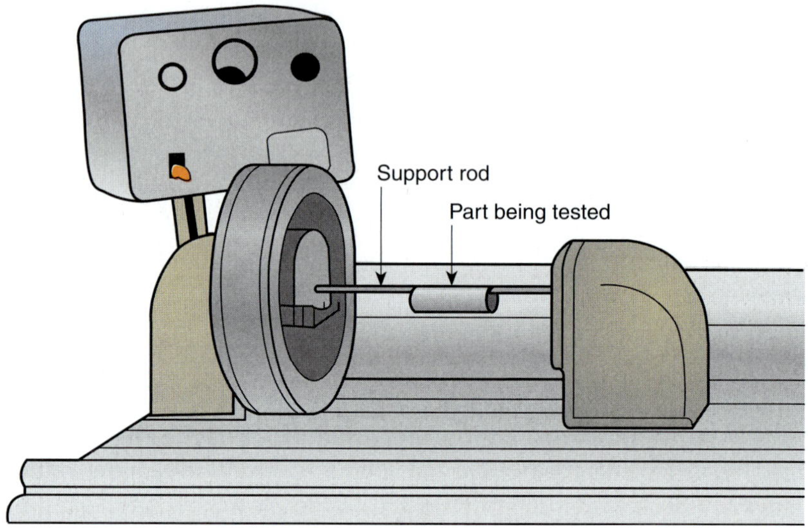

Figure 9-7. *Small parts such as wrist pins can be circularly magnetized by placing them over a nonmagnetic rod through which the magnetizing current flows.*

longitudinal magnetism. A method of magnetizing a part for magnetic particle inspection. Current flows through a solenoid, or coil, that encircles the part so the lines of magnetic flux pass lengthwise through the part. Longitudinal magnetism makes it possible to detect faults that extend across the part.

Longitudinal Magnetization

A part is magnetized longitudinally by supporting it between the heads of the magnetizing machine, and passing current through the windings of a coil, or solenoid, that surrounds the part. *See* Figure 9-8. Lines of magnetic flux caused by current flowing through the coil will magnetize the part. Faults extending across the part show the best, and those that extend along the length of the part show the least.

The Actual Inspection

The overhaul manual for an engine specifies which parts should be inspected by the magnetic particle method and explains the way the part should be inspected. Figure 9-9 is an excerpt from an overhaul manual.

For example, the crankshaft should be inspected twice by the continuous process. The shaft is circularly magnetized by clamping it firmly between the heads of the machine and passing 2,500 amps through it. Flow the oxide-carrying fluid over the shaft while the current is flowing. The journals, fillets, oil holes, thrust flanges, and propeller flange must be carefully examined for indication of fatigue or heat cracks.

Part is longitudinally magnetized by mounting it in the magnetizing machine and passing direct current through a solenoid, or coil, that encircles the part.

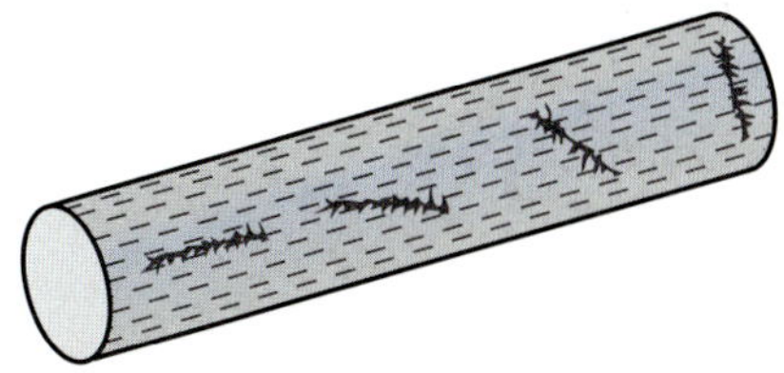

Lines of flux extend lengthwise through the magnetized part.

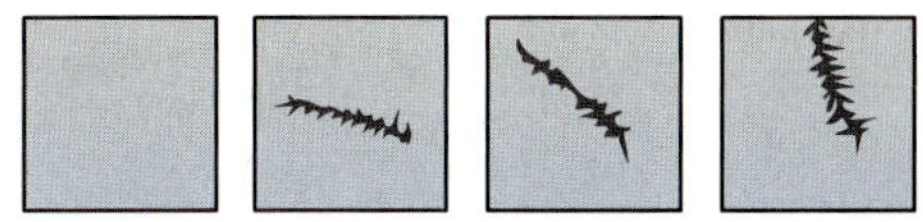

Longitudinal magnetism causes faults that are perpendicular to the length of the part to show up best.

Figure 9-8. *Longitudinal magnetization*

Magnetic Particle Inspection
Fluorescent Method Preferred, Wet Continuous Procedure Required

Part	Method of Magnetization*	D.C. Amperes	Critical Areas	Possible Defects
Crankshaft	Circular and Longitudinal	2500	Journals, fillets, oil holes, thrust flanges, prop flange	Fatigue cracks, heat cracks
Connecting rod	Circular and Longitudinal	1800	All areas	Fatigue cracks
Camshaft	Circular and Longitudinal	1500	Lobes, journals	Heat cracks
Piston pin	Circular and Longitudinal	1000	Shear planes, ends, center	Fatigue cracks
Rocker arms	Circular and Longitudinal	800	Pad, socket under side arms and boss	Fatigue cracks
Gears to 6-inch diameter	Circular or on Center Conductor	1000 to 1500	Teeth, splines, keyways	Fatigue cracks
Gears over 6-inch diameter	Shaft Circular teeth between heads two times 90°	1000 to 1500	Teeth, splines	Fatigue cracks
Shafts	Circular and Longitudinal	1000 to 1500	Splines, keyways, change of section	Fatigue cracks, heat cracks
Thru bolts Rod bolts	Circular and Longitudinal	500	Threads under head	Fatigue cracks

** NOTE: Longitudinal Magnetism: Current applied to solenoid coil surrounding the work.*
Circular Magnetism: Current passed through work or through non-magnetic conductor bar inserted through

Figure 9-9. *Excerpts from an overhaul manual describing the required magnetic particle inspection*

The crankshaft is longitudinally magnetized by clamping it firmly between the heads to support it, but do not flow current through it. Adjust the machine so 2,500 amps flows through the windings of the solenoid to produce the magnetic field. Flow the oxide-carrying fluid over the shaft while the current is flowing. Carefully examine the journals, fillets, oil holes, thrust flanges, and propeller flange for indication of fatigue or heat cracks.

Fluorescent Penetrant Inspection

Nonferrous components, such as the crankcase, accessory case, oil sump, and cylinder heads, are inspected for cracks that extend to the surface by fluorescent penetrant inspection.

For a successful inspection, the part must be thoroughly cleaned. It may be degreased and cleaned with hot trichlorethylene vapors. The hot part is then immersed in a vat of penetrant solution for a specified dwell time. This allows the thin liquid to seep into all cracks and faults that extend to the surface.

The part is removed from the vat and allowed to drain. A coarse spray of warm water removes the bulk of the penetrant from the surface but because the penetrant is not soluble in water, an emulsifier must be used to remove it completely. Tiny globules of the emulsifier form a suspension with the penetrant on the surface, and make it possible for a warm water rinse to remove all of the surface penetrant, without affecting that which has seeped into the cracks or faults.

The part is thoroughly dried, then immersed in a vat of developer and allowed to remain for a specified dwell time. The penetrant has a high affinity for the developer and will be pulled out from any faults that extend to the surface. The part is inspected in a darkened booth under an ultraviolet or black light, and the faults will show as green lines.

Here are some thoughts pertaining penetrant inspection:

- The most important rule regarding penetrant inspection is that the part must be thoroughly cleaned before the inspection begins. Dirt or grease can fill a crack or fault and prevent the penetrant from getting into it.
- A part must not be abrasive blasted before inspected. The blasting is likely to peen over the edges of a fault and prevent the penetrant from getting into it.
- The temperature and dwell time relate to the size of the fault. If the part is hot and the dwell time is long, the penetrant will seep into tiny faults.
- Penetrant that remains on the surface will be picked up by the developer and mask any faults. It is for this reason that rough surfaces require special care and skill for a meaningful inspection.

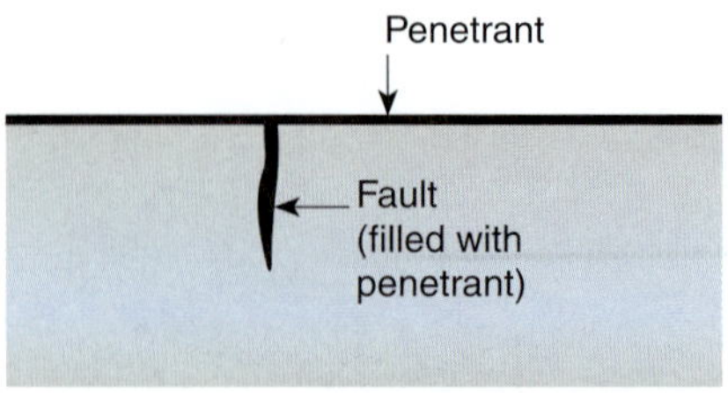

Penetrant seeps into all faults that extend to the surface of the part.

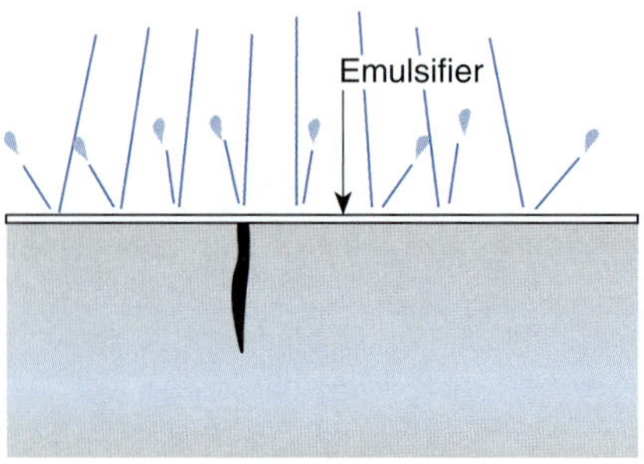

An emulsifier makes it possible for a hot water rinse to remove all of the penetrant that remains on the surface.

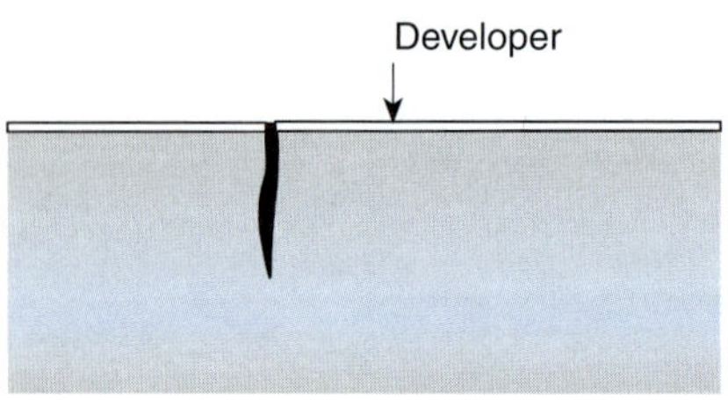

A developer pulls some of the penetrant from the fault and makes a line on the surface that is visible when viewed under ultraviolet light.

Figure 9-10. *Principle of a fluorescent penetrant inspection*

penetrant dwell time. The length of time a part is left in the penetrant when preparing it for inspection by the fluorescent or dye penetrant method. The hotter the part and the longer the penetrant dwell time, the smaller the fault that will be detected.

Dimensional Inspection

A careful dimensional inspection with instruments whose accuracy can be traced to the National Institute of Standards and Technology (NIST) is important to ensure that the parts meet the dimensional requirements listed in the engine overhaul manual.

Refer back to Figure 9-4 on Page 306, the excerpt from the table of limits. Discussed below is how certain measurements from this table are made. Chapter 7 of the *General* textbook of this *Aviation Maintenance Technician Series* explains the way many of the precision measuring instruments are read.

Cylinder and Head Assembly

The cylinder bore in the lower 4¼ inches of the barrel is measured with an inside micrometer caliper or with telescoping gages and outside micrometer calipers. The dimensions are taken twice, with the measurements made at 90° to each other—one measurement parallel to the wrist pin, and the other perpendicular to it. Measurements taken with these two orientations will show the maximum amount of difference caused by wear. The new-parts dimension of the bore is between 5.2510 and 5.2530 inches, with the serviceable limit being 5.2560 inches. The difference between the two measurements is the amount the cylinder is out of round. The difference allowed for new parts is 0.0020 inch, with the serviceable limit being 0.0030 inch.

serviceable limits. Limits included in a reciprocating engine overhaul manual. If a part measures outside of the new-parts limits, but within the serviceable limits, it will not likely wear to the point of causing engine failure within the next TBO interval.

The diameter is measured at a distance of 5.75 inches from the open end of the barrel. The difference between this and the original dimension is the choke of the cylinder. This dimension should be between 0.0030 and 0.0050-inch less than the dimension measured in the lower part of the barrel. The serviceable limit is only 0.0010-inch difference between the two measurements.

choke of a cylinder. The difference in the bore diameter of a reciprocating engine cylinder in the area of the head and in the center of the barrel.

The cylinder barrel is required to fit into the crankcase with a fit between 0.0040L (Loose) and 0.0100L. This means that the outside diameter of the cylinder skirt should be between four and ten thousandths of an inch smaller than the diameter of the hole in the crankcase into which the cylinder fits.

If the intake valve guides are to be replaced, their fit in the cylinder head must between 0.0010T (Tight) and 0.0025T. The diameter of the valve guide must be between one thousandth of an inch and two and a half thousandths of an inch larger than the diameter of the hole in the cylinder head into which it is installed. This is called an interference fit, and to install the guide, heat the cylinder in an oven and chill the guide with dry ice. The hole expands when the cylinder head is heated, and the valve guide shrinks when chilled. The temperature difference allows the parts to be easily assembled, and when they operate at the same temperature, the guide will not loosen.

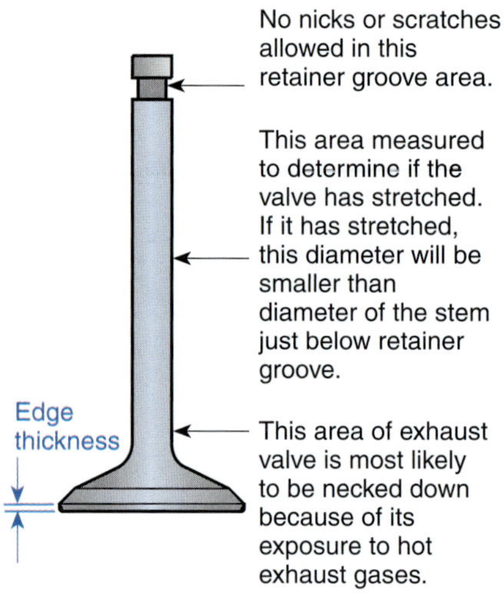

Figure 9-11. *Points to inspect on a poppet valve*

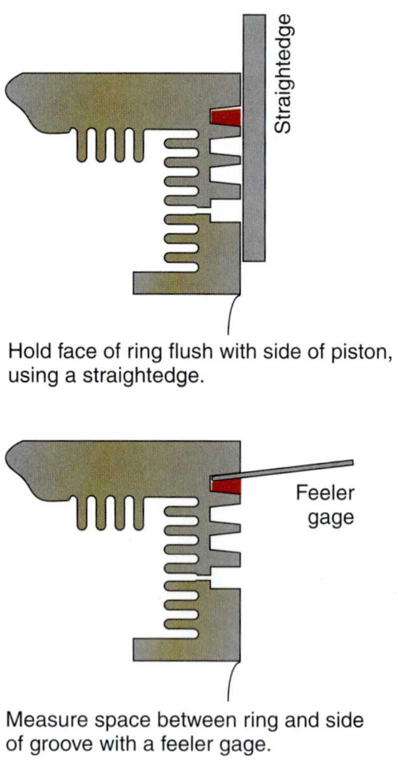

Figure 9-12. *Measurement of the side clearance of a piston ring*

Valves and Valve Springs

During the overhaul of high-performance engines, the exhaust valves are usually replaced, but the intake valves, which operate much cooler, may be reused if they are in good condition. When determining that the used intake and new exhaust valves are dimensionally acceptable, the outside diameter of the valve stem and the inside diameter of the valve guide are measured. The difference between the two measurements must be within the limits specified in the table of limits: 0.0012L to 0.0027L for the intake valve, with the service limit being 0.0050L, and 0.0030L to 0.0050L for the exhaust valve, with its service limit being 0.0070L. The measurement of the valve stem should be made near its center where it receives the most wear, and the inside diameter of the valve guide should be measured parallel to the axis of the rocker arm (perpendicular to the rocker arm shaft).

The face of the valve should be ground to assure its sealing with the seat, and the head thickness at the edge must be great enough to prevent it becoming incandescent and causing preignition.

Valve springs are checked for their compression strength by measuring the amount of force required to compress the spring to a specified length. For an example, a typical outer valve spring requires between 49 and 55 pounds to compress it to a length of 1.791 inches. The serviceable limit for this compression is 46 pounds. Further compression to 1.275 inches must require between 117 and 133 pounds, with 109 pounds being the serviceable limit.

Piston Rings

Three measurements are made on piston rings: end gap, side clearance in the ring groove, and tension.

The end gap is measured by placing the ring in the cylinder, even with the mounting flange and square in the bore. The gap between the two ends is measured with feeler gages. The gap for the top ring should be between 0.0330 and 0.0490 inch, with a service limit for this ring at 0.0570 inch.

The ring is placed in the tester with the gap horizontal. The tester measures the amount of force needed to close the gap to the specified dimension. For example, the top ring should require between 13 and 17 pounds in order to close the gap to between 0.0330 and 0.0490 of an inch. Serviceable limits would allow the ring to compress to the same gap with only 12 pounds of force.

The side clearance in the ring groove is measured by installing the ring in the groove and measuring the distance between the side of the ring and the side of the groove with a feeler gage. The top ring should be loose in the groove with a clearance of between 0.0040 and 0.0060 inch, with a service limit of 0.0080 inch.

Crankshaft

Install new main bearing inserts in the crankcase halves, bolt the halves together, and tighten all bolts to the correct torque. For the bolts that would normally extend through the cylinder mounting flange, protect the surface with washers having a large bearing area to prevent damage to the case. Measure the inside diameter of the bearings with an inside micrometer or a telescoping gage and an outside micrometer. Measure the diameter of the bearing journal with an outside micrometer caliper and compare the two measurements. The diameter of the bearing journal on the crankshaft should be between 0.0018 and 0.0047-inch smaller than the inside diameter of the bearing. The serviceable limit for this measurement is 0.0061-inch difference.

Check the crankshaft runout at the center main bearing journal with the shaft supported in V-blocks at the thrust and rear journals. Measure the propeller flange runout with the shaft supported at the front and rear journals.

Support the crankshaft in a pair of roller-type V-blocks at the specified journals, and measure the amount of runout with a dial indicator mounted on a stand, so the spindle rests on the journal at the location specified in the overhaul manual. Rotate the shaft until the gage deflects the maximum amount, and turn the dial face until the gage reads zero. Rotate the crankshaft until the gage again reads maximum. The reading at this point is the amount the shaft is run out. The maximum allowable runout at the center journal is 0.0150 inch, and at the propeller flange is 0.0050 inch.

feeler gages. A type of measuring tool consisting of strips of precision-ground steel of accurately measured thickness.

Feeler gages are used to measure the distance between close-fitting parts, such as the clearances of a mechanical system or the distance by which moving contacts are separated.

runout. A measure of the amount a shaft, flange, or disk is bent or fails to run true. Runout is normally measured with a dial indicator.

V-blocks. A fixture that allows a shaft to be centered and rotated to measure any out-of-round condition.

Connecting Rods

There are four important measurements on a connecting rod:

- Clearance between the big end bearing and the crankpin on the crankshaft
- End clearance between the connecting rod and the crankpin cheek
- Clearance between the piston pin and the small end bushing
- Twist and convergence, or bending, of the rod

Install new bearing inserts in the big end of the connecting rod and the connecting rod cap. Assemble the cap to the rod with the cylinder numbers on the two parts matching and on the same side, and tighten the nuts to the recommended torque. Measure the inside diameter with an inside micro-meter or with a telescoping gage and an outside caliper. The inside diameter of the bearing should be between 0.0009 and 0.0034-inch greater than the outside diameter of the crankpin journal. The serviceable limit is 0.0060 loose.

The end clearance between the edge of the connecting rod and the side of the crank cheek is measured with a feeler gage. It should be between 0.0060 and 0.0100 inch, with 0.0160 inch as the serviceable limit.

Measure the inside diameter of the bushing in the small end of the rod with a telescoping gage and an outside micrometer. Measure the outside diameter of the center of the wrist pin that goes with the rod. The inside diameter of the bushing should be between 0.0022 and 0.0026 inch larger than the diameter of the wrist pin. The serviceable limit for this measurement is 0.0040 inch loose.

The connecting rod twist and convergence, or bend, is measured on a fixture such as the one in Figure 9-13.

The correct mandrels with arbors on each end are inserted into the small-end bushing and the big-end bearing, and the assembly is placed on a pair of parallel bars on a surface plate. Twist is measured by placing a feeler gage between the end of the arbor and the parallel bar. Convergence is measured between particular points on the two arbors on each side of the rod. The difference between the measurements on the two sides is the amount of convergence, or the amount the rod is bent. The new-parts maximum amount of twist or convergence allowed is 0.0005 inch for each inch of distance between the two mandrels. The serviceable limit is 0.0010 inch for each inch of length.

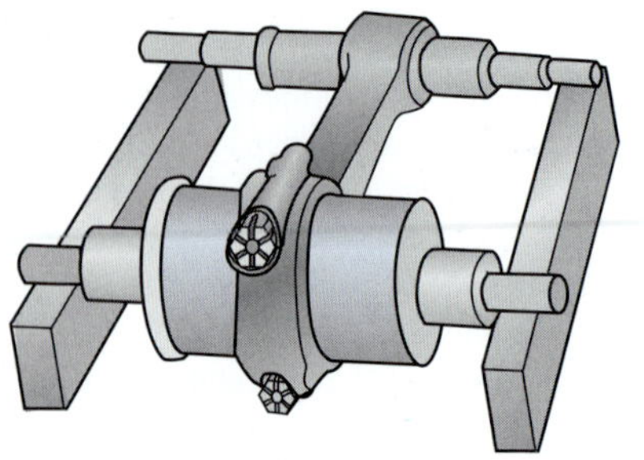

Fixture for measuring connecting rod twist.

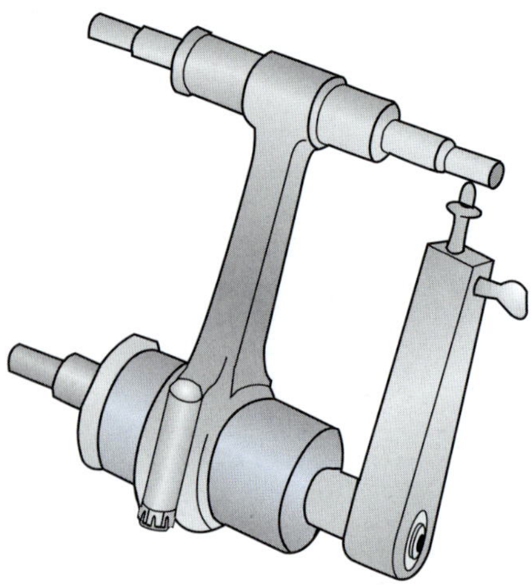

Fixture for measuring connecting rod convergence.

Figure 9-13. *A fixture such as this is used to measure the twist and convergence of a connecting rod.*

Repair

When all parts have been inspected dimensionally and visually, using the appropriate nondestructive inspection method, the faults are corrected and the parts made ready for reassembly.

Crankcase

The crankcase is the central part of an aircraft engine, and even though it is made as light as possible, it must withstand tremendous stresses. Crankcases do crack, and certain FAA-certificated repair stations are approved to repair them by welding.

A complete crankcase repair consists of welding any cracks with one of the inert-gas arc welding procedures and peening the repair to relieve stresses that are left in the metal by welding. When the weld is complete, the repair is machined to match the rest of the surface. Bearing cavities that have worn excessively can be built up by welding, and all of them line-bored to be sure they are in perfect alignment. The camshaft does not normally have bearing inserts, but it rides in bearings cut in the crankcase. If these have worn, they can be line-bored to an oversize dimension and a camshaft with oversize journals can be installed.

line boring. A method of boring several holes in such a way that they all have a common center.

If any studs are bent or otherwise damaged, they can be removed and new ones installed. Loose studs can be removed and the hole cleaned up and tapped for the installation of an oversize stud. Some damaged stud holes can be repaired by tapping the hole and installing Heli-Coil inserts.

Crankshaft

The crankshaft is the most highly stressed component in the engine, and any repairs that are made to it must be done by a facility that is fully equipped, experienced, and approved by the FAA.

The bearing journals on most crankshafts are case hardened by nitriding. When a shaft is nitrided, it is heated in an atmosphere of ammonia gas (NH_3) until some nitrogen combines with the alloys in the steel to form extremely hard nitrides in the surface of the metal. This hardening process is done at a lower temperature than other case hardening procedures, so it does not warp the metal. Unlike hard plating, nitriding causes the metal to grow only a few ten thousandths of an inch. Nitrided surfaces are not only hard, but also brittle; therefore, nitrided crankshafts should never be straightened to correct excessive runout.

Some manufacturers allow crankshafts whose journals have worn out of round to be reground to an undersize dimension and the shaft renitrided. Crankshaft grinding is a highly specialized operation, because the original radius between the bearing journal and the cheeks must be preserved to prevent stress concentrations that would cause crankshaft failure.

Dynamic dampers installed on a crankshaft to compensate for torsional vibration are extremely critical, and if not properly installed, they can cause severe vibrations and stresses. Visual and dimensional inspection of the attachment blades and the attaching hardware are extremely critical.

Almost all aircraft engines have hollow crankshafts that allow engine oil to be carried to the connecting rod bearings. Some of these engines are equipped with sludge plugs in the crankpins. These thin metal spool-shaped plugs are pressed into the hole in the crankpin. Oil flows through transfer tubes from the main bearing journal into the crankpin and out through a drilled hole to lubricate the connecting rod bearing. Any sludge in the oil is slung out by centrifugal action and is trapped around the outside of the sludge plug. When the engine is overhauled, the sludge plugs are removed and when the crankshaft is reassembled, new plugs are pressed into place. The fit of the plug is important because it must not allow oil to leak past its ends, as this would starve the connecting rod bearing.

sludge. A heavy contaminant that forms in an aircraft engine lubricating oil because of oxidation and chemical decomposition of the oil.

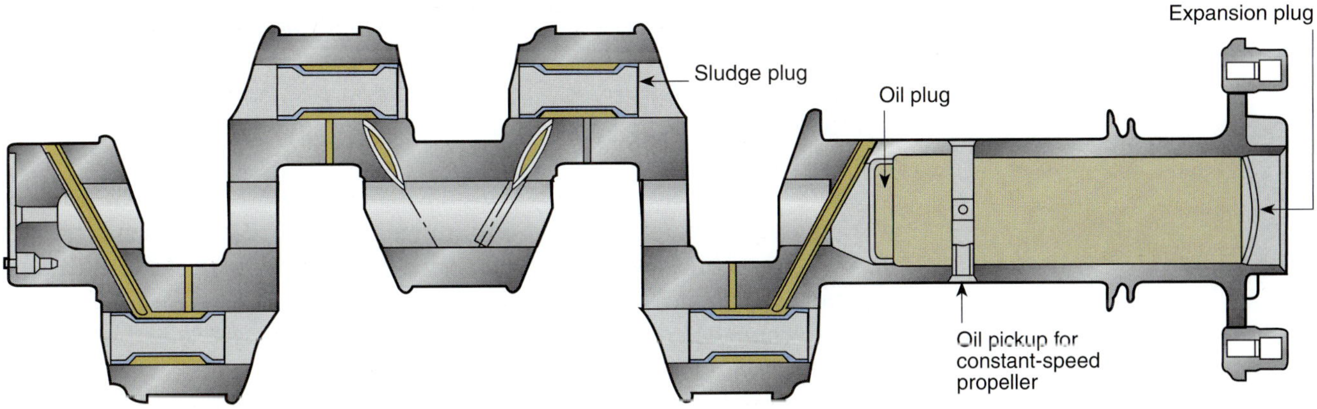

Figure 9-14. *Sludge plugs are pressed into hollow crankpins of some crankshafts to trap and hold sludge in the lubricating oil.*

Engines equipped with hydraulic constant-speed propellers route the control oil through the forward end of the crankshaft. *See* Figure 9-14. When the engine is fitted with a fixed-pitch propeller, a steel oil plug is pressed into the hollow shaft just ahead of the first throw, and an expansion plug is installed in the end of the crankshaft. The expansion plug must be removed, the accumulation of sludge cleaned out, and a new plug installed.

Cylinders

Cylinder reconditioning consists of thoroughly cleaning them and inspecting the barrel visually and dimensionally, inspecting the head for cracks using the fluorescent penetrant method, replacing the valve guides and seats, honing the cylinder walls, and replacing the rocker shaft bushings.

Cracks in certain areas that do not exceed specific limits in some cast aluminum alloy cylinder heads may be repaired by welding.

The cylinder is placed in a temperature-controlled oven and heated to expand the head. Special water-injected tools are used to cool the valve guides and seats so they will contract and can be removed. Bosses for the valve guides and seats are line-bored so when new ones are installed, they will be perfectly concentric. The head is again heated to expand the holes and new valve guides, and seats are chilled with dry ice to shrink them, then they are installed.

Great strides have been made in the treatment of engine cylinder barrels. For many years there were three types of cylinder bores found in aircraft engines: plain steel, porous chrome-plated steel, and nitrided steel.

Porous chrome plating has been used to reclaim worn cylinders since World War II. Cylinder barrels that have worn out-of-round can be ground to shape, including the choke, and the walls electrolytically plated with a special hard chromium to restore the desired dimensions and provide a hard, wear-resistant surface. Oil does not naturally adhere to the chrome plating, but the surface contains a spider web pattern of thousands of tiny cracks. When the plating is completed, the current is reversed for a controlled length of time to enlarge these cracks into channels of specific dimensions that hold oil for lubrication.

The walls of nitrided steel cylinder barrels have been treated by heating the cylinder in an atmosphere of ammonia (NH_3) gas. Nitrogen from the ammonia combines with certain alloying elements in the steel to produce an extremely hard and wear-resistant surface. Nitriding is not a plating, but is rather a change in the chemical structure of the steel surface, and it diffuses into the base metal a few thousandths of an inch.

One problem with a nitrided steel cylinder barrel is its susceptibility to rusting. If an engine with nitrided cylinders sits idle for any length of time with the cylinders inadequately preserved, the walls will rust.

The surface of the walls of plain steel and nitrided steel cylinders should be honed to a specified roughness with a special cylinder hone. Notice in Figure 9-4 on Page 306 that the new-parts cylinder bore surface roughness must be between 15 and 25 RMS. This means that cylinder walls are roughened with hones that make grooves in the surface that are between 15 and 25 millionths of an inch (0.000 015 and 0.000 025-inch) deep. The actual depth is measured with a profilometer that draws a diamond-tipped stylus across the surface of the cylinder wall. The stylus produces an electrical signal that is proportional to the depth of the grooves in the surface.

The honing pattern is critical. The notation in Figure 9-15 specifies that the cylinder wall be honed with a 45° crosshatch, and the surface roughness should be 20 microinches RMS.

There are three new cylinder bore treatments that have been developed by Engine Components, Inc., of San Antonio, Texas that are eliminating many problems inherent with both chrome plating and nitriding. These proprietary treatments are Cermicrome®, Cermisteel™, and CermiNil™.

In the Cermicrome® process, the prepared cylinder bore is plated with hard chromium, and then silicon carbide particles are locked into the surface. The silicon carbide particles are four times as hard as chromium, and unlike chromium, oil will adhere to them. This treatment has the advantages of porous chrome plating but provides more rapid ring break-in and more predictable oil consumption.

The Cermisteel™ process impregnates silicon carbide particles into plain steel or nitrided steel cylinder bore surfaces. This process increases the surface hardness by more than 400% and significantly reduces friction between the rings and the cylinder bore. It also reduces ring wear and gives the engine better oil consumption.

The CermiNil™ process is similar to chrome plating except that a special proprietary nickel alloy that contains carbide particles throughout the plating matrix is deposited on the prepared cylinder bore. This process has the advantage of a rapid break-in and predictable oil consumption. The bore does not rust, as does a nitrided bore, and the plating process is more environmentally safe than chrome plating. The cross hatching used on a CermiNil™ cylinder bore is only about eight microinches deep compared with the 15 to 25 microinches for plain steel or nitrided bores.

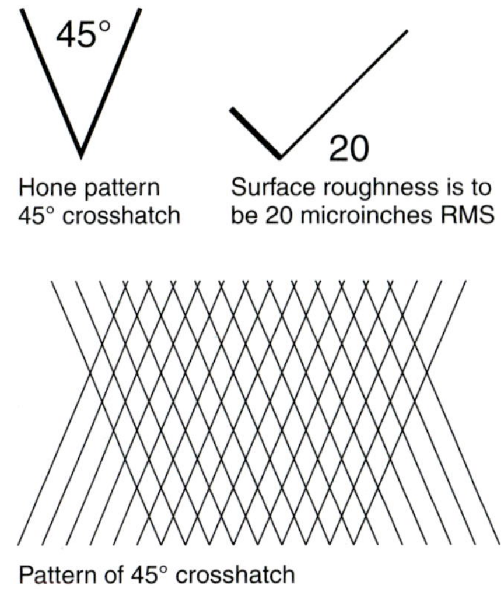

Figure 9-15. *Symbols used to specify the honing pattern and surface roughness for a cylinder wall. A 45° crosshatch with a surface roughness of 20 microinches is to be used.*

surface roughness. The condition of the surface of a reciprocating engine cylinder wall that has been honed to make it hold lubricating oil. Surface roughness is measured in microinches RMS.

Figure 9-16 shows the markings used on cylinders to identify their characteristics. The type of cylinder bore is identified either by a band of the proper color between the flange and the lower fin, or by the outer tip of the cylinder head fins between the spark plug and the cylinder barrel. Painted fin tips between the spark plug and the rocker box indicate the reach of the spark plug that must be used.

Cylinder barrel configuration	**Color**
Chromed bore	Orange
Nitrided bore	Blue
Cermisteel™ bore	2 White stripes
Cermicrome® bore	2 Orange stripes
CermiNil™ bore	2 Silver stripes
Rebarreled cylinder	White
+0.010 oversize barrel (Lyc.)	Green
+0.015 oversize barrel (TCM)	Black
+0.020 oversize barrel (Lyc.)	Yellow
Painted fin tips between spark plug and rocker box identifies required spark plug reach	**Color**
Long reach spark plugs	Yellow
Short reach spark plugs	Gray or Blue

Figure 9-16. *Cylinder color codes*

Valves, Guides, and Seats

Normal overhaul practices on many high-performance engines require the exhaust valves be replaced at each overhaul. The less hostile environment in which the intake valves operate does not always require their replacement if they pass all non-destructive testing and dimensional inspections. Regardless of whether new or reconditioned valves are installed, it is extremely important that the valves seal perfectly in their seats and allow absolutely no leakage.

The face of most intake valves is ground with an angle of 30° relative to the plane of the valve seat, while exhaust valves are ground with an angle of 45°. The flatter angle for the intake valves allows better flow of the fuel-air charge into the cylinder and greater valve life. The steeper angle for the exhaust valve allows the valve to seat tighter and prevents carbon buildup on the seat.

One key factor of valve reconditioning is that the valve face and seat must be perfectly concentric. To achieve this, the hole for the valve guide and the recess for the valve seat are line-bored so they have the same center. The cylinder head is uniformly heated in a oven to expand the holes, and the guides and seats are chilled with dry ice to shrink them. The guides and seats are installed with this temperature difference, and when they are at the same temperature, the fit is so tight that they will not loosen.

The valve is placed in a valve grinding machine and any minor pitting on the face is removed with a fine grinding wheel. The face of the valve is ground so that it runs true within the allowable limits, normally 0.0020 inch or less. The valve must not be ground so much that the edge becomes sharp or featheredged. The thickness of the edge must be no less than that specified by the engine manufacturer. If the edge is sharp, it will get so hot that it will become incandescent and prematurely ignite the fuel-air mixture, causing preignition, which can easily lead to detonation. *See* Figure 9-17.

The face angle of the valve seat is critical, and it must be ground exactly as is specified by the engine manufacturer. The seat is ground so that it provides an interference angle with the valve. The engine manufacturers specify that there be between 0.5° and 1° difference in the angle of the valve and the valve seat. *See* Figure 9-18.

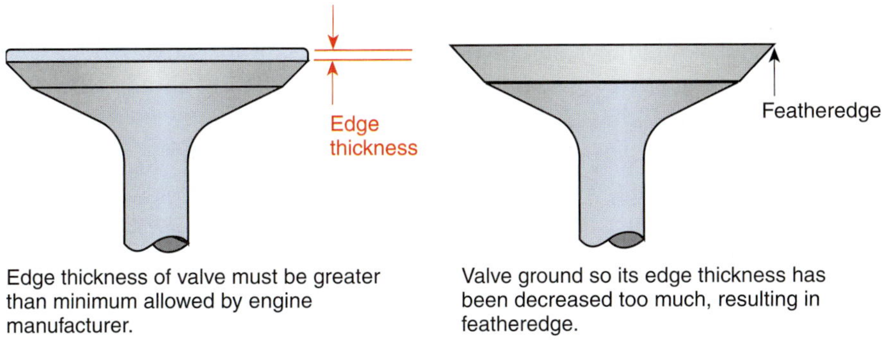

Figure 9-17. *The valve face must not be ground so much that the edge becomes sharp. A featheredged valve can overheat and cause preignition.*

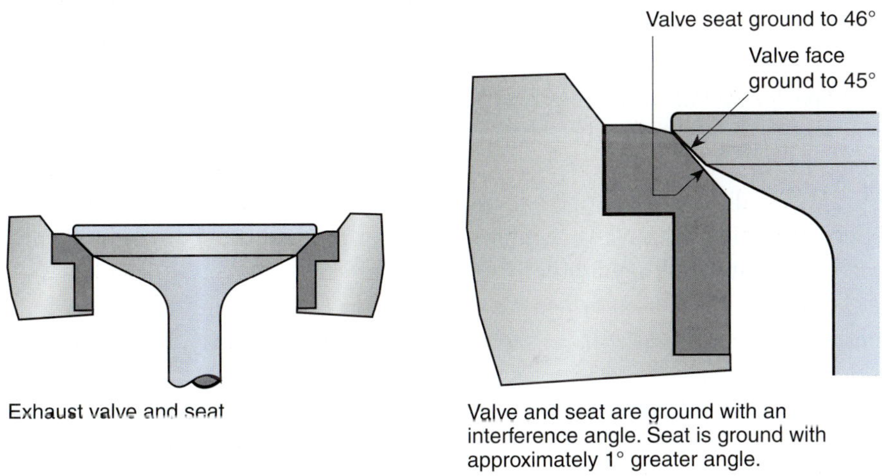

Figure 9-18. *Valves are ground with an interference angle. The seat is ground with a 0.5° to 1° greater angle than the valve face.*

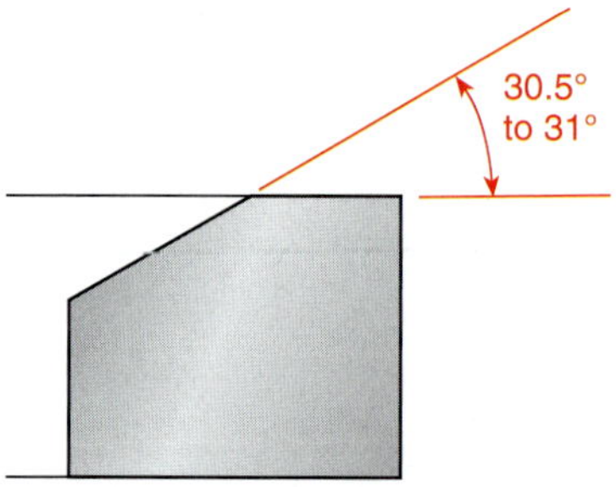

First cut is the valve seat angle plus the amount needed for the interference angle.

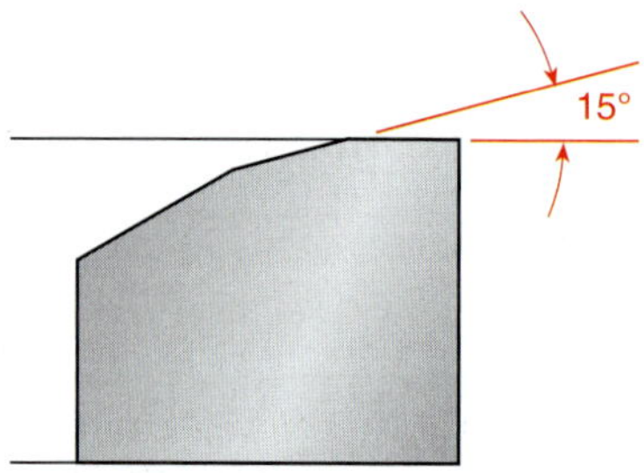

Second cut uses a 15° stone, establishing the diameter of the seat.

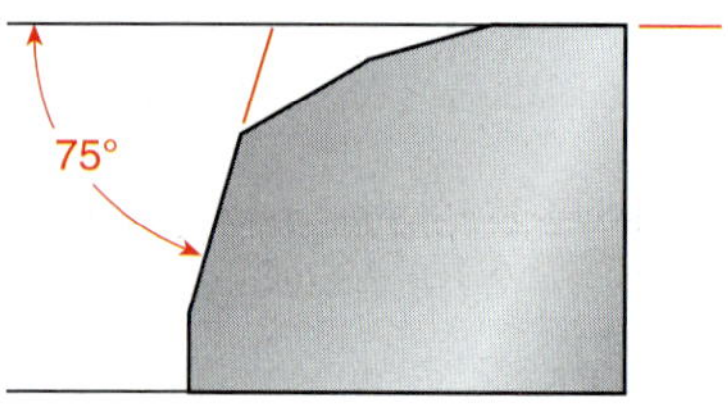

Third cut uses a 75° stone, cutting the bottom of the seat to establish the seat width.

Figure 9-19. *Procedure used to grind the seat for a typical intake valve*

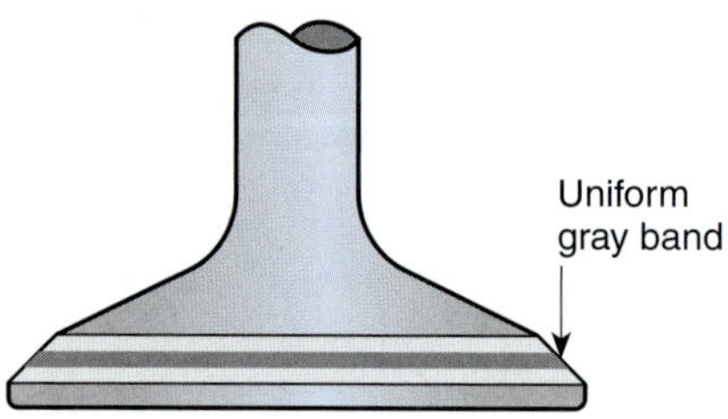

Figure 9-20. *A properly lapped valve will have a thin uniform gray band in the center of the ground face. This band is the actual sealing area between the valve and its seat.*

The valve seat is shrunk into place, and the valve guide is reamed to its final dimension. A valve-seat grinding tool piloted from the valve guide is used to grind the seat so it is concentric with the valve guide and has the correct face angle.

The first stone used to grind the seat has an angle that will give the seat the proper interference angle with the valve. This is normally one-half to one degree greater than the valve face angle. After the seat angle is ground, a 15° stone is used to grind the top of the seat to the correct diameter as specified in the overhaul manual. A third stone, this one having a 75° angle, grinds the bottom to narrow the seat to the specified dimension. *See* Figure 9-19.

Exhaust valve seats are ground in the same way as the intake, but they are not normally narrowed by using the third stone.

After the valves and valve seats have been ground to the proper angles, the valves are lapped into their seats. A small amount of valve-lapping compound is placed on the valve face and the valve is inserted in the valve guide. The valve is rotated against the seat and the lapping compound abrades the surfaces to form a gastight fit.

When the lapping is completed, every trace of the compound is removed from both the seat and the valve, and the valve is identified so it will be returned to the cylinder into which it has been fitted. *See* Figure 9-20.

Valve Operating Mechanism

The camshaft and tappet bodies have been responsible for a number of engine failures, and they must be carefully examined for any indication of pitting or spalling. Any indication of spalling on the face of any tappet bodies is a signal that the cam lobes may also have problems, and they should be carefully examined and a new camshaft installed if there are any indications of surface irregularities or feathering at the edges of the lobe.

It is important that the tappet bodies be kept with the same lobe on the camshaft when the engine is reassembled, and any time a new or reground camshaft is installed in the engine, all tappet bodies must be replaced with new ones.

The hydraulic assemblies that ride inside the tappet bodies deserve special consideration. The two parts, the plunger and the plunger body, are matched units and they are not interchangeable.

The assemblies should never be magnetized, as magnetization will prevent the check valve from seating.

After the hydraulic assembly is thoroughly cleaned and visually inspected for signs of damage, it is tested by assembling the plunger in the plunger body and quickly depressing it with your finger. If it bounces back, the check valve is seating properly and the assembly is satisfactory. If it does not bounce back, the valve is not seating and the assembly must be replaced. *See* Figure 9-22.

Reassembly

When all parts have been inspected and repaired, the engine is ready to be reassembled. Here, we discuss some typical reassembly procedures, but it is extremely important to follow the manufacturer's overhaul manual in detail.

It is critical that you use only the lubricants specified in the overhaul manual when assembling the engine. The use of an improper lubricant can prevent the piston rings from seating, and the improper application of lubricant can cause bearings to be damaged before the lubricating oil reaches them when the engine is started for the first time.

Crankshaft

Check to be sure that the crankshaft is ready for installation. The sludge plugs, oil plug, and expansion plug must be properly installed, and if any gear is attached to the crankshaft, it must be installed, properly torqued, and safetied. All counterweights and dynamic dampers must be properly assembled and the retaining rings correctly installed.

Mount the crankshaft vertically on an assembly stand. Lubricate the connecting rod bearings and install the connecting rods, making sure that the numbers stamped on the rods and caps are on the correct side of the crankshaft. Use new bolts and nuts, and tighten the nuts to the specified torque, and safety them as shown in the overhaul manual. If new rods are being installed, match them weightwise to the rods on the opposite side of the engine.

Crankcase

Place the crankcase halves on a clean, flat surface and install the main bearing inserts, paying careful attention to the locking tangs or dowels. If the engine uses mushroom-type tappet bodies such as those in Figure 9-21, lubricate and install the tappet bodies in the crankcase halves, and then place the camshaft in its bearing journals.

Lubricate the main bearing inserts, and install the crankshaft in the crankcase half. Install the front oil seal in the recess around the crankshaft.

Apply a thin layer of nonhardening gasket compound to the mating surfaces of both crankcase halves, and embed in this compound, on one of the case halves, lengths of very fine silk thread on both sides of the bolt holes.

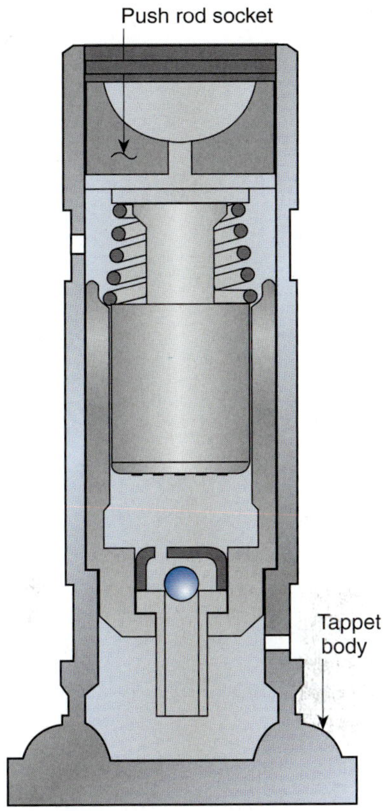

Figure 9-21. *The tappet bodies should be kept with the same lobe on the camshaft.*

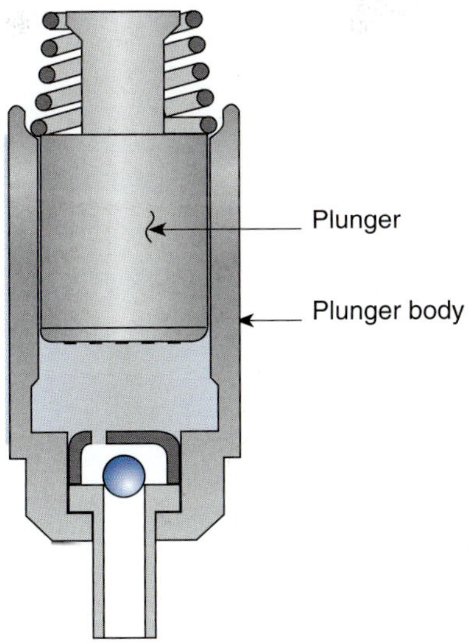

Figure 9-22. *The plunger and plunger body have a precision fit and are matched units.*

Hold the tappet bodies in place, and very carefully lower the crankcase half over the one in which the crankshaft and camshaft are installed. Keep the halves parallel as they are being assembled so they will not bind. Bolt the two halves together, and install torque hold-down plates over the cylinder pads. Tighten all nuts to the specified torque and in the sequence specified in the overhaul manual.

Place the assembled crankcase on the assembly stand in a vertical position for the installation of the cylinders.

Pistons and Rings

Lubricate the pistons, rings, and wrist pins with the proper lubricant, and using the correct ring expander, expand the rings and slip them into their grooves. Pay particular attention to the way the ring is installed, with the side of the ring specified in the overhaul manual toward the top of the piston. Stagger the ring gaps as directed in the overhaul manual.

It is extremely important that only the type of ring approved for the particular type of cylinder bore be used. Chrome-plated piston rings should never be installed in cylinders with chrome-plated walls. If this is done, the rings will not seat and the engine will pump oil.

Cylinders

Lubricate the valve stems and install the valves in their guides. Place the cylinder over a wooden post with a rounded end to hold the valves in place while the springs are installed.

Install the spring retainers and valve springs over the valve stems, and using the correct valve spring compressor, compress the valve springs and install the split keys. Install valve rotators on the end of the stems if they are used.

Install any required fin stabilizers and intercylinder baffles. Place the cylinder base seal around the cylinder skirt.

Put the wrist pin in one side of the piston, and using the correct piston ring compressor, compress the piston rings and install the piston in the cylinder, pushing it almost to the wrist pin.

Install the hydraulic valve lifters in the mushroom-type tappet bodies, or install the complete valve lifter if they are of the non-mushroom type. Remove the torque hold-down plates and install the cylinders.

Rotate the engine so the connecting rod is fully extended, and push the wrist pin through the small end of the connecting rod and through the other side of the piston. Check that the piston pin plugs are securely in place. Compress the oil wiper ring, if one is installed, and carefully move the cylinder down over the piston until the skirt enters the crankcase. Pushrod housings with proper seals must be installed on some engines as the cylinder is moved into place. Put new nuts on the cylinder studs and tighten them to the correct torque in the sequence specified in the overhaul manual.

Install the pushrods, rocker arms and their shafts. Press on the pushrod to completely collapse the hydraulic unit and measure the clearance between the rocker arm and the end of the valve stem. If this is not within the limits specified in the overhaul manual, replace the pushrod with one of a different length that will allow the correct clearance.

Final Assembly

Install the oil sump, the induction pipes and manifold, and the carburetor or fuel injection system. Install and time the magnetos to the engine and install new spark plugs and a new ignition harness. Install new or overhauled accessories such as pumps, propeller governor, and generator or alternator. Replace all baffles, and install new hoses, air ducts, and belts.

Testing

When the engine is completely assembled, it is installed in a test cell and connected to a dynamometer or fitted with a test club. An air scoop is installed to force cooling air through the cylinder fins. A clean induction air filter is installed and an exhaust system is connected to the cylinders. The engine is serviced with the proper quantity of break-in oil, and is connected to a fuel system that provides an adequate flow of the proper grade of fuel.

test club. A wide-blade, short-diameter propeller used on a reciprocating engine when it is run in a test cell. A test club applies a specific load to the engine and forces the maximum amount of air through the engine cooling fins.

The test stand should be equipped with accurately calibrated test instruments. These should include:

- A tachometer
- An oil pressure gage
- An oil temperature gage
- A cylinder head temperature gage installed on the cylinder specified by the engine manufacturer
- A water manometer for measuring the pressure inside the crankcase
- A voltmeter and ammeter

Preoil the engine by forcing oil through the engine with a pressure oiler until some runs out of the connection for the oil pressure gage. If no pressure oiler is available, one spark plug may be removed from each cylinder and the engine turned over with the starter, until pressure is indicated on the oil pressure gage.

When the engine is preoiled and a final check of the entire installation shows that all controls are properly connected and have the correct travel, the engine is started and run according to a recommended test schedule. A typical test schedule is seen in Figure 9-23.

Recommended Run-In Schedule		
RPM	**Time (Minutes)**	**Remarks**
1200	10	
1500	10	
1800	10	Check magneto drop.
2000	10	Do not exceed 125 RPM on either magneto or
2200	10	35 RPM between magnetos.
2400	10	
Full throttle RPM	15	
Full throttle RPM	60	Oil consumption run

Figure 9-23. *Typical run-in schedule for an overhauled engine*

STUDY QUESTIONS: RECIPROCATING ENGINE OVERHAUL

Answers begin on Page 341. *Page numbers refer to chapter text.*

38. Aircraft reciprocating engines _______________ (do or do not) have a specific time they must be given a top overhaul. *Page 304*

39. A top overhaul __________________ (may or may not) be used to extend the TBO of an engine. *Page 304*

40. When an aircraft engine is overhauled in the field with all parts conforming to new parts limits, the engine operating time ________________ (may or may not) be set back to zero hours. *Page 305*

41. An engine may be assembled with parts that are outside of the new parts dimensions but within the serviceable limits. Assembling an engine in this way is generally _____________________ (illegal or uneconomical). *Page 304*

42. An engine that is rebuilt by the manufacturer or an agency approved by the manufacturer is considered to have _____________ hours of operating time. *Page 305*

43. All usable original parts of an engine _____________ (are or are not) kept together and assembled in a rebuilt engine. *Page 305*

44. The carburetor and magnetos _______________ (are or are not) normally installed on the engine when it is sent to an overhaul facility for major overhaul. *Page 305*

45. Three abrasives that are usually recommended for removing carbon deposits from engine components are ____________________, ______________________, or ____________________. *Page 308*

46. Softened carbon deposits may be removed from piston ring grooves with a/an ________________________. *Page 308*

47. After steel parts have been degreased and steam cleaned, they should be protected from rust by coating them with _________________________. *Page 309*

48. The two most widely used types of nondestructive inspection for aircraft engine components are ____________________ and _________________________. *Page 309*

49. Magnetic particle inspection __________________ (can or cannot) detect subsurface faults. *Page 309*

50. Two types of magnetic particle inspection are _________________ and _________________. *Page 311*

51. Two ways a part can be magnetized are ____________________ and ___________________. *Page 311*

52. When inspecting a part for cracks parallel to its length, the part should be magnetized ___________________ (circularly or longitudinally). *Page 311*

53. When inspecting a part for cracks perpendicular to its length, the part should be magnetized ___________________ (circularly or longitudinally). *Page 312*

54. When the magnetizing current flows through a part, the part is magnetized ___________________ (circularly or longitudinally). *Page 311*

55. For a fluorescent penetrant inspection to detect a fault, the fault must extend to the surface. This statement is ____________________ (true or false). *Page 314*

56. The penetrant used in a fluorescent penetrant inspection ____________________ (is or is not) soluble in water. *Page 314*

57. After a part has been in the penetrant vat for the required length of time, it is rinsed with water to remove the bulk of the penetrant from the surface. Then the surface is covered with a/an ____________________ that makes it possible for a hot water rinse to remove all the penetrant from the surface. *Page 314*

Continued

STUDY QUESTIONS: RECIPROCATING ENGINE OVERHAUL *Continued*

58. Abrasive blasting __________________ (is or is not) a good method to use for cleaning a part that is to be inspected by the fluorescent penetrant method. *Page 314*

59. All precision measuring instruments used to overhaul an FAA-certificated aircraft engine must have their accuracy traceable back to the ___________________________ . *Page 315*

60. The bore diameter of a cylinder may be measured with a/an __________________________ or by an outside micrometer and a/an _________________________ . *Page 315*

61. A cylinder barrel is measured twice, with the measurements taken at 90° to each other. The difference between the two dimensions is the amount the cylinder is ____________________ (tapered or out-of-round). *Page 315*

62. The cylinder diameter is measured near the bottom, and again near the top. The difference in these two readings is the _________________ of the cylinder. *Page 315*

63. If a part is larger than the hole into which it is to be installed, the fit is called a/an ______________________ fit. *Page 315*

64. If the diameter of a valve stem, measured in the center, is less than it is near the spring retainer groove, the valve has been ____________________ . *Page 316*

65. Three measurements that are important for a piston ring are:
    a. ____________________________
    b. ____________________________
    c. ____________________________
    Page 316

66. Four dimensional measurements that are made to a connecting rod are:
    a. ____________________________
    b. ____________________________
    c. ____________________________
    d. ____________________________
    Page 317

67. Damaged stud holes can be repaired by tapping them with a special tap and installing __________________ inserts. *Page 318*

68. The bearing journals of a crankshaft are case hardened by ______________________ (carburizing or nitriding). *Page 319*

69. It __________________ (is or is not) generally recommended that nitrided crankshafts be straightened to correct excessive runout. *Page 319*

70. Sludge plugs are removed, and the sludge is cleaned from the crankpin cavity of a crankshaft during ___________________ (annual inspection or major overhaul). *Page 319*

71. Cylinder heads are inspected for cracks by _________________ (fluorescent penetrant or magnetic particle) inspection. *Page 320*

72. A table of limits specifies the fit of a valve guide in the cylinder head be between 0.0010T and 0.0025T. The diameter of the valve guide should be ________________________ (larger or smaller) than the diameter of the hole into which it is to be installed. *Page 322*

73. Perfect alignment between the valve guide and the valve seat is assured by ________________________ the holes into which the new guides and seats are installed. *Page 322*

74. Valve guides and valve seats are installed in the cylinder heads with a/an ___________________ fit. *Page 315*

75. Two methods used for making cylinder walls hard and wear resistant are:
   a. _____________________________
   b. _____________________________
   Page 320

76. Cylinder wall surface roughness is measured in _______________________ . *Page 321*

77. A cylinder that has an orange band of paint around its base has cylinder walls that are ____________________ (chrome-plated or nitrided). *Page 322*

78. A cylinder that has the tips of the cylinder head fins between the spark plug and the cylinder barrel painted blue has ____________________ (chrome-plated or nitrided) cylinder walls. *Page 322*

79. An engine that has the tip of the cylinder head fins between the spark plug and the rocker box painted yellow requires a __________________ (long or short) -reach spark plug. *Page 322*

80. Intake valves are ground to an angle of __________ degrees relative to the plane of the seat. *Page 322*

Continued

STUDY QUESTIONS: RECIPROCATING ENGINE OVERHAUL *Continued*

81. Exhaust valves are ground to an angle of __________ degrees relative to the plane of the seat. *Page 322*

82. If the edge of a valve face is ground to a featheredge, it can become so hot that it can cause __________________ . *Page 323*

83. Valve seats are ground so that their angle is 0.5° to 1° ____________________ (greater or less) than the angle of the valve face. *Page 323*

84. When grinding a valve seat, one stone grinds the seat to the correct angle, and another stone is used to grind the top of the seat to get the correct ___________________ of the seat. *Page 324*

85. The bottom of the valve seat is ground to provide the correct ___________________ of the seat. *Page 324*

86. After grinding the valve face and seat, the final gastight seal is ensured by ____________________ the valve into the seat. *Page 324*

87. Any indication of spalling on the tappet body is cause for a careful examination of the ________________ on the camshaft. *Page 324*

88. The plunger and plunger body of a hydraulic valve lifter ______________________ (should or should not) be inspected by the magnetic particle method. *Page 324*

89. Chrome-plated piston rings ___________________ (should or should not) be installed in an engine that has chrome-plated cylinder walls. *Page 326*

90. If when installing a cylinder on the crankcase, the hydraulic valve lifter is completely collapsed and the clearance between the rocker arm and the valve stem is too great, the ___________________ should be replaced with one having a different length. *Page 327*

91. An overhauled engine should be run in the test cell either connected to a dynamometer or with a ____________________ installed. *Page 327*

Powerplant Troubleshooting

Troubleshooting is one of the most important skills an aviation maintenance technician must develop. Systematic troubleshooting can save hours of time and prevent components being needlessly replaced. It can be learned if you apply a few very simple rules.

Rules for Systematic Troubleshooting

1. *Know how the system should operate.*

   This sounds absurdly simple, but it is the secret of successful troubleshooting. You must know the way the components work separately, and the way they work together.

2. *Observe how the system is operating.*

   Any difference between the way a system is operating and the way it should operate is an indication of trouble, and you can concentrate your efforts on these differences to find the problem.

3. *Divide the system to find the trouble.*

   Time is valuable in aviation maintenance and you must find the trouble in the shortest length of time. To do this, you must have a minimum of lost motion.

   When you have isolated the system that is malfunctioning, you must then find which subsystem or component is not working as it should.

4. *Look for the obvious problem first.*

   Popped circuit breakers, and switches or valves in the wrong position have been the cause of wasted troubleshooting time.

An Example of Systematic Troubleshooting

Let's examine a simple troubleshooting problem that is likely to occur. A pilot comes into your shop and reports that the engine on his airplane will not start. The airplane is a low-wing model with a six-cylinder, naturally aspirated engine equipped with an up-draft float carburetor, a fixed-pitch propeller, and magnetos with an induction vibrator system for starting.

Begin your troubleshooting process by asking the pilot some pointed questions:

- Does the starter turn the engine over? *Yes, but the battery is now run down.*
- Did the engine fire at all? *No.*
- Do you smell gasoline? *No.*

By now you know that you will need a ground power unit to supply power to crank the engine since the battery has been run down. Go to the airplane, check to be sure that the ignition switch and the master switch are OFF, then plug in the GPU.

Mentally, divide the engine into six basic systems and analyze each of them:

1. *Mechanical*

   Pull the propeller through by hand for several revolutions. There should be a relatively uniform compression on all cylinders, and no sound of scraping, binding, or knocking.

2. *Induction*

   The inlet air filter should be free of obstruction, and there should be no indication of fuel dripping from the carburetor air inlet.

3. *Fuel*

   Visually check the fuel quantity in one of the tanks, and drain some fuel from the sump of that tank. The fuel should be clean and free from water, and it should be the correct color for the required fuel grade.

   With the fuel selector valve on the tank with good fuel in it, drain some fuel from the engine fuel strainer. This fuel should also be clean and free from water.

4. *Lubrication*

   This system will not normally be the cause of an engine's failure to start, but you should check the quantity and appearance of the engine oil to be sure it is sufficient to run the engine.

5. *Cooling*

   This system will not prevent the engine from starting, but look into the inlet air scoop to be sure there are no bird nests or other obstructions to the flow of cooling air.

6. *Electrical*

   The starter should turn the engine over.

   Since this engine has an induction vibrator, you may hear it buzz when the master switch is on and the ignition switch is in the START position.

   If the engine had an impulse coupling, you would hear it snap when the propeller was rotated.

When everything on the outside of the engine and aircraft appears to be normal, you may attempt to start the engine.

- With the ground power unit connected and the master switch ON, and the fuel selector to a tank with good fuel, turn the auxiliary fuel pump ON. There should be an indication on the fuel pressure gage. Since the fuel pressure is measured at the carburetor, this means there is fuel all the way to the carburetor.
- Since there is no excessive fuel odor, and there was no indication of fuel dripping from the bottom of the carburetor, you know that the engine is not flooded. Prime the engine with the hand primer for the number of strokes recommended in the Pilot's Operating Handbook for the particular airplane.

- Open the throttle slightly, check to be sure there is no one near the propeller, and call "Clear." If the propeller is clear, move the ignition switch to the START position.
- The starter cranks the engine vigorously, yet the engine does not start.

The pilot's complaint has been verified, and you are ready to find the problem. The engine must have three things to start: air, fuel, and a spark to ignite the fuel-air mixture. These are the things you know:

- The engine has air because the air filter is unobstructed.
- The engine has fuel because there is fuel pressure indicated at the carburetor.

The problem is most likely the lack of a spark for starting. Consider a wiring diagram of the ignition system such as that in Figure 9-24:

- The instrument lights circuit breaker is good because there is power to the starter solenoid coil. The starter cranks the engine.
- The ignition switch is probably good because the starter works and the right magneto appears to be grounded.
- The left magneto is apparently not getting the pulsating DC it needs for the starting spark.

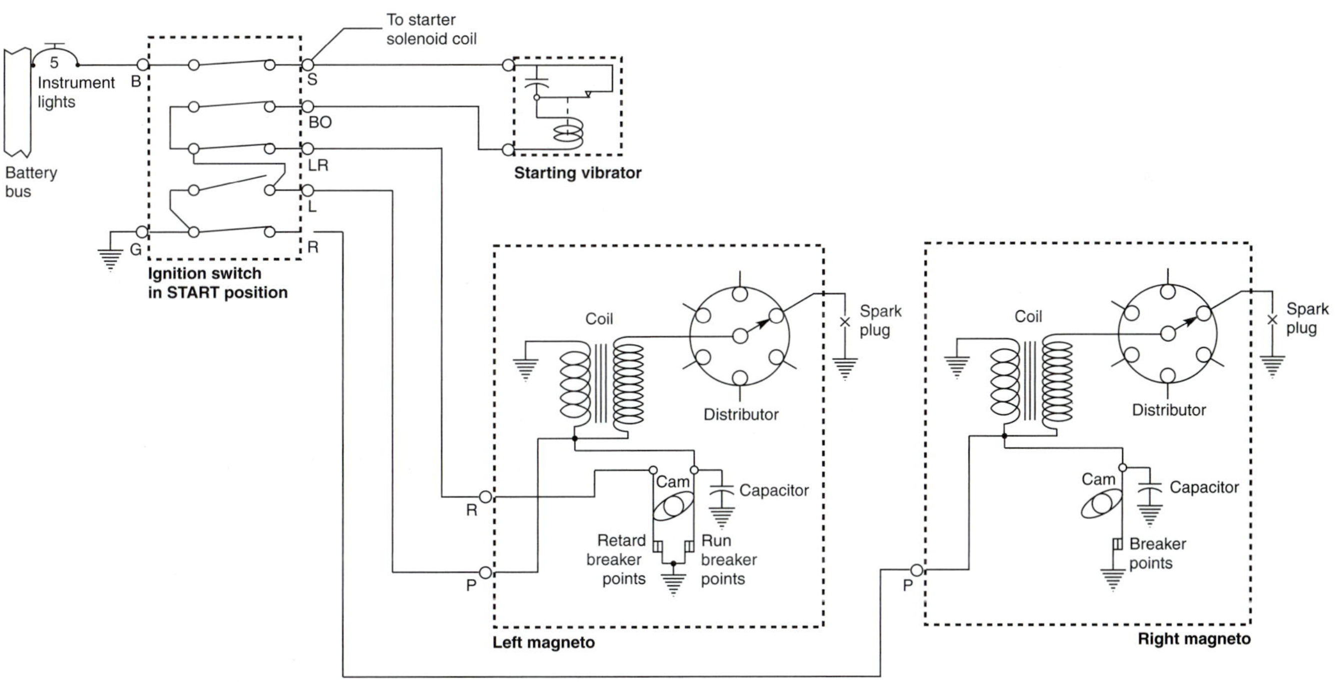

Figure 9-24. *Circuit for induction vibrator and magneto ignition system*

Remove the cowling to reach the starting vibrator and feel it while a qualified person in the cockpit holds the ignition switch in the START position. You should be able to feel the vibrator buzzing. If it does not vibrate, remove the plug and check for voltage at the input terminal of the plug. If there is voltage there when the ignition switch is in the START position, the vibrator is most likely faulty. Make a final check of the vibrator with it disconnected from the aircraft circuit. Ground the coil terminal and apply voltage to the input terminal. If the vibrator does not buzz, it is faulty and should be replaced. If it buzzes, the problem is in one of three places:

- The wiring between the coil terminal and the ignition switch
- The ignition switch itself
- The wiring between the ignition switch and the P or R terminals of the left magneto

This example shows the way to systematically analyze a typical problem. To review this procedure:

- Know how the system should operate.
- Observe how the system is operating.
- Divide the system to find the trouble.
- Look for the obvious problems first.

Figure 9-25 is a troubleshooting chart, typical of those furnished by the engine manufacturers.

Engine Troubleshooting Chart

Engine is Hard to Start

Cause	Solution
Flooded	Clear engine.
Throttle open too far	Close to position for approximately 800 - 1000 RPM.
Impulse coupling not operating properly	Remove and check.
Defective spark plugs or ignition lead	Inspect and replace if necessary.
Low voltage at vibrator	Replace battery.
Inoperative vibrator	Replace vibrator.
Magneto improperly timed to engine	Check magneto timing.
Magneto internal timing wrong	Check internal timing.

Engine Idles Rough

Cause	Solution
Mixture too rich or too lean	Adjust idle mixture.
Plugged injector nozzles	Clean nozzles.
Induction air leak	Check hoses, gaskets, intake pipes and flange bolts.
Cracked engine mounts	Replace mounts.
Uneven cylinder compression	Perform differential compression check and make necessary repairs.
Faulty ignition system	Check mag. drop and condition of plugs and leads.
Fuel pressure too low	Adjust to pressure specified in Operator's Manual.

Figure 9-25. *Engine troubleshooting chart*

Engine Troubleshooting Chart *(continued)*

Engine Will Not Produce Static RPM

Cause	Solution
Restriction in induction air system	Inspect and remove restriction.
Injector too rich or too lean	Replace or recalibrate injector.
Propeller out of adjustment (low pitch)	Adjust low-pitch stop.
Propeller governor not adjusted correctly	Adjust governor.
Crankshaft to camshaft timing off	Remove accessory housing and time correctly.
Exhaust muffler baffles broken and blocking exhaust outlet	Remove muffler, replace if necessary.
Excessively dirty air filter	Clean or replace filter element.

Low Oil Pressure

Cause	Solution
Insufficient oil	Fill oil sump to correct level.
Pressure relief valve out of adjustment	Adjust oil pressure to proper value.
Dirt or metal chips under oil pressure relief valve	Remove, disassemble, and clean.
Damaged oil pressure relief valve seat	Repair or replace seat.
High oil temperature	Check all areas that involve lubrication system or oil cooling system.
Restriction at inlet of oil pump	Remove and clean oil suction screen and passage to inlet side of oil pump.

High Oil Pressure

Cause	Solution
Oil pressure relief valve improperly adjusted	Adjust to correct pressure.
Improper weight of oil	Drain oil and replace with correct weight.
Oil passage from pressure relief valve to sump plugged	Remove blockage from oil passage.
Incorrect spring in pressure relief valve	Replace with correct spring.
Oil temperature too low	Allow oil temperature to increase before opening throttle.

Excessive Oil Consumption

Cause	Solution
Improper grade of oil	Drain oil and replace with correct grade.
Failure of new rings to seat properly	Use correct grade of oil and proper operating procedures.
Failed or failing bearings	Replace affected bearing.
Worn piston rings and/or cylinder	Remove cylinders, deglaze barrels, and replace piston rings.
Worn valve guides	Replace worn guides.

High Cylinder Head Temperature

Cause	Solution
Spark plug of improper heat rating installed in engine	Replace with proper spark plugs.
Cooling baffles missing or broken	Repair or replace affected baffles.
Partially plugged fuel nozzle	Clean nozzle.
Fuel line with too small an I.D.	Replace with line having correct I.D.
Magnetos improperly timed	Check magneto-to-engine timing and adjust as necessary.
Mixture control improperly rigged	Rig for complete travel.

High Manifold Pressure at Idle

Cause	Solution
Improperly adjusted fuel injector or carburetor	Adjust idle mixture.
Incorrect hydraulic valve lifters installed	Replace with correct lifters.
Hydraulic lifters bleeding down too rapidly	Replace with new lifters.
Air leak in induction system	Check for leak and repair.

Figure 9-25. *Engine troubleshooting chart (continued)*

Reciprocating Engine Removal and Installation

Most modern engine overhaul is done in facilities that specialize in this work, but the engine must be properly removed from the aircraft and prepared for shipping to the overhaul facility. When a new, overhauled, or rebuilt engine is received in the shop, it must be properly installed in the aircraft and the aircraft prepared for flight. In this portion of the text, we consider the basic procedure for removing and installing the engine in a typical general aviation single-engine airplane.

Removal

The procedures to follow when removing the engine from an aircraft differ greatly from one aircraft to another. When removing an engine, it is extremely important that the instructions in the aircraft maintenance manual be followed in detail.

Park the aircraft in the hangar where it will be out of the way and not have to be moved during the time the engine is off of the airframe.

Most small tricycle landing gear airplanes are extremely tail heavy without the engine, so you must support the tail on a suitable stand to prevent it dropping to the floor as the engine is being removed.

Use a hoist with sufficient capacity and special slings or attachments specified in the aircraft service manual. Have the proper stands or racks ready to receive the propeller and the engine when they are removed, and if either is to be shipped to an outside overhaul facility, have their shipping containers beside the aircraft so they can be prepared for shipping with a minimum of handling.

Disconnect or remove the battery, and place the fuel selector valve in the OFF position. Remove the cowling and store it on suitable racks where it will not be damaged while it is off the aircraft.

Drain the fuel strainer and disconnect the fuel line from the carburetor or fuel injector, and drain all of these lines. Drain the oil sump and oil cooler. Disconnect the oil lines and plug them with the proper plug or cap. Do not wrap the open ends of fuel or oil lines with masking tape or duct tape.

Remove the propeller and put it in the shipping container or on a propeller rack. Tape vinyl sheeting over the hub of constant-speed propellers to keep dirt or other foreign matter out. Cover the end of the propeller shaft to keep foreign matter out of it.

Remove the exhaust system and the induction air inlet system.

Disconnect the electrical leads, instrument lines, and controls that connect the engine to the cockpit. These items include:

- Starter cable
- Cylinder head temperature and exhaust gas temperature thermocouples
- Oil temperature capillary tube or electrical wires to the temperature probe
- Oil pressure gage line
- Electrical wires to the generator or alternator
- Throttle and mixture controls
- Vacuum hoses to the vacuum pump
- Manifold pressure line
- Fuel flowmeter line to the fuel injection manifold
- Engine primer lines
- Flexible ducts from the heater shroud to the cabin heater valve

When the engine is completely disconnected from the airframe, attach the cable from the hoist to the lifting eye on the crankcase. Loosen the engine mount bolts and carefully lift the engine from the airframe. Place it on the proper engine stand, or in the shipping container if it is to be shipped.

Installation

When a new or overhauled engine is received from the manufacturer or overhaul facility, it normally has the fuel metering and ignition systems installed, but all accessories specific to the airframe must be installed.

The vacuum pump, hydraulic pump, propeller governor, starter, alternator or generator, and the necessary wiring, plumbing, and baffles must be installed. The serial numbers of all the components must be recorded, and any paperwork that has accompanied them must be placed with the aircraft maintenance records.

Attach a hoist of adequate capacity to the engine lifting eye, and raise it into the proper position. Install new engine shock mounts and attach the engine to the airframe. Torque and safety the engine mount bolts as directed in the aircraft service manual.

Install the exhaust system and the inlet induction air system. Connect all fluid lines, electrical wires, and controls to the engine. Install the propeller. Adjust all controls to be sure that the stops on the engine component are reached first, and that the cockpit control will move a small amount more before it reaches its stop, and that the control springs back a small amount.

The largest amount of engine wear occurs when the engine is first started, before the bearings are adequately covered with oil. It is extremely important before starting an engine for the first time after installation, that the bearings have immediate access to plenty of fresh oil. Put the correct amount of the recommended grade of oil in the sump and preoil the engine by forcing it through the engine with a pressure oiler until some runs out of the connection for the oil pressure gage. If no pressure oiler is available, remove one spark plug from each cylinder and turn the engine over with the starter until pressure is indicated on the oil pressure gage.

Install the cowling, take the aircraft out of the hangar, and run the engine until the oil is warm. Check for any oil leaks. If everything checks out as it should, complete all necessary paperwork, and approve the aircraft for a test flight.

STUDY QUESTIONS: RECIPROCATING ENGINE REMOVAL AND INSTALLATION

Answers begin on Page 341. Page numbers refer to chapter text.

92. Disconnected fluid lines should have their ends protected with the proper _______________ or _______________. *Page 338*

93. A properly adjusted engine control should contact the stop at the component _______________ (before or after) the stop in the cockpit is reached. *Page 339*

Answers to Chapter 9 Study Questions

1. false
2. poorer
3. FULL RICH
4. COLD
5. low
6. FULL RICH
7. IDLE CUTOFF
8. keep cranking the engine
9. CO_2
10. low
11. remaining the same, increasing
12. a. annual inspection
    b. 100-hour inspection
13. 12
14. 100
15. is not
16. is
17. is
18. is
19. differential
20. compression
21. 80
22. 60
23. intake valve
24. exhaust valve
25. piston rings
26. after the flow has started
27. growth trend
28. bearing inserts
29. next to the end
30. internal
31. fuel dye
32. springback
33. air bleed
34. filtered
35. cracks
36. corrosion
37. owner, operator
38. do not
39. may not
40. may not
41. uneconomical
42. zero
43. are not
44. are
45. plastic pellets, glass beads, crushed walnut shells
46. cotton string
47. clean oil
48. magnetic particle inspection, fluorescent penetrant inspection
49. can
50. continuous, residual
51. circularly, longitudinally
52. circularly
53. longitudinally
54. circularly
55. true
56. is not
57. emulsifier
58. is not
59. National Institute of Standards and Technology or National Bureau of Standards
60. inside micrometer caliper, telescoping gage
61. out-of-round
62. choke
63. interference
64. stretched
65. a. end gap
    b. side clearance
    c. tension
66. a. clearance between small-end bushing and wrist pin
    b. clearance between big-end bearing and the crankpin
    c. twist
    d. convergence
67. Heli-Coil
68. nitriding
69. is not
70. major overhaul

Continued

Answers to Chapter 9 Study Questions *(Continued)*

71. fluorescent penetrant
72. larger
73. line boring
74. interference
75. a. chrome plating
    b. nitriding
76. microinches RMS
77. chrome-plated
78. nitrided
79. long
80. 30
81. 45
82. preignition
83. greater
84. diameter
85. width
86. lapping
87. lobes
88. should not
89. should not
90. pushrod
91. test club
92. plugs, caps
93. before

THEORY & CONSTRUCTION

10

Continued

Continued

Theory & Construction

10

Propulsion Principles

Newton's third law of motion, which states that every action has an equal and opposite reaction, is the basic explanation for all aircraft propulsion. An aircraft powerplant, whether it be a reciprocating engine and a propeller, or a turbojet or rocket engine, produces thrust in basically the same way, by changing the momentum of a mass of air.

There are two basic methods of producing thrust to propel an aircraft: aerodynamic action and jet reaction (or jet propulsion).

1. Aerodynamic action produces thrust when an engine turns a propeller, a rotating airfoil that produces lift along a horizontal plane. The wash from this lift accelerates a large mass of air rearward through a small change in velocity.
2. Jet reaction engines, commonly called jet engines, produce thrust by heating a mass of air inside the engine and discharging it at a high velocity through a specially shaped nozzle. The amount of thrust is determined by the mass of the air and by the amount it is accelerated. Jet engines accelerate a small mass of air through a large change in velocity.

thrust. The forward-directed force produced by the reaction from a rearward-acceleration of a mass of fluid.

velocity. A vector quantity that expresses both the speed an object is moving and the direction in which it is moving.

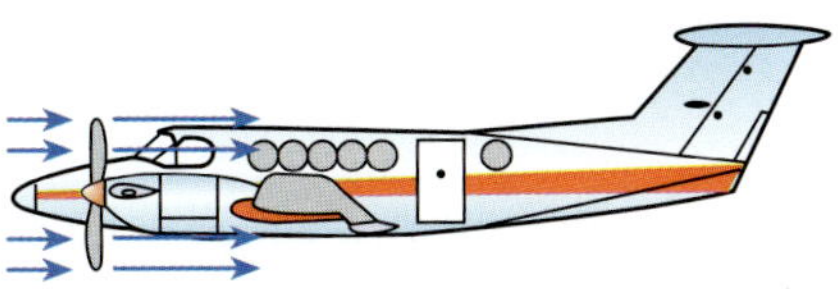

Figure 10-1. *An engine-propeller combination produces thrust as a long propeller accelerates a large mass of air rearward through a small change in velocity.*

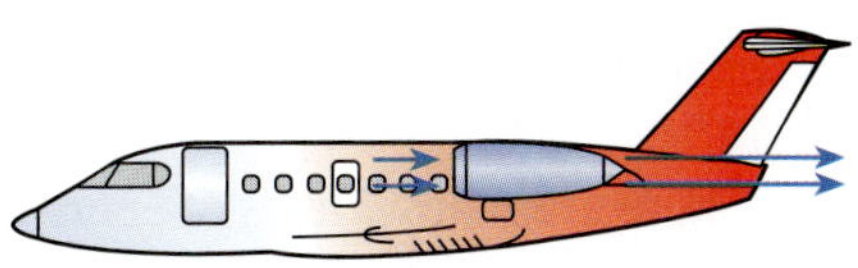

Figure 10-2. *A jet engine produces thrust by accelerating a small mass of air through a large change in velocity.*

The principle of thrust produced by jet propulsion was known as early as the first century A.D. Hero, or Heron, a mathematician of Alexandria in Egypt, is credited with inventing a steam engine called an aeolipile. Steam generated in a boiler was directed into a hollow sphere mounted above the boiler on an axle. Bent nozzles protruded from the periphery of the sphere in such a way that steam escaping from the nozzles produced a reaction force that rotated the sphere. The principle is much the same as that of the rotating lawn sprinkler, which is spun by the reaction of water as it leaves the nozzles.

Figure 10-3. *The principle of Hero's aeolipile was described in the first century A.D. Steam generated in the body was directed into the hollow sphere and exited through bent nozzles. The reaction from the escaping steam rotated the sphere.*

The Chinese made the first practical applications of jet propulsion when they devised solid-fuel rockets.

A rocket engine contains chemicals that, when ignited, release gases and a tremendous amount of heat. The heat expands the gases, which are ejected at a high velocity from a nozzle at the rear of the rocket.

A Practical Review of Physics

gas turbine engine. A form of heat engine that produces torque and thrust by converting the chemical energy in the fuel into heat to drive a turbine.

The technical aspects of aviation are all practical applications of physics. The *General* textbook in this *Aviation Maintenance Technician Series* contains a coverage of basic physics. In this chapter of *Powerplant*, we review the aspects of practical physics that apply directly to the production of thrust and power by gas turbine engines.

The Physics of Energy and Matter

When considered in their very basic nature, matter and energy are expressions of the same thing. Matter is defined as something that has mass and exists as a solid, liquid, or gas. Energy is thought of as a source of usable power. A gallon of jet fuel has mass, and when it is burned, it changes most of its mass into heat energy.

Energy

Energy is defined as the capacity of a physical system to do work. The law of conservation of energy states that energy can neither be created nor destroyed, but it can be changed from one form into another.

acceleration. The amount the velocity of an object is increased by a force during each second it is acted upon by that force. Acceleration is normally expressed in terms of feet per second, per second (fps^2).

It is by changing the forms of energy that flight by jet propulsion is possible. Essentially, this is done by changing the chemical energy in the fuel into heat energy, and then into mechanical energy to accelerate air through the engine. It is this acceleration that is primarily responsible for moving the aircraft.

There are many forms of energy: chemical, mechanical, electrical, heat, light, and nuclear. But there are only two basic classifications into which all energy fits: potential and kinetic. As air passes through a gas turbine engine and energy is added to and taken from it, there is continual change between its potential and kinetic energies. The total energy in air passing through the engine is always the sum of its kinetic and potential energies.

Potential Energy

Potential energy is the energy in an object derived from its position or condition. When a fluid, such as the air flowing through a jet engine, is considered, the potential energy in the fluid relates to its pressure.

Kinetic Energy

Kinetic energy is the energy possessed by an object because of its motion. Kinetic energy in the air flowing through a jet engine relates to its velocity. The kinetic energy in moving air is equal to one half of its mass times the square of its velocity.

$$E_k = \frac{M \cdot V^2}{2}$$

Matter

Matter is something that has mass, takes up space, and exists as a solid, liquid, or gas. When we think of matter, mass and weight must be taken into consideration. They are often expressed in the same terms, but are actually quite different.

matter. Something that has mass, takes up space, and exists as a solid, liquid, or gas.

Mass

Mass is a measure of the amount of matter an object contains. The mass of the object is not dependent upon gravity and is therefore different from, but relates to, the weight of the object. It is common practice to express both mass and weight in terms of pounds. The thrust produced by a gas turbine engine is determined by the mass flow of air through the engine.

mass. A measure of the amount of matter in an object.

Mass is technically measured in slugs, with one slug being the amount of mass accelerated one foot per second, per second (fps^2) when it is acted upon by a force of one pound.

slug. The unit of mass which experiences an acceleration of one foot per second per second when a force of one pound acts on it.

One slug of air under standard conditions has a volume of approximately 421 cubic feet, and the gravity effect on this air causes it to weigh 32.2 pounds.

In practical formulas used to find the mass of the air flowing through a gas turbine engine, the weight of the airflow in pounds per second is divided by the acceleration due to gravity, 32.2 feet per second, per second.

$$\text{Mass flow of air} = \frac{\text{Weight of airflow in pounds per second}}{\text{Acceleration due to gravity}}$$

$$= \frac{W_a}{g}$$

$$= \frac{W_a}{32.2}$$

Weight

The weight of an object is the force with which it is attracted to the earth. Weight is equal to the product of the object's mass times the acceleration due to gravity.

$$\text{Weight} = \text{Mass} \cdot \text{Acceleration}$$

Force

Force is a vector quantity (a quantity that has both direction and magnitude) that causes an object to accelerate in the direction of its application. In practical mechanics such as this study of gas turbine engines, force is normally expressed in units of pounds.

Work

Work is the transfer of energy to a body by the application of a force that moves the body in the direction of the force. It is measured as the product of the force, in pounds, and the distance through which the body moves, in feet or inches, and is expressed in terms of foot-pounds or inch-pounds.

$$\text{Work} = \text{Force} \cdot \text{Distance}$$

Power

Power is the rate at which work is done. It is the number of foot-pounds of work done in a specified length of time. Power is expressed in units such as foot-pounds per minute or foot-pounds per second.

$$\text{Power} = \frac{\text{Work}}{\text{Time}}$$

Horsepower

The horsepower is a commonly used measure of power. One horsepower is equal to 33,000 foot-pounds of work done in one minute, or 550 foot-pounds of work done in one second. One horsepower is also equal to 746 watts of power.

The Physics of Motion

Many important aspects of the practical physics of gas turbine engines relate to motion and the forces which cause or change it.

Speed

Speed is a measure of the rate of motion and is normally considered to be the distance traveled divided by the time used to travel. Speed does not take direction of travel into consideration and is expressed in such terms as feet per second.

Velocity

Velocity is a vector quantity whose magnitude is the speed of the object and whose direction is the direction in which the object is moving.

Momentum

momentum. A measure of the motion of a body equal to the product of its mass times its velocity.

Momentum is a measure of the motion of an object equal to the product of its mass times its velocity. In the study of gas turbine engines, the momentum of air flowing through the engine is found by dividing the weight of the airflow, in pounds per second, by the acceleration due to gravity (32.2 fps^2). This is then multiplied by the velocity of the air in feet per second.

$$\text{Momentum} = \frac{W_a}{g} \cdot V$$

Acceleration

Acceleration is the amount the velocity of an object (measured in feet per second) is increased by a force during each second it is acted upon by that force. Acceleration is normally expressed in terms of feet per second, per second (fps^2).

Deceleration is similar to acceleration, but is a measure of the decrease in an object's velocity.

deceleration. The amount the velocity of an object is decreased by a force during each second it is acted upon by that force.

Newton's Laws of Motion

Many events that take place in a gas turbine engine may be explained by one or more of Newton's laws of motion.

Newton's first law is called the law of inertia. It states that an object at rest will remain at rest, and an object in motion will continue in motion in a straight line and will not change its velocity unless it is acted upon by an outside force.

inertia. The tendency of a body to resist acceleration.

Newton's second law is called the law of acceleration. It states that the amount an object will accelerate is inversely related to its mass and directly related to the amount of force acting on it.

Newton's third law is the law of action and reaction. It states that for every action there is an opposite and equal reaction.

The Physics of Gas Flow

The principle upon which a gas turbine engine operates is that of accelerating a mass of air as it passes through the engine by the addition of energy from burning fuel. Understanding the things that happen to the air as it passes through the engine helps us understand the way thrust is developed.

Bernoulli's Principle

We can best understand the important relationship between the two basic forms of energy (kinetic and potential) by studying Bernoulli's principle. It states that the total energy in a column of moving fluid remains constant throughout the column. This total energy is made up of potential energy in the form of pressure, and kinetic energy in the form of velocity.

Figure 10-4 shows a duct carrying a flow of fluid. At point 1, the fluid has the velocity V_1 and pressure P_1. As the fluid moves through the duct, it reaches the restrictor at point 2, where the velocity increases to V_2, and the pressure decreases to P_2. The product of V_1P_1 is equal to that of V_2P_2. In other words, the total energy is the same at both locations. As the flow continues, the duct widens to its original dimensions. The fluid slows to its original velocity and its pressure rises to its original value. The total energy has not changed.

The relationship stated by Bernoulli's principle is that when the total energy in moving fluid is not changed, any increase in the fluid's velocity will result in a corresponding decrease in its pressure.

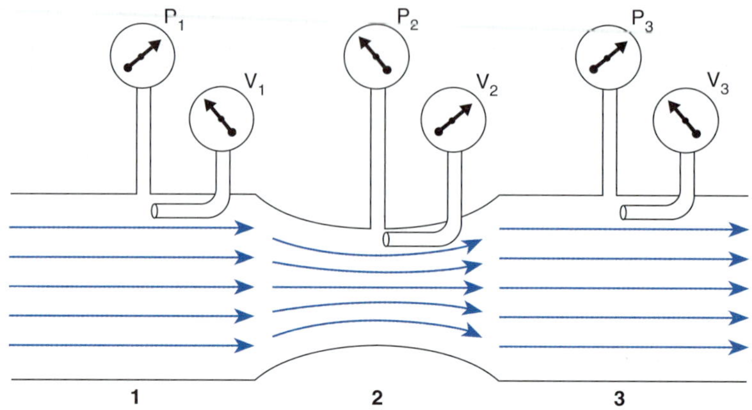

Figure 10-4. *When energy is neither added to nor taken from fluid in motion, any restriction causes the fluid to speed up and its pressure to decrease.*

Subsonic Flow Through a Duct

When air flows through a duct at a speed below the speed of sound, it behaves as just explained in Bernoulli's principle.

density. A measure of the amount of mass in a unit volume.

When the duct converges or becomes smaller in the direction of the flow, the air speeds up and its pressure decreases. When it flows through a diverging duct—one whose size increases in the direction of flow—its velocity decreases and its pressure increases. The changes in pressure and velocity do not change the density of the air as long as the flow remains subsonic. *See* Figure 10-5.

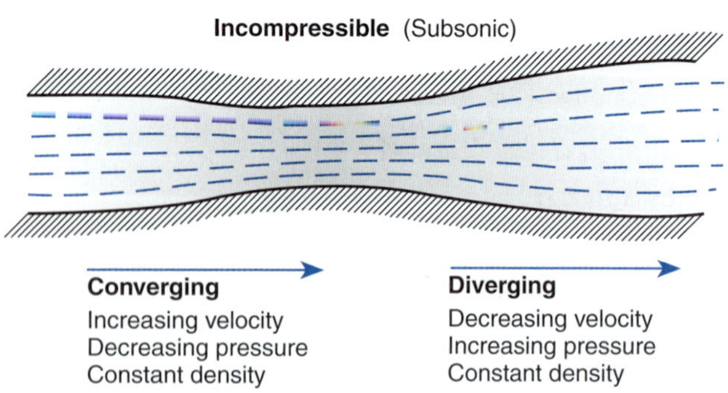

Figure 10-5. *When air flows through a converging duct at speeds below the speed of sound, its velocity increases and its pressure decreases. As it leaves the restriction and enters the diverging portion of the duct, its velocity decreases and its pressure increases.*

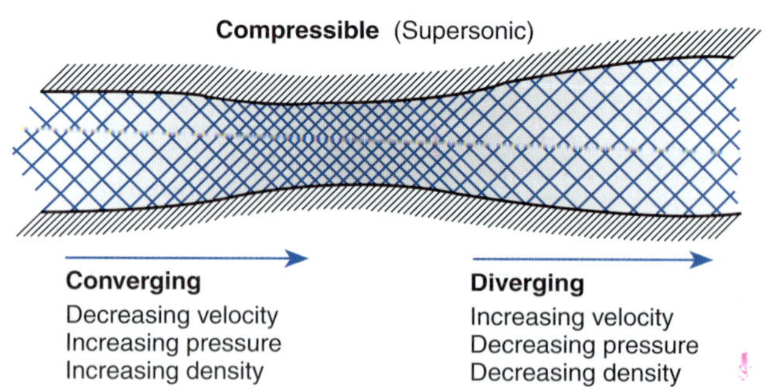

Figure 10-6. *As supersonic air flows through the converging portion of a duct, it compresses and its density increases. Its velocity decreases and its pressure increases. As it passes through the diverging portion of the duct, it expands and its density decreases, its velocity increases, and its pressure decreases.*

Supersonic Flow Through a Duct

When air flows through the duct at a speed greater than the speed of sound, the pressure-velocity relationship is quite different from that at a subsonic speed.

When supersonic air flows through a converging duct, it cannot speed up, but rather, it compresses. Its speed decreases and its pressure and density increase. When it flows through a diverging duct, it expands and its density and pressure decrease, and its velocity increases. *See* Figure 10-6.

Pressure Waves

When air flows through a duct or over a surface at a supersonic speed, pressure waves form. There are three types of pressure waves: normal shock waves, oblique shock waves, and expansion waves.

normal shock wave. A pressure wave that forms at right angles to a surface when air moves at the speed of sound.

oblique shock wave. A pressure wave that forms on a sharp-pointed object when air flows past it at a supersonic speed.

expansion wave. The change in pressure and velocity of supersonic air as it passes over a surface that drops away from the flow.

Normal Shock Waves

When the velocity of moving air reaches the speed of sound, pressure disturbances in the air cannot propagate back against the flow, and they accumulate at the point at which the speed of sound is reached, forming a normal shock wave.

When a supersonic airstream passes through a normal shock wave, five things happen:

1. The airstream slows to subsonic.
2. The airflow direction immediately behind the wave remains unchanged.
3. The static pressure of the airstream behind the shock wave increases.
4. The density of the airstream behind the wave is greatly increased.
5. The energy in the airstream is greatly reduced.

static pressure. The pressure of an unmoving fluid.

Gas turbine engines require the velocity of air at the inlet to the compressor to be subsonic. Some supersonic airplanes have a specially shaped inlet duct that causes air entering the engine to form one or more shock waves that slow the air to a speed below the speed of sound.

Oblique Shock Waves

An oblique shock wave forms when a supersonic airstream strikes a sharp object.

When a supersonic airstream passes through an oblique shock wave, five things occur:

1. The airstream slows down, but remains supersonic.
2. The flow direction changes to follow the surface.
3. The static pressure of the airstream behind the shock wave increases.
4. The density of the airstream behind the shock wave increases.
5. Some energy in the airstream is converted into heat and wasted.

Expansion Waves

When the surface over which air is flowing at a supersonic speed drops away from this air flow, an expansion wave forms.

When a supersonic airstream passes through an expansion wave, five things result:

1. The airstream accelerates, and the air behind the expansion wave has a higher supersonic velocity.
2. The flow direction changes to follow the surface.
3. The static pressure of the airstream behind the expansion wave decreases.
4. The density of the airstream behind the expansion wave decreases.
5. No loss of energy in the airstream occurs.

Airflow through a Choked Nozzle

choked nozzle. A nozzle that limits the speed of the gases flowing through it by the formation of a normal shock wave.

When subsonic air flows into a converging duct or nozzle, its velocity increases. If the passage is designed in such a way that the airflow reaches the speed of sound at the narrowest point, a normal shock wave forms, and the air passing through the shock wave is slowed to a subsonic velocity. This principle is used with a choked exhaust nozzle. The exhaust gases leaving the engine accelerate to the speed of sound where they form a shock wave. They can no longer accelerate, so the energy that would normally cause acceleration is now converted into pressure energy that produces a pressure differential across the nozzle opening that contributes to the total thrust.

Pressure

The pressure of a fluid is the amount of force it applies to a given area. Pressure is normally measured in pounds per square inch or in inches of mercury. A pressure of one inch of mercury is the pressure that will support a column of mercury one inch high.

There are three types of pressure commonly used in gas turbine engine technology: absolute pressure, gage pressure, and differential pressure.

Pressure must be measured from some reference level. Absolute pressure is referenced from zero pressure, or a vacuum. Gage pressure is referenced from the existing atmospheric pressure, and differential pressure is the difference between two pressures.

total pressure. The pressure a column of moving fluid would have if it were stopped.

ram pressure. Pressure produced when a moving fluid is stopped.

Two additional types of pressure important in the study of gas turbine engines are static pressure and total pressure. Static pressure is the pressure of still air, but when air is in motion, its pressure increases because of the ram effect. The total pressure is therefore the sum of the static pressure and the pressure that results from ram effect, or ram pressure.

$$P_{TOTAL} = P_{STATIC} + P_{RAM}$$

Heat and Temperature

A gas turbine engine is a heat engine. Its purpose is to convert the chemical energy in the fuel into heat energy, and then to convert this heat energy into useful work or thrust.

Temperature is a measure of the effect of heat energy, and it is expressed in terms of degrees as measured on four different scales. The two most commonly used scales are Celsius (°C) and Fahrenheit (°F). As their reference point, both scales use the temperatures at which pure water freezes and boils. The Celsius scale divides the temperature spread between these points into 100 equal degrees. Water freezes at 0°C and boils at 100°C. The Fahrenheit scale divides the spread into 180 equal degrees. Water freezes at 32°F and boils at 212°F. Absolute zero is -273°C and -460°F. Absolute temperature is measured in Kelvin and Rankine degrees. The Kelvin scale uses the same divisions as the Celsius scale. All molecular motion stops at 0°K; water freezes at 273°K and boils at 373°K. The Rankine scale uses the same divisions as the Fahrenheit scale. All molecular motion stops at 0°R; water freezes at 492°R and boils at 672°R.

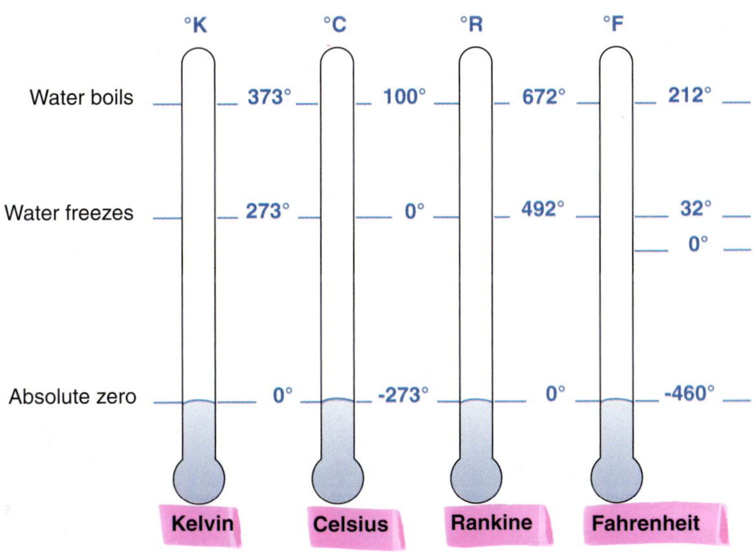

Figure 10-7. *The four temperature scales use the same three reference points: the temperature at which all molecular motion ceases, the temperature at which pure water changes from a liquid into a solid, and the temperature at which pure water changes from a liquid into a gas.*

Two types of temperature are important in the study of gas turbine engines. These are static temperature and total temperature. Static temperature is the temperature of the ambient or surrounding air that is still, or not moving. Total temperature is the temperature a column of moving air has when stopped. Total temperature is the sum of the static temperature and the temperature rise caused by the ram effect.

total temperature. The temperature of moving fluid that has been stopped from its motion.

Standard Atmospheric Conditions

Gas turbine engine operation is affected by changes in air temperature and pressure. To determine whether or not an engine is operating as the manufacturer designed it, engineers have agreed upon conditions that constitute a standard day. All engine performance can be reduced to the performance it would have under these standard conditions. By converting all data to this standard, the effect of variables such as pressure and temperature can be eliminated and engine performance meaningfully compared.

The standard day conditions that have been accepted by the International Civil Aviation Organization (ICAO) are seen in Figure 10-8.

Figure 10-9 shows how pressure and temperature change with altitude. Notice that the pressure continues to drop as altitude increases, but the temperature stabilizes at the beginning of the stratosphere, which is 36,089 feet. The speed of sound, which is affected only by the temperature, stabilizes at the same altitude as the temperature.

stratosphere. The upper part of the earth's atmosphere, extending upward from about 36,000 feet above the surface of the earth, to about 85,000 feet.

Parameter	U.S. Units	Metric Units
Pressure, P_0	2116.22 lb/ft^2 29.92 in. Hg	$1.013250 \cdot 10^5$ N/m^2 760 mm Hg
Temperature, T_0	518.67°R 59.0°F	288.15°K 15.0°C
Acceleration due to gravity, g_0	32.1741 ft/sec^2	9.80665 m/sec^2
Specific weight, $g_0\rho_0$	0.076464 lb/ft^3	1.2250 kg/m^3
Density, ρ_0	0.0023769 lb-sec^2/ft^4	0.12492 kg-sec^2/m^4

Figure 10-8. *Parameters of the ICAO standard day*

standard day. A set of atmospheric conditions that have been agreed upon by scientists and engineers who work with aircraft and engines that allow them to correct all measurements to the same conditions.

geopotential of the tropopause. The point in the standard atmosphere where the temperature stops dropping and becomes constant.

Altitude Feet	Temperature °F	°C	Pressure In. Hg	Speed of Sound Knots
0	59.00	15.0	29.92	661.7
1,000	55.43	13.0	28.86	659.5
2,000	51.87	11.0	27.82	657.2
3,000	48.30	9.1	26.82	654.9
4,000	44.74	7.1	25.84	652.6
5,000	41.17	5.1	24.90	650.3
6,000	37.60	3.1	23.98	647.9
7,000	34.04	1.1	23.09	645.6
8,000	30.47	-0.8	22.23	643.3
9,000	26.90	-2.8	21.39	640.9
10,000	23.34	-4.8	20.58	638.6
15,000	5.51	-14.7	16.89	626.7
20,000	-12.32	-24.6	13.75	614.6
25,000	-30.15	-34.5	11.12	602.2
30,000	-47.90	-44.4	8.885	589.5
35,000	-65.82	-54.2	7.041	576.6
*36,089	-69.70	-56.5	6.683	573.8
40,000	-69.70	-56.5	5.558	573.8
45,000	-69.70	-56.5	4.355	573.8
50,000	-69.70	-56.5	3.425	573.8
55,000	-69.70	-56.5	2.693	573.8
60,000	-69.70	-56.5	2.118	573.8
65,000	-69.70	-56.5	1.665	573.8
70,000	-69.70	-56.5	1.310	573.8
75,000	-69.70	-56.5	1.030	573.8
80,000	-69.70	-56.5	0.810	573.8
85,000	-64.80	-53.8	0.637	577.4
90,000	-56.57	-49.2	0.504	583.4
95,000	-48.34	-44.6	0.400	589.3
100,000	-40.11	-40.1	0.320	595.2

*Geopotential of the tropopause

Figure 10-9. *Conditions within the ICAO standard atmosphere*

STUDY QUESTIONS: PROPULSION PRINCIPLES

Answers begin on Page 420. Page numbers refer to chapter text.

1. Two methods of producing thrust to drive an aircraft are:
   a. Aerodynamic Action
   b. Jet Reaction
   Page 347

2. A rocket engine produces its thrust by pushing against the air behind the rocket. This statement is false (true or false). *Page 347*

3. When a gallon of jet fuel is burned, most of its mass is changed into ~~mechanical~~ heat energy. *Page 348*

4. Energy can neither be created nor destroyed, but it can be changed from one form to another. This statement is True (true or false). *Page 348*

5. The pressure of moving gas is a measure of its potential (potential or kinetic) energy. *Page 348*

6. The velocity of moving gas is a measure of its kinetic (potential or kinetic) energy. *Page 349*

7. The mass of the air flowing through a gas turbine engine is found by dividing its weight by a constant value that represents the acceleration caused by gravity. *Page 349*

8. Force is a vector quantity that has both direction and magnitude. *Page 350*

9. Work is the product of force and distance. *Page 350*

10. Power is the rate at which work is done. *Page 350*

11. One horsepower is equal to 33,000 foot-pounds of work done in one minute, 550 foot-pounds of work done in one second, and 746 watts. *Page 350*

12. The rate at which an object is traveling in a given direction is the velocity (speed or velocity) of the object. *Page 350*

13. The product of the mass of an object times its velocity is the ____________ of the object. *Page 350*

Continued

STUDY QUESTIONS: PROPULSION PRINCIPLES *Continued*

14. Acceleration is a measure of the increase in the velocity of an object each second it is acted on by a force. Acceleration is measured in ______________________ . *Page 351*

15. Bernoulli's principle explains that if energy is neither added to nor taken from a body of moving fluid, any increase in its velocity will result in a corresponding __________________ (increase or decrease) in its pressure. *Page 351*

16. When air flows through a converging nozzle at a subsonic speed, its velocity will __________________ (increase or decrease) and its pressure will __________________ (increase or decrease). *Page 352*

17. When air flows through a converging nozzle at a supersonic speed, its velocity will __________________ (increase or decrease) and its pressure will __________________ (increase or decrease). *Page 352*

18. When air flows through a diverging nozzle at a subsonic speed, its velocity will __________________ (increase or decrease) and its pressure will __________________ (increase or decrease). *Page 352*

19. When air flows through a diverging nozzle at a supersonic speed, its velocity will __________________ (increase or decrease) and its pressure will __________________ (increase or decrease). *Page 352*

20. After a supersonic airstream passes through a normal shock wave, its velocity is ________________________ (subsonic or supersonic). *Page 353*

21. After a supersonic airstream passes through an oblique shock wave, its velocity is ________________________ (subsonic or supersonic). *Page 353*

22. An exhaust nozzle that prevents gases from flowing through it faster than the speed of sound is said to be a __________________ nozzle. *Page 354*

23. The pressure of still air is called ________________ pressure. *Page 354*

24. The pressure of a column of moving air that has been stopped is the __________________ pressure of the air. *Page 354*

25. The absolute temperature scale that uses the same size graduations as is used by the Celsius scale is the ___________________ (Kelvin or Rankine) scale. *Page 355*

26. The ICAO standard day conditions for sea level temperature and pressure are:
    a. ____________________ °C
    b. ____________________ °F
    c. ____________________ inches of mercury
    Page 356

Aircraft Turbine Engines

The principle of jet propulsion may be understood by visualizing a tin can with a firecracker in it. The can is suspended by a string, as shown in Figure 10-10. When the firecracker, which acts as a reaction engine, is ignited, the air inside the can is heated, and it expands uniformly and produces a force in all directions. The forces that act on the can's side walls cancel each other out, but since the gases are free to escape from the open end, a force is produced that acts on the closed end of the can. This force is called thrust, and it is caused by the acceleration of air leaving the can. This thrust causes the can to swing violently.

Notice that the reaction engine does not push against anything outside of the engine. Its reaction is within, which is the reason it can function beyond the limits of the earth's atmosphere.

jet propulsion. A method of propulsion by accelerating a relatively small mass of air through a large change in velocity.

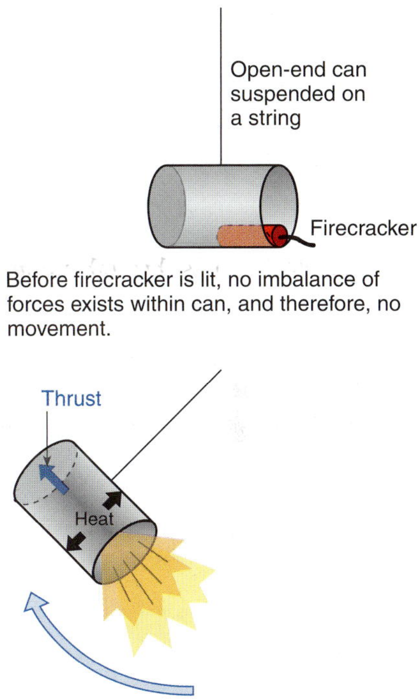

Figure 10-10. *Principle of jet propulsion*

Non-Air-Breathing (Rocket) Engines

A firecracker in a can has no practical application in transportation, but if the firecracker were replaced with a specially shaped powder charge, and the open end of the can were replaced with a nozzle, the powder would burn, rather than explode, and sustained thrust would result.

Since rockets carry with their fuel all of the oxygen needed for combustion, they can operate in outer space where there is no air. The largest rocket engines do not use solid propellant, but instead generate heat by the chemical reaction between two liquids, a fuel and an oxidizer. *See* Figure 10-11 on the next page.

rocket engine. A form of reaction engine whose fuel and oxidizer contain all of the oxygen needed for the release of heat energy.

Air-Breathing Reaction Engines

There are several types of air-breathing reaction engines, with the gas turbine engine being the most popular and useful. Engines with limited usefulness are pulse-jets and ramjets.

reaction engine. A form of heat engine that produces thrust by heating a mass of air inside the engine and discharging it at a high velocity through a specially shaped nozzle. The amount of thrust is determined by the mass of the air and the amount it is accelerated.

Pulse-Jet Engines

One of the early air-breathing reaction engines was a pulse-jet engine. Fuel is sprayed into the combustion chamber and ignited. The heated air expands and closes the one-way shutter valve at the front of the combustion chamber and exits through the open jet nozzle at the rear. As soon as the pressure inside the combustion chamber decreases, the shutters open, and air flows in, mixes with the fuel, and is ignited. Thrust is produced in a series of pulses, and the sound produced by the pulse-jet-powered missiles used during World War II caused them to be called "buzz bombs." *See* Figure 10-12.

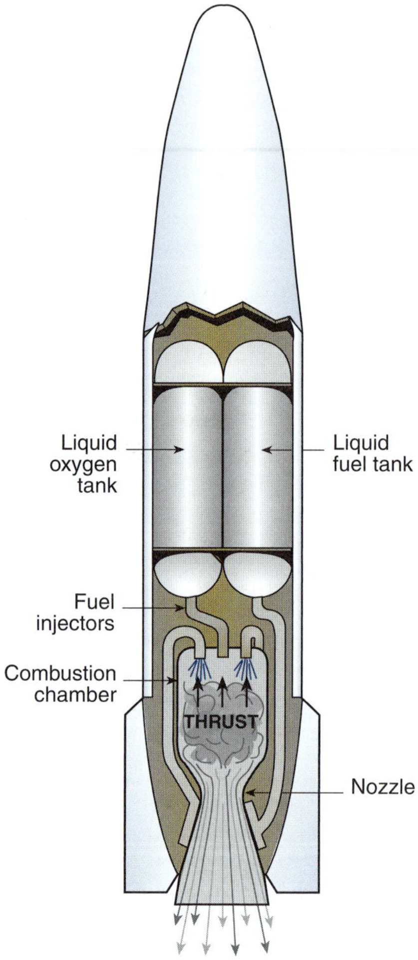

Figure 10-11. *When liquid fuel and liquid oxygen combine, the intense reaction releases a tremendous amount of heat. The resultant gases expand inside the combustion chamber and produce equal forces in all directions.*

The side forces cancel, but the open nozzle does not produce a force to cancel the one at the front of the combustion chamber, and thrust results.

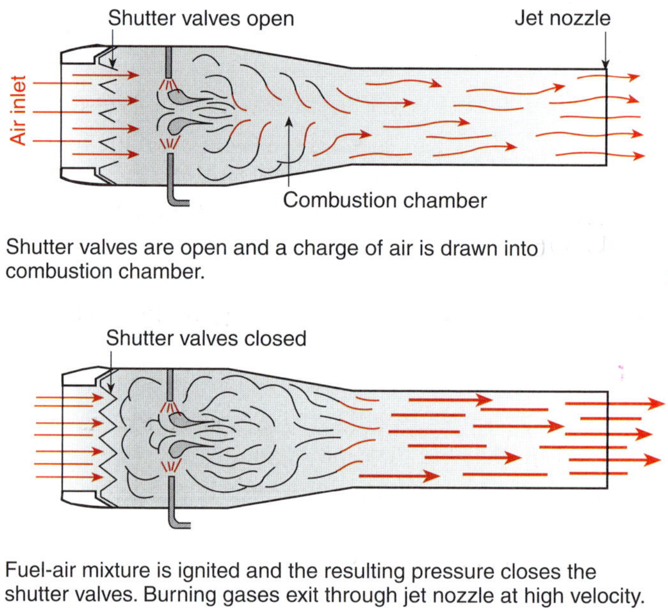

Figure 10-12. *A pulse-jet engine burns the fuel in a series of rapid explosions and produces thrust in pulses.*

pulse-jet engine. A type of simple air-breathing reaction engine used during World War II to power jet-propelled missiles.

Ramjet Engines

A ramjet is the simplest type of air-breathing reaction engine. It depends upon high-velocity air entering the front of the engine and forming a barrier. When fuel is sprayed into the engine and ignited, the air expands, and the barrier forces it to leave the engine at a high speed through the nozzle in the rear. The acceleration of the air produces forward thrust.

Ramjet engines are used in some military unmanned aircraft that are boosted to a speed high enough for the engine to function.

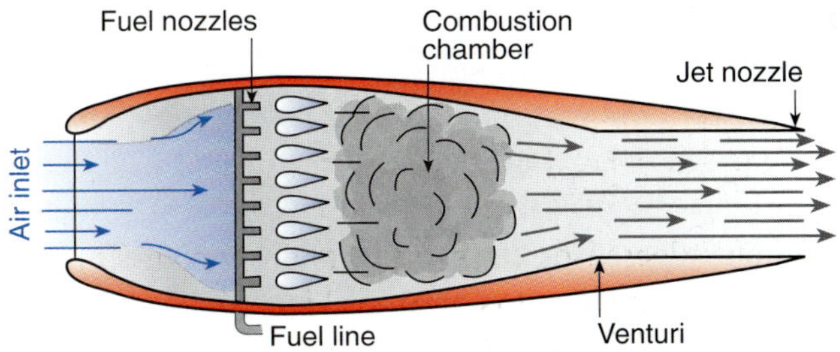

Figure 10-13. *A ramjet engine depends upon high-velocity air entering the engine to produce a barrier that forces the expanding gases to exit the rear of the engine and generate a forward thrust.*

One limitation of a ramjet engine is that the airflow inside the engine must be slowed to a subsonic velocity in order to sustain combustion of the fuel-air mixture. The scramjet (supersonic combustion ramjet) engine has overcome this limitation by its design which allows the fuel to be ignited and burn in the supersonic air traveling through the engine.

Afterburners used to augment the thrust of some turbojet and turbofan engines are a form of ramjet and are discussed beginning on Page 375.

ramjet engine. The simplest type of air-breathing reaction engine that depends on high-velocity air entering the front of the engine and forming a barrier that forces the expanding gases to leave through the nozzle at the rear.

scramjet (supersonic combustion ramjet). A special type of ramjet engine whose fuel can be ignited while the vehicle is moving at a supersonic speed.

afterburner. A component, similar to a ramjet engine, in the exhaust system of a turbojet or turbofan engine that is used to increase the thrust produced by the engine for takeoff and for special flight conditions.

turbine. A wheel with vanes radiating out from its circumference. Kinetic energy in a fluid is converted into mechanical power by the impulse or reaction of the fluid with the vanes.

Turbojet Engines

The most important breakthrough in air-breathing reaction engines has been the turbojet. A turbojet engine takes air in through an air inlet, compresses it, and mixes it with a liquid fuel that is ignited and burned. The hot gases escape through a turbine that extracts part of the energy and uses it to drive the compressor. The remaining energy forces the expanded air out through a specially shaped exhaust nozzle in the form of a high-speed jet.

The term "jet propulsion" normally means flight powered by a turbojet engine, but this definition is not always accurate. In Italy the Caproni-Campini CC-2 flew in 1940. This airplane was technically jet-propelled, but not as we think today. It used a 900-horsepower reciprocating engine to turn a three-stage compressor to compress the air into which fuel was injected and burned. Although the airplane flew, it was no faster than one powered by a conventional reciprocating engine and propeller, and further development along this line ceased.

Gas Turbine Engines

Water and steam turbines have been used for years to generate electricity and to power ships and trains. The first practical aeronautical application of a turbine was made in 1918 when one was used in a turbosupercharger. Hot exhaust gases from the engine spun a turbine that drove an air compressor and increased the mass of the air being fed into the induction system of the engine.

Whittle, Sir Frank. The British Royal Air Force flying officer who in 1929 filed a patent application for a turbojet engine.

axial turbine. A turbine that is turned by a fluid flowing through it in a direction that is approximately parallel to the shaft on which the turbine wheel is mounted.

axial-flow compressor. A type of compressor used in gas turbine engines in which the air passes through the compressor in essentially a straight line, parallel to the axis of the compressor.

von Ohain, Dr. Hans Pabst. The designer and developer of the first turbojet engine to actually power an airplane.

In 1929 Frank Whittle, a British Royal Air Force flying officer, filed a patent application for a turbojet engine. The engine used an axial turbine to drive a centrifugal compressor. Fuel was added to the compressed air and burned, and the expanding gases as they left the engine, flowed through the turbine to drive the compressor. Energy that was not thus used was ducted out of the engine through an exhaust nozzle in a high-velocity stream that produced thrust.

At about the same time Whittle was developing his engine in England, Dr. Hans von Ohain was working independently in Germany on a similar principle. His first engine used a centrifugal compressor, but his later engines used axial-flow compressors. Both types of compressors are discussed in detail later in this chapter.

Two basic problems faced Whittle and von Ohain. There were no compressors that could supply the tremendous amount of air needed, and there were no combustion chambers that could withstand the release of so much energy in such a short time, and in such a small space without being destroyed. It was only after years of research and development that these two problems, along with others that cropped up, were solved, and truly practical jet engines were built.

Von Ohain's engine first flew in a Heinkel He 178 on August 27, 1939, and Whittle's engine first flew in a Gloster E.28 on May 15, 1941. The first jet flight in America was made on October 2, 1942, in a Bell XP-59A that was powered by two General Electric I-A engines. These were versions of a Whittle engine built in America by General Electric under a special agreement between England and the United States.

Types of Gas Turbine Engines

Reciprocating engines have several built-in limitations that have been overcome by the gas turbine:

- The reciprocating action of the pistons sliding back and forth in the cylinders causes vibrations and significant internal stresses for which the engine must be designed. As a result, they are heavy for the power they produce.
- Reciprocating engines have a large number of moving parts that require the engine to be overhauled when a relatively few hours of operation have been accumulated.
- The only way a reciprocating engine can produce thrust is by driving a propeller.

TBO (time between overhauls). A time period specified by the manufacturer of an aircraft engine as the maximum length of time the engine can be operated between overhauls without normal wear causing critical parts to be worn beyond safe limits.

A gas turbine engine, on the other hand, is extremely light for the amount of power it produces. It has few moving parts, and with the development of new materials, has achieved an exceptionally high TBO (time between overhauls). Its high versatility lies in the fact that it can be used to produce either thrust or torque to power aircraft.

Turbojet

The original study by Frank Whittle that led to the successful gas turbine engine was directed toward jet propulsion. Whittle's engine was designed to propel an aircraft by imparting a tremendous amount of acceleration to the mass of air flowing through it by the addition of energy from burning fuel. There was a lack of interest in Whittle's theory of such a radical method of propulsion, because it was inefficient at the low altitude and low speeds flown by airplanes in the 1930s.

A turbojet engine consists of a compressor that takes air in through the aircraft air inlet duct and increases its pressure. It then sends this compressed air through a diffuser into a burner, or combustor. Fuel is then added and burned, and the resulting expanded and heated air flows through a turbine that extracts some of the energy to drive the compressor. A high-velocity stream of exhaust gases leaves the engine through a specially shaped exhaust nozzle where it produces thrust. The amount of thrust is proportional to the increase in the momentum of air that passes through the engine.

turbojet engine. A gas turbine engine that produces thrust by accelerating the air flowing through it by burning fuel in the air.

diffuser. A component in a gas turbine engine that decreases the velocity and increases the pressure of the air flowing through.

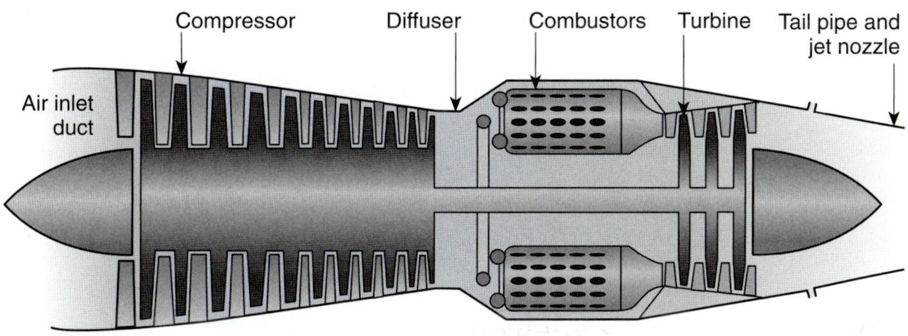

Figure 10-14. *Basic components of a turbojet engine*

Turboprop

A turbine engine can drive a propeller by extracting some of the energy that remains in the exhaust gases after they have driven the compressor. This can be done by connecting the propeller to the compressor through a set of reduction gears. But the propeller can be more efficiently driven through appropriate reduction gears by a turbine separate from the core engine, the portion of the engine that drives the compressor. An engine that uses a separate turbine to drive the propeller is called a free-turbine engine.

core engine. The gas generator portion of a turboshaft, turboprop, or turbofan engine.

free-turbine engine. A gas turbine engine that drives a fan, propeller, or output shaft by a turbine other than that which drives the core-engine compressor or compressors.

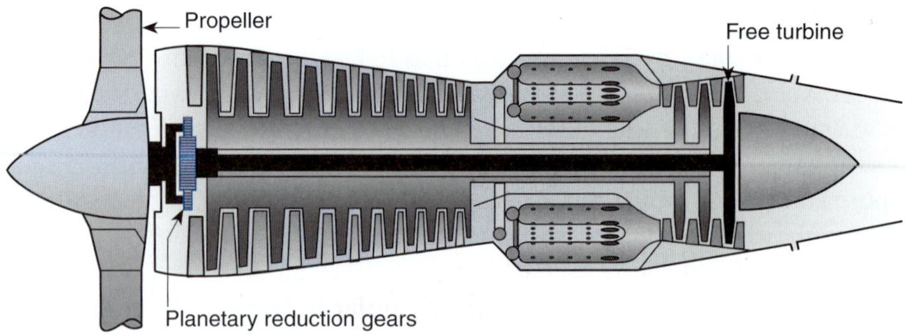

Figure 10-15. *A turboprop engine such as this uses a free turbine to drive the propeller through a series of planetary reduction gears.*

Turboshaft

Turbine engines are ideally suited for powering helicopters, because they operate most efficiently at the constant RPM required by a helicopter. Most turboshaft engines drive their output shaft with a multistage free turbine that extracts as much energy as possible from the exhaust gases.

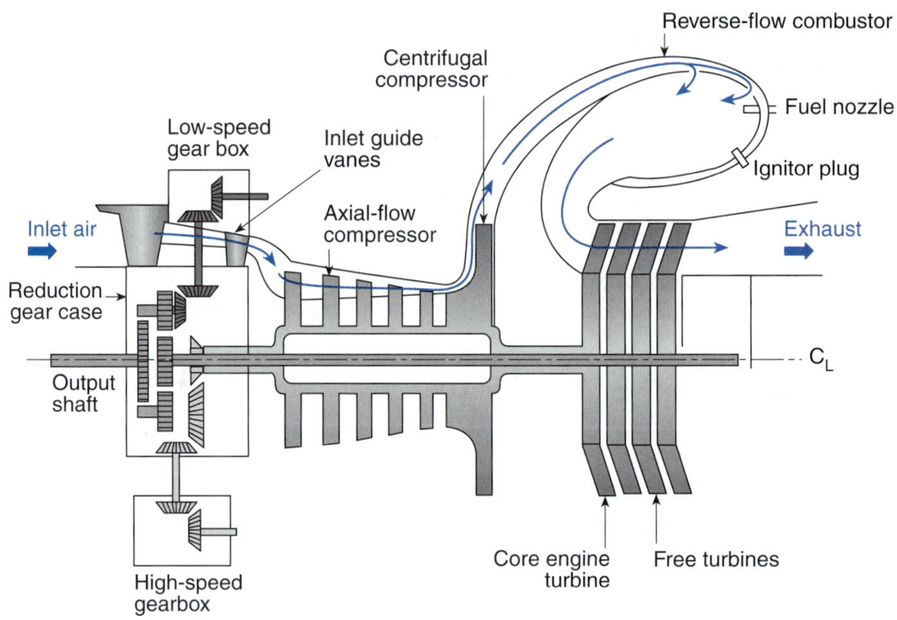

Figure 10-16. *A turboshaft engine uses a multistage free turbine to extract as much energy as possible from the exhaust gases before they leave the engine.*

Turbofan

Turbojet engines produce thrust by accelerating a smaller mass of air by a larger amount, and turboprop engines accelerate a larger mass by a smaller amount. A turbofan engine has some characteristics of both.

Turbojet engines were the principal type of thrust-producing turbine engines used for the first few decades of jet-propelled flight, but turbofan engines, with their advantages of increased propulsive efficiency, lower noise level, and lower specific fuel consumption, have almost replaced them.

A turbofan engine produces thrust similar to that produced by a combination of turbojet and turboprop engines. A turbofan engine, also called a bypass engine, has a set of lengthened blades in the first stage or stages of the low-pressure compressor. Air that flows through the fan section of the engine bypasses, or flows around the outside of the core engine. The amount of thrust produced by the fan varies between 30% and 75% of the total thrust, depending upon the bypass ratio.

The bypass ratio, which is the ratio of the mass of air moved by the fan to the mass of air moved by the core engine, varies from less than 1:1 for low-bypass engines, up to about 8:1 for some high-bypass engines.

Some low-bypass-ratio turbofan engines have long annular fan-discharge ducts that surround the full length of the engine. The fan-discharge air and the exhaust from the core engine are discharged separately in some long-duct engines, and in others, the two discharges are mixed before leaving the engine.

turbofan engine. A gas turbine engine that has a set of lengthened blades on the low-pressure compressor or low-pressure turbine. Air moved by these special blades bypasses the core engine and produces between 30% and 75% of the total thrust.

propulsive efficiency. A measure of the effectiveness with which an aircraft engine converts the fuel it burns into useful thrust.

bypass engine. Another name for a turbofan engine.

bypass ratio. The ratio of the mass of air moved by the fan to the mass of air moved by the core engine.

low-bypass engine. A turbofan engine whose bypass ratio is less than 2:1.

high-bypass engine. A turbofan engine whose bypass ratio is 4:1 or greater.

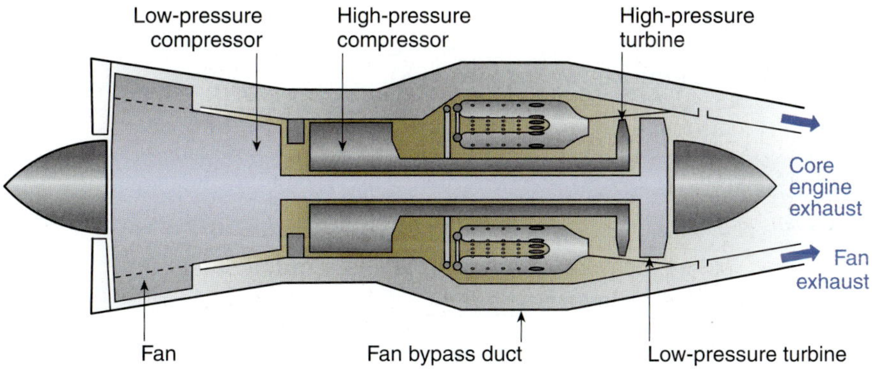

Figure 10-17. *This dual axial-flow, low-bypass-ratio turbofan engine has long ducts and separate exhausts for the fan discharge and core engine.*

Most of the high-bypass-ratio turbofan engines use short fan-discharge ducts like those in Figure 10-18.

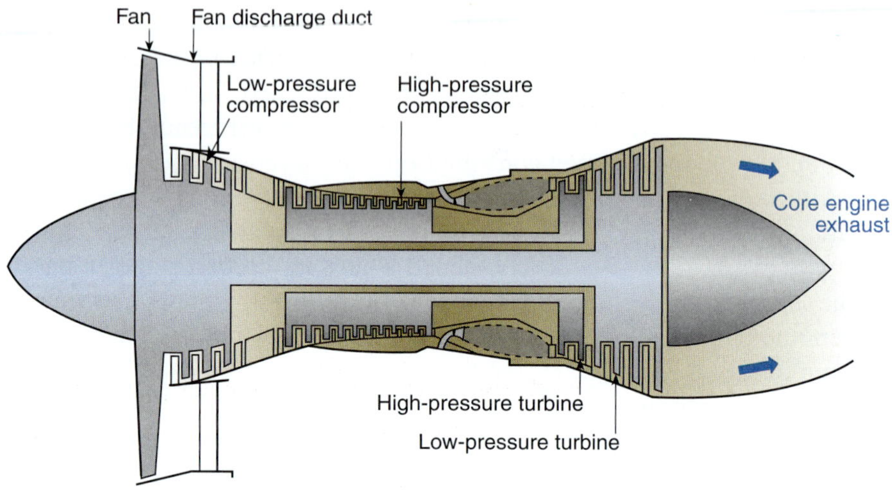

Figure 10-18. *A high-bypass-ratio turbofan engine with short fan discharge ducts*

UHB (ultrahigh-bypass) engine. A turbine engine that drives a pair of multiblade, contrarotating, variable-pitch propellers. UHB engines have a bypass ratio in excess of 30:1.

Unducted Fan™ engine. The registered trade name by General Electric of an ultrahigh-bypass engine.

PropFan™ engine. The registered trade name by Hamilton Standard of an ultra-high-bypass turbine engine.

scimitar shape. The curved shape of the blades of the propellers mounted on UHB engines.

contrarotating. Two components rotating in opposite directions.

There is a special type of ultra-high-bypass (UHB) engine with characteristics more like a turboprop than a turbofan. The Unducted Fan™ and the Prop-Fan™ drive scimitar-shaped, multiblade, contrarotating, variable-pitch propellers. Their bypass ratios exceed 30:1 and they are capable of powering airplanes in the speed range of Mach 0.8. These engines are discussed in more detail in Chapter 19, beginning on Page 687.

STUDY QUESTIONS: AIRCRAFT TURBINE ENGINES

Answers begin on Page 420. Page numbers refer to chapter text.

27. Two types of rocket engines are ____________________ and ____________________. *Page 359*

28. An afterburner used to augment the thrust of a turbojet or turbofan engine is a form of ________________ engine. *Page 361*

29. The function of the turbine in a turbojet engine is to drive the ______________________. *Page 361*

30. The two components that presented the greatest challenges in the development of the gas turbine engine were the ________________ and the _________________. *Page 362*

31. The first successful turbojet engine built in America was built by the ________________________ Company. *Page 362*

32. The six basic components in a turbojet engine are the:
   a. ____________________________
   b. ____________________________
   c. ____________________________
   d. ____________________________
   e. ____________________________
   f. ____________________________
   Page 363

33. The portion of a turboprop engine that produces the hot gases that drive the turbines is called the ______________________ engine. *Page 363*

34. A turboprop engine that uses a separate turbine to drive the propeller reduction gearing is called a/an _________________________ engine. *Page 363*

35. A gas turbine engine used to drive the transmission of a helicopter is called a/an ____________________ engine. *Page 364*

36. Three advantages of a turbofan engine over a turbojet engine are:
   a. __________________________
   b. __________________________
   c. __________________________
   Page 365

37. The ratio of the amount of air that passes through the fan to the amount of air that flows through the core engine is called the __________________ ratio of the engine. *Page 365*

Thrust

A gas turbine engine releases the energy in the fuel in a constant-pressure thermodynamic cycle called the Brayton cycle. The same events—intake, compression, expansion, power, and exhaust—that occur in the constant-volume Otto cycle used by reciprocating engines, take place in a gas turbine engine. The basic difference between the two cycles is that in an Otto-cycle engine, the events take place in the same location, in the engine cylinder, but at different times. In the Brayton cycle, the events occur at the same time, but at different locations within the engine.

Figures 10-19 and 10-20 explain how the Brayton cycle operates:

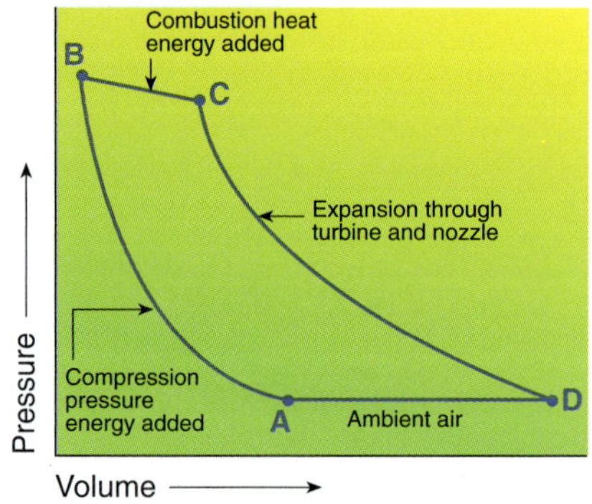

Figure 10-19. *In the Brayton thermodynamic cycle, energy from the fuel is released at a relatively constant pressure. The addition of heat energy causes the volume of the air to increase.*

1. Air enters the inlet duct and flows to the inlet of the compressor, point A, at ambient pressure.
2. The air passes through the compressor, which raises its pressure and decreases its volume to that represented by point B.
3. The air then passes into the combustion chamber where fuel is injected into it and burned. The pressure between points B and C remains relatively constant as the heat energy is added, but the volume and temperature increase.
4. The hot gases leave the combustion chamber and pass through the turbine where energy is extracted and the pressure drops to near ambient at point D.

The pressures between points A and D are nearly the same, but at D the volume and thus the velocity are much greater. The air has expanded at a relatively constant pressure.

The Production of Thrust

Horsepower, which is commonly used as a measure the output of a reciprocating engine, is not an appropriate measure for the output of a turbojet or turbofan engine, because time and distance are not always involved.

When a turbojet engine is running but the airplane is stopped on the runway, the time and distance values required for measuring power are not applicable. Torque and RPM are produced by the turbine, but the power developed is being used within the engine itself, primarily to turn the compressor.

The engine is developing thrust and is pushing against the aircraft, even when it is not producing external power. When the brakes are released, the airplane moves forward. Thrust in pounds, rather than horsepower, is used as a measure of the propulsive force developed by a turbojet or turbofan engine.

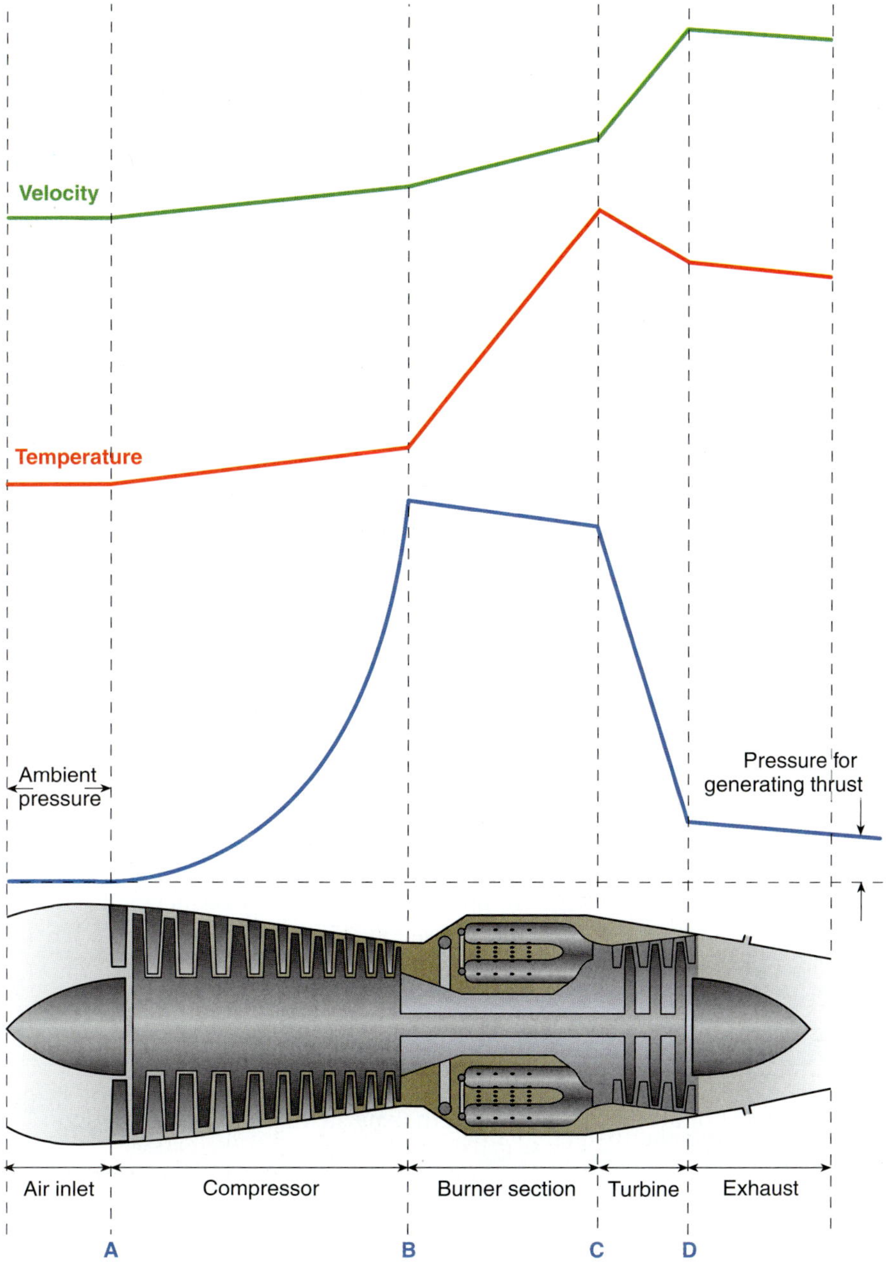

Figure 10-20. *The events that take place in the Brayton thermodynamic cycle occur at the same time but at different locations within the engine. The changes in pressure, temperature, and velocity are shown as they relate to the Brayton cycle.*

net thrust. Thrust produced by a turbojet or turbofan engine in flight.

gross thrust. Thrust produced by a turbojet or turbofan engine when the engine is static, or not moving.

Thrust produced by a turbojet or turbofan engine is determined by three things:

1. The change in momentum experienced by air flowing through the engine
2. The momentum of the fuel
3. The force caused by the difference in pressure across the exhaust nozzle multiplied by the area of the nozzle

There are two types of thrust: net thrust and gross thrust. Net thrust considers the momentum of air entering and leaving the engine, and gross thrust considers the inlet momentum to be zero.

Mass

Both weight and mass may be expressed in terms of pounds, but they are not the same. Mass is the amount of matter in an object, and is not affected by the gravitational pull of the earth. Weight, on the other hand, is the effect of gravity on mass.

$$\text{Mass} = \frac{\text{Weight (pounds)}}{\text{Acceleration due to gravity}}$$

$$= \frac{\text{Weight (pounds)}}{32.2 \text{ fps}^2}$$

Momentum Change

Momentum is the product of the mass of an object multiplied by its velocity. The momentum of air and fuel passing through a gas turbine engine is found by dividing their weight in pounds per second by the acceleration due to gravity, or 32.2 feet per second, per second. This is then multiplied by the velocity of gases leaving the engine.

$$\text{Momentum of air and fuel leaving engine} = \frac{W_a + W_f}{g} \cdot V_2$$

W_a = Weight of air flowing through the engine in pounds per second
W_f = Fuel flow in pounds per second
g = Constant for acceleration due to gravity of 32.2 feet per second^2
V_2 = Velocity of exhaust gases in feet per second

The fuel is carried in the aircraft, so it has no initial velocity relative to the engine. The momentum of air and fuel entering the engine is:

$$\text{Momentum of air and fuel entering engine} = \frac{W_a}{g} \cdot V_1 + 0$$

V_1 = Velocity of incoming air in feet per second. This is the speed of the aircraft.

The portion of thrust produced by the change in momentum of air and fuel passing through the engine is the sum of the change in momentum of the air and the change in the momentum of the fuel.

The formula for this portion of the thrust is:

$$F = \frac{W_a}{g}(V_2 - V_1) + \frac{W_f}{g}(V_2)$$

In some practical problems, the mass of the fuel flow is omitted, as it is quite small in comparison with the mass of the airflow.

Air Pressure at Nozzle

Many subsonic jet nozzles operate in a choked condition. This means that air flowing through the convergent section of the exhaust duct reaches the speed of sound and can no longer accelerate. The energy that would have caused acceleration now increases the pressure and creates a component of thrust by the difference between the exhaust pressure and the pressure of air surrounding the exhaust nozzle.

This component of thrust may be found by this formula:

$$F = A_j (P_2 - P_{am})$$

A_j = Area of the exhaust jet nozzle in square inches
P_2 = Static air pressure at the jet nozzle discharge in pounds per square inch
P_{am} = Static pressure of ambient air at the jet nozzle in pounds per square inch

The total net thrust produced by a turbojet or turbofan engine may be found by combining the two types of thrust just discussed into a single formula:

Net thrust (F_n) = Thrust caused by change in momentum of the air
+ Thrust caused by change in momentum of the fuel
+ Thrust caused by pressure drop across the exhaust jet nozzle

$$F_n = \underbrace{\left[\frac{W_a}{g}(V_2 - V_1)\right]}_{\text{Momentum of the air}} + \underbrace{\left[\frac{W_f}{g}(V_2)\right]}_{\text{Momentum of the fuel}} + \underbrace{\left[A_j (P_2 - P_{am})\right]}_{\text{Pressure drop across exhaust jet nozzle}}$$

Example: Find the net thrust produced by a turbojet engine that has these specifications:

Airflow through the engine W_a	=	85 pounds per second
Airspeed	=	325 mph (476 feet per second)
Jet nozzle velocity V_2	=	1,475 feet per second
Area of jet nozzle A_j	=	275 in^2
Pressure at jet nozzle P_2	=	76 in Hg (37.33 psi)
Ambient pressure P_{am}	=	28.75 in Hg (14.12 psi)
Fuel flow	=	5,250 pounds/hour (1.46 pounds/sec.)

$$F_n = \frac{W_a}{g}(V_2 - V_1) + \frac{W_f}{g}(V_2) + A_j(P_2 - P_{am})$$

$$= \frac{85}{32.2}(1475 - 476) + \frac{1.46}{32.2}(1475) + 275(37.33 - 14.12)$$

$$= 2.64(999) + 0.045(1475) + 275(23.21)$$

$$= 2637.4 + 66.8 + 6382.8$$

$$= 9{,}087 \text{ pounds}$$

Turbofan Engine Thrust

When computing the thrust produced by a turbofan engine, compute the thrust of the core engine as just described, and then compute the thrust produced by the momentum change of air passing through the fan. The total thrust is the sum of the thrust from the core engine plus the thrust from the fan.

$$\text{Thrust}_{\text{TOTAL}} = \text{Thrust}_{\text{CORE ENGINE}} + \text{Thrust}_{\text{FAN}}$$

Distribution of Thrust

The net thrust in an engine is made up of several components, as seen in Figure 10-21. When momentum is added to the mass of air flowing through the engine, the thrust is forward (+), and when momentum is lost, the thrust is rearward (–).

The air flowing into the engine through the intake duct produces no thrust of either type, but as the air is compressed, its momentum is increased by energy taken from the turbine, and a forward thrust is produced. As air passes through the combustion section, energy is added by the burning fuel and another increment of forward thrust is added. When the hot air leaves the combustion section, it flows through the turbine where much of its energy is extracted to turn the compressor. This results in a rearward thrust. As exhaust gases expand through the convergent exhaust duct, more rearward thrust is produced. When all rearward components of thrust are subtracted from the forward components, the resultant thrust left is the thrust available for propulsion. *See* Figure 10-21.

convergent duct. A duct that has a decreasing cross section in the direction of flow.

Measurement of Thrust

Thrust produced by a turbojet or turbofan engine may be measured on the ground in an engine test stand. The engine is mounted on a floating mount that pushes against a calibrated scale that has a remote indicator in the control booth. The indicator is calibrated in pounds of thrust.

It is necessary to know the amount of thrust an engine develops when operating in the aircraft. Some smaller jet engines, especially those with centrifugal compressors, use the compressor speed in % RPM as an indication of the thrust. RPM can be used because it is approximately proportional to the thrust the engine is producing.

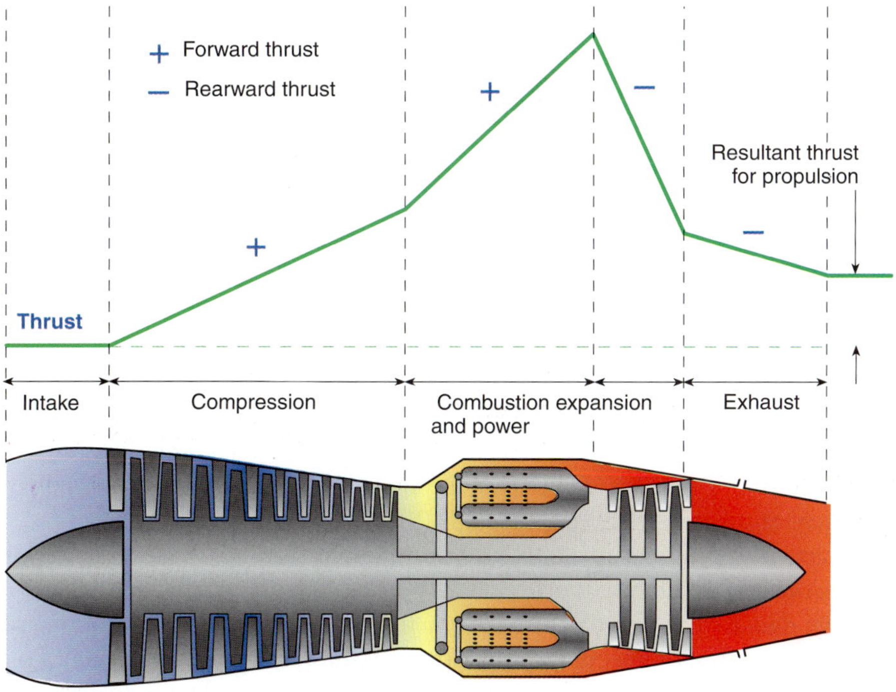

Figure 10-21. *Distribution of thrust through a turbojet engine*

Almost all larger engines indicate the amount of thrust produced either by measuring the turbine discharge total pressure or more generally, by measuring the engine pressure ratio (EPR).

The total pressure is the static pressure plus the rise in pressure caused by the ram effect. The engine pressure ratio is the ratio of the turbine discharge total pressure to the compressor inlet total pressure.

Factors Affecting Thrust

Gas turbine engines operate under widely varying conditions that affect the amount of thrust the engine produces. These conditions affect either the momentum of the air and fuel flowing through the engine or the pressure differential across the exhaust nozzle.

During most normal high-speed operations, the exhaust nozzle operates in a choked condition. This means that the gases flowing out of the engine have been accelerated to the speed of sound, and they cannot be accelerated any further. The value V_2 in the thrust formula remains relatively constant, changing only as the temperature of the exhaust gas changes the speed of sound.

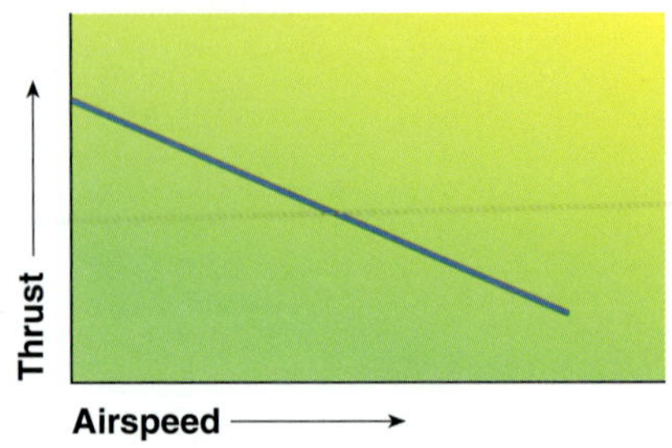

Figure 10-22. *As the airspeed of a turbojet- or turbofan-powered aircraft increases, the net thrust decreases.*

Airspeed

When a turbojet or turbofan engine is operating at high power and the aircraft in which it is mounted is not moving, the value of V_1 in the thrust formula is zero and thrust is maximum.

$$F_n = \frac{W_a}{g}(V_2 - V_1) + \frac{W_f}{g}(V_2) + A_j(P_2 - P_{am})$$

When the aircraft is moving forward, the velocity, V_1, of the air entering the engine increases, but at high power, the exhaust nozzle is choked, and the value V_2 does not increase proportionally. Therefore, as airspeed increases, net thrust decreases, as shown in the curve in Figure 10-22.

Mass Airflow

The basis of the thrust produced by a turbojet or turbofan engine is the change in momentum of air flowing through the engine. Anything that increases the mass of air increases the thrust. Two factors that affect the mass of air are its density and the ram effect.

Air density has a profound effect on the thrust produced. The volume of air flowing through the engine is relatively fixed for any particular RPM by the size and geometry of the inlet duct system. But since thrust is determined by mass, not the volume of air, any increase in its density increases its mass and thus the thrust.

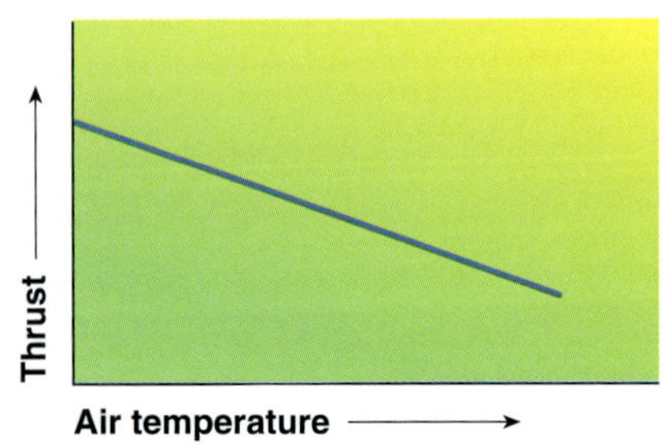

Figure 10-23. *As the temperature of the air increases, its density decreases, causing the thrust produced by the engine to decrease.*

As the temperature of the air increases, its density decreases, and the thrust produced by the engine decreases. *See* Figure 10-23.

As the air pressure increases, its density increases, causing the thrust produced by the engine to increase. *See* Figure 10-24.

Altitude has a double effect on thrust. As altitude increases, the air becomes colder and denser, up to the beginning of the stratosphere. This causes the thrust to increase. But, at the same time, an increase in altitude causes a decrease in pressure, thus a decrease in density and a corresponding decrease in thrust. Since the loss of thrust caused by decreasing pressure is greater than the increase caused by decreasing temperature, the thrust decreases as the aircraft ascends.

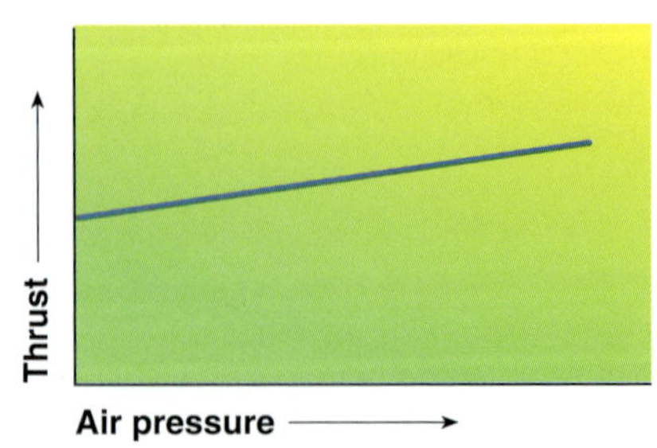

Figure 10-24. *As the pressure of the air increases, its density increases, and the thrust produced by the engine increases.*

At the beginning of the stratosphere at approximately 36,000 feet, the temperature stabilizes at -56.5°C (-69.7°F) and remains at this temperature up to around 85,000 feet. The pressure continues to fall above 36,000 feet, and the thrust therefore drops off at a faster rate than it does at the lower altitude. This increased drop-off in thrust makes 36,000 feet a chosen altitude for long-range cruise in jet-powered aircraft. *See* Figure 10-25.

Ram Effect

As a turbine-powered aircraft begins to move forward for takeoff, air is rammed into the inlet duct, and the mass flow through the engine increases. The faster the aircraft moves, the greater the increase in thrust. This is shown in Figure 10-26, curve A.

But, as was shown in Figure 10-22, an increase in forward speed decreases the amount the air is accelerated, and the thrust decreases. This is seen in Figure 10-26, curve B.

The thrust increase caused by the ram effect is greater than the decrease caused by the increasing airspeed, and the net increase is seen in curve C. Notice that the thrust increase becomes greater as the airspeed increases.

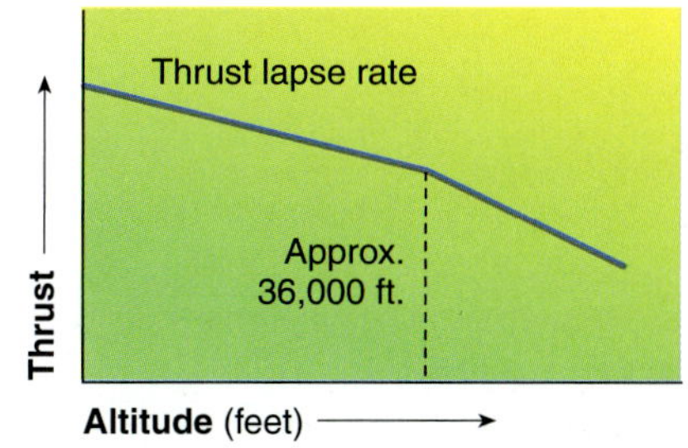

Figure 10-25. *Air temperature stabilizes at about 36,000 feet, but pressure continues to drop. The thrust produced by a jet engine decreases at a faster rate above this altitude.*

Thrust Augmentation

The amount of thrust a jet engine can produce may be increased in two ways: by water injection and by the use of afterburners.

Water Injection

On a hot day, the temperature of the air entering a turbine engine is high and its density is low; consequently, the thrust the engine can produce is correspondingly low. To increase the takeoff thrust, some engines have used water injection. Demineralized water or a mixture of demineralized water and methyl alcohol is sprayed into the air inlet, and sometimes also into the diffuser, after the air has passed through the compressor. The water evaporates and lowers the temperature of the air flowing through the engine. This increases the mass airflow and reduces the temperature at the turbine inlet. Increasing the mass airflow and decreasing the turbine inlet temperature allows more fuel to be burned without the turbine inlet temperature reaching its limit. Takeoff thrust can be increased by between 10% and 30% by the use of water injection.

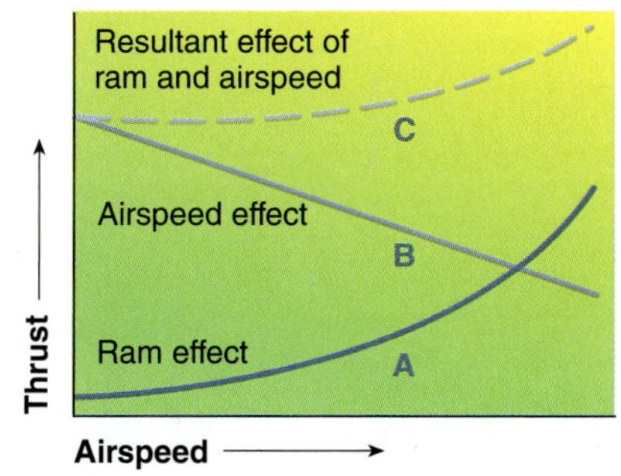

Figure 10-26. *As the airspeed of an aircraft increases, ram effect increases the thrust, but the airspeed effect decreases the thrust. The ram effect is greater, so the net thrust increases with an increase in airspeed.*

Afterburners

Only about 25% of the air passing through the core engine is used for combustion; the remaining air is used for cooling. This unburned air leaving the engine contains a large amount of oxygen.

An afterburner is essentially a ramjet attached to the rear of a turbojet or turbofan engine. When additional thrust is required, fuel is sprayed into the extremely hot exhaust gases and ignited. The additional heat further accelerates the air and produces a thrust increase in the neighborhood of 50%. This additional thrust is obtained at the cost of a threefold increase in the normal fuel consumption.

Afterburners are used on supersonic military fighter airplanes for bursts of extra speed, and on supersonic commercial airplanes to allow them to reach an efficient cruising altitude in much less time.

An afterburner consists of the afterburner duct, fuel spray nozzles or spray bars, and flame holders. An adjustable exhaust nozzle is opened to increase the nozzle area to maintain the proper discharge pressure when the afterburner is in use, and closed to decrease the area when it is not in use. Afterburners are discussed in more detail beginning on Page 513.

duct heater. A thrust augmentation system similar to an afterburner in which fuel is added to the fan-discharge air and burned.

Some turbofan engines use a duct heater for thrust augmentation. A duct heater is similar to an afterburner, but the fuel is added to the fan-discharge air and burned.

Thrust Horsepower

When a jet-powered aircraft is moving, time and distance are relevant, and the engine produces power. Thrust horsepower can be computed when the thrust produced by the engine in pounds and the speed of the airplane in miles per hour are known.

One horsepower is equal to 33,000 foot-pounds of work per minute, which is the same as 375 mile-pounds per hour. When an airplane is flying at 375 miles per hour, each pound of thrust is the equivalent of one horsepower. When the speed of the airplane is doubled, the amount of horsepower produced by each pound of thrust is also doubled.

$$\text{Thrust horsepower} = \frac{\text{Net thrust (pounds)} \cdot \text{miles per hour}}{\text{375 mile-pounds per hour}}$$

The thrust horsepower produced by an engine and propeller can be approximated by multiplying the brake horsepower the engine is producing by an industry-accepted standard for propeller efficiency (η) of 80%. If a reciprocating engine is producing 1,000 brake horsepower, it is producing 800 thrust horsepower.

$$\begin{aligned}\text{Thrust horsepower} &= \text{Brake horsepower} \cdot \text{Propeller efficiency}\\ &= 1{,}000 \cdot 80\%\\ &= 800\end{aligned}$$

Equivalent Shaft Horsepower

ESHP (equivalent shaft horsepower). A measure of the power produced by a turboprop engine that combines the shaft horsepower being delivered to the propeller and the thrust being developed at the engine exhaust.

Turboprop engines are rated in their equivalent shaft horsepower (ESHP). This rating takes into consideration both the shaft horsepower delivered to the propeller and the thrust developed at the engine exhaust. Under static conditions, one shaft horsepower is equal to approximately 2.5 pounds of thrust.

The formula for ESHP is:

$$\text{ESHP (static)} = \text{SHP} + \frac{F_n}{2.5}$$

ESHP = equivalent shaft horsepower
SHP = shaft horsepower as determined by the torquemeter and tachometer
F_n = net thrust in pounds
2.5 = a constant that relates horsepower to static thrust.
One hp = 2.5 pounds of thrust

In flight, the ESHP considers the thrust produced by the propeller, which is found by multiplying the net thrust in pounds by the speed of the aircraft in miles per hour. Divide this by 375 times the propeller efficiency, which is considered to be 80%.

$$\text{ESHP (flight)} = \text{SHP} + \frac{F_n \cdot V}{375 \cdot \eta}$$

V = airspeed in miles per hour
375 = a constant; mile-pounds per hour for one horsepower
η = propeller efficiency; an industry standard of 80%.

Example: Find the equivalent shaft horsepower produced by a turboprop airplane that has these specifications:

Airspeed = 260 mph
Shaft horsepower indicated on the cockpit indicator = 525 SHP
Net thrust = 195 pounds

$$\begin{aligned}\text{ESHP (flight)} &= \text{SHP} + \frac{F_n \cdot V}{375 \cdot \eta} \\ &= 525 + \frac{195 \cdot 260}{375 \cdot 0.8} \\ &= 525 + 169 \\ &= 694\end{aligned}$$

Under these flight conditions, the engine is producing 694 ESHP.

STUDY QUESTIONS: THRUST

Answers begin on Page 420. *Page numbers refer to chapter text.*

38. The thermodynamic cycle of a gas turbine engine is the ____________________ cycle. *Page 368*

39. The Brayton thermodynamic cycle is the constant ____________________ (pressure or volume) cycle. *Page 368*

40. The thrust a turbojet or turbofan engine produces when the aircraft is not moving is _________________ (gross or net) thrust. *Page 370*

41. The thrust produced by a turbofan engine is the sum of the thrust produced by the _______________ and that produced by the __________________ . *Page 370*

42. When momentum is added to the air passing through the engine, the thrust is _________________ (forward or rearward). *Page 370*

43. When all of the rearward components of thrust have been subtracted from the forward components, the remaining thrust is that which is available for _______________________ . *Page 372*

44. Three parameters that may be measured to determine the amount of thrust a turbojet engine is developing in flight are:
    a. __________________________
    b. __________________________
    c. __________________________
    Page 373

45. Three things that affect the thrust produced by a turbojet or turbofan engine are:
    a. __________________________
    b. __________________________
    c. __________________________
    Pages 374, 375

46. An increase in the airspeed, with all else remaining the same, will cause turbine engine thrust to ______________________ (increase or decrease). *Page 374*

47. An increase in the temperature of the inlet air, with all else remaining the same, will cause turbine engine thrust to ______________________ (increase or decrease). *Page 374*

48. An increase in the pressure of inlet air, with all else remaining the same, will cause turbine engine thrust to ______________________ (increase or decrease). *Page 374*

49. An increase in altitude will cause turbine engine thrust to ______________________ (increase or decrease). *Page 375*

50. An efficient cruising altitude for a turbojet- or turbofan-powered aircraft is near 36,000 feet. Above this altitude, the thrust produced by the engine ___________________ (increases or drops off more rapidly). *Page 375*

51. An increase in ram effect, with all else remaining the same, will cause turbine engine thrust to ______________________ (increase or decrease). *Page 375*

52. Airspeed change and ram effect both cause a change in the thrust a turbine engine produces. The ________________________ (airspeed change or ram effect) has the greater effect on the thrust. *Page 375*

53. Two ways the thrust can be augmented on a turbojet or turbofan engine are:
   a. __________________________
   b. __________________________
   Page 375

54. One pound of thrust is equal to one thrust horsepower when the aircraft is flying at ___________ miles per hour. *Page 376*

55. When computing the static thrust produced by a turboprop engine, one shaft horsepower produces approximately ______________ pounds of thrust. *Page 376*

56. When computing the equivalent shaft horsepower of a turboprop engine, the propeller efficiency is considered to be _____________ percent. *Page 376*

cold section. The portion of a gas turbine engine ahead of the combustion section.

hot section. The portion of a gas turbine engine that contains the burner section, the turbines, and the exhaust section.

Turbine Engine Terms and Definitions

An aircraft gas turbine engine is divided into two sections: the cold section and the hot section. The cold section includes the inlet air duct, the compressor, and the diffuser. The hot section includes the combustion section, the turbine, and the exhaust system.

The air inlet duct is technically a part of the airframe, but it is so important in the development of thrust that it is included with the engine as a part of the cold section.

In order to compare gas turbine engines, the industry has agreed upon certain standard abbreviations and symbols. Some manufacturers have proprietary terms and names, but the ones in Figure 10-28 are widely accepted and used.

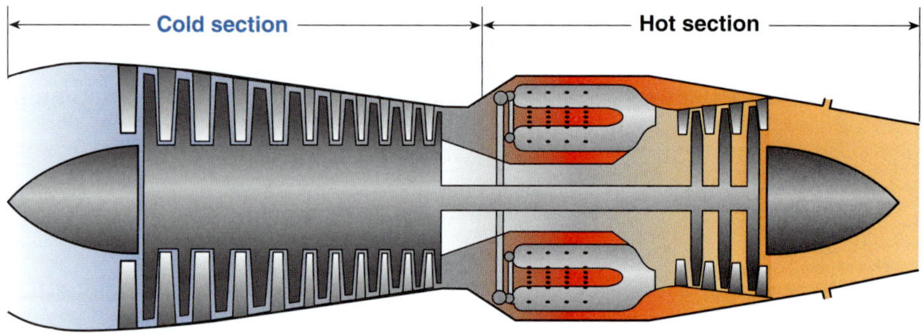

Figure 10-27. *A gas turbine engine is divided into two sections: the cold section, which includes the air inlet duct, compressor, and diffuser, and the hot section which includes the combustion section, turbine, and exhaust system.*

A	Cross-sectional area usually in square inches or square feet
am	Ambient
c	Velocity of sound in air in feet per second
Δ **(Delta)**	Difference or change
ESHP	Equivalent shaft horsepower (turboprop)
ESFC	Equivalent specific fuel consumption (turboprop)
η **(eta)**	Efficiency
F_g	Gross thrust
F_n	Net thrust
g	Acceleration due to gravity, a constant 32.2 fps^2
M	Mach number
N	Compressor speed in RPM or in percent of maximum RPM
N_1	Speed of the low-pressure compressor
N_2	Speed of the high-pressure compressor
N_3	Speed of the free turbine
P or psia	Absolute pressure
p or psig	Gage pressure
ρ **(rho)**	Density
T	Absolute temperature °R or °K
t	Temperature °F or °C
V	Velocity in feet per second
W	Weight in pounds
w	Flow rate of a gas in terms of pounds per second or pounds per hour

Subscripts used to modify the letters and symbols:

a	air
am	ambient
av	average
b	burner
bl	bleed
c	compressor
d	diffuser or duct
e	exhaust or exit
f	fuel
F	turbofan stations
g	gross
i	initial conditions
j	jet
n	net
o	sea level standard conditions
p	propulsion or propeller
r	ram
s	static
t	total or turbine
th	thermal
v	volume

Figure 10-28. *Gas turbine engine abbreviations and symbols*

Engine Station Designations

To standardize the locations in gas turbine engines, a numbering system has been devised that allows a person to identify a location by numbers. The locations are shown in Figures 10-29 and 10-30, and typical numbers for a dual-compressor turbofan engine with afterburner are identified in Figure 10-31.

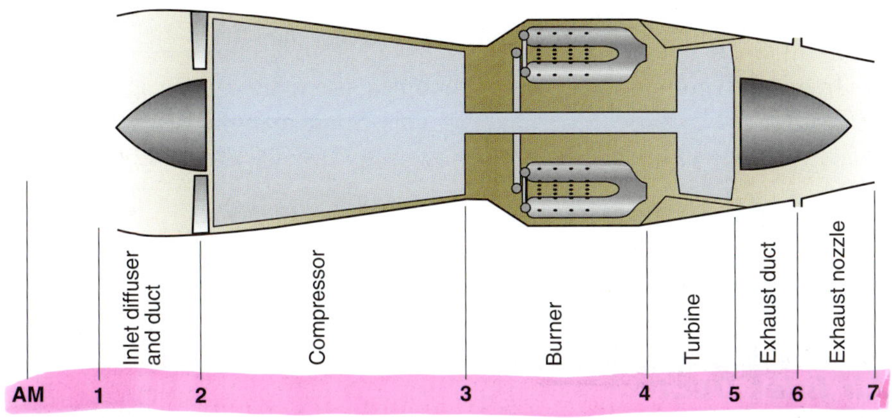

Figure 10-29. *Station location designators for a single-compressor turbojet engine without an afterburner*

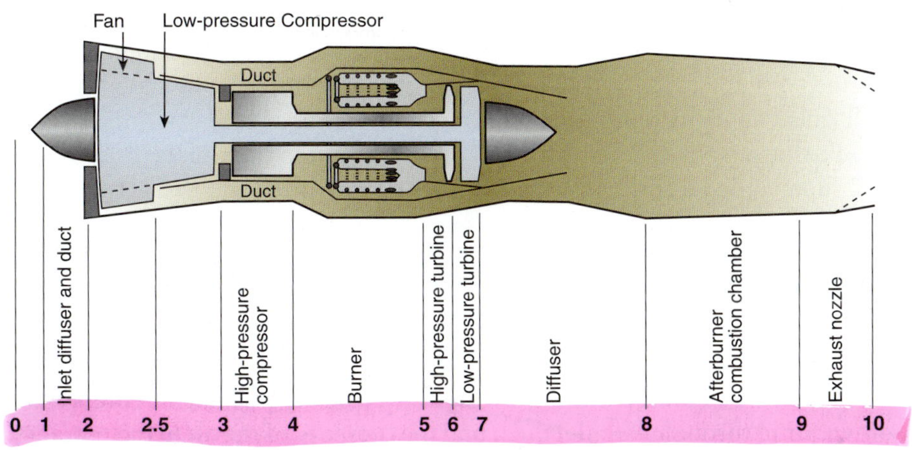

Figure 10-30. *Station location designators for a dual-compressor turbofan engine with an afterburner*

AM or 0	Ahead of the engine
1	Entrance to the aircraft air inlet duct
2	Fan (N_1 compressor) inlet
3	N_2 compressor inlet
4	Burner inlet
5	N_2 turbine inlet
6	N_1 turbine inlet
7	Diffuser
8	Afterburner combustion chamber
9	Afterburner duct
10	Exhaust nozzle discharge

Figure 10-31. *Typical station location designations for a dual-compressor turbofan engine with afterburner*

STUDY QUESTIONS: TURBINE ENGINE TERMS AND DEFINITIONS

Answers begin on Page 420. Page numbers refer to chapter text.

57. The speed of the high-pressure compressor in a dual-compressor turbofan engine is called ___________ (N_1 or N_2). *Page 380*

58. A pressure identified as P_{t2} is _____________ (static or total) pressure measured at the _________________ (compressor or turbine) ________________ (inlet or outlet). *Pages 380, 381*

The Cold Section

The cold section of a gas turbine engine extends from the inlet ducts ahead of the compressor to the rear flange of the diffuser where the combustion section attaches.

Air Inlet Ducts

The air inlet ducts are technically part of the airframe, but because of their extreme importance to the operation of the engine, we will include them in the study of the cold section.

For a gas turbine engine to operate effectively, the compressor must be supplied with a constant flow of subsonic air with a minimum of disturbance. The design and fabrication of the inlet ducts are carefully matched to the engine, and any repairs to these critical components must be made so they will cause no distortion of the airflow entering the engine.

For the purpose of study, we will divide them into inlet ducts for subsonic and supersonic aircraft.

divergent duct. A duct that has an increased cross-sectional area in the direction of flow.

ram-recovery speed. The speed at which the ram effect caused by the forward movement of an aircraft increases the air pressure at the compressor inlet until it is the same as that of the ambient air.

Subsonic Inlet Ducts

Most inlet ducts on subsonic airplanes are of the divergent type shown in Figure 10-32. Air flowing into a divergent duct expands slightly and converts some of its velocity energy into pressure.

When the engine is running at a high speed on the ground, the compressor draws air in through the inlet duct, and the pressure of the air at the entrance to the compressor is slightly lower than that of the surrounding or ambient air. When the airplane moves down the runway for takeoff, air is rammed into the duct until its pressure becomes the same as that of the ambient air. The speed at which this happens is called the ram-recovery speed. As the airplane continues to increase its forward speed, the ram effect becomes greater, and even though there is some loss due to the increased velocity of the air entering the engine, the thrust increases. This is seen in Figure 10-26 on Page 375.

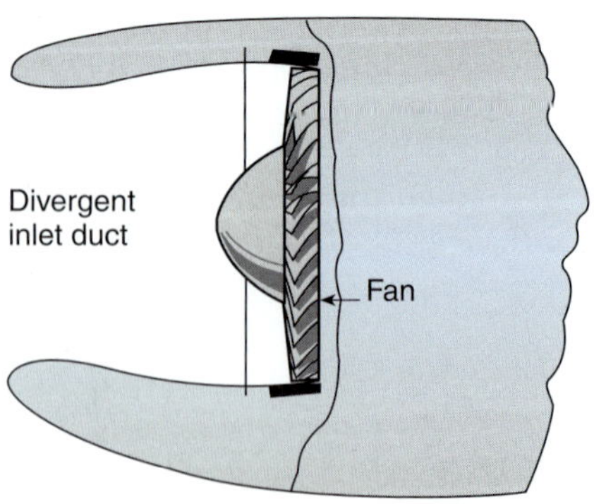

Figure 10-32. *The subsonic inlet duct for this turbofan engine is divergent. It decreases the velocity and increases the pressure of air entering the fan.*

Figure 10-33 shows the type of inlet duct used by a typical high-bypass-ratio turbofan. The air flows through the divergent duct into the fan, and although part of the air passes out through the fan discharge, part of it also flows into the low-pressure compressor and supercharges the core engine.

The air inlet for turboprop engines has caused some unique problems for engineers. The propeller reduction gears are located at the front of the engine and thus interfere with a smooth flow of air entering the compressor. Figure 10-34 illustrates three types of inlet ducts for turboprop engines. The ducted spinner inlet is the most efficient, but it is complex to maintain and to anti-ice. The conical spinner with the inlet duct around it is an effective alternative. For engines that have offset reduction gearing, the air scoop can be located below or above the reduction gears, depending upon which way the gears are offset.

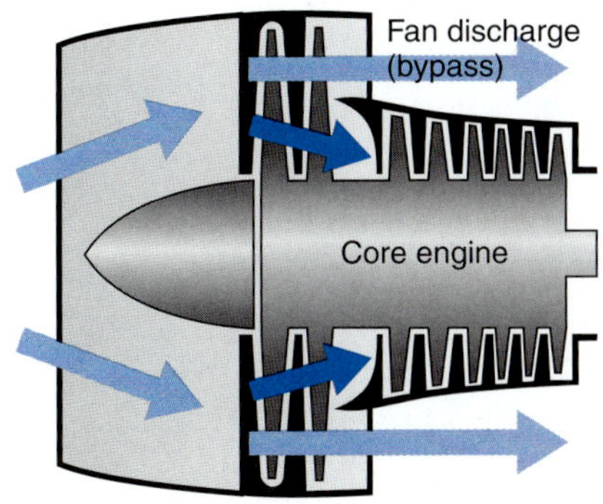

Figure 10-33. *Divergent inlet duct used on a high-bypass-ratio turbofan engine. Part of the fan discharge air is used as the inlet air for the core engine.*

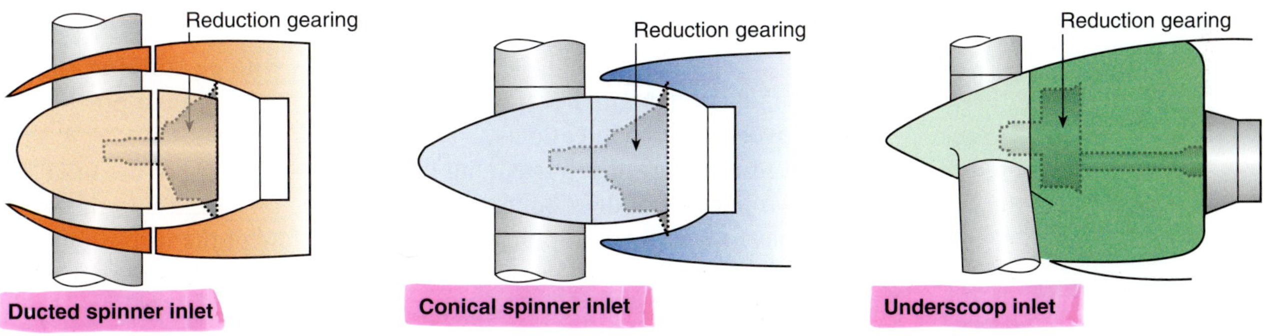

Figure 10-34. *Typical inlet ducts for turboprop engines*

The compressor on the popular Pratt & Whitney of Canada PT6 engine is located at the rear of the engine, and the air flows forward. The induction air flows through an inlet duct, normally located below the engine, into a plenum chamber that surrounds the engine and acts as a diffuser. This increases the pressure of the air entering the compressor.

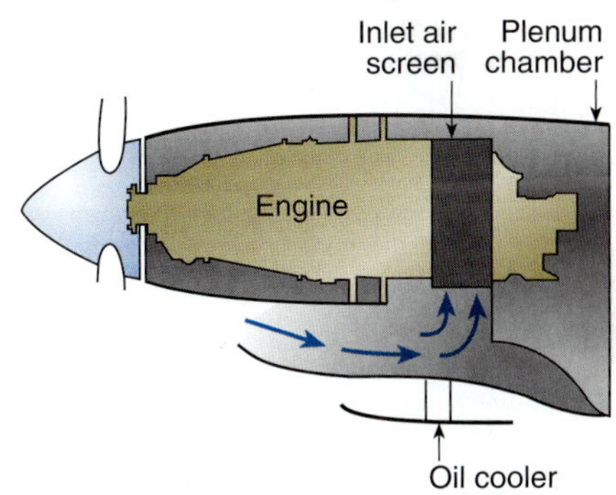

Figure 10-35. *The Pratt & Whitney of Canada PT6 turboprop engine has the compressor inlet at the rear. Air flows through the inlet duct into a plenum chamber, and then through an inlet screen into the compressor.*

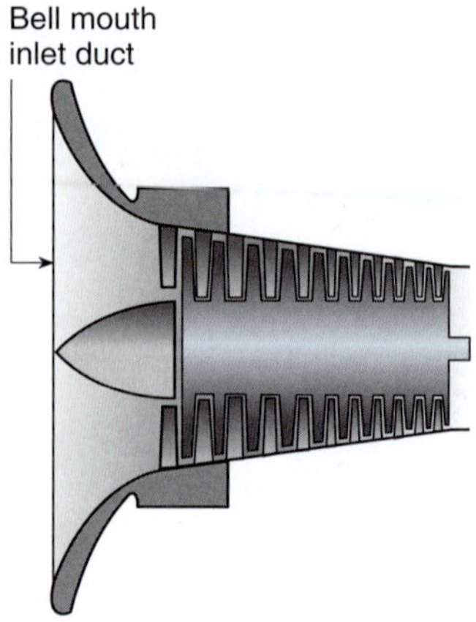

Figure 10-36. *A bell mouth inlet has almost no duct losses and is used when calibrating a gas turbine engine on a test stand.*

bell mouth inlet duct. A wide, smooth, convergent inlet-air duct used to direct air into the compressor of a gas turbine engine.

duct losses. A decrease in pressure of the air flowing into a gas turbine engine caused by friction.

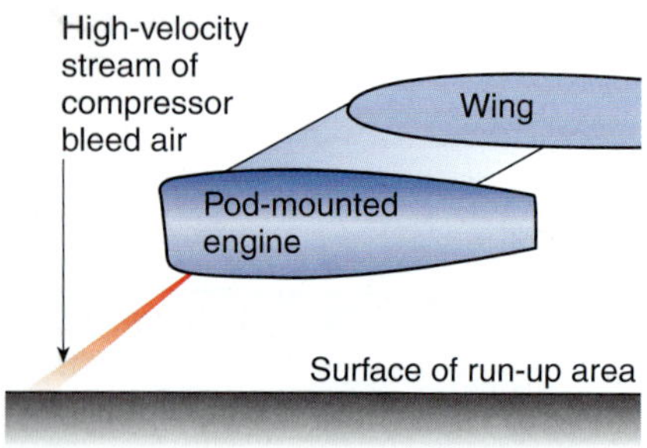

Figure 10-37. *Vortex dissipaters are used on some engines that are mounted on pods below the wing. A stream of high-velocity compressor bleed air is directed into the area where vortices would normally form, and dissipates them as they form.*

vortex dissipater. A high-velocity stream of compressor bleed air blown from a nozzle into an area where vortices are likely to form.

Turbine engines installed in most run-in stands and on some helicopters have bell mouth inlets. These inlets, like that in Figure 10-36, provide a smooth curve for the induction air to follow when flowing into the compressor. The duct losses with a bell mouth are extremely low, and this type of inlet is used when calibrating the performance of gas turbine engines.

Blow-In Doors

Under some conditions of airflow in a gas turbine engine, the inlet duct does not furnish sufficient air to prevent a compressor stall. Installations in which this is a possibility are often equipped with blow-in doors.

Blow-in doors are installed in the side of the inlet duct and are spring-loaded to hold them closed. But when the inlet air pressure becomes a specified amount lower than that of the ambient air, the pressure differential forces them open and furnishes additional air to the compressor inlet.

Foreign-Object Damage

Foreign-object damage (FOD) is a major problem with gas turbine engine operation, and when an aircraft is not operating, it is common practice to install covers over the inlet ducts.

Inlet screens have been tried, but the amount they impede the inlet air, and the probability of their being covered with ice or sucked into the compressor have made them impractical for all but a few of the turboshaft and turboprop engines.

Ice that forms on the lip of the inlet duct can break off and cause damage to the compressor. This problem is minimized by routing warm compressor bleed air between the skins of the inlet duct to prevent the formation of ice.

When air is drawn into the inlet duct, a high-energy vortex is often formed, which creates a strong suction that reaches to the ground from engines mounted low in pods beneath the wing. These vortices can pick up rocks and other debris that, if allowed to enter the engine, will damage the compressor. To prevent this type of damage, vortex dissipaters may be installed below and in front of the inlet duct. These are nozzles that blow a stream of high-velocity compressor bleed air into the area where the vortices form. They force the vortex to dissipate before it can gain enough energy to pick up debris.

Supersonic Inlet Ducts

The air approaching the compressor inlet must always be at a speed below the speed of sound. When the aircraft is flying at supersonic speed, the inlet air must be slowed to subsonic speed before it reaches the compressor. This is done by using a convergent-divergent, or CD, inlet duct such as the one in Figure 10-38. Air enters the convergent portion of the duct at a supersonic speed, and the velocity decreases until the narrowest part of the duct is reached. At this point the air velocity has been reduced to the speed of sound and a normal shock wave forms. Beyond this point, the duct becomes larger. The air, which has passed through the shock wave, is now flowing at a subsonic speed, and it is further slowed down as it flows through the divergent portion of the duct. By the time it reaches the compressor, its speed is well below the speed of sound, and its pressure has been increased.

Aircraft that fly at very high speeds normally have variable inlet ducts that change their shape as the airspeed changes. This is done either by lowering or raising a wedge, or by moving a tapered plug in or out of the duct. One or more oblique shock waves form in the duct to slow the air to near sonic, and then a normal shock wave forms to complete the transition from supersonic to subsonic. *See* Figure 10-39.

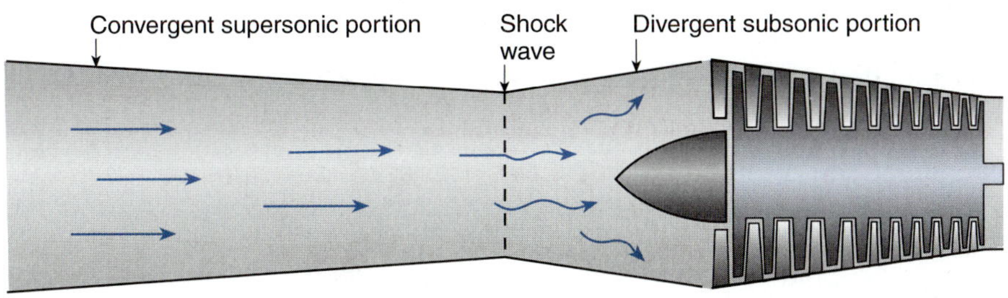

Figure 10-38. *Supersonic inlet air is slowed to Mach 1.0 in the convergent portion of the duct and a shock wave is formed. When the air flows through the shock wave, it slows to a subsonic speed and is further slowed in the divergent portion of the duct before it enters the compressor.*

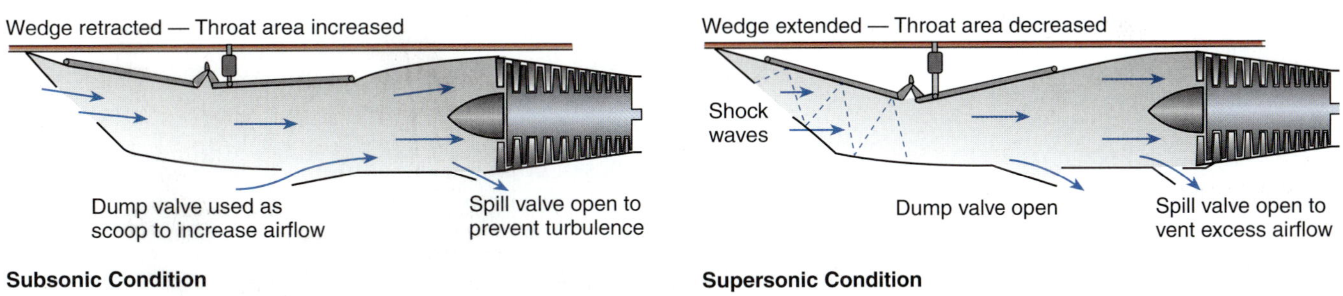

Figure 10-39. *This movable-wedge variable inlet duct slows supersonic airflow to a subsonic speed by forcing it to flow through a series of oblique shock waves, and then finally through a normal shock wave.*

blow-in doors. Spring-loaded doors in the inlet duct of some turbojet or turbofan engine installations that allow additional air to flow during conditions of low ram pressure.

compressor stall. A condition in a turbine engine axial-flow compressor in which the angle of attack of one or more blades becomes excessive, and the smooth airflow through the compressor is disrupted.

FOD (foreign-object damage). Damage to the parts of a gas turbine in the gas path by the ingestion of articles such as runway debris, ice, birds, mechanic's tools, and loose hardware.

vortex. A whirling mass of air that sucks everything near it toward its center.

convergent-divergent duct. A duct that has a decreasing cross section in the direction of flow (convergent) until a minimum area is reached. After this point, the cross section increases (divergent).

Compressors

As mentioned earlier, two components that caused the most problems in gas turbine engine development were the compressor and the combustion chambers.

The primary function of a gas turbine engine is to add the maximum amount of energy to the air as it passes through. This energy is released by burning a mixture of fuel and air. When fuel is burned in still air, it releases energy, but the rate of release is quite low. If a stream of compressed air is blown through the burning fuel, it releases far more energy. This action can be visualized by considering the old-fashioned blacksmith forge. The smith builds a fire in a small pile of coal. This fire releases just enough heat energy to keep the coal burning. But when the smith wants a piece of steel red hot, he blows a stream of air from a bellows or compressor through the fire. The added air contains enough oxygen to allow the fuel to release more heat.

The function of the compressor in a gas turbine engine is to supply the correct amount of air for the fuel, and to supply it with the correct pressure and velocity.

The basic principle of all compressors used in gas turbine engines is the same. The compressor converts mechanical energy from the turbine into kinetic energy in the air. The compressor accelerates the air, which then flows through a diffuser, slowing it down and converting most of the kinetic energy (velocity) into potential energy (pressure) and some into heat.

compressor bleed air. Air that is tapped off from a turbine engine compressor.

customer bleed air. Air that is tapped off a turbine engine compressor and used for such airframe functions as air conditioning and pressurization.

The majority of air flows from the compressor into the combustion section, but some of it, called compressor bleed air, is used for anti-icing the inlet ducts and for cooling parts of the hot section. Other bleed air is used for cabin pressurization, air conditioning, fuel system anti-icing, and pneumatic engine starting. Bleed air used for purposes other than engine operation is called customer bleed air.

Types of Compressors

There are two basic types of compressors used on gas turbine engines: centrifugal and axial-flow. Some engines use both types.

Centrifugal Compressors

centrifugal compressor. A compressor that uses a vaned plate-like impeller.

Centrifugal compressors were used on many of the earliest gas turbine engines because of their ruggedness, light weight, ease of construction, and high pressure ratio for each stage of compression. A typical centrifugal compressor consists of three components: the impeller, the diffuser, and the manifold.

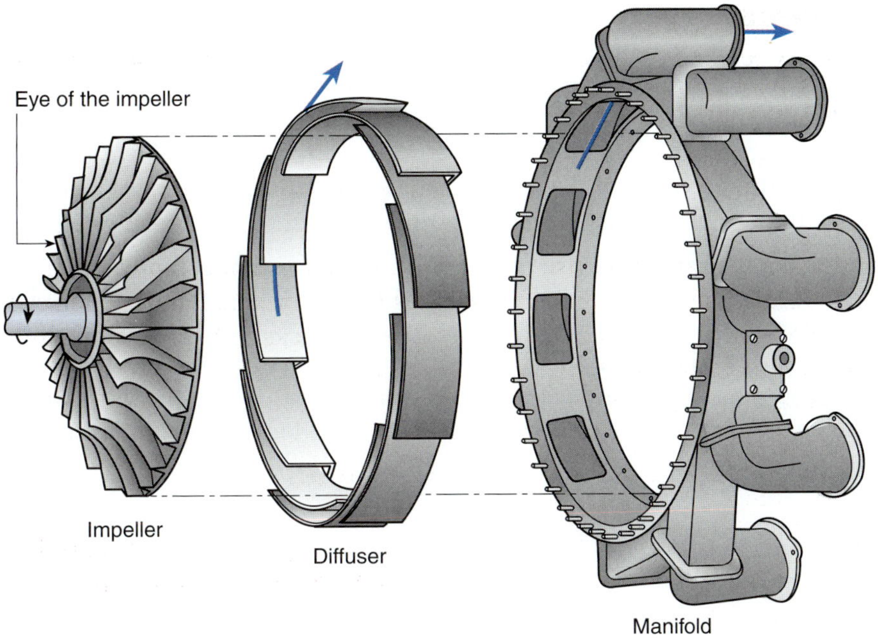

Figure 10-40. *A typical single-entry centrifugal compressor.*

Principle of Operation

Air enters the eye, or the center, of the fast-rotating impeller and is accelerated to a high velocity as it is slung to the outer edge by centrifugal force. The high-velocity air then flows into the diffuser which fits closely around the periphery of the impeller. There it flows through divergent ducts where some of the velocity energy is changed into pressure energy. The air, which has slowed down and has had its pressure increased, flows into the manifold through a series of turning vanes. From the manifold, the air flows into the combustion section of the engine. *See* Figure 10-41.

The compression pressure ratio of a single-stage centrifugal compressor is normally in the range of 6:1 to 7:1. The volume of air that can be moved by a single-entry centrifugal compressor is determined by the compressor's diameter. If the diameter is too large, the tip speed will become too high, and efficiency will decrease. Large-diameter compressors also require the engine to be large and therefore difficult to streamline.

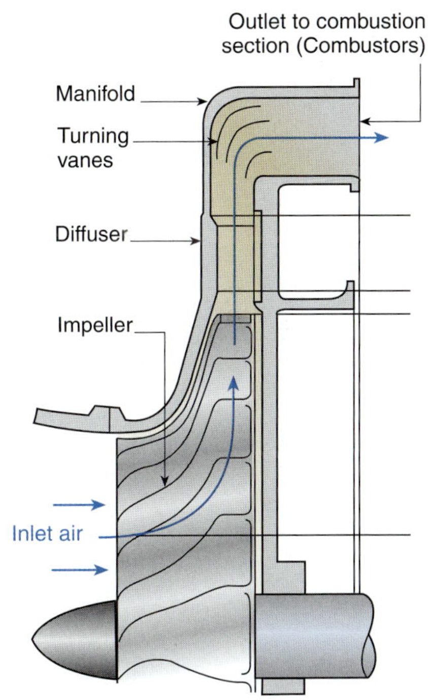

Figure 10-41. *Air is accelerated as it is slung outward through the vanes of the impeller. In the diffuser, part of the velocity energy is changed into pressure. The air then flows through the manifold into the combustion section of the engine.*

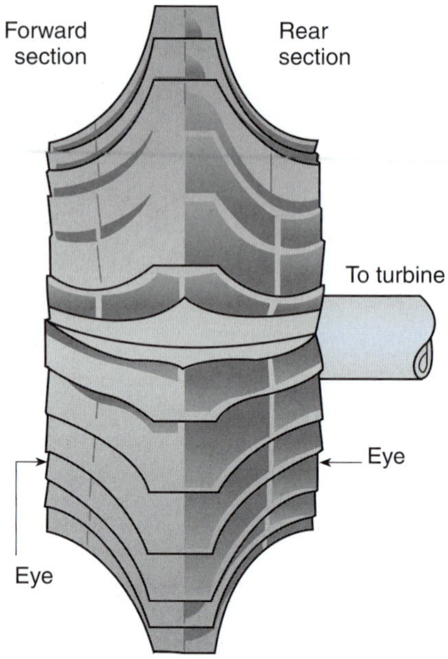

Figure 10-42. *A double-entry centrifugal compressor can move a large volume of air without its outside diameter being excessively large.*

When a large volume of air is needed, a double-entry compressor, as shown in Figure 10-42, can be used. This is the type of compressor that was used in the first Whittle flight engine in England and in the first turbojet engine built in the United States. One major problem with double-entry compressors is the difficulty in designing effective inlet ducts to supply air to the rear section. This is normally done by bringing the air into a plenum chamber where the velocity of incoming air is converted into pressure and fed into both the front and rear sections of the compressor.

The pressure rise per stage is high for centrifugal compressors, but the number of stages is limited. On some engines, the turbine turns two single-entry centrifugal compressors. This configuration is used on a popular series of turboprop engines like the one shown in Figure 10-43. One of the main problems with this type of engine is the pressure losses experienced by the air flowing between the stages.

plenum chamber. An enclosed chamber in which air can be held at a pressure slightly higher than that of the surrounding air.

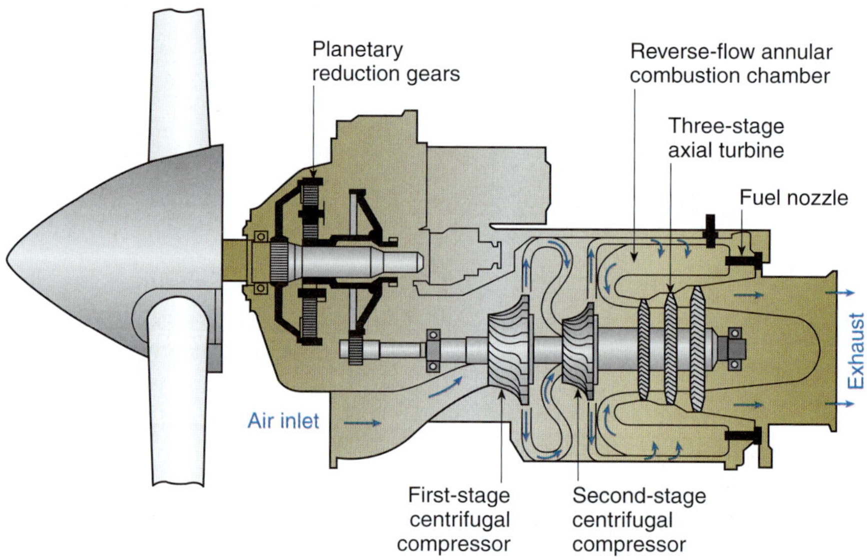

Figure 10-43. *The two-stage single-entry centrifugal compressor in this turboprop engine is driven by a three-stage axial turbine.*

Axial-Flow Compressors

compression ratio. The ratio of the pressure of the air at the discharge of a turbine engine compressor to the pressure of the air at its inlet.

Axial-flow compressors are, as their name implies, compressors in which the air passes axially or straight through the compressor. They are heavier than a centrifugal compressor and are much more costly to manufacture, but they are capable of a much higher overall compression ratio, and they have a smaller cross-sectional area which makes them easier to streamline. Axial-flow compressors have therefore become the standard for large gas turbine engines and are also used on many small engines.

An axial-flow compressor may have as few as two stages, when it is used in conjunction with a centrifugal compressor, or as many as 18 stages in some of the largest dual-spool turbofan engines.

dual-spool gas-turbine engine. An axial-flow turbine engine that has two compressors, each driven by its own stage or stages of turbines.

Principle of Operation

Axial-flow compressors are made up of a number of stages of rotor blades that are driven by the turbine, and that rotate between stages of fixed stator vanes. Both the rotor blades and stator vanes have airfoil shapes and are mounted so that they form a series of divergent ducts through which the air flows.

Figure 10-44 shows the way an axial-flow compressor changes the pressure and velocity of the air flowing through it. The incoming air passes through a set of inlet guide vanes that change its direction of flow so that it enters the first stage of rotor blades at the correct angle. These vanes cause a very slight increase in air velocity and a corresponding small decrease in air pressure.

inlet guide vanes. A set of stator vanes in front of the first stage of compression in a gas turbine engine.

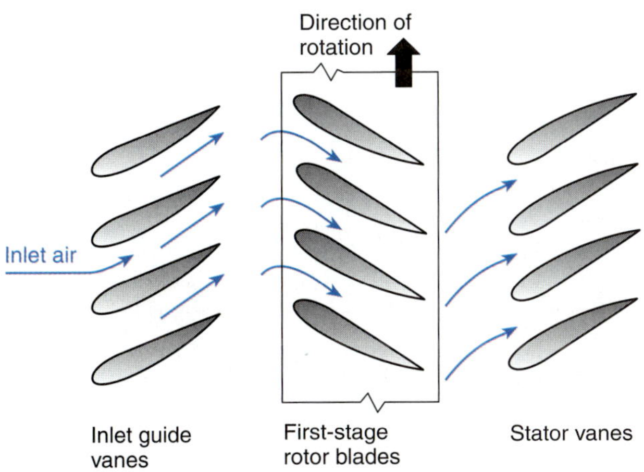

Figure 10-44. *Airflow through the inlet guide vanes and one stage of axial-flow compression*

After leaving the guide vanes, the air enters the first stage of compression. The rotor blades, turned at a high speed by the turbine, pick up the air and force it back across their airfoil shape.

Energy taken from the turbine is added to the air as it passes through the rotor, and the air is accelerated. The air leaves the trailing edges of the rotor blades and flows between the stator vanes. These vanes form a series of divergent ducts, and as the air flows through them, its pressure increases and its velocity drops back to its original value. The air passes through all stages of compression and then leaves the compressor through a set of exit guide vanes. As it moves through the compressor, it flows in a more or less straight line, usually swirling less than 180°. The exit guide vanes remove this swirl and direct the air into the diffuser, where it is prepared for the combustion section. *See* Figure 10-45.

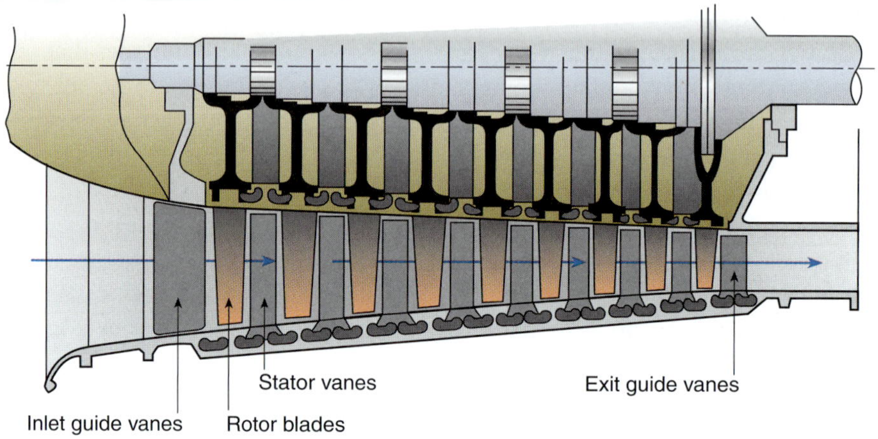

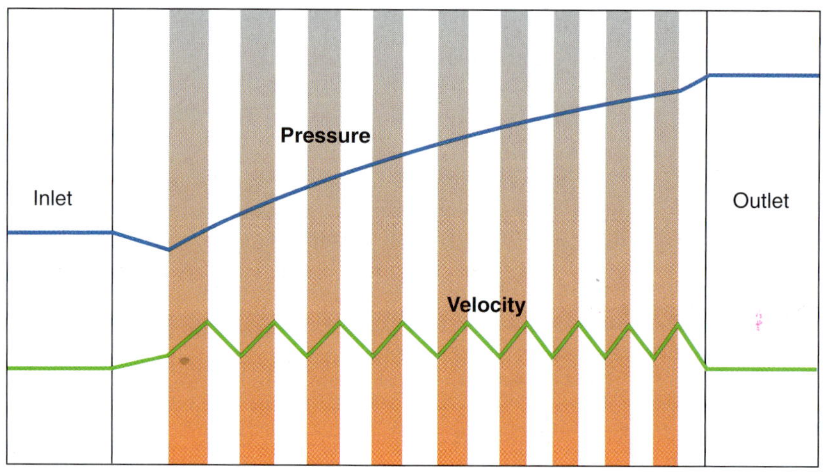

Figure 10-45. *Pressure and velocity changes in the air flowing through an axial-flow compressor*

Stage	Inlet Pressure (psi)	Pressure Ratio	Discharge Pressure (psi)
1	14.7	1.2	17.64
2	17.64	1.2	21.17
3	21.17	1.2	25.40
4	25.40	1.2	30.48
5	30.48	1.2	36.58
6	36.58	1.2	43.90
7	43.90	1.2	52.68
8	52.68	1.2	63.22
9	63.22	1.2	75.86
The total compression ratio is 75.86 ÷ 14.7 = 5.16			

Figure 10-46. *Finding the total compression ratio for a nine-stage axial-flow compressor with a compression pressure rise of 1.2 for each stage*

Compression Ratio

As mentioned earlier, the compression ratio of an axial-flow compressor is high. The pressure ratio for each stage is low, but by using a large number of stages, the overall ratio can be very high.

If the pressure ratio across each stage is only about 1.2:1, the pressure of the air at the discharge of each stage is 1.2 times the pressure of the air at its inlet. The total pressure ratio may be determined by finding the pressure rise in each stage and then dividing the final pressure by the inlet pressure. For example, if there are nine stages, and each stage has a pressure ratio of 1.2, the total pressure ratio is 5.16. *See* Figure 10-46.

If the discharge pressure, inlet pressure, and number of stages of compression are known, the pressure rise per stage can be found by taking the root of the compression ratio that represents the number of compression stages. The pressure rise per stage of the compressor in Figure 10-46 may be found by using this formula:

$$\text{Pressure rise per stage} = \sqrt[n]{CR} = CR^{\frac{1}{n}} = 5.16^{\frac{1}{9}} = 1.2$$

CR = compression ratio of the total compressor
n = number of stages

Surges and Stalls

The rotor blades in an axial-flow compressor are airfoils similar to those used on an airplane wing or a helicopter rotor. The "lift" they produce is the aerodynamic force that moves air through the various stages of the compressor. As airfoils, they are subject to stalls when their angle of attack becomes excessive.

Angle of Attack

The angle of attack of any airfoil is the acute angle formed between its chord line and the relative wind. In an axial-flow compressor, the angle of attack is determined by two parameters: the velocity of air flowing through the compressor, and the rotational speed of the compressor.

Figure 10-47 shows how the angle of attack of a compressor blade is determined. The stator vanes change the direction of air, and the air leaving the stator vanes can be represented as a vector quantity that has both direction and length and is identified as vector A. The rotor movement can also be represented as a vector quantity whose length relates to its rotational speed and whose direction is opposite the direction of rotation. This is seen as vector B.

Joining the nose of vector A with the tail of vector B produces the resultant vector C, which represents the direction and velocity of the resultant wind. Placing the resultant wind vector so that it points to the center of pressure of the rotor blade, allows us to visualize the blade's angle of attack (α).

Notice the two variables in Figure 10-47 and how they affect the angle of attack. Any decrease in the inlet air velocity shortens vector A, and with the compressor RPM remaining the same, increases the angle of attack.

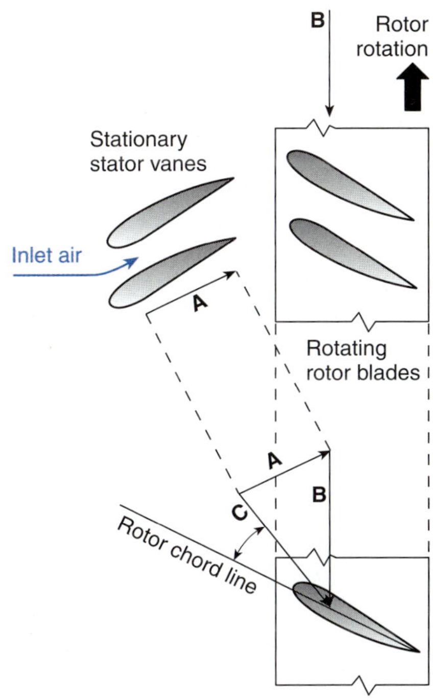

Figure 10-47. *Determination of the angle of attack of an axial-flow compressor blade*

When the angle of attack becomes excessive, the rotor blade stalls and disrupts the smooth airflow through the compressor. When only a few blades stall, the effect is minimal and is noticed by a fluttering or rumbling sound. When the entire compressor disk stalls, the effect can be a drastic slowing down of the airflow through the engine. This can result in a loud explosive noise, with a resulting RPM fluctuation and a serious increase in exhaust gas temperature. A stall that affects the entire compressor and restricts the airflow through the engine is called a surge.

compressor surge. A stall that affects the entire compressor.

Some causes of compressor stall and surge are:

- Obstruction to the inlet airflow
- Excessive pressure in the burner section
- Abrupt flight maneuvers that prevent air flowing directly into the inlet air ducts
- Choking of the airflow through the engine
- High crosswind component, especially at takeoff and low airspeed

Two design features that may be used to minimize or prevent compressor stall are variable guide vanes and interstage air bleeds.

Some engines have variable inlet guide vanes, and others also have variable-angle stators in the first few stages. The angle of these vanes is changed by the engine fuel control, and it is continuously adjusted so that the air leaving the vanes strikes the rotor blades of the following stage in a direction that produces a low angle of attack.

The other way of decreasing the angle of attack is to bleed some air from the last stages of compression. This decreases the pressure buildup and allows a freer flow of air through the compressor. The bleed valves are automatically controlled by the engine fuel control.

Cascade Effect

cascade effect. The cumulative effect that occurs when the output of one series of components serves as the input to the next series.

The comparison of rotor blades and stator vanes with airplane wings and helicopter rotors has its limits. The close proximity of many compressor blades and the nearness of the leading edge of the stator vanes to the trailing edge of the rotor blades cause a rather complex cascade-effect change in the pressure pattern. The camber of the rotor blades and stator vanes are opposite each other. This allows the high-pressure air from a stator vane to be pulled into the low pressure produced by the following rotor blade. This series of pressure differentials across all of the stages causes a pumping action that moves the air through the compressor.

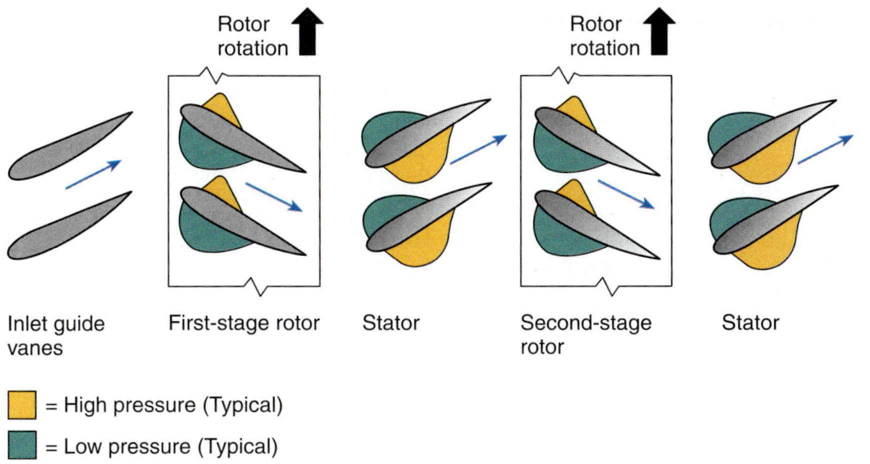

Figure 10-48. *The cascade effect due to the close proximity of the compressor blades and the stator vanes causes a pumping action that increases airflow through the engine*

As the air flows through the compressor, its pressure increases and, if the area through which it flowed were uniform, its velocity would decrease. To maintain constant velocity through the compressor, the blades in the rear sections must be shorter than those in the front, causing the passage through which the air flows to become smaller as the pressure increases.

Types of Axial-Flow Compressors

Up to this point, the axial-flow compressors discussed have been single-spool compressors. This means there is only one rotating element. Additional efficiency is gained in the larger engines by using two compressors. This type of engine is called a split-spool, or dual-spool, engine. Additional propulsive efficiency is gained by the turbofan engine, which is essentially a third stage of axial-flow compression.

single-spool gas-turbine engine. A type of axial-flow-compressor gas turbine engine that has only one rotating element.

Single-Spool Axial-Flow Compressors

There is a limit to the number of stages of compression that can be built into an axial-flow compressor. If there are too many, the rearmost stages will be operating inefficiently while the front stages will be overloaded. This restricts the airflow through the compressor and can lead to compressor surge. This overloading can be prevented by bleeding off some of the interstage air during part-throttle operation, but this wastes power. Practically all modern high power turbojet and turbofan engines have split compressors in which there are two or more compressors, each driven by its own stage, or stages, of turbines.

dual-spool gas-turbine engine. An axial-flow turbine engine that has two compressors, each driven by its own stage or stages of turbines.

high-pressure compressor. The second-stage compressor in a dual-spool gas turbine engine.

N_1. A symbol representing the rotational speed of the low-pressure compressor in a dual-spool gas turbine engine.

N_2. A symbol representing the rotational speed of the high-pressure compressor in a dual-spool gas turbine engine.

low-pressure compressor. The first-stage compressor in a dual-spool gas turbine engine.

Dual-Spool Axial-Flow Compressors

Figure 10-49 illustrates a modern two-spool turbojet engine. The rearmost, or high-pressure, compressor is connected to and driven through a hollow shaft by the forward stage of turbines. This is called the N_2 compressor, and its speed is governed by the fuel control. The N_2 compressor, because of its lighter weight, is the one rotated by the starter for starting the engine.

The larger, or N_1, compressor at the front of the engine is the low-pressure compressor, and it is driven by the rearmost stages of turbines. Notice that this engine uses two stages of turbines to drive the N_1 compressor.

The total compression ratio of a dual-spool engine is found by multiplying the compression ratio of the N_1 compressor by the compression ratio of the N_2 compressor. If the compression ratio of N_1 is 3.2:1 and the compression ratio of N_2 is 4.1:1, the total compression ratio is $3.2 \cdot 4.1 = 13.12:1$.

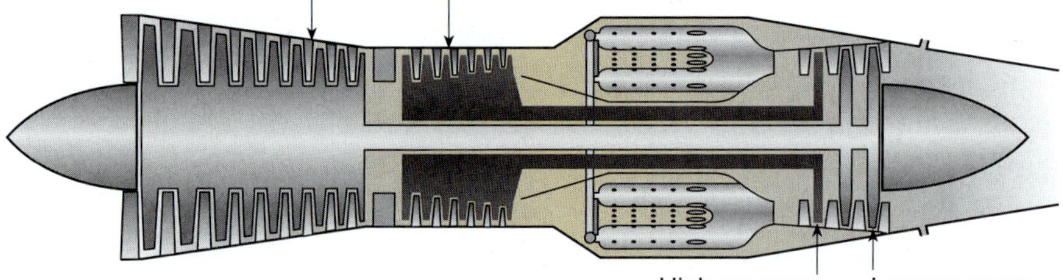

Figure 10-49. *Configuration of a typical dual-spool axial-flow turbojet engine*

For any given power setting, the speed of the N_2 compressor is held steady by the governor in the fuel control. The N_1, or low-pressure compressor, is not governed, and it changes its RPM in flight as the atmospheric and inlet duct conditions change. As the air becomes less dense, the compressor speeds up and moves a corresponding larger volume of air, and as the air becomes more dense, it slows down and moves a smaller volume of air. This varying speed allows the N_1 compressor to supply the N_2 compressor with a relatively constant mass of air for any given power setting.

Some engines, notably the new generation of large Rolls-Royce turbofans, have three spools. The fan, which is referred to as the low-pressure (LP) compressor, the intermediate-pressure (IP) compressor, and the high-pressure (HP) compressor are all driven by separate turbines. The fan which turns at a relatively low speed requires a great deal of torque, therefore its turbine has three stages.

Turbofans

The advantages of lower specific fuel consumption, lower noise level, and greatly improved propulsive efficiency have caused turbofan engines to just about replace turbojets for most commercial, airline, and military operations. These engines have an additional stage of compression ahead of the normal low-pressure compressor.

Turbofan Engine Configurations

The fan may be connected directly to the low-pressure compressor (Figure 10-50); or it may be driven by its own stages of free turbines (Figure 10-51); or it may be gear-driven from the low-pressure compressor (Figure 10-52). Some turbofan engines, like that in Figure 10-53, have the fan mounted behind the

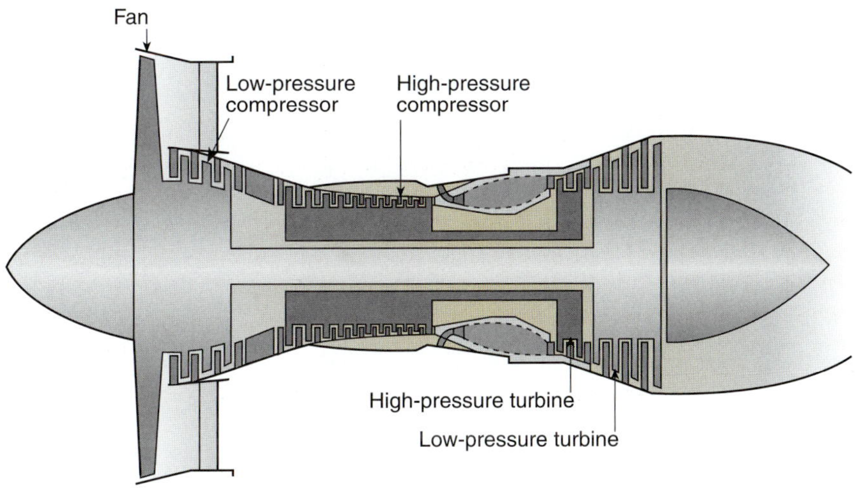

Figure 10-50. *Turbofan engine with the fan being a part of the low-pressure compressor*

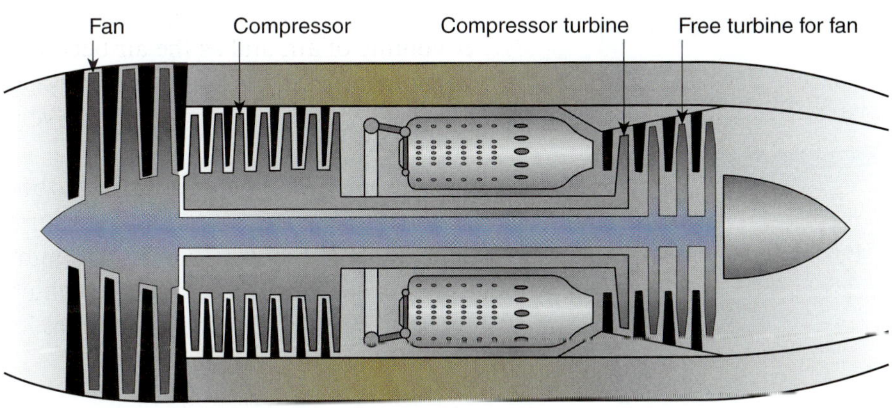

Figure 10-51. *Turbofan engine with the fan driven by its own stages of turbines*

bluckets. The portions of aft-fan blades in the exhaust of the core engine used to drive the fan.

engine and driven by a free turbine. These aft fan blades are actually part of the turbine blades. The portion of the blades that drive the fan by the energy received from the gases leaving the engine is called a blucket.

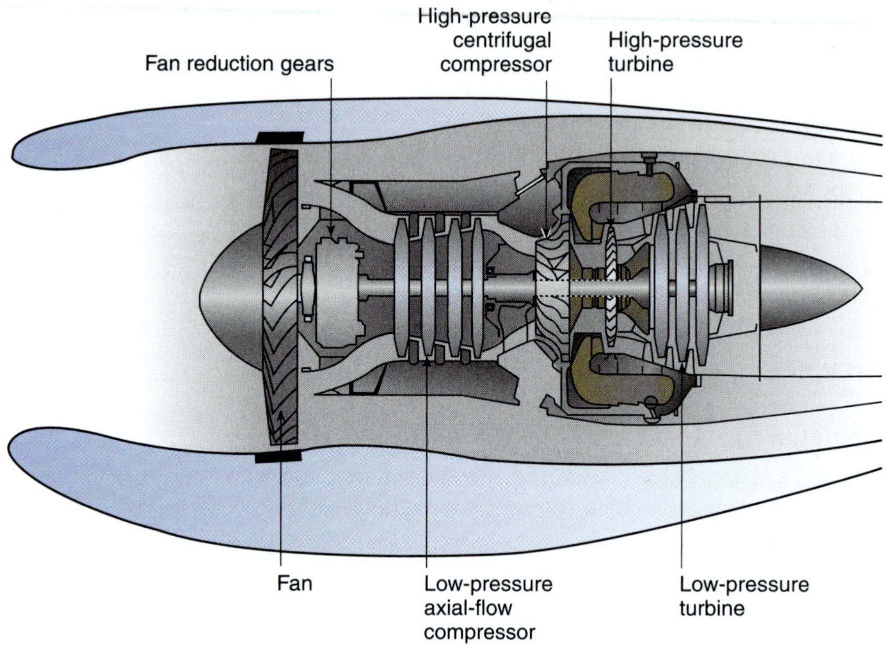

Figure 10-52. *Turbofan engine with the fan driven through a set of planetary reduction gears from the low-pressure compressor*

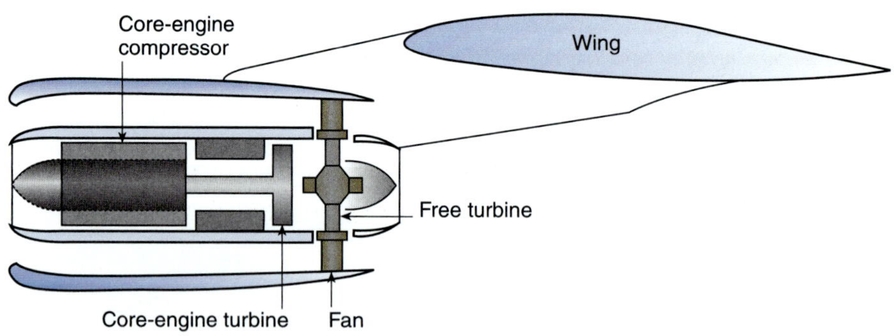

Figure 10-53. *Aft-fan turbofan engine with the fan blades an actual part of the free turbine blades*

aft-fan engine. A turbofan engine in which the fan blades are an extension of the free turbine blades.

annular duct. A duct, or passage, that surrounds an object.

fan pressure ratio. The ratio of the fan-discharge pressure to the fan inlet pressure.

Turbofan Engine Bypass Ratio

The bypass ratio of a turbofan engine is the ratio of the mass of air moved by the fan to the mass of air moved by the core engine. Turbofan engines may be classified as low-bypass, medium-bypass, or high-bypass engines. Low-bypass engines have a bypass ratio of 1:1 or less and normally pass their discharge air through annular ducts around the engine. The fan-discharge air may mix with the exhaust from the core engine, or it may be discharged through separate ducts as is shown in Figure 10-17 on Page 365.

Medium-bypass turbofan engines have a bypass ratio in the range of 2:1 to 3:1.

High-bypass turbofan engines are the engines on most large transport aircraft. Their bypass ratio is in the range of 4:1 to as high as 9:1. Almost all of these engines discharge the fan bypass air through a set of fan exit vanes to remove the swirl from the air as it leaves the fan case. Figure 10-54 shows this type of fan.

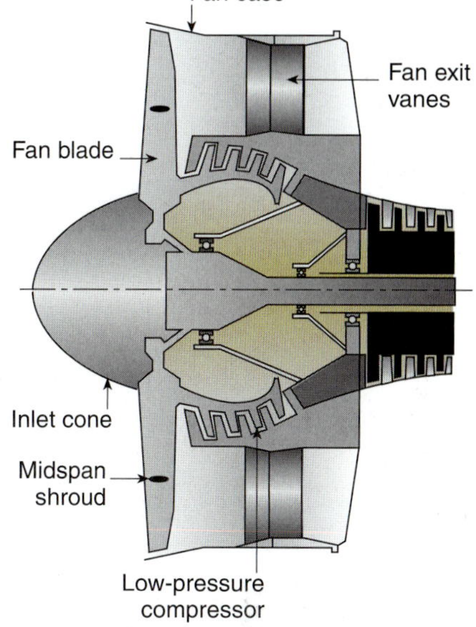

Figure 10-54. *High-bypass turbofan engine in which the fan discharge air exits the fan case through a set of fan exit guide vanes*

Turbofan Pressure Ratio

The turbofan engine has gained popularity because it has the characteristic of an engine/propeller combination that accelerates a large mass of air and passes it around the outside of the engine. The pressure ratio of a fan is a measure of the change in pressure of the air as it flows through the fan. The fan pressure ratios range from about 1.4:1 to as high as 2.6:1.

The fan blade design has a great effect on the fan pressure ratio. Many blades used on the large, high-bypass-ratio engines have a high-aspect-ratio, but some of the more modern engine designs use wider or lower-aspect-ratio blades made of titanium or some extremely strong composite materials.

Some high-aspect-ratio blades have midspan shrouds, or braces, that form a ring around the fan at the mid portion of the blade to stiffen it and prevent flutter. *See* Figure 10-55.

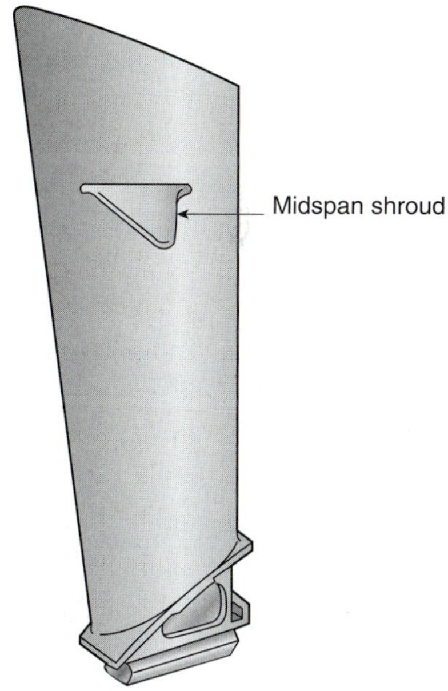

Figure 10-55. *Turbofan blade with a midspan shroud, or snubber. The snubbers of adjacent blades touch to form a shroud ring that stiffens the blades and prevents flutter.*

Compressor Design Features

The function of the compressor is to increase the pressure of air flowing through the engine with a minimum change in its velocity. *See* Figure 10-45 on Page 390.

Convergent Airflow Path

If the compressor blades were all the same length, and the air flowed through a constant-area duct, its velocity would decrease as its pressure increased. To keep the air velocity relatively constant as its pressure is increased, the rear blades of the compressor are shorter than those at the front, and the passage through which the air flows become smaller as the pressure increases. There are two ways of decreasing the size of the airflow passage: by holding the outside of the compressor housing constant and increasing the diameter of the drum or disks on which each stage of rotor blades are mounted, or by keeping the disks or drum the same diameter and decreasing the outside diameter of the compressor case. Both methods are used. *See* Figure 10-56.

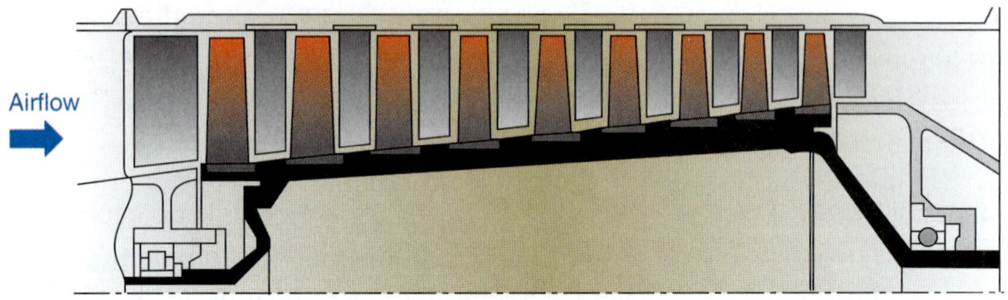

Outside diameter of compressor housing is constant, and the drum on which blades are mounted increases its diameter toward rear of compressor.

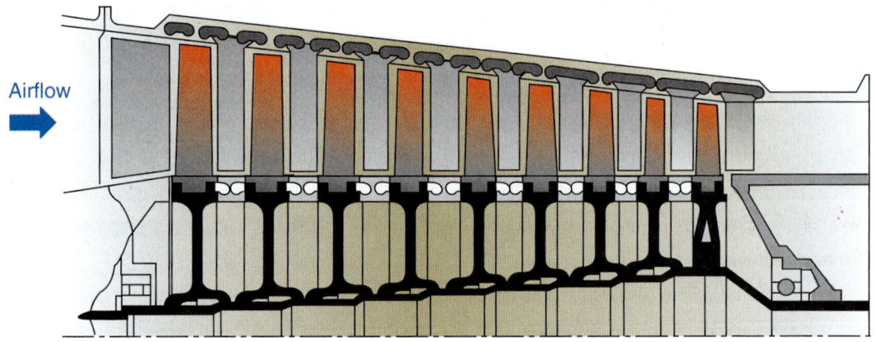

Disks on which blades are mounted are all same diameter, but outside of compressor case decreases toward rear of compressor.

Figure 10-56. *Two methods of decreasing the area of the airflow path through an axial-flow compressor*

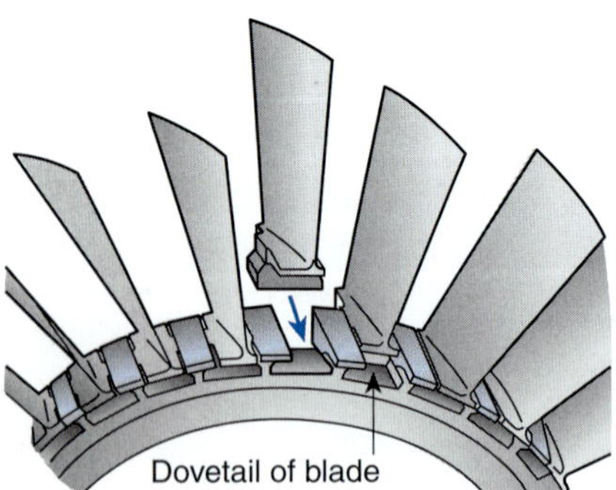

Figure 10-57. *The dovetail method of securing compressor blades to the disk allows them to rock slightly and thus minimize stresses at the root.*

Blade Attachment

The compressor blades are not rigidly attached to the compressor drum or disks, but are loose in their mountings so that they are free to rock. When the engine is running, centrifugal force holds the blades in their correct position. Allowing the blades to be loose prevents stresses at the root. One popular way of attaching the blades to the disks is by the dovetail method shown in Figure 10-57.

The stator vanes are mounted in retaining rings that are attached to the compressor housing.

Blade Design

Almost all compressor blades are designed with some twist to give them the correct pressure gradient along their length as their velocity changes from root to tip.

The tip of a compressor blade is most important. Some blade tips are squared off, and others have the tip thickness reduced. These tips with reduced thickness are called profile tips. The thinner tips have a high natural resonant frequency and are therefore not subject to the vibrations that would affect a blade with a squared tip. The profile tip also provides a more aerodynamically efficient shape for the high-velocity air moved by the blade. These profile tips often touch the housing and make a squealing noise as the engine is shut down. For this reason profile tips are often called squealer tips.

profile tip (compressor blade tip). The tip of an axial-flow compressor blade whose thickness is reduced to give it a higher resonant frequency.

Air leakage around the tips of compressor blades causes a loss of compressor efficiency. There are several methods used to prevent this loss.

In some engines, the compressor blades have their profile tips so close to the case that they actually touch an abradable strip in the compressor housing. When the tip touches the strip, it abrades part of it away and provides a fit that minimizes the air losses. Other blades have abradable tips that are designed to wear away when they touch the housing, which provides an extremely close fit.

abradable strip. A strip of material in the compressor housing in which a groove is abraded by the compressor blade tip. The groove ensures the minimum blade tip clearance.

abradable tip (compressor blade tip). The tip of an axial-flow compressor blade that will abrade, or wear away, when it contacts the compressor housing.

Guide Vanes and Stator Vanes

Air entering the first stage of the compressor is turned by the inlet guide vanes so that it flows in the correct direction to be picked up by the rotor blades. Inlet guide vanes are similar to the stator vanes, but they are designed to have a minimum effect on the velocity or pressure of the incoming air.

The inlet guide vanes on most engines are fixed, but on some they are variable and are controlled hydraulically with fuel from the fuel control.

stage of a compressor. One disk of rotor blades and the following set of stator vanes in an axial-flow compressor.

Stator vanes are airfoils similar to rotor blades, but are mounted so they form a series of divergent ducts to decrease the velocity and increase the pressure of air flowing through them.

Like the inlet guide vanes, most stator vanes are fixed, but in some engines the vanes in the first few stages of the stators may have their angle varied by the fuel control. Variable vanes and interstage bleed air are used to prevent compressor stall and surge.

Hybrid Compressor Engine

Some of the popular smaller gas turbine engines have both an axial-flow and a centrifugal compressor.

The Pratt & Whitney of Canada PT6 turboprop engine in Figure 10-58 has a three-stage axial-flow compressor followed immediately by a centrifugal compressor on the same shaft. These compressors are driven by a single-stage compressor turbine. Notice that the inlet air is taken in at the rear of the engine through the air inlet screen. It flows forward through the two compressors and then reverses its direction and flows to the rear through the combustor. Here, energy is added, and the hot gases reverse their direction

hybrid compressor engine. A gas turbine engine that has both centrifugal and axial-flow compressors.

and flow forward through the compressor turbine and the free turbine that drives the propeller reduction gears. It leaves the free turbine and exits the engine through the exhaust ducts at the front of the engine. *See* Figure 10-58.

The Garrett TFE731 turbofan engine in Figure 10-59 also has a hybrid compressor arrangement, but it works differently from that in the PT6 engine.

The fan is driven through a series of gears from the four-stage axial-flow, low-pressure compressor. This is, in turn, driven by a three-stage axial turbine. A single-stage centrifugal high-pressure compressor is driven by its own single-stage axial turbine.

Air enters the engine through a divergent inlet duct and passes through the fan. Part of the fan discharge flows into the first stage of the low-pressure compressor, and the remainder flows around the engine through the fan duct and leaves through the fan-duct discharge.

The air moved by the central part of the fan flows through the four stages of the low-pressure axial-flow compressor and then into the eye of the centrifugal high-pressure compressor. From there the air is slung outward, and passed through a diffuser and into the reverse-flow combustor where

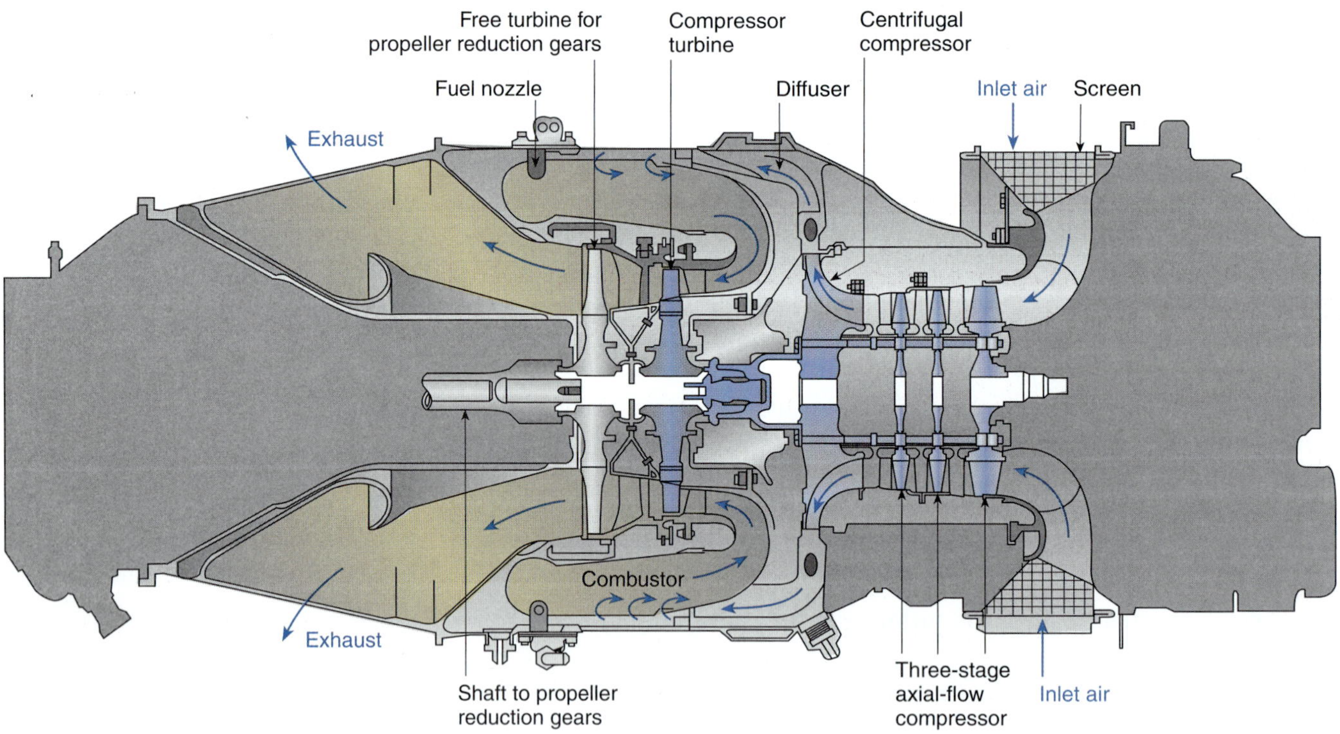

Figure 10-58. *The Pratt & Whitney of Canada PT6 engine has a three-stage axial-flow compressor on the same shaft as a single-entry centrifugal compressor. Air enters the engine at the rear, flows forward through the compressors, reverses its direction as it flows through the combustor, and reverses again before it flows forward through the compressor turbine and the free turbine that drives the propeller reduction gears.*

fuel is added and burned. The hot gases again reverse their direction and flow through the single stage of the high-pressure turbine and then through the three stages of the low-pressure turbine. From there they pass out of the engine through the exhaust duct. *See* Figure 10-59.

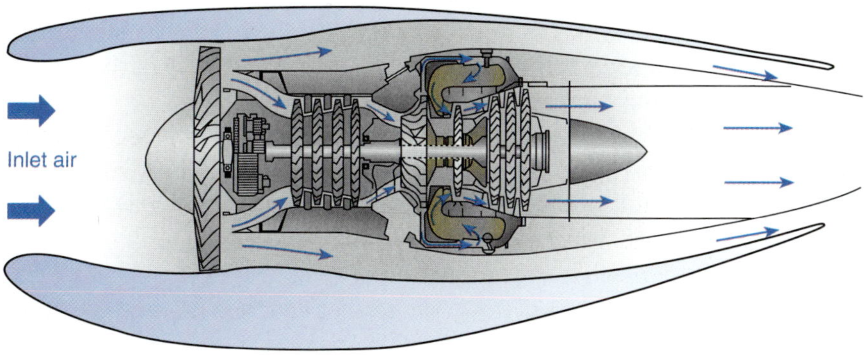

Figure 10-59. *The Garrett TFE731 turbofan engine uses a multistage axial-flow, low-pressure compressor and a centrifugal high-pressure compressor. The two compressors are driven at different speeds by different turbines.*

Diffuser Section

The diffuser section is normally bolted to the rear of the compressor and furnishes a diverging area into which the air leaving the compressor spreads out and decreases its velocity while increasing its pressure.

The diffuser, or compressor-discharge, pressure is the highest pressure of any location within the engine. This high pressure gives the expanding gases in the combustion section a force to push against to accelerate the gases out the rear of the engine.

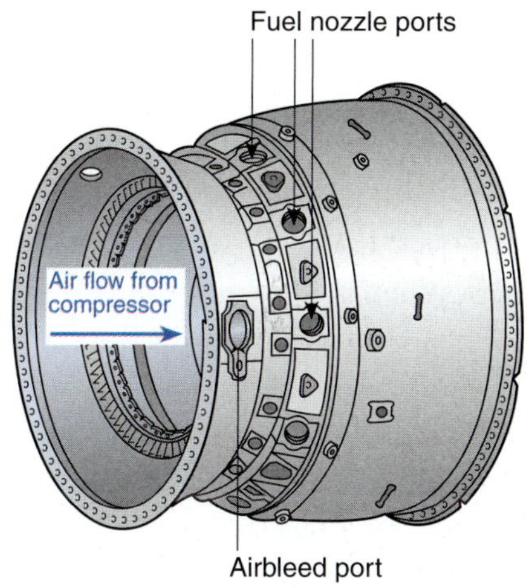

Figure 10-60. *A typical diffuser section used with an axial-flow compressor. The air is at its highest pressure in the diffuser. Much of the compressor bleed air is taken from the diffuser.*

STUDY QUESTIONS: THE COLD SECTION

Answers begin on Page 420. Page numbers refer to chapter text.

59. The two sections into which a gas turbine can be divided are the __________________ section and the __________________ section. *Page 380*

60. The inlet ducts are actually part of the __________________ (airframe or engine). *Page 382*

61. Inlet ducts for turbine engines may be divided into two classifications according to the speed of the aircraft on which they are installed. These are:
    a. __________________
    b. __________________
    Page 382

62. Most subsonic inlet air ducts are __________________ (convergent or divergent). *Page 382*

63. The airspeed at which the ram effect of the air flowing into a divergent inlet air duct causes the pressure at the front of the compressor to be the same as the ambient pressure is called the __________________ speed. *Page 382*

64. The air that flows into compressors mounted at the rear of the engine first flow into a/an __________________ chamber where part of the velocity energy is changed into pressure energy. *Page 383*

65. When gas turbine engines are calibrated on a run-in stand, they are normally fitted with a __________________ -type inlet duct. *Page 384*

66. Some gas turbine engine inlet ducts have spring-loaded doors that automatically open when the engine demands more air than can be supplied through the normal duct. These are called __________________ doors. *Page 384*

67. Dirt and objects on the runway are prevented from being sucked into low-mounted turbine engines by destroying the vortex with a vortex __________________ . *Page 384*

68. Supersonic inlet ducts are of the __________________ (convergent, divergent, or convergent-divergent) type. *Page 384*

69. When supersonic air flows through the convergent portion of a CD duct, its velocity __________________ (increases or decreases). *Pages 384, 385*

70. When supersonic air flows through a CD inlet duct, its velocity at the compressor inlet is ______________________ (subsonic or supersonic). *Page 385*

71. Supersonic air that has passed through a normal shock wave is ______________________ (subsonic or supersonic). *Page 385*

72. The compressor of a gas turbine engine must supply the correct amount of air for the fuel, and it must have the correct ________________ and ________________ . *Page 386*

73. Energy added to the air by a compressor initially increases the ___________________ (pressure or velocity) of the air. *Page 386*

74. The velocity of the air passing through a compressor is a measure of its ________________ (kinetic or potential) energy. *Page 386*

75. The pressure of the air passing through a compressor is a measure of its ________________ (kinetic or potential) energy. *Page 386*

76. After passing through the compressor, the air flows through a diffuser, which converts some of the __________________ energy into pressure energy. *Page 386*

77. In addition to increasing the pressure of the air, some of the energy used to drive the compressor is converted into ______________ . *Page 386*

78. Air taken from the compressor to anti-ice the inlet ducts is called compressor ______________________ . *Page 386*

79. Two types of compressors used in gas turbine engines are ____________________ and _____________________ compressors. *Page 386*

80. The three components of a centrifugal compressor are:
    a. ____________________________
    b. ____________________________
    c. ____________________________
    Page 386

81. The center of a centrifugal compressor where the air enters is called the __________________ . *Page 387*

82. The ducts in the diffuser are ________________ (convergent or divergent). *Page 387*

Continued

STUDY QUESTIONS: THE COLD SECTION *Continued*

83. A centrifugal compressor that has vanes on both sides is called a/an ____________________ compressor. *Page 388*

84. The air taken into a double-entry centrifugal compressor is normally taken from a/an __________________ chamber that gets its air from the inlet duct. *Page 388*

85. The stators used with an axial-flow compressor cause the air to __________________ (speed up or slow down). *Page 389*

86. After the air leaves the compressor, it flows through a/an _______________________ before it enters the combustion section. *Page 389*

87. The overall compression ratio of an axial-flow compressor is generally ___________________ (higher or lower) than that of a centrifugal compressor. *Page 390*

88. The pressure ratio per stage of an axial-flow compressor is ___________________ (higher or lower) than that of a centrifugal compressor. *Page 390*

89. The blades in an axial-flow compressor can stall if the angle of attack becomes too ________________ (high or low). *Page 391*

90. Two variables that affect the angle of attack of a compressor rotor blade are:
    a. _____________________________
    b. _____________________________
    Page 391

91. The direction of the air entering the first stage of an axial-flow compressor is changed by the ________________________ so that it enters at the correct angle. *Page 389*

92. If the airflow entering an axial-flow compressor is interrupted, the angle of attack of the compressor blades will ______________________ (increase or decrease). *Pages 391, 392*

93. A stall that affects the entire compressor and seriously restricts the airflow through the engine is normally called a/an ____________________ . *Page 392*

94. Two design features that may be used to prevent or minimize compressor stalls are:
    a. ______________________________
    b. ______________________________
    Page 392

95. Variable inlet guide vanes and interstage air bleeds are controlled automatically by the ________________________. *Page 392*

96. An axial-flow compressor that has two separate compressors, each driven by its own turbine, is called a/an ________________________ compressor. *Page 393*

97. The rearmost compressor in a dual-spool engine is the ____________________ (high or low)-pressure compressor. *Page 394*

98. The low-pressure compressor is called the __________________ (N_1 or N_2) compressor. *Page 394*

99. The speed of the low-pressure compressor ________________ (is or is not) controlled by a governor. *Page 394*

100. The compressor rotated by the starter for starting the engine is the __________________ (N_1 or N_2). *Page 394*

101. As the air becomes less dense, the speed of the N_1 compressor in a dual-spool engine _____________________ (increases, decreases, or remains the same). *Page 394*

102. Four ways of driving the fan in a turbofan engine are:
   a. ____
   b. ____
   c. ____
   d. ____
   Pages 395, 396

103. The ratio of the mass of air moved by the fan to the mass of air moved by the core engine is called the _____________________ of a turbofan engine. *Page 397*

104. The ratio of the fan-discharge air pressure to the pressure of the ambient air ahead of the fan is the fan __________________ ratio. *Page 397*

105. The compressor blades at the rear of an axial-flow compressor are __________________ (longer or shorter) than the blades at the front of the compressor. *Page 398*

106. A compressor blade with a tip having a decreased thickness is called a ___________________ tip or a ____________________ tip. *Page 399*

107. Some compressor blades actually touch and wear a groove in a special ___________________ strip in the compressor housing. This keeps the blade tip clearance to an absolute minimum. *Page 399*

Continued

STUDY QUESTIONS: THE COLD SECTION *Continued*

108. An engine that has both a centrifugal and an axial-flow compressor is called a/an ____________________ compressor engine. *Page 399*

109. The angle of the inlet guide vanes is variable on some engines. The angles are changed hydraulically by the ____________________ . *Page 399*

110. Some engines have variable-angle stator vanes in the first few stages. These vanes are variable to prevent ____________________ . *Page 399*

111. The highest air pressure in a gas turbine engine is that of the air in the ____________________ . This is the same as the ____________________ pressure. *Page 401*

The Hot Section

The hot section of a gas turbine begins with the combustion section and includes the turbine and the exhaust system components. The hot section is subjected to the most severe stresses in the engine and requires careful inspection and maintenance. A major type of damage is cracks caused by the extreme heat to which these parts are exposed.

Combustion Section

combustor (combustion chamber). The section of a gas turbine engine in which the fuel is injected and burns.

The design of a successful combustion section, or combustors, was one of the major obstacles in building the first successful gas turbine engines. For even in a medium-power engine, the heat energy released per cubic foot of combustor volume is several thousand times that released in a typical home heating system, and the pressures in the extremely thin-wall combustors are about ten times as high as those in an industrial furnace with its thick walls of firebrick.

The combustors used in a turbine engine have several very stringent requirements. Some of these are:

- Minimum pressure loss in gases as they pass through the combustor
- High combustion efficiency, therefore low smoke emission
- Low risk of flame blowout
- Combustion occurring entirely within the combustor
- Uniform temperature distribution throughout the gases
- Low enough temperature of the gases leaving the combustor to prevent damage to the turbine
- Combustor design providing easy starting

Combustors are made of thin sheets of corrosion-resistant metal that can withstand high temperatures. The walls of the combustors have a series of carefully designed holes or slots that allow air to enter.

Part of the air discharged from the compressor flows into the combustors through the openings, and part of it flows around the outside of the combustor. About 60 times as many pounds of air flow through the combustion section as pounds of fuel, but only about one fourth of this air mixes with the fuel and becomes part of the combustion process. The remainder of the air is used to cool the combustor walls and to mix with the burned gases to cool them before they flow through the turbine.

There are a number of combustor designs used in gas turbine engines. The most common of these are the multiple-can, can-annular, and annular types. Combustors may also be of the straight-through or reverse-flow type.

straight-through combustor. A combustor in a gas turbine engine through which the air flows in an essentially straight line.

Multiple-Can Combustors

Many engines having centrifugal compressors use multiple-can combustors such as the one in Figure 10-61. Air enters the left end of the combustor and divides into two streams. The primary air, which makes up about one fourth of the mass and enters into the combustion process, flows into the inner liner where its velocity is slowed down enough that it will not blow out the flame of the burning fuel. The remainder of the air, the secondary air, flows between the liners. Part of it enters the inner liner downstream of the combustion in a turbulent fashion so it will mix with the primary air to lower its temperature enough that it will not damage the turbine.

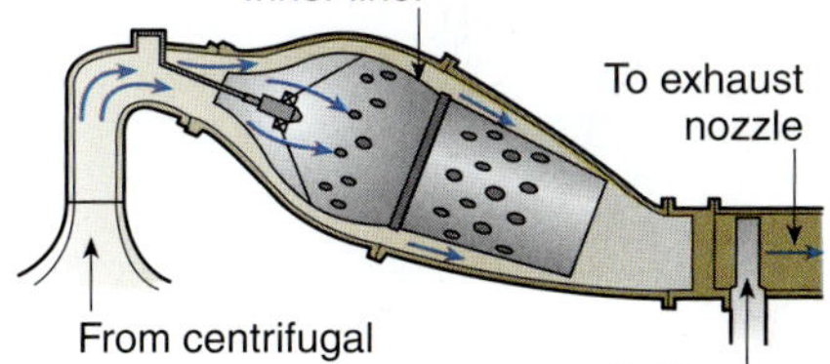

Figure 10-61. *A typical can-type combustor*

Usually between eight and ten can-type combustors are arranged around the outside of the turbine shaft between the compressor manifold and the turbine.

multiple-can combustor. A combustor that consists of a series of individual burner cans arranged around the periphery of a centrifugal compressor.

Igniters are normally installed in only two of the cans to provide the spark for starting the engine. The cans are connected by small flame propagation tubes, also called crossover tubes, so that the flames from the cans containing the igniters are carried to the other cans for ignition.

The advantage can-type combustors have is that individual cans may be removed for inspection and replacement without disturbing the others. But they have the disadvantage that the turbine is subjected to uneven temperatures, and if one can fails, the resulting extreme temperature difference can cause turbine failure.

Can-Annular Combustors

Can-annular combustors are used on many large turbojet and turbofan engines. They consist of individual cans into which fuel is sprayed and ignited. They mount on an annular duct through which hot gases from individual cans are collected and directed uniformly into the turbine.

can-annular combustor. A combustor that consists of individual cans mounted on an annular duct which collects the hot gases and directs them uniformly into the turbine.

Figure 10-63 shows a typical can-annular combustor. Fuel is sprayed into the combustor from a fuel manifold that discharges through the holes in the end of the cans. It mixes with air from the compressor and is burned. Secondary air flows through the holes in the can and cools the air before it passes through the turbine. The cutaway can in Figure 10-63 shows the perforated tube that carries additional air for cooling.

Can-annular combustors have the advantage that individual combustors may be removed for inspection and replacement. The cans are shorter than those used in a multiple-can combustor and therefore have a lower pressure drop through them. Combining the gases from all of the cans provides a uniform temperature at the turbine, even if one of the fuel nozzles becomes clogged.

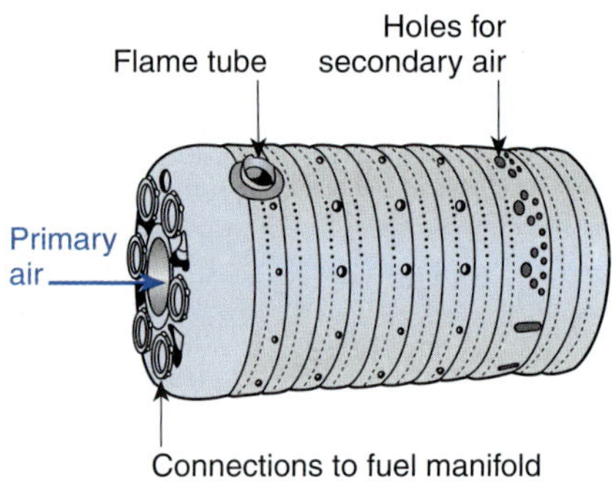

Figure 10-62. *An individual can used in a can-annular combustor*

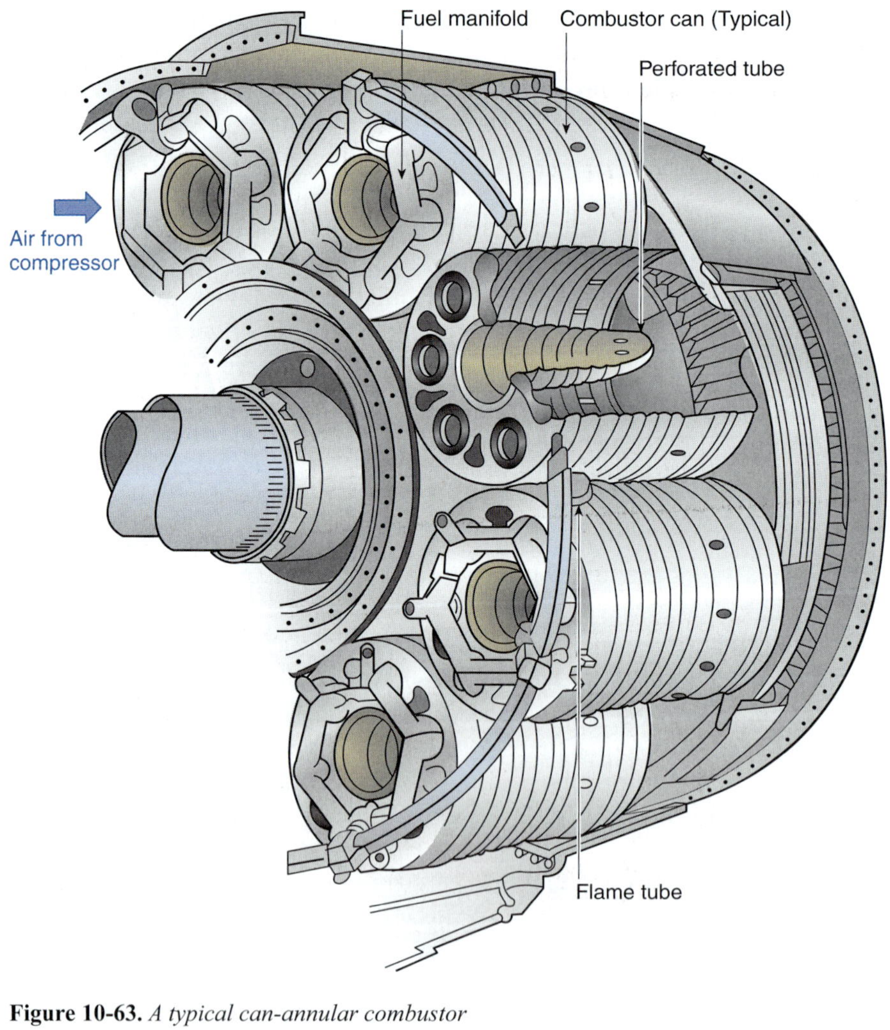

Figure 10-63. *A typical can-annular combustor*

Annular Combustors

Many modern axial-flow-compressor engines use a single annular combustor. Annular combustors make the most efficient use of the limited space available, and they provide exceptionally efficient mixing of the fuel vapor with the air. They require a minimum amount of cooling air to keep the temperature low enough to prevent damage to the turbine, and they provide an even temperature distribution of the air leaving the combustor. The main disadvantage of an annular combustion chamber is that it cannot be replaced without removing the engine from the aircraft.

Reverse-Flow Annular Combustors

In engines where the overall length is critical, some manufacturers have chosen to use reverse-flow combustors such as the one used in the Pratt & Whitney of Canada PT6. *See* Figure 10-58 on Page 400 and Figure 10-64. The air enters the engine from the rear and flows forward through the axial-flow compressor, then through the centrifugal compressor and diffuser. It flows around the outside of the combustor and reverses direction to enter the combustor and flow rearward. Fuel is sprayed in at the forward end of the combustor and burned. Secondary air flows in through the slots and holes in the inner liner of the combustor to dilute and cool the air before it leaves. The air then makes its final reverse in direction and flows forward through the compressor turbine and the propeller turbine, which remove most of its energy. The hot air, with the greater part of its energy removed, leaves the engine near the front through the exhaust stacks.

reverse-flow combustor. A turbine engine combustor in which the air enters the outer case and reverses its direction as it flows into the inner liner. It again reverses its direction before it flows through the turbine.

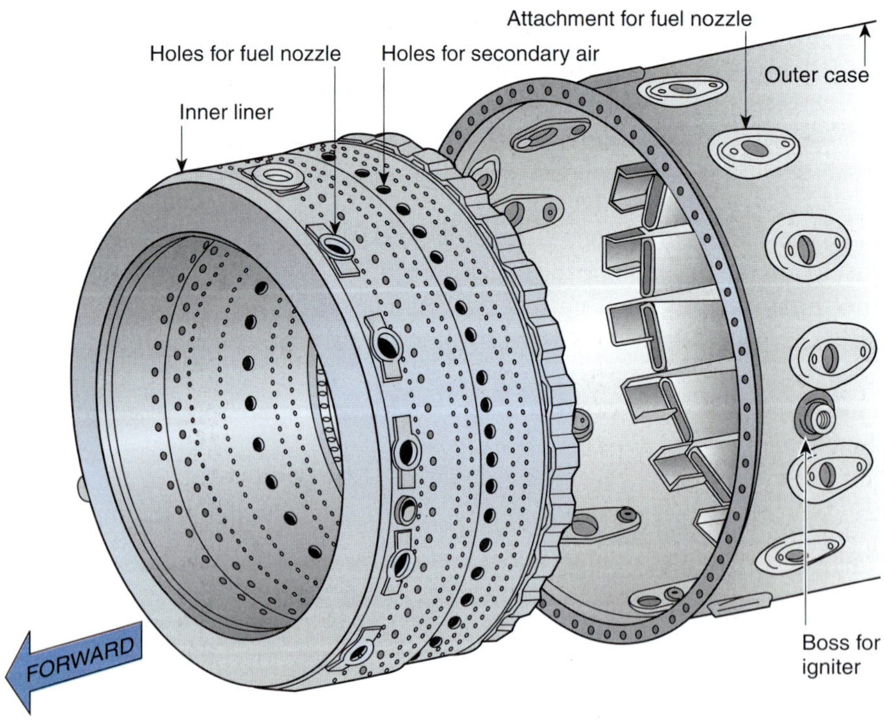

Figure 10-64. *A reverse-flow annular combustor for a small turboprop engine*

Turbine Section

The turbine is the power-producing component in a gas turbine engine. About three-fourths of the energy in the gases leaving the combustion section is converted into shaft horsepower and is used to drive the compressors and fan. The remaining energy drives the accessories and accelerates the gases to produce thrust.

Turboprop and turboshaft engines require a great deal of shaft horsepower other than that used to drive the compressor, so these engines have additional turbine stages to extract the maximum amount of energy from hot gases.

radial-inflow turbine. A turbine similar in appearance to a centrifugal compressor rotor. The hot gases flow into the turbine from its outside rim, then radially inward through the vanes, and out at its center.

There are two basic types of turbines used in aircraft engines: axial and radial-inflow. All gas turbine engines, except some of the smallest used to drive auxiliary power units, use axial turbines. The small APU compressors and turbochargers used with reciprocating engines are driven by radial-inflow turbines.

The turbine section operates in the most hostile environment in the engine. The temperature of the gases is at its highest as they enter the first stage of the turbine through the turbine inlet guide vanes, and the velocity of these gases is the highest as they leave the turbine. *See* Figure 10-20 on Page 369.

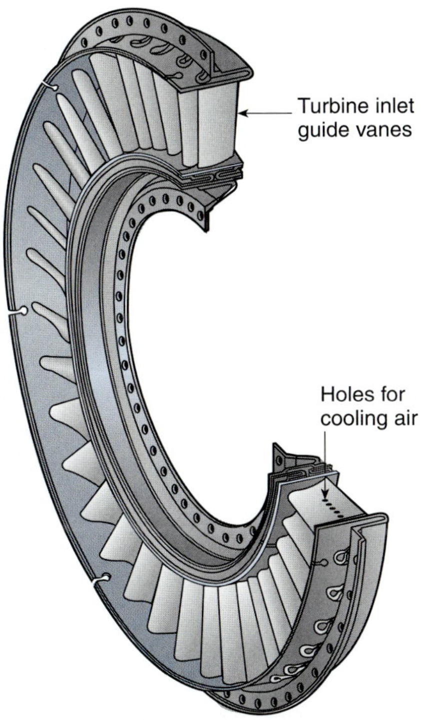

Figure 10-65. *Section of a typical ring of turbine nozzles, or turbine inlet guide vanes*

turbine inlet guide vanes. A series of stator vanes immediately ahead of the first-stage turbine.

Turbine Section Elements

The turbine section consists of a ring of turbine nozzles, or guide vanes, and one or more stages of turbines. Each stage consists of a set of guide vanes followed by a turbine disk with its attached blades.

Turbine Inlet Guide Vanes

A ring of turbine inlet guide vanes, or nozzles, mounts between the discharge of the combustion section and the first stage of turbines. It forms a series of convergent ducts that increase the velocity and decrease the pressure of the gases, and directs them so they will enter the turbine blades at the optimum angle for maximum efficiency.

When the engine is producing between cruise and takeoff power, the gases flowing through the turbine nozzles are moving at the speed of sound, and the nozzle is operating in a choked condition, preventing any further increase in the velocity of the gases. Since the gases can no longer accelerate, they produce a back pressure greater than the pressure behind the turbine. This choked condition causes a relatively constant flow of gases through the turbine.

Since the speed of sound depends upon the temperature of the air, the velocity of gases flowing through a choked nozzle depends upon their temperature.

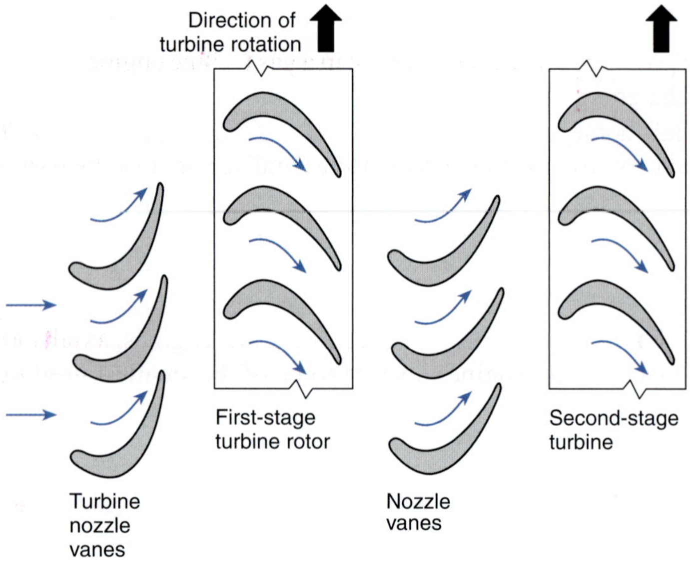

Figure 10-66. *Air flowing through the convergent turbine nozzles is accelerated, and its direction is changed so that it enters the following turbine blades at the angle that provides maximum efficiency.*

turbine nozzle. Another name for turbine inlet guide vanes.

Turbine Design and Construction

The turbine must absorb enough energy from the hot gases flowing from the combustion section to drive the compressor fast enough to move a tremendous mass of air through the engine. The mass of air flowing through the engine is not changed as the energy is absorbed by the turbine, but the pressure of the air is decreased, and its velocity is increased, as is seen in Figure 10-20 on Page 369. Much of the energy is converted from that available for reactive thrust into torque used to drive the compressor. Turboprop and turboshaft engines convert so much of the energy into torque that very little thrust remains at the tail pipe.

Turbine disks are normally made of a heavy nickel alloy forging. The blades are most generally attached by the fir-tree method which allows them to be loose when the engine is cold, but tighten as the engine reaches its operating temperature.

One recent development in turbine construction is to machine an entire turbine wheel from a single slab of steel, making the disk and the blades an integral unit called a blisk.

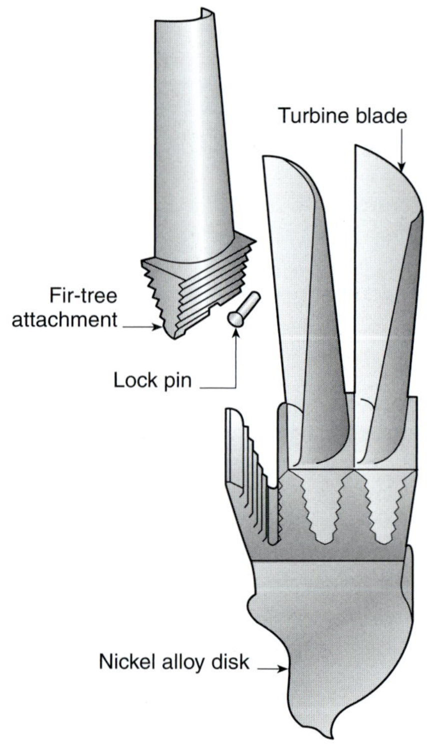

Figure 10-67. *Turbine disks are made of heavy nickel alloy forgings with the blades attached by the fir-tree method.*

blisk. A turbine wheel machined from a single slab of steel. The disk and blades are an integral unit.

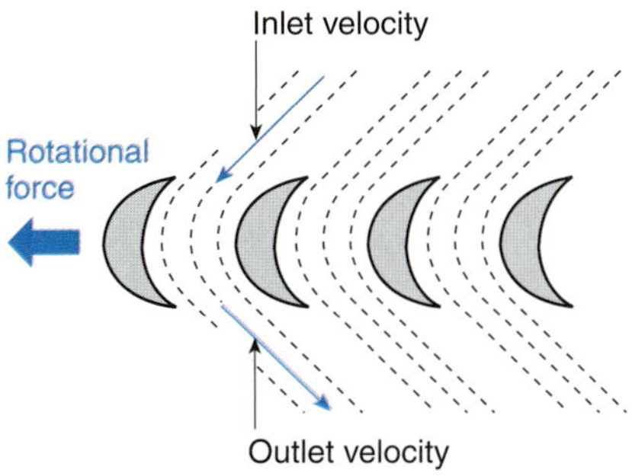

Figure 10-68. *An impulse turbine rotor blade absorbs energy from the change in the direction of air flowing through it.*

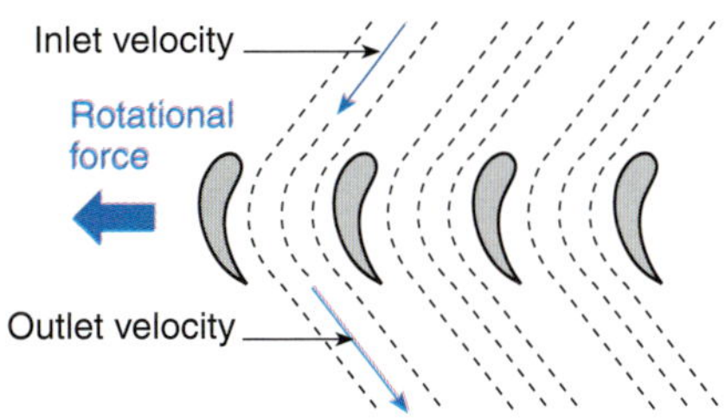

Figure 10-69. *A reaction turbine blade produces rotation by the reaction to the force that causes air flowing between the blades to accelerate.*

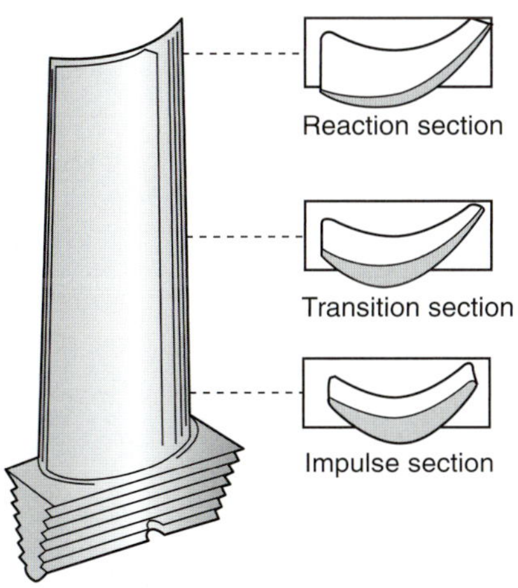

Figure 10-70. *A typical reaction-impulse turbine blade that has an impulse section near the root which changes to a reaction section at the tip.*

Turbine Blade Design

There are two basic designs of turbine blades: reaction and impulse. A third design, a reaction-impulse blade, is a combination of the two.

The impulse turbine blade in Figure 10-68 is driven by the energy absorbed from the fast-flowing gases as their direction of flow is changed. The air experiences neither a pressure drop nor speed change as it flows through an impulse turbine, but the direction of flow is drastically changed. The energy required to change the direction produces the force that causes the turbine to rotate.

Reaction turbine blades produce their turning force by an aerodynamic action of the air as it accelerates through the converging ducts formed between the blades. The velocity vectors in Figure 10-69 show that the outlet velocity is greater than the inlet velocity. The reaction from energy used to accelerate the air produces a force that rotates the turbine.

A reaction-impulse turbine blade combines the action of both types. Notice that the root section has an impulse shape, and the blade progressively changes to a reaction shape with a considerable twist, which with the other blades, forms converging ducts. *See* Figure 10-70.

Shrouded Turbine Blades

Some early turbines used shrouded blades to reduce vibration and minimize tip losses. These blades had a basic T-shape, with the cross bar of the T forming a ring around the turbine as shown in Figure 10-71. Most modern engines use thinner, unshrouded turbine blades that are more efficient than the older type.

Blade-Tip Clearance Control

Some modern high-power turbofan engines control the tip losses by active clearance control (ACC). This is done by blowing a stream of cold fan-discharge air from nozzles around the turbine cases when the engine is developing takeoff power. This cold air shrinks the case around the turbine, minimizing tip losses, and increasing engine fuel efficiency. The ACC is controlled by the full-authority, digital electronic control (FADEC), discussed in Chapter 12, beginning on Page 476.

Turbine Cooling

The primary factor that limits the amount of power a gas turbine engine can produce is the maximum temperature that can be tolerated at the turbine inlet. This is called the turbine inlet temperature (TIT).

Some modern engines have increased fuel efficiency because of increased allowable TIT, and one way of doing this is by cooling the turbine inlet guide vanes and the first-stage rotor blades.

Turbine cooling is done by flowing compressor bleed air through hollow guide vanes and rotor blades. Air leaves the blade's surface through specially shaped holes in such a way that it forms a film of air over the blade to insulate the surface from the hot gases. Figure 10-72 shows the cross section of two modern turbine blades through which cooling air flows to increase the allowable TIT.

Air used for turbine system cooling is bled from one of the higher stages of the compressor, and while its temperature is greater than 1,000°F, it is far cooler than the gases that drive the turbine. This air flows through the hollow blades and exits with the exhaust gases.

It is necessary to cool only the turbine inlet guide vanes and the first stage turbine blades. The gases lose enough of their energy when passing the first stage that their temperature drops to within the range allowed for succeeding stages.

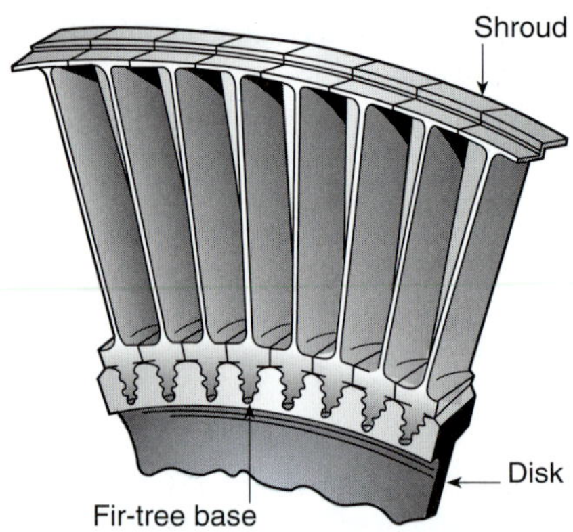

Figure 10-71. *Some older engines used shrouded turbines to reduce vibration and minimize airflow losses around the tips of the blades.*

Turbine Blade Construction

Along with great strides in increasing the allowable TIT by cooling the turbine blades, metallurgical advances in new materials and procedures for manufacturing the blades have resulted in additional allowable temperature increases. Ceramics offer many potential improvements.

One advancement that has increased the allowable temperatures is a blade made of a single crystal of metal. It has been proven that many of the present-day turbine blade failures have been along the grain boundaries within the metal of the blade. New technologies have allowed turbine blades to be made of a single crystal of the metal. This avoids all of the grain boundaries and allows a much higher inlet temperature with its accompanying increased efficiencies.

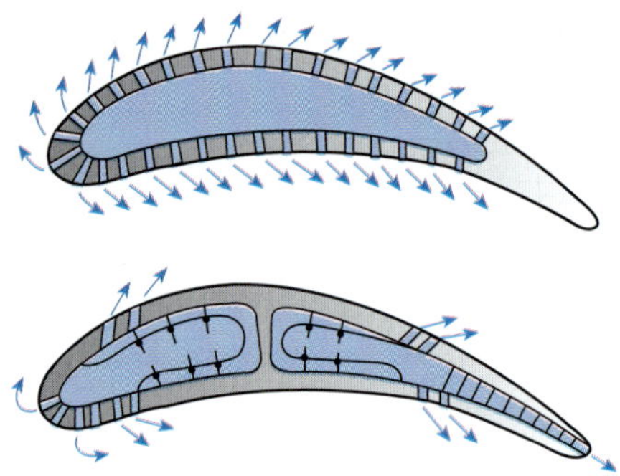

Figure 10-72. *Cross sectional views of two typical air-cooled turbine blades*

Turbine Failures

The turbine section of a gas turbine engine operates in a most hostile environment and is the portion of the engine that requires the most careful inspection. These inspections are discussed with the hot-section inspections in Chapter 15, beginning on Page 530.

Three main causes of turbine failure are creep, metal fatigue, and corrosion.

Creep is the deformation of a metal part that is continually exposed to high centrifugal loads and high temperature. When the metal is operated within its elastic limit, it will always return to its original configuration when the stress is removed; but if the load is increased, part of the deformation will remain after the stress is removed. Some metals show a permanent deformation even after exposure to a lower amount of stress, if the load is maintained over a long period of time. This type of deformation is called creep.

ACC (active clearance control). A system for controlling the clearance between tips of the compressor and turbine blades and the case of high-performance turbofan engines by shrinking the engine case with a flow of cool fan discharge air.

TIT (turbine inlet temperature). The temperature of the gases from the combustion section of a gas turbine engine as they enter the turbine inlet guide vanes or the first stage of the turbine.

creep. The deformation of a metal part that is continually exposed to high centrifugal loads and high temperature.

elastic limit. The maximum amount of tensile load, in pounds per square inch, a material is able to withstand without being permanently deformed.

Metal fatigue is a weakening of the metal that is subjected to repeated cyclic loading. Each time a gas turbine engine is started, operated through its flight routine, and shut down, it goes through a severe heat cycle. The numerous heat cycles that an engine goes through makes the metal parts subject to metal fatigue.

Corrosion is the electrolytic action that occurs when the alloying agents in the metal combine with elements in the air to form salts that have no strength. This corrosive action is accelerated by exposure to extremely high temperatures.

STUDY QUESTIONS: THE HOT SECTION

Answers begin on Page 420. Page numbers refer to chapter text.

112. Approximately __________________ (what part) of the air that flows into the combustion section is actually used in the combustion process. *Page 407*

113. The air that flows through the combustion section but is not used in the combustion process is used for __________________ . *Page 407*

114. Three types of combustors are:
    a. ____________________________
    b. ____________________________
    c. ____________________________
    Page 407

115. A combustor in which the air leaving the compressor makes a 180° turn, flows through the combustor, and makes another 180° turn before it flows through the turbine, is a/an ____________________________ combustor. *Page 409*

116. Multiple-can combustors are most generally used on engines with ____________________ (axial-flow or centrifugal) compressors. *Page 407*

117. The air used in the combustion process is called ____________________ (primary or secondary) air. *Page 407*

118. There are normally ________________ (how many) igniters used to ignite the fuel-air mixture in a gas turbine engine. *Page 407*

119. The fuel-air mixture in cans not having an ignitor is ignited by flames from the other cans that travel through __________________________ . *Page 407*

120. If a single can-type combustor fails because of a clogged fuel nozzle, the turbine is likely to be damaged because of the extreme _____________________ differences. *Page 407*

121. A combustion section that has individual cans, but in which the hot gases from all of the cans are collected before they go to the turbine, is called a/an ______________________ combustor. *Page 407*

122. The cans used in a can-annular combustor are ______________________ (longer or shorter) than those used in an multiple-can combustor. *Page 408*

123. When the length of an engine must be kept as short as possible, a/an ____________________ combustor can be used. *Page 409*

124. The greater portion of the energy released from the fuel in a turbojet engine is used ______________________ (for thrust or to drive the compressor). *Page 410*

125. Turboprop and turboshaft engines use additional stages of ________________ to produce the shaft horsepower needed to drive the propeller or helicopter rotors. *Page 410*

126. The type of turbine used in gas turbine engines is the ____________________ (axial or radial-inflow) type. *Page 410*

127. The stationary vanes in a turbine section are located in the airflow path __________________ (before or after) the turbine rotors. *Page 410*

128. The component in the turbine section that directs hot gases from the combustion section into the turbine is the ring of turbine _____________________ or ______________________ . *Page 410*

129. When the gases flowing through a turbine nozzle can no longer accelerate, the nozzle is said to be operating in a/an ___________________ condition. *Page 410*

130. The speed of sound depends upon the _________________ (temperature or pressure) of the air. *Page 410*

131. The point in a gas turbine engine at which the temperature is the highest is the ____________________ . *Page 410*

132. A set of guide vanes followed by a turbine disk with its blades is called a/an _______________ . *Page 410*

133. The turbine inlet guide vanes form a series of _________________ (convergent or divergent) nozzles. *Page 410*

134. Three things the turbine inlet guide vanes do to the air as it passes through them are:

a. ________________________

b. ________________________

c. ________________________

Page 410

Continued

STUDY QUESTIONS: THE HOT SECTION *Continued*

135. Turbine disks are normally made of a __________________ alloy. *Page 411*

136. Turbine blades are normally attached to the disk by the __________________ method. *Page 411*

137. Three types of turbine blades are:
 a. __________________
 b. __________________
 c. __________________
 Page 412

138. When the hot gases drive an impulse turbine, their direction is changed, and their speed __________________ (increases, decreases, or remains the same). *Page 412*

139. When the hot gases drive a reaction turbine, their direction is changed, and their speed __________________ (increases, decreases, or remains the same). *Page 412*

140. In a reaction-impulse turbine blade, the root section of the blade acts as a/an __________________ blade. *Page 412*

141. Some turbine engines have used shrouded turbine blades to reduce __________________ and minimize tip losses. *Page 412*

142. Some modern high-performance turbofan engines reduce turbine blade tip losses by shrinking the turbine case when the engine is producing its maximum power. The case is shrunk by blowing a stream of __________________ over it. *Page 412*

143. The primary factor that limits the amount of power a gas turbine engine can produce is the maximum temperature that can be tolerated at the __________________ . *Page 412*

144. Air used to cool the turbine blades is taken from the __________________ (higher or lower) stages of the compressor. *Page 413*

145. Three factors that cause turbine failure are:
 a. __________________
 b. __________________
 c. __________________
 Page 413

Accessory Systems

The accessory systems of a gas turbine engine are covered in their own chapters in this text. Here we will consider only the locations of these systems on the engine itself.

Accessory Drives

On almost all reciprocating engines, the accessories such as magnetos, pumps, alternators, and starters are mounted behind the cylinders, but on a turbojet or turboprop engine, the exhaust nozzle takes up this valuable space. For this reason, other locations must be chosen for the accessories.

Gas turbine engines have an abundance of compressed air available to operate many of the components, which on an aircraft powered by a reciprocating engine, are driven by gears in the engine. Some of these accessories are hydraulic pumps, air conditioning units, and various actuators.

Certain other components, such as oil pumps, fuel pumps, fuel controls, and starters, are driven by bevel gears from the turbine-compressor drive shaft. There are two basic locations for mounting gear-driven accessories: below the low-pressure compressor and below the high-pressure compressor behind the fan. Both of these locations are seen in Figure 10-73.

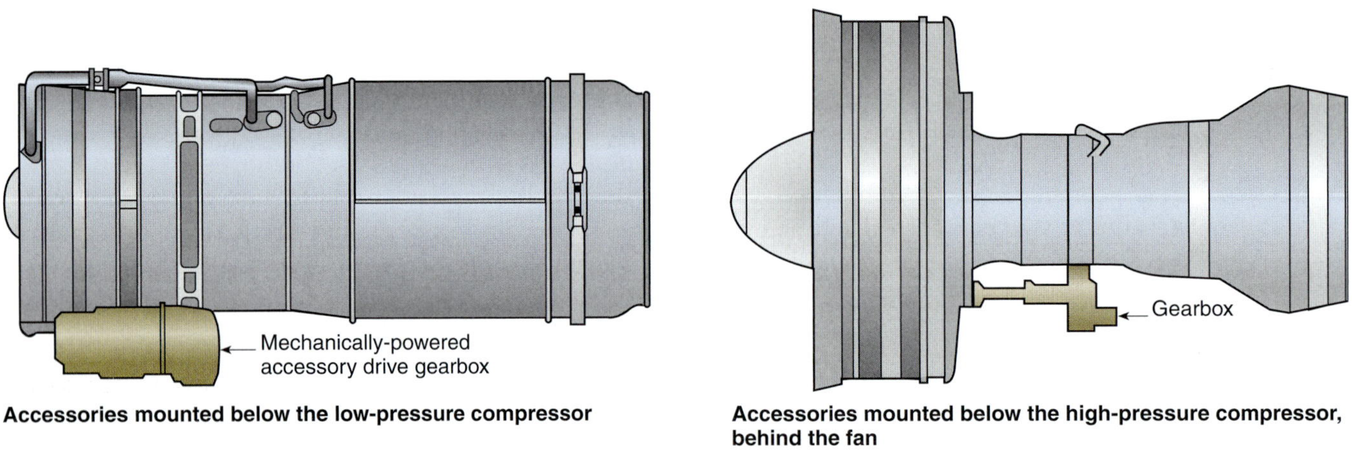

Figure 10-73. *Locations for gear-driven accessories on turbojet and turbofan engines*

Propeller Reduction Gear Systems

Reciprocating engines turn slow enough that some of the smaller ones drive the propeller directly from the crankshaft. Even the larger engines that drive the propeller through a series of reduction gears seldom use a gear reduction ratio of more than 1:2. Turboprop engines have an entirely different problem: since the turbine turns at such a high speed, multistage planetary reduction gears are normally used, with ratios in the range of 1:10 (0.10) to 1:15 (0.0667) considered common.

There are several ways to mount the reduction gears on a gas turbine engine. Figure 10-74 shows how the reduction gears are mounted on the Pratt & Whitney of Canada PT6 engine. The air inlet is around the rear of the engine, and most of the accessories are mounted on the rear. The exhaust leaves the engine near the front. The planetary propeller reduction gears are at the front, in line with the turbine-compressor shaft.

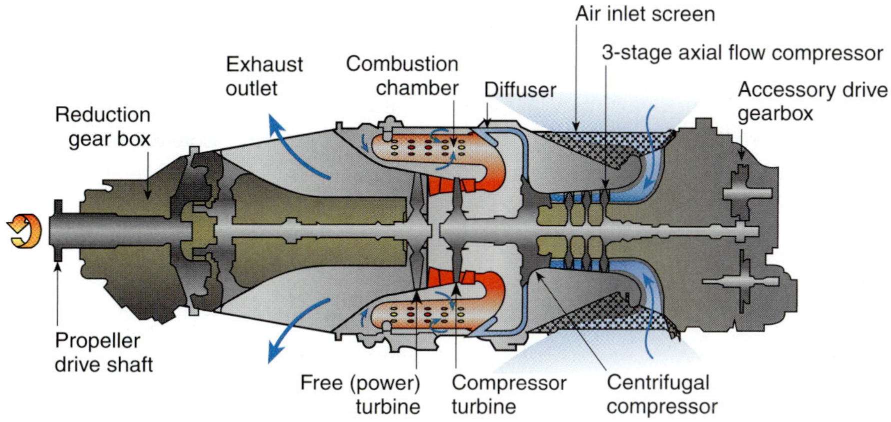

Figure 10-74. *The propeller reduction gears for the PT6 engine are at the front of the engine with the propeller shaft in line with the turbine-compressor shaft.*

The large Allison 501 engine extends the reduction gearing out ahead of the engine and supports it by struts. The reduction gears may be located either above or below the center line of the engine. This allows the aircraft manufacturer a degree of freedom in designing the air inlet ducts for the engine. *See* Figure 10-75.

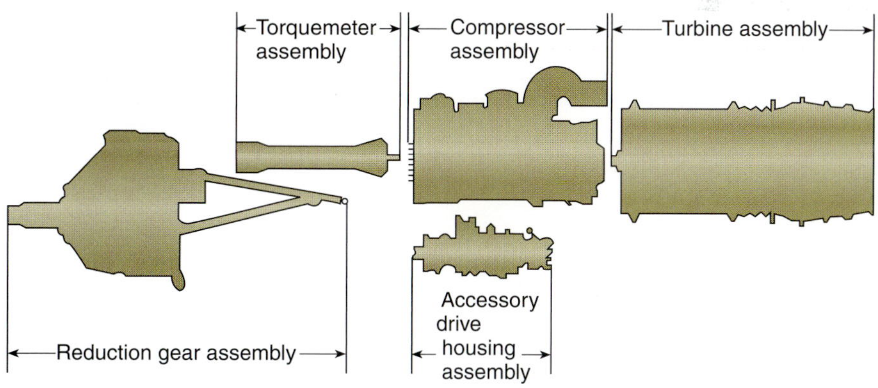

Figure 10-75. *The propeller reduction gears for the Allison 501 engine are separate from the engine and supported by struts. The reduction gearing may be above or below the engine center line.*

The small Garrett TPE331 engine has the reduction gearing outside of it, and like the Allison 501, can be mounted either above or below the engine center line. *See* Figure 10-76.

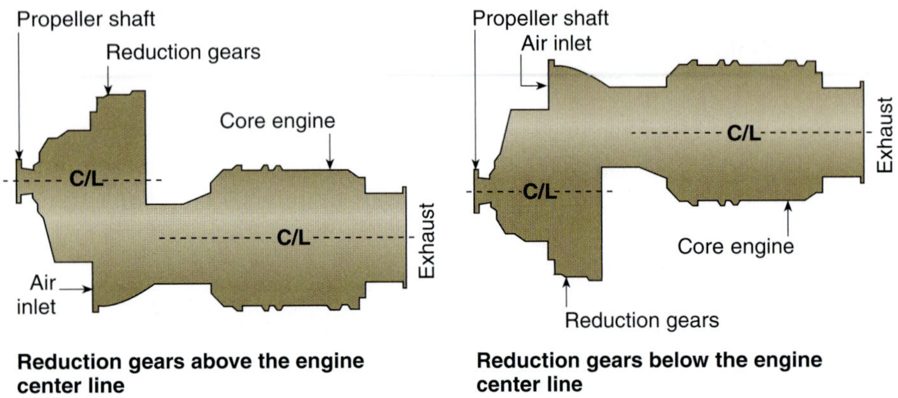

Figure 10-76. *The propeller reduction gears for the Garrett TPE331 engine are separate from the engine and may be above or below the engine center line.*

STUDY QUESTIONS: ACCESSORY SYSTEMS

Answers begin on Page 420. Page numbers refer to chapter text.

146. Two ways gas turbine engines accessories may be driven are:

a. ________________________________

b. ________________________________

Page 417

147. The propeller reduction gearing used by a turboprop engine is normally of the multistage ____________________ type. *Page 417*

Answers to Chapter 10 Study Questions

1. a. aerodynamic action
   b. jet reaction
2. false
3. heat
4. true
5. potential
6. kinetic
7. acceleration
8. direction, magnitude
9. force, distance
10. work
11. 33,000, 550, 746
12. velocity
13. momentum
14. feet per second, per second
15. decrease
16. increase, decrease
17. decrease, increase
18. decrease, increase
19. increase, decrease
20. subsonic
21. supersonic
22. choked
23. static
24. total
25. Kelvin
26. a. 15
    b. 59
    c. 29.92
27. solid propellant, liquid propellant
28. ramjet
29. compressor
30. compressor, combustion chambers
31. General Electric
32. a. air inlet
    b. compressor
    c. diffuser
    d. combustors or burners
    e. turbine
    f. tail pipe and jet nozzle
33. core
34. free-turbine
35. turboshaft
36. a. increased propulsive efficiency
    b. lower noise level
    c. lower specific fuel consumption
37. bypass
38. Brayton
39. pressure
40. gross
41. core engine, fan
42. forward
43. propulsion
44. a. engine RPM
    b. turbine discharge total pressure
    c. EPR (engine pressure ratio)
45. a. airspeed
    b. mass airflow
    c. ram effect
46. decrease
47. decrease
48. increase
49. decrease
50. drops off more rapidly
51. increase
52. ram effect
53. a. water injection
    b. afterburners
54. 375
55. 2.5
56. 80
57. N_2
58. total, compressor, inlet
59. cold, hot
60. airframe
61. a. subsonic
    b. supersonic
62. divergent
63. ram recovery
64. plenum
65. bell mouth
66. blow-in
67. dissipater
68. convergent-divergent
69. decreases
70. subsonic
71. subsonic
72. pressure, velocity
73. velocity
74. kinetic
75. potential
76. velocity
77. heat
78. bleed air
79. centrifugal, axial-flow
80. a. impeller
    b. diffuser
    c. manifold
81. eye
82. divergent

83. double-entry
84. plenum
85. slow down
86. diffuser
87. higher
88. lower
89. high
90. a. velocity of air through compressor
    b. compressor rotational speed
91. inlet guide vanes
92. increase
93. surge
94. a. variable inlet guide vanes
    b. interstage air bleeds
95. fuel control
96. dual-spool
97. high
98. N_1
99. is not
100. N_2
101. increases
102. a. connect directly to the low-pressure compressor
    b. drive with its own free turbine
    c. drive from low-pressure compressor through gears
    d. fan blades mounted on blades of an aft free-turbine
103. bypass ratio
104. pressure
105. shorter
106. profile, squealer
107. abradable
108. hybrid
109. fuel control
110. compressor stall
111. diffuser, compressor-discharge
112. one fourth
113. cooling
114. a. multiple-can
    b. can-annular
    c. annular
115. reverse-flow
116. centrifugal
117. primary
118. two
119. flame propagation tubes
120. temperature
121. can-annular
122. shorter
123. reverse-flow
124. to drive the compressor
125. turbines
126. axial
127. before
128. guide vanes, nozzles
129. choked
130. temperature
131. turbine inlet
132. stage
133. convergent
134. a. increase its velocity
    b. decrease its pressure
    c. change its direction
135. nickel
136. fir-tree
137. a. reaction
    b. impulse
    c. reaction-impulse
138. remains the same
139. increases
140. impulse
141. vibration
142. fan-discharge air
143. turbine inlet
144. higher
145. a. creep
    b. metal fatigue
    c. corrosion
146. a. by compressor bleed air
    b. gear-driven
147. planetary

LUBRICATION & COOLING SYSTEMS

11

Continued

Lubrication & Cooling Systems

11

Lubrication is a vital function in both reciprocating and gas turbine engines, and while it performs many of the same functions in both types of engines, the systems are different. It is especially important to note that the lubricants differ and are not compatible.

Reciprocating engines have an abundance of moving parts, such as the pistons, connecting rods, crankshaft, valve operating mechanism, and accessory drive gears, and their lubrication system absorbs much heat from the cylinder walls and from the undersides of the pistons. For this reason, they carry a large quantity of oil and have a high oil consumption rate. It is not uncommon for some large radial engines to carry twenty or thirty gallons of relatively high-viscosity mineral-base oil and use as much as four or five gallons per hour.

viscosity. The resistance of a fluid to flow.

Gas turbine engines, on the other hand, have only one basic moving part, plus the accessory gears. The lubrication system must absorb a great amount of heat, the majority of which comes from the turbine shaft bearings. Large turbine engines carry between five and eight gallons of low-viscosity synthetic-base oil. In order to absorb the heat, oil circulates through the engine at a high flow rate several times each minute. Since the oil has no contact with the combustion area, and oil seals are used around the turbine/compressor shaft, very little is lost out the exhaust. As a result, a turbine engine does not consume nearly as much oil as a reciprocating engine, normally less than a pint an hour.

Requirements for Turbine Engine Lubricants

Lubricants in a turbine engine function like those in a reciprocating engine. They reduce friction in the gears and bearings, absorb heat from the bearings, protect metal parts of the engine against corrosion, and pick up and carry contaminants into filters where they are trapped and held. In addition to this, in many turbine engines, hot engine oil is routed through an oil-to-fuel heat exchanger where part of its heat warms the fuel to prevent formation of ice crystals.

viscosity index. A measure of the change in the viscosity of a fluid with a change of temperature.

synthetic oil. Oil made by chemical synthesis (changing) of a mineral, animal, or vegetable base.

kinematic viscosity. The ratio of the absolute viscosity of a fluid to its density.

For an oil to perform properly, it must have these characteristics:

- Low viscosity—It must be able to flow readily under all temperature conditions.
- High viscosity index—It must not change its viscosity appreciably with changes in its temperature.
- High film strength—It must not break down under the loads imposed by the accessory gear tooth pressures.
- Low volatility—It must not evaporate at the low air pressures of high altitude.
- High flash point—It must not release vapors that would ignite at the temperatures inside the engine.
- Low pour point—It must be pourable at low temperatures.
- Good antifoaming characteristics—It must be able to easily release all the air it accumulates as it passes through the engine.
- Low carbon deposit formation—It must not deposit coke or hard carbon deposits on hot engine parts. The carbon that forms must be held in the oil until removed by the filters.

Synthetic Lubricating Oil

Oil that meets the requirements for turbine engine lubrication is a synthetic product made by synthesizing, or changing, an animal-, vegetable-, or mineral-base material into a more chemically complex compound.

The viscosity of turbine engine synthetic lubricating oil is not expressed in terms of SAE numbers, as is reciprocating engine oil, but rather in centistokes, which is a measure of its kinematic viscosity, or the ratio of its absolute viscosity to its density.

There are two grades of synthetic oil used for turbine engine lubrication. These conform to Military Specifications MIL-L-7808, known as Type I oil, and MIL-L-25699, known as Type II oil. Type I, a 3-centistoke oil, is an older, lighter weight oil. Type II, or 5-centistoke oil meets the more stringent requirements of the newer, higher-performance engines.

Synthetic oils are definitely not interchangeable with straight mineral oil or with the AD oils used in reciprocating engines. And Type I oil should not be added to Type II, because it will degrade its superior characteristics. Some manufacturers recommend that different brands of oil not be mixed, as some of the proprietary additives may not be compatible.

Synthetic oils have some toxic additives that can cause skin irritation. They can also soften certain types of aircraft finishes, causing cadmium and zinc plating to flake off and contaminate the lubrication system.

Turbine Engine Lubrication Systems

There are two basic classifications of turbine engine lubrication systems: wet-sump and dry-sump. There is another type used on some of the smaller engines designed for special short-duration operation. This is a nonreturn system, in which the bearings are lubricated by a pressure spray and the oil is then collected and disposed of.

Wet-Sump Lubrication System

The wet-sump lubrication system was used in some early turbine engines, but today is found only on the smaller engines such as those used to drive auxiliary power units (APU).

wet-sump lubrication system. A lubrication system in which the oil supply is carried within the engine itself.

APU (auxiliary power unit). A self-contained motor-generator carried in an aircraft to generate power for ground operations and to start the main engines.

In a wet-sump system, pressurized oil is used to lubricate the turbine rotor coupling and rotor-shaft bearings, but the accessory drive gears are splash-lubricated by oil carried in the gear housing that serves as the oil reservoir. Oil that has lubricated the bearings drains by gravity and is collected and directed back to the gear housing, where it is stored until it recirculates through the system.

Some modern engines, notably the Pratt & Whitney of Canada JT15D turbofan and the PT6 turboprop, carry their oil supply in a tank that is an integral part of the engine, but all of the lubrication is done by pressure and the oil is returned to the tank by scavenger pumps. Therefore, these are not wet-sump engines.

Dry-Sump Lubrication System

By far the more popular lubrication system is the dry-sump type, in which the oil, after serving its lubrication and cooling functions, is returned by scavenger pumps to a tank outside of the engine proper. There are two types of dry-sump lubrication systems: the hot-tank system (Figure 11-1), and the cold-tank system (Figure 11-2).

dry-sump lubrication system. A lubrication system that carries the oil supply in a tank not normally part of the engine.

Hot-Tank Lubrication System

hot-tank lubricating system. A turbine engine lubricating system in which the oil cooler is located in the pressure subsystem.

In a hot-tank lubrication system, the oil cooler is in the pressure subsystem, and the scavenged oil is not cooled before it is returned to the tank.

In the lubrication system in Figure 11-1, the oil is carried in the oil tank (1) and flows by gravity to the main pressure pump (2). From there, the pressurized oil flows through the filter (3) and through the air/oil heat exchanger, or oil cooler (6), to the bearings through the four bearing filters (9). The pump pressure is maintained at the correct value by the pressure relief valve (4). If the pressure exceeds that for which the relief valve is set, the valve offseats and returns the excess oil back to the inlet side of the pump. If the filter should clog, the filter bypass valve (5) will offseat and allow unfiltered oil to flow through the system.

When the oil temperature is low, the temperature bypass valve (7) is open and the oil flows directly to the bearing filters and the bearings. When the temperature is high enough to require cooling, the valve restricts the flow of oil and forces it through the cooler. If, for any reason, the cooler should clog, the cooler bypass valve (8) will open and allow oil to flow to the bearings.

After leaving the oil cooler, the oil flows through the bearing filters (9) to jets which spray the oil into the bearings.

The oil drains from the bearings and is collected by scavenger pumps (10) and returned to the oil tank.

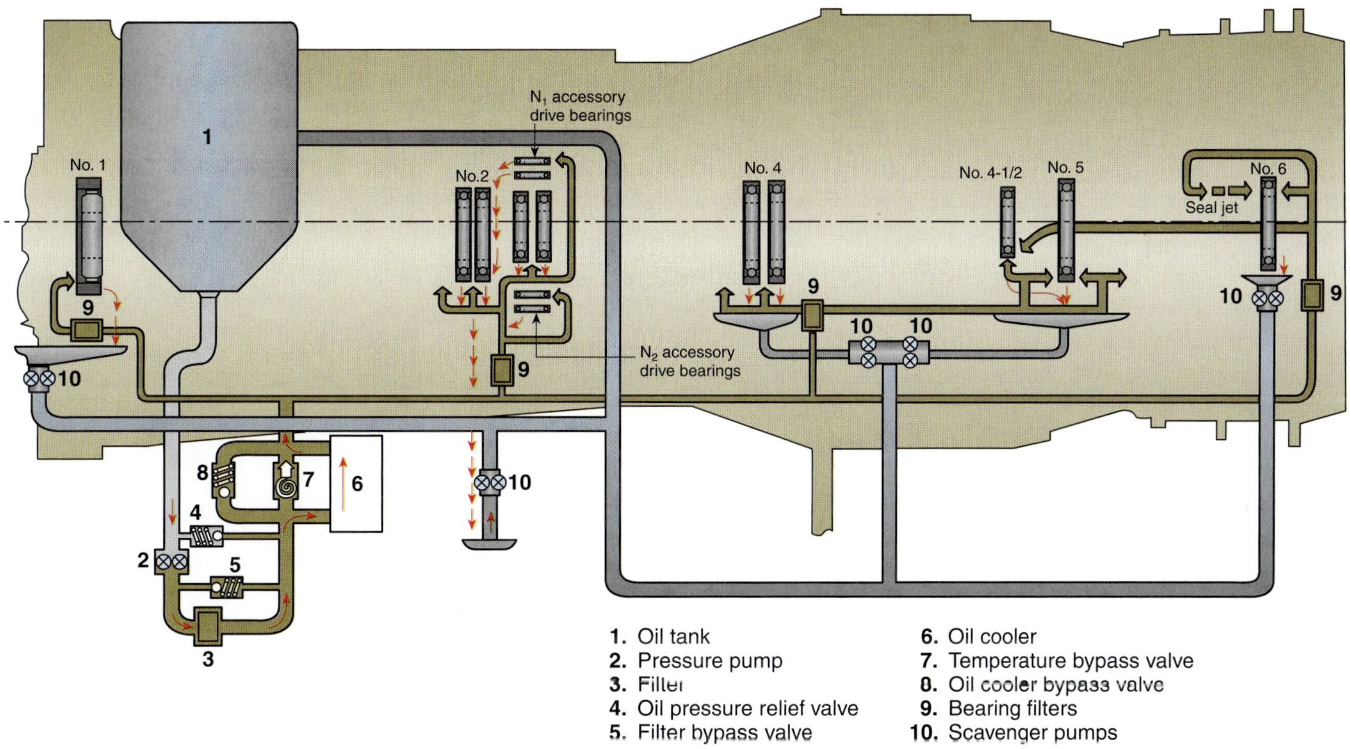

Figure 11-1. *A typical hot-tank dry-sump lubrication system for a dual-spool axial-flow turbojet engine*

Cold-Tank Lubrication System

The cold-tank system in Figure 11-2 is the same as the hot-tank system, except for the placement of the oil cooler and the temperature bypass and cooler bypass valves.

cold-tank lubrication system. A turbine engine lubricating system in which the oil cooler is in the scavenge subsystem.

The oil goes directly from the pressure pump (2) through filters (3) and (9) to the bearings. From the bearings, it drains down and is collected and returned to the oil tank by the scavenger pumps (10). If the oil is cool enough, it returns to the tank through the temperature bypass valve (7), but if it is too hot, this valve closes, forcing the oil to flow through the cooler where it gives up the excess heat.

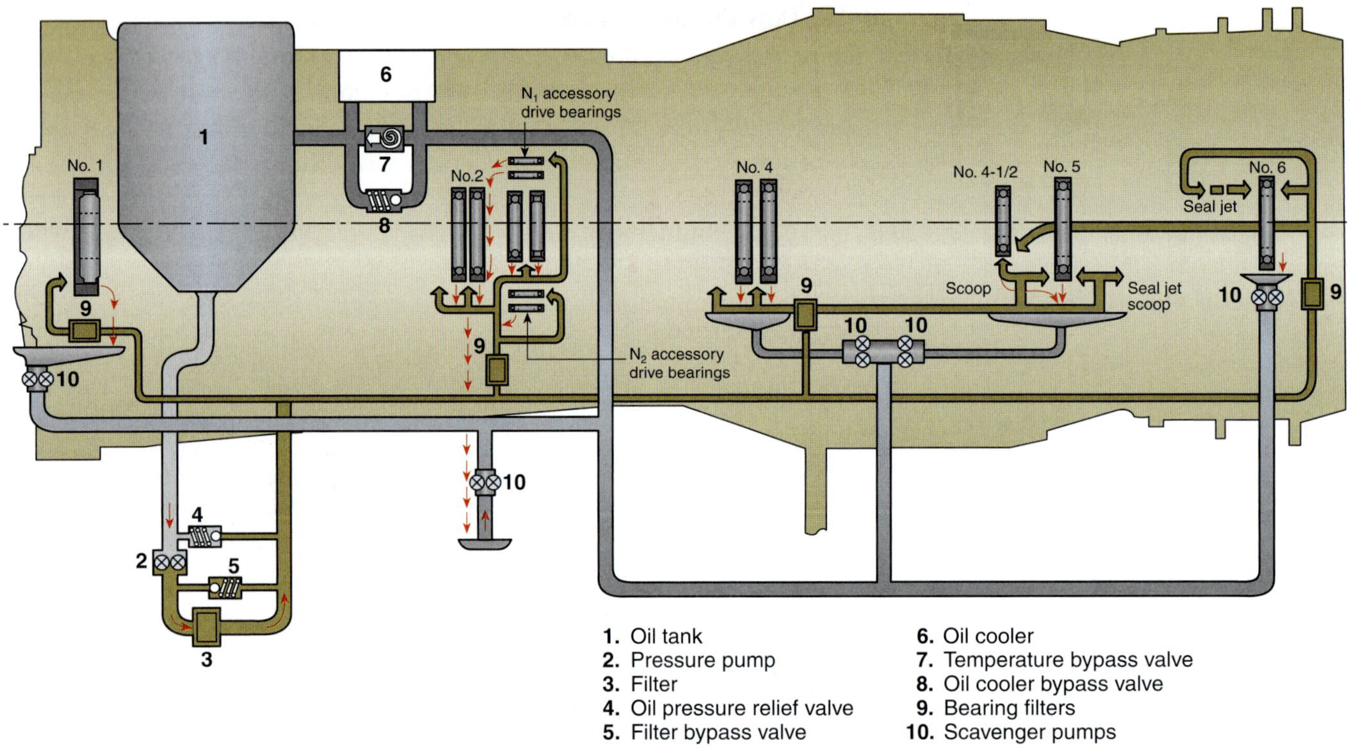

Figure 11-2. *A typical cold-tank dry-sump lubrication system for a dual-spool axial-flow turbojet engine*

Lubrication System Subsystems

Turbine engine lubrication systems are logically divided into three basic subsystems: pressure, scavenge, and vent.

Pressure Subsystem

The pressure subsystem supplies the correct amount of clean lubricating oil at the proper pressure and temperature to all of the bearings and gears. It consists of the oil tank, the pressure pump, the pressure relief valve, the main oil filter, the oil cooler (for hot-tank systems), the bearing (last-chance) filters, and the oil jets, or nozzles.

scavenge oil system. The subsystem in the lubrication system of a gas turbine engine that collects the oil after it has lubricated the bearings and gears and returns it to the oil tank.

air-oil separator. A component in a turbine engine lubrication system that removes the air from the scavenged oil before it is returned the oil tank.

dwell chamber. A chamber in a turbine engine oil tank in which entrained air is allowed to escape from the scavenged oil.

Scavenge Subsystem

The scavenge subsystem picks up the oil after it has performed its lubrication and cooling functions and returns it to the oil tank where it can be recirculated through the system. The scavenge subsystem consists of the sumps, in which oil from the bearings and gears collects, the various scavenger pumps, and the air-oil separator, or dwell chamber, in the oil tank. The oil cooler with its bypass valve and temperature bypass valve are in the scavenge subsystem of cold-tank systems.

Vent Subsystem

The vent subsystem provides a slight pressure on the oil in the oil tank to assure a positive flow of oil to the pump inlet to prevent pump cavitation. It also vents the various bearing compartments and gear boxes to the expansion space above the oil in the oil tank to maintain a uniform air pressure at the oil jets. This ensures the proper flow of oil to the bearings.

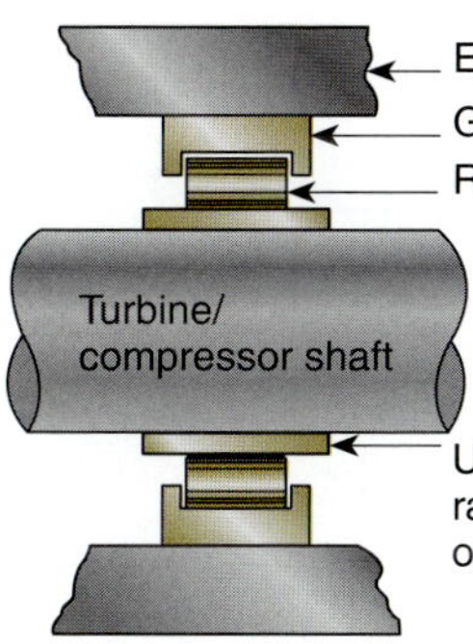

Figure 11-3. *The inner race of a roller bearing is shrunk onto the shaft, and in this bearing, it is not grooved. The ungrooved race allows the rollers to move axially as the engine expands and contracts with temperature changes.*

Bearings and Seals

The turbine/compressor shaft is the single main rotating element in a gas turbine engine and is supported by roller and ball bearings. The rotating shaft must be sealed to prevent oil that lubricates the bearings from getting into the airstream as it passes through the engine.

Bearings

Roller bearings are used to support radial loads because of their large contact area. The rollers in these bearings are not mounted in cages, and the races are not tapered. The ungrooved inner race is shrunk onto the shaft to hold it securely in place, and the rollers in the grooved outer race are free to move axially on the inner race when the engine expands and contracts with changes in temperature.

radial bearing load. The load on a bearing perpendicular to the shaft on which the bearing is mounted.

It is important that the races for roller bearings be in perfect alignment so the load will be applied to the rollers uniformly. Some roller bearing installations are oil damped. The outer race of these bearings is installed in a cavity a few thousandths of an inch larger than the outside diameter of the race, and pressurized engine oil is directed into this cavity. This film of oil allows the race to shift slightly to compensate for minor misalignment between the shaft and the housing and to cushion vibrations in the rotor assembly. Oil-damped bearings are also called squeeze film bearings. *See* Figure 11-4.

Ball bearings are used when there is an axial, or thrust, load; they can support radial loads as well. The balls are normally spaced around the races by thin sheet-metal cages. Some ball bearings use two-piece, or split, inner races to facilitate bearing replacement. *See* Figure 11-5.

oil-damped bearing. A type of roller bearing installation in which the outer race is installed in an oil damper compartment. Oil damped bearings compensate for sight misalignment and absorb vibrations of the shaft.

squeeze film bearings. Another name for oil-damped bearing.

axial bearing load. The load on a bearing parallel to the shaft on which the bearing is mounted.

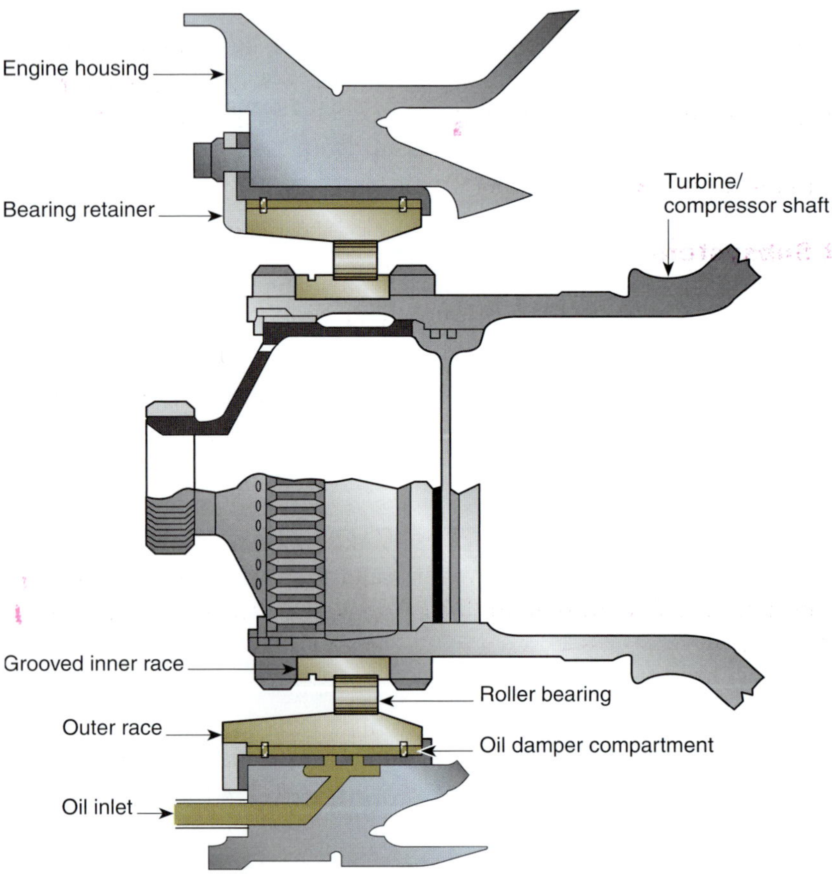

Figure 11-4. *Roller bearings that support the turbine/compressor shaft are oil damped by being installed in a cavity a few thousandths of an inch larger than the outside of the outer race. A cushion of oil absorbs vibrations and compensates for any slight misalignment.*

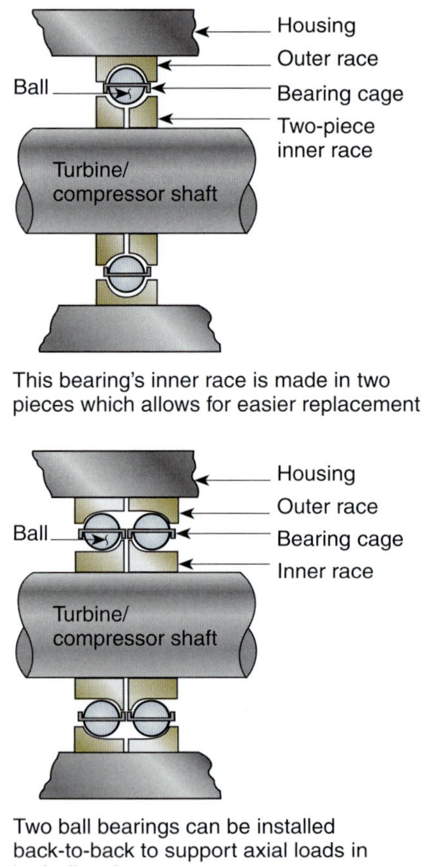

Figure 11-5. *Ball bearings can support both axial and radial loads, and the balls are normally spaced around the races in a thin sheet-metal cage.*

Seals

High rotational speeds and high temperatures inside a gas turbine engine require different seals from those used in reciprocating engines. Two types of commonly used seals are carbon seals and labyrinth seals.

Carbon Seals

Carbon seals are used for applications where an absolute minimum of leakage is allowed. These are rubbing-type seals in which contact is maintained between the stationary carbon element and a polished steel surface, or race.

circumferential coil spring (garter spring). A coil spring formed into a ring.

The seal in Figure 11-6 is a typical ring-type, or circumferential, seal. A highly polished, hardened steel seal race rotates with the shaft, and carbon segments held in a seal housing rub against the race. A circumferential, or garter-type, coil spring encircles the seal and holds the segments in contact with the race.

A face-type carbon seal, shown in Figure 11-7, seals by pressing the carbon against a polished steel race mounted on and rotating with the shaft. Its polished surface is perpendicular to the shaft, and the carbon seal element, mounted in a seal holder, is pressed firmly against the race by a series of coil springs.

Labyrinth Seals

labyrinth seal. A type of nonrubbing air and/or oil seal used around the main-shaft bearings in a gas turbine engine.

Labyrinth seals are non-rubbing seals, because the rotating portion of the seal does not perform its sealing action by pressing against the stationary race. Labyrinth seals consist of a series of knife edges that come very near, but do not touch the fixed land. Figure 11-8 shows a cross-sectional view of a labyrinth seal.

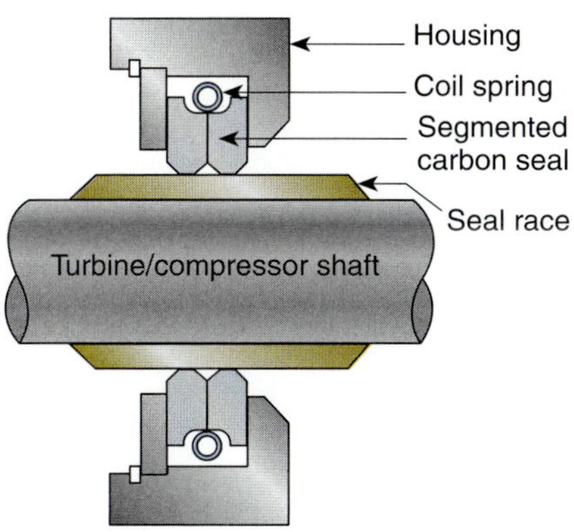

Figure 11-6. *Ring-type carbon oil seal*

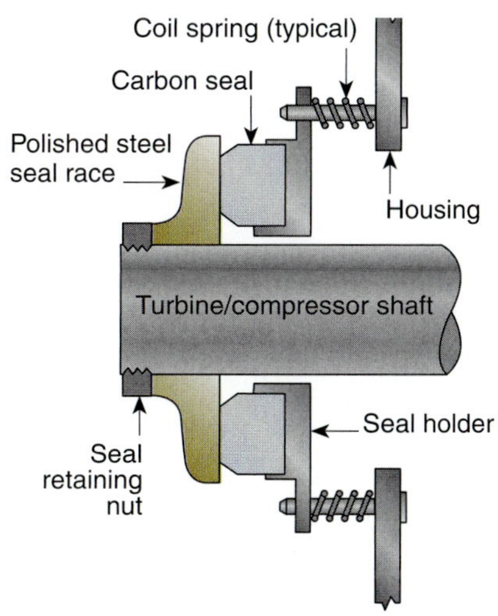

Figure 11-7. *Face-type carbon oil seal*

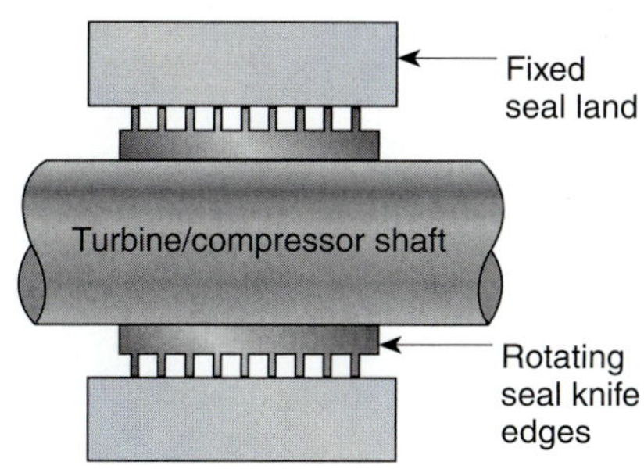

Figure 11-8. *Labyrinth-type seal*

Labyrinth seals may be placed on both sides of a bearing sump, as shown in Figure 11-9, to prevent the loss of lubricating oil into the engine airstream. Pressurizing air bled from one of the compressor stages flows into the outer chamber of the seal. Some of this air leaks through both sets of labyrinth-type seals; the remainder flows out of the compartment through the overboard drain.

Bearing Lubrication

The bearing in Figure 11-9 is lubricated by a spray of oil from a jet, or nozzle. This nozzle incorporates a calibrated orifice that assures the correct amount of oil is supplied to the bearing at all engine operating speeds.

After lubricating the bearing, the oil drains out of the inner compartment and is returned to the oil tank by a scavenger pump. This pump has a capacity considerably greater than the amount of oil used to lubricate the bearing, and air is pulled from the bearing chamber along with the oil. The low pressure in the inner chamber and the higher pressure in the outer chamber causes a small flow of air through the labyrinth seal. This inward flow of air prevents any oil flowing outward through the oil seal.

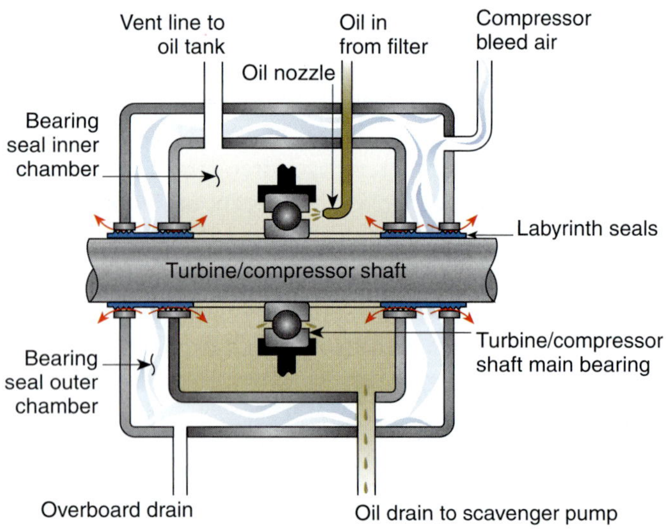

Figure 11-9. *Labyrinth-type oil and air seals around a turbine-shaft main bearing*

Lubrication System Components

The components described in this portion of the *Aviation Maintenance Technician Series* are generic in nature and typical of those used in gas turbine engines. However, some components are unique to a specific engine and are identified by the name of the engine.

Oil Tanks

The oil tanks used with a turbine engine are normally mounted either on the engine or near it. These tanks are required by Federal Aviation Regulations to have an oiltight filler cap and an expansion space of 10% of the tank capacity. The tank must be designed so that it is impossible to inadvertently fill the expansion space.

Figure 11-10 shows a representative oil tank that contains most of the features typically found.

deaerator. A component in a turbine engine lubrication system that removes air from the scavenged oil before it is returned to the tank.

cavitating. The creation of a low pressure in an oil pump when the inlet system is not able to supply all of the oil the pump requires.

The oil returned to this tank by the scavenger pumps contains a large amount of air, and it enters the tank through the deaerator in the oil return inlet where it is swirled to release as much of the air as possible. This air is used to pressurize the tank to ensure a positive supply of oil to the inlet of the main pressure pump to prevent the pump cavitating. A pressuring valve maintains the pressure in the tank at approximately 4 psi.

In normal operation, the oil level is above the horizontal baffle, and the swivel assembly on the oil pickup tube is submerged in oil that contains no entrained air. The two flapper doors in the horizontal baffle are normally open, but if any abrupt maneuver should attempt to force oil out of the lower chamber, the flapper doors automatically close to prevent oil from flowing away from the pickup tube.

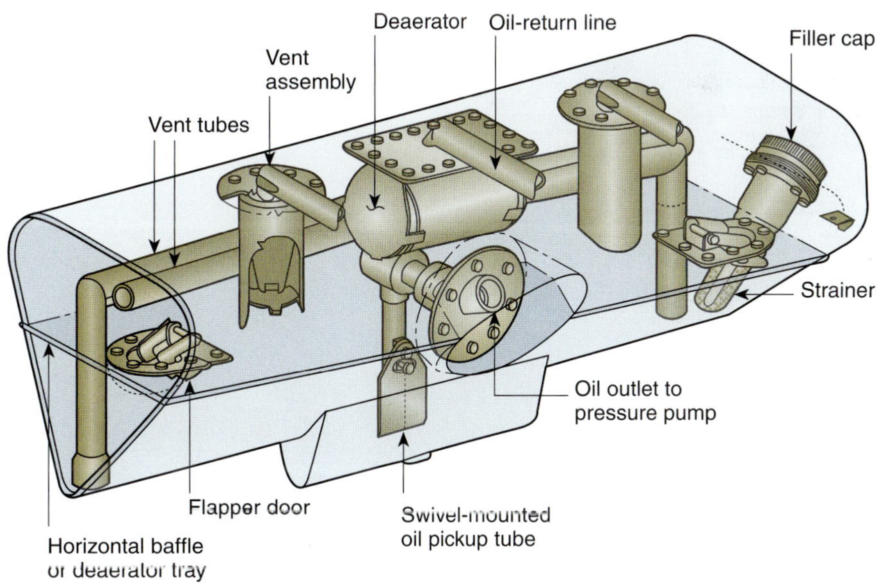

Figure 11-10. *Representative turbine engine oil tank*

The two vent tubes and the vent assemblies with their check valves ensure that, in any maneuver, the space above the oil will always be vented, and oil cannot flow out of the vent lines.

The popular Pratt & Whitney of Canada PT6 turboprop engine uses as its oil tank a part of the engine between the accessory gear box and the compressor inlet section. While the oil tank is physically inside the engine, the lubrication system is of the dry-sump type. The oil pump and the oil screen are both mounted inside the oil tank, with the strainer being serviceable from the outside.

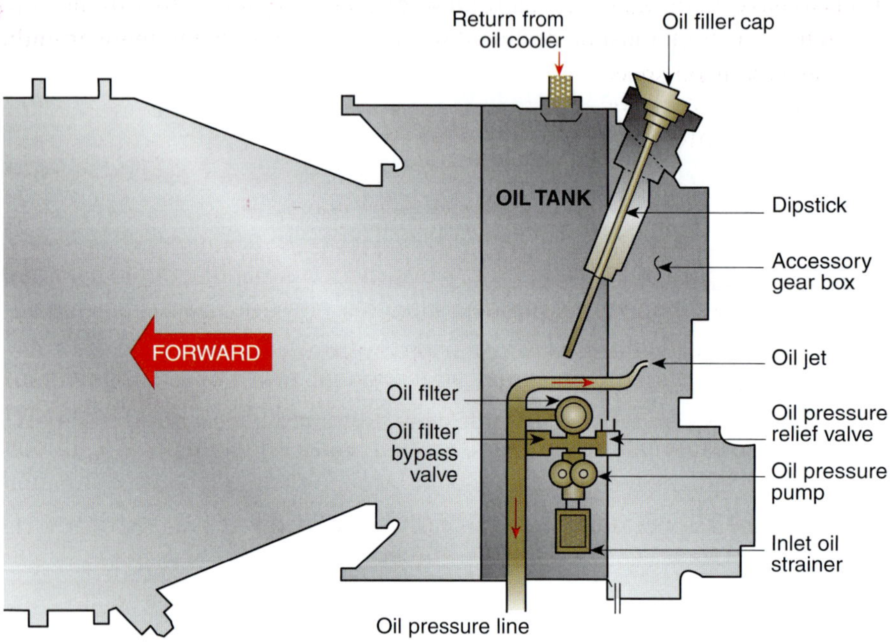

Figure 11-11. *The oil tank of the Pratt & Whitney of Canada PT6 turboprop engine is built into the engine itself.*

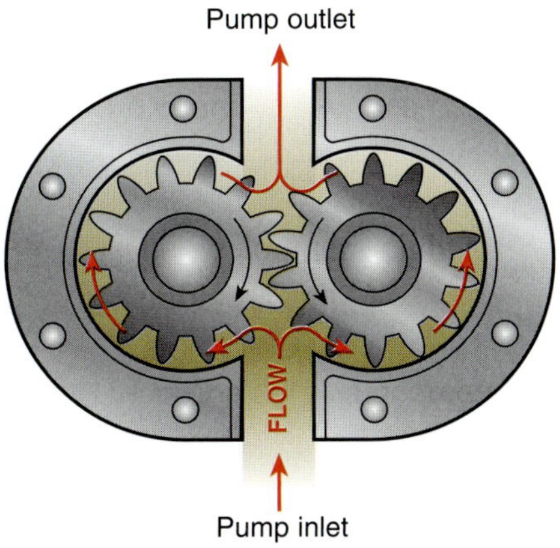

Figure 11-12. *Spur-gear pumps are often used in turbine-engine lubrication systems for both pressure and scavenger functions. Oil picked up on the inlet side, as the teeth become unmeshed, is carried around the housing in the spaces between the teeth and discharged as the teeth come into mesh.*

Oil Pumps

The oil pumps used in a turbine engine lubricating system are all positive-displacement pumps because they move a specific quantity of oil each time they rotate. There are two basic functions of these pumps in a gas turbine engine: pressure pumps produce oil pressure to lubricate bearings and gears, and scavenger pumps pick up the oil after it has performed its duties and return it to the oil tank.

The most generally used pressure pumps are spur-gear and gerotor types. Scavenger pumps with their high-volume and low-pressure requirements may be of the spur-gear, gerotor, or vane type.

It is common practice to use several pump sections in a single housing and to drive them all with the same drive shaft. Figure 11-15 shows a typical spur-gear pump that has one stage of pressure pump and one stage of scavenger pump. Notice that the gears for the scavenger pump are much larger than those for the pressure pump. This is because the hot oil returning to the tank has expanded and contains a great deal of air. Another interesting feature of this pump is the necked-down shear section of the drive shaft. If the gears should bind, the shaft will shear at this weakened point rather than force the pump to continue to turn, which could cause serious damage to the pump or the engine.

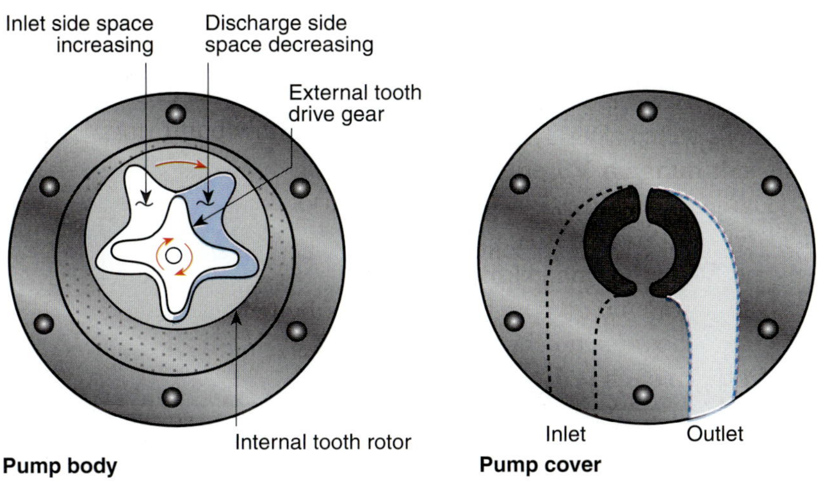

Figure 11-13. *Gerotor pumps pick up oil as the space between the internal-tooth-driven-rotor gear and the external-tooth drive gear becomes larger and discharge it as the space becomes smaller.*

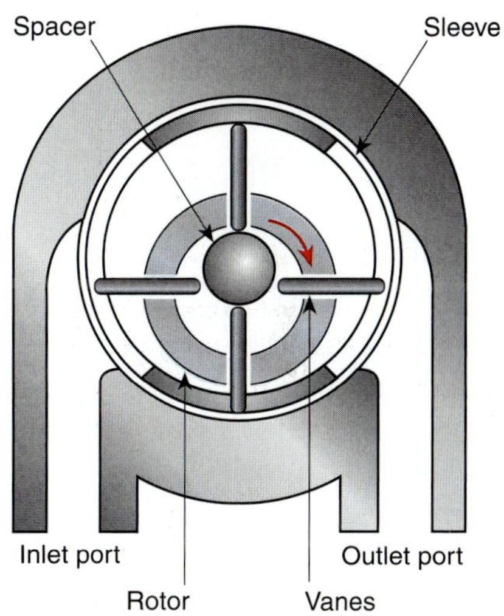

Figure 11-14. *Vane-type pumps are sometimes used as scavenger pumps. They have a high volume and produce a relatively low pressure.*

positive-displacement pump. A fluid pump that moves a specific volume of fluid each time it rotates.

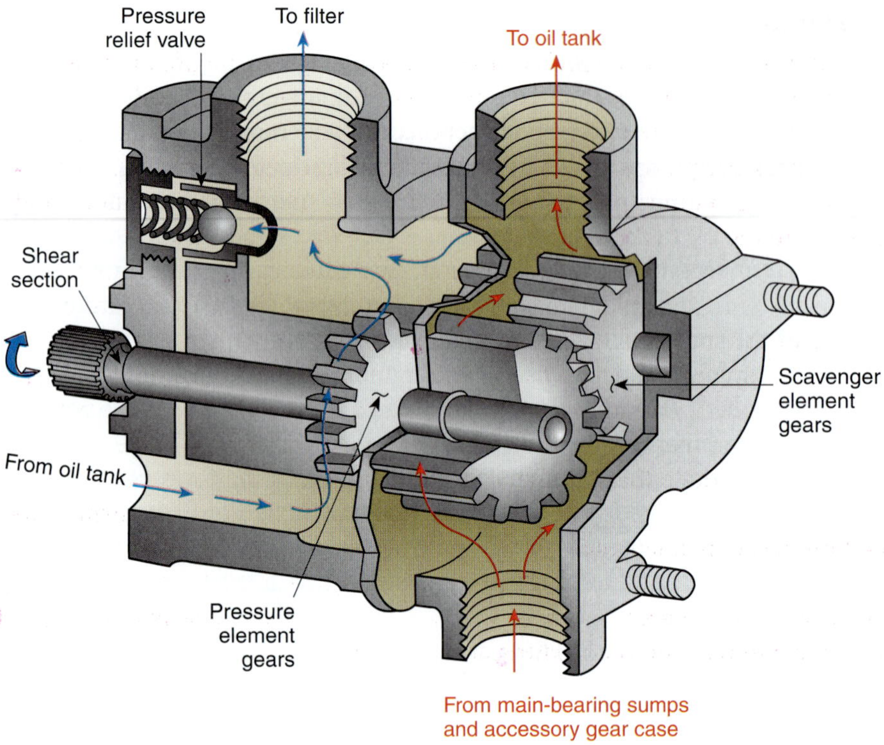

Figure 11-15. *A typical two-stage turbine-engine oil pump. The scavenger element is much larger than the pressure element, because the scavenged oil is hot and contains air.*

Oil-Pressure Relief Valves

All oil pumps used on turbine engines are of the positive-displacement type, and as a result, require a pressure relief valve to maintain constant output pressure as the engine speed changes.

A typical relief valve is in the discharge side of the pump and is spring loaded to lift off its seat when the pressure of the oil is in excess of the valve setting. The oil that passes through the valve returns to the inlet side of the pump. *See* Figure 11-16.

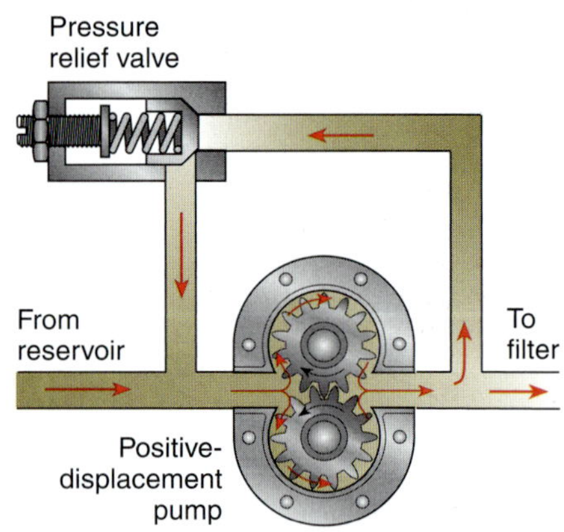

Figure 11-16. *A typical oil pressure relief valve maintains the pump output pressure at a constant value as the speed of the pump changes.*

Oil Filters

It is extremely important that the oil circulating through a gas turbine engine be kept as clean as possible. To do this, the oil is filtered after it leaves the pressure pump and again just before it sprays out from the jet nozzles.

The effectiveness of an oil filter is measured in microns, with one micron being one millionth of a meter, or approximately 39 millionths of an inch (0.000 039 inch). To visualize the effectiveness of a filter, the normal unaided human eye can detect an object having a diameter of approximately 40 microns; a typical human hair has a diameter of about 100 microns.

Three popular types of filters used in turbine engine lubrication systems are wire-mesh screen filters, screen-disk filters, and pleated-fiber filters.

A wire-mesh screen such as the one in Figure 11-17 is able to remove contaminants larger than 40 microns.

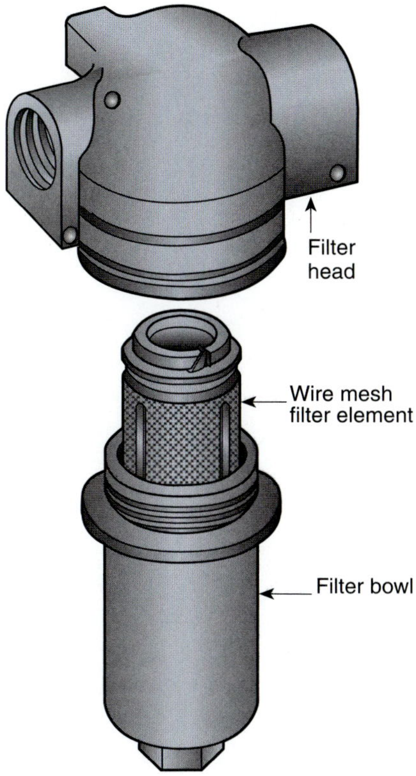

Figure 11-17. *A wire-mesh filter is usually capable of removing contaminants larger than 40 microns.*

micron. A micro meter, or one millionth of a meter. One micron is 0.000 039 inch.

The screen-disk filter in Figure 11-18 is made in the form of a series of wire-screen disks stacked on a perforated tube inside the filter bowl. Oil enters from the outside of the disks and flows through the screen which traps the contaminants, and the filtered oil leaves through the tube in the center of the disks.

The pleated resin-impregnated fiber filter, as shown in Figure 11-19, can normally remove contaminants in the 15-micron range.

Some engine oil filters have provisions for bypassing the filter element if it should clog. For example, if the element in the filter in Figure 11-20 should clog, the ball-type bypass valve will be forced off its seat, and unfiltered oil will flow through the engine.

If cold oil is too viscous to flow through the filter, it unseats the ball and flows through the bypass valve until it warms up and its viscosity decreases enough for it to flow through the filter. The bypass valve then closes, and the filtering action becomes normal.

Other filters are designed with sufficient capacity to furnish enough filtered oil for the engine to operate satisfactorily when the filter is partially clogged. These filters have a red indicator button on the housing, which pops up to inform the maintenance technician that the filter is partially clogged so that the appropriate maintenance action can be taken.

Most metal wire-mesh filters can be cleaned by soaking the elements in a solvent and blowing them out with compressed air. Some of the pleated fiber filters are cleaned by plugging the openings in both ends of the filter with rubber plugs and placing the element in a special cleaning machine with the specified amount of the proper solvent. Cleaning machines like the one shown in Figure 11-21 apply just the correct amount of vibration to loosen the contaminants. After the element has been in the machine for the specified length of time, it is removed and allowed to dry without being blown with compressed air.

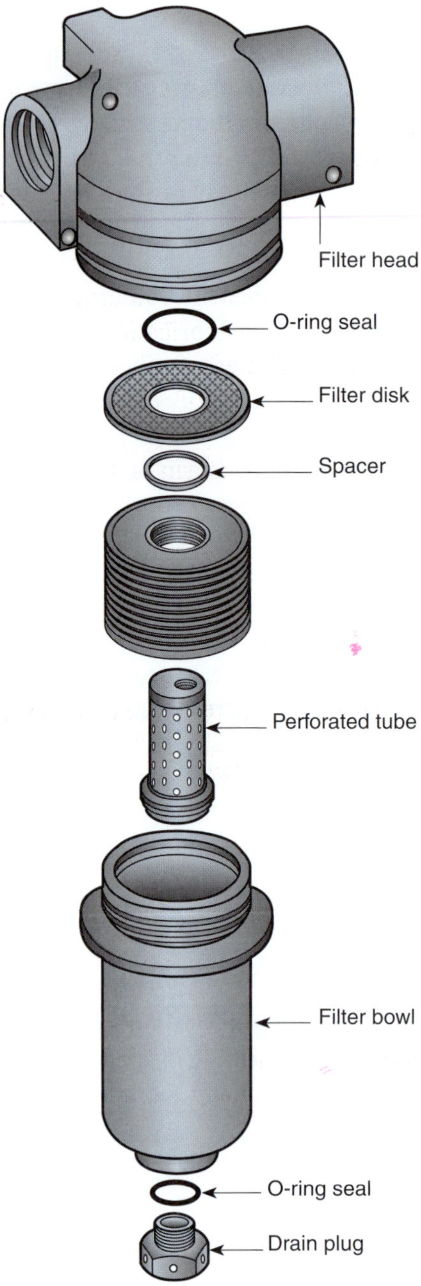

Figure 11-18. *Disk-type oil filter*

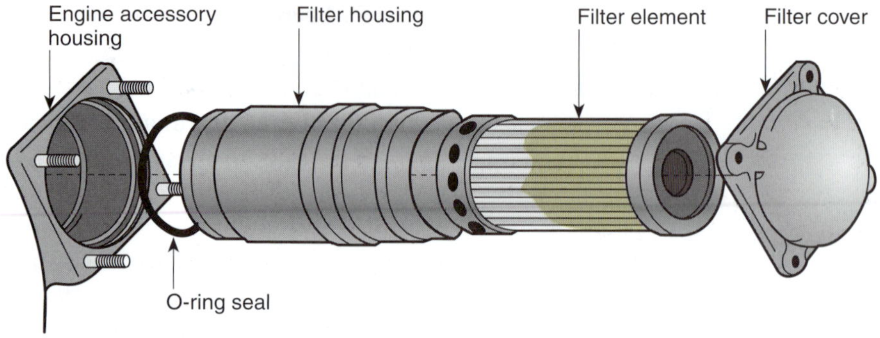

Figure 11-19. *Pleated resin-impregnated fiber oil filter*

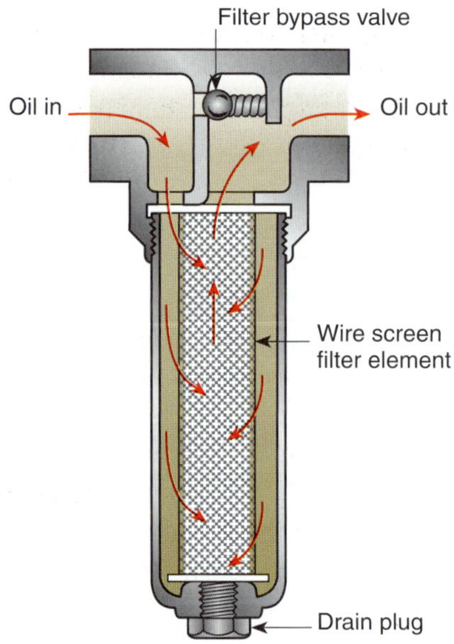

Figure 11-20. *If the oil is too viscous, or if the filter element becomes clogged, the oil will offseat the bypass valve and flow directly into the engine oil passages.*

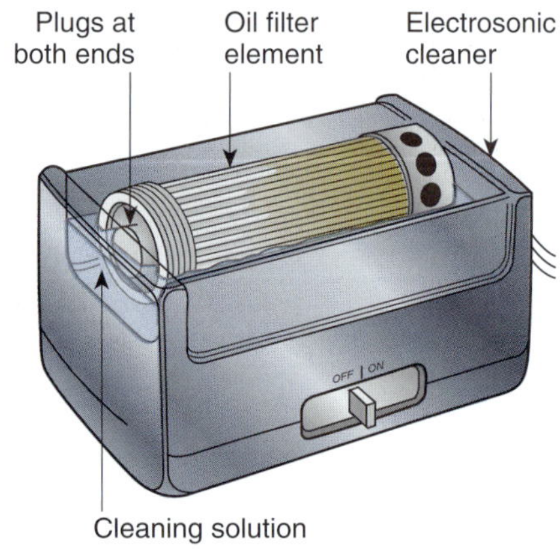

Figure 11-21. *Electrosonic oil filter cleaner used for certain types of pleated filter elements*

last-chance oil filter. A small filter installed in the oil line to the bearing jet in a gas turbine engine.

Last-Chance Filters

To ensure that the bearings receive only clean oil, many engines have screen-type in-line filters installed just before the oil jets. These are often called last-chance filters and they can be cleaned only when the engine is disassembled for overhaul.

Oil Coolers

The lubrication system of a gas turbine engine picks up a great deal of heat from the turbine shaft bearings, and this heat may be transferred into either the air surrounding the aircraft or into the fuel.

Most of the early oil coolers were of the oil-to-air type, similar in operation to those used on reciprocating engines, but almost all modern coolers are of the oil-to-fuel type. This type of cooler serves a dual purpose: it removes heat from the oil and uses it to warm the fuel, preventing the formation of ice crystals.

Figure 11-22 shows a typical oil-to-fuel heat exchanger. The fuel flows into the cooler from the left side and through a series of passages and leaves from its right side. The oil enters the temperature control valve housing and

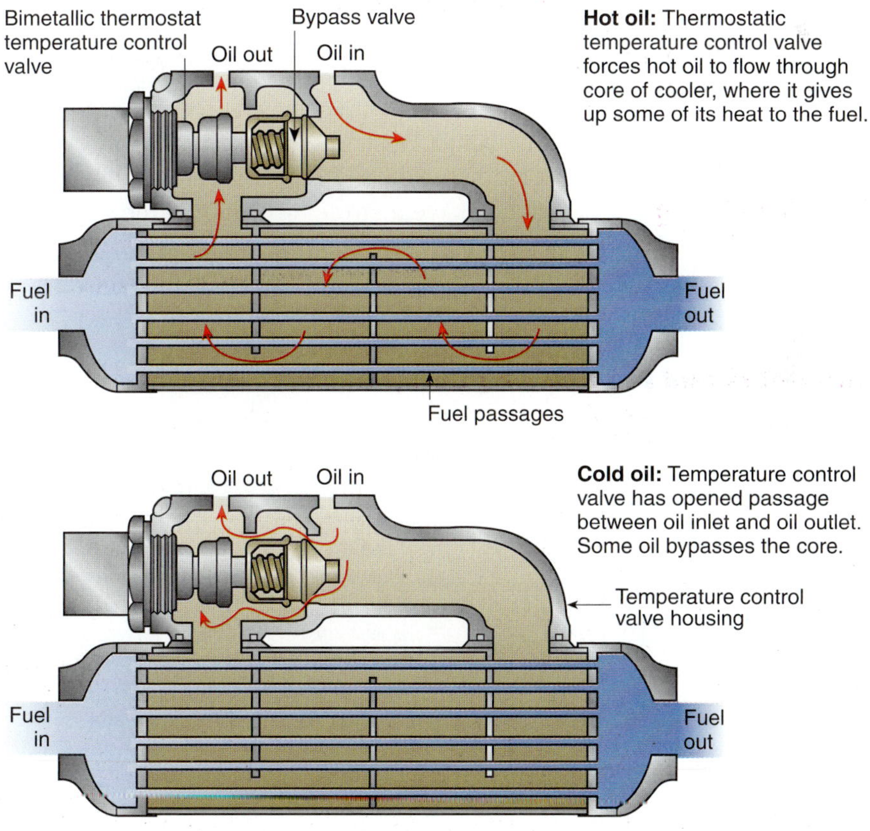

Figure 11-22. *Typical oil-to-fuel heat exchanger*

flows through the cooler, passing around the fuel passages four times. It leaves the cooler and flows across a bimetallic thermostat valve. If the oil is cooler than the valve calls for, the valve moves to the left and allows some of the oil to bypass the cooler and flow directly to the outlet. In operation, the valve assumes a position that maintains the correct oil temperature. Heat from the oil transferred into the fuel warms it enough to prevent the formation of ice crystals on the fuel filter element.

The oil cooler in Figure 11-22 has a spring-loaded bypass valve that remains seated for normal operation, but if the oil in the cooler should congeal and block the flow, the valve will move off its seat and allow the oil to bypass the cooler until it warms up and resumes its normal flow.

Air-Oil Separator

Since the oil picked up by the scavenger pumps from the bearing compartments contains a large amount of air, it is returned to the tank through a deaerator that swirls the oil to separate the air. *See* Figure 11-10 on Page 434. The returned oil flows into the deaerator tray, which forms a dwell chamber in which any remaining air will separate before the oil is picked up by the pressure pump.

Some engines have a gear-driven air-oil separator that removes oil from the vent air taken from the bearing compartments. This oil-laden air is mechanically slung outward to the walls of the separator chamber where the oil collects and drains back into the sump. The oil-free air then goes to the pressurizing and vent valve.

Other engines use a flow-through type separator that contains a series of Teflon ribbons. These ribbons have a strong affinity for the oil, and as the air and oil vapors are forced through the separator, the oil collects on the ribbons and is returned to the engine sump. The oil-free air flows out the vent line.

Breathers and Pressurizing Components

The vent subsystem in some engines uses a pressurizing valve to maintain the air in the bearing compartments and gear boxes at the same pressure as the air above the oil in the oil tank. A system of this type is seen in Figure 11-23.

After lubricating the bearing (1), the oil drains down and collects in the sump (2) and is returned to the tank by the scavenger pump (3). This pump has a large capacity and pulls not only all of the oil from the bearing compartment, but air as well. This air returns to the tank with the oil and is removed by the deaerator (4) and used to pressurize the tank. This air pressure assures a positive flow of oil to the inlet of the main pressure pump (8) and prevents the pump cavitating.

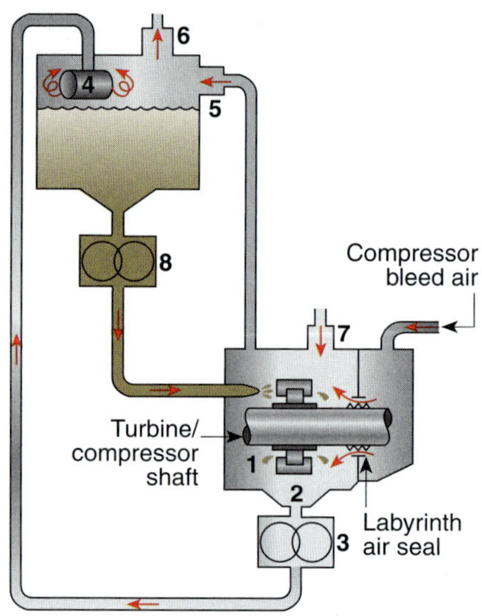

1. Bearing
2. Sump
3. Scavenger pump
4. Deaerator
5. Vent line & check valve
6. Pressurizing valve
7. Vacuum valve
8. Pressure pump

Figure 11-23. *Pressurizing system that maintains the same air pressure in the bearing compartment as that in the air above the oil in the oil tank*

The bearing compartment is vented to the air space above the oil in the tank through the vent line and check valve (5). The tank is also vented to the ambient air through the spring-loaded pressurizing valve (6), which maintains the pressure of the air in the tank at approximately three or four psi. If this pressure rises too high, this valve offseats and vents the excess air.

If the scavenger pump lowers the pressure in the bearing compartment too much, air is drawn into the system through the vacuum valve (7).

Chip Detector

chip detector. A component in a lubrication system that attracts and holds ferrous metal chips circulating with the engine oil.

Magnetic chip detectors are installed in the scavenger systems of many turbine engines to attract and hold ferrous metal flakes or chips that may have been picked up by the oil and circulating through the system.

The simplest chip detectors are easily removed so they can be inspected for contaminants. Many detectors are connected into an electrical circuit that illuminates a warning light on the instrument panel when metal particles bridge the gap in the detector.

Lubrication System Instrumentation

Because of the importance of a turbine engine lubrication system, it is imperative that certain conditions are monitored and the information displayed for the flight crew. The following is a discussion of these conditions; the way this information is displayed is discussed in Chapter 16.

Oil Pressure

It is important to recognize that in a turbine engine lubrication system, it is the pressure of the oil at the oil jet, or nozzle, that determines the amount of oil sprayed into the bearings. In bearing compartments that are pressurized to prevent the loss of oil through the seals, the air pressure in the compartment influences the amount of oil that leaves the nozzle. In engines where this condition exists, the oil pressure indication is actually a differential pressure, the difference between the pressure produced by the oil pump and the pressure of the air in the vent system.

Low-Pressure Warning Light

EICAS (engine indication and crew alerting system). An electronic instrument system installed in modern turbine-powered aircraft that senses engine parameters and displays them on one of two multicolor display units on the instrument panel.

In addition to the oil pressure gage on the instrument panel, or on the EICAS display, many turbine-powered aircraft also have an oil-pressure warning light that illuminates if the pressure ever drops to a predetermined low value.

Oil Temperature

The high rate of circulation allows the oil temperature to stabilize. For this reason the temperature is measured in some engines in the pressure subsystem as the oil leaves the main filter, and in other engines, in the scavenger subsystem just before it returns to the tank.

Oil Quantity

The quantity of oil in the tank of large aircraft is measured electrically and displayed on the flight engineer's instrument panel, or in aircraft that do not have a flight engineer station, on the EICAS display.

Many large engines have provisions for filling the tank remotely by using a service cart with a hand pump. These tanks are filled until there is an indication of oil flowing from the overflow line.

The quantity of the oil in the smaller engine tanks is determined by a dipstick, similar to the one in Figure 11-24, or by a sight gage on the side of the oil tank. The dipstick is mounted on the oil tank filler cap. To determine the quantity of oil, remove the cap, wipe all of the oil from the dipstick, and reinstall it. Then remove it and check to see the amount of the stick that is wet with the oil. Markings on a dipstick as in Figure 11-24 are opposite to those on an automobile engine dipstick. These marks indicate the number of U.S. quarts of oil needed to fill the tank to its maximum level, not the amount of oil in the tank. Notice that the maximum level of the oil when it is cold is at the one quart mark, but when it is hot, it has expanded until it is even with the small circle between the notations for MAX COLD and MAX HOT. The numbers beside the illustration of the dipstick show the relationship between U.S. quarts, Imperial quarts, and liters.

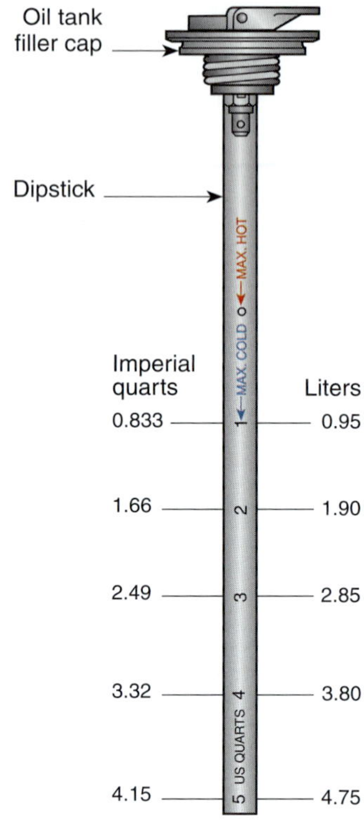

Figure 11-24. *Typical oil quantity dipstick used in a turbine engine oil tank*

dipstick. A quantity-indicating gage made in the form of a thin metal rod.

annunciator panel. A panel of warning lights visible to the flight crew.

Filter Bypass Warning Light

As mentioned in the section on filters, some filters have a button that pops up to warn the technician that the filter element is clogging. Some aircraft have a pressure-actuated electrical switch that senses the pressure on both sides of the filter. If the element should begin to clog, and the pressure drop across the filter rises to a predetermined level, a warning light on an annunciator panel will illuminate.

Lubrication System Servicing

Servicing the lubrication system of a gas turbine engine is an important maintenance task. Different engines have different requirements, and it is important, before attempting any servicing, to be totally familiar with the instructions for that engine and its installation in the particular aircraft.

Use only the oil specified in the operations or service manual, and fill the tank to the level appropriate for the existing oil temperature. It is always a good policy to service the lubrication system as soon as possible after the engine has been shut down. The reason is that when the engine sits idle for a time, some oil from the tank may seep down into the engine. Then when the tank is filled to the correct level, there will be too much oil in the system.

If, when checking the oil before starting the engine, the level is below the normal "add oil" mark, motor the engine with the starter with the ignition off to allow the scavenger system to return the oil to the tank. Afterwards check the oil level and service it accordingly.

The number of hours a turbine engine can operate between oil changes is much greater than for a reciprocating engine. In fact, the operators of some large engines do not have a specified change interval, but rather depend upon effective filter servicing and the normal replenishing of the oil to take the place of a scheduled oil replacement.

Oil Analysis

SOAP (spectrometric oil analysis program). A computer based oil analysis program that measures the number of parts of various chemical element per million parts of oil.

The discussion of reciprocating engine lubrication systems mentions that a very useful maintenance tool is the spectrometric oil analysis program (SOAP). The principle of this program is discussed in Chapter 3 beginning on Page 121.

The oil analysis program for a turbine engine consists of the same two areas used for reciprocating engines: spectrometric analysis of the oil and an evaluation of the contents of the filter element.

The laboratories used for the oil analysis program should be approved by the engine manufacturer. This assures recognition of any abnormal growth trends of a particular metal in the oil. The kit furnished by the lab includes containers for the oil taken from the oil tank and from the filter element, instructions for taking the samples, and forms for recording the results of the tests.

Normally, the sample of oil should be taken shortly after the engine has been run. A tube is inserted into the oil tank to get a sample of oil from the middle of the tank, and this oil is placed in the sample bottle furnished in the kit. The filter is back-flushed to remove entrapped metal particles, and any that are found are examined to determine where they came from.

The sample sent to the laboratory must be identified with the type and serial number of the aircraft and engine, the number of hours on the filter since the last oil change, the number of hours since the last sample was taken, and the amount of oil added since the last sample. This information allows the laboratory to make a meaningful analysis.

A normal engine should show a slow but steady growth of metal in the oil as hours of engine operation build up. At each oil change, the metal contents drop and the growth trend begins at a slightly higher rate. Any abrupt increase of metals in the oil indicates abnormal wear of a component. The engine manufacturer's service representative can suggest the appropriate action to take.

STUDY QUESTIONS: TURBINE ENGINE LUBRICATION SYSTEMS

Answers are on Page 452. Page numbers refer to chapter text.

1. Turbine engines carry a ____________________ (larger or smaller) quantity of oil than reciprocating engines. *Page 425*

2. Most of the heat absorbed by oil in a turbine engine lubricating system comes from the ______________________________ . *Page 425*

3. Five functions of the oil in a turbine engine lubrication system are:
   a. ______________________________
   b. ______________________________
   c. ______________________________
   d. ______________________________
   e. ______________________________
   Page 425

4. Most turbine engine oil has a ____________________ (mineral or synthetic) base. *Page 426*

5. Turbine engine oil has a ________________ (lower or higher) viscosity than that used in a reciprocating engine. *Page 426*

6. The viscosity of a turbine engine oil is expressed in ______________________ . *Page 426*

7. Synthetic oils ____________________ (are or are not) interchangeable with straight mineral oils. *Page 426*

8. A turbine engine lubrication system that carries all of the oil in the engine itself is a ______________ (wet or dry)-sump system. *Page 427*

9. The most popular type of lubrication system used on modern gas turbine engines is the _____________ (wet or dry)-sump system. *Page 427*

10. In a hot-tank, dry-sump lubrication system, the oil cooler is located in the ______________________ (pressure or scavenger) subsystem. *Page 428*

11. The component that maintains the oil pump pressure at the correct value is the ____________________ valve. *Page 428*

12. If the oil filter should clog, the __________________ valve will unseat and allow unfiltered oil to flow through the system. *Page 428*

Continued

STUDY QUESTIONS: TURBINE ENGINE LUBRICATION SYSTEMS *Continued*

13. After lubricating the bearings, oil is returned to the tank by the ____________________ pumps. *Page 428*

14. In a cold-tank, dry-sump lubrication system, the oil cooler is located in the ____________________ (pressure or scavenger) subsystem. *Page 429*

15. The three basic subsystems of a turbine engine lubrication system are:
    a. ____________________
    b. ____________________
    c. ____________________
    Page 430

16. Two types of bearings that are used to support the turbine/compressor shaft in a turbine engine are:
    a. ____________________
    b. ____________________
    Page 430

17. The rollers used in roller bearings ______________ (are or are not) normally mounted in cages. *Page 430*

18. Ball bearings _________________ (can or cannot) be used to support a thrust load. *Page 431*

19. Oil-damped roller bearings accomplish two functions. These are:
    a. ____________________
    b. ____________________
    Page 431

20. The balls used in ball bearings _________________ (are or are not) normally mounted in cages. *Page 431*

21. Two types of air and oil seals used in a turbine engine are:
    a. ____________________
    b. ____________________
    Page 432

22. The type of seal that allows the minimum amount of leakage is the ____________________ (carbon or labyrinth) seal. *Page 432*

23. A labyrinth seal _________________ (does or does not) seal by rubbing against its race. *Page 432*

24. Turbine engine main bearings are lubricated by oil sprayed out through a ______________ or ____________. *Page 433*

25. Oil returned to the oil tank by the scavenger pump contains a large amount of air. This air is removed by a/an ______________________ . *Page 434*

26. The air returned to the oil tank with the scavenged oil is used to __________________________ the oil tank. *Page 434*

27. Turbine engine oil tanks are pressurized to prevent the pressure pump from _____________________ . *Page 434*

28. The two most generally used types of oil pressure pumps are:
    a. _____________________________
    b. _____________________________
    Page 436

29. A scavenger pump has a ________________ (larger or smaller) capacity than a pressure pump. *Page 436*

30. Vane-type pumps are sometimes used as __________________ (pressure or scavenger) pumps. *Page 436*

31. Spur-gear, gerotor, and vane-type pumps are all ___________________ (positive or variable) -displacement pumps. *Page 436*

32. An oil pressure relief valve is installed in __________________________ (series or parallel) with the pump. *Page 437*

33. The weakened section of the drive shaft of an oil pump designed to break if the pump should seize is called a/an ____________________ section. *Page 436*

34. The effectiveness of oil filters is measured in terms of _________________________ . *Page 438*

35. One micron is one millionth of a/an ______________ . This is approximately 0.000 039 inch. *Page 438*

36. If the oil is too viscous to flow through an oil filter, the ____________________ valve will offseat and allow unfiltered oil to flow through the system. *Page 438*

37. If a technician finds an oil filter with the red indicator button popped out, it means that the filter is partially ___________________ . *Page 438*

38. Last-chance, or bearing, filters are cleaned when the engine is given a/an __________________ (overhaul or 100-hour inspection). *Page 440*

Continued

STUDY QUESTIONS: TURBINE ENGINE LUBRICATION SYSTEMS *Continued*

39. Two types of oil coolers used on a turbine engine are:
    a. ______________________________
    b. ______________________________
    Page 440

40. The temperature control in a turbine engine lubrication system is normally a thermostatic valve inside or in parallel with the ________________________ . *Pages 440, 441*

41. The oil-tank pressurizing system uses air returned to the tank with the oil by the ______________________ pump to maintain a positive air pressure above the oil in the tank. *Page 441*

42. An early warning of metal particle contamination in the lubricating oil is provided to the flight crew by an indicator light actuated by a/an _____________________ . *Page 442*

43. The oil pressure indication that relates to the flow of oil available at the bearing is a differential pressure. This pressure is the difference between these two pressures:
    a. ______________________________
    b. ______________________________
    Page 442

44. In a turbine-engine lubrication system, there is normally a ______________________ (large or small) difference between the temperature of the oil leaving the main filter and the oil entering the tank. *Page 442*

45. Three types of oil quantity indicators used on turbine engines are:
    a. ______________________________
    b. ______________________________
    c. ______________________________
    Page 443

46. If a turbine engine lubrication system is not serviced soon after it is shut down, it is possible that the oil level in the tank will _____________________ (rise or lower). *Page 443*

47. Two areas of an oil analysis program for turbine engines are:
    a. ____
    b. ____
    Page 444

48. The oil sample taken for oil analysis should be taken from the __________________ (top, bottom, or middle) of the oil in the tank. *Page 444*

Turbine Engine Cooling Systems

As with other systems in heat engines, the requirements of the cooling systems differ greatly between a reciprocating engine and a gas turbine engine.

The minimum amount of air is taken into the cylinder of a reciprocating engine to mix with the fuel and form a combustible mixture. When the fuel burns, it releases its heat energy and about a third of the energy is converted into useful work. The rest is wasted, much of it carried away by the exhaust system. The remainder is removed by air flowing through the oil cooler or between the fins on the cylinders, or by a liquid coolant flowing around the cylinder walls.

In a turbine engine, fuel is burned continually, and it releases so much heat that it can destroy the engine if not removed.

Of the huge volume of air flowing through a turbine engine, only about 25% is actually involved in the combustion process, with the remainder used for cooling. The cooling air flows through louvered slots or holes in the combustion chamber liners and forms a high-velocity film on either side of the metal of which the liners are made. This air carries the heat away before it can be absorbed into the metal.

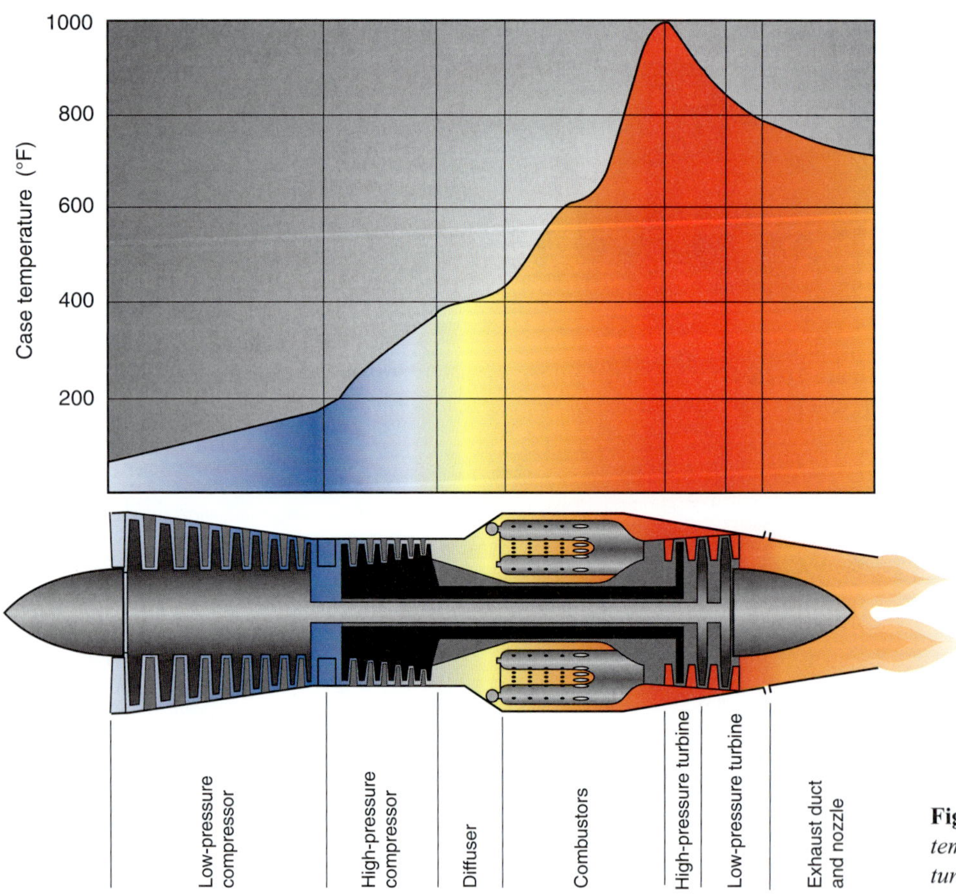

Figure 11-25. *Typical outer-case temperature distribution in a dual-spool turbojet engine*

Figure 11-25 on the previous page shows the temperature distribution along the outer case of a representative dual-spool turbojet engine. The compression of the air raises its temperature as it passes through the low- and high-pressure compressors, and the temperature rises precipitously when the fuel-air mixture is burned in the combustors. Once burning is completed, the hot concentrated gases flow through the turbine inlet guide vanes. This is the point at which the temperature is at its highest. The turbines absorb energy from the hot stream of gas, and the gas temperature drops rapidly. After leaving the last stage of turbine, the hot gases are concentrated into the thrust-producing jet.

Turbine engine installations are designed to prevent the hot engine case from causing a fire in the aircraft structure. One way of keeping the temperature within the engine pod low enough to prevent damage is by routing cooling air from the front of the engine or the fan discharge into both the forward compartment, or cold section, and the aft compartment, or hot section.

After passing around the cold-section case, the cooling air leaves the engine through exit ducts, and the air used to cool the hot section exits through a discharge around the exhaust nozzle. *See* Figure 11-26.

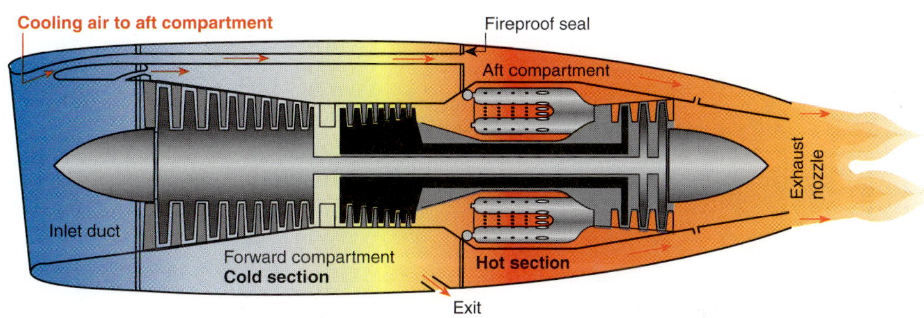

Figure 11-26. *Cool air flows between the engine case and the outer skin of the pod to prevent the high temperatures of the case from damaging the pod.*

Turbine-engine designers work under conflicting requirements. The engine parts must be kept cool enough to avoid destruction, but at the same time, the amount of power the engine can produce is determined by the air temperature as it passes through the turbine inlet guide vanes. The higher this temperature, the more power. One way of allowing the turbine inlet temperature to increase without destroying the engine is to direct compressor bleed air through hollow inlet guide vanes and hollow first-stage rotor blades. Figure 10-72 on Page 413 shows how cooling air flowing through the hollow turbine blades forms an air film that insulates the metal of the blades from the hot gases.

The air used for cooling the blades and guide vanes is bled from one of the higher stages of the compressor, and while its temperature is high, it is far cooler than the gases flowing through the turbine. After passing through the hollow vanes and blades, this air exits the engine with the exhaust gases.

It is necessary to cool only the turbine inlet guide vanes and the first stage turbine blades, because the gases lose enough energy when passing the first stage that their temperature falls within the range allowed for the stages that follow.

STUDY QUESTIONS: TURBINE ENGINE COOLING SYSTEMS

Answers are on Page 452. Page numbers refer to chapter text.

49. About 75% of the air that passes through a gas turbine engine is used for ____________________ . *Page 449*

50. The burning gases inside the combustors are prevented from contact with the metal by a film of ____________________ . *Page 449*

51. The hottest point in a gas turbine engine is found at the ______________________________ . *Page 450*

52. The turbine blades in some high-performance turbine engines are hollow and are cooled by a flow of ______________________ . *Page 451*

Answers to Chapter 11 Study Questions

1. smaller
2. turbine shaft bearings
3. a. reduces friction
   b. absorbs heat from the bearings
   c. protects against corrosion
   d. picks up and carries contaminants into the filters
   e. warms fuel to prevent formation of ice crystals
4. synthetic
5. lower
6. centistokes
7. are not
8. wet
9. dry
10. pressure
11. pressure relief
12. filter bypass
13. scavenger
14. scavenger
15. a. pressure
    b. scavenge
    c. vent
16. a. roller
    b. ball
17. are not
18. can
19. a. compensate for slight misalignment between shaft and housing
    b. cushion vibrations in the rotor assembly
20. are
21. a. carbon seals
    b. labyrinth seals
22. carbon
23. does not
24. jet, nozzle
25. deaerator
26. pressurize
27. cavitating
28. a. spur gear type
    b. gerotor type
29. larger
30. scavenger
31. positive
32. parallel
33. shear
34. microns
35. meter
36. bypass
37. clogged
38. overhaul
39. a. oil-to-air
    b. oil-to-fuel
40. oil cooler
41. scavenger
42. chip detector
43. a. pressure produced by oil pump
    b. air pressure in the vent system
44. small
45. a. dipstick
    b. sight gage
    c. electrical oil quantity indicator
46. lower
47. a. spectrometric oil analysis
    b. evaluation of the filter element contents
48. middle
49. cooling
50. air
51. turbine inlet guide vanes
52. compressor bleed air

Fuel Metering Systems

12

Fuel Metering Systems

12

Turbine Engine Fuels

Aviation gasoline reached its peak of development and production by the end of World War II. After the war, reciprocating engine-powered airliners required large volumes of aviation gasoline, or avgas. However, since turbine engines have almost completely replaced reciprocating engines for military and airline flying, the demand for avgas has dropped so precipitously that the petroleum industry is considering its phase-out. As the demand for avgas decreased, the demand for turbine engine fuel, or jet fuel, increased.

jet fuel. Fuel designed and produced to be used in aircraft gas turbine engines.

In the early days of turbine engine development, kerosine was used as the principal fuel, but the military services—the main users of turbine engines—needed a fuel that had better cold-weather starting and air-starting capability. This brought about blended fuels. Today there are two basic types of jet fuel: kerosine-type and blended.

kerosine. A light, almost colorless hydrocarbon liquid obtained from crude oil by the process of fractional distillation.

The grade numbers used for reciprocating engine fuels relate to their characteristics, but the identification numbers or letters used for turbine engine fuels have no relationship to the fuel's characteristics.

Jet A and A-1 are similar to commercial kerosine and have characteristics similar to those of military JP-8. Their flash points are between 110°F and 150°F. The difference between Jet A and Jet A-1 is their freezing points. Solid crystals begin to form in Jet A at -40°F and at -58°F in Jet A-1.

flash point. The temperature to which a liquid must be raised for it to ignite, but not continue to burn, when a flame is passed above it.

Jet B is called a wide-cut fuel, because it is a blend of gasoline and kerosine fractions, and it is similar to military JP-4. Jet B has a freezing point of around -60°F.

freezing point. The temperature at which solids, such as wax crystals, separate from a hydrocarbon fuel as it is cooled.

Kerosine has approximately 18,500 Btu of heat energy per pound, and it nominally weighs 6.7 pounds per gallon. Aviation gasoline has 20,000 Btu per pound, and its nominal weight is 6 pounds per gallon. Therefore, while kerosine has a lower heat energy per pound, it has a higher heat energy per gallon than gasoline.

Many aviation gas turbine engine manufacturers allow some aviation gasoline to be used in their engines when turbine fuel is not available. The amount of time aviation gasoline can be used is limited for two reasons: the TEL, or tetraethyl lead, in the aviation gasoline causes deposits to form on the turbine blades, and aviation gasoline does not have the lubricating properties of kerosine. The use of too much gasoline can cause excessive wear on the fuel control.

Jet Fuel Volatility

volatility. The characteristic of a liquid that relates to its ability to vaporize, or change into a gas.

Volatility of jet fuel is important because it is a compromise between conflicting factors. Its volatility should be high enough for good cold-weather starting and aerial restarting, but low enough to prevent vapor lock and reduce fuel losses by evaporation.

Under normal temperature conditions, gasoline, with its 7-psi vapor pressure, gives off so much vapor in a closed container or tank that the fuel-air mixture is too rich to burn. Under these same conditions, the vapor given off by Jet B with its 2- to 3-psi vapor pressure may produce a fuel-air mixture that is explosive. Jet A, with its extremely low vapor pressure of around 0.125 psi, has such a low volatility that under normal conditions it does not give off enough vapor to form an explosive fuel-air mixture.

Jet Fuel Viscosity

viscosity. The resistance of a fluid to flow.

sump. The low point in a fuel tank in which water or contaminants collect and are held until they can be drained out.

Jet fuel is more viscous than gasoline, and because of this, it holds contaminants in suspension and slows their settling out in the tank sumps. Water is held in the fuel in a suspended state, and at high altitude and low temperature, it can condense out, collect on the strainer elements, and freeze.

Microbial Growth in Jet Fuel Tanks

Jet fuel absorbs water from the atmosphere, and the amount absorbed is determined by the humidity of the atmosphere and the temperature of the fuel. As the fuel temperature increases, it can hold more water, and as the temperature decreases, the water in the fuel precipitates out and settles in the bottom of the tanks. Here, it collects around the sealant used to fuel-proof the tank seams. Microscopic organisms, such as microbes and bacteria, live in this water and multiply at the interface between the water and the fuel to form a scum that holds the water in contact with the tank structure. The water acts as an electrolyte, and corrosion forms in the aluminum alloy of which the tank is made.

To prevent the formation of scum in the tanks, jet fuel may be treated with an additive such as ethylene glycol monomethyl ether or proprietary chemicals such as PFA 55MB, or Prist. These additives, which kill the microbes and bacteria, may be mixed into the fuel at the refinery, or injected into the fuel as it is pumped into the aircraft tanks.

Fuel Anti-Icing

When water that has precipitated from the fuel freezes on the fuel filters, it can shut off the flow of fuel to the engine. To prevent this, some aircraft are equipped with fuel heaters, a form of air-to-fuel or oil-to-fuel heat exchanger. Compressor bleed air or hot engine oil flows through one part of the heater, and fuel flows through another. The air or oil gives up some of its heat to the fuel and raises its temperature enough to prevent ice crystals forming on the filter element.

The additive that prevents formation of the microbial scum in the fuel tanks also acts as an anti-icing, or antifreeze, agent. It mixes with the water that condenses out of the fuel and lowers its freezing point so it cannot freeze on the filters.

Fuel Handling

The cleanliness, safety, and fire precautions required for handling any liquid fuel must be observed when handling jet fuel, but additional precautions must be observed due to its unique nature.

Elaborate precautions are taken at fuel storage facilities and on fuel delivery trucks or underground fuel hydrants to protect against water contamination. Also, a fire hazard exists when static electricity is produced as fuel flows at a high rate through the fuel hose. The fuel truck or delivery hose must be grounded, and the aircraft structure must be connected to the fuel hose with a good low-resistance electrical ground wire.

Fueling jet aircraft requires careful attention to details specified by the aircraft manufacturer. Many jet aircraft are fueled from a single-point pressure fueling port, and adjacent to the fueling port is a fueling control panel which contains fuel quantity gages for each tank, fueling valve switches that activate the fueling valves, lights to show the position of the fueling valves, a fueling power switch, and a fuel gage test switch. The maximum permissible fueling pressure and the maximum permissible defueling pressure are marked on a placard at the fueling control panel.

If the selected tank is to be completely filled, the fueling valve will automatically close when the tank is full, but if the tank is to be partially filled, the valve can be closed by the fueling operator when the fuel quantity gage shows the appropriate amount of fuel is in the tank.

When it is necessary to remove fuel from an aircraft tank, the same procedures used for fueling should be followed. The defueling process must never be conducted in a hangar, but should be done in an open area where there is adequate ventilation. The aircraft should be properly grounded to protect against a static electricity buildup. The fuel removed from the aircraft must be protected against contamination and identified so it can be returned to the proper storage facility.

STUDY QUESTIONS: TURBINE ENGINE FUELS

Answers are on Page 484. Page numbers below refer to chapter text.

1. The first fuel used for aviation gas turbine engines was ______________________ . *Page 455*

2. The two basic types of modern turbine engine fuel are:
   a. ______________________________
   b. ______________________________
   Page 455

3. The identification numbers or letters used with a turbine engine fuel ____________________ (do or do not) relate to the fuel's characteristics. *Page 455*

4. The civilian jet fuel that is a kerosine-type fuel is Jet _____________ (A or B). *Page 455*

5. The difference between Jet A and Jet A-1 is their ___________________ (freezing or flash) points. *Page 455*

6. The most widely used military jet fuel is JP- ______________ (4 or 8). *Page 455*

7. One gallon of Jet A fuel has _______________ (more or fewer) Btu of heat energy than one gallon of aviation gasoline. *Page 455*

8. Two reasons for allowing only a limited amount of aviation gasoline to be used in a turbine engine are:
   a. _________________________________
   b. _________________________________
   Page 455

9. The vapor pressure of Jet A is _________________ (higher or lower) than the vapor pressure of aviation gasoline. *Page 456*

10. Jet fuel is ____________________ (more or less) viscous than aviation gasoline. *Page 456*

11. The amount of water absorbed in jet fuel is determined by the temperature of the fuel. Warm fuel will absorb and hold ___________________ (more or less) water than cold fuel. *Page 456*

12. Microscopic organisms that live in water that collects in a jet aircraft fuel tank form a scum that holds the water in contact with the metal of which the tank is made and causes _____________________ . *Page 456*

13. The fuel additive that may be added to jet fuel as it is pumped into the aircraft tank serves two purposes. These are:

    a. ________________________________

    b. ________________________________

    Pages 456, 457

14. The maximum allowable pressure for fueling a jet aircraft is displayed on a placard at the __________________________ . *Page 457*

Turbine Engine Fuel Systems

The function of a turbine engine fuel system is to furnish the correct amount of clean fuel, free from vapor and at the correct pressure, to the fuel control under all operating conditions. Chapter 8 of the *Airframe* textbook discusses the airframe portion of the fuel system for a turbine engine transport aircraft. Figure 12-1 shows the engine portion of the fuel system.

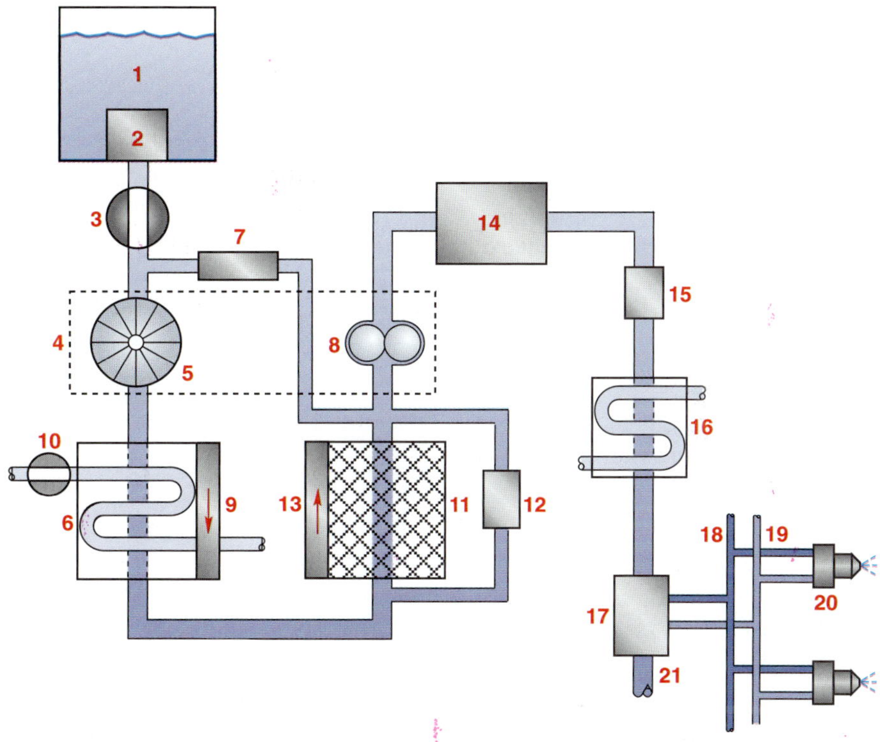

1 Fuel tank
2 Boost pump
3 Engine shutoff valve
4 Main multistage fuel pump
5 Low-pressure pump element
6 Air-fuel heat exchanger
7 Impeller bypass valve
8 High-pressure pump element
9 Heat exchanger bypass valve
10 Bleed air valve
11 Filter
12 Pressure differential switch
13 Filter bypass valve
14 Fuel control
15 Flowmeter transmitter
16 Oil-fuel heat exchanger
17 Flow divider and dump valve
18 Primary fuel manifold
19 Secondary fuel manifold
20 Fuel nozzles
21 Overboard drain

Figure 12-1. *Typical fuel system for a turbofan engine*

The fuel is delivered from the fuel tank (1) by the submerged boost pump (2). It flows through the engine shutoff valve (3) into the main multistage fuel pump (4). It flows first through a low-pressure centrifugal primary-stage pump (5), and then through an air-fuel heat exchanger (6). An external impeller-bypass valve (7) will bypass fuel directly from the boost pump in the fuel tank to the spur-gear-type high-pressure secondary-stage pump (8) if the impeller should block the flow of fuel.

The fuel temperature is maintained high enough by hot compressor bleed air flowing through the air-fuel heat exchanger to prevent the formation of ice crystals. A built-in spring-loaded bypass valve (9) in the heat exchanger will open and allow fuel to flow around the core if the core should, for any reason, become clogged. The temperature of the fuel is displayed on the flight engineer's instrument panel, and the valve (10) that controls the flow of bleed air is electrically actuated from the FE panel.

The fuel flows from the heat exchanger through the filter (11). If the filter should become partially clogged with ice crystals, a pressure-differential switch (12) across the filter element will close and turn on a fuel-ice warning light on the FE panel. The flight engineer can then open the bleed air valve, which allows hot air to flow through the heat exchanger and warm the fuel to melt the ice and prevent further ice formation. If the filter element should clog completely, the fuel will flow around it through a built-in filter-bypass valve (13).

From the filter, the fuel flows through the high-pressure stage (8) of the pump (4), and then into the fuel control (14). The fuel control is mounted on the engine and contains pressure and temperature sensors and the flyweight-type N_2 governor. The fuel shutoff and fuel control levers in the cockpit actuate valves in the fuel control.

N_2. The speed of the high-pressure turbine and compressor in a dual-spool gas turbine engine.

From the fuel control, the fuel passes through the fuel flow transmitter (15) and the oil-fuel heat exchanger (16). From there it goes through the pressurization and dump valve (17). Fuel for starting and low-speed operation flows into the primary manifold (18), and the greater part of the fuel flows through the secondary manifold (19). From the two manifolds, the fuel flows to the nozzles (20), and from there it is sprayed out into the combustors and burned.

When the engine is shut down, the fuel in the manifolds drains back through the dump portion of the pressurization and dump valve and overboard through the drain line (21).

Fuel System Components

A turbine engine fuel system uses some of the same basic types of components as a reciprocating-engine system, but the components are quite different due to the large volume of fuel that must be moved and the extremely low temperatures encountered at the altitudes flown by turbine-powered aircraft.

Fuel Pumps

Submerged centrifugal boost pumps move fuel from the tanks. These pumps are controlled from the flight engineer's panel and are used for three purposes:

1. To pressurize the fuel in the line between the tank and the engine-driven pump, thus ensuring a positive feed of fuel to the pump;
2. To transfer fuel from one tank to another to balance the fuel load; and
3. To pump fuel from the tank into the dump chute when fuel is being dumped.

Engine-driven pumps are normally of the multistage type like that in Figure 12-3. The fuel flows from the tank under boost pump pressure into the centrifugal impeller. The pressure is boosted by this element, and the fuel leaves the pump and flows through the fuel heater and filter and returns to the dual-element, spur-gear-type high-pressure pump.

Notice that both stages of the high-pressure pump are protected by shear sections. If the gears in either section should jam, the shear section will break

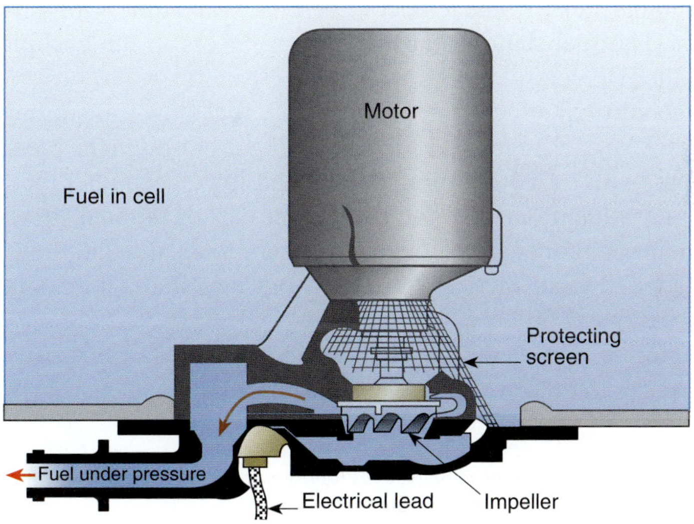

Figure 12-2. *Submerged centrifugal boost pumps move the fuel from the tank to the inlet of the engine-driven pump.*

and disable that section while the other will continue to function. The check valves in the discharge lines from the gears prevent fuel flowing back into an inoperative section.

After the fuel leaves the gears, it flows through the check valves and out to the fuel control. The pressure relief valve in the pump outlet maintains the pump discharge pressure at its set value. If the pressure rises above this value, the valve will offseat, allowing the fuel that caused the excess pressure to flow back to the inlet side of the gears.

Fuel Strainers

Turbine engine fuel controls contain such close-tolerance components that even the smallest contaminants can cause serious problems. For this reason, turbine engine fuel systems often have a microfilter which uses a replaceable cellulose filter element capable of removing foreign particles as small as 10 to 25 microns. To visualize this size, a human hair has a diameter of approximately 100 microns. Such a filter is seen in Figure 12-4 on the next page.

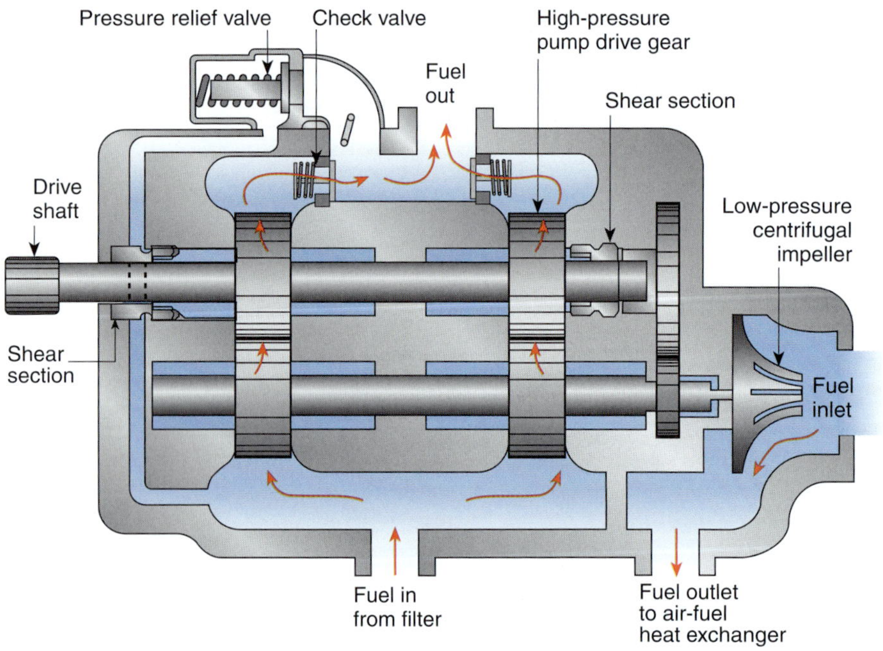

Figure 12-3. *A multistage turbine engine fuel pump. This pump has a centrifugal impeller for the first stage pressure and a dual spur-gear-type high-pressure stage. The pressure relief valve controls the output pressure from the pump.*

Another type of filter widely used for turbine engine fuel systems is the wafer screen filter seen in Figure 12-5. The filtering element consists of a stack of wafer-type screen disks made of a 200-mesh bronze, brass, or stainless steel wire screen. This type of filter has the ability to remove very tiny particles from the fuel, and at the same time, to withstand the high pressures found in a turbine engine fuel system.

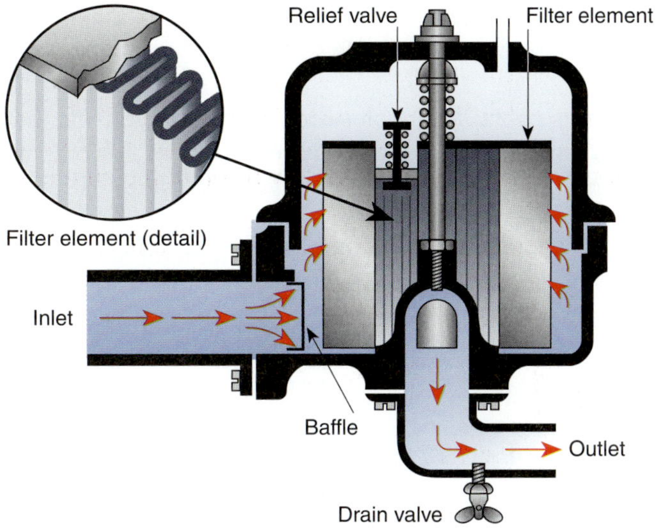

Figure 12-4. *Microfilter used in a turbine engine fuel system*

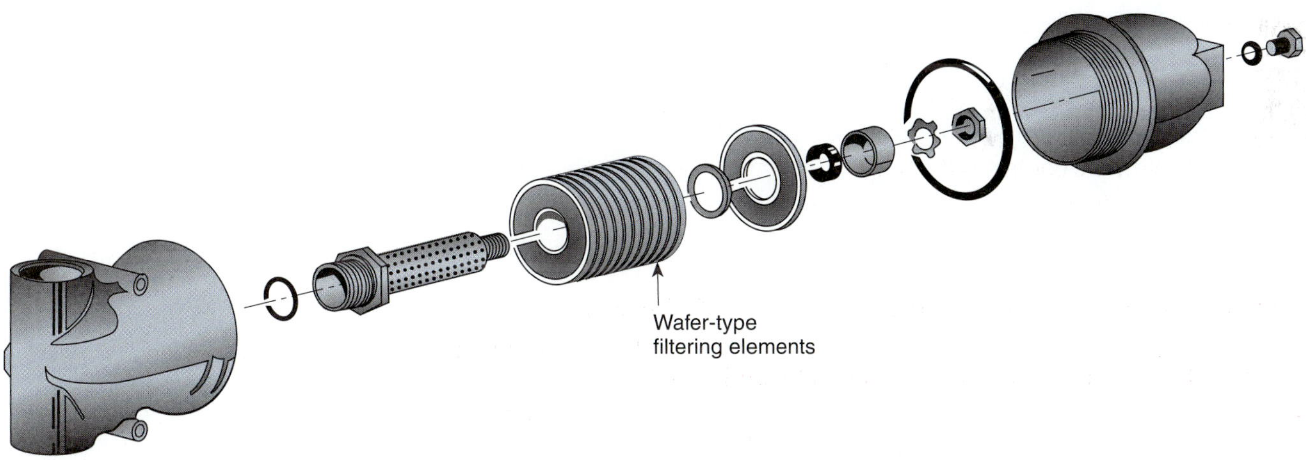

Figure 12-5. *Wafer-type fuel filter*

Fuel Controls

The carburetor or continuous-flow fuel injector system installed on a reciprocating engine gives the pilot direct control over the amount of fuel-air mixture delivered to the engine. The more the throttle is opened, the more air and fuel flow into the engine. On the other hand, the turbine engine is far more demanding of its fuel metering system, and the pilot does not have direct control over the amount of air or fuel the engine receives.

throttle. The control in an aircraft that regulates the power or thrust the pilot wants the engine to produce.

Most turbojet and turbofan engines control the engine RPM to get the desired thrust. The pilot has a throttle control, or more correctly, a power lever or thrust lever, used to program the fuel control so the engine will give the amount of power or thrust needed for the existing flight condition.

The fuel control senses a number of parameters and meters the correct amount of fuel to the burners only when the airflow through the engine is adequate and when other conditions are right.

To best understand the operation of the complete fuel system in Figure 12-1, we will temporarily leave the fuel control and consider the other components in the system. Then, beginning on Page 468, we will return to the discussion of fuel control and consider some of its complexities.

Fuel Flowmeter

After leaving the fuel control, the fuel flows through the fuel flowmeter transmitter. This device senses the mass of the fuel flowing to the nozzles and indicates the number of pounds per hour on the instrument panel. The mechanism used to measure the fuel flow is described in Chapter 16, beginning on Page 563.

Fuel-Oil Heat Exchanger

vaporize. The changing of a liquid into a vapor.

atomize. The process of breaking a liquid down into tiny droplets or a fine spray.

The next component in the typical turbofan fuel system is the fuel-oil heat exchanger. This is of the type described in Chapter 11 on turbine engine lubrication systems, illustrated in Figure 11-22 on Page 440. Fuel flows directly through the tubes in the cooler, and engine oil flows around the tubes. Heat from the oil warms the fuel, and the fuel cools the oil.

Fuel Nozzles

The end point of the turbine engine fuel system is the fuel nozzles. Liquid fuel will not burn, and in order for it to release its energy, it must be vaporized so it will mix with the air to form a combustible mixture. There are two types of nozzles used to discharge the fuel: the widely used atomizing nozzles and the more seldom used vaporizing nozzles. Only the atomizing nozzles are discussed in this text.

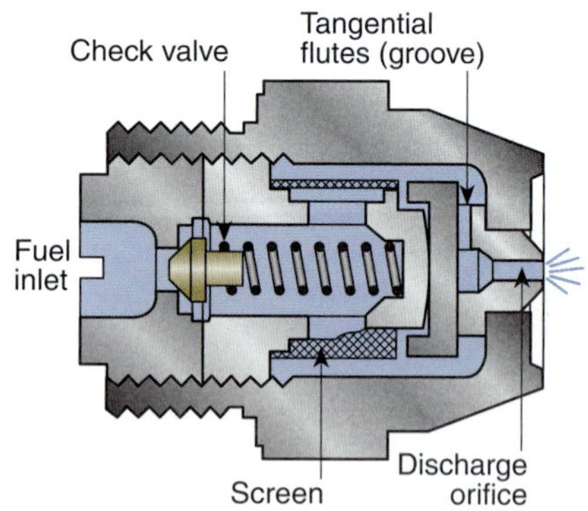

Figure 12-6. *A simplex fuel nozzle*

Two types of atomizing nozzles are simplex and duplex. The simplex nozzle in Figure 12-6 was one of the first successful nozzles. This nozzle screws directly into the fuel manifold inside the combustor, and fuel under pressure from the fuel control forces the check valve off its seat and enters

the nozzle. This fuel then passes through a series of tangential flutes, or grooves, and sprays out through the single discharge orifice in very fine droplets in a cone-shaped pattern. As soon as the engine is shut down and the fuel pressure drops to below the value for which the check valve is set, the check valve closes and shuts off all flow to the discharge orifice. This prevents fuel from dribbling out and continuing to burn.

The basic problems with simplex nozzles are their inadequate atomization and improper spray pattern at low speed and low pressure. They are designed to best operate at high pressures.

Two types of duplex nozzles are used on modern engines: single-line and dual-line nozzles.

Single-line duplex nozzles are supplied with fuel from the fuel control through a single fuel line or manifold. These nozzles incorporate a flow divider valve that allows the fuel to spray from a central orifice in a wide spray pattern for starting and idling. When the fuel control meters sufficient pressure to open the flow divider, fuel flows to the secondary orifice. The large volume of the secondary fuel and the high pressure at which it leaves the nozzle narrow the spray pattern and force the fuel farther down the combustor. *See* Figure 12-7.

Some engines are equipped with dual-line duplex fuel nozzles. These nozzles have separate passages through which the primary and secondary

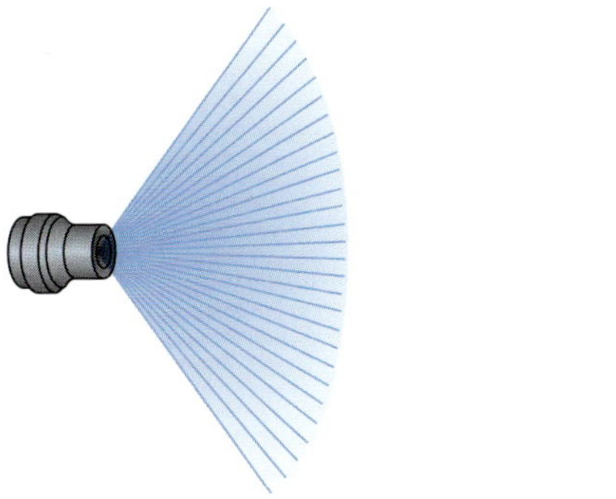

Spray pattern from the primary orifice alone

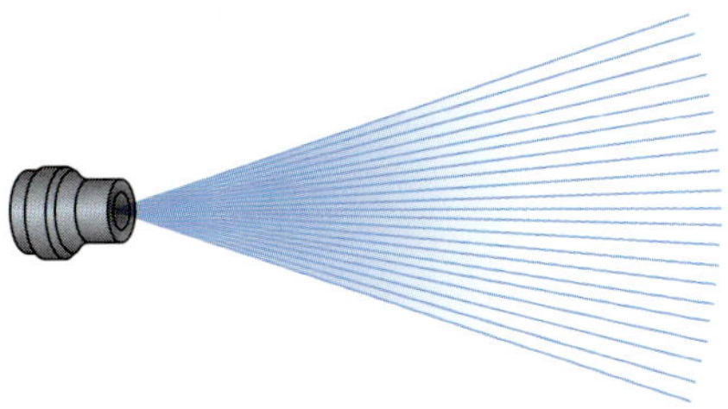

Spray pattern from both the primary and secondary orifices

Figure 12-7. *Spray pattern from a duplex fuel nozzle*

flow divider. A component in a turbine engine fuel system that routes all of the fuel to the primary nozzles or primary orifices when starting the engine or when the RPM is low.

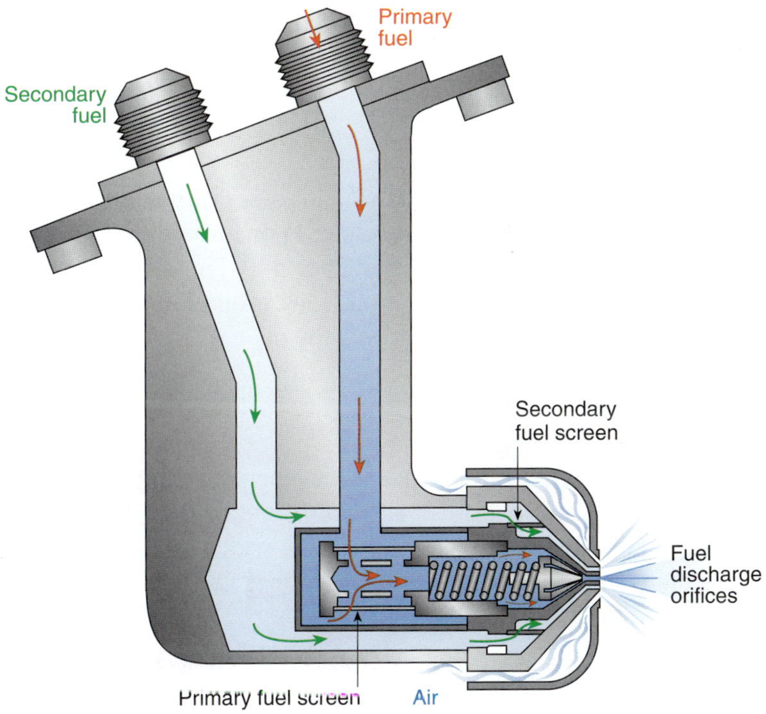

Figure 12-8. *Dual-line duplex fuel nozzle*

fuels flow. For starting and low-flow conditions, only primary fuel comes from the pressurizing valve, and it sprays out from the center primary orifice in a wide spray. *See* Figure 12-8.

Pressurizing and Dump Valve

Dual-line duplex nozzles must have a flow divider between the fuel control and the nozzles. This valve is often called a pressurizing and dump valve. When the engine is being started, the fuel from the fuel control builds up pressure high enough to open the valve to the primary fuel line. This fuel flows into the nozzle and sprays out through the primary discharge orifice in a wide spray pattern. When the engine accelerates and the fuel pressure from the fuel control increases, the spring-loaded pressurizing valve opens farther and allows fuel to also flow out through the secondary fuel line into the nozzle and out through the secondary discharge orifice. This orifice surrounds the primary orifice, and fuel spraying from it narrows the spray pattern and sends it farther down the combustor.

When the engine is shut down and there is no more fuel pressure from the fuel control, a spring-loaded valve inside the pressurization and dump valve opens and allows all the fuel in the lines to the nozzles to drain.

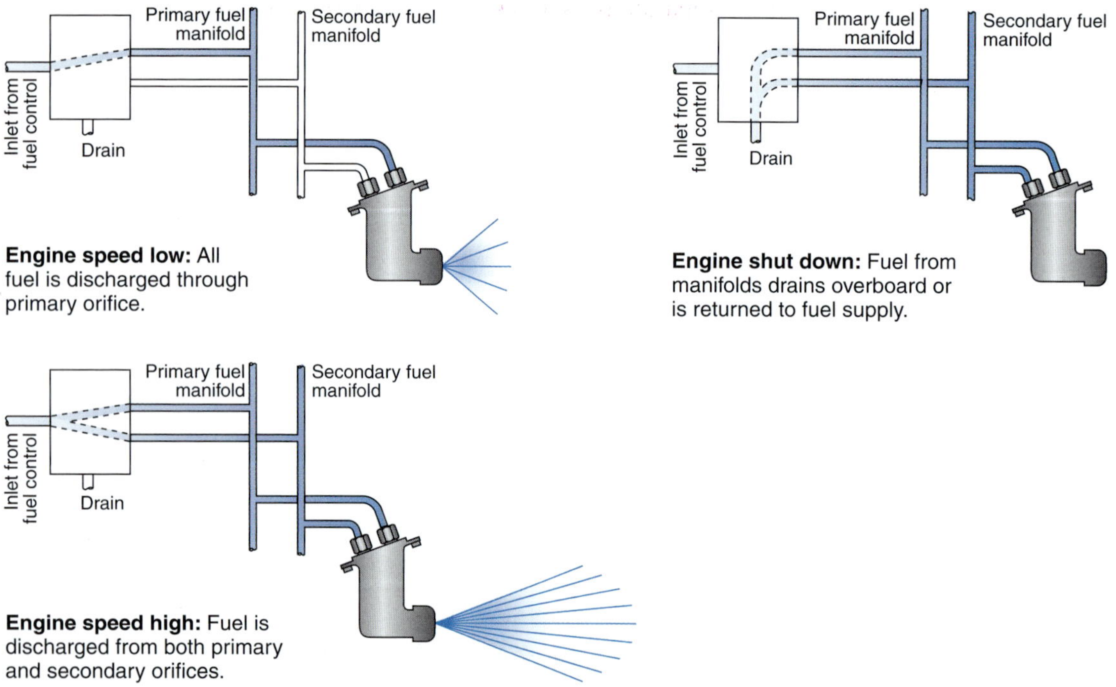

Figure 12-9. *A pressurizing and dump valve acts as a flow divider to send fuel to the secondary discharge line when the fuel pressure is sufficiently high. When the engine is shut down, the fuel in the manifolds drains overboard.*

Turbine Engine Fuel Control

Turbine inlet temperature, or TIT, is the most critical temperature in a gas turbine engine and is normally the factor that limits the amount of thrust an engine can produce.

A turbine engine fuel control is designed to sense a number of parameters and integrate them to meter the fuel to the burners for any set of conditions, so that the engine will produce the desired power or thrust while ensuring the TIT does not exceed its allowable limits.

The fuel control meters fuel to the burners only when the airflow through the engine is adequate and other conditions are right.

The basic parameters that are sensed by a typical fuel control are:

- Power lever angle (PLA)—This is the input by the pilot that specifies the amount of power or thrust desired.
- Compressor inlet total temperature (T_{t2})—This relates to the density of the air entering the engine.
- Compressor RPM (N or N_2)—This is important for controlling the fuel for steady-state operation and for limiting the rate of acceleration and deceleration.
- Burner pressure (Pb)—This relates to the weight of the air flowing through the engine.

Fuel controls for many turbojet and turbofan engines vary the thrust by controlling rotor RPM. On some turbojets, thrust and TIT are allowed to vary inversely with compressor inlet temperature. As the inlet air temperature rises and the air becomes less dense, the compressor load decreases and the RPM increases. The fuel control sends less fuel to the burners, and the thrust and the TIT decrease.

On cold days when the inlet air is dense, the compressor load is high, and additional fuel is metered to bring the compressor up to the desired speed. This increases the thrust and TIT.

The fuel controls used on modern axial-flow engines sense the compressor inlet pressure and the burner pressure and protect the engine from excessive internal pressures, particularly during takeoff on a cold day at low altitude. At the same time they allow the engine to produce as much thrust as possible on hot days by maintaining the TIT at a constant value and by allowing the RPM to vary.

As long as the throttle remains in a certain position, the fuel control will vary fuel flow as compressor inlet conditions change, and the approximate percentage of full engine thrust requested by the pilot will be maintained. The TIT for any given increment of thrust will remain approximately constant.

Maximum thermal efficiency and thrust are obtained when the TIT is held close to its allowable limit. This is done by calculating the TIT that will be produced under various conditions. The fuel control is programmed to vary the fuel flow according to a predetermined schedule that ensures the engine will always operate within safe temperature limits. This complex scheduling is done in most hydromechanical fuel controls by a three-dimensional cam in the fuel control computer.

hydromechanical. Any device that combines fluid pressures with mechanical actions to achieve a desired result.

Combustion temperature varies directly with the fuel flow. When the fuel flow increases, the temperature of the air leaving the burners increases, and this increases the pressure acting on both the turbine and the compressor. This gives the turbine additional heat and pressure energy, but at the same time, back pressure acts on the compressor. The compressor accelerates in spite of this back pressure, and the increased compressor speed forces additional air through the engine, and as a result, the thrust increases. Reducing the fuel flow lowers the combustion temperature and decreases the thrust.

The fuel control automatically changes the fuel flow as compressor inlet conditions change. The fuel flow decreases when the compressor inlet pressure decreases as altitude is gained, and it increases during descent.

To accelerate the engine, energy must be supplied to the turbine in excess of that needed to maintain a constant RPM. However, if the fuel flow is increased too rapidly, an overly rich mixture may result, which will cause excessive TIT or produce a compressor surge. The fire may even go out because of a condition known as a rich blowout.

surge. A condition of unstable airflow through the compressor of a gas turbine engine in which the compressor blades have an excessive angle of attack.

It is also possible to reduce the fuel flow at a rate faster than the compressor can reduce the airflow to the burners. If the fuel flow is reduced too quickly during deceleration, the engine may experience a lean die-out. The fuel control must supply the correct amount of fuel to maintain a fuel-air ratio that will prevent the engine from flaming out during either acceleration or deceleration.

rich blowout. A condition in which the fire in a gas turbine engine goes out because the fuel-air mixture ratio is too rich to sustain combustion.

lean die-out. A condition in which the fire in a gas turbine engine goes out because the fuel-air mixture ratio is too lean to sustain combustion.

Surge is a condition of unstable airflow through the compressor in which the compressor blades have an excessive angle of attack. The susceptibility to surge increases with the low density of high altitude.

At certain compressor speeds and inlet air temperatures, care must be taken not to meter fuel into the engine too rapidly, as it can produce a high burner pressure before the RPM and airflow can increase proportionately. If the burner pressure increases too rapidly, the flow of air through the compressor will slow down, the angle of attack of the compressor blades will increase and the blades will stall, causing the engine to surge. To avoid this, the fuel control must limit acceleration fuel flow.

There are many variations of fuel controls used on gas turbine engines, but they can be divided into two basic groups: hydromechanical and electronic. Most of the basic controls are of the hydromechanical type, with some sensing more engine parameters than others. The large airliners and many of the high-performance military aircraft use electronic controls because of their ability to sense more parameters and accomplish more control functions than is possible with a hydromechanical control.

servo system. A type of automatic control system in which a small input force is used to control a large output force.

three-dimensional cam. A drum-shaped cam in a hydromechanical fuel control whose outer surface is ground in such a way that followers riding on the surface, as the cam is moved up and down and rotated, can move mechanical linkages to control the fuel according to a preprogrammed schedule.

Turbine engine fuel controls are extremely complicated devices. The hydromechanical types contain many components such as speed governors, servo systems, three-dimensional cams, sleeve and pilot valves, feedback or follow-up devices, and metering systems. Electronic fuel controls are a maze of printed circuit boards, thermocouples, amplifiers, relays, electrical servo systems, switches, and solenoids.

In this text, we will not attempt to trace the operation of any particular fuel control, but will consider the basic functions of a typical hydromechanical fuel control and two categories of electronic fuel controls.

Hydromechanical Fuel Control

The basic principle of turbine engine fuel metering is to vary the flow of fuel to the discharge nozzles by maintaining a constant pressure drop across a metering orifice whose area is varied by the pilot's throttle position and by the engine operating conditions.

A basic hydromechanical fuel control consists of these basic components:

- A pump to increase the pressure of the fuel
- A shutoff valve to stop the flow of fuel to the engine
- A relief valve to protect the fuel control when the shutoff valve is closed
- A metering valve to regulate the flow of fuel to the engine
- A minimum-flow adjustment to prevent the metering valve from completely stopping the flow of fuel

In Figure 12-10, the fuel to the fuel control is supplied from the tank (1), by the boost pump (2), through the fuel valve (3). Its pressure is increased by the pump (4), and its volume is controlled by the up or down movement of the throttle valve (5). A shutoff valve (6) stops all flow to the nozzles (7) when the engine is shut down.

This basic fuel control can be improved by adding a pressure relief valve (8) around the pump to limit the amount of pressure the pump can produce when the throttle valve is closed. A minimum flow adjustment (9) ensures that the nozzles will always receive at least the minimum amount of fuel needed to keep the engine operating under idle conditions. *See* Figure 12-11.

Two ways of controlling the amount of fuel delivered to the nozzles are by varying the pressure drop across a fixed-size metering orifice and by maintaining a constant pressure drop across a variable-size orifice. In Figure 12-12 the size of the metering orifice is determined by the position of the throttle valve, which is controlled by the parameters sensed by the fuel control. A differential relief valve (10) is installed across the throttle valve to maintain a constant pressure drop as the size of the orifice changes. *See* Figure 12-12 on the next page.

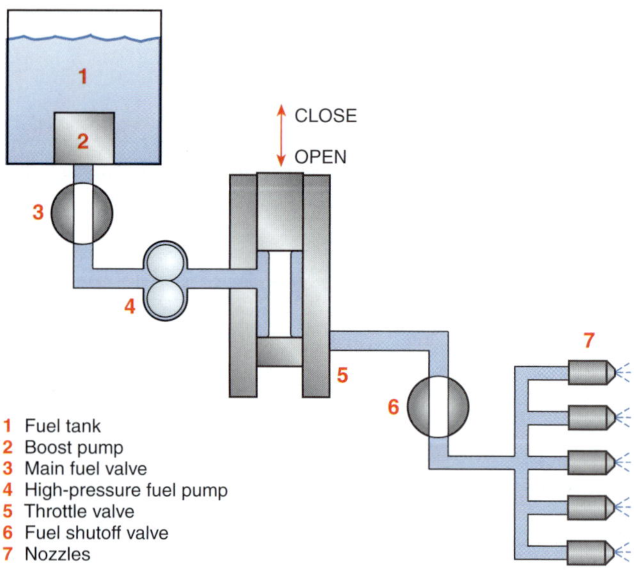

Figure 12-10. *The basic fuel control*

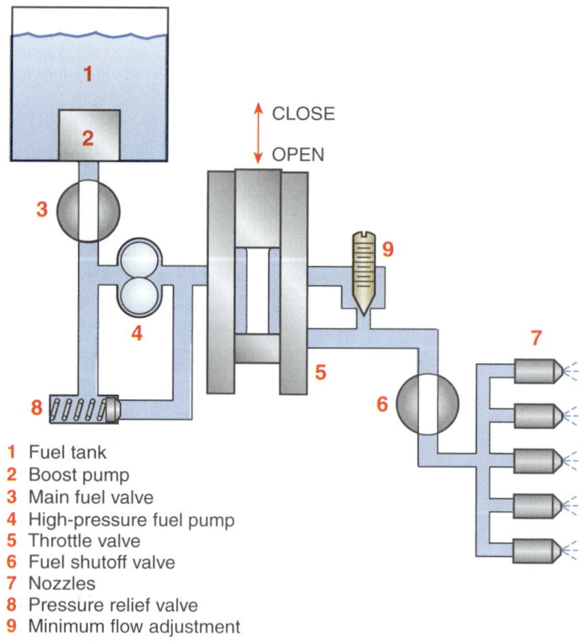

Figure 12-11. *The basic fuel control with the pressure relief valve (8) and minimum flow adjustment (9)*

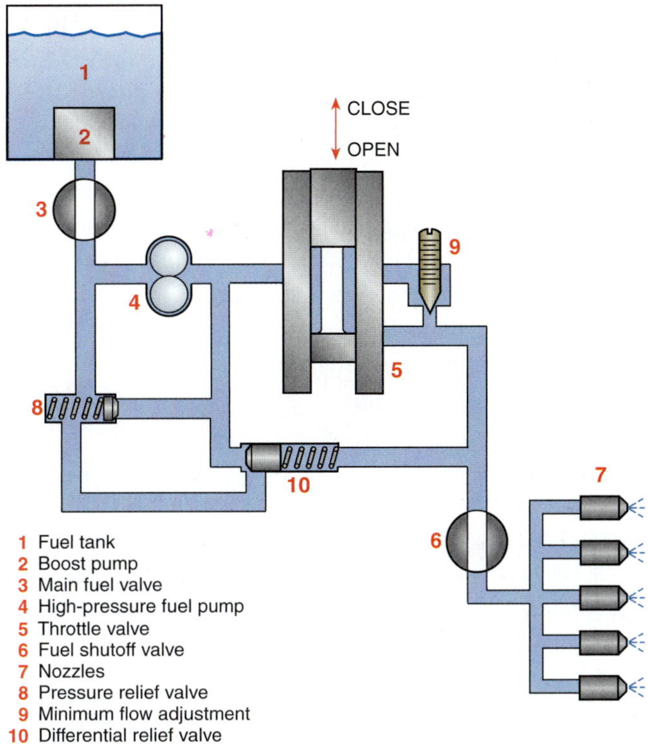

Figure 12-12. *The pressure drop across the throttle valve is held constant by the differential relief valve (10).*

When the throttle valve is almost closed, the flow through the valve is low. The pressure is low on the outlet side and high on the inlet side. The high inlet-side pressure moves the differential relief valve to the right, compressing its spring and allowing some of the fuel to flow back to the inlet side of the pump.

When the throttle valve is opened to increase the flow of fuel to the nozzles, the pressure on the inlet side of the throttle drops, and the spring behind the differential relief valve moves it to the left to restrict the fuel returning to the pump inlet. The inlet-side pressure then rises to maintain a constant pressure drop across the throttle valve.

The basic fuel control senses the pilot's throttle position, the engine RPM, and the burner pressure to control the amount of fuel metered to the nozzles. In Figure 12-13, a flyweight governor (11) has been added to the basic system to control the position of the throttle valve. The flyweights (12) are driven by the engine at a speed proportional to the engine RPM, and the compression of the speeder spring (13) is controlled by the pilot's throttle lever.

A bellows-actuated burner-pressure sensor valve (14) between the discharge side of the throttle valve and the inlet side of the fuel pump senses the difference between the burner pressure and ambient pressure so the fuel pressure at the nozzles can vary as the burner, or compressor discharge, pressure changes.

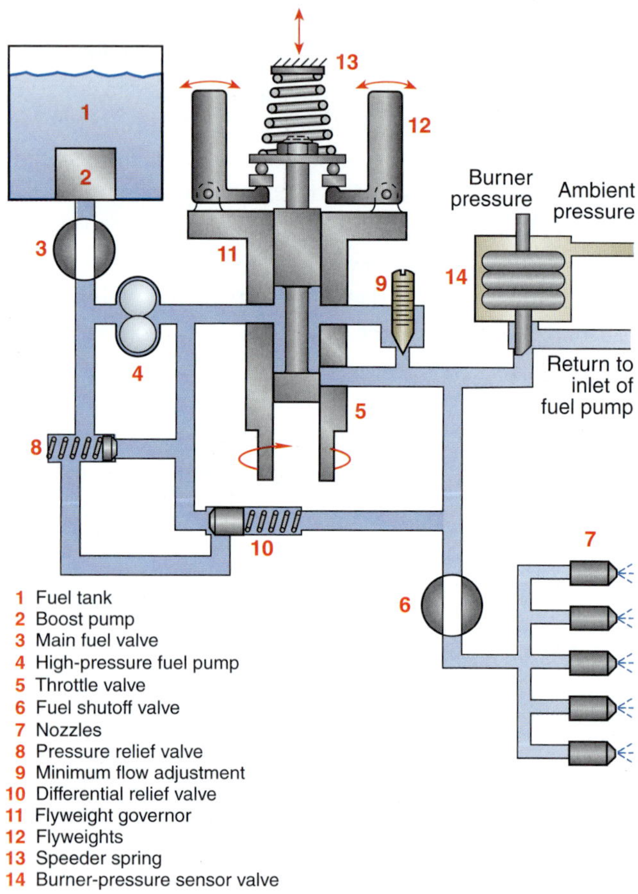

Figure 12-13. *The position of the throttle valve is controlled by the flyweight governor (11) maintaining a balance between the compression force on the speeder spring (13) (controlled by the pilot's throttle) and the centrifugal force acting on the flyweights (12). The difference between burner pressure and ambient air pressure is sensed by the burner-pressure sensor valve (14). This valve controls the pressure downstream of the throttle valve.*

The pilot moves the throttle forward to increase thrust from the engine. This movement increases the compression of the speeder spring. The flyweights move inward, and the throttle valve moves down, sending more fuel to the engine. The engine speeds up, and centrifugal force causes the flyweights to sling outward until they return to their upright, or on-speed, condition. The throttle valve has increased the size of the metering orifice, and the differential relief valve maintains the constant pressure drop across the throttle valve in its new position.

When the fuel flow is initially increased, it is possible that the fuel-air mixture will enrich enough to cause a rich blowout before the engine speeds up enough to move sufficient air through the compressor and provide the correct fuel-air mixture. But this is prevented by the burner-pressure sensor.

Fuel flow through the throttle valve increases before the compressor discharge pressure, or burner pressure, increases enough to prevent a rich blowout. The bellows in the burner-pressure sensor opens the bypass valve, and some of the fuel bypasses back to the inlet side of the pump instead of going to the nozzles. As soon as the engine accelerates to the desired speed, the burner pressure increases enough to shut off the fuel bypass, and all the metered fuel goes to the nozzles.

droop. A progressive decrease in RPM with load in an engine whose speed is governed with a flyweight-type governor.

Almost all hydromechanical fuel controls use some type of flyweight governor to maintain conditions called for by the pilot. These governors have a characteristic known as droop.

When the engine operates under a constant load, the centrifugal force acting on the flyweights is balanced by the force from the compression of the speeder spring. The throttle valve is in a position to meter the correct amount of fuel.

If the load increases, the engine slows down and the flyweights pull inward. This allows the throttle valve to drop down and meter more fuel to the nozzles. The engine speed increases again, but now the throttle valve is more open than it originally was (the valve is farther down), and in this on-speed condition, the flyweights are slightly more inward than they were when there was no load on the engine. The speeder spring exerts a little less force, and the engine speed is slightly lower than when the engine was operating with no load. As the load increases, the stabilized speed progressively decreases; and this condition is called droop.

Emergency Fuel Control

Some hydromechanical fuel controls have provisions for operating the engine in emergency mode should the automatic control features fail. The pilot is still able to control the engine's speed, but because there is no automatic scheduling of the fuel, extreme care must be taken not to change the throttle position too rapidly, as this can cause the engine to flame out.

Turboprop Engine Fuel Control

The hydromechanical fuel control for a turboprop or turboshaft engine receives the pilot's signal for a given level of power. The control then takes certain variables into consideration and adjusts the engine fuel flow to provide the desired power, yet not exceed the RPM and TIT limitations of the engine.

The control system of a turboprop or turboshaft engine has an additional job not shared by its turbojet and turbofan counterparts: it must control the speed of the propeller or the free turbine, and it usually governs the pitch angle of the propeller blades.

More details on the turboprop fuel control are covered in Chapter 19, "Propellers," beginning on Page 693.

Electronic Engine Control Systems

Advances in gas turbine technology have demanded more precise control of engine parameters than can be provided by hydromechanical fuel controls alone. These demands are met by electronic engine controls, or EEC, of which there are two types: supervisory and full-authority.

Supervisory Electronic Engine Control

The first type of EEC is a supervisory control that works with a proven hydromechanical fuel control.

The major components in the supervisory control system include the electronic control itself, the hydromechanical fuel control on the engine, and the bleed air and variable stator vane control. The hydromechanical element controls the basic operation of the engine including starting, acceleration, deceleration, and shutdown. High-pressure rotor speed (N_2), compressor stator vane angles, and engine bleed system are also controlled hydromechanically. The EEC, acting in a supervisory capacity, modulates the engine fuel flow to maintain the designated thrust. The pilot simply moves the throttle lever to a desired thrust setting position, such as full takeoff thrust, or maximum climb. The control adjusts engine pressure ratio (EPR) as required to maintain the thrust rating, compensating for changes in flight and environmental conditions. The control also limits engine operating speed and temperature, ensuring safe operation throughout the flight envelope.

EPR (engine pressure ratio). The ratio of the turbine discharge total pressure to the compressor inlet total pressure.

If a problem develops, control automatically reverts to the hydromechanical system, with no discontinuity in thrust. A warning signal is displayed in the cockpit, but no immediate action is required by the pilot. The pilot can also revert to the hydromechanical control at any time.

Full-Authority Digital Electronic Control (FADEC)

The supervisory control was a step toward the full-authority, fully redundant EEC. It controls all engine functions and eliminates the need for the backup hydromechanical control used in the supervisory system. The modern full-authority EEC is a digital electronic device called a full-authority digital electronic control, or FADEC.

One of the basic purposes of the FADEC is to reduce flight crew workload. This is achieved by the FADEC's control logic, which simplifies power settings for all engine operating conditions. The throttle position is used to achieve consistent engine settings regardless of flight or environmental conditions.

closed-loop control. A type of control in which part of the output is fed back to the input.

ADC (air data computer). An electronic computer in an aircraft that senses pitot pressure, static pressure, and total air temperature.

FADEC (full-authority digital electronic control). A highly precision digital electronic fuel control that functions during all engine operations. It includes the electronic engine control and functions with the flight management computer to meter the fuel to the nozzles in such a way that prevents overshooting of power changes and over-temperature conditions. FADEC furnishes information to the engine indication crew alerting system.

The FADEC establishes engine power through direct closed-loop control of the engine pressure ratio (EPR), which is the thrust-rating parameter. The required EPR is calculated as a function of throttle lever angle, altitude, Mach number, and total air temperature. The air data computer supplies altitude, Mach number, and total air temperature information, and sensors provide measurements of engine temperatures, pressures, and speeds. This data is used to provide automatic thrust control, engine limit protection, transient control, and engine starting.

FADEC uses a preprogrammed schedule to obtain the correct EPR for the various throttle lever angles, and it provides the correct EPR for any chosen angle during changing flight or environmental conditions.

To get the desired thrust, the pilot has only to set the throttle lever to a position which aligns the EPR command from the control with the reference indicator from the aircraft thrust management computer. The control system automatically accelerates or decelerates the engine to the desired EPR level without the pilot having to continually monitor the EPR gage. Once a power setting has been selected, the FADEC maintains it until the throttle lever position is changed.

A constant throttle lever angle setting can be used for takeoff and climb. In addition, since the pilot sets engine thrust according to EPR, and the system controls the EPR by using a given throttle lever angle, the same thrust rating will be obtained on each engine at the same throttle position. This eliminates throttle stagger.

The FADEC has many advantages over both the hydromechanical and supervisory EEC. Some of these are:

- Requires no engine trimming
- Ensures improved engine starts
- Provides a constant idle speed with changes in atmospheric conditions and changing service bleed air requirements
- Saves fuel by providing improved engine bleed air management
- Fully modulates the active clearance control (ACC) system in place of the more conventional stepped modulating system
- Ensures more repeatable engine transients due to the higher precision of its digital computer
- Provides engine limit protection by automatically limiting critical engine pressures and speeds

ACC (active clearance control). A system for controlling the clearance between tips of the compressor and turbine blades and the case of high-performance turbofan engines. Cool fan discharge air shrinks the case around the compressor and turbine.

The FADEC has dual electronic channels, each with its own processor, power supply, program memory, selected input sensors, and output actuators. Power to each electronic control channel is provided by a dedicated, engine-gearbox-driven alternator. This redundancy provides high operational reliability. No single electronic malfunction will cause an engine operational problem. Each control channel incorporates fault identification, isolation, and accommodation logic.

While electronic controls are highly reliable, malfunctions can occur. A hierarchy of fault-tolerance logic will take care of any single or multiple faults. The logic also identifies the controlling channel, and if computational capability is lost in the primary channel, the FADEC automatically switches to the secondary channel. If a sensor is lost in the primary channel, the secondary channel will supply the information. If data from the secondary channel is lost, the FADEC will produce usable synthesized information from the parameters that are available. If there is not enough data available for synthesizing, the control modes switch. For example, if EPR is lost, the engine will be run on its N_1 ratings.

N_1. The speed of the low-pressure turbine and compressor in a dual-spool gas turbine engine.

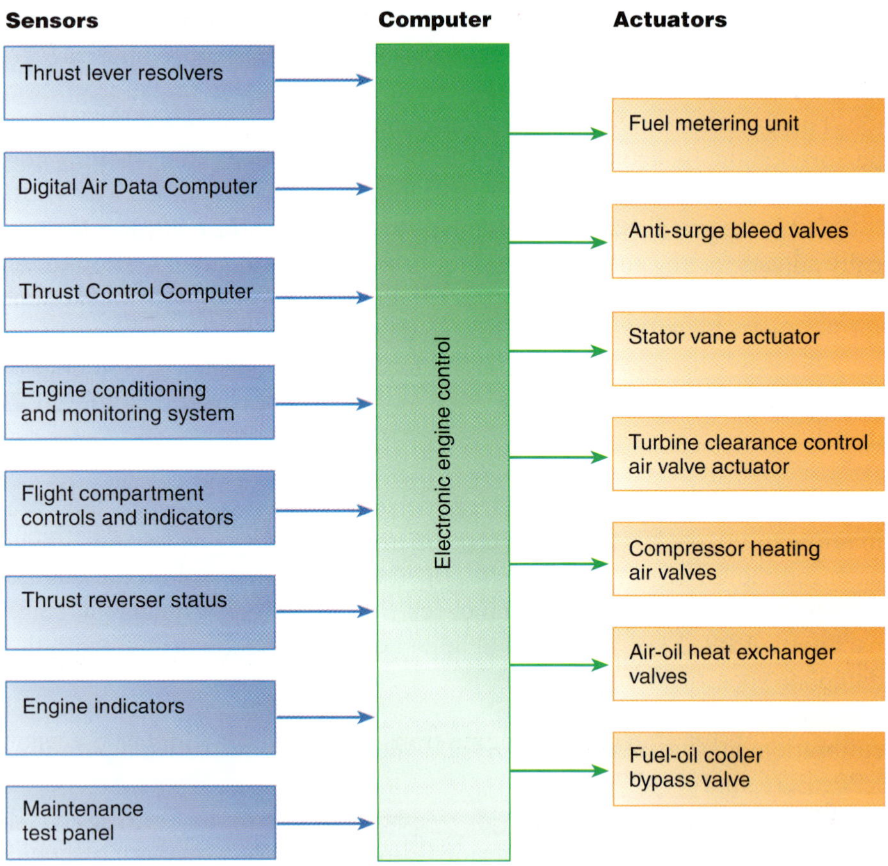

Figure 12-14. *Parameters sensed and controls actuated by an Electronic Engine Control*

In the unlikely event both channels of electronic control are lost, the torque motors are spring-loaded to their fail-safe positions. The fuel flow will go to minimum flow, the stator vanes will move to fully open, the air-oil cooler will open wide, and the ACC will shut off.

The FADEC includes extensive self-test routines which are continuously actuated. BITE, or built-in test equipment, can detect and isolate faults within the EEC and its input and output devices. The fault words of the control are decoded into English messages by a maintenance monitor, and they identify the faulty line-replaceable unit (LRU). Inflight fault data is recorded so it can be recalled during shop repair. The FADEC is able to isolate problems and indicate whether the fault is within itself or in a sensor or actuator. In the shop, computer-aided troubleshooting can identify a fault at the circuit-board level.

BITE (built-in test equipment). A computerized troubleshooting system that isolates a problem and identifies the LRU that contains the fault.

LRU (line replaceable unit). Components in an aircraft that are designed so they can be replaced as a unit while the aircraft is on the flight line.

Fuel Control Adjustments

The EGT, and thus the thrust, produced by some of the first turbojet engines was adjusted by varying the area of the exhaust nozzle. This was done by trimming the end of the tail pipe to increase the nozzle area or installing small metal tabs called "mice" to decrease it. The adjustment of the engine conditions is still called trimming, even though these procedures are no longer followed.

There is not much maintenance an aviation maintenance technician can perform on the fuel control of a gas turbine engine other than removing and replacing the units and making the adjustments the manufacturer specifies.

There are no adjustments to be made to an EEC, as the control automatically adjusts for any performance deterioration. The adjustments allowed for a hydromechanical fuel control are normally the specific gravity adjustment for the fuel, idle RPM, and full throttle RPM.

specific gravity. The ratio of the density of a material to the density of pure water.

The procedure for making these adjustments is outlined in the engine service manual. The latest information and procedures must be followed in detail.

Turbojet and turbofan engines are precision devices, and when they are manufactured, each is run in the test cell and calibrated. The RPM at which each individual engine produces its rated thrust is stamped on the engine data plate, and the technician can check the engine performance by comparing the EPR, EGT, and RPM obtained during an engine runup with the dataplate specifications.

dataplate specifications. Specifications of each gas turbine engine that were determined in the manufacturer's test cell when the engine was calibrated.

The engine is prepared as the manufacturer specifies, and accurate calibration instruments are connected to the engine to measure temperature, EPR, EGT, and RPM.

The calibration should be made when there is no wind blowing, but if there is, the aircraft should be headed directly into it. The engine should be operated as specified in the engine manufacturer's instructions and all adjustments made according to these instructions.

STUDY QUESTIONS: TURBINE ENGINE FUEL SYSTEMS

Answers are on Page 484. Page numbers below refer to chapter text.

15. Turbine engine fuel pumps often have two stages; the low-pressure stage uses a ____________________ (centrifugal or spur-gear-type) element, and the high-pressure stage uses a ____________________ (centrifugal or spur-gear-type) element. *Page 462*

16. Two sources of heat used to prevent the formation of ice crystals in jet fuel are:
    a. ______________________________
    b. ______________________________
    Page 461

17. If the fuel filter should clog, fuel will flow to the engine through the ____________________ valve. *Page 460*

18. If the gears should jam in an engine-driven fuel pump, extensive damage is prevented by a/an ____________________ in the drive shaft. *Page 462*

19. The two high-pressure sections of a spur-gear-type fuel pump are in ____________________ (series or parallel). *Page 463*

20. The pressure produced by an engine-driven fuel pump is maintained by a/an ____________________ . *Page 463*

21. A jet engine fuel control, in its normal mode of operation, ____________________ (does or does not) give the pilot direct control of the amount of fuel metered into the engine. *Page 465*

22. The fuel flowmeter used with a turbine engine measures the ____________________ (mass or volume) of the fuel and displays the information on the instrument panel in units of ____________________ (pounds or gallons) per hour. *Page 465*

23. Two basic types of fuel nozzles used in turbine engines are:
    a. ______________________________
    b. ______________________________
    Page 465

24. Two types of atomizing fuel nozzles are:
    a. ______________________________
    b. ______________________________
    Page 465

Continued

STUDY QUESTIONS: TURBINE ENGINE FUEL SYSTEMS *Continued*

25. An immediate shutoff of fuel from the nozzle when the engine is shut down is provided in a simplex fuel nozzle by the ________________________ . *Page 466*

26. In a duplex fuel nozzle, the fuel used to start the engine and for idling is the ____________________ (primary or secondary) fuel. *Page 467*

27. The spray pattern of the primary fuel is ____________________ (wider or narrower) than the pattern of both the primary and secondary fuel flowing together. *Page 466*

28. The component in a single-line duplex fuel nozzle that opens the passage to the secondary discharge orifices when the fuel pressure is sufficiently high is the ______________________ valve. *Page 466*

29. When an engine equipped with a pressurizing and dump valve is shut down, the fuel in the nozzle manifolds ________________________________ . *Page 467*

30. The most critical temperature in a gas turbine engine is the _____________________ temperature. *Page 468*

31. Four of the most basic parameters sensed by a turbine engine fuel control are:
    a. ______________________________
    b. ______________________________
    c. ______________________________
    d. ______________________________
    Page 468

32. When the engine RPM is held constant by the fuel control, an increase in ambient or compressor inlet air temperature will cause the thrust to _____________________ (increase or decrease). *Page 468*

33. When the engine RPM is held constant by the fuel control, a decrease in ambient or compressor inlet air temperature will cause the TIT to _____________________ (increase or decrease). *Page 468*

34. Modern turbine engine fuel controls maintain the _____________________ (TIT or RPM) constant and allow the ______________________ (TIT or RPM) to vary with changing atmospheric conditions. *Page 468*

35. Hydromechanical fuel controls hold the TIT near its allowable limit by providing a predetermined fuel schedule that compensates for varying conditions. This is normally done by using a/an ______________________ cam in the fuel control computer. *Page 469*

36. If the fuel flow to the engine is increased too rapidly during acceleration, the fire will likely go out because the mixture will be too ________________ (rich or lean). *Page 469*

37. The tendency of a turbine engine to surge ____________________ (increases or decreases) with an increase in altitude. *Page 470*

38. When the burner pressure increases too rapidly, the mass airflow through the engine will _____________ (increase or decrease). *Page 470*

39. The two basic classifications of turbine engine fuel controls are:
   a. _____________________________
   b. _____________________________
   Page 470

40. The adjustment in a hydromechanical fuel control that assures the nozzles will always receive the amount of fuel needed to keep the engine operating under idle conditions is the _____________________ adjustment. *Page 471*

41. Most hydromechanical fuel controls vary the amount of fuel delivered to the nozzles by ______________ _________________________ (holding a constant pressure drop across a variable orifice, or varying the pressure drop across a fixed orifice). *Page 471*

42. Droop is a condition in a flyweight governor that causes the stabilized engine RPM to _______________ (increase or decrease) slightly as the load increases. *Page 474*

43. If a hydromechanical fuel control is operated in its emergency mode, the fuel schedule that causes the engine to change its speed is ____________________ (manual or automatic). *Page 474*

44. The pitch angle of the propeller blades installed on a turboprop engine is controlled by the _____________________. *Page 475*

45. Two types of electronic engine controls are:
   a. __________________________
   b. __________________________
   Page 475

46. A supervisory EEC adjusts the ________________________ to maintain the required fuel flow for the thrust called for by the position of the throttle lever. *Page 475*

Continued

STUDY QUESTIONS: TURBINE ENGINE FUEL SYSTEMS *Continued*

47. If the electronic circuitry in a supervisory EEC should fail, control automatically reverts to the ______________________________ . *Page 475*

48. A full-authority digital electronic control (FADEC) _______________________ (does or does not) control a hydromechanical fuel control. *Page 475*

49. The thrust setting parameter for a FADEC is the ____________________________ (EPR or RPM). *Page 476*

50. When a turbine engine is controlled by a FADEC, the EPR is maintained constant for any position of the ____________________________ as the flight and environmental conditions change. *Page 476*

51. Adjusting the fuel control of a gas turbine engine is called ______________________ the engine. *Page 478*

52. The RPM at which a specific engine develops its rated thrust is recorded in/on the engine ______________________ . *Page 478*

Answers to Chapter 12 Study Questions

1. kerosine
2. a. kerosine-type
   b. blended
3. do not
4. A
5. freezing
6. 4
7. more
8. a. TEL forms deposits on the turbine blades
   b. avgas does not have the lubricating properties of kerosine
9. lower
10. more
11. more
12. corrosion
13. a. prevents formation of microbial scum
    b. lowers the freezing point of water condensed from the fuel
14. fueling control panel
15. centrifugal, spur-gear type
16. a. compressor bleed air
    b. engine lubricating oil
17. filter bypass
18. shear section
19. parallel
20. pressure relief valve
21. does not
22. mass, pounds
23. a. atomizing
    b. vaporizing
24. a. simplex
    b. duplex
25. check valve
26. primary
27. wider
28. flow divider
29. drains
30. turbine inlet
31. a. power lever angle
    b. compressor inlet total temperature
    c. compressor RPM
    d. burner pressure
32. decrease
33. increase
34. TIT, RPM
35. three-dimensional
36. rich
37. increases
38. decrease
39. a. hydromechanical
    b. electronic
40. minimum flow
41. holding a constant pressure drop across a variable orifice
42. decrease
43. manual
44. fuel control
45. a. supervisory
    b. full-authority
46. hydromechanical fuel control
47. hydromechanical system
48. does not
49. EPR
50. throttle lever
51. trimming
52. data plate

Ignition & Starting Systems

13

Ignition & Starting Systems

13

Turbine Engine Ignition Systems

Turbine engine fuel requires a much hotter spark for ignition than does aviation gasoline. As a result, turbine engine ignition systems are entirely different from the magneto systems installed on reciprocating engines.

Combustion in a reciprocating engine is intermittent, and a timed high-voltage spark is required in each cylinder to ignite the fuel-air mixture to begin the power strokes. Reciprocating engines require two magnetos with high-voltage distributors and two spark plugs in each cylinder. The distributors and spark plugs are connected with high-voltage, shielded leads.

Turbine engines, on the other hand, use continuous combustion and require an electrical spark only during the start sequence to ignite the fuel-air mixture. Once the fire is started, the ignition system can be turned off. Turbine engines, even those with multiple-can combustors, normally use only two igniters installed on opposite sides of the engine. Flame tubes, such as those in Figure 10-63 on Page 408, connect the cans to ensure that all the fuel-air mixture is ignited.

flame tubes. Small-diameter metal tubes that connect the combustors to carry the ignition flame to all combustion chambers.

The conditions for ignition in a reciprocating engine are ideal because the fuel-air mixture ratio is correct, warm, and turbulent. In a turbine engine, there is a large volume of cold air flowing through the combustion chamber, and the fuel for starting is at best atomized rather than vaporized. The spark needed to ignite the fuel must be intense. Starting a turbine engine after a flameout in flight is further complicated by the low density and low temperature of the air.

flameout. A condition of turbine engine operation when the fire unintentionally goes out. Improper fuel-air mixture or interruption of the air flow through the engine can cause a flameout.

High-Energy Ignition Systems

Turbine engine ignition systems are rated in joules, or watt-seconds, with four-joule systems used on some of the smaller engines and 20-joule systems commonly used on larger engines.

joule. A measure of energy. In terms of electrical energy, one joule is equal to one watt-second.

An extremely short time is required to ignite the fuel-air mixture, normally only a few microseconds (millionths of a second). But a high rate of current is needed to produce a spark intense enough to vaporize and ignite the fuel. Because the duration of the spark is so short, the ignition system can furnish this intense current without being excessively large or heavy.

Some early turbine-engine ignition systems produced their high voltage by changing direct current into pulsating DC with an electromagnetic vibrator or a set of contact points opened by a motor-driven cam. This pulsating

DC was converted into high-voltage AC by a step-up transformer, or coil, similar to that used in a magneto. The resulting high-voltage AC produced the needed sparks in the combustion chamber by arcing across the electrodes in the igniters.

Almost all modern ignition systems produce high-energy sparks by storing high-voltage electrical energy in a capacitor and discharging it across the electrodes in the igniters.

There are two types of high-energy, or capacitor-discharge, ignition systems: low-voltage systems with about 1,000 volts at the igniters, and high-voltage systems with an output of more than 5,000 volts. Both may be powered with AC or DC.

igniter. The component in a turbine-engine ignition system that provides a high-energy spark for igniting the fuel-air mixture in the combustion chamber for starting.

A typical turbine engine ignition system consists of two ignition exciters, two high-tension leads, and two igniters. The exciters, which convert the 28-volt DC or 115-volt AC into the high voltage, are normally sealed units and cannot be serviced in the field. In some installations, the two exciters are built into a single unit like the one in Figure 13-1.

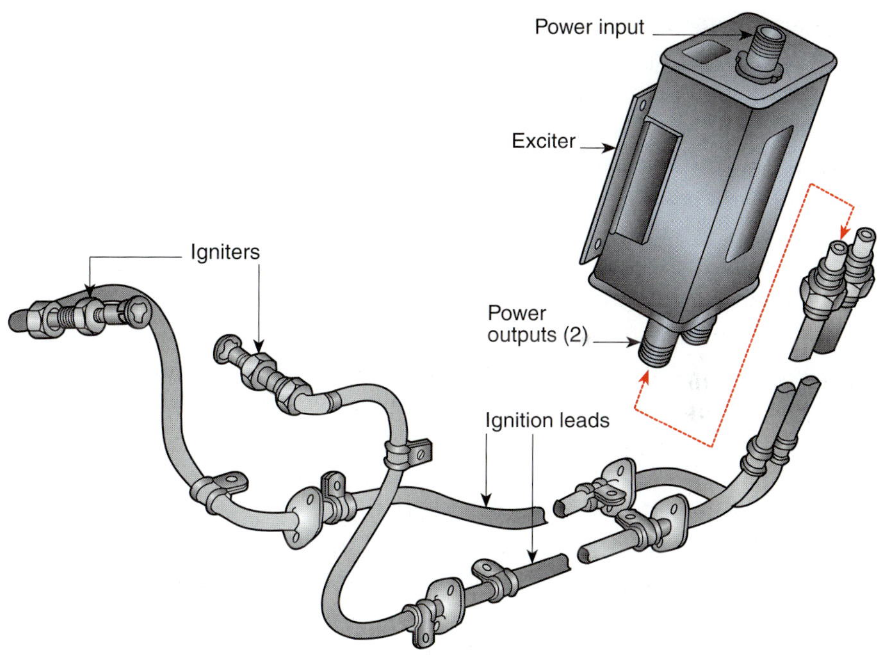

Figure 13-1. *Typical ignition system components for a gas turbine engine*

There is a danger of a turbine engine flaming out during certain flight conditions, and for this reason, the ignition system is normally turned on for takeoff, landing, or flight into turbulent conditions, and when the engine anti-ice system is actuated. The tremendous amount of energy released by these ignition systems limits the length of time they can be used without a cooling-down period. This cooling down is done by observing a duty cycle.

duty cycle. A schedule that allows a device to operate for a given period of time, followed by a cooling down period before the device can be operated again.

Intermittent-duty systems, those normally used for the main ignition, have a duty cycle which allows operation for two minutes; then they must be cooled down for about three minutes. After a second two-minutes of operation, about 20 minutes must be allowed for the system to cool down before it is used a third time. The specific duty cycle for each system is found in the operators manual for the engine, and it must be carefully observed.

Some engines have, in addition to the intermittent-duty exciter, a continuous-duty ignition system that has no operating time limit. Continuous-duty systems normally provide energy to only one igniter and may be kept energized for much of the flight.

Ignition Exciters

The exciters, the heart of the ignition system, change the input power into pulsating DC that charges a capacitor to a specific high voltage. When this voltage is reached, the capacitor discharges across the electrodes of the igniter.

If your knowledge of electrical circuits is weak, review Chapter 4 of the *General* textbook and Chapter 7 of the *Airframe* textbook of this *Aviation Maintenance Technician Series.* These two chapters cover basic electricity and electrical circuits.

The circuit in Figure 13-2 illustrates the principle of an intermittent-duty DC ignition system. The exciter receives its power from the ignition power circuit breaker in the aircraft. Current flows into the unit through a π (pi) filter (1) that consists of an inductor (2) and two capacitors (3). Radio-frequency electromagnetic energy is produced inside the exciter and is prevented from entering the aircraft electrical system by the inductor. It is then directed to ground through the capacitors. The entire exciter unit is housed inside a hermetically-sealed metal case (17) that is electrically grounded to the engine.

pi (π) filter. An electronic filter used to prevent radio frequency energy produced in the ignition exciter from feeding back into the aircraft electrical system. The filter is made of an inductor with a capacitor on its input and output. The name is derived from the resemblance of the three components on a schematic diagram to the Greek letter pi (π).

Current flows from the filter into the vibrator coil (4) that also serves as the primary winding of a step-up transformer, then to ground through the relay contacts (6). As it flows, a magnetic field builds up around the primary winding and cuts across the turns in the secondary winding (5). The strength of the magnetic field produced by the primary winding increases until it pulls the contacts open. The current immediately stops flowing, and the magnetic field collapses. The capacitor (7) across the contacts aids the

RF (radio frequency) energy. Electromagnetic energy with a frequency high enough to radiate from any conductor through which it is flowing.

hermetically sealed. A complete seal, especially against the escape or entry of air.

collapse of the field and prevents arcing across them. The collapsing magnetic field cuts across the many turns in the secondary winding, inducing a high voltage in it.

half-wave rectifier. An electrical rectifier circuit that converts AC into pulsating DC. Only one alternation of each cycle is present in the output.

The high voltage produced in the secondary coil (5) causes current to flow through the half-wave rectifier diode (8) into the storage capacitor (9). Each pulse of current into the storage capacitor causes the voltage across it to build up until the ionization voltage of the discharge tube (10) is reached. At this voltage, the gas in the tube becomes conductive and the storage capacitor discharges through it.

As the discharge current flows through the primary winding of the trigger transformer (11) and into the trigger capacitor (12), it induces a high voltage into the secondary winding of the trigger transformer, which is connected through the ignition lead (16) to the igniter (13). The voltage rises across the electrodes in the igniter until the air in the gap ionizes and becomes conductive. Then, the energy in the storage and trigger capacitors discharges to ground across the ionized gap. This current produces the short-duration, high-energy spark needed to ignite the fuel-air mixture in the combustion chamber.

The safety resistor (15) bleeds off any energy stored in the trigger capacitor when the ignition system is shut down, and prevents voltage building up to a damaging level if the system is operated with the lead to the igniter plug disconnected. The bleeder resistor (14) bleeds off energy in the storage capacitor when the system is not operating. The bleeder and safety resistors are extremely important for these reasons.

To avoid a severe electrical shock, no component in a turbine engine ignition system should be touched for at least five minutes after the system is turned off.

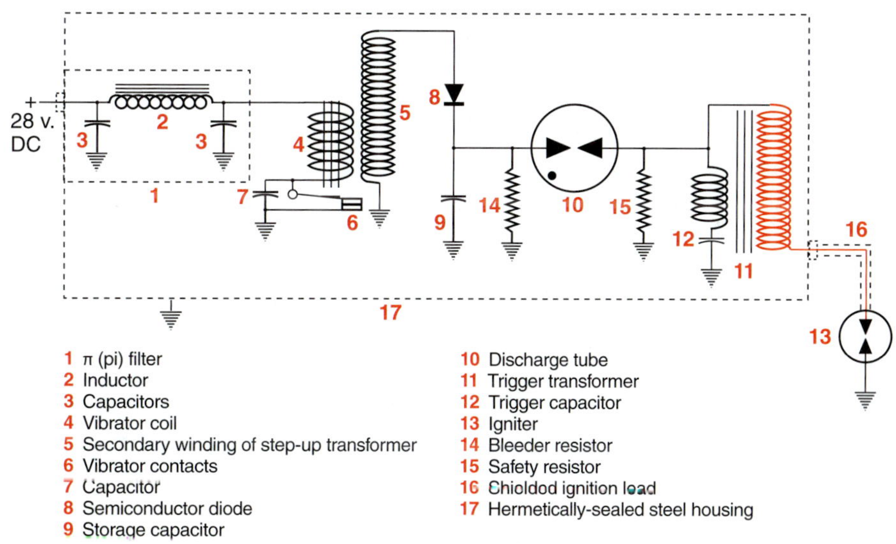

Figure 13-2. *Circuit diagram of a DC-input turbine-engine ignition exciter*

A typical high-voltage, intermittent-duty AC igniter uses a voltage-doubler circuit like the one in Figure 13-3.

This system is powered by 115-volt, 400-hertz AC from the aircraft electrical system through π filter (1). This filter passes the 400-Hz frequency but prevents radio-frequency energy produced in the exciter from feeding back into the aircraft electrical system where it might interfere with installed electronic equipment. The 400-Hz AC then goes to the power transformer (2) where its voltage is increased.

The high voltage from the secondary of the transformer goes into the voltage doubler composed of the doubler capacitor (3), resistor (4), diode (5), resistor (6), diode (7), and the storage capacitor (8).

During the half cycle when the top of the power transformer secondary winding is negative and the ground is positive, electrons charge the doubler capacitor to the transformer secondary peak voltage through diode (5) and resistor (4).

peak voltage. The voltage of AC electricity that is measured from zero voltage to the peak of either alternation.

During the next half cycle when the top of the secondary winding is positive, its voltage is the same as that across the charged doubler capacitor, and the two are in series across the storage capacitor through resistor (6) and diode (7).

Continued

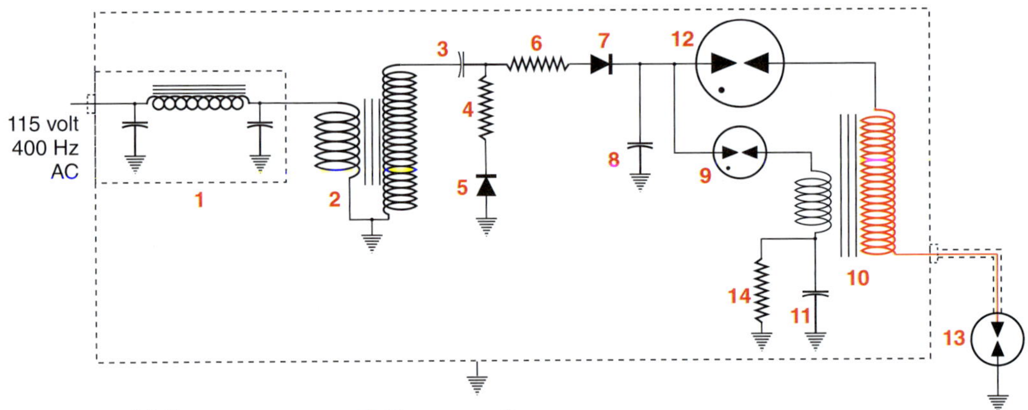

1 π (pi) filter
2 Power transformer
3 Doubler capacitor
4 Resistor
5 Diode
6 Resistor
7 Diode
8 Storage capacitor
9 Discharge tube
10 High-voltage trigger transformer
11 Trigger capacitor
12 Discharge tube
13 Igniter plug
14 Bleeder resistor

Figure 13-3. *Circuit diagram of an AC-input turbine-engine ignition exciter*

The charge on the storage capacitor increases with each cycle of the AC until its voltage rises to the ionization voltage of the gas in the discharge tube (9). At this point, the tube conducts and allows some of the electrons in the storage capacitor to flow through the primary winding of the high-voltage trigger transformer (10). This current charges the trigger capacitor (11) and induces a high voltage into the secondary winding of the transformer. The voltage across the secondary winding rises to the ionization voltage of the gas in the discharge tube (12), and when the gas ionizes, the tube conducts and the storage capacitor discharges to ground across the electrodes in the igniter (13). The bleeder resistor (14) completely discharges the trigger capacitor so it will accept the charge from the next cycle.

Autoignition System

While a turbine-engine ignition system is used only for starting the engine, it is important that it be energized so it can relight the engine if it should flame out. An autoignition system such as the one in Figure 13-4 is installed on some turboprop engines to serve as a backup for takeoff and landing and for flight conditions in which the engine could flame out.

The engine-start switch has three positions. In the ENGINE START AND IGNITION position, current flows to the generator control, the coil of the starter relay, and the coil of the ignition power relay. Current flowing in the generator control opens the generator-field circuit for the starter-generator and connects the series starter winding to the electrical system. (*See* Figure 13-9 on Page 502). Current flowing in the coil of the ignition power relay closes its contacts, allowing current to flow from the bus to the ignition exciter unit and to the IGNITION ON light on the annunciator panel.

motor. (*vb*) The act of rotating a turbine engine using the starter, with the ignition system deactivated. An engine is motored to force air through it to purge fuel fumes.

When the engine-start switch is placed in the STARTER ONLY position to motor or rotate the engine without starting it, current flows to the coil of the starter-only relay and moves its contacts so it no longer supplies a ground for the ignition-power relay. Since the ignition-power relay cannot be actuated, no current flows to the ignition exciters nor to the IGNITION ON light on the annunciator panel. However, current does flow to the generator control and the coil of the starter relay.

When the autoignition control switch is placed in the ON or ARMED position, current flows from the bus through the 5-amp starter-control circuit breaker to the compressor-discharge-pressure switch. When the engine is producing a specified compressor-discharge pressure, the pressure switch moves its contacts so that current can flow to the AUTOIGNITION ARMED light on the annunciator panel.

If the engine should lose power and the compressor-discharge pressure drops below the specified value, the pressure-switch contacts shift and send current to the coil of the ignition-power relay. When this relay shifts its position, current flows from the bus through the ignition-power circuit breaker to the ignition exciter and the IGNITION ON light on the annunciator panel.

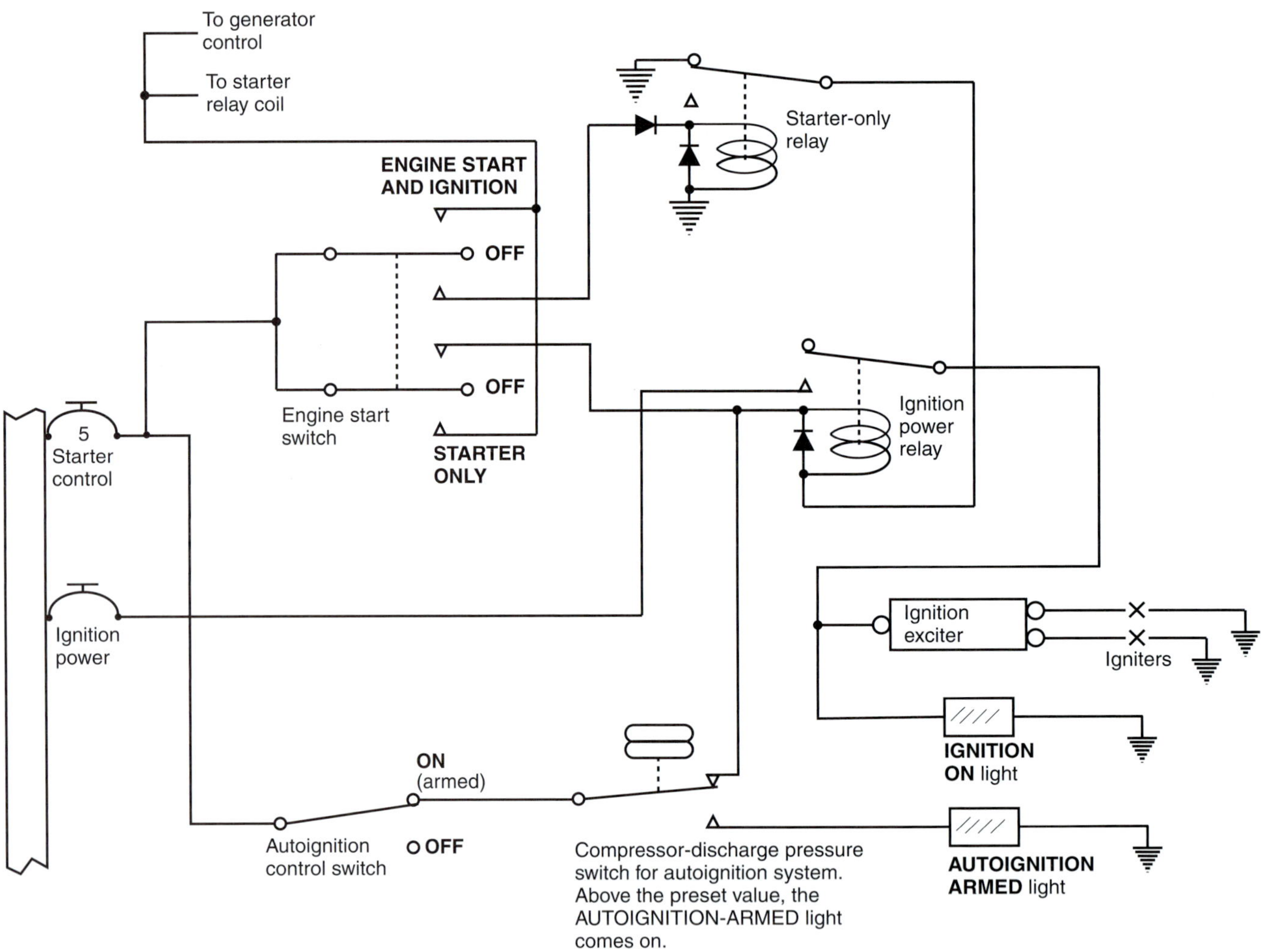

Figure 13-4. *Ignition system for a turboprop engine incorporating an autoignition system that activates the exciter if the compressor-discharge pressure drops to a predetermined low value*

Igniters

Turbine-engine igniters serve the same purpose as spark plugs in a reciprocating engine: they ignite the fuel-air mixture, but their operating conditions are entirely different. In a reciprocating engine, a relatively low-energy spark must jump across the electrodes of a spark plug every other revolution of the crankshaft. But in a turbine engine, high-energy sparks are required, but only when the engine is being started or during flight conditions when there is danger of the engine flaming out. Igniters are not prone to carbon fouling as are spark plugs, because the high-energy spark blasts off any deposits that form on the firing end.

There are a few spark plug designs that are usable in all aircraft reciprocating engines, but because of the difficulty in igniting the fuel-air mixture in a turbine engine, igniters are tailored to the engine. Generally, an igniter designed for one engine will not function properly in another model.

There are two basic types of igniters: spark igniters (by far the more popular) and glow plug igniters, which are used in some of the smaller engines.

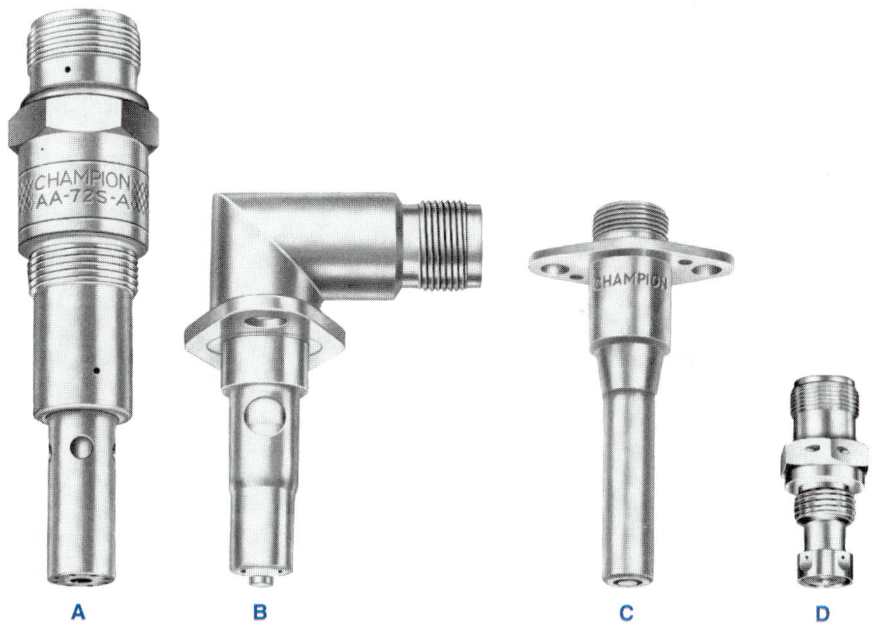

A High-voltage air-cooled recessed gap igniter
B High-voltage air-cooled surface gap igniter
C Low-voltage shunted surface gap igniter
D Low-voltage glow plug igniter

Figure 13-5. *Typical types of turbine-engine igniters*

The firing-end configurations of the different types of igniters in Figure 13-6 show some of their interesting characteristics.

The tip of the surface-gap igniter A in Figure 13-6 protrudes slightly into the combustion chamber and the spark roughly follows the surface of the insulator between the electrodes.

The tip of igniter B also protrudes into the combustion chamber, but is air cooled. Air that passes around the outside of the combustion chamber flows into the igniter through the hole in its side. This hole is also seen in Figure 13-5B.

Igniter C has a recessed gap, and its tip does not protrude into the combustor. The spark does not follow the surface of the insulator, but extends out beyond the tip.

The low-voltage shunted-gap igniter in D operates on a different principle from the others. There is a ceramic semiconductor material between the center electrode and the shell. The resistance of this semiconductor is low when it is relatively cool, but increases as it heats up. When the storage capacitor discharges through the igniter, the current initially flows to ground through the semiconductor, which gets so hot it becomes incandescent and its resistance increases. The air gap between the electrodes becomes ionized, and its resistance becomes lower than that of the semiconductor, so the remainder of the current discharges across the air gap in a surge as a high-energy spark.

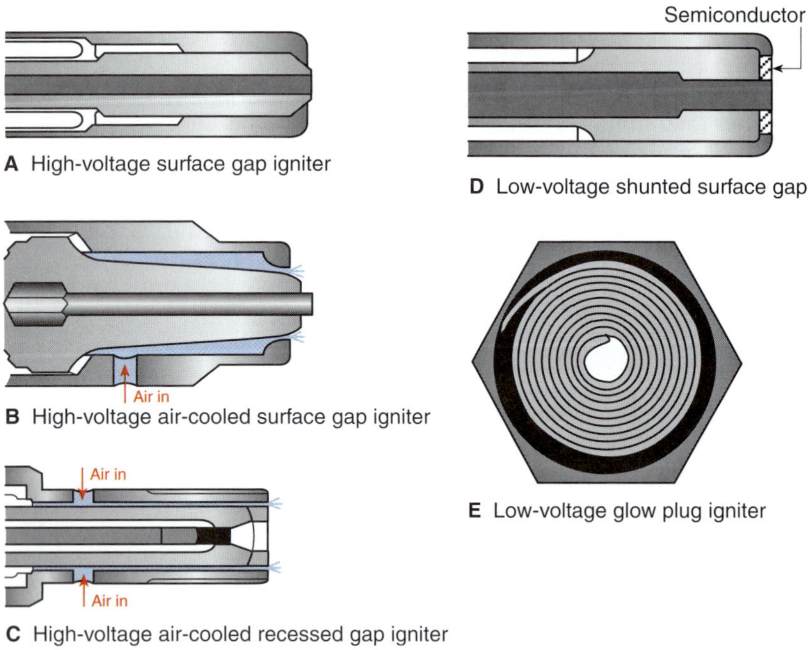

Figure 13-6. *Firing-end configuration of typical gas turbine igniters*

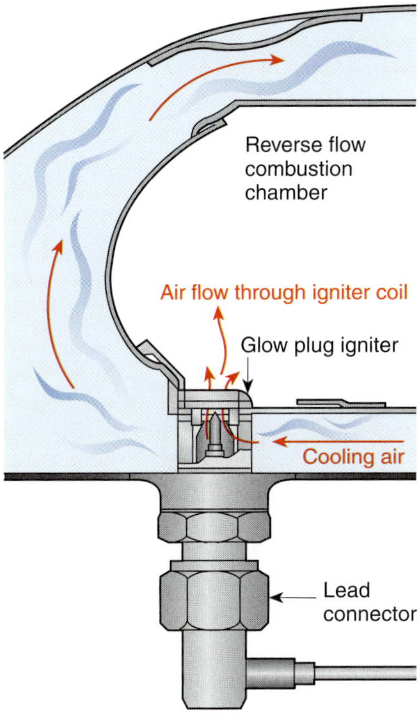

Figure 13-7. *Installation of a glow plug in a turbine engine combustor*

shielding. A metal braid that encloses wires carrying high-frequency alternating current or high-voltage DC that has radio-frequency energy superimposed on it. The shielding intercepts any electromagnetic radiation and conducts it to the engine structure so it will not interfere with any installed electronic equipment.

The glow plug igniter in E is only used in some of the smaller engines and is installed as seen in Figure 13-7. Some of the cooling air flowing around the outside of the combustor flows into it through the coils of the glow plug. The coils are heated with current from the ignition exciter until they glow orange-yellow, and fuel from the fuel nozzle spraying on the coil is ignited. Air blowing through the coil produces a streak of flame that ignites the mixture in the combustor. Glow plug igniters are especially suited for cold-weather starting.

Turbine Engine Ignition System Servicing

Turbine engine ignition systems must be handled with extreme care because the high voltage can be lethal.

Before disconnecting the lead from an exciter or igniter, be sure to pull the ignition power circuit breaker. Disconnect the power lead to the exciter and observe the time specified in the engine maintenance manual before removing the igniter lead. This time, normally about five minutes, allows energy stored in the capacitors to bleed safely to ground through the bleeder and safety resistors. As soon as the lead is removed from the igniter, ground the center conductor to the engine to ensure that the capacitors are completely discharged.

Igniters should be removed from the engine with extreme care and examined according to the instructions in the engine service manual. Some igniters can be cleaned, and when servicing them, be sure to follow the igniter manufacturer's instructions in detail.

Exciters are sealed units and cannot be opened for servicing, but should be carefully inspected for secure mounting on the engine and all electrical and ground connections should be clean and tight. The high-voltage leads between the exciter and igniters should be supported as shown in the engine maintenance manual, and there should be no broken shielding braid. Some exciters contain radioactive material, and when disposed of, it must be done in accordance with local environmental requirements.

STUDY QUESTIONS: TURBINE ENGINE IGNITION SYSTEMS

Answers are on Page 506. Page numbers refer to chapter text.

1. Turbine engines normally have _____________ (how many) igniter plugs. *Page 487*

2. The fuel-air mixture in combustors that have no igniter is ignited by flames carried from other combustors through ___________________ . *Page 487*

3. The conditions for igniting the fuel-air mixture in a turbine engine is ____________________ (more or less) severe than they are in a reciprocating engine. *Page 487*

4. Turbine engine ignition systems are rated according to the energy in the spark in units of _____________ (joules or watts). *Page 487*

5. One joule is one _________________ (watt or watt-second). *Page 487*

6. The duration of the spark across the gaps of a turbine engine igniter is __________________ (short or long). *Page 487*

7. By limiting the duration of the spark, the current flow for each spark can be ___________________ (increased or decreased). *Page 487*

8. A high-energy ignition system produces the high current in each spark by storing electrons in a/an _____________________ . *Page 488*

9. Turbine-engine exciters __________________ (can or cannot) be serviced in the field. *Page 488*

10. Intermittent-duty ignition systems must be operated according to a ________________ , which allows them to cool down between the times they are operated. *Page 489*

11. Continuous-duty exciters _________________ (are or are not) normally used for initial engine starting. *Page 489*

12. Interference with electronic equipment installed in the aircraft by the electromagnetic energy generated in the ignition exciter is prevented by the __________________ in the exciter unit filter. *Page 489*

13. The vibrator in an ignition exciter changes DC into ____________________ (AC or pulsating DC). *Page 489*

14. The electrons that flow across the gap in the igniter are those that have been stored in the ________________ and the ________________ capacitors. *Page 490*

Continued

STUDY QUESTIONS: TURBINE ENGINE IGNITION SYSTEMS *Continued*

15. Energy that remains stored in the capacitors when the ignition system is shut down is bled off through the _______________ and _________________ resistors. *Page 490*

16. The high-voltage AC exciter uses a _____________________ circuit to increase the voltage from the power transformer. *Page 491*

17. Two basic types of turbine engine igniters are:
    a. ___________________________
    b. ___________________________
    Page 494

18. Shunted-gap igniters are normally used in ___________________ (high or low) -voltage systems. *Page 495*

19. After removing the power lead from an exciter, you should wait for a specified length of time before removing the lead from the igniter. This period of time allows the energy that is stored in the ____________________ to bleed off to ground. *Page 496*

20. After removing the lead from the igniter, ground the center conductor to the ________________________ . *Page 496*

Turbine Engine Starting Systems

Gas turbine engines are normally started by first rotating the compressor with a starter to move air through the engine and purge it of any fuel vapors that may have collected; then the ignition system is energized. When the engine has reached a specified RPM, fuel is sprayed into the combustors and the engine starts. As soon as it accelerates to a self-sustaining idle speed, the starter and ignition systems are deactivated. The start sequence is programmed so the pilot has only to initiate the action; the rest follows automatically.

In dual-spool axial-flow engines, the starter rotates the high-pressure (N_2) compressor because it has less mass than the N_1 compressor and requires less energy from the starter.

There are several types of starters in use, but the two most commonly used are air starters for larger engines and electrical starters for smaller ones.

Air Turbine Starter

Air turbine starters are popular for jet transport aircraft because of their light weight for the torque produced. They do, however, require a large volume of low-pressure compressed air, but for airline operations, this is no problem because most jet transport airplanes have auxiliary power units (APU) that provide all the necessary air for starting. Ground power units (GPU), normally available at the airline gates, can also furnish the needed air, as can a cross-flow of compressor bleed air from an already running engine.

Figure 13-8 on the next page shows the operating principle of an air turbine starter. Air at a pressure of approximately 35 to 40 psi enters the starter housing at its aft end and flows through the turbine nozzle where its direction is altered so that it enters the blades of the axial turbine at the proper angle. Air flowing through the turbine spins it at a speed of around 50,000 RPM, and the high speed of the turbine shaft is reduced to approximately 2,400 RPM by the double set of reduction gears. The output of the reduction gears is splined to the ratchet in the sprag clutch and drives it in a clockwise direction.

The pawls in the clutch housing are held against the teeth of the ratchet by a series of leaf springs. When the pawls are engaged, the clutch housing rotates with the ratchet. When the engine starts, the housing rotates faster than the ratchet, and the pawls slide over the ratchet teeth until speed builds up high enough to produce sufficient centrifugal force to hold the pawls away from the ratchet against the force of the springs.

APU (auxiliary power unit). A self-contained motor-generator compressor carried in an aircraft to generate electrical power and compressed air for ground operations and for starting the main engines.

GPU (ground power unit). A service component used to supply electrical power and compressed air to an aircraft when it is operating on the ground.

sprag clutch. A freewheeling, non-reversable clutch that allows torque to be applied to a driven unit in one direction only.

When the engine is shut down, the housing slows enough that centrifugal force can no longer hold the pawls back, and they ride over the ratchet, making a clicking noise as the turbine and compressor spin down.

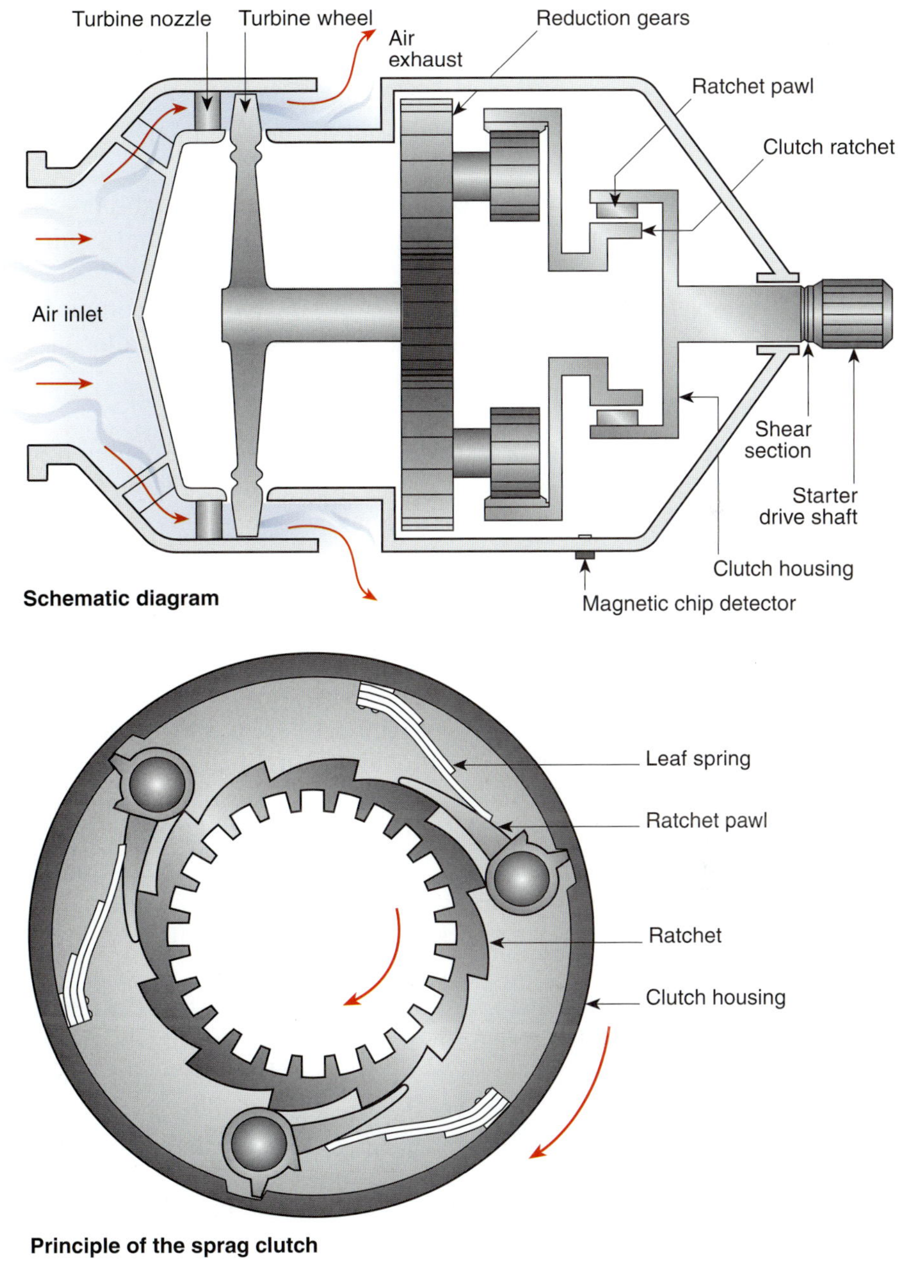

Figure 13-8. *An axial-flow air starter*

Air turbine starters have a shear section on the drive shaft. If the ratchet mechanism should fail to release, the engine will try to drive the starter to a higher speed than it is designed for. When the torque becomes sufficiently high, the shear section will break, protecting the starter from further damage.

Air from the APU flows through a solenoid-actuated butterfly-type shutoff and regulating valve controlled by the start switch in the cockpit. The regulator function of the valve protects the starter from too high an air pressure.

When the engine starts and reaches a self-sustaining speed, a centrifugal switch opens and de-energizes the solenoid. A spring then closes the air valve.

There is a secondary safety feature in this starter that prevents the unloaded turbine from reaching its burst speed if, for any reason, the air supply should not shut off after the engine starts. When the airflow through the nozzle reaches the speed of sound, the nozzle becomes choked, which prevents further increase in the flow. The turbine stabilizes in an overspeed condition.

choked nozzle. A nozzle that limits the speed of gases flowing through it. The gases speed up until they reach the speed of sound. At this point, a normal shock wave forms that prevents any further acceleration.

Air turbine starters have their own self-contained lubrication system with the oil held in the starter housing. A magnetic chip detector is incorporated into the drain plug to warn the technician of any gear distress that produces flakes or shavings of metal.

Electric Starter

Some turbine engines use series-wound electric starters. These starters are normally connected to the engine through a ratcheting starter jaw similar to that used in a reciprocating engine. When the automatic start sequence begins, the jaw moves out and engages a mating jaw on the engine starter drive. When the engine starts and accelerates to its self-sustaining speed, there is no longer a load on the jaws, and a spring moves the starter jaw away from the jaw in the engine. The automatic start sequence of events then removes the electrical power from the starter.

Starter Generator

Many smaller turbine engines installed in business jet airplanes have a combination starter-generator rather than two separate components. These units resemble heavy-duty, compound-wound DC generators, but have an extra set of series windings. These series motor windings are switched into the circuit when the engine is being started, but as soon as it is running, they are automatically switched out.

starter-generator. A single-component starter and generator used on many smaller gas-turbine engines. It is used to start the engine, and when the engine is running, its circuitry is shifted so that it acts as a generator.

Figure 13-9 shows a typical starter-generator circuit. When the start switch is placed in the START position, current flows through the start/ignition circuit breaker and the upper contacts of the start switch, to the coil of the starter relay. This current produces a magnetic pull that closes the relay and allows current to flow from the bus to the starter-generator through its C+ terminal, the series motor windings, the armature, and the

starter-generator series windings, and then to ground. At the same time, current flows through the ignition-cutoff switch into the ignition exciter to provide the intense heat at the ignitors needed to ignite the fuel-air mixture.

When light-off is achieved, the start switch is moved to the RUN position. Current flows from the bus through the generator-field circuit breaker and the coil of the generator-field relay. This current produces a magnetic pull that closes the field relay contacts to connect the generator field to the voltage regulator. When generator-field current flows, the generator produces current, and as soon as the voltage builds up to the specified value, the contacts inside the reverse-current cutout relay close. The generator output current flows from the B+ terminal through the reverse-current relay, to the bus through the generator circuit breaker.

When the start switch is placed in the OFF position, current is shut off to the generator-field relay, and it opens, disconnecting the generator field from the voltage regulator, and the generator stops producing load current.

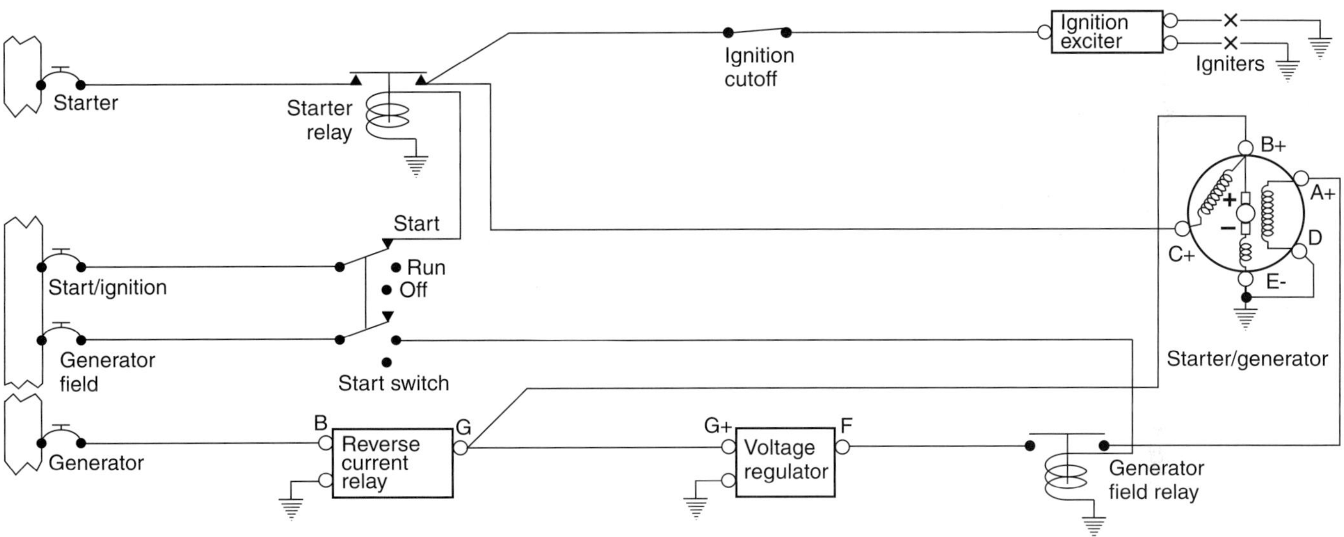

Figure 13-9. *Basic starter-generator circuit*

Combustion Starters

There are two types of combustion starters that may be used on aircraft that do not have access to a GPU (ground power unit) or an installed APU (auxiliary power unit).

The cartridge starter uses an electrically-ignited solid powder charge to produce a large volume of hot gas that spins a turbine that is connected to the engine through a set of reduction gears and a clutch. In the start sequence, the powder charge is ignited, and the gases produced as it burns drive the turbine, which rotates the engine compressor.

Some engines may be started with a gas-turbine starter that consists of a small, totally self-contained gas turbine engine. This engine has a centrifugal compressor, a reverse-flow combustor, and a free turbine that drives the engine compressor through a clutch and a series of reduction gears.

The clutches in both types of combustion starters automatically disconnect the starter from the drive shaft when the starter is no longer operating.

Air Impingement Starter

Some engines are started by a stream of low-pressure air that passes through a check valve and is blown through the turbine blades. The air source is turned off when the engine reaches a speed that allows it to accelerate without further assistance. No components are required for this type of starter except an air line, a check valve, and the nozzle that directs the air into the turbine.

air impingement starter. A turbine engine starter that basically consists of a nozzle that blows a stream of compressed air against the turbine blades to rotate the compressor for starting the engine.

Auxiliary Power Units

An auxiliary power unit, or APU, for a modern jet transport airplane is a compact, self-contained gas-turbine-powered unit that provides electric power and compressed air when the main engines are not operating, thus freeing the aircraft from dependence on ground power equipment. The APU can also be operated in flight to serve as a backup for electrical power.

The APU with its own automatic controls and safety devices, including a complete fire protection system, is most often installed in the tail cone of the airplane. Air flows in from the outside of the airplane through an APU air-inlet door into an air-inlet plenum chamber, a compartment in which the pressure of the air is stabilized before it enters the APU compressor.

A typical APU drives one or two AC generators of the same type as those installed on the main engines. These generators can normally supply all of the airplane's electrical load. The APU is rotated for starting by a 28-volt DC motor which is controlled either by switches on the cockpit APU control panel, or from a ground-accessible control panel. Fuel is normally supplied from one of the airplane's main fuel tanks, and the APU lubrication system is self-contained, consisting of an independent oil supply and the necessary pump, regulator, cooler, filters, and instruments.

Most of the modern APU fuel controls are of the FADEC, or "full-authority digital engine control" type. They automatically control the fuel for starting and then bring the engine up smoothly to its rated RPM. This RPM is automatically maintained as the air or electrical load conditions change. The FADEC makes troubleshooting simple by displaying fault codes for any problem the APU encounters.

The APU supplies the aircraft pneumatic system with compressed air and provides an ample supply for operating the main engine air-turbine starters. This air is obtained either as bleed air from the APU turbine compressor, or in some installations from a load compressor driven by a free turbine in the engine.

Opening the bleed air valve places the maximum load on the APU, and when maximum power is being used, the pneumatic load is automatically modulated, or regulated, to prevent the EGT from exceeding its safe limits.

The APU fuel control incorporates an automatic time-delay feature that closes the bleed air valve to remove most of the load, and thus reduce the APU temperature before it is shut down.

STUDY QUESTIONS: TURBINE ENGINE STARTING SYSTEMS

Answers are on Page 506. *Page numbers refer to chapter text.*

21. In the process of starting a turbine engine, the ignition is turned on ______________________ (before or after) the engine has started rotating. *Page 499*

22. In the process of starting a turbine engine, the fuel is turned on ______________________ (before or after) the ignition. *Page 499*

23. In a dual-spool turbine engine, the starter rotates the ____________________ (high or low)-pressure compressor. *Page 499*

24. The engines on most jet transport aircraft are started with _________________ (air or electric) starters. *Page 499*

25. Air for starting the engines on a jet transport aircraft is normally supplied by the __________________ . *Page 499*

26. The primary advantage of an air turbine starter is its low weight-to-________________ ratio. *Page 499*

27. An air turbine starter requires a _______________ (large or small) volume of _______________ (high or low)-pressure compressed air. *Page 499*

28. Three sources of compressed air to operate an air turbine starter are:
    a. ___________________________
    b. ___________________________
    c. ___________________________
    Page 499

29. The air turbine starter described in this text is disconnected from the engine when the engine starts by a/an ________________________ . *Page 499*

30. In the air turbine starter described in this text, the pawls are held away from the ratchet when the engine is running by ________________________ . *Page 499*

31. The air turbine starter described in this text is protected from attaining too high a speed if the ratchet should fail to release the clutch by a/an ____ . *Page 501*

32. The pressure of the air supplied to the air turbine starter is controlled by a shutoff and ______________________ valve. *Page 501*

33. If the air valve does not shut off after the engine has started, the turbine will be protected from reaching its burst speed by the turbine nozzle operating in a/an ______________________ condition. *Page 501*

34. An electric starter uses a ____________________ (series or shunt)-wound motor. *Page 501*

35. The type of starter used by most small business jet turbine engines is a/an _________________________ . *Page 501*

36. Two types of combustion starters are:
    a. ___________________________
    b. ___________________________
    Page 502

37. The APU in most modern jet transports are located in the ___________________ (airplane tail cone or in the landing gear wheel well). *Page 503*

38. The two things supplied by an APU are:
    a. _______________________________________
    b. _______________________________________
    Page 503

39. The electrical generator installed on an APU is generally __________________________ (identical to or different from) the generators on the engines. *Page 503*

40. Before the air enters the APU, it is stabilized in an air inlet ___________________________ . *Page 503*

41. Fuel for the APU comes from ________________________________ (the aircraft fuel tanks or an APU fuel tank). *Page 503*

42. The type of fuel control used on most APUs is a/an ______________________________ (hydromechanical control or FADEC). *Page 503*

43. Before shutting the APU down, an automatic ____________________________ closes the bleed air valve to remove most of the load and reduce the APU temperature. *Page 504*

44. APU troubleshooting is done by fault codes displayed by the _____________________ . *Page 503*

Answers to Chapter 13 Study Questions

1. two
2. flame tubes
3. more
4. joules
5. watt-second
6. short
7. increased
8. capacitor
9. cannot
10. duty cycle
11. are not
12. inductor
13. pulsating DC
14. storage, trigger
15. safety, bleeder
16. voltage doubler
17. a. spark type
    b. glow plug
18. low
19. capacitors
20. engine
21. after
22. after
23. high
24. air
25. APU
26. torque
27. large, low
28. a. APU
    b. GPU
    c. cross-flow from an operating engine
29. sprag clutch
30. centrifugal force
31. shear section in the drive shaft
32. regulating
33. choked
34. series
35. starter generator
36. a. cartridge starters
    b. gas-turbine starters
37. airplane tail cone
38. a. electrical power
    b. compressed air
39. identical to
40. plenum chamber
41. the aircraft fuel tanks
42. FADEC
43. time delay
44. FADEC

Exhaust Systems

14

Exhaust Systems

14

Turbine Engine Exhaust

Energy remaining in the exhaust gases after they leave the turbine can be utilized to produce thrust. To do this, components in the exhaust system must straighten and accelerate the stream of gases.

After the gases leave the turbine, they flow through a convergent duct formed between the exhaust cone and the exhaust, or tail, pipe as shown in Figure 14-1. The exhaust cone is an engine component, but the exhaust pipe is an airframe component.

Some engines have an exhaust collector with struts between the forward end of the exhaust cone and the tail pipe to support the rear turbine bearing and straighten the gas flow. Such a collector is seen in Figure 14-2. As gases flow through the convergent duct between the exhaust cone and the exhaust pipe, they are accelerated and leave the exhaust nozzle at the highest practical velocity.

The opening at the end of the tail pipe is called the exhaust nozzle, or jet nozzle. Its outlet area is critical because it determines the velocity of the gases as they leave the engine. The nozzles on most turbojet and low-bypass turbofan engines have an area that usually causes them to operate in a choked condition. By the time the gases reach the end of the tail pipe, they have accelerated to the speed of sound and can accelerate no further. The remaining energy that would otherwise be converted into velocity is now converted into a pressure differential across the nozzle. This differential produces additional thrust.

The RPM and exhaust gas temperature of some of the early gas turbine engines were adjusted by changing the size of the exhaust nozzle; the area was increased by trimming the end of the nozzle or decreased by installing around the edges small metal tabs called "mice."

Engines with afterburners have variable-area nozzles that modulate, or open or close automatically, as the fuel flow changes. On the early afterburner engines, these nozzles had two positions, but the area of the nozzles installed on modern high-performance engines is continuously varied by the EEC (electronic engine control) to maintain the proper back pressure on the turbine.

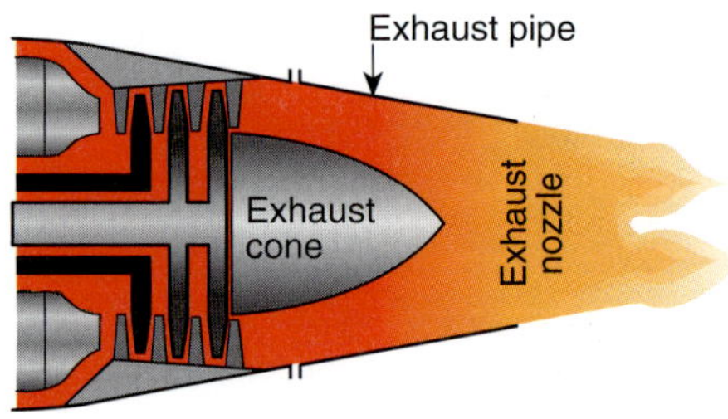

Figure 14-1. *A convergent exhaust duct is formed between the exhaust cone and the exhaust pipe.*

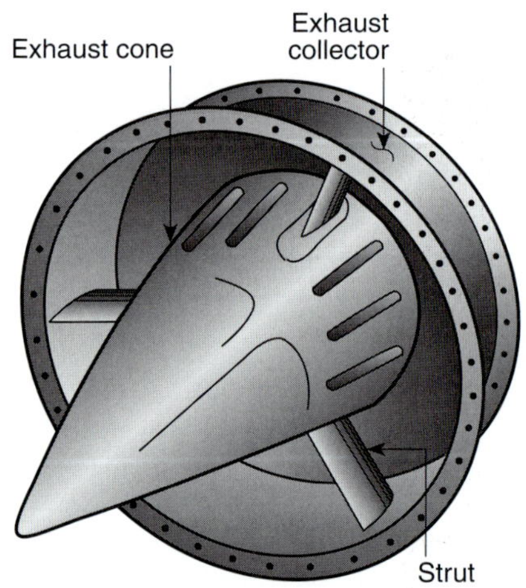

Figure 14-2. *The exhaust collector bolts to the engine housing just aft of the turbine. The tail pipe bolts to the ring around the collector.*

exhaust cone. The fixed conical fairing centered in the exhaust stream immediately aft of the last-stage turbine wheel.

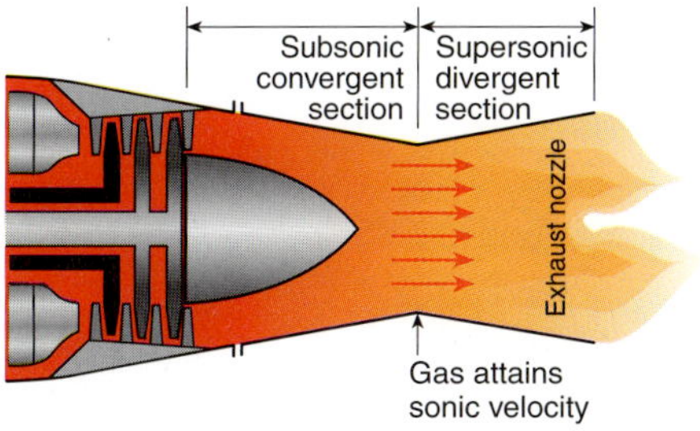

Figure 14-3. *A convergent-divergent exhaust nozzle causes an increase in thrust by increased acceleration of the exhaust gases.*

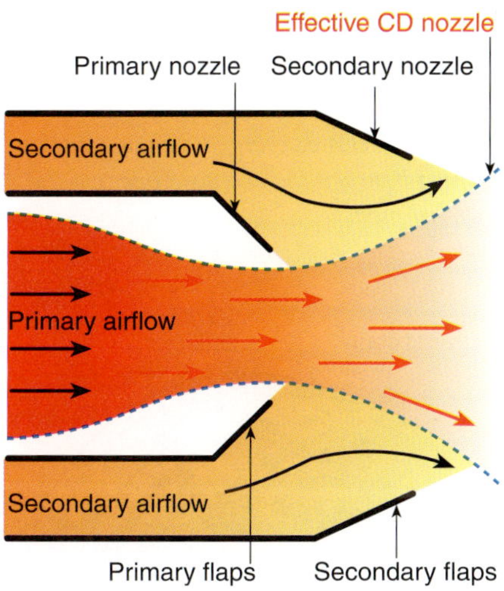

Figure 14-4. *The CD exhaust nozzle is formed in some high-performance engines with an afterburner using two sets of exit flaps. The primary flaps form a converging nozzle, and as the gases leave, their radial expansion is controlled by the flow of secondary gases which are, in turn, controlled by the position of the secondary flaps.*

tail pipe. The portion of the exhaust system of a gas turbine engine through which the gases leave.

exhaust nozzle. The opening at the rear of the exhaust pipe.

The thrust produced by the engines of supersonic airplanes is increased by using convergent-divergent, or CD, exhaust nozzles whose cross-sectional area changes as is shown in Figure 14-3.

The gases leave the turbine section and enter the convergent portion of the nozzle at a subsonic speed. Their speed increases as the duct gets smaller until they reach the speed of sound at the narrowest point where a shock wave forms and prevents further acceleration. The gases leave the narrowest point at the speed of sound, and as the duct area increases, they accelerate to a higher supersonic speed. The benefits of a CD nozzle increase as the flight Mach number of the aircraft increases.

For the most efficient operation, the rate of change of the divergent portion of the nozzle cannot be fixed, but must vary automatically as the airflow through the engine changes. Some high-performance engines have a CD nozzle whose boundary is formed by a wall of air. Figure 14-4 shows the way such a nozzle works.

The afterburner duct has two sets of overlapping flaps, one set at the end of the primary airflow passage and the other end at the end of the secondary airflow passage. The size of the openings formed by these flaps is varied with hydraulic actuators controlled by the EEC. The primary flaps adjust the opening to achieve sonic velocity at the primary nozzle, and as the gases leave, they try to expand in all directions, but are restrained from expanding radially by the flow of secondary air controlled by the secondary flaps. Therefore, all of their expansion is in an axial direction, which increases the velocity of the gas. The shape of an effective CD nozzle is controlled by the relative position of the two sets of flaps.

Some high-speed jet aircraft have CD inlet ducts, or diffusers, as well as CD exhaust nozzles. The principle upon which these ducts operate is discussed in Chapter 10, beginning on Page 382. Their function is exactly opposite to that of the CD exhaust nozzles discussed here. Figure 14-5 compares the purpose and function of a CD inlet duct and a CD exhaust nozzle.

CD Inlet Duct
Purpose: To slow supersonic air to a subsonic speed before it enters the compressor.

The air enters the convergent portion of the CD duct at supersonic velocity.
The air slows to the speed of sound at the narrowest point of the duct.
The subsonic air flowing through the divergent portion of the duct further slows down.

CD Exhaust Nozzle
Purpose: To increase the thrust produced by an engine by accelerating the exhaust gases to a high supersonic speed.

The gases enter the convergent portion of the nozzle at a subsonic speed.
Their speed increases to the speed of sound at the narrowest point of the nozzle.
The supersonic flow in the divergent portion of the nozzle accelerates the air to a higher supersonic speed.

Figure 14-5. *Comparison between the function of a CD inlet duct and a CD exhaust nozzle*

Noise Suppressors

Noise from high-powered turbojet engines around commercial airports has caused complaints from area residents. However, turbofan engines have reduced the noise problem considerably.

The amount of noise produced by a turbojet engine relates to the velocity of the exhaust gases, but the distance the noise can be heard relates to the frequency of the sound. You have perhaps noticed the way a band is heard as it approaches from a distance. The first sound heard is the low-frequency of the drum, and it is only when the band is much nearer that the high-pitch of the horns can be heard. The same principle applies to turbine engine noise suppressors. The total amount of sound cannot be decreased without sacrificing power. But, if its frequency is increased, the sound will not be audible at a long distance. A sound suppressor, as shown in Figure 14-6, may be used to replace the normal exhaust nozzle. This type of nozzle is called a corrugated-perimeter noise suppressor, and it works by breaking the normal exhaust stream into a number of smaller streams. The frequency of some of the sound produced by these smaller streams is above the audible range of the human ear, and these high frequencies are absorbed by the atmosphere before they can travel a great distance from the aircraft.

Turbofan engines extract much more energy from the exhaust gases to drive the fan, and their exhaust gas velocities are lower than those of a turbojet engine of comparable power. For these reasons, turbofan engines do not produce enough noise to require noise suppressors.

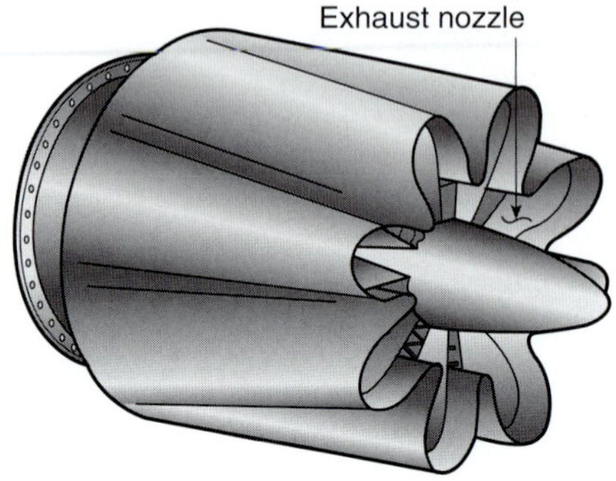

Figure 14-6. *A corrugated-perimeter noise suppressor*

sound suppressor. The airframe component that replaces the turbine engine tail pipe. It reduces the distance the sounds made by the exhaust gases propagate by converting low-frequency vibrations into high-frequency vibrations.

Thrust Reversers

Modern turbine-powered aircraft are normally so heavy and land at such a high speed that the aircraft brakes cannot be depended upon for complete speed control while the aircraft is in its landing roll.

Propellers with reverse-pitch capability can be used to slow airplanes equipped with reciprocating and turboprop engines. Turbojet and turbofan engines do not have reversing propellers, but they do have provisions for reversing some of their thrust.

Thrust reversers, which produce a rearward thrust of between 40% and 50% of the engine's rated forward thrust, can be installed on these engines. These reversers are used on the ground to decrease the landing roll, and in many aircraft they can be used in flight to allow descent at a steep angle without the aircraft building up excessive speed.

A thrust reverser operates by deflecting part of the exhaust gases, or fan discharge air, forward. Two ways of doing this are with a set of clamshell doors or a series of cascade vanes.

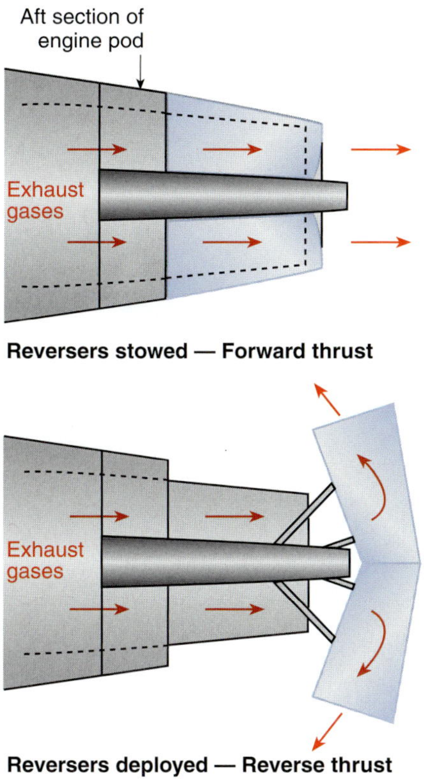

Figure 14-7. *A clamshell-type thrust reverser*

A clamshell thrust reverser is often called a mechanical-blockage reverser. A pair of scoop-shaped doors that normally lie alongside the exhaust nozzle may be deployed by sliding them rearward and opening them so they block the normal nozzle and deflect the exhaust gases forward. This is shown in Figure 14-7.

A cascade thrust reverser is sometimes called an aerodynamic-blockage reverser, and it is used on many large commercial engines to provide reverse thrust by deflecting the fan discharge, as well as the exhaust from the core engine.

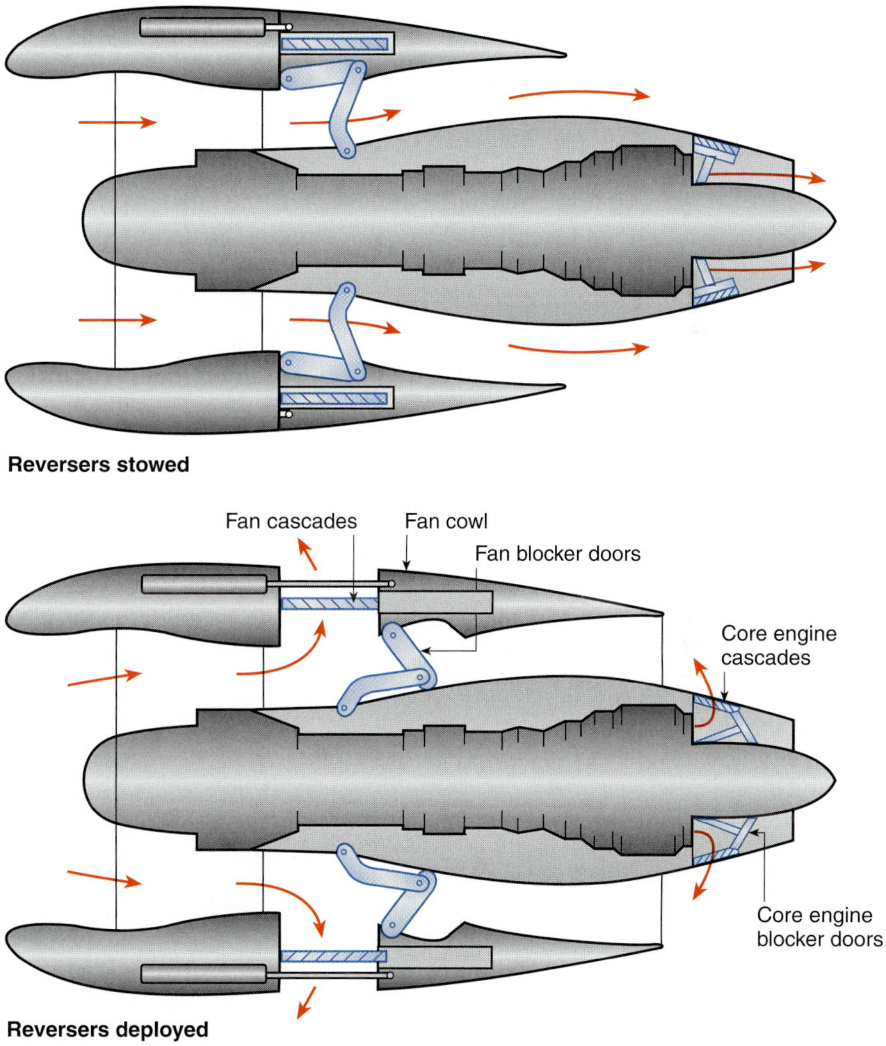

Figure 14-8. *The top view of a cascade-type thrust reverser installed on a high-bypass turbofan engine*

When the reverser is stowed, both the fan discharge duct and the core engine exhaust are unobstructed. But when deployed, the rear portion of the fan cowl moves aft, and blocker doors block the fan duct. The fan discharge air flows out through a series of reverser vanes which deflect the fan discharge air forward. At the same time, blocker doors cover the exit to the core engine exhaust nozzle, and its exhaust is directed forward through a series of cascade vanes. *See* Figure 14-8.

Thrust reversers are actuated by the pilot using a control mounted on the engine throttle. The throttle is retarded to the idle position and the reverse thrust control is moved. The reversers deploy, and further aft movement of the control increases the engine speed and thus the amount of reverse thrust.

Thrust reversers are not normally used on the ground when the speed is less than approximately 60 knots because of the danger of recirculating exhaust gases and ingesting foreign objects stirred up by the high-velocity gases.

Afterburners

Only about one-fourth of the air flowing through a gas turbine engine is used in the combustion process. The air flowing out the exhaust system is hot and contains a relatively large amount of oxygen.

afterburner. A component in the exhaust system of a turbojet or turbofan engine used to increase the thrust for takeoff and for special flight conditions.

Engine manufacturers have long recognized that by introducing additional fuel into this hot, fast-moving stream of gases, the thrust can be increased by as much as 50%. But this increase comes about only with a tremendous increase in fuel consumption, which is acceptable for military aircraft because of their critical need for sudden bursts of speed.

Afterburners were not used on civilian aircraft until the supersonic Concorde, but on this aircraft they are fuel efficient, allowing it to quickly reach a high cruising altitude where fuel consumption is economically low.

Even though a tremendous amount of fuel is consumed by the afterburners, cruising altitude is reached far sooner than possible with the engines alone, and the longer time spent at high altitude saves more fuel than is used by the afterburners.

An afterburner, called a reheat system in the United Kingdom, is essentially a ramjet engine in the exhaust of a turbojet or turbofan engine. Fuel is sprayed into the oxygen-rich, extremely hot gases flowing from the engine and is ignited by either a spark igniter or a spurt of flame through the turbine. As this fuel burns, it further heats and expands these gases. The tremendous increase in thrust is caused by the addition of the mass of fuel and the increase in momentum of the air flowing through the engine.

Jet engine fuel will not ignite in the fast-moving stream of exhaust gases, so the fuel is injected just ahead of a series of flame holders that reduce the velocity of the gases and create turbulence to allow the fuel to be properly vaporized and ignited.

When the afterburner is operating, the additional mass of gases leaving the exhaust system would increase the back pressure on the turbine if the jet nozzle area were not increased. This is done by using an adjustable nozzle that has a given area when the afterburner is not operating, but automatically increases when it is operating. Some older installations used two-position nozzles, but the more modern high-performance engines have nozzles whose area is continuously varied by the EEC (electronic engine control). *See* Figure 14-9.

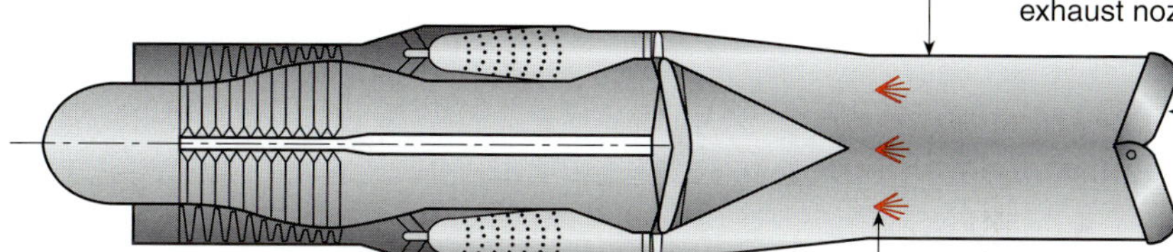

Figure 14-9. *An afterburner installed in the exhaust of a turbojet engine*

Vectored Thrust Engines

Much research has been done on decreasing the length of runway needed for turbojet aircraft. Some design studies used lift turbine engines that provided vertical thrust for takeoff. This approach is impractical because of the additional weight needed to produce thrust used only during takeoff. Other approaches have been to mount the turbine engine so it could be swiveled, pointing down for takeoff and horizontally for forward flight.

The British-designed Harrier solved the problem in a simpler way. Instead of a single exhaust nozzle, the Rolls-Royce vectored-thrust Pegasus turbofan engine has four nozzles that can be rotated in the order of 100°. *See* Figure 14-10. The two rear, or hot, nozzles deflect the exhaust from the core engine, and the two forward, or cold, nozzles deflect the fan exhaust. The four nozzles are linked together and chain-driven by an air turbine motor so that their exhaust can be directed rearward to produce forward thrust, downward to produce lift, or forward for in-flight braking. At the low speeds possible with this unique aircraft, there is not enough air flowing over the control surfaces for normal aerodynamic control, so reaction control valves located near the wing tips and fore and aft on the fuselage control a flow of compressor bleed air to produce reaction forces.

For takeoff, the nozzles are rotated so they point downward, and generate enough upward thrust to appreciably shorten the takeoff run. Once airborne, the nozzles can be pointed straight back to produce enough forward thrust to propel the Harrier in the Mach .85 to .90 range.

It is possible to increase the thrust produced by vectored-thrust engines by using plenum-chamber burning, or PCB, a form of thrust augmentation. Fuel injected into the fan-discharge air in the two forward nozzles and burned has proven to increase the overall engine thrust by as much as 50%.

PCB (plenum chamber burning). A method of thrust augmentation used on engines with vectored nozzles. Fuel injected into the fan-discharge air is burned to increase thrust.

The flexibility of the vectored thrust engine was discovered by the pilots flying the Harrier. By vectoring the thrust forward, the Harrier is able to decelerate far faster than any conventional aircraft. This feature has caused much study, and some of the new generation of fighter aircraft, notably the Lockheed Martin-Boeing F-22 and the Russian SU-35, have improved maneuverability by vectoring the thrust from their single exhaust nozzles in flight. Aircraft equipped as such are said to have the capability of VIFF (vectoring in forward flight).

VIFF (vectoring in forward flight). A method of enhancing the maneuverability of an airplane by vectoring the exhaust gases and/or fan-discharge air to produce thrust components not parallel to the longitudinal axis of the aircraft.

Research is being conducted on a thrust vector control system for transport aircraft in which the thrust of the engines can be deflected by about 20°. This deflected thrust would provide a method of control that would be independent of the aerodynamic controls.

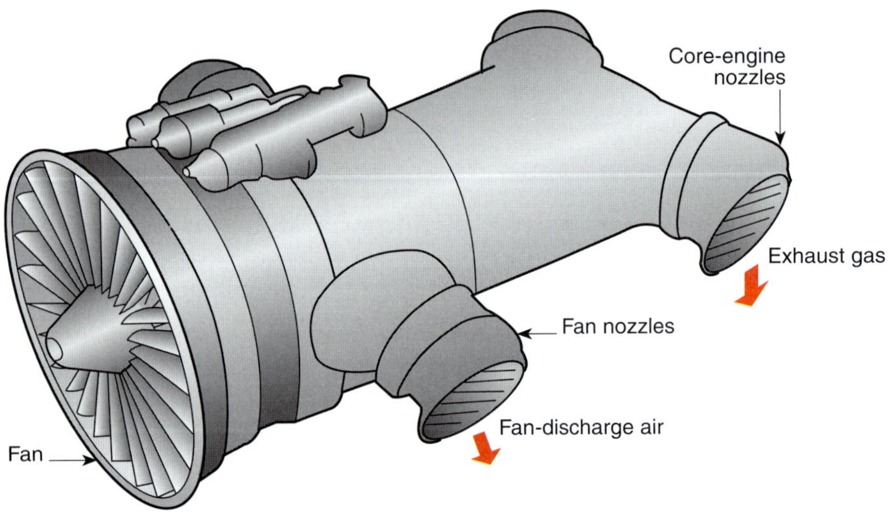

Figure 14-10. *The four controllable nozzles allow the thrust produced by this turbofan engine to be vectored downward for takeoff and aft for forward flight.*

STUDY QUESTIONS: TURBINE ENGINE EXHAUST

Answers are on Page 518. Page numbers refer to chapter text.

1. The most common type of exhaust duct for the smaller turbojet and turbofan engines is a ____________ (convergent or divergent) duct. *Page 509*

2. The exhaust cone is considered to be an ____________ (engine or airframe) component. *Page 509*

3. Exhaust gases leave the exhaust nozzle in most turbojet and low-bypass turbofan engines at the speed of ____________. *Page 509*

4. A convergent-divergent exhaust nozzle increases the thrust produced at supersonic flights by increasing the ____________ (pressure or velocity) of the exhaust gases. *Page 510*

5. Exhaust gases enter a CD nozzle at a ____________ (subsonic or supersonic) speed. *Page 510*

6. In a CD nozzle that uses two sets of overlapping flaps to control the gases as they leave the afterburner duct, the flaps are actuated hydraulically and they are controlled by the ____________. *Page 510*

7. A CD inlet duct and a CD exhaust nozzle serve ____________ (the same or a different) function. *Page 510*

8. A noise suppressor used on a turbojet engine exhaust system decreases the total amount of sound the engine produces. This statement is ____________ (true or false). *Page 511*

9. The amount of noise a turbojet engine produces is related to the ____________ of the exhaust gases. *Page 511*

10. Low-frequency sound waves travel a ____________ (greater or shorter) distance through the air than high-frequency waves. *Page 511*

11. Two reasons a turbofan engine produces less noise than a turbojet engine are:
    a. ____________
    b. ____________
    Page 511

12. A thrust reverser installed on a turbojet or turbofan engine produces a forward thrust that is between ____________ and ____________ percent of the engines rated forward thrust. *Page 511*

13. Two types of thrust reversers are those that use _______________ doors and _______________ vanes. *Page 511*

14. Thrust reversers are most effective on the ground when the speed is __________________ (low or high). *Page 513*

15. An afterburner is used on a supersonic commercial transport to decrease the time needed for the aircraft to reach its cruising ______________________ (airspeed or altitude). *Page 513*

16. When the afterburner is being used, the area of the exhaust nozzle is _____________________ (increased or decreased). *Page 514*

17. The engine in a Harrier aircraft produces a component of vertical thrust by vectoring the ______________________ (engine or exhaust). *Page 514*

Answers to Chapter 14 Study Questions

1. convergent
2. engine
3. sound
4. velocity
5. subsonic
6. Electronic Engine Control
7. a different
8. false
9. velocity
10. greater
11. a. the fan extracts additional energy from the exhaust gases.
    b. the exhaust gas velocities are lower.
12. 40, 50
13. clamshell, cascade
14. high
15. altitude
16. increased
17. exhaust

Operation & Maintenance

15

Operation & Maintenance

15

Turbine Engine Operation

Turbine engine operation is quite different from that of reciprocating engines because turbine engines produce so much more power or thrust for their weight. A reciprocating engine may tolerate an improper start-up or shut-down procedure, but it is possible to destroy a turbine engine in a matter of seconds if the wrong procedure is followed when starting it.

Turbine engines and turbine-powered aircraft are extremely expensive, and the only way for them to be cost effective is to keep them operational. Inspections and maintenance must be done with a minimum of downtime.

Starting Gas Turbine Engines

Because an improper start can easily damage a turbine engine, most maintenance operators require all technicians who are responsible for running the engines to be fully instructed in the procedures for normal starting and possible problems that could arise.

This text will not furnish instructions for starting a turbine engine, but will outline some of the more important safety considerations and describe the events that occur in a normal engine start sequence.

Before a turbine engine is started, the aircraft must be moved to a location that is approved for engine run-up, and it must be positioned so that the high-velocity wake from the engine is not directed toward other aircraft, ground servicing equipment, or any nearby buildings. A rule of thumb is that a hazard area extends approximately 100 feet behind an idling turbojet engine and 200 feet behind the engine when it is developing takeoff thrust. Figure 15-1 shows the hazard areas around operating turbojet engines.

wake. The high-velocity stream of turbulent air behind an operating aircraft engine.

Pay particular attention to the area in front of the aircraft. Before starting the engine, check the area ahead of the inlet duct for loose objects that could possibly be ingested when the engine sucks in the tremendous amount of air that flows through it when it is operating. Rocks and loose bits of concrete can cause expensive damage. No one should approach within 25 feet of an inlet duct when the engine is operating, because the low-pressure area ahead of the engine is strong enough that a person could be sucked into the engine.

The sound made by a turbine engine can permanently damage your hearing, so an approved type of hearing protector should be worn at all times when working near operating engines.

If the start is to be made from the aircraft battery, be sure it is fully charged. If a GPU is to be used, it must be appropriate for the aircraft and must be properly connected.

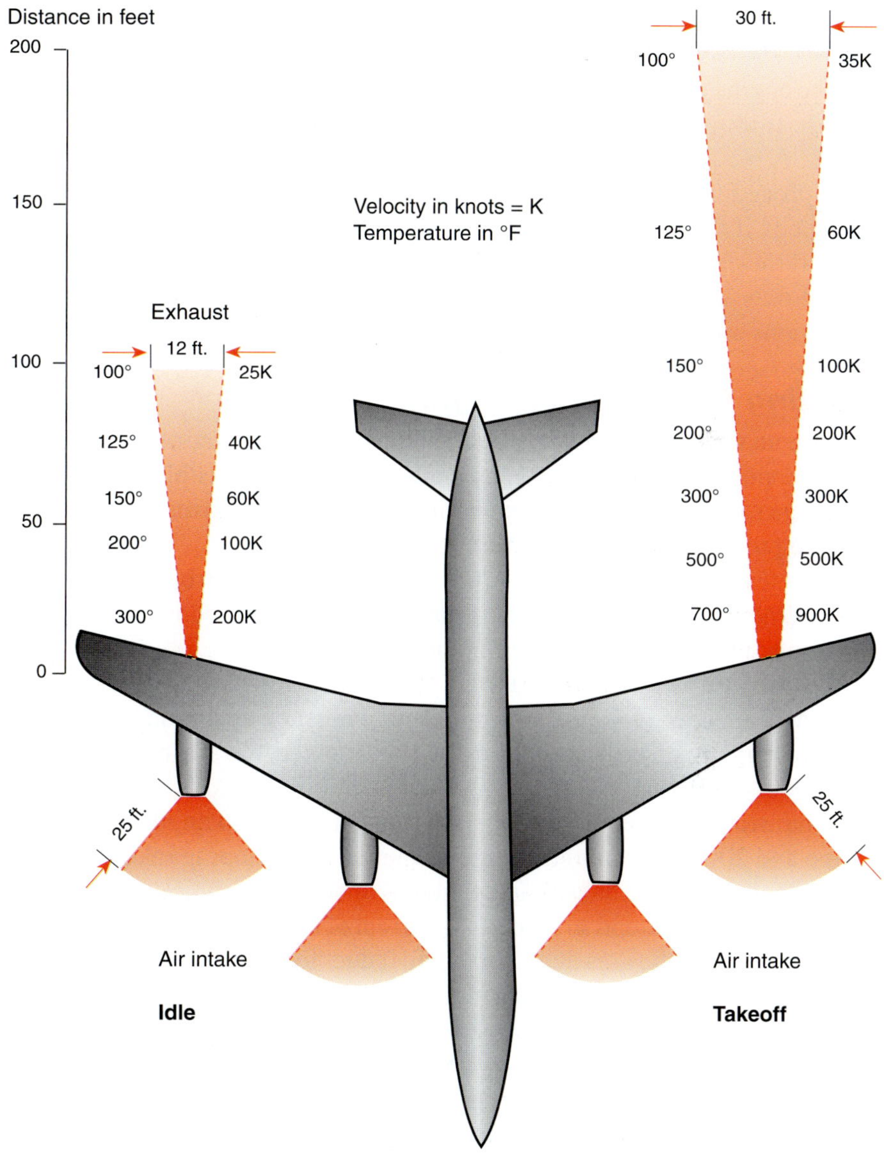

Figure 15-1. *Hazard areas ahead of and behind operating turbojet engines*

The aircraft wheels should be chocked and all controls set according to the operations manual.

- Turn on the electrical power.
- Turn on the fuel and the boost pumps.
- Move the start switch to the START position.

At this point the automatic start sequence begins. A typical turbine engine start consists of these steps:

- The starter begins to rotate the compressor to produce a flow of air through the engine that clears out any fuel vapor left from a previous operation.
- The ignition exciters are then energized and begin producing high-energy sparks.
- When the engine reaches a speed of about 10% to 20% RPM, move the start or throttle lever to the IDLE position.
- The engine speed continues to pick up, and at the correct RPM, fuel sprays into the combustion chambers and is ignited by the sparks.

A proper light up is indicated by a rapid rise in EGT that stabilizes at the correct temperature, an indication of oil pressure and a stabilized fuel flow. When the RPM reaches the correct value, the starter and the ignition system are automatically or manually de-energized. *See* Figure 15-2.

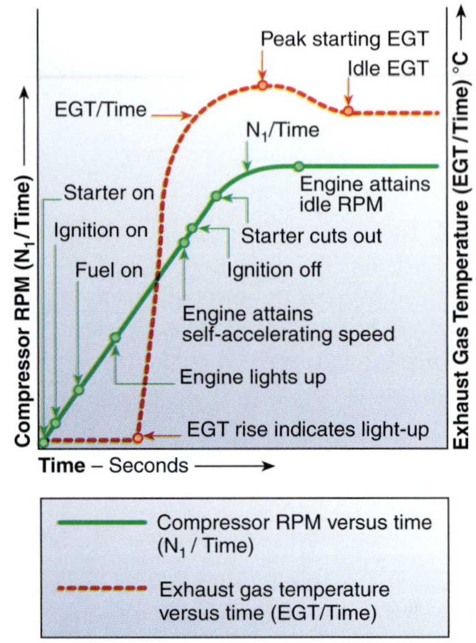

Figure 15-2. *Events in a typical starting sequence of a gas turbine engine*

RPM (revolutions per minute). The rotational speed of an engine.

EGT (exhaust gas temperature). The temperature of the gases as they leave the last stage of the turbine.

Improper Starts

There are three major instruments to monitor when starting a turbine engine: the exhaust gas temperature indicator, the tachometer, and the oil pressure gage. Abnormal readings from these instruments indicate an improper start, and require immediate action to prevent engine damage.

No Oil Pressure

If no oil pressure is indicated after a turbine engine has reached a specified speed, turn the fuel and ignition OFF, discontinue the start, and thoroughly investigate the cause of the problem.

Hot Start

If the exhaust gas temperature (EGT) rises above its allowable limit, the engine is experiencing a hot start. Turn the fuel and ignition OFF and discontinue the start. A hot start can cause serious damage, so investigate the cause and determine what damage, if any, has been done. Hot starts are usually

caused by too rich a fuel-air mixture, a result of too much fuel metered into the burners for the amount of air moved through the engine by the compressor.

Hot starts may cause damage that is not immediately apparent, so any occurrences should be recorded in the engine maintenance record.

Hung Start

A hung, or false, start is one in which the engine lights up normally, but does not reach a speed that allows continued acceleration without help from the starter. If a hung start occurs, shut the engine down and determine the reason it failed to accelerate properly.

self-accelerating speed. The speed attained by a gas turbine engine during start-up that allows it to accelerate to its normal idling speed without any assistance from the starter.

A hung start is often caused by insufficient power to the starter or by the starter cutting off before the engine reaches its self-accelerating speed.

Turbine Engine Maintenance

Turbine engine maintenance is considerably different from reciprocating engine maintenance. Under normal circumstances, reciprocating engines are overhauled on an hours-of-operation basis. When they have accumulated the number of hours determined by the manufacturer to be the economical limit for operation between overhauls, they are removed from the aircraft and completely disassembled, inspected, and repaired as necessary. They are then reassembled, tested, and reinstalled in the aircraft.

module (modular engine construction). The method of turbine engine construction in which the engine is made up of several modules, or units, that can be removed and replaced independent of the rest of the engine.

Compared to reciprocating engines, turbine engines have few moving parts that wear, and they are built in modules that can be exchanged without having to remove the engine from the aircraft.

operating cycle. One complete series of events in the operation of a turbine engine that consists of starting the engine, taking off, landing, and shutting the engine down.

Operating hours are not the only criteria used to determine when an overhaul is needed. Operating cycles are also important. One operating cycle consists of starting the engine, taking off, landing, and shutting the engine down. Engines installed on commuter airliners that make many short-duration flights will need to be overhauled with fewer total hours than engines on aircraft whose flights are all of long-duration.

On-Condition Maintenance

on-condition maintenance. A maintenance program that monitors the operating condition of an engine and allows major repairs or replacement to be made when the engine performance deteriorates to a specific level rather than on a predetermined time schedule.

Turbine engines are not necessarily removed from the aircraft and overhauled when a specified number of operating hours or operating cycles have been reached. Some engines are maintained according to an on-condition maintenance program.

On-condition maintenance is described in detail in the operations manual for the particular engine. It consists primarily of monitoring the engine performance at regular intervals and determining when maintenance is required, based on the deterioration of certain operating parameters.

A typical on-condition maintenance program includes these items:

- Check the operation of the chip detector daily, and inspect it visually for trapped particles every 50 operating hours.
- Record and maintain a log of every engine start. This record must include outside air temperature, pressure altitude, and maximum EGT.
- Listen for unusual noises during engine coast-down and if any are heard, determine their cause.
- Change engine oil at 100-hour intervals, and perform a spectrometric analysis of the oil, recording its results. Perform an acidity test of the oil at 25-hour intervals.
- Record and maintain a log of all power-assurance checks as prescribed in the aircraft flight manual.

power-assurance check. A test run made of a gas turbine engine to determine the way its performance compares with its performance when it was new or freshly overhauled.

If a problem shows up in any particular module, the module can be replaced without having to remove the engine.

Engines operated according to an on-condition schedule are allowed to remain in service as long as the approved inspections and checks indicate they are serviceable.

Trend Monitoring

Graphs or curves are used to show trends in changing conditions, and trend monitoring curves reveal much about the internal condition of a gas turbine engine.

The engine manufacturer or overhauler collects data such as N_1, N_2, EGT, and fuel flow when the engine is run in the test cell. This data is reduced to standard day conditions and used to create a series of standard reference baselines. Routinely, checks are made to compare the current performance of the engine with its test-cell performance. The same parameters are measured and reduced to standard day conditions, and the differences between the original and the new readings are plotted on a graph. One or two deviations from the baseline do not necessarily indicate an abnormal condition, but when the deviations in all the parameters are plotted over a number of operating hours or a given period of time, trends become apparent. These trends, when properly interpreted, are important maintenance tools that warn of impending problems before they could be detected by any other method.

trend monitoring. A system of routine comparison of engine performance parameters with a base line of these same parameters established when the engine was new or newly overhauled.

N_1. The rotational speed of the low-pressure compressor.

N_2. The rotational speed of the high-pressure compressor.

Types of Maintenance

Most modern turbine engines are built in a modular form to decrease downtime. Some smaller turboprop engines are made in two modules: the gas generator, or core engine, and the reduction gearing. A large turbofan engine may have a number of modules, such as the inlet case, including the inlet guide vanes, the compressor stator, compressor rotor, diffuser, accessory gear box, combustion section, turbine stator, turbine rotors, and exhaust cone.

gas generator. Another name for the core engine.

core engine. The gas generator portion of a turbofan, turboprop, or turboshaft engine.

When damage to a single module is discovered, or a component reaches its allowable service life, the module is removed from the engine and an exchange module is installed. The removed module is then returned to the manufacturer or to an approved facility to be overhauled. Many modules can be replaced without removing the engine from the aircraft.

Turbine Engine Inspections and Repair

As mentioned before, turbine engines are designed for efficient maintenance with as little downtime as possible. One procedure that has improved efficiency is the built-in provision for inspecting the inside of the engine without disassembling it. This is done with a borescope or with one of its modern counterparts.

Borescope, Fiberscope, Electronic Imaging

borescope. An inspection tool for viewing the inside of a turbine engine without having to disassemble it.

It has long been the practice when inspecting reciprocating engines to disassemble them and examine the component parts. As engine output increased over the years, the susceptibility to detonation became a serious problem, and borescope inspection of the inside of installed cylinders became an important maintenance tool.

Turbine engines are lightweight for the amount of power or thrust they produce and are expensive to disassemble. Because of this, engine manufacturers have placed borescope ports at strategic locations, so that technicians can examine critical internal areas without disassembling the engine.

fiber optics. The technique of transmitting light or images through long, thin, flexible fibers of plastic or glass.

There are three types of internal visual inspection instruments commonly used in turbine engine maintenance: rigid-tube borescopes, flexible fiber-optic scopes, and video-imaging scopes.

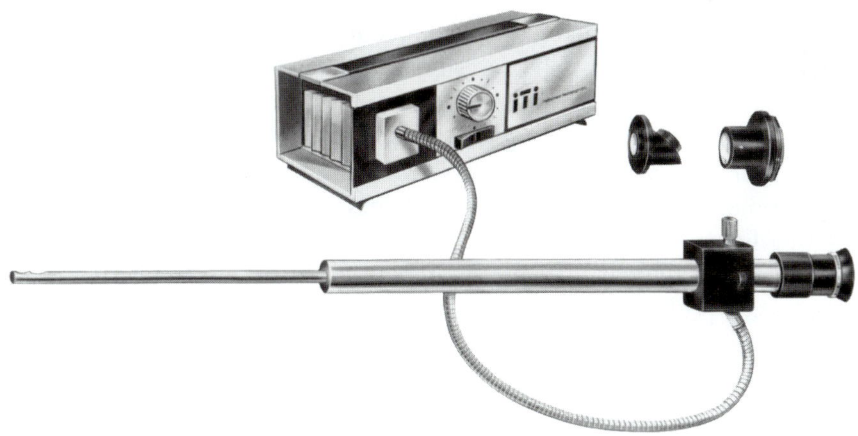

Figure 15-3. *Rigid-tube borescope*

A rigid-tube borescope can be inserted into the engine through an inspection port, and a controllable power source allows you to regulate the intensity of the light produced by the lamp at the end of the scope tube. Insert the tube into the appropriate port and adjust the light. Aim the instrument at the area to be inspected and focus to get the sharpest image.

Flexible-tube fiber optic scopes are more versatile than the rigid-tube scope. These instruments consist of a light guide and an image guide made of bundles of optical fibers enclosed inside a protective sheath. A power supply with a controllable light source is connected to the light guide, and an eyepiece lens is situated so it can view the end of the image guide. Bending and focusing controls on the instrument housing allow you to guide the probe inside the

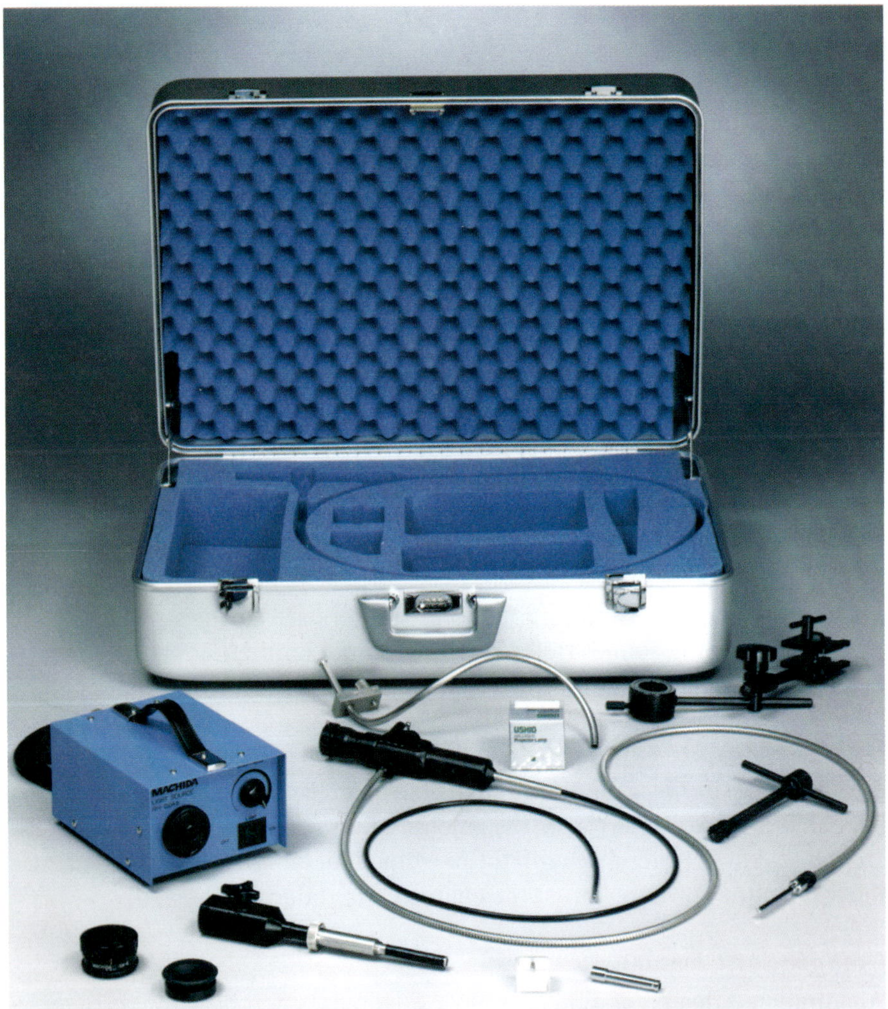

Figure 15-4. *Flexible-tube fiber-optic inspection scope.*

Figure 15-5. *Video display for a fiber-optic inspection scope*

digitized image. A modified image picked up by the miniature TV camera in the end of a fiber optic probe.

motor the engine. To rotate a turbine engine with the starter with the fuel and ignition turned off.

cold section. The portion of a gas turbine engine ahead of the combustion section.

engine and focus to get the clearest image of the area. Adapters are normally included that allow attachment of a still or video camera to the eyepiece, providing a permanent record of the interior of the engine. This state-of-the-art inspection tool is the video imaging system shown in Figure 15-4.

The probe is inserted into the engine through one of the inspection ports, and the tip is guided to the area to be inspected. The sensor in the tip of the probe acts as a miniature TV camera and picks up an image of the area illuminated by the probe. This image is digitized, enhanced, and displayed on a video monitor. It can also be recorded on video tape. *See* Figure 15-5.

Routine Inspections

There are three basic routine inspections to which every aircraft turbine engine is subjected: preflight inspections, cold-section inspections, and hot-section inspections.

Preflight Inspection

Typical routine inspections of a turbine engine require the following to be checked before each flight:

- Exterior of the cowling, including the inlet area and attachment to the airframe.
- Inlet guide vanes and the first stage compressor blades of the fan. When the compressor is rotated, there should be no unusual noise.
- Exhaust area including the rear turbine and the thrust reversers. Check for cracks.
- Lubricating oil quantity. This should be checked as soon after engine shutdown as possible. Record the amount of oil added.
- CSD (constant-speed drive) oil quantity.
- Starter oil quantity.
- Ignition system. Determine that the igniters are firing by listening for their snapping sound. Before turning on the ignition, motor the engine with the starter to pull through enough air to remove fuel vapors.

Cold Section Inspection

The cold section of a turbofan engine, which includes the fan, compressors, and diffuser, determines the amount of air supplied to the combustion chamber. Any interference with the movement of air will seriously decrease the amount of power or thrust the engine can produce.

Compressor Cleaning

Accumulated dirt on compressor blades decreases aerodynamic efficiency and deteriorates engine performance. Dirty compressor blades normally cause high EGT and unsatisfactory acceleration.

Compressors may be cleaned in either of two ways: grit blasting or liquid washing.

Grit blasting is done by blanking off certain orifices and valves in the air path and injecting an approved soft grit material into the inlet duct while the engine is being rotated with the starter. Exercise extreme care when using the grit material to prevent damage to the corrosion-protected surfaces inside the engine.

Liquid washing is by far the more widely used method of compressor cleaning. There are two types of liquid washing: desalination washing and performance-recovery washing.

When an engine is operated in a heavily salt-laden environment, it is usually recommended that the compressor be washed on a regular schedule by spraying demineralized water through the engine while it is being motored by the starter at about 15% to 20% RPM. This washing removes the salt deposits before they build up enough to cause performance deterioration. When using the starter to rotate the engine, be sure not to operate it longer than the manufacturer allows, as it can overheat and become permanently damaged.

If performance has deteriorated, the engine may be given a performance-recovery washing. The engine is rotated with the starter or is run at idle speed, and a stream of demineralized water and cleaning fluid is injected into the inlet duct through either a hand-held spray nozzle or a built-in ring of spray nozzles. The active agent in the cleaning fluid breaks down the contaminants so they can pass harmlessly through the engine. The agent leaves a corrosion-inhibiting film on the compressor to prevent the accumulation of further deposits. The cleaning agent is then rinsed off by spraying clean demineralized water through the engine. Shut off the water and run the engine up to a specified RPM, to completely dry it out.

blending. A method of repairing damaged compressor and turbine blades by removing the damage with a fine file to form a shallow depression with generous radii.

jeweler's file. A small, fine-cut metalworking file used by jewelers to manufacture and repair jewelry.

riffle file. A hand file with its teeth formed on a curved surface that resembles a spoon.

Compressor Repair

The extent of repairs to compressor blades that the engine manufacturer allows is described in detail in the engine service manual.

The blades are inspected, and damage is evaluated to determine if it falls within the repairable limits allowed. If it is excessive, the compressor module must be returned to the manufacturer or sent to an overhauler for repair.

Damage of the type in Figure 15-6 that is within allowable limits may be repaired by blending. The damage is removed with a fine-cut jeweler's file, moved parallel to the length of the blade, and the file marks are then removed with a fine-grit abrasive stone. A properly blended area must be as smooth as the undamaged portion of the blade.

Corrosion pits can be removed with a fine riffle file and the file marks removed with a fine abrasive cloth. If the blades are bent, do not straighten them; instead, send the compressor module out for major repair. *See* Figure 15-7 on the next page.

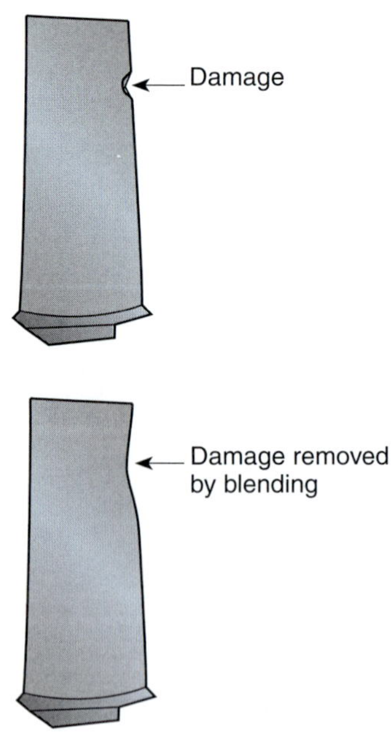

Figure 15-6. *Compressor blade repair*

Term	Appearance	Cause
Bow	Bent blade	Foreign objects
Burn	Injury to surface, evidence of discoloration	Excessive heat
Burr	A ragged or turned out edge	Grinding or cutting action
Corrosion	Pits and breakdown of surface	Corrosive environment
Cracks	A partial fracture or separation	Excessive stress due to shock, overloading, or faulty processing
Creep	Elongation of a blade	Continued and/or excessive heat and centrifugal force
Dent	Small smoothly rounded hollow	Striking part with a dull object
Gall	Transfer of metal from one surface to another	Severe rubbing
Gouging	Displacement of material from a surface; a cutting or tearing	Presence of a comparatively large foreign body between moving parts
Score	Deep scratches	Presence of chips between surfaces
Scratch	Narrow, shallow marks	Sand or fine foreign particles between surfaces

Figure 15-7. *Typical damage to compressor elements*

hot section. The portion of a gas turbine engine that operates at a high temperature.

Hot Section Inspection

The hot section of a turbine engine is the heart of the system. It is this section and its components that extract energy from the flow of extremely hot, high-pressure gas.

The hot section consists of the fuel nozzles, combustion chambers, turbine inlet guide vanes, turbine wheels and blades, and exhaust system. These components are designed and constructed to extract the maximum amount of energy from the fuel, and any damage or distortion will lower the component and system efficiency enough to impair engine performance.

Hot section inspections are normally required on a time or cycle basis, but anytime the EGT becomes excessive, a hot section inspection should be performed.

Hot section inspections can usually be conducted with the engine on the airframe. They consist essentially of the following items:

- Review the engine operating records for the period just before the inspection for indications of hot starts, hung starts, takeoffs involving overtemperature or overspeed conditions, or fluctuations in the oil or fuel pressure.
- Connect accurately calibrated test instruments, and run the engine to check its actual performance. Record the results.
- Follow the engine manufacturer's instructions in detail, and use the required tools to remove the power section from the engine.

- Remove the igniters and fuel nozzles. Check the nozzle flow pattern and clean them if the pattern is not as specified in the engine maintenance manual. A partially clogged fuel nozzle will cause localized hot spots in the combustion liner.
- Remove the combustion liner and check for cracks and indication of localized hot spots. If there are cracks that exceed the allowable limit, the liner will have to be repaired at a facility equipped and approved by the engine manufacturer.
- Remove the inlet guide vane ring, and carefully examine the condition of all vanes.
- Measure the clearance between the tips of the turbine blades and the housing, and the clearance between the turbine wheel and the exhaust cone; both measurements are critical.
- Remove the turbine wheel, and examine all the blades for indication of damage and tip erosion. Stress-rupture cracks that appear perpendicular to the leading and trailing edges are one of the more prevalent types of damage. Waviness of the leading or trailing edges is an indication of overheating. Damaged turbine wheels cannot be repaired in the field, but can be repaired at facilities which specialize in this type of work.

When all the damaged parts have been repaired or a replacement module is to be installed, check the replacement parts carefully, and reassemble the engine. Make a calibration run-up with the same test instrumentation used on the preinspection run-up. This new information compares the present performance with that before the inspection and provides a new baseline to use in trend monitoring.

The following pointers are important when working on the hot section of an engine:

- When disassembling the engine, mark the location of the parts with the type of marker approved for this purpose. Felt-tip markers are acceptable, but never use a lead pencil or a wax or grease pencil. The carbon from the graphite or the grease or wax will infuse into the hot metal and make it brittle, causing it to crack.
- Replace all items, such as seals and locks, specified by the engine manufacturer. Use only new components, prepared and installed according to the instructions in the service manual.
- Follow the engine manufacturer's service manual in detail when making any inspection. The allowable damage varies from one engine model to the next. Only repairs specifically allowed by the manufacturer should be made.

Continued

- When shipping turbine wheels to the engine manufacturer or to an overhaul facility, be sure to use the correct shipping container for that particular wheel. These wheels are extremely expensive and can be easily damaged by careless handling in transit.

Nonroutine Inspections

In addition to routine inspections, other inspections must be conducted when certain abnormal conditions arise. These conditions, specified in the operations manual, may be as obvious as foreign object damage (FOD), or as subtle as changes in parameters noted by trend monitoring or oil analysis.

Foreign Object Damage (FOD)

FOD (foreign object damage). Damage to the components in the gas path of a turbine engine caused by ingesting objects that are not part of the engine.

EPR (engine pressure ratio). The ratio of the turbine discharge total pressure to the compressor inlet total pressure.

FOD is one of the major conditions to guard against in turbine engine operation. The air inlet acts as a huge vacuum cleaner and ingests small debris such as concrete chips or rocks from the ramp, or tools that have been carelessly left within range. Birds and ice can be ingested during flight.

FOD to the fan or inlet guide vanes or to the visible stages of the compressor is justification for removal of the engine and a complete over-haul. It is impossible to determine the full extent of the damage without disassembling the engine.

Less obvious damage that causes performance deterioration calls for an examination of the engine's interior with borescopes or videoscopes. Performance deterioration that could be caused by FOD may be detected by an increase in EGT with a corresponding decrease in EPR, or by a change in the normal relationship between the RPM of the low- and high-pressure compressors (N_1 and N_2).

FOD to the blades in the fan can cause nicked or scratched blades or shingled blades.

Nicks and scratches on some blades and in certain locations on the blades can be repaired by careful filing and blending. If the damage is within the limits specified in the service manual, it can be carefully removed with a jeweler's file by filing parallel to the length of the blade. Radius the cleaned out area as specified in the service manual, and remove all the file marks with a 150-grit abrasive stone. If the damage is beyond the limits allowed for field repair, the fan must be removed and sent to a repair facility to be disassembled, the blades repaired or replaced and the fan rebalanced.

Many fans have a mid- or partial-span shroud, or snubber, formed by projections on both sides of the blade that touch the projections on adjacent blades, to form a shroud around the central portion of the fan. *See* Figure 10-55 on Page 397. This shroud stiffens the blades and prevents flutter. If there is any sudden deceleration of the fan, such as is caused by a bird strike,

or if the engine should overspeed, the blades can shift so that the shroud segment on one blade overlaps the segment on the adjacent blade in much the same way as shingles on a roof. This type of damage is called shingling. Any shingled blades must be examined for indication of damage.

One cause of FOD that we technicians do not like to think about is damage caused by our own carelessness. Turbine inlet ducts make convenient shelves for holding tools and parts, but placing anything in the duct is bad practice. A bolt, nut, or wrench left in the duct when the engine is started can destroy the engine.

Be sure your tools are all accounted for when your work is finished, and properly dispose of every used piece of safety wire and packing material.

Overtemperature or Overspeed Operation

The rapid response of the FADEC used on many modern high-performance turbine engines has virtually eliminated overtemperature and overspeed conditions, but engine manufacturers do have limits that, if exceeded, require specialized inspections.

Overtemperature operations require careful inspection of the hot section, and overspeed operations require a careful inspection of the fan and compressors. The engine service manual details the limits and describes the inspections that must be performed when limits of either parameter are exceeded.

Repair Considerations

The high precision with which turbine engines are built and the high temperatures at which they operate demand the highest quality repairs. Technicians are not allowed to deviate from the procedures detailed in the maintenance manuals furnished by the engine manufacturer and approved by the FAA.

It is extremely important that all threaded fasteners on a turbine engine be properly torqued and safetied. Here, we will consider the importance of torque wrenches and notice some requirements for proper safety wiring.

Torque Wrenches

The precision with which a turbine engine is built and the tremendous amount of power developed by such a lightweight device make it imperative that every fastener be tightened to the exact torque specified by the manufacturer. To obtain the maximum strength from a threaded fastener, the load applied to the threads must be greater than the maximum load applied to the joint.

If a threaded fastener does not fail while being torqued, it will not fail in service. While being torqued, the fastener is subjected to both torsional and tensile stresses. After installation, it is subjected only to tensile stress.

Torque wrenches are precision tools and must be handled with care. Check their calibration on a regular schedule against a master, the accuracy of which can be traced back to the National Institute of Standards and Technology (NIST). This was formerly known as the National Bureau of Standards.

Unless otherwise stated in the service manual, specified torque values are for clean, dry threads. When lubricated threads are called for, the type and amount of lubricant will be specified.

There are some applications in which a series of fasteners must be torqued to a specified low value in a certain sequence, and then retorqued to a higher final value. Follow the engine service manual carefully when special instructions such as this are specified.

When using a snap-action, or micrometer-type, torque wrench, stop the pull as soon as the wrench snaps. The snapping action is an indication that the desired torque has been reached, but it does not prevent you from applying further torque.

prevailing torque. The torque required to turn a threaded fastener before it contacts the surface it is intended to hold.

Some applications measure the torque applied to a threaded fastener by measuring the amount the nut is turned after it contacts its bearing surface. Torque measured by this method is not affected by the prevailing torque of the nut, or by lubrication or lack of lubrication of the threads.

There are times when you must use special adapters on a torque wrench. When adapters such as those in Figures 15-8 and 15-9 are used, you may determine the wrench indication needed to obtain a specific torque at the adapter by using the appropriate formula.

When an adapter is used that adds length to the wrench arm, the wrench indication needed for a specific torque at the adapter is found by using the formula:

$$T_W = \frac{T_A \cdot L}{L + E}$$

T_W = Torque indicated on the wrench
T_A = Torque applied at the adapter
L = Lever length of torque wrench
E = Arm of the adapter

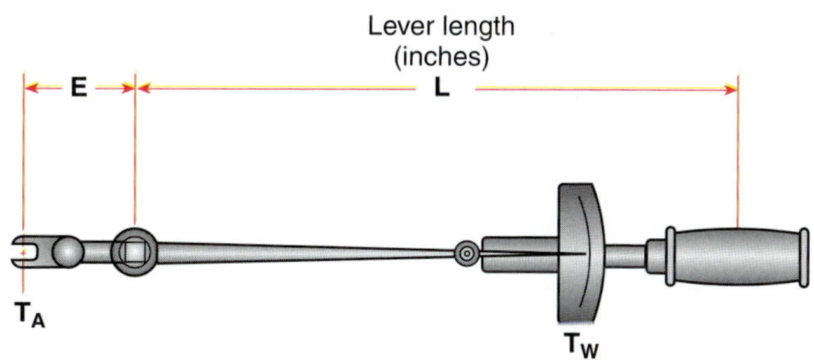

Figure 15-8. *This adapter lengthens the wrench arm.*

When the adapter shortens the wrench arm, the wrench indication needed for a specific torque at the adapter is found by using the formula:

$$T_W = \frac{T_A \cdot L}{L - E}$$

T_W = Torque indicated on the wrench
T_A = Torque applied at the adapter
L = Lever length of torque wrench
E = Arm of the adapter

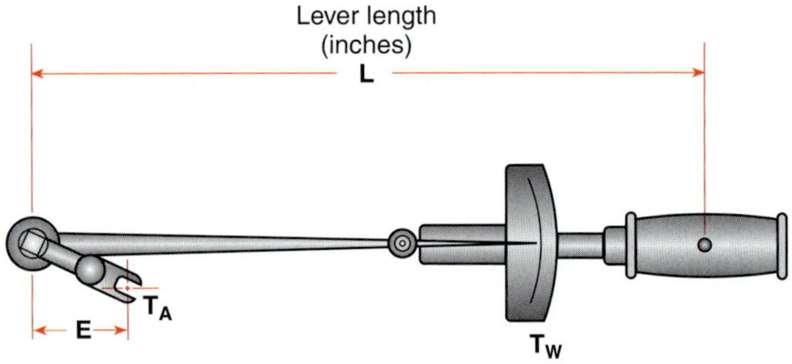

Figure 15-9. *This adapter shortens the wrench arm.*

Torque wrenches are available with calibrations in several different increments. Figure 15-10 allows you to convert between the indications of six of the most widely used increments.

Inch Grams	Inch Ounces	Inch Pounds	Foot Pounds	Centimeter Kilograms	Meter Kilograms
7.09	0.25				
14.17	0.5				
21.26	0.75				
28.35	1.0				
113.40	4.0	0.25			
226.80	8.0	0.50			
453.59	16.0	1.00	0.08	1.11	
	96.0	6.00	0.50	6.92	
	192.0	12.00	1.00	13.83	0.138
	384.0	24.00	2.00	27.66	0.277
	576.0	36.00	3.00	41.49	0.415
	768.0	48.00	4.00	55.32	0.553
	960.0	60.00	5.00	69.15	0.692
		72.00	6.00	82.98	0.830
		84.00	7.00	96.81	0.968
		96.00	8.00	110.64	1.106
		108.00	9.00	124.47	1.245
		120.00	10.00	138.31	1.383

Figure 15-10. *Torque conversion table*

Safety Wiring

The effect of vibration on the lightweight structure used in an aircraft and its engines makes safetying all fasteners an important function of aviation maintenance. The principles of safety wiring are quite simple, but skill is developed only by practice. Figure 15-12 shows one of the most commonly used procedures.

The following are some basic rules of safetying that must be observed:

- Before safety wiring a fastener, be sure that it is properly torqued.
- Be sure to use the method of safetying specified in the engine maintenance manual. When safety wiring is specified, the size and type of wire called for must be used.
- Install the wire so that it always pulls the fastener in the direction of tightening. This will prevent the fastener from backing off if it should loosen.
- Loop the wire around the outside of the fastener so that it is routed under the wire protruding from the hole. This causes the loop to stay down and prevents slackening. The direction of twist should reverse from run to run, or from run to pigtail.
- Be sure the twists are tight and even, and the twisted wire between the fasteners is taut but not too tight. The recommended number of twists per inch depends upon the diameter of the wire. *See* Figure 15-11. A twist is considered to be one half of a complete turn.
- Be sure the pigtail at the end of the wire is no more than ¾-inch long and has a minimum of 4 twists. Double the cutoff end back and bend it under so it will not snag or cut anything that rubs across it.

Safety wire twisting can be done with a pair of duckbill pliers, but using a safety wire twisting tool like that in Figure 15-13 will make the job much faster and the results more uniform.

Wire diameter (inch)	**Twists per inch**
0.020 – 0.025	8 – 14
0.032 – 0.041	6 – 11
0.051 – 0.060	4 – 9

Figure 15-11. *The recommended number of twists per inch is determined by the diameter of the safety wire.*

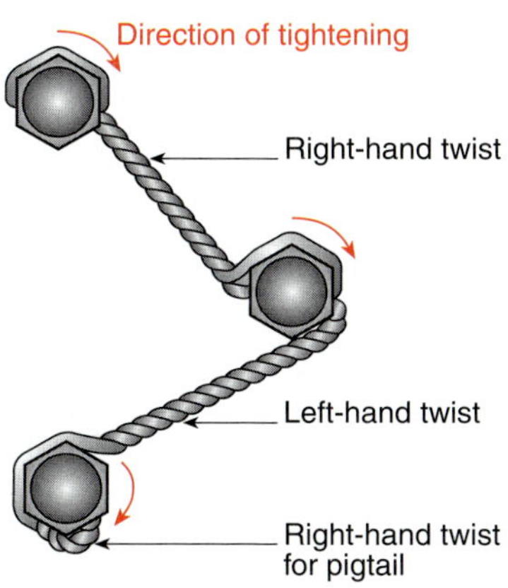

Figure 15-12. *Safety wire should always pull fastener in the direction of tightening.*

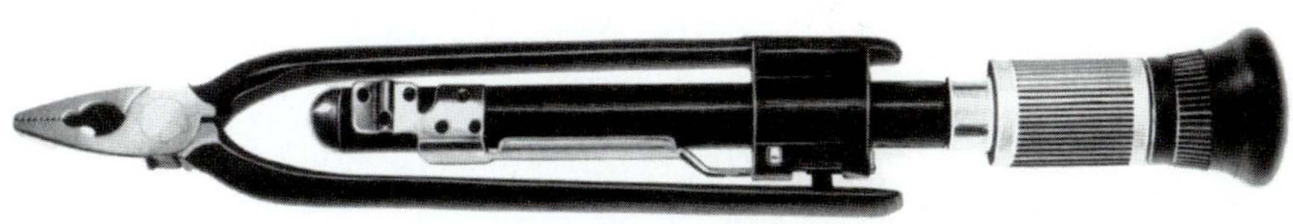

Figure 15-13. *Reversible safety wire twisting tool allows the same tool to be used for twisting left- and right-hand runs and pigtails.*

Turbine Engine Testing

It is important that maintenance technicians and flight crews have accurate information about the operating conditions of a turbine engine. The Jet-Cal Analyzer/Trimmer, a handy test instrument manufactured by Howell Instruments, Inc. of Fort Worth, Texas, is one of the standard instruments for turbine engine maintenance. This portable instrument has a master EGT, EPR, and tachometer, and it is connected into the aircraft instrument system with leads that are part of the tester. The EGT, EPR, and tachometer in the aircraft can be checked for accuracy without having to run the engine.

The JetCal is able to check the EGT system for continuity and resistance. Electrical heaters are placed over the thermocouple probes, and the cockpit instrument is compared with the actual temperature of the probes. A master AC signal is sent to the tachometer, and the cockpit RPM indication is compared with that of the master indicator. The EPR indicator is also checked by comparing its reading with that of the master instrument.

Instruments on the Analyzer/Trimmer can also be used when running the engine for trimming. This JetCal has a self-contained computer that reduces the instrument indications to standard day conditions so that the data from the actual engine performance can be compared with the manufacturer's dataplate performance. *See* Figure 15-14 on the next page.

standard day conditions. Conditions that have been decided upon by the ICAO (International Civil Aeronautics Organization) for relating all aircraft and engine performance to a given reference.

dataplate performance. The performance specifications of a turbine engine that were observed and recorded by the engine manufacturer or overhauler and recorded on the engine dataplate.

Engine Trimming

Engine trimming is the adjustment of the fuel control to obtain the correct idle and trim speed RPM. Every engine has its own trimming instructions and these must be followed in detail. To conduct a typical trimming procedure, follow these steps:

1. Move the aircraft to a location where there is no debris on the ramp that can be sucked into the engine. Head the aircraft directly into the wind.
2. Connect accurately calibrated test instruments into the system.
3. Install the fuel control trim stop recommended by the engine manufacturer.
4. Accurately measure the ambient temperature and the existing barometric pressure at the engine air inlet. This information allows the data collected on the trimming run to be corrected to standard-day, sea-level conditions.
5. Run the engine as specified in the engine maintenance manual, and if any adjustments are needed, make them in accordance with the manufacturer's instructions.

engine trimming. A maintenance procedure in which the fuel control is adjusted to cause the engine to produce the required EGT or EPR at a specified RPM.

Figure 15-14. *JetCal Analyzer/Trimmer*

Turbine Engine Troubleshooting

The technique of troubleshooting a turbine engine is no different from that used for other aircraft components or systems. All systematic troubleshooting must follow these basic rules:

- *Know the way the system should operate.* Be familiar with how the components work together and separately.
- *Observe how the system is operating.* Any difference between the way a system is operating and the way it should operate is an indication of trouble.
- *Divide the system to find the trouble.* Isolate the system that is malfunctioning, then find which subsystem or component is not working as it should.
- *Look for the obvious problem first.*

Every engine design is different in its operation, and therefore no troubleshooting guide can be comprehensive. Figure 15-15 lists some of the more common troubles and their basic causes found on a wide range of engines.

Problem	Possible Cause
Compressor fails to rotate during attempted start	Inadequate or no air supply to starter; Internal rotating parts binding; Sheared starter drive shaft or jaws
Engine fails to light up when power lever is advanced to idle	Main power switch off; Defective ignition system; No fuel to engine
Engine lights up but fails to accelerate to idle	Starter cut out too soon; Inadequate air supply; Loose or broken burner pressure sensor line; Burner pressure limiter stuck open; Fuel control acceleration schedule
out of limits; Fuel control improperly rigged	
Engine fails to decelerate properly	Fuel control improperly rigged; Malfunctioning fuel control
Fluctuating RPM	Contaminated fuel; Defective tachometer; Malfunctioning compressor bleed valves; Fuel tank vent obstructed; Clogged fuel filter
Inability to obtain maximum fuel flow	Improper trim; Incorrect travel of power lever linkage; Malfunctioning fuel control; Clogged fuel filters
Engine slow in accelerating	Defective fuel control
Hot start	Defective ignition; Insufficient cranking speed; Accumulation of fuel in engine
Low EGT	Improper trim; Malfunctioning fuel control; Defective thermocouple or EGT gage
Engine roughness or vibration	Interference between turbine rotor inner air seals and inner seal rings; Main bearing failure; Wrong combustion flame pattern
High EGT	Improper trim; Insufficient air; Defective thermocouple or EGT gage
High oil temperature	Insufficient oil in tank; Clogged oil strainer; Bearing failure; Defective oil pump
Low oil temperature	Defective oil temperature gage; Thermostatic valve in fuel-oil cooler not functioning
High or low oil pressure	Oil pressure relief valve malfunctioning; Main oil pump malfunctioning; Defective oil pressure gage
Fluctuating oil pressure	Insufficient oil supply; Clogged or dirty strainer; Malfunctioning oil pressure relief valve
Excessive oil consumption	Oil leakage; Loose oil tank cap; Malfunctioning breather pressurizing valve

Figure 15-15. *Basic turbine engine troubleshooting*

STUDY QUESTIONS: TURBINE ENGINE OPERATION & MAINTENANCE

Answers are on Page 546. *Page numbers refer to chapter text.*

1. A turbine engine is ____________________ (more or less) tolerant of improper starting procedures than a reciprocating engine. *Page 521*

2. The hazard area is considered to extend ____________ feet behind an idling turbojet engine and _______________ feet behind one that is developing takeoff power. *Page 522*

3. It is dangerous to be closer than ________________ feet to the inlet ducts of any operating turbojet or turbofan engine. *Page 522*

4. When a turbine engine is being started, the ignition sparks occur _________________ (before or after) the fuel is sprayed into the combustion chambers. *Page 523*

5. A proper light-up is indicated by a rapid rise in _______________ . *Page 523*

6. Three instruments that should be monitored when starting a turbine engine are:
   a. ___________________________
   b. ___________________________
   c. ___________________________
   Page 523

7. A hot start is normally caused by a fuel-air mixture that is too ______________ (rich or lean). *Pages 523, 524*

8. A hung, or false, start is usually caused by a malfunction of the ____________________ (fuel control or starter). *Page 524*

9. Two methods of determining when a turbine engine should be given a routine inspection are:
   a. ____________________________
   b. ____________________________
   Page 524

10. A sequence of events that consists of an engine start, takeoff, landing, and shutdown is called a/an ________________________ . *Page 524*

11. Engines operated on an approved on-condition maintenance schedule ________________ (are or are not) required to be overhauled when a specific number of operational cycles have been reached. *Page 524*

12. Trend monitoring compares the current operating conditions of an engine with standard base line conditions that were recorded by the engine ________________ or __________________ when the engine was run in the test cell. *Page 525*

13. Inspection of the inside of a turbine engine installed on the aircraft is possible by using a/an ____________________. *Page 526*

14. Inspection of the inside of a turbine engine is done through ____________________ provided by the engine manufacturer. *Page 526*

15. Before the ignition system is checked on a preflight inspection, the engine must be motored with the starter to move air through the engine and completely purge it of any ___________________. *Page 528*

16. Three components of the cold section of a turbofan engine are:
    a. ___________________________
    b. ___________________________
    c. ___________________________
    Page 528

17. Two indications of a dirty compressor are:
    a. ___________________________
    b. ___________________________
    Page 528

18. Two types of field cleaning of a turbine engine compressor are:
    a. ___________________________
    b. ___________________________
    Page 529

19. Two types of liquid washing of a turbine engine compressor are:
    a. ___________________________
    b. ___________________________
    Page 529

20. The most frequently used method of field cleaning a compressor is to wash it with __________________ water and a cleaning fluid. *Page 529*

21. If the engine is run for a performance-recovery compressor washing, it is operated at _______________ (idle or takeoff) RPM. *Page 529*

Continued

STUDY QUESTIONS: TURBINE ENGINE OPERATION & MAINTENANCE *Continued*

22. The amount of repair that can be made to a compressor blade is detailed in the ______________________ . *Page 529*

23. When the edges of a fan or compressor blade are repaired by blending, the damaged area is removed with a fine file, and the file marks are removed with a fine-grit ______________________ . *Page 529*

24. It __________________ (is or is not) recommended that bent compressor blades be straightened. *Page 529*

25. Five components in the hot section of a turbine engine are:
    a. ________________________________
    b. ________________________________
    c. ________________________________
    d. ________________________________
    e. ________________________________
    Page 530

26. A hot section inspection is normally required after a hot start or if, for any reason, the _____________ has exceeded a specified limit. *Page 530*

27. A hot section inspection normally requires that the engine be removed from the aircraft. This statement is ___________________ (true or false). *Page 530*

28. A localized hot spot in a combustion chamber can be caused by a partially clogged _________________________ . *Page 531*

29. Stress-rupture cracks normally appear ___________________ (perpendicular or parallel) to the leading and trailing edges of the turbine blades. *Page 531*

30. Waviness of the leading or trailing edges of a turbine blade is normally an indication of ___________________ . *Page 531*

31. Two critical measurements to be made of a turbine wheel during a hot section inspection are:
    a. ______________________________
    b. ______________________________
    Page 531

32. Instruments installed in the aircraft _________________ (are or are not) normally considered accurate enough for a preinspection and postinspection runup for a hot section inspection. *Page 530*

33. The information gained in a postinspection run-up after a hot section inspection is used as a base line for _______________________ . *Page 531*

34. A lead pencil _______________________ (is or is not) an acceptable marker to use on hot section components. *Page 531*

35. When removing damage to the leading or trailing edge of a compressor blade, the file should be moved _______________________ (parallel to or across) the length of the blade. *Page 529*

36. When the segment of a midspan shroud of a turbofan blade overlaps the segment on the adjacent blade, the fan blade is said to be _______________________ . *Pages 532, 533*

37. Fan blade shingling is normally caused by a ___________________ or by a/an ___________________ condition. *Pages 532, 533*

38. An engine that has had an overtemperature operation should be given a _______________________ inspection. *Page 533*

39. For the maximum strength to be provided by a threaded fastener, the load applied when the fastener is torqued must be ___________________ (greater or less) than the maximum load applied to the joint. *Page 533*

40. Torque wrenches used in aviation maintenance must be calibrated using masters that have their accuracy traced to the _______________________________ . *Page 534*

41. Unless otherwise stated, the torque specified in torque tables is for clean and __________________ (dry or lubricated) threads. *Page 534*

42. When a torque wrench with a length of 15 inches is used with an adapter that increases the arm length by 3 inches, a torque of 200 inch-pounds will be produced at the adapter when the torque wrench indicates ___________________ inch-pounds. *Page 534*

43. When a torque wrench with a length of 15 inches is used with an adapter that shortens the arm length by 3 inches, a torque of 200 inch-pounds will be produced at the adapter when the torque wrench indicates ___________________ inch-pounds. *Page 535*

44. A JetCal Analyzer/Trimmer has three accurate master instruments. These are:
    a. ______________________________
    b. ______________________________
    c. ______________________________
    Page 537

Continued

STUDY QUESTIONS: TURBINE ENGINE OPERATION & MAINTENANCE *Continued*

45. When a turbine engine is trimmed, the ambient air temperature should be measured adjacent to the ____________________ . *Page 537*

46. The ambient air temperature and barometric pressure must be corrected to ____________________ conditions, in order to compare the instrument indications with the data furnished by the engine manufacturer. *Page 537*

47. When a turbine engine is trimmed, the aircraft should be headed ____________________ (into or across) the wind. *Page 537*

48. The four basic rules for systematic troubleshooting are:

    a. ____________________
    b. ____________________
    c. ____________________
    d. ____________________

    Page 539

Answers to Chapter 15 Study Questions

1. less
2. 100, 200
3. 25
4. before
5. EGT
6. a. EGT indicator
   b. tachometer
   c. oil pressure gage
7. rich
8. starter
9. a. operating hours
   b. operating cycles
10. operating cycle
11. are not
12. manufacturer, overhauler
13. borescope
14. borescope ports
15. fuel vapors
16. a. fan
    b. compressors
    c. diffuser
17. a. high EGT
    b. unsatisfactory acceleration
18. a. grit blasting
    b. liquid washing
19. a. desalination washing
    b. performance-recovery washing
20. demineralized
21. idle
22. engine service manual
23. abrasive stone
24. is not
25. a. fuel nozzles
    b. combustion chambers
    c. turbine inlet guide vanes
    d. turbine wheels and blades
    e. exhaust system
26. EGT
27. false
28. fuel nozzle
29. perpendicular
30. overheating
31. a. blade tip clearance
    b. clearance between turbine wheel and exhaust cone
32. are not
33. trend monitoring
34. is not
35. parallel to
36. shingled
37. bird strike, overspeed
38. hot section
39. greater
40. National Bureau of Standards or National Institute of Standards and Technology (NIST)
41. dry
42. 166.7
43. 250
44. a. tachometer
    b. EGT indicator
    c. EPR indicator
45. engine air inlet
46. standard day, sea level
47. into
48. a. Know the way the system should operate.
    b. Observe how the system is operating.
    c. Divide the system to find the trouble.
    d. Look for the obvious problem first.

INSTRUMENT SYSTEMS

16

Continued

Instrument Systems

16

The Evolution of Powerplant Instruments

The first aircraft did not have any powerplant instruments, as their engines ran for a short time and the pilot could tell by the sound when the engine was developing sufficient RPM for flight. As airplanes and engines evolved, tachometers, oil pressure, and water temperature gages were installed.

Even today, the minimum powerplant instruments required by 14 CFR §91.205 are simply:

- a tachometer for each engine
- an oil pressure gage for each engine using a pressure system
- a coolant temperature gage for each liquid-cooled engine
- an oil temperature gage for each air-cooled engine
- a manifold pressure gage for each altitude engine

Most powerplant instruments have been of the conventional round-dial analog type, in which a pointer, driven by the parameter being measured, moves across a graduated dial. As the number of engines and the need for data increased, instrument panels became so crowded that vertical tape instruments became popular. Data for four engines can be displayed by vertical instruments in the space that would be required for two engines using circular instruments.

More precise data is displayed using drum-type instruments. In these instruments, several narrow drums with numbers on their outer rim are mounted in the instrument case so a portion of the rim shows through the instrument dial. The same mechanism that drives the pointer drives the drums, but the indication is in digital rather than analog form.

When large airplanes with four reciprocating engines became the standard for long-range airline flying, the more than one hundred instruments and controls relating to the powerplants were installed on a special panel so they could be monitored by a flight engineer. Keeping track of instrument indications and recording the data kept the flight engineer busy on the long flights.

Jet transports with a flight engineer's station have far fewer engine instruments than planes with reciprocating engines, and the most modern jet transports do not even have a flight engineer's station. All necessary powerplant data is gathered by solid-state transducers or optoelectronic devices

powerplant. The complete installation of an aircraft engine, propeller, and all accessories needed for its proper functioning.

altitude engine. A reciprocating engine whose rated takeoff power can be produced to an established higher altitude.

analog indicator. An indicator that shows the value of the parameter being measured by a number marked on a graduated dial aligned with a movable pointer.

vertical tape instrument. A tall rectangular instrument that displays the quantity of the parameter being measured by a movable strip of colored tape. The presentation resembles a vertical bar graph.

Figure 16-1. *The drums in this instrument, similar in appearance to the odometer in an automobile speedometer, are driven by the same mechanism that drives the pointer. These drums show the precise engine pressure ratio, and movement of the analog pointer makes trends easy to visualize.*

pressure. A measure of force applied uniformly over a given unit of surface area.

kilopascal. The unit of pressure produced when one newton of force acts uniformly over an area of one square meter.

gage pressure. Pressure referenced from the existing atmospheric pressure.

differential pressure. A single pressure that is the difference between two opposing pressures.

absolute pressure. Pressure referenced from zero pressure, or a vacuum.

millibar. A unit of pressure in the metric system. One millibar is one thousandth of a bar, 0.014 69 psi, or 0.029 52 in. Hg.

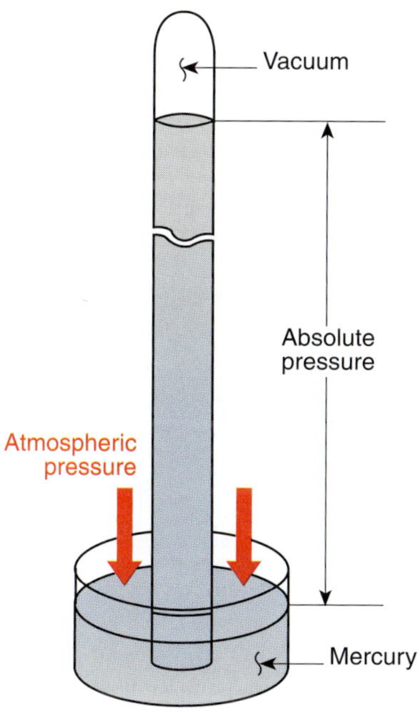

Figure 16-2. *The absolute pressure of the atmosphere is measured by the height of a column of mercury the atmospheric pressure supports.*

and fed into computers. The flight crew is provided with all the necessary data on video displays. This type of electronic instrumentation is discussed as the EICAS, or Engine Indication and Crew Alerting System, beginning on Page 580.

Types of Powerplant Instruments

There are three types of data important for monitoring the operation and performance of an aircraft powerplant: pressure, temperature, and mechanical movement.

Pressure Measurement

Pressure is a measure of force acting on a given unit of area. In the English system it is normally expressed in terms of pounds per square inch (psi). In the metric system, it is expressed in kilopascals (kPa) with one kPa = 0.14503 psi.

Types of Pressures

Because pressure relates to force, it must have a reference from which it is measured, and in powerplant instrumentation, three different references are used. There are, therefore, three types of pressure: absolute pressure, gage pressure, and differential pressure.

Absolute Pressure

Absolute pressure is referenced from zero pressure, or a vacuum, and is generally expressed in inches or millimeters of mercury (in. Hg or mm Hg), millibars (mb), or pounds per square inch absolute (psia).

Absolute pressure is normally measured with some type of barometer. A mercury barometer like that in Figure 16-2 consists of a closed glass tube about an inch in diameter and about 30 inches long. The tube is filled with mercury and inverted with its open end submerged in a bowl of mercury. The mercury in the tube drops down and leaves a vacuum above it. Atmospheric pressure pressing down on the mercury in the bowl holds it up in the tube to a height determined by the weight of the atmosphere, or the atmospheric pressure. At sea level under standard-day conditions, the column of mercury will be 29.92 inches, or 760 millimeters, high.

Manifold pressure in a reciprocating engine is an often used application of absolute pressure. *See* MAP (Manifold Absolute Pressure), beginning on Page 554.

Gage Pressure

Gage pressure is the pressure added to a fluid by a pump or other device. It is referenced from the existing atmospheric pressure and is normally expressed in units of pounds per square inch gage (psig).

Engine oil pressure is a familiar application of gage pressure. Atmospheric pressure acts on the oil at both the inlet and outlet of the pump and is thus canceled. Only the pressure added to the oil by the pump is indicated on the gage.

Differential Pressure

Differential pressure, or ΔP, is the difference between two pressures and is normally expressed in pounds per square inch differential (psid).

Many fuel filters used with turbine engines have differential-pressure switches or transducers across their filtering element. These devices sense an excessive pressure drop across the filter and turn on a warning light that is visible to the flight crew. In some installations they open a valve that allows warm compressor bleed air to flow through the air-fuel heat exchanger if ice crystals are blocking the filter.

Total Pressure

Total pressure, P_T, is the pressure a moving fluid would have if it were stopped without any losses. Total pressure is the sum of dynamic pressure (q) and static pressure (P_S).

Dynamic Pressure

Dynamic pressure, q, is the pressure of a moving fluid that results from its motion and is equal to one half of the fluid density (ρ) times the square of its velocity. Dynamic pressure is the difference between the total pressure of a fluid and its static pressure.

Static Pressure

Static pressure, P_S, is the pressure of a fluid that is still, or not in motion, and is measured perpendicular, or normal, to the surface on which it acts.

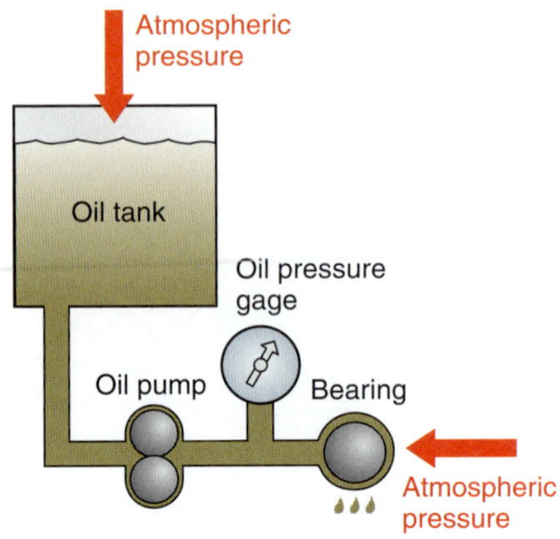

Figure 16-3. *Gage pressure is the amount of pressure added to a fluid by a pump.*

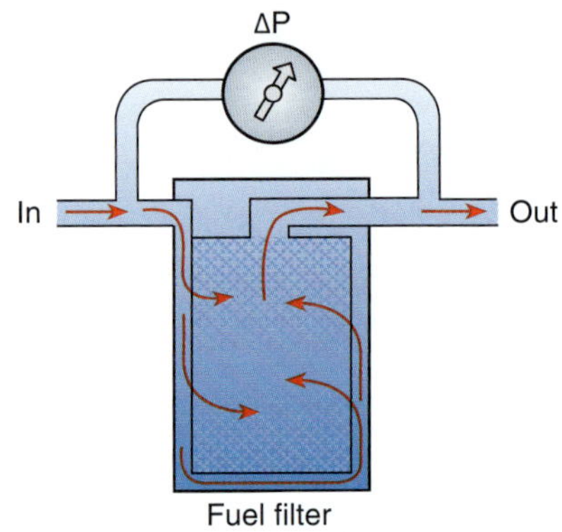

Figure 16-4. *Differential pressure, or ΔP, is the difference between two pressures.*

standard-day conditions. A set of atmospheric conditions agreed upon by scientists and engineers who work with aircraft and engines, which allow all measurements to be corrected to the same conditions.

transducer. A device that changes energy from one form to another. Commonly used transducers change mechanical movement or pressures into electrical signals.

total pressure. The pressure a column of moving fluid would have if it were stopped from its motion.

static pressure. The pressure of an unmoving fluid.

Pressure-Measuring Instruments

All the pressures just mentioned are measured to provide important data for the operation of aircraft engines. Some of the more commonly used pressure-measuring instruments are described below.

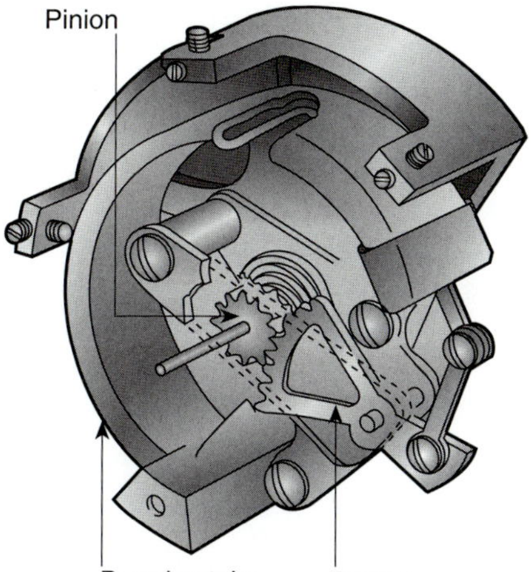

Figure 16-5. *A Bourdon tube instrument is used to measure such gage pressures as engine lubricating oil pressure.*

Engine Lubricating Oil Pressure

The pressure that forces the lubricating oil through the engine is a gage pressure and is most generally measured with a Bourdon tube instrument such as that in Figure 16-5.

The Bourdon tube has a flat or elliptical cross section and is formed into a curve. One end of the curved tube is sealed and attached to the tail of the sector gear. The other end is fixed in the case where the pressure is introduced. The pressure forces the tube to assume a round cross section, and as it rounds out, it straightens the curve and rotates the sector gear. The sector meshes with and rotates the pinion on whose shaft the instrument pointer is mounted. The pointer moves across a graduated dial an amount proportional to the pressure inside the tube.

The line between the oil pressure gage and the engine is filled with a lightweight oil such as kerosine, and the fitting in the engine to which the line is attached has a very small orifice, one made with a number 60 drill (0.040 inch). The restricted fitting serves two functions: it smooths the pulsations in the oil pressure, and it prevents a major loss of engine oil if the instrument line should break. The lightweight oil in the line speeds up the instrument indication when it is cold because it does not become thick or viscous.

Bourdon tube. The major component in a gage-pressure measuring instrument.

capillary tube. A glass or metal tube with a tiny inside diameter.

Pressure Gage—Measuring Oil Temperature

Oil temperature on most small general aviation aircraft engines is measured with a simple Bourdon tube instrument. The Bourdon tube is connected through a small-diameter capillary tube to the temperature-sensing bulb that is normally mounted inside the engine oil pressure screen. The bulb, capillary, and Bourdon tube are filled with a highly volatile liquid such as methyl chloride and sealed as a unit. As the temperature of the oil increases, the temperature of the liquid inside the instrument increases and part of it evaporates. Since the system is sealed, vapors are trapped, and their pressure increases proportionally to the temperature rise. The Bourdon tube begins to straighten, and as it does, it moves the pointer over the graduated dial indicating the temperature of the oil.

Fuel Pressure

Fuel pressure for small reciprocating engines is measured with a capsule-type instrument mechanism like the one in Figure 16-6. Pressurized fuel flows into the instrument capsule through a small-diameter tube. The pressure expands the capsule and rotates the sector gear which, in turn, rotates the pinion. The pointer attached to the pinion shaft moves over a graduated dial an amount proportional to the pressure inside the capsule.

The inherent danger of bringing gasoline into the aircraft cockpit has caused a change in many aircraft in which the fuel pressure is converted into electrical signals at the engine and the pressure read on electrical instruments.

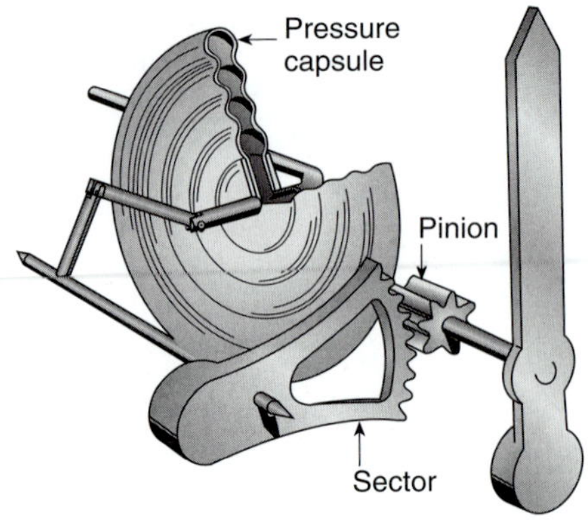

Figure 16-6. *A capsule-type mechanism is used to measure low-gage pressures such as the fuel pressure for reciprocating engines.*

Fuel Pressure Warning System

The fuel pressure indicating systems used for large reciprocating engines measure the difference between the fuel pressure and the induction air pressure at the carburetor. These pressure gages are sensitive enough for normal operation, but an indicator with a more rapid response and positive indication is needed to warn the pilot of a dangerous drop in fuel pressure.

The warning system senses the fuel pressure at the carburetor inlet and generates an electrical signal that gives the first indication when a tank becomes empty and the selector valve should be switched to a full tank. As soon as the pump begins to draw air from the tank, the fuel pressure drops, and the pressure warning system contacts close, sending an electrical signal that turns on a warning light or flashes a warning on an annunciator panel.

The pressure-sensitive mechanism is generally a bellows that can be adjusted to change the pressures at which it actuates. Maintenance and troubleshooting procedures for fuel pressure warning systems are found in the manufacturer's maintenance manuals.

Pressure Gage—Measuring Fuel Flow

Fuel flow in fuel-injected horizontally opposed reciprocating engines is measured with a capsule-type pressure gage installed as shown in Figure 16-7 on the next page. This flowmeter works on the principle that the pressure drop across a fixed orifice is directly proportional to the amount of fluid flowing through it.

A sample of fuel at the same pressure as that being sent to the injector nozzles is directed inside the capsule. The capsule expands and moves the pointer over the graduated dial an amount proportional to the pressure drop across the injector nozzles.

The capsule of a fuel flowmeter used on a naturally aspirated engine is mounted inside an instrument case that is vented to the atmosphere. The fuel pressure tries to expand the capsule and the atmospheric pressure opposes the expansion. The pressure shown on the instrument dial is therefore gage pressure. On turbocharged engines, the instrument case is sealed and connected to the air inlet of the fuel control unit, so the pressure inside the case

naturally aspirated engine. A reciprocating engine that depends upon the atmospheric pressure to force the fuel-air mixture into the cylinders.

upper-deck pressure. The pressure in the induction system of a reciprocating engine between the turbocharger compressor discharge and the fuel control throttle valve.

MAP (manifold absolute pressure). The absolute pressure that exists within the induction system of a reciprocating engine.

ΔP (delta P). Differential pressure.

is turbocharger discharge, or upper-deck, pressure. This same pressure is directed into the shrouds around the injector nozzles. The pressure shown on the instrument dial is therefore a differential pressure, i.e., the difference between the fuel pressure delivered to the nozzles and the upper-deck pressure. *See* Figure 16-7.

One of the more versatile types of pressure measuring instruments is the differential bellows. This instrument uses two sets of bellows, a measuring bellows and a reference bellows, connected through a linkage to the sector gear. The sector rotates a pinion to which the instrument pointer is attached. *See* Figure 16-8.

To measure gage pressure, the reference bellows is open to the atmosphere. For absolute pressure, the reference bellows is evacuated and sealed, and for differential pressure, one pressure is directed into the measuring bellows and the other into the reference bellows.

MAP (Manifold Absolute Pressure)

The power produced by a reciprocating engine is determined by two variables: the amount of pressure forcing the piston down, and the number of power strokes per minute.

The rotational speed of the crankshaft, measured by the tachometer, is directly related to the number of power strokes per minute, but the pressure inside the cylinder is too difficult to measure directly. The absolute pressure

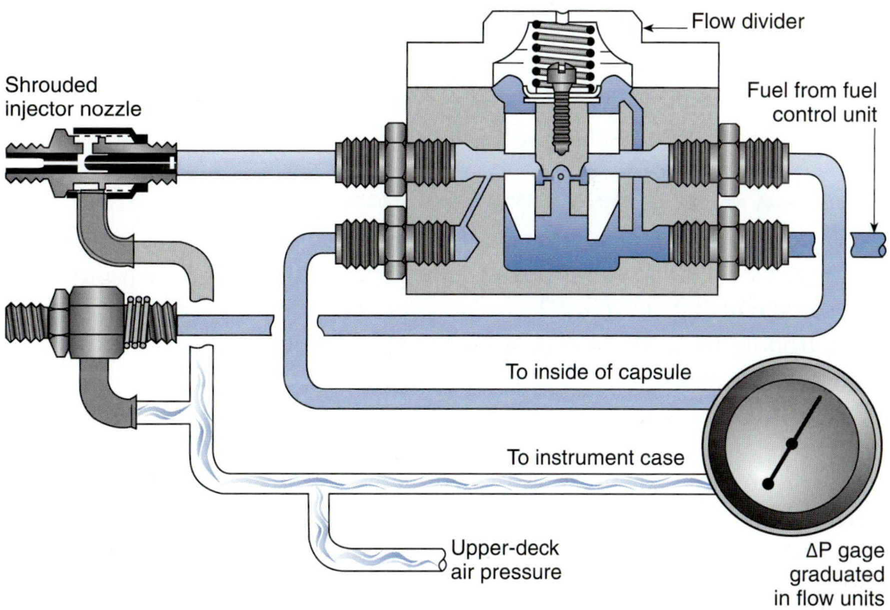

Figure 16-7. *The fuel flowmeter used on a turbocharged horizontally opposed fuel-injected engine is a differential pressure gage that measures the difference between the fuel pressure delivered to the nozzle and the upper-deck air pressure.*

of the air in the induction system, just before the intake valve, has a definite relationship to the pressure developed inside the cylinder. It is this pressure, the manifold absolute pressure (MAP), that is easy to measure and is used with the RPM to show the technician or flight crew the amount of power the engine is developing.

When the engine is not running, the pointer indicates the existing barometric pressure in inches of mercury. It should read the same as the barometric scale on the altimeter when the altimeter pointers are adjusted to the existing airport elevation.

When the engine starts and is idling, the manifold pressure lowers to typically around 15 inches. A naturally aspirated engine can never develop a manifold pressure greater than that of the atmosphere, but supercharged and turbocharged engines can. *See* Figure 16-9.

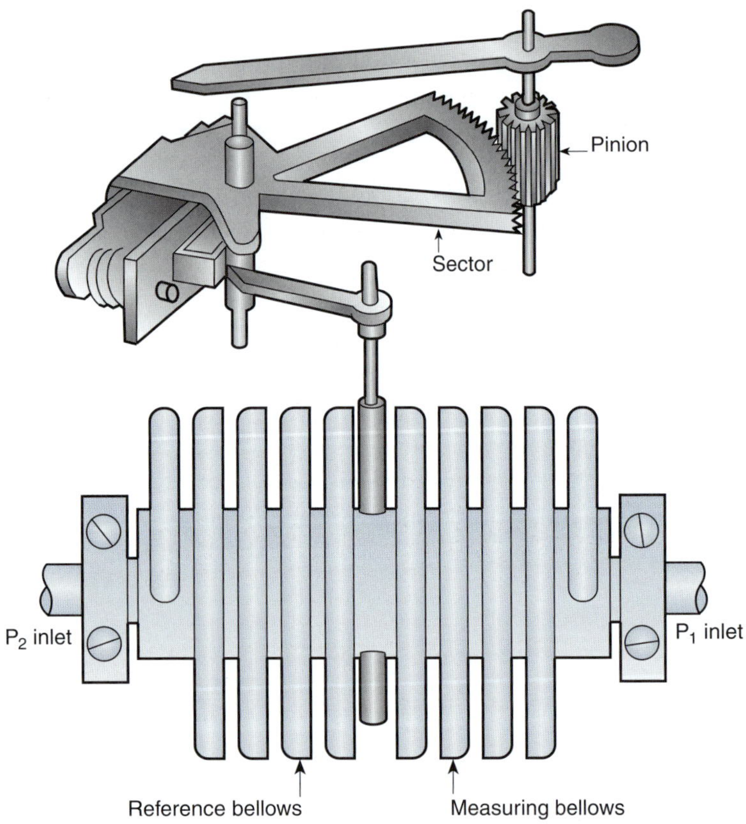

Figure 16-8. *A differential bellows mechanism measures the difference between two pressures, P_1 and P_2. This versatile mechanism can measure absolute, differential, or gage pressure.*

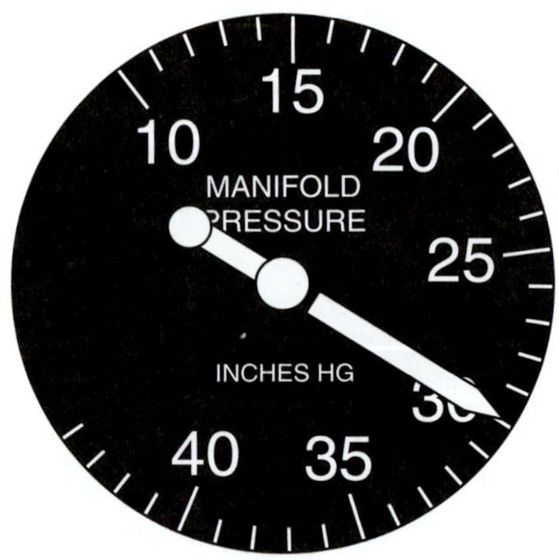

Figure 16-9. *Manifold pressure gage for a turbocharged horizontally opposed general aviation engine*

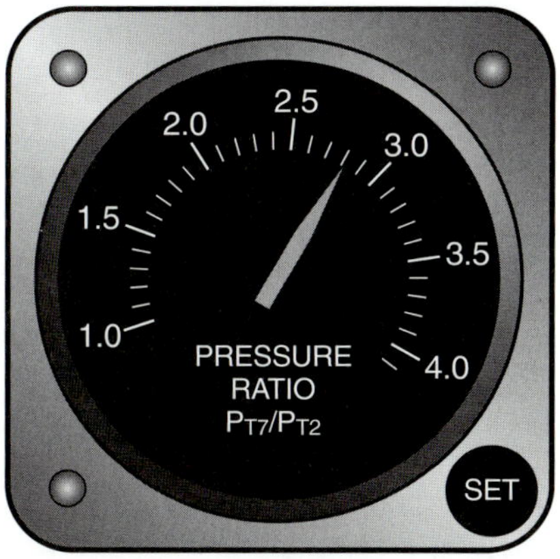

Figure 16-10. *The EPR (engine pressure ratio) indicator measures the ratio of the turbine discharge total pressure to the compressor inlet total pressure (P_{T7}/P_{T2}) to determine the amount of thrust the engine is developing.*

EPR (Engine Pressure Ratio)

The amount of thrust produced by a gas turbine engine is determined by the change in the momentum of air flowing through the engine. In the early days of turbine engine operation, EGT and RPM were used to measure thrust.

Modern axial-flow engines indicate the thrust by measuring the EPR, or engine pressure ratio. A pressure pickup measures the total pressure at the compressor inlet (P_{T2}), and another pickup measures the total pressure at the discharge of the last turbine stage (P_{T5} or P_{T7}).

Temperature Measurement

Temperature is one of the most important measurements in aircraft engine operation. Operational temperatures range from well below freezing for carburetor air temperature and fuel temperature gages to around 1,000°C for exhaust gas temperature gages.

The oil temperature of some smaller engines is measured with a pressure gage that measures the vapor pressure of a liquid in a sealed tube. This type of measurement is described beginning on Page 552. Almost all other measurements are made electrically by either a ratiometer or thermocouple instrument.

EPR (engine pressure ratio). The ratio of the turbine discharge total pressure to the compressor inlet total pressure.

momentum. A force caused by the inertia of a moving body as it tries to keep the object moving in the same direction, at the same speed.

Nichrome. An alloy of nickel and chromium.

ratiometer indicator. An analog temperature measuring instrument in which the pointer deflection is proportional to the ratio between the current flowing in an internal reference circuit and that flowing through the temperature-sensing probe.

Ratiometer Instruments

Carburetor air temperature, fuel temperature, and oil temperature are typically measured with ratiometer instruments. Figure 16-11 illustrates the operation of a typical moving-magnet ratiometer. The temperature-sensing bulb consists of a coil of fine Nichrome wire encased and sealed in a thin stainless steel tube. This bulb is immersed in the fluid whose temperature is being measured. The resistance of the Nichrome wire varies directly with its temperature. At the low-end temperature, the bulb resistance is approximately 20 ohms, and at the high end, its resistance is about 200 ohms.

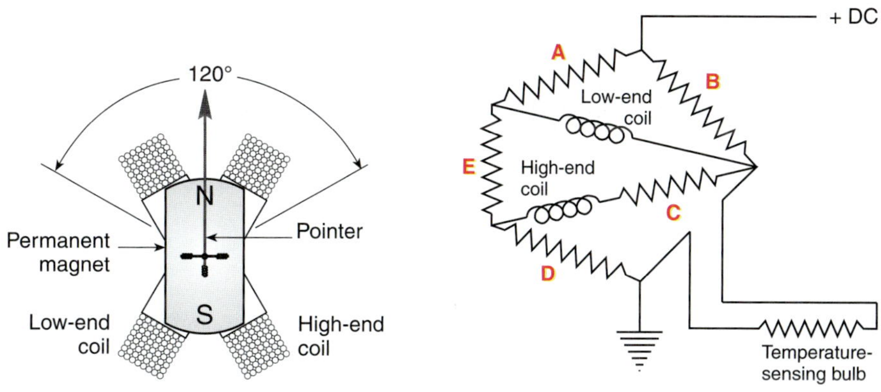

Figure 16-11. *Principle of ratiometer operation for measurement of relatively low temperatures*

Follow the circuit in Figure 16-11 to see the way this instrument operates. When the temperature and bulb resistance are low, most of the current flows to ground through resistor A, the low-end coil, and the temperature-sensing bulb. The magnetic field from the low-end coil pulls the permanent magnet with its attached pointer toward the low-temperature side of the dial. When the temperature and resistance of the bulb increase, current flows to ground through resistors B, C, the high-end coil, and resistor D. The resulting magnetic field from the high-end coil moves the pointer toward the high-temperature side of the dial.

Thermocouple Instruments

Cylinder head temperature for reciprocating engines and exhaust gas temperature for both reciprocating and turbine engines are measured with thermocouple instruments. These instruments do not require external power, as a thermocouple is an electrical generator.

A thermocouple is a loop made of two different types of wire, as shown in Figure 16-12. The temperature is measured at one of the points at which the two wires are joined—called the measuring junction. The other point is the reference junction and is located inside the instrument case.

A voltage is produced between the two junctions proportional to the temperature difference between them. This voltage causes current to flow, and the current is measured by a milliammeter whose dial is calibrated in terms of °C or °F.

Since the indicator is a current-measuring instrument, the resistance of the conductors between the junctions is critical. Thermocouple leads usually have a resistance of either two or eight ohms, and their length must not be altered to suit the installation, as this would change their resistance. If they are too long, they may be neatly coiled and secured so that they will not interfere with the controls and will not become overheated. If resistance is too low, a special resistor may be installed in the negative lead.

For the temperature indication of the measuring junction to be meaningful, the reference junction temperature must be held constant. This is actually done in a laboratory with a container of melting ice, but this is not practical in an aircraft instrument, so the indicator needle is mounted on a bimetallic hairspring that causes it to move back as the cockpit temperature increases. Moving the pointer in this way compensates for reference junction temperature changes.

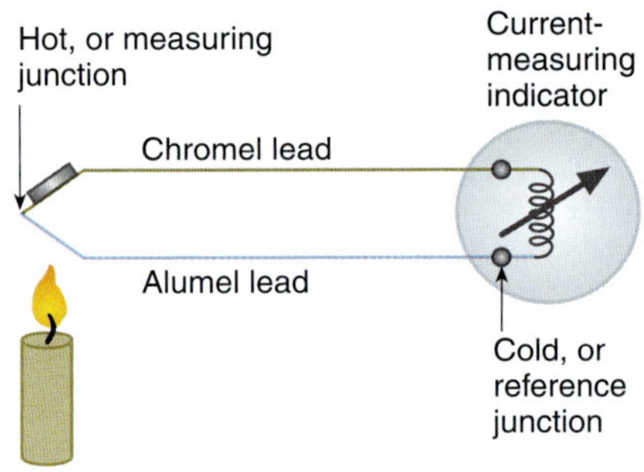

Figure 16-12. *Operation principle of a thermocouple used to measure high temperatures*

milliammeter. An instrument that measures electrical current in units of thousandths of an ampere.

bimetallic hairspring. A flat, spiral-wound spring made of two strips of metal laid side-by-side and welded together. The two metals have different coefficients of expansion, and as the temperature changes, the spiral either tightens or loosens.

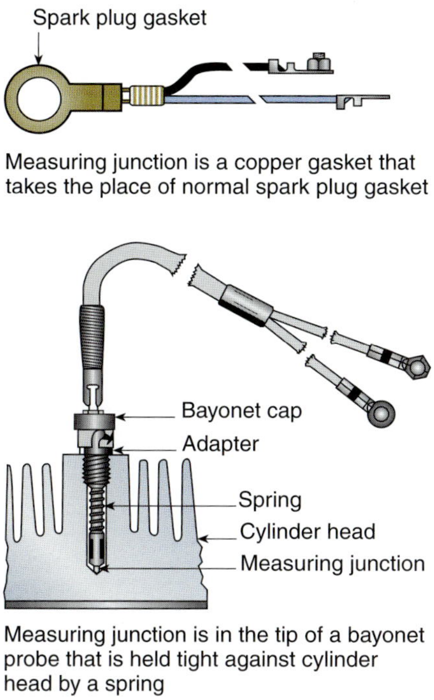

Figure 16-13. *Cylinder head temperature measurement*

CHT (Reciprocating Engine Cylinder Head Temperature)

The cylinder head temperature of a reciprocating engine is measured with thermocouples made of copper or iron wire for the positive lead and constantan, a copper-nickel alloy, for the negative lead. One end of each wire is embedded in a special copper spark plug gasket or is joined inside a bayonet probe like that in Figure 16-13. This is called the hot, or measuring, junction. The other ends are connected to the instrument movement inside the case and form the cold, or reference, junction.

EGT (Reciprocating Engine Exhaust Gas Temperature)

Higher temperatures, such as that from exhaust gas of both reciprocating and turbine engines, are measured with thermocouples made of chromel and alumel wires. Chromel, an alloy of nickel and chromium, is used as the positive lead, and alumel, an alloy of nickel, aluminum, manganese, and silicon, is used as the negative lead. Figure 16-14 shows a typical EGT system for a reciprocating engine. The thermocouple is mounted in the exhaust stack, usually within six inches of the cylinder head. The indicator is a current-measuring instrument similar to that used for measuring cylinder head temperature.

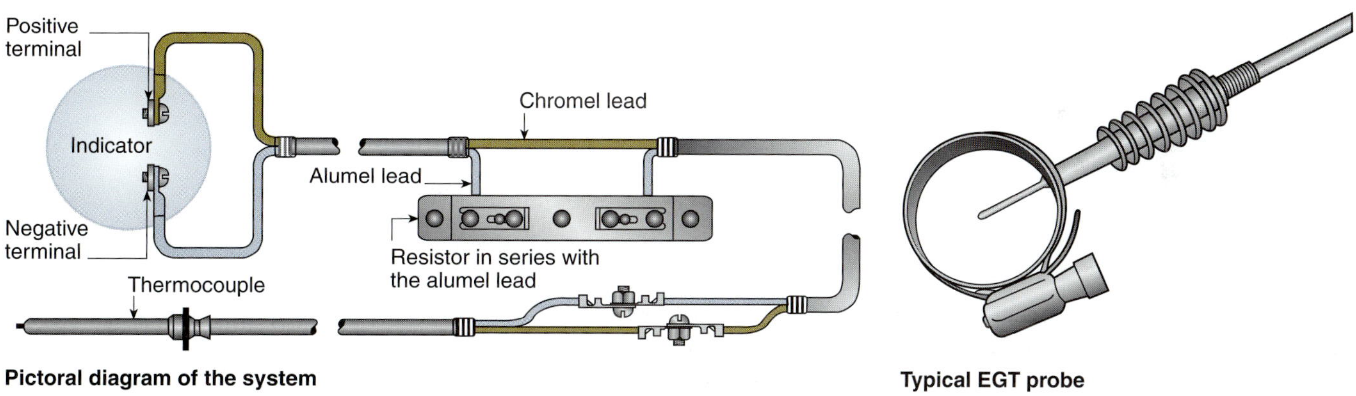

Figure 16-14. *Exhaust gas temperature measuring system for a reciprocating engine*

EGT (Turbine Engine Exhaust Gas Temperature)

The most critical temperature in a turbine engine is the turbine inlet temperature (TIT). This high temperature is extremely difficult to measure, as any probe placed just ahead of the first stage turbine would interfere with the critical flow of gases entering the turbine, and any probe sensitive enough to efficiently measure changes in the temperature would be easily damaged by the high-velocity flow of hot gases. Engineering tests show a definite relationship between the temperature of the gases as they leave the last stage of the turbine and their temperature as they enter the first stage. The temperature leaving the turbine is easy to measure, and this measurement, the EGT, is the temperature displayed to the flight crew. As long as the EGT remains within its allowable range, the TIT is not excessive.

EGT (exhaust gas temperature). The temperature of the gases as they leave the cylinder of a reciprocating engine or the turbine of a gas turbine engine.

A turbine engine EGT system is similar to that used for a reciprocating engine, except that several thermocouples are connected in parallel and arranged around the tail cone so they can sample the temperature at several locations. The parallel connections give one indication that is the average temperature of the gases leaving the turbine. Figure 16-15 shows a typical circuit for a turbine engine EGT system.

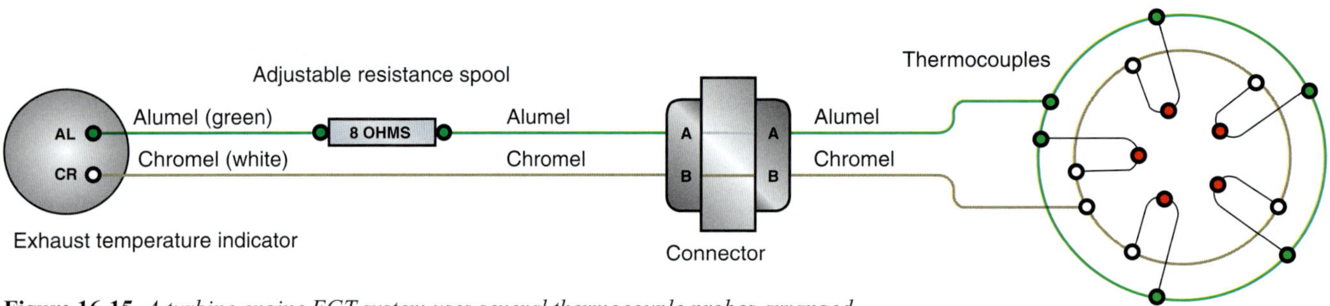

Figure 16-15. *A turbine engine EGT system uses several thermocouple probes arranged around the periphery of the tail pipe, just behind the last stage of the turbine. The thermocouples are connected in parallel to obtain an average temperature.*

Mechanical Movement

The instruments that measure mechanical movement discussed in this chapter are various types of tachometers for measuring engine RPM, fuel flow indicators for reciprocating and turbine engines, and torquemeters that measure the amount of torque produced by reciprocating and turboprop engines.

Tachometers

One of the earliest aircraft instruments was a tachometer that told the pilot the RPM of the engine. Today, tachometers are still one of the required instruments in all powered aircraft, and their indications allow the pilots to monitor engine performance. Tachometers for reciprocating engines indicate the engine speed in RPM times 100. Turbine-engine tachometers indicate the compressor speed in percent of the rated RPM.

tachometer. An instrument that measures the rotational speed of an object.

optoelectronic device. An electronic device that produces, modulates, or senses electromagnetic radiation in the ultraviolet, visible light, or infrared portions of the energy spectrum.

eddy current. Current induced into a conductor because of a moving or nonuniform magnetic field.

odometer. The portion of an automobile speedometer that indicates the distance traveled.

Some of the earliest tachometers were centrifugal instruments that used the same principle as early steam-engine governors. During World War II, one popular tachometer used a rather complicated clockwork mechanism that momentarily coupled a shaft from the engine directly to the indicating needle approximately once per second.

The mechanism in most modern mechanical tachometers resembles that used in automobile speedometers; the most popular electrical tachometer is based on a three-phase synchronous motor. Digital electronic tachometers count pulses from an optoelectronic device or from the breaker points in the ignition system of a reciprocating engine.

Mechanical Tachometers

The most widely used tachometer for the smaller reciprocating engines is of the magnetic drag type. (*See* Figure 16-16.) A relatively small permanent magnet inside the instrument case is driven by a steel cable from the engine at one half of the crankshaft speed. Riding over but not touching the outside of this magnet is an aluminum drag cup. A steel shaft attached to the outside center of this cup rides in bearings so it is free to rotate. The instrument pointer is attached to this shaft whose rotation is restrained by a calibrated hairspring.

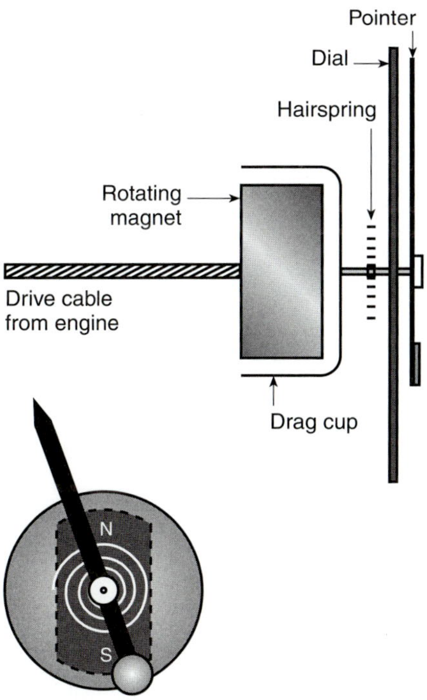

Figure 16-16. *Simplified diagram of a magnetic drag tachometer*

When the engine is operating, the magnet spins inside the drag cup, and as it spins, its lines of flux cut across the aluminum cup and induce an eddy current in it. This current produces a magnetic field, which interacts with that of the rotating magnet and tries to magnetically lock the drag cup to the rotating magnet. But the calibrated hairspring restrains rotation of the cup, so it turns only a portion of a revolution. The pointer on the shaft moves in front of a dial marked in RPM times 100 to indicate the speed of the crankshaft.

This type of tachometer is mounted inside a steel case to prevent the magnetic flux from interfering with other instruments in the panel.

Most magnetic drag tachometers have an hourmeter built into them that is the counterpart to the odometer on an automobile speedometer. A worm gear driven by the magnet drive shaft turns a series of narrow drums with numbers on their outer surfaces. These numbers show the hours the engine has operated. The hours indication is a function of the engine speed and it is accurate only at the cruise RPM of the engine. When replacing a magnetic drag tachometer, the replacement must be one designed for the cruise RPM of the engine. This RPM is stamped on the instrument case. The hours indication on this type of tachometer is normally considered by the FAA to be sufficiently accurate for measuring inspection intervals and total engine operating time.

worm gear. An irreversible gear system in which a helical worm gear meshes with and turns a spur gear whose teeth are cut at an angle to its face.

Magnetic drag tachometers are not known for their accuracy, and since engine RPM is extremely important, the tachometer indication should be checked with a stroboscopic or laser tachometer any time its accuracy is in question.

Electric Tachometers

Some of the older electric tachometers were simply voltmeters that measured the voltage produced by an AC or DC permanent-magnet tachometer generator. This system has the inherent limitation that its accuracy is dependent upon the strength of the permanent magnet in the generator, and this strength deteriorates with time.

The much more accurate and widely used system employs a three-phase AC permanent-magnet generator on the engine to drive a small synchronous motor inside the tachometer case. This motor in turn drives a magnet assembly and a drag disk as shown in Figure 16-17. The drag disk with its calibrated hairspring and pointer operates in exactly the same way as the drag cup in the mechanical tachometer. This instrument is inherently accurate, as the frequency of the generator is determined only by the RPM of the engine, and variations of the strength of the generator magnet have little or no effect on its accuracy.

stroboscopic tachometer. A tachometer used to measure the speed of any rotating device without physical contact.

laser tachometer. A highly accurate tachometer that shines a laser beam on the rotating element that has a reflective tape or contrasting mark. The reflected laser beam is converted into electrical pulses which are counted and displayed on a monitoring instrument.

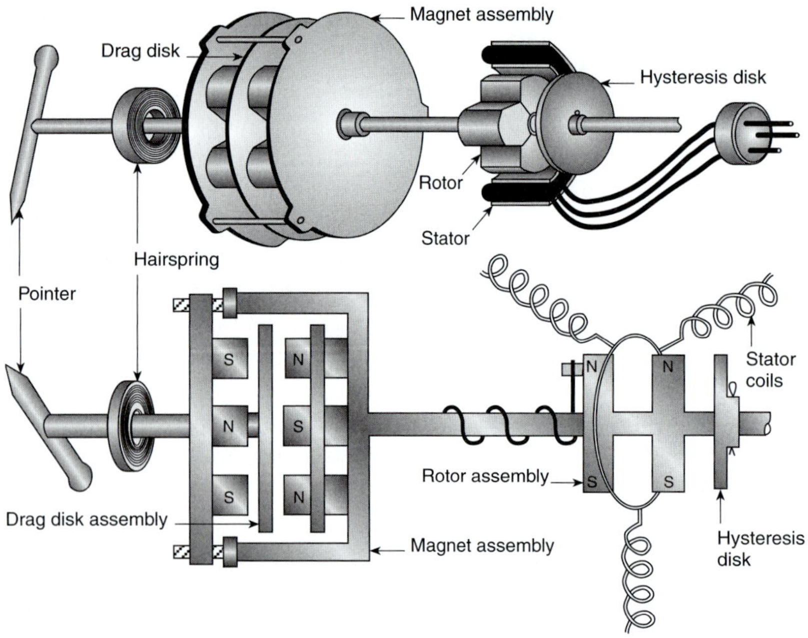

Figure 16-17. *Simplified diagram of a three-phase AC tachometer*

Synchroscopes

Some tachometers designed for multiengine aircraft have synchroscopes built into them. These instruments simply show a small disk in a cutout on the instrument dial. This disk is marked with light- and dark-colored segments so it is easy to see when it is turning. It is driven by two synchronous motor windings on its single shaft, and these windings are excited by the output of the two engine-driven AC tachometer generators. When the engines turn at the same speed, the torque produced by the two windings cancel, and the disk remains still. But when one engine turns faster than the other, one set of windings delivers more torque than the other, and the disk rotates in the direction of the faster engine. The speed of rotation is one half of the difference in the speed of the two engines.

Fuel Flowmeters for Large Reciprocating Engines

Autosyn system. A remote indicating instrument system that uses an electromagnet rotor, excited with 400-hertz AC, and a three-phase distributed-pole stator.

Many large reciprocating engines use a vane-type flowmeter in the fuel line between the engine-driven pump and the carburetor. The measuring vane, shown in Figure 16-18, is moved by the flow of fuel, and its movement is measured with an Autosyn transmitter. The pointer on the Autosyn fuel flow indicator in the cockpit follows the movement of the vane and shows the fuel flow in gallons per hour. This type of instrument is accurate and reliable, but it shows only the volume of fuel flowing to the engine, not its mass. Some indicators used with this type of flowmeter have two scales on the dial. One scale is calibrated in gallons and the other in pounds. The pounds indication is only approximate, as it is based on the nominal weight of gasoline being six pounds per gallon. It does not take into consideration changes in fuel density with changes in temperature.

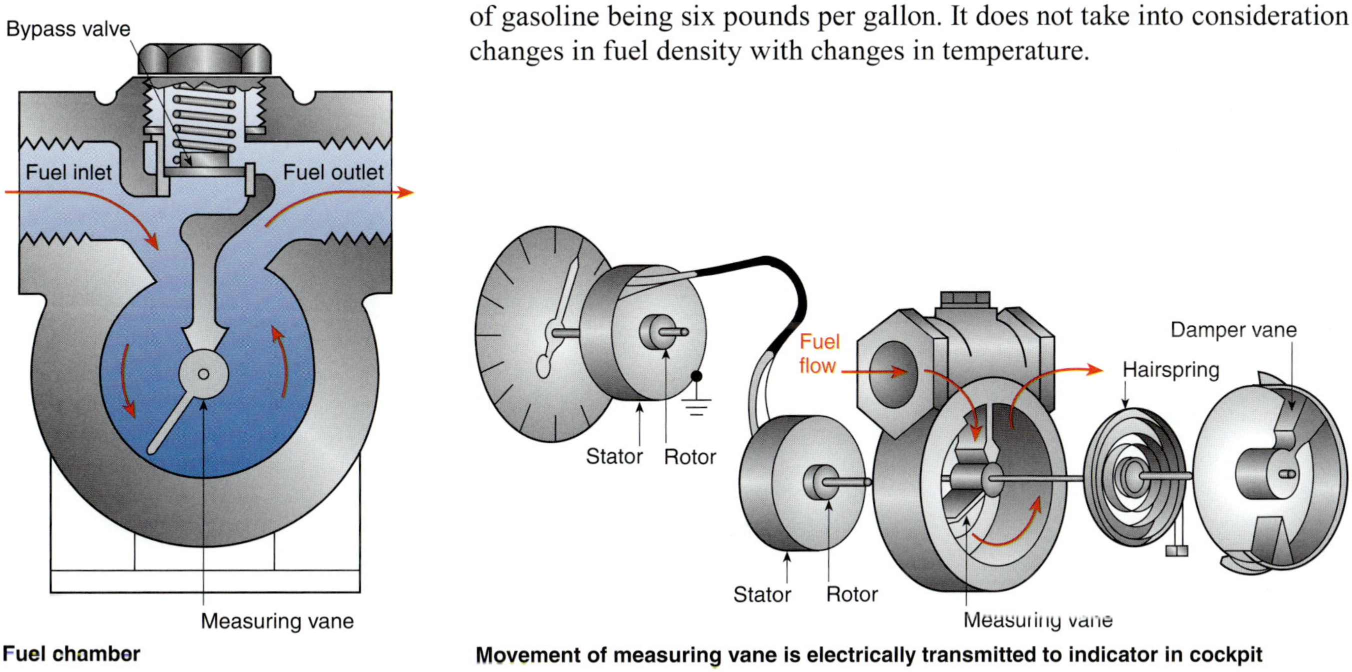

Figure 16-18. *Volume-type fuel flowmeter used with a large reciprocating engine*

Flowmeters for Turbine Engines

Turbine engine fuel flow is measured as a mass flow rather than volume flow, and the transmitting system is illustrated in Figure 16-19.

The impeller and the turbine are mounted in the main fuel line leading to the engine. The impeller, driven at a constant speed by a special three-phase AC motor, imparts a swirling motion to the fuel passing through it, and this swirling fuel deflects the turbine. The turbine is restrained by two calibrated restraining springs, and the amount it deflects is affected by both the volume and the density of the fuel.

The amount of turbine deflection is transmitted to an electrical indicator in the cockpit by a Magnesyn transmitter built into the flowmeter.

Magnesyn system. A remote indicating system that uses a permanent magnet as its rotor and a toroidal coil excited by 400-hertz AC as its stator.

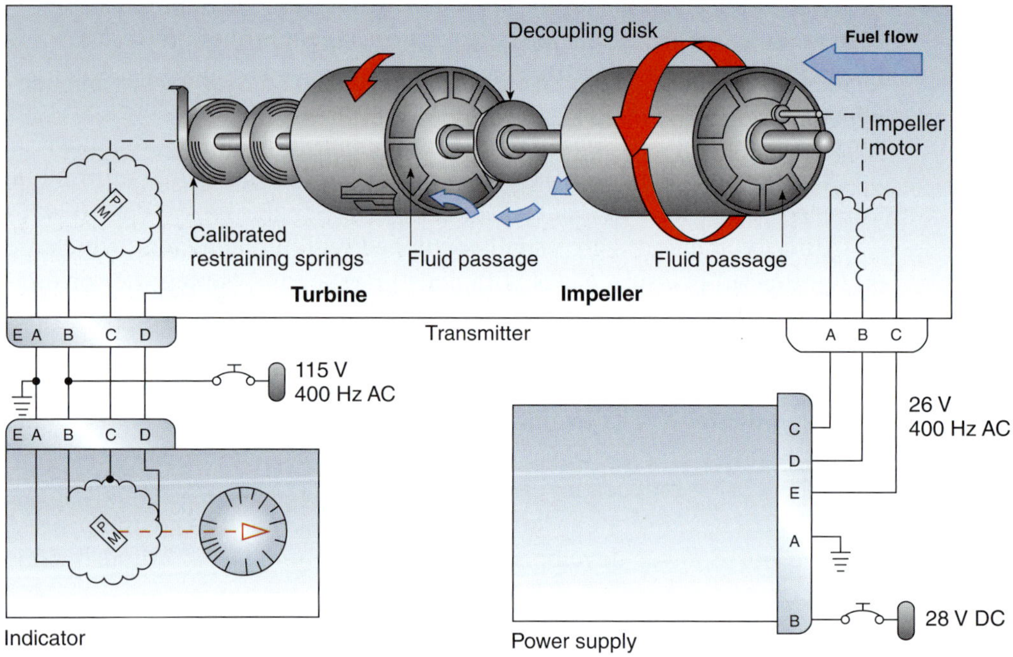

Figure 16-19. *The rotation of the turbine in this mass fuel flowmeter is affected by both the volume and density of the fuel.*

Torquemeters

It is important that a flight crew know the power an engine is developing in flight, but there are no mechanical instruments that provide this information directly. However, by knowing the RPM, manifold pressure, pressure altitude, and outside air temperature, and using a chart like the one in Chapter 2, Figure 2-12 on Page 36, a pilot can determine this information.

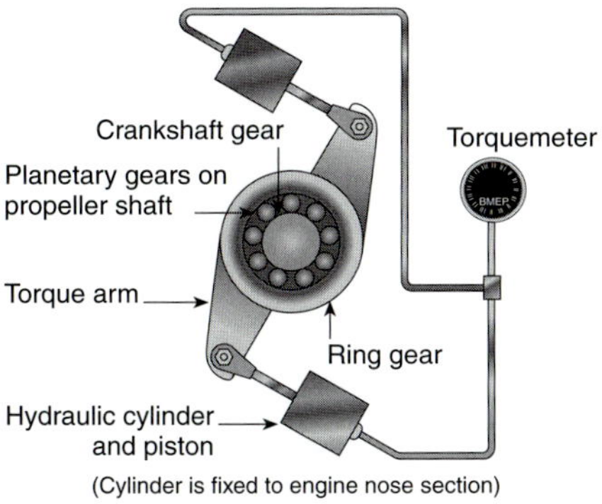

Figure 16-20. *The reaction between the ring gear of the propeller reduction gearing system and the nose section of the engine is proportional to the amount of torque the engine is producing.*

BMEP (brake mean effective pressure). The average pressure inside the cylinder of a reciprocating engine during the power stroke.

RPM (revolutions per minute). A measure of rotational speed. One RPM is one revolution made in one minute.

brake horsepower. The horsepower actually delivered to the output shaft of an engine.

POH (Pilot's Operating Handbook). A document published by an airframe manufacturer and approved by the FAA that lists the operating conditions for a particular model of aircraft.

Some engines installed on reciprocating-engine-powered airliners and bombers built near the end of World War II were equipped with torque noses like that in Figure 16-20. The ring gear in the propeller reduction gear system is allowed a small freedom of rotary movement, which is restrained by hydraulic cylinders connected between this gear and the nose section of the engine.

A reaction force from the torque of the propeller tries to rotate the ring gear, and the pressure produced in the hydraulic cylinders is therefore proportional to the torque the engine is producing. This pressure is measured on a pressure gage calibrated in pounds per square inch of BMEP (brake mean effective pressure). By using the chart in Figure 2-9 on Page 29, and knowing the BMEP and RPM, the flight crew can determine the brake horsepower the engine is developing.

The amount of torque produced by a modern turboprop engine is extremely important, because these engines are capable of producing more power than the airframe in which they are installed is designed to handle. The airframe manufacturer therefore establishes a flat-rate limit for the torque the engine is allowed to produce. This limit holds until a pressure altitude is reached at which the power is limited by the exhaust gas temperature. Flat rating is explained in more detail in Chapter 19, beginning on Page 694.

Turboprop engines are usually equipped with torquemeters, which are quite often oil pressure gages with dials graduated in horsepower or foot-pounds of torque.

When a torquemeter is calibrated in foot-pounds of torque, the horsepower the engine is developing can be determined by the relationship be-tween the RPM and the torque. The limiting horsepower is reached at a given torque when operating at 100% RPM for takeoff, but a higher torque may be allowed when the RPM is reduced for cruise. The POH (Pilot's Operating Handbook) for the particular aircraft specifies the RPM, torque, and EGT limits.

The horsepower developed by a turboprop engine is found by this formula:

$$HP = K \cdot RPM \cdot \text{Torque (foot-pounds)}$$

The K factor (typically in the neighborhood of 0.0001) in this formula is a constant for the particular engine. Notice in this formula that for any given horsepower, the torque varies inversely as the RPM. The higher the RPM, the lower the torque for any given power.

The operation of a torquemeter for a turboprop engine is described and illustrated in Figures 16-21 and 16-22.

The power extracted from the three stages of turbines in Figure 16-21 is delivered to the reduction gears through a long, thin torsion shaft splined at both ends. One end is splined into the rear of the hollow turbine/compressor main shaft, and the other is splined into a gear on the same shaft as the pinion gear for the propeller reduction gear system.

When the turbine and compressor are turning at 100%, which is approximately 42,000 RPM, the torque applied to the torsion shaft is only slightly more than 100 foot-pounds, which can easily be carried by the small-diameter shaft. But when this speed is reduced to 1,600 RPM by the 26:1 propeller reduction gear ratio, the torque at the propeller shaft is approximately 2,600 foot-pounds, which is sufficient to produce 416 horsepower at the propeller shaft.

$$\begin{aligned} HP &= K \cdot RPM \cdot \text{Torque (foot-pounds)} \\ &= 0.0001 \cdot 1{,}600 \cdot 2{,}600 \\ &= 416 \text{ horsepower} \end{aligned}$$

The torque sensor in Figure 16-22 on the next page uses a helical gear set that moves a pilot valve. This gear set consists of an internal-tooth female gear and an external-tooth male gear with the teeth cut at an angle. The female gear is driven by the high-speed pinion of the propeller reduction gear system. This gear is in turn driven by the turbines through the torsion shaft. The male gear is driven by the turbine/compressor main shaft. The two drive gears turn at the same speed, but any twisting of the torsion shaft caused by the torque reaction of the propeller produces an angular difference between the two gears. This angular difference causes the female helical gear to move in over the male gear and move the pilot valve to the right an amount proportional to the torque reaction.

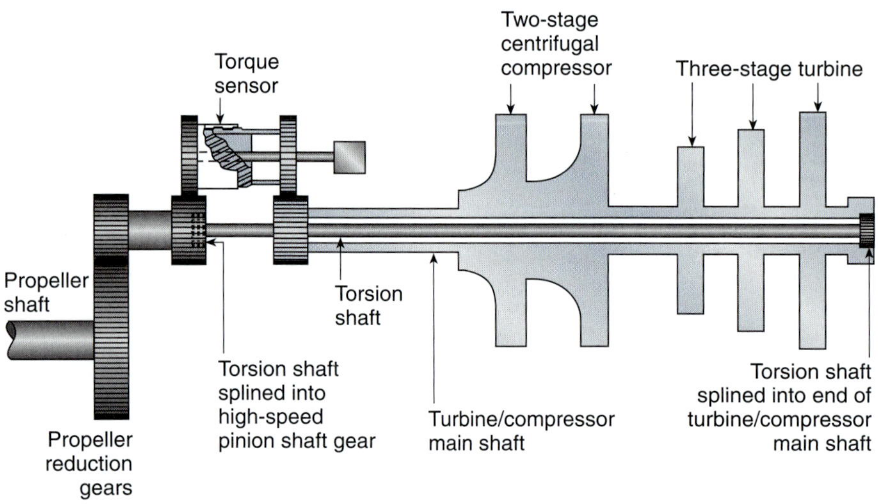

Figure 16-21. *The torque sensor is an oil valve driven by the angular difference between two gears that turn at the same speed. One is driven directly by the main turbine/compressor shaft, and the other is driven by this shaft through a long, thin torsion shaft.*

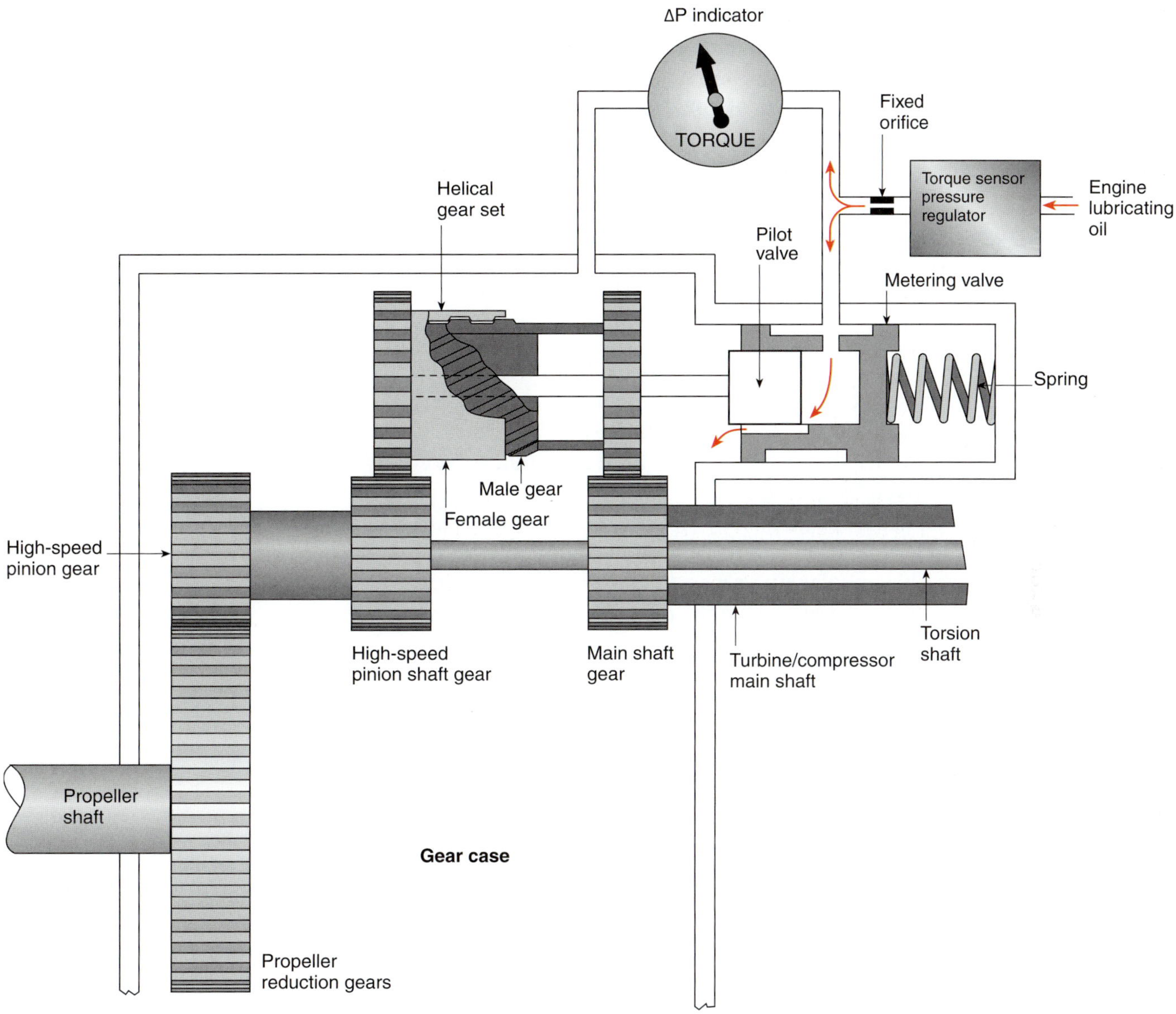

Figure 16-22. *Operating principle of the torque sensor*

Engine lubricating oil, under pressure from the torque-sensor pressure regulator, flows through a fixed orifice and into a metering valve. When the torque is low, there is little angular difference between the two drive gears, and the pilot valve is moved to the left, allowing the oil to flow through a channel in the metering valve body and into the gear case. The differential pressure indicator, calibrated in foot-pounds of torque, measures the pressure drop across the metering valve.

When the propeller load increases and the fuel control sends more fuel to the engine to keep the propeller RPM constant, the torque increases and twists the torsion shaft. This twisting movement causes the female gear and the pilot valve to move to the right. The pilot valve restricts the flow of oil into the gear case, and the torque indication on the ΔP indicator increases. To prevent the pressure indication from continuing to rise, the pressure acting on the metering valve moves it to the right, compressing the spring which holds the metering valve constant for any given torque, and the torquemeter indication remains steady.

Almost all torquemeter sensors can sense a negative torque. If the engine fails, and the aerodynamic forces on the propeller drive the engine, the negative torque sensor will move the propeller feather valve and automatically feather the propeller. There is more on this system in Chapter 19 on Page 700.

STUDY QUESTIONS: TYPES OF POWERPLANT INSTRUMENTS

Answers are on Page 586. *Page numbers refer to chapter text.*

1. Four powerplant instruments required for an air-cooled, supercharged reciprocating engine are:
   a. ______________________________
   b. ______________________________
   c. ______________________________
   d. ______________________________
   Page 549

2. An instrument whose indication is given by a pointer moving across a graduated circular dial is called a/an ______________________________ indicator. *Page 549*

3. Three types of data important for monitoring the operation and performance of an aircraft powerplant are:
   a. ______________________________
   b. ______________________________
   c. ______________________________
   Page 550

4. Three types of pressure commonly measured in powerplant instrumentation are:
   a. ______________________________
   b. ______________________________
   c. ______________________________
   Page 550

5. Pressure referenced from a vacuum is ______________________________ pressure. *Page 550*

6. The absolute pressure measuring instrument most widely used with a reciprocating engine is the ______________________________ gage. *Page 550*

7. Pressure referenced from the existing atmospheric pressure is ______________________________ pressure. *Page 551*

8. The pressure a moving fluid would have if it were stopped without any losses is called ______________________________ pressure. *Page 551*

9. The pressure that results from the motion of a fluid is called ______________________________ pressure. *Page 551*

10. The pressure of a fluid not in motion is called ______________________________ pressure. *Page 551*

11. The type of instrument mechanism normally used to measure gage pressure is a/an ______________________. *Page 552*

12. An oil pressure gage is connected to the engine through a/an ______________________ fitting. *Page 552*

13. The line between the engine and the oil pressure gage is filled with a/an ______________________. *Page 552*

14. When oil temperature is measured with a pressure gage, the Bourdon tube, capillary tube, and temperature-sensing bulb are all filled with a/an ______________________ liquid and sealed as a unit. *Page 552*

15. A fuel pressure warning light is ______________________ (more or less) sensitive than a fuel pressure gage. *Page 553*

16. The fuel flowmeter used with a fuel-injected horizontally opposed engine is a/an ______________________ gage. *Page 553*

17. The fuel flowmeter used on a turbocharged fuel-injected horizontally opposed engine measures ______________________ (gage or differential) pressure. *Page 553, 554*

18. The two instruments used together to show the power a reciprocating engine is producing are the:
    a. ______________________
    b. ______________________
    Page 554

19. When the engine is not running, the manifold pressure gage indicates the existing ______________________ pressure. *Page 555*

20. It ______________________ (is or is not) possible for a supercharged reciprocating engine to have a manifold pressure higher than the existing atmospheric pressure. *Page 555*

21. The amount of thrust produced by a gas turbine engine is determined by the change in the ______________________ of air flowing through the engine. *Page 556*

22. The thrust produced by an axial-flow turbine engine is found by measuring the ______________________. *Page 556*

Continued

STUDY QUESTIONS: TYPES OF POWERPLANT INSTRUMENTS *Continued*

23. EPR is the ratio of what two pressures?
    a. ______________________________
    b. ______________________________
    Page 556

24. Two types of electrically operated temperature measuring instruments are:
    a. ______________________________
    b. ______________________________
    Page 556

25. The sensitive element in the temperature bulb used with a ratiometer indicator is made of ______________________ wire. *Page 556*

26. The two junctions in a thermocouple are the:
    a. ______________________________
    b. ______________________________
    Page 557

27. A thermocouple is useful for measuring temperature, because the voltage difference between the two junctions is proportional to the ____________________________ between the two junctions. *Page 557*

28. The reference junction in an aircraft thermocouple instrument is compensated for temperature changes by mounting the indicator needle on a/an _______________________ . *Page 557*

29. Two combinations of metals used for thermocouples to measure cylinder head temperature are:
    a. _______________ and ________________
    b. _______________ and ________________
    Page 558

30. The metals used in thermocouples for measuring exhaust gas temperature in both reciprocating and turbine engines are ___________________ and __________________ . *Page 558*

31. The most critical temperature in a turbine engine is the _______________________ temperature. *Page 559*

32. On most turbine engines, the turbine inlet temperature _________________ (is or is not) measured directly. *Page 559*

33. The temperature most generally measured in a turbine engine that relates to the turbine inlet temperature is the ________________________ temperature. *Page 559*

34. When several thermocouples are used to average the temperature of the gases leaving the tail cone of a turbine engine, they are connected in ____________________ (series or parallel). *Page 559*

35. The dials of reciprocating engine tachometers are graduated in RPM times ____________________ . *Page 559*

36. The dials of tachometers used for turbine engines are graduated in ____________________ of the rated RPM. *Page 559*

37. The tachometer for a reciprocating engine turns at __________________ (what fraction) of the crankshaft speed. *Page 560*

38. Magnetic drag tachometers used in most small general aviation aircraft must be matched to the cruise RPM of the engine so the _________________________ will be accurate. *Page 560*

39. The accuracy of a tachometer installed in an aircraft can be checked with a/an _____________________ , or _____________________ tachometer. *Page 560*

40. Three-phase AC tachometers use the ____________________ (voltage or frequency) of the AC produced by the tachometer generator to indicate the RPM of the engine. *Page 561*

41. An instrument that shows the difference in RPM between two engines installed on a multiengine airplane is called a/an _______________________ . *Page 562*

42. A vane-type flowmeter measures the _____________________ (mass or volume) of fuel flowing to the engine. *Page 562*

43. Flowmeters used on turbine engines measure the ________________________ (mass or volume) of the fuel flowing to the engine. *Page 563*

44. Hydraulic cylinders in the nose section of certain large radial engines oppose the reaction between the propeller reduction gears and the nose section of the engine. This reaction is proportional to the _______________________ the engine is producing. *Page 564*

45. A flat-rate limit to the amount of torque a turboprop engine is allowed to produce at low pressure altitudes is normally established by the _______________ (airframe or powerplant) manufacturer. *Page 564*

46. The amount of torque a turboprop engine is allowed to produce at high pressure altitude is limited by the ________________________ . *Page 564*

47. For a given amount of power, the torque varies _______________________ (directly or inversely) as the RPM. *Page 564*

Powerplant Instrument Marking, Installation, and Maintenance

The repair and installation of aircraft instruments differs from maintenance for other aircraft components. 14 CFR §65.81(a) allows a certificated mechanic to "perform or supervise the maintenance, preventive maintenance or alteration of an aircraft or appliance, or a part thereof, for which he is rated (*but excluding* major repairs to, and major alterations of propellers, and *any repair to, or alteration of, instruments*)..."

An aviation maintenance technician with a powerplant rating is authorized to perform the required 100-hour inspections on powerplant instruments and their systems. This rating also authorizes the removal and replacement of the instruments and the replacement of range markings if these marks are on the outside of the glass and do not require opening the instrument case.

Any actual repair or calibration to an instrument must be done by the instrument manufacturer or by an FAA-certificated repair station approved for the particular repair to that instrument.

Instrument Range Marking

Some powerplant instruments have colored range marks that enable the pilot to see at a glance whether a particular system or component is or is not operating in a safe and desirable range. The colored marks point out approaching operating difficulties. Figure 16-23 lists the colors, types, and their meanings. *See* Figures 16-23 and 16-24.

The technician who installs an instrument in an aircraft and the technician who conducts an inspection are responsible for determining that the instruments are properly marked for the aircraft in which they are installed. Type Certificate Data Sheets for the aircraft and the engine specify the range marks required.

bezel. The rim which holds the glass cover in the case of an aircraft instrument.

Some instruments have the range marks on the glass rather than on the dial, and must have a slip mark that shows when the glass has slipped and the marks are no longer properly aligned with the numbers on the dial. The slip mark is a white line painted across the instrument bezel and onto the glass at the bottom of the instrument. If the glass should slip, the mark will be broken, warning the flight crew that the range markings are no longer correct.

Color and Type of Mark	Meaning
Green arc	Normal operating range
Yellow arc	Caution range
White arc	Special operations range
Red arc	Prohibited range
Red radial line	Do not exceed indication
Blue radial line	Special operating condition
Red triangle, dot or diamond	Maximum limit for high transients such as starting

Figure 16-23. *The color, type, and meaning of range marks on powerplant instruments*

Carburetor air temperature	
Yellow arc	Range where carburetor ice is most likely to form
Green arc	Normal operating range
Red radial line	Maximum allowable inlet air temperature
Cylinder head temperature	
Green arc	Normal operating range
Yellow arc	Operation approved for limited time
Red radial line	Never-exceed temperature
Manifold pressure gage	
Green arc	Normal operating range
Yellow arc	Precautionary range
Red radial line	Maximum permissible manifold absolute pressure
Fuel pressure gage	
Green arc	Normal operating range
Yellow arc	Precautionary range
Red radial line	Maximum and/or minimum permissible fuel pressure
Oil pressure gage	
Green arc	Normal operating range
Yellow arc	Precautionary range
Red radial line	Maximum and/or minimum permissible oil pressure
Oil temperature gage	
Green arc	Normal operating range
Yellow arc	Precautionary range
Red radial line	Maximum and/or minimum permissible oil temperature
Tachometer (Reciprocating engine)	
Green arc	Normal operating range
Yellow arc	Precautionary range
Red arc	Restricted operating range
Red radial line	Maximum permissible rotational speed
Tachometer (Turbine engine)	
Green arc	Normal operating range
Yellow arc	Precautionary range
Red radial line	Maximum permissible rotational speed
Tachometer (Helicopter)	
Engine tachometer	
Green arc	Normal operating range
Yellow arc	Precautionary range
Red radial line	Maximum permissible rotational speed
Rotor tachometer	
Green arc	Normal operating range
Red radial line	Maximum and minimum rotor speed for power-off operational conditions
Torque indicator	
Green arc	Normal operating range
Yellow arc	Precautionary range
Red radial line	Maximum permissible torque pressure
Exhaust gas temperature indicator (Turbine engine)	
Green arc	Normal operating range
Yellow arc	Precautionary range
Red radial line	Maximum permissible gas temperature

Figure 16-24. *Required range marking for powerplant instruments*

Instrument Installation

A powerplant technician is authorized to install powerplant instruments in an instrument panel, and it is his or her responsibility during an inspection to be sure that the instruments are secure in their mounting, and that all the hoses and wires attached to them are in good condition and do not interfere with any control.

Many electrical instruments, such as tachometers and cylinder head temperature indicators, have iron or steel cases that prevent their magnets from interfering with nearby instruments or electronic equipment. Lines of magnetic flux cannot cross iron or steel, and these cases entrap the flux. But as a precaution, electrical instruments should not be mounted near a magnetic compass.

Most instruments are installed in the panel with brass machine screws and nuts mounted in holes in the instrument case or nut plates installed in the panel. Because panel space is limited, many modern powerplant instrument cases are flangeless and are held in clamps attached to the back of the panel. The instrument is connected to its electrical harness or hose and slipped through the hole in the panel until it is flush and properly aligned; then the clamp-tightening screw in the panel is turned to tighten the clamp around the instrument case. *See* Figure 16-25.

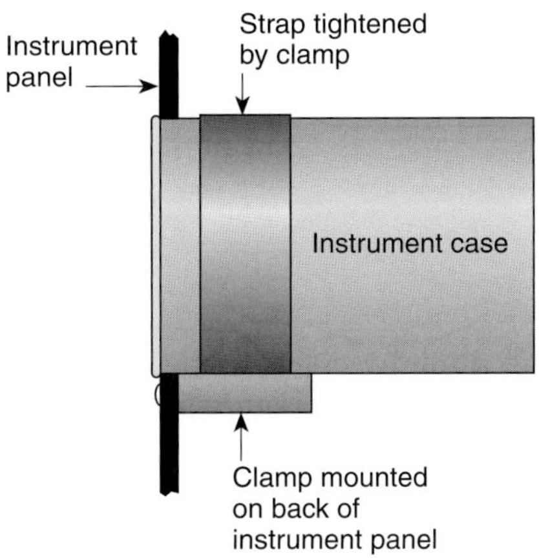

Figure 16-25. *Flange-mounted instrument case*

Instrument Handling

Aircraft instruments are delicate and sensitive and require special handling. The cases of such instruments as oil pressure, fuel pressure, and manifold pressure are made of thermosetting plastic and can be cracked if the fittings are overtightened.

Cylinder head temperature and exhaust gas temperature gages are thermocouple-type instruments whose moving coils are damped through the thermocouple. When the thermocouple is not connected, the instrument is not damped, and the pointer can swing violently enough to knock it out of balance and cause inaccurate indications. Any time a thermocouple instrument is not connected to its thermocouple, a piece of uninsulated wire, such as safety wire, should be wrapped around the terminals to short-circuit them and allow damping current to flow. Be sure, however, that this wire is removed before the thermocouple is connected.

STUDY QUESTIONS: POWERPLANT INSTRUMENT MARKING, INSTALLATION, AND MAINTENANCE

Answers are on Page 586. Page numbers refer to chapter text.

48. An aviation maintenance technician with a powerplant rating ________________ (is or is not) authorized to perform the required 100-hour inspections on powerplant instruments and their systems. *Page 572*

49. Any repair or calibration to an instrument must be done by the instrument manufacturer or by a/an ________________-certificated repair station approved for the particular repair to that instrument. *Page 572*

50. The maximum limit for a high-temperature transient, such as could occur during starting, is marked on a powerplant instrument dial with a/an ______________________________________ . *Page 572*

51. The proper range markings required for a powerplant instrument may be found in the _________________________ for the aircraft in which the instrument is installed. *Page 572*

52. If an instrument has range marks on the glass, there must be a ___________________ (what color) slip mark painted between the glass and the instrument case bezel. *Page 572*

53. Certain instruments that contain magnets are prevented from interfering with nearby instruments or electronic equipment by installing them in cases made of __________________ or __________________ . *Page 574*

54. When a thermocouple is not attached to a cylinder head temperature indicator, a piece of _________________________ wire should be wrapped around the terminals of the instrument. *Page 574*

microprocessor. A single silicon chip that contains the arithmetic and logic functions of a computer.

semiconductor transducer. A piezoelectric crystal that converts input energy of one form, such as pressure, into output energy of another, such as an electrical signal.

data. Input for computer processing in the form of numerical information that represents characters or analog quantities.

CRT (cathode-ray tube). An electronic display tube in which a controlled stream of electrons causes a portion of the phosphor-coated tube face to emit light.

LCD (liquid crystal display). A digital display in which alphanumeric or graphic data is displayed on a glass screen, as voltages applied to segments etched into the glass change the transparency of a liquid crystal material between the glass sheets of which the screen is made.

MFD (multifunction display). A liquid crystal or CRT display that shows a number of parameters and replaces several analog-type indicators.

glass cockpit. An aircraft instrument system that uses a few color cathode-ray-tube displays to replace a large number of mechanically-actuated instruments.

Electronic Instrumentation

One of the fundamental axioms of aircraft instrumentation is "If a parameter can be measured, it can be controlled, and if it can be controlled, it can be controlled automatically." But automatic control was difficult before the advent of digital electronics and the microprocessor.

Microprocessors have revolutionized powerplant instrumentation, and tiny semiconductor transducers and optoelectronic devices have replaced mechanical sensors for measuring pressure, temperature, and mechanical movement. Rather than sending data from these transducers to analog-type indicators for the flight crew to interpret, the data is directed into a computer where it is analyzed and corrected as needed and then displayed on a video screen.

The most elaborate electronic powerplant instrument system is the EICAS (Engine Indication and Crew Alerting System) discussed beginning on Page 580. The EICAS does not display all the powerplant parameters when they are within their normal operating range, but when any condition exceeds its allowable limit, the flight crew is alerted of an impending problem. There are simpler systems available for general aviation aircraft and we will discuss those first.

Digital Indicating and Control Systems

Color cathode-ray tubes (CRT) and color liquid-crystal displays (LCD) are used as multifunction displays (MFD) in the modern "glass cockpits." A single MFD replaces a number of mechanical analog-type indicators and has the added advantage that only those indications that show abnormal conditions need to be displayed. In addition to displaying instrument indications, CRTs are used to display checklists and operational history for the portions of a system that are showing trouble. They can also display performance deterioration caused by the malfunction and display suggested corrective action.

Digital systems lend themselves to self-examination of their operating condition and the diagnosis of faults. This is done by the portion of the system known as BITE, or built-in test equipment. BITE checks the system, and when a malfunction is detected, traces it to the nearest line replaceable unit, or LRU, and informs the flight crew of the action that should be taken.

Microcomputers

Analog computers, developed during World War II, opened tremendous possibilities for automation, but they were too large and required too much electrical power to be used for aircraft powerplant control. This problem was solved with the advent of the microcomputer.

Electronic computers operate with numbers by assigning a value of voltage to each digit. Using the decimal number system requires the computer to manipulate ten different values, or conditions, and the precision required

complicates the process of computing. But the binary number system does everything the decimal system does, and it uses only two conditions: 0 and 1, or electrically, voltage and no voltage.

Dedicated digital computers are the heart of almost all modern aircraft electronic control and indicating systems. Figure 16-26 is a simple diagram of a digital computer that contains the three essential components: a central processing unit, or CPU, a memory, and input/output devices.

The CPU is the heart of the computer, and it contains a clock, a control unit, and an arithmetic/logic unit (ALU). The memory contains all the instructions and data stored in the form of binary numbers called words. The input devices receive signals from temperature, position, or pressure sensors, and commands from the pilot. The output devices can range from video displays to electric motors or other types of actuators.

The memory section contains two types of memory, ROM and RAM. ROM, or Read-Only Memory, is permanent memory built into the computer and contains the instructions the computer follows to boot up (start up) and perform a number of diagnostic tests. It also contains all the steps the computer should follow to process signals from the input devices and give the desired results in the output.

The RAM, or Random-Access Memory, is a read-write memory in which data can be held until needed, and then called out and manipulated by the ALU until returned to storage. RAM is called volatile memory because it loses its data when the power is turned off.

BITE (built-in test equipment). A troubleshooting system installed in many modern jet aircraft that monitors engine and airframe systems. When a fault is found, the system identifies the LRU (line replaceable unit) that contains the fault.

LRU (line replaceable unit). Components in an aircraft that are designed so they can be replaced as a unit while the aircraft is on the flight line.

volatile memory. Computer memory that is lost when the power to the computer is turned off.

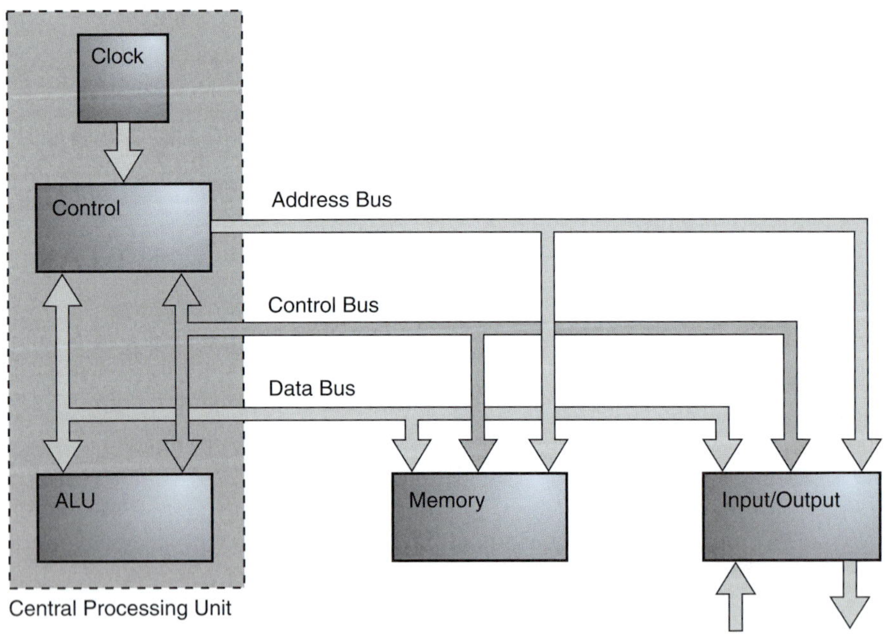

Figure 16-26. *A simplified block diagram of a dedicated digital computer used in electronic aircraft powerplant controls*

Dedicated digital computers work on a principle that can be understood by following the operation of a computer designed to maintain cylinder head temperature within an optimum range, and to warn the pilot if it gets out of this range. The aircraft has a digital cylinder head temperature indicator with a two-line remarks display, as in Figure 16-27. The light bars on the indicator are green up to a temperature of 100°C. They are blue between 100°C and 230°C, and red above 230°C. The bars flash above 260°C.

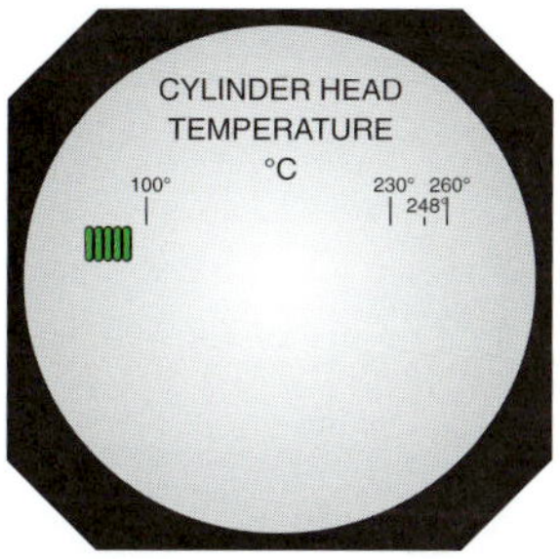

CHT is 90°C. Bars are green, and no remarks are displayed.

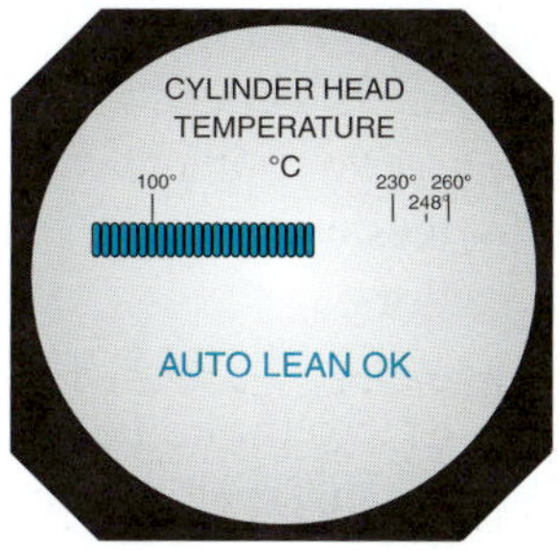

CHT is 185°C. Bars are blue, and notation shows that engine can be operated with carburetor mixture set to AUTO LEAN.

CHT is 235°C. Bars are red, and notation shows AUTO RICH must be used.

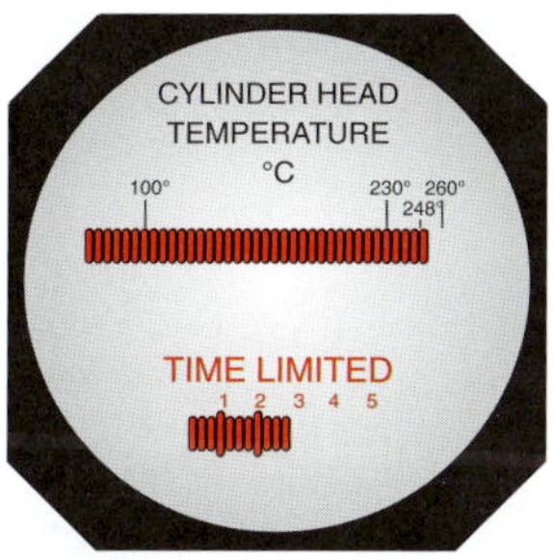

CHT is 250°C. Bars are red, and engine is in time limited range. It has been operating in this range for almost 3 minutes out of the allowable 5 minutes.

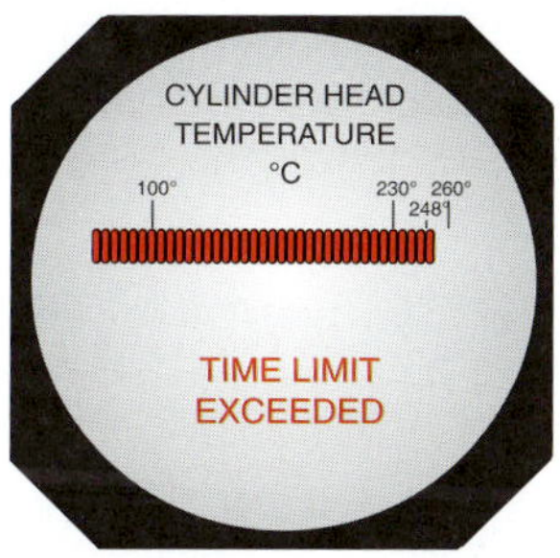

CHT is still 250°C and has been above 248°C limit for more than the allowable 5 minutes. TIME LIMIT EXCEEDED warning is flashing.

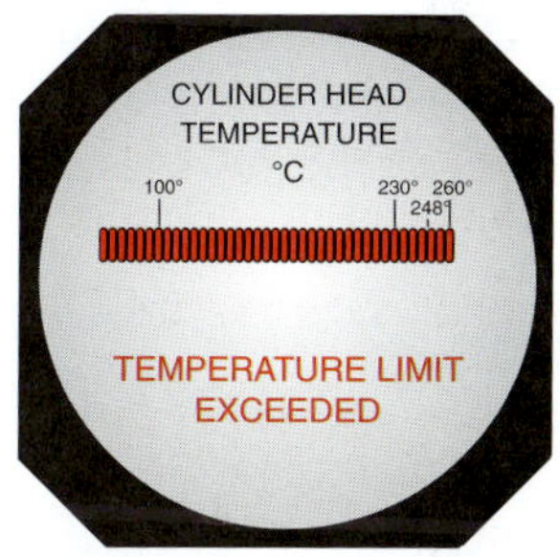

CHT is 260°C and light bars and TEMPERATURE LIMIT EXCEEDED warning is flashing.

Figure 16-27. *This type of digital cylinder head temperature indicator is driven by a dedicated digital computer. Each light bar in the display represents 5°C.*

Listed below are basic steps the computer accomplishes, and while it requires some time for us to read them, the computer moves through them continuously, with each step taking only a few thousandths of a second.

1. When the master switch is turned on, the first signal from the clock tells the control to clear all storage areas in the computer and perform all necessary diagnostic tests.
2. The next signal from the clock tells the control to fetch the first instruction. The program in ROM tells the control to determine from the landing gear squat switch (an input device) if weight is on the landing gear.

squat switch. An electrical switch actuated by the landing gear scissors on the oleo strut.

3. If the squat switch indicates that weight is on the landing gear, an instruction is sent to the controller signaling the cowl flap motor to open the cowl flaps.
4. The clock continually tells the control to fetch instructions from the memory. These steps are a loop of instructions that tell the controller to sense the input from the CHT thermocouple and illuminate the correct light bars on the indicator display.
5. The loop of instructions also continually monitors the landing gear squat switch, and when it signals that the weight is off the landing gear, the cowl flap motor (an output device) is instructed to close the cowl flaps.
6. The loop continues to search all the input devices until the temperature sensor indicates that the CHT has reached 100°C. At this time the display at the bottom of the indicator reads AUTO-LEAN OK.
7. The loop continues to search all the input devices until the temperature sensor indicates that the CHT has reached 230°C. This time the display reads AUTO-RICH ONLY.
8. The loop continues to search all the input devices until the temperature sensor shows that the CHT has reached 248°C. A signal is sent to the cowl flap motor to open the cowl flaps enough to keep the temperature below 248°C. Any time the loop detects that the CHT is above 248°C, the display reads TIME-LIMITED. The control directs a display on the CHT indicator to light up a bar graph showing the length of time the engine is allowed to operate in this temperature range. A red bar lights up every 15 seconds.
9. If the loop sampling the temperature finds that it remains above 248°C for the full five minutes that are allowed, the display flashes TIME LIMIT EXCEEDED.
10. If the temperature ever reaches 260°C, the light bars and the words TEMPERATURE LIMIT EXCEEDED flash.

Engine sensors
- Compressor speeds
  - N_1, N_2, and N_3
- Engine Pressure Ratio (EPR)
- Exhaust Gas Temperature (EGT)
- Fuel flow
- Oil pressure
- Oil quantity
- Oil temperature
- Vibration

System sensors
- Hydraulic quantity
- Hydraulic pressure
- Control surface positions
- Electrical system voltage
- Electrical system current
- Electrical system frequency
- Generator drive temperature
- Environmental control system temperatures
- APU exhaust gas temperature
- APU speed
- Brake temperature

Figure 16-28. *Parameters sensed by a typical EICAS*

EICAS (engine indication and crew alerting system). An electronic instrumentation system that monitors airframe and engine parameters and displays the essential information on a video display on the instrument panel.

EEC (electronic engine control). An electronic fuel control for a gas turbine engine that senses the power-lever angle (PLA), engine RPM, bleed valve and variable stator vane position, and the various engine pressures and temperatures. It meters the correct amount of fuel to the nozzles for all flight conditions to prevent turbine overspeed and overtemperature.

Computerized Fuel System

Another application for a dedicated computer is the computerized fuel system, or CFS. This versatile instrument uses a small turbine rotor mounted in the fuel line between the fuel injection unit and the flow divider to which the fuel injection nozzles attach. All the fuel that flows to the cylinders must pass through the turbine, which spins at a speed proportional to the rate of fuel flow. As it spins, it interrupts a beam of light between a light-emitting diode and a phototransistor. The resulting pulses of light are converted into pulses of electricity and are directed into the input of the computer.

When the instrument is properly programmed with the amount of fuel on board at engine startup, it gives the pilot the following information:

- The number of pounds or gallons of fuel on board
- The fuel flow in gallons or pounds per hour
- The fuel time remaining at the present rate of flow
- The number of gallons of fuel used since the engine was initially started

EICAS

The EICAS (engine indication and crew alerting system) replaces a myriad of individual instruments and furnishes the flight crew with the necessary information. A typical EICAS senses the parameters listed in Figure 16-28, and in addition, interfaces with such systems as the maintenance control display panel (MCDP) of the flight control computer (FCC), the thrust management system (TMS), the electronic engine control (EEC), the flight management computer (FMC), the radio altimeter, and the air data computer (ADC).

The EICAS consists of two multicolor CRT display units, two computers, two control panels, a pilot's display select panel, and a maintenance panel. There are also two display-switching modules, the cancel/recall switches, and the captain's and first officer's caution lights.

The two EICAS display units are mounted one above the other as shown in Figure 16-29. The primary engine parameters, such as EPR, EGT, and N_1 speed, are continually displayed in the upper right-hand side of the unit. These parameters are shown in the form of an analog display along with the actual value in digits. The left-hand side of the upper display shows crew-alerting warnings and cautions. The lower display shows engine secondary parameters such as N_2 speed, fuel flow, oil quantity, oil pressure, oil temperature, and engine vibration. The status of airframe systems may be displayed as well as maintenance data. The master caution lights as well as aural signals back up the displays on the EICAS.

Alert messages appear on the left side of the upper display in one of three colors. If white, it is an advisory message, and action should be taken when time is available. If yellow, it is a caution message, which requires immediate crew awareness and future compensatory action. If red, it is a warning message, which requires immediate corrective or compensatory action by the crew. Any time a warning message appears on the EICAS, an aural and visual warning are also initiated. The cancel/recall switches are used to review and control the caution and advisory messages.

EICAS is not only useful to the flight crew, but is an essential tool for ground maintenance. When any subsystem malfunctions, the EICAS automatically records the parameters that identify the malfunction as soon as they are detected. Also, if the flight crew want to record all the parameters that exist at any specific time, they can press a single button, and all data will be automatically recorded for the maintenance technicians to analyze and evaluate.

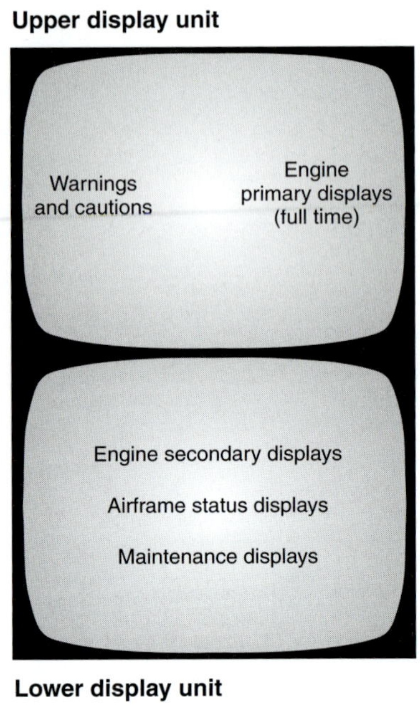

Figure 16-29. *Location of information on the EICAS multifunction displays*

FMC (flight management computer). An electronic flight instrumentation system that enables the flight crew to initiate and implement a given flight plan and monitor its execution.

ADC (air data computer). An electronic computer in an aircraft that senses pitot pressure, static pressure, and total air temperature and produces an indication of altitude, indicated airspeed, true airspeed, and Mach number.

Notice in Figure 16-30 that the EICAS has two computers that receive data from the airplane system and engine sensors. One computer at a time processes and displays all the information required by the flight crew, who select the computer from the pilot's display select panel.

The three primary engine parameters needed to set and monitor engine thrust, EPR, N_1, and EGT, are shown on the upper display at all times by both an analog pointer and a digital readout.

The secondary parameters, N_2, fuel flow, oil pressure, oil temperature, oil quantity, and vibration, are displayed on the lower display at power up and when manually selected. At other times, the lower display is blank. These secondary parameters are displayed by both a digital readout and an analog pointer.

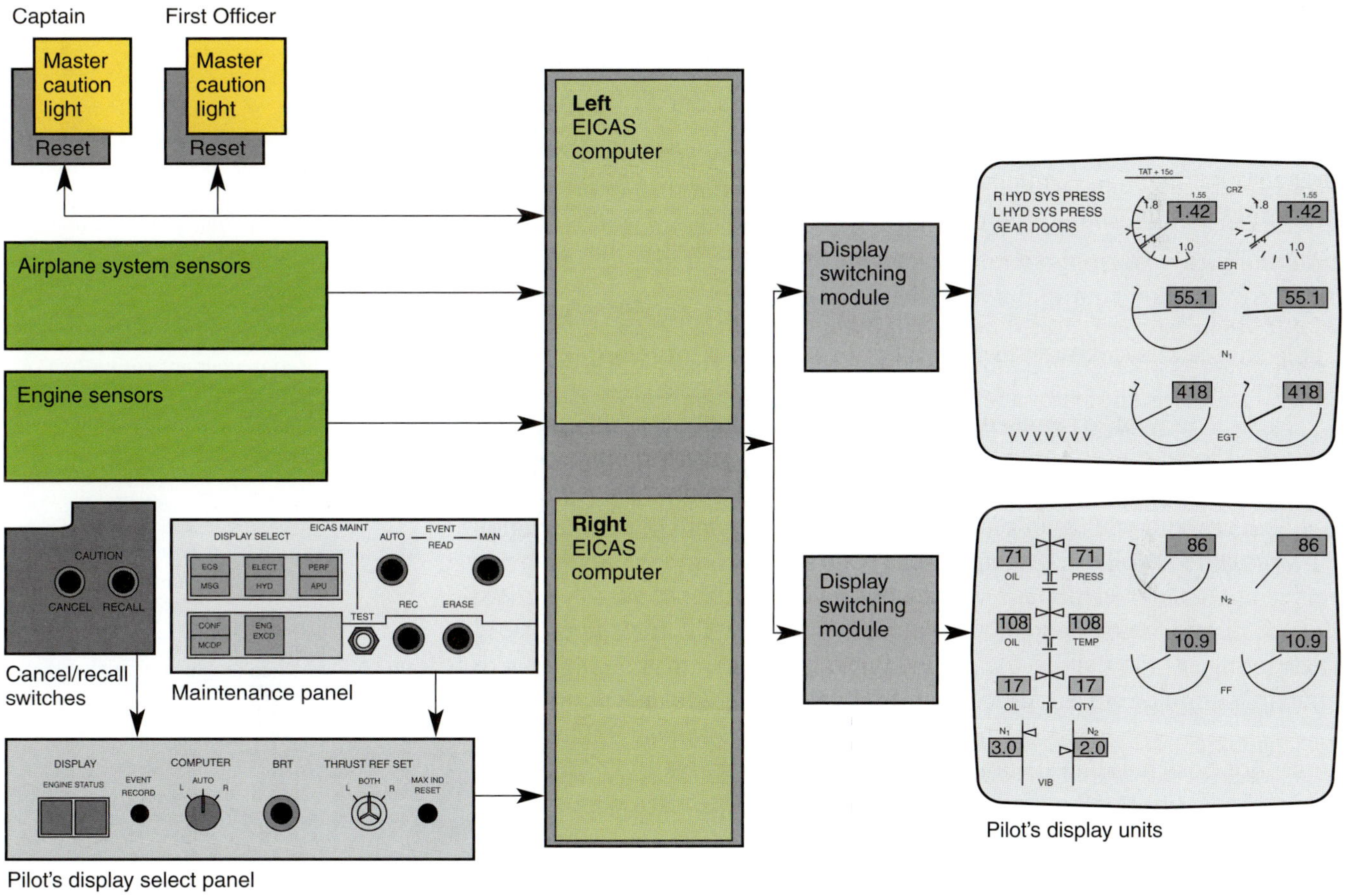

Figure 16-30. *Block diagram of an EICAS*

If any parameter exceeds the limits for which it is programmed (an exceedance condition), the EICAS automatically displays this parameter for both engines in the appropriate color. This alerts the flight crew to an impending problem and indicates its seriousness.

exceedance condition. A condition in which a parameter sensed by the EICAS exceeds the limits for which it is programmed.

The colors used on the multicolor CRT displays and their meanings are shown in Figure 16-31.

White: General color used for all scales, normal operating range of pointers, digital readouts and digital readout boxes, advisory messages

Red: Warning messages, redline limits on scales, exceedance condition for pointers, digital readouts, and digital readout boxes

Green: Thrust mode and reference readout, EPR target cursor, TAI readout, selected temperature readout, and thrust reverser messages

Blue: Not used for EICAS displays, it is displayed during EICAS test only

Yellow: Caution messages. Yellow band on scales, yellow band condition for pointer, digital readouts and digital boxes, maximum limit marker, and thrust reverser readout

Magenta (Pink): Inflight start envelope, fuel-on command index, crossbleed messages, and EPR/FMC target cursor

Cyan (Blue): All parameter names, status cue and secondary engine data cue

Black: Background color

Figure 16-31. *Colors and their meanings used on an EICAS display*

STUDY QUESTIONS: ELECTRONIC INSTRUMENT SYSTEMS

Answers are on Page 586. Page numbers refer to chapter text.

55. When the BITE detects a malfunction, it traces it to the nearest ____________________ . *Page 576*

56. Dedicated microcomputers used in aircraft engine control systems are ____________________ (analog or digital) computers. *Page 577*

57. The three basic components in a digital computer are:
    a. ______________________________
    b. ______________________________
    c. ______________________________
    Page 577

58. Two types of memory in a digital computer are:
    a. ______________________________
    b. ______________________________
    Page 577

59. RAM ________________________ (does or does not) allow data to be written into the memory as well as read from it. *Page 577*

60. RAM is normally ___________________________ (volatile or nonvolatile). *Page 577*

61. Three engine primary parameters displayed on the upper display of an EICAS are:
    a. ______________________________
    b. ______________________________
    c. ______________________________
    Page 581

62. Three types of information displayed on the lower EICAS display are:
    a. ______________________________
    b. ______________________________
    c. ______________________________
    Page 581

63. Alert messages that appear on the upper EICAS panel are colored. An advisory message is ________________ , a caution message is ________________ , and a warning message is ________________ . *Page 581*

Answers to Chapter 16 Study Questions

1. a. tachometer
   b. oil pressure gage
   c. oil temperature gage
   d. manifold pressure gage
2. analog
3. a. pressure
   b. temperature
   c. mechanical movement
4. a. absolute pressure
   b. gage pressure
   c. differential pressure
5. absolute
6. manifold pressure
7. gage
8. total
9. dynamic
10. static
11. Bourdon tube
12. restricted
13. lightweight oil
14. volatile
15. more
16. pressure
17. differential
18. a. tachometer
    b. manifold pressure gage
19. barometric
20. is
21. momentum
22. EPR (engine pressure ratio)
23. a. turbine discharge total pressure
    b. compressor inlet total pressure
24. a. ratiometer instruments
    b. thermocouple instruments
25. Nichrome
26. a. measuring junction
    b. reference junction
27. temperature difference
28. bimetallic hairspring
29. a. copper, constantan
    b. iron, constantan
30. chromel, alumel
31. turbine inlet
32. is not
33. exhaust gas
34. parallel
35. 100
36. percent
37. one half
38. hourmeter
39. stroboscopic, laser
40. frequency
41. synchroscope
42. volume
43. mass
44. torque
45. airframe
46. exhaust gas temperature
47. inversely
48. is
49. FAA
50. red triangle, dot, or diamond
51. Type Certificate Data Sheets
52. white
53. iron, steel
54. uninsulated
55. LRU (Line Replaceable Unit)
56. digital
57. a. central processing unit
    b. memory
    c. input/output devices
58. a. ROM
    b. RAM
59. does
60. volatile
61. a. EPR
    b. EGT
    c. N_1 speed
62. a. engine secondary parameters
    b. airframe system status
    c. maintenance data
63. white, yellow, red

ELECTRICAL SYSTEMS

17

Continued

Electrical Systems

17

The importance of the electrical system of a modern aircraft cannot be overemphasized. No technician can consider his or her technical education complete without a good understanding of electricity, electronics, and the fundamentals of digital electronics.

The *General* textbook of this *Aviation Maintenance Technician Series* includes an introduction to both DC and AC electricity and solid-state electronic components. This chapter assumes a knowledge of basic electrical theory, and a review of Chapter 4 in the *General* textbook is in order before beginning the study of this chapter.

The *Airframe* textbook of this series provides examples and explanations of various electrical systems and circuits found in modern aircraft. This *Powerplant* textbook considers the generation and control of DC and AC electricity and the portions of the electrical systems that are usually maintained by a powerplant technician.

DC (direct current). Electrical current in which the electrons always flow in the same direction.

AC (alternating current). Electrical current in which the electrons continually change their rate of flow and periodically reverse their direction.

Electrical systems and components are serviced by aviation maintenance technicians in the same way as other systems and components. But many airlines and some larger fixed base operators have technicians who specialize in electrical systems.

Some operators consider any electrical component on the engine side of the firewall to be within the domain of a powerplant technician, and components or portions of the system on the airframe side of the firewall to be the responsibility of the airframe technician.

Method of Generation	Type of Electricity
Chemical energy	DC
Heat	DC
Light	DC
Pressure	AC
Magnetism	AC

Figure 17-1. *Methods of generating electrical energy*

Generation of Current Electricity

In the *General* textbook, it was shown that there are two types of electricity: static and current. Static electricity produces electrical fields caused by an accumulation of electrons that are not moving. Current electricity, a more useful type, is caused by the movement of electrons. In DC (direct current) electricity, the electrons move through the circuit in one direction. In AC (alternating current), the electrons periodically reverse their direction of flow.

There are five ways current electricity can be generated. Three methods—the conversion of chemical energy, heat, and light—produce DC, while pressure and magnetism produce AC. *See* Figure 17-1.

All electrical energy used in an aircraft comes from two sources: the conversion of chemical energy and magnetism. Storage batteries convert chemical energy into electricity for starting the engine and for use in emergencies, but the electricity for normal operation is produced by magnetism in a generator or alternator.

Electricity Produced by Magnetism

An electron is an invisible subatomic particle that has a negative electrical charge. Electrons spin in orbital shells around the nucleus of the atoms in all materials in the same way planets spin around the sun. Not only do electrons orbit the nucleus of the atoms, but they also spin at an extremely high speed about their own axes, and each spinning electron has a magnetic polarity.

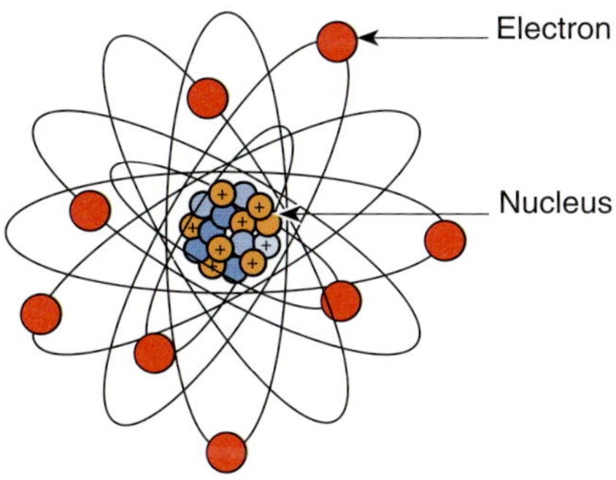

Figure 17-2. *All matter is made of atoms that contain a nucleus surrounded by electrons. The electrons orbit the nucleus in rings, or shells, and they spin about their own axes.*

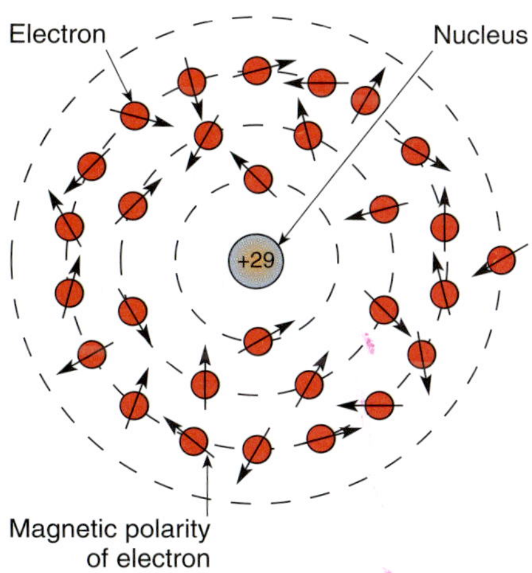

Figure 17-3. *The electrons surrounding the nucleus of a copper atom have a random orientation of their spin axes, and there is no resultant magnetic field.*

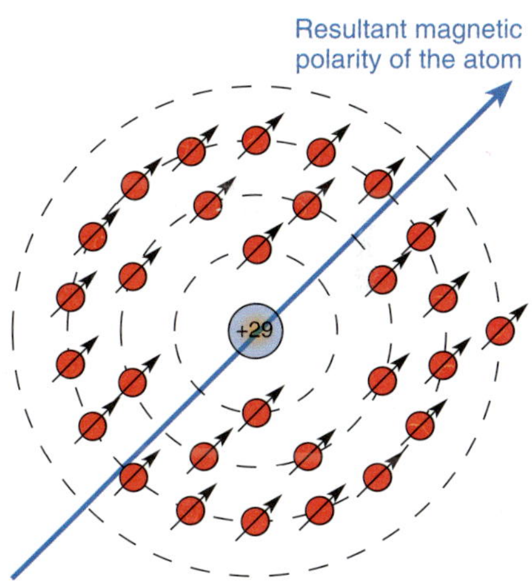

Figure 17-4. *When a copper wire has an electrical potential difference between its ends, the spin axes of all of the electrons are aligned, and each atom has a resultant magnetic field. Electrons move from the valence shell of one atom to another, producing current flow.*

electrical potential. The electrical force caused by a deficiency of electrons in one location and an excess of electrons in another.

In a piece of copper wire, there are 29 electrons spinning around the nucleus of each copper atom, and the magnetic polarities of these electrons have no common orientation. Therefore, there is no resultant magnetic field around the wire.

There is a direct relationship between current electricity and magnetism. When a copper wire has an electrical potential difference between its ends, the spin axes of all electrons are forced into alignment, and electrons from the outer shell (valence shell) of the atoms move from one atom to another. This movement of electrons is called electrical current. When current flows in a conductor, a magnetic field surrounds the conductor.

magnetic flux. Lines used to represent a magnetic field.

magnetic field. The invisible, but measurable, force surrounding a permanent magnet or current-carrying conductor.

The resultant magnetic field of the atoms in the copper wire causes invisible but detectable lines of magnetic flux to surround the wire. These lines of flux surround all current-carrying conductors and are of extreme importance in the generation of electricity.

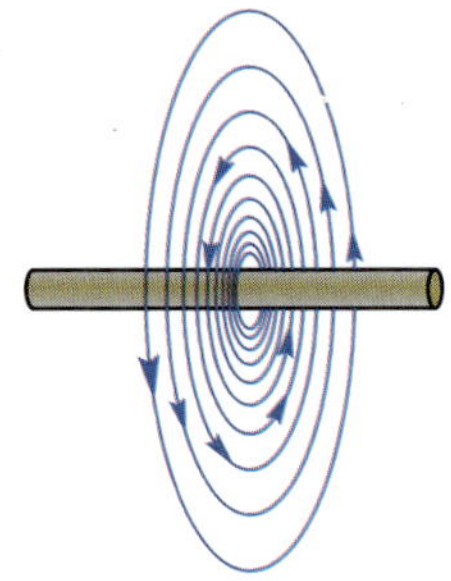

Figure 17-5. *Invisible but detectable lines of magnetic flux surround any conductor through which electrons are flowing.*

Electricity is generated any time there is relative movement between a conductor and a magnetic field that causes the conductor to cut across or move through the field. This is illustrated in Figure 17-6. A magnetic field surrounds the wire carrying current from one end of the battery to the other. A second conductor, this one containing a current-measuring instrument, is moved back and forth, toward, and away from the current-carrying wire. As this wire moves through the magnetic field, its electrons change their spin axes, and some are forced from one atom to the next, creating a flow of current.

The amount of current is determined by three things:

- The strength of the magnetic field
- The number of conductors passing through the field
- The speed with which the magnetic field is crossed

The direction the current flows or the polarity of the voltage generated is determined by the direction the conductor moves relative to the magnetic field.

There are three basic ways of obtaining relative motion between the conductor and the magnetic field:

- The magnetic field remains stationary and the conductor moves through it.
- The conductor remains stationary and the magnetic field moves across it.
- The conductor and the magnetic field remain stationary, and the strength of the magnetic field changes.

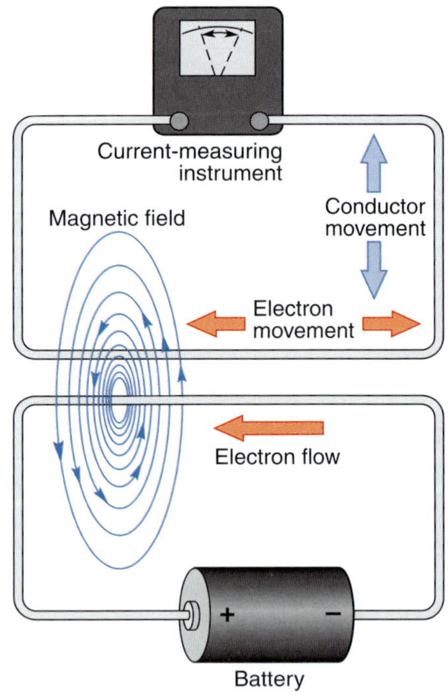

Figure 17-6. *Any time a conductor cuts across or moves through a magnetic field, electrons are forced to move through it.*

AC Generator Principles

slip ring. A smooth, continuous ring of brass or copper mounted on the rotor shaft of an electrical generator or alternator. Brushes riding on the smooth surface of the slip ring carry current into and out of the rotor coil.

Because AC is easier to produce than DC, we will discuss the basic AC generator in Figure 17-7 first. A loop of wire rotates in the magnetic field, which is caused by lines of magnetic flux between the two pole pieces. The two ends of the loop terminate in slip rings connected to the external circuit through carbon brushes.

When the loop is in position A, both sides are moving parallel to the lines of flux, and no lines are being cut. No current is produced, and there is no voltage difference between the two slip rings.

As the loop rotates with its light-colored side moving upward in front of the north pole, and the dark side moving downward in front of the south pole, electrons leave the light slip ring and return from the external circuit through the dark slip ring. The current starts at zero and peaks at the point the maximum number of lines of flux are being cut, position B. Then as the loop continues to rotate, the current drops back to zero.

In position C, the loop again moves parallel to the lines of flux and no current is produced. At position D, the maximum number of lines of flux are again cut, but this time, the dark side moves upward in front of the north pole, and the light side moves downward in front of the south pole. Electrons leave through the dark slip ring and return from the external circuit through the light slip ring.

The generator shown in Figure 17-7 illustrates the principle, but it is not practical. Practical generators have many turns of wire in the rotating coil, and the coil is wound around a laminated soft iron core to concentrate the magnetic field. Additional discussion of AC generators begins on Page 606.

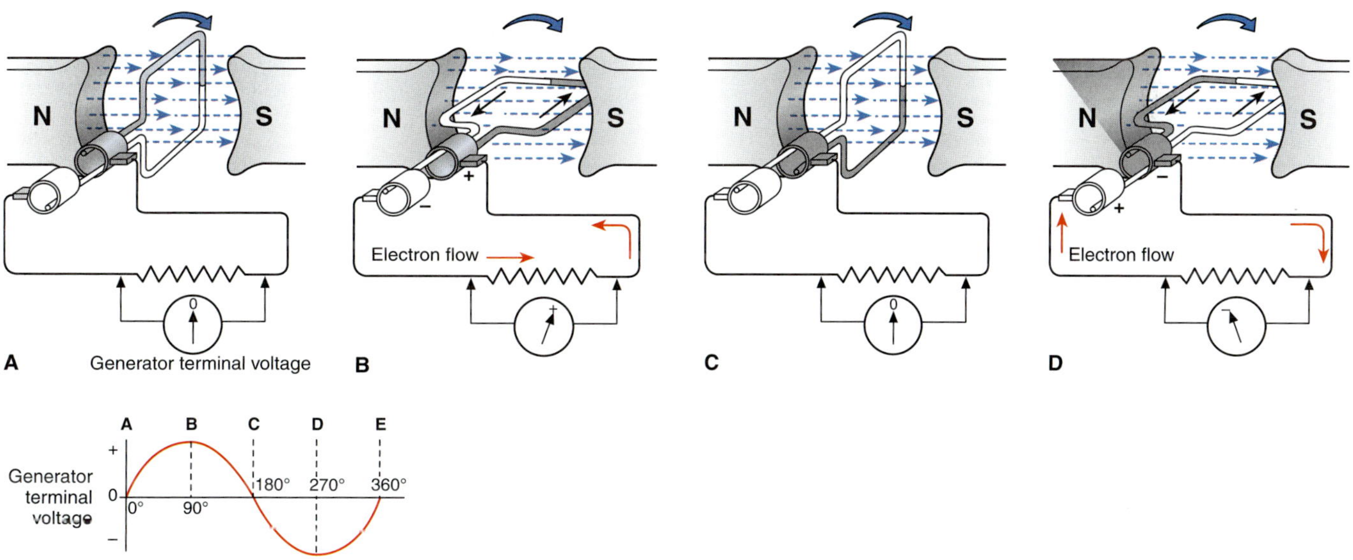

Figure 17-7. *The voltage produced by a coil of wire turning in a magnetic field varies with the number of flux lines being cut.*

DC Generators and Alternators

All electricity produced by magnetic generators is AC, but before it is taken from the generator it is rectified, or changed into DC. There are two types of rectifiers used for making this change: mechanical and semiconductor. DC generators use mechanical rectifiers, and DC alternators of the type installed on the smaller aircraft use semiconductor rectifiers.

rectifier. A device that allows electrons to flow in one direction while preventing their flow in the opposite direction.

Figure 17-8 illustrates the principle of a DC generator. The pole pieces and rotating coil are similar to those in the description of the AC generator, but the ends of the coil are terminated in the two segments of a split-ring commutator. Carbon brushes ride on the commutator to connect the coil to the external circuit.

commutator. A mechanical rectifier mounted on the armature shaft of a DC generator. Carbon brushes riding on the commutator carry current into and out of the armature.

In position A, the conductors are moving parallel to the lines of flux and no lines are being cut. Therefore, no electrons are forced into the external circuit. As the light-colored side moves upward in front of the north pole, position B, electrons are forced out through the light-colored commutator segment and the light brush, and from left to right through the external load back into the dark side of the loop. Another 90° of rotation brings the loop into the neutral position C with no lines of flux being cut and no current produced. At this point, the brushes switch commutator segments with the dark brush riding on the light segment.

As the loop continues to rotate, the current builds up to the maximum at point D as the maximum number of lines of flux are cut. The dark side of the loop is moving upward in front of the north pole, but since this side is connected to the light brush, the current continues to flow through the external load from left to right.

pulsating DC. Direct current whose voltage periodically changes, but whose electrons flow in the same direction all of the time.

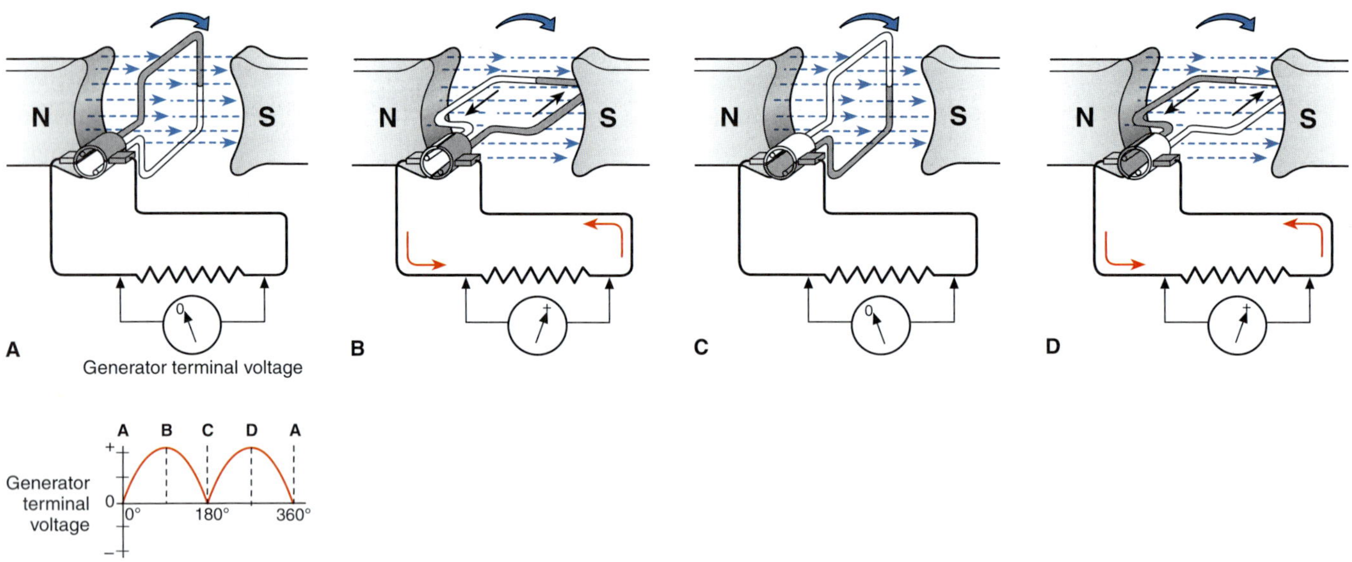

Figure 17-8. *A DC generator produces alternating current, but it is rectified by the commutator and brushes before leaving the generator.*

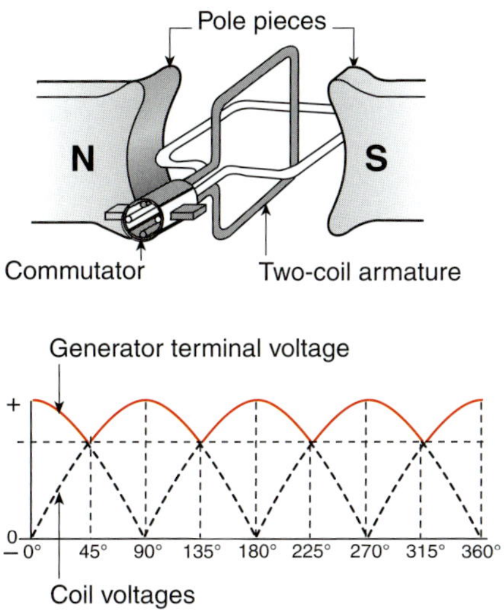

Figure 17-9. *Adding a second coil and additional commutator segments increases the ripple frequency and decreases the ripple amplitude.*

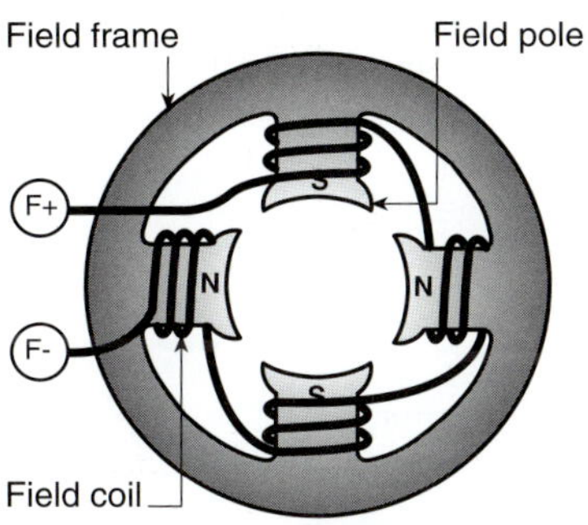

Figure 17-10. *The magnetic field for practical DC generators is produced by electromagnets. The strength of the field is determined by the amount of current flowing in the field coils.*

permanent magnet. A piece of hardened steel that has been exposed to a strong magnetizing force which has aligned the spin axes of the electrons surrounding its atoms.

electromagnet. A magnet produced by an electrical current flowing through a coil of wire.

The generator in Figure 17-8 is not practical because there is only one loop of wire, and the output waveform is low-frequency pulsating DC. But by adding a second coil at right angles to the first, the frequency of the pulsations, called the ripple frequency, is doubled, and the terminal voltage is much smoother than that produced by a single coil. By adding even more coils, the ripple frequency is further increased and the output voltage becomes smoother. Practical generators have many coils of wire in the armature, and the larger generators have more than one pair of magnets that produce the magnetic field.

Field Magnets

Up to this point in the description of the DC generator, the magnetic field was assumed to be produced by permanent magnets. In practical generators, the magnetic field is produced by electromagnets. The frame of a typical aircraft DC generator has four electromagnetic poles arranged as shown in Figure 17-10. The strength of the magnetic field, and thus the voltage produced by the generator, are determined by the voltage regulator which controls the amount of current flowing in the field coils.

DC generators are said to be self-excited. This means that the current that flows through the field coils is taken from the generator armature. For the generator to begin producing this excitation current, the field frame must retain a small amount of residual magnetism. When the generator starts turning, the residual magnetism produces a small amount of field excitation current. This current intensifies the magnetic strength of the field, and the output current increases further. The generator output voltage and thus the field current continue to increase until the voltage regulator restricts the field current and holds it at a value that produces the desired output voltage.

If the field circuit is opened by either the generator field switch or by a break in the field circuit, the generator output voltage will drop to residual voltage, usually two volts or less.

When a generator is disassembled, the residual magnetism is normally lost, but can be restored when the generator is reassembled by flashing the field. A pulse of current from the battery is sent through the field coils in the direction in which current normally flows. The field produced by this current magnetizes the steel generator field frame.

If the field coil is in series with the armature, as in Figure 17-11A, the output voltage will increase with the generator load, and there is no practical way to control it. This arrangement is not used in aircraft generators.

Shunt-connected fields, as in 17-11B, tend to decrease the generator output voltage as the load increases. A voltage regulator, which acts as a variable resistor in the field circuit, holds the output voltage constant as the load varies. This arrangement is used for low-output aircraft generators.

Compound-wound generators, like that in Figure 17-11C, have both series and shunt fields, with the voltage regulator controlling the current in the shunt field. This arrangement is used for high-output aircraft generators.

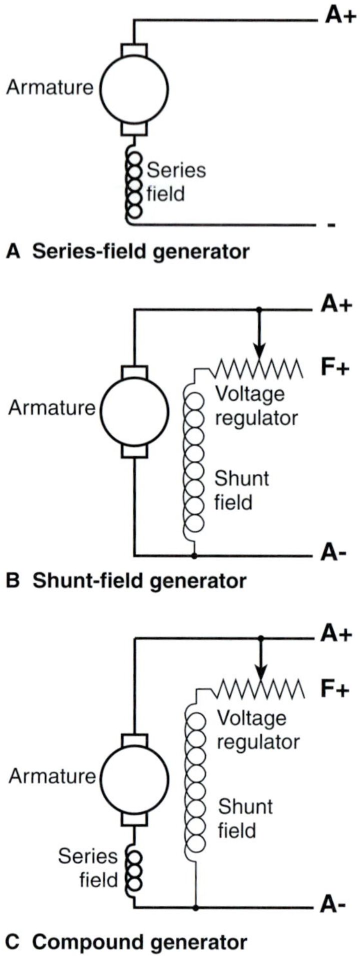

Figure 17-11. *Generator field connections*

Armature

Practical generators have a number of coils wound in slots around a drum-like core of laminated soft iron on the armature shaft. The ends of the coils are soldered into copper commutator segments, which are insulated from each other and from the armature shaft. *See* Figures 17-12 and 17-13.

The coils in positions A and C of Figure 17-8 on Page 593 are moving parallel to the lines of magnetic flux. This is called the neutral position, or neutral plane, and there is no voltage difference between the ends of the coil, and no current is flowing from it. When the coil is in its neutral plane, the brushes can short-circuit the ends without sparking.

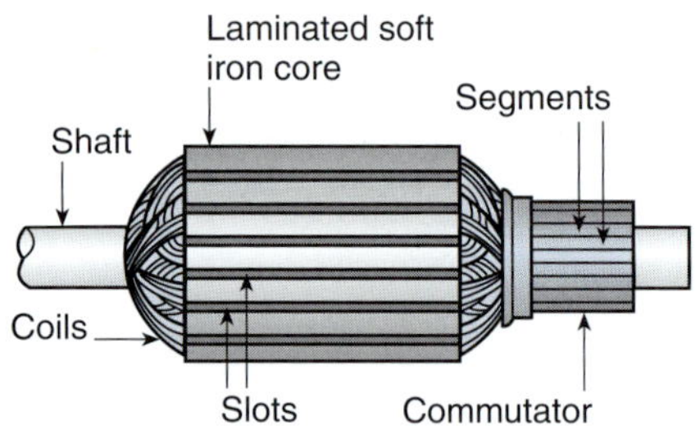

Figure 17-12. *Typical armature used on aircraft DC generators*

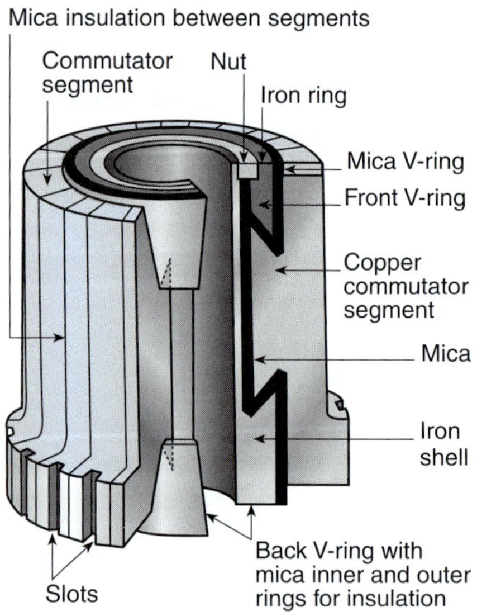

Figure 17-13. *The commutator used on a DC generator is made up of a number of copper segments insulated from each other and from the armature shaft on which it is mounted. The ends of the armature coils are soldered into slots in the copper segments.*

residual magnetism. The magnetism that remains in the field frame of a generator when no current is flowing in the field coils.

residual voltage. The voltage produced in a generator armature when the armature is rotated in the residual magnetism.

flashing the field. A maintenance procedure that restores residual magnetism to a DC generator field frame.

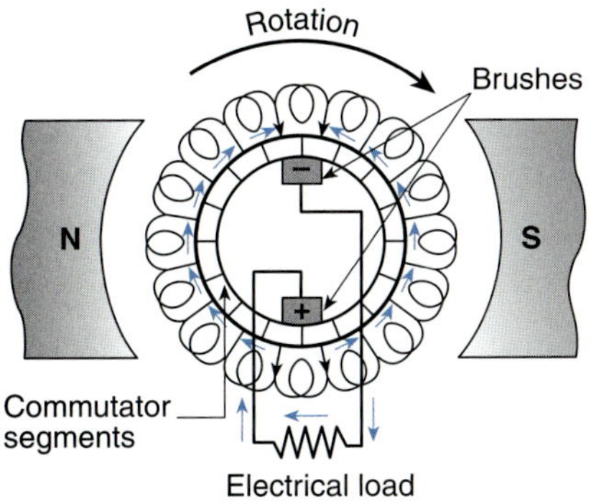

Figure 17-14. *The brushes contact the commutator of a multicoil armature when there is no voltage difference between the ends of the coils. In this position, there is no sparking between the brushes and the commutator segments.*

In armatures having many coils, the ends of two of the coils are soldered into each of the commutator segments as seen in Figure 17-14. The brushes that pick up the current are arranged so they short across the segments when the coils are passing through their neutral position.

Armature Reaction

There is no sparking between the brushes and the commutator when the brushes short-circuit the ends of the coils moving through the neutral plane. But the neutral plane of an uncompensated generator does not remain fixed; it moves as the armature, or load, current changes.

Figure 17-15 shows how the neutral plane shifts. The magnetic flux produced by the field coils extends straight across the space between the poles as seen in view A, and the neutral plane is perpendicular to this flux.

The armature coils rotate between the poles in the direction shown by the arrow in view B. A magnetic field encircles the conductor in each coil, and the orientation of this field is determined by the electron movement's direction in the conductor. The strength of the magnetic field is determined by the amount of current flowing. The dot represents the head of the current arrow and indicates that the electrons are leaving the coil, so the lines of flux encircle the conductor in a clockwise direction. The cross represents the tail of the current arrow and shows that the electrons are moving into the coil. Therefore, the lines of flux encircle the conductor in a counterclockwise direction.

The magnetic field surrounding the armature coils reacts with the stationary magnetic field between the poles and distorts it, as in view C. The neutral plane is perpendicular to the resulting lines of flux and has therefore shifted in the direction of armature rotation.

short circuit. A low-resistance connection between two points in an electric circuit.

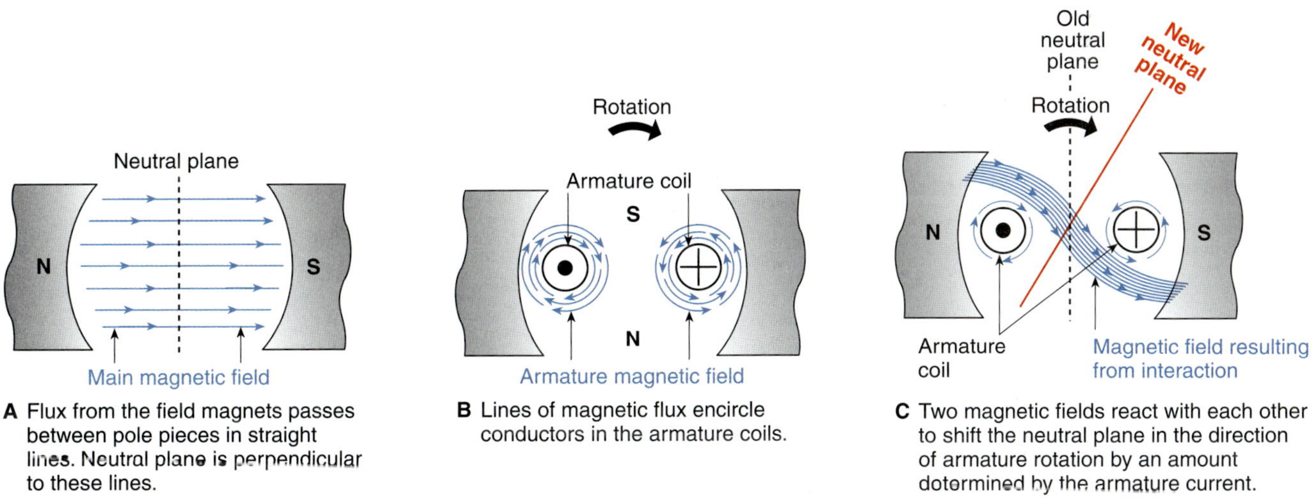

Figure 17-15. *The production of armature reaction*

If the brushes are installed in a fixed location, and there are no provisions for counteracting armature reaction, the neutral plane will shift away from the brush location as the load current increases, and there will be arcing between each brush and the commutator. This arcing damages the commutator and wears the brushes; therefore, it must be minimized.

Most small shunt-wound generators are designed so that the brushes are aligned with the neutral plane when the generator is producing its rated current, and there is some arcing when less current flows.

Compensating Windings and Interpoles

High-output compound-wound DC generators have two provisions for minimizing armature reaction: compensating windings and interpoles.

Figure 17-16 shows the field frame of a four-pole, high-output generator. Between each field pole there is a smaller pole shoe over which the interpole coils are wound. Figure 17-17 shows the wiring diagram of this generator. There are two positive and two negative brushes. The positive brushes are connected by a heavy conductor to the generator B terminal, the positive output terminal. The negative brushes connect to the series compensating windings and the windings around the interpoles. Both of these windings are of heavy wire and have relatively few turns.

The coils around the interpoles are wound so that their polarity is the same as the next field pole in the direction of rotation. The intensity of their magnetic field varies with the armature current.

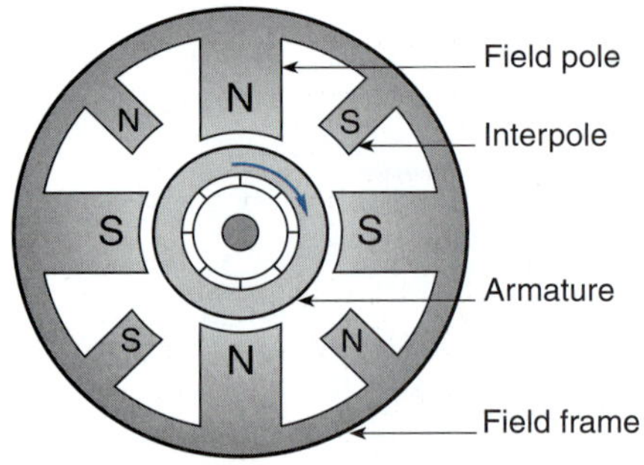

Figure 17-16. *The field frame of a high-output DC generator has an interpole between each of the field poles. The magnetic fields from the interpoles oppose the magnetic fields from the armature windings and minimize armature reaction.*

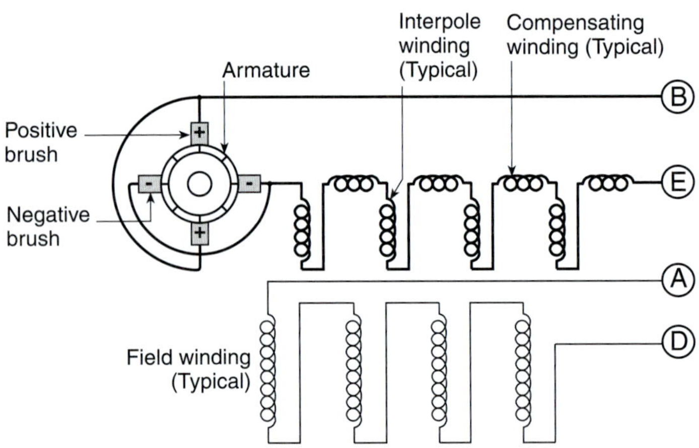

Figure 17-17. *The interpole and compensating windings are in series with the armature coils. The magnetic fields produced by current flowing in these windings vary with the armature current and minimize armature reaction.*

compensating winding. A series winding in a compound-wound DC generator. The magnetic field from the compensating windings helps cancel the field distortion caused by armature current.

interpole. A field pole in a compound-wound DC generator used to minimize armature reaction. Interpoles are located between each of the regular field poles, and their coils are in series with the armature winding.

pole shoe. Inward extensions from the field frame of a generator around which the field coils are wound.

The compensating windings are embedded in the faces of the field poles, and their varying magnetic field works with the fields from the interpoles to effectively cancel the distortion of flux between the field poles. The result minimizes armature reaction, and the neutral plane remains aligned with the brushes as the armature current changes. Arcing between the brushes and the commutator is thus minimized.

DC Generator Controls

Figure 17-18 is a simplified diagram of a high-output compound-wound generator. Notice that the interpole and compensating windings have been reduced to a single symbol for the series fields. Only two brushes and a single shunt field are included in this diagram.

DC generators require three basic controls or protection devices: a current limiter, a voltage regulator, and a reverse-current cutout and switch. These controls are seen in the circuit diagram of a twin-engine aircraft electrical power system in Figure 17-19.

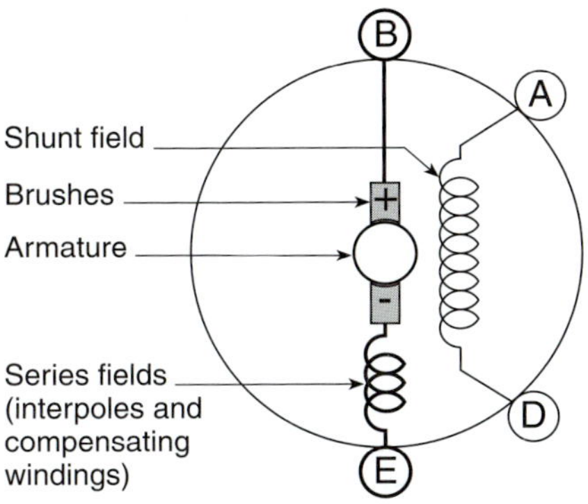

Figure 17-18. *Simplified diagram of the internal circuit of a high-output DC generator*

slow-blow fuse. A special type of circuit protection device that allows a momentary flow of excess current, but opens the circuit if the excessive flow is sustained.

Current Limiter

High-output generators use a form of slow-blow fuse as the current limiter. This fuse is located in the engine compartment and is inaccessible in flight. The current limiter will pass a momentary excess of current, but if the circuit uses more than the generator's rated current for an extended period of time, a fusible link will melt and open the circuit to protect the generator.

Small generators are protected by automatic current limiters that monitor the output current, and when it exceeds the generator rating, the current limiter reduces the field current and thus the output voltage, which in turn, reduces the output current.

Reverse-Current Cutout and Switch

A DC generator acts as a motor if voltage is supplied to its output terminal. To prevent battery discharge and generator damage, all DC generators must have an automatic switch to disconnect the generator from the electrical system when the generator voltage is lower than that of the battery.

The reverse-current cutout relay and switch in Figure 17-19 is a differential-voltage cutout that opens the main generator output circuit any time the generator voltage is a specific amount lower than the battery voltage. A switch mounted on the pilot's or flight engineer's panel can open this circuit at any time.

The reverse-current cutout relay used with small generators is an electromagnetic switch that closes and connects the generator to the electrical system when the generator voltage is higher than the battery voltage. When the generator voltage drops below the battery voltage, current flows from the battery into the generator. This reverse flow of current opens the relay and disconnects the generator from the battery.

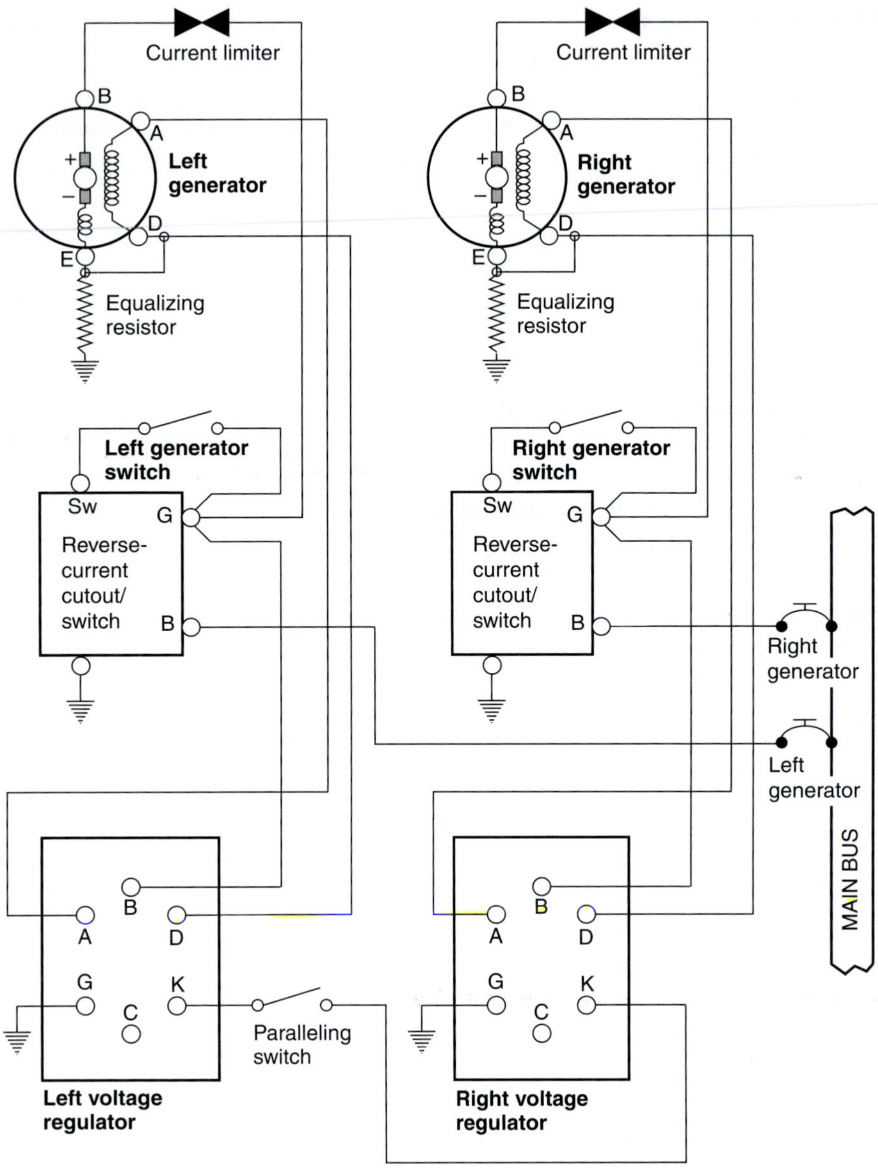

Figure 17-19. *Electrical power system for a twin-engine aircraft using DC generators and their controls*

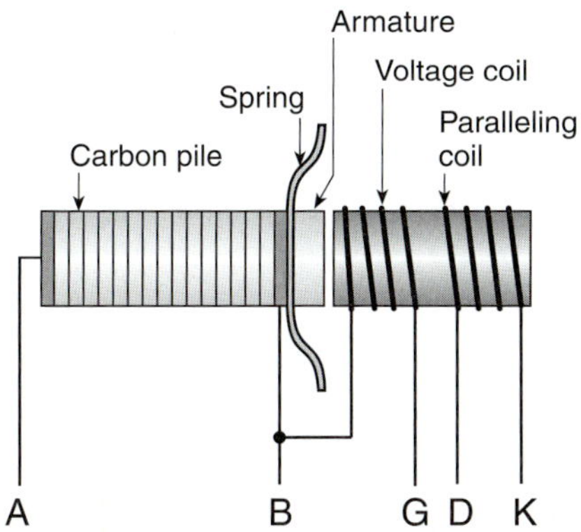

Figure 17-20. *Basic schematic diagram of a carbon pile voltage regulator*

carbon pile voltage regulator. A voltage regulator for a high-output DC generator that uses a stack of pure carbon disks for the variable resistance element.

Voltage Regulator

The output voltage of a DC generator is regulated by controlling the amount of field current. Most large generators are equipped with carbon pile voltage regulators in which a stack of carbon disks acts as a variable resistor between the generator output, terminal B, and the shunt field, terminal A. Figure 17-20 shows how this regulator operates.

When the voltage is low, the spring in the regulator compresses the carbon pile, reducing its resistance. Maximum current flows through the generator shunt field, and the output voltage rises.

As voltage builds up, current flowing through the voltage coil in the regulator increases and produces a magnetic pull on the armature. This pull opposes the spring force and loosens the carbon pile, increasing its resistance. The field current and therefore the generator output voltage decrease.

In a multiengine aircraft, the generators must share the electrical load equally. This is done using paralleling circuits in the voltage regulators, as shown in Figures 17-19 and 17-20.

Terminal E of each generator does not connect directly to ground, but rather to equalizing resistors. These are low-resistance, high-current precision resistors that produce a 0.5-volt drop when the rated current of the generator flows through them. The paralleling coils in both voltage regulators are connected in series between the tops of the two equalizing resistors. The paralleling switch is between the two paralleling coils.

When the paralleling switch is closed, and both generators are producing the same amount of current, the voltage drops across the two equalizing resistors are the same, and no current flows through the paralleling coils. But if, for any reason, the right generator begins to produce more voltage than the left, its current increases and the voltage drop across its equalizing resistor increases. Current flows through the paralleling coils in a direction that increases the magnetic pull on the armature in the right voltage regulator and decreases the pull on the armature in the left regulator. The carbon pile in the right regulator loosens and increases its resistance. The field current to the right generator decreases, and so does its output voltage. The magnetic pull on the armature in the left regulator decreases, and the spring tightens the carbon pile. Its resistance decreases and the field current in the left generator increases. Its voltage rises until it produces the same current as the right generator. Opening the paralleling switch disables the paralleling circuit, and each generator operates independently.

Small Aircraft DC Generator Systems

For many years small aircraft used a DC generator and a three-element electromagnetic control similar to those used on older automobiles. Some small generator systems are still used, but most have been replaced by DC alternators.

Low-output DC generators have two field poles and are controlled and protected by vibrator-type controls. There are two basic types of small generators that look and act alike, but the voltage regulators are connected into the circuits differently. Figure 17-21 shows the two types.

The generator commonly known as an A-circuit generator has the shunt field connected to the insulated brush, and the vibrating-type voltage regulator is in the circuit between the field winding and ground. The B-circuit generator has the field winding connected to the uninsulated, or grounded, brush, and the voltage regulator is between the field winding and the generator armature.

The controls for a low-output DC generator consist of a vibrating-type voltage regulator and current limiter and an electromagnetic reverse-current cutout relay, all mounted in a single metal housing.

The reverse-current cutout relay in Figure 17-22 is an electromagnetic switch with normally open contacts between the generator output and the battery bus. There are two coils wound around the relay core. One coil consists of many turns of fine wire between the generator terminal and ground, and the other coil has a few turns of heavy wire between the generator terminal and the contacts.

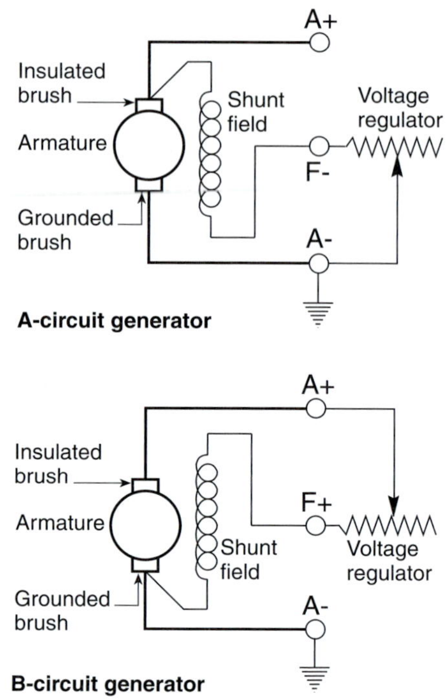

Figure 17-21. *Two types of low-output DC generators for light aircraft*

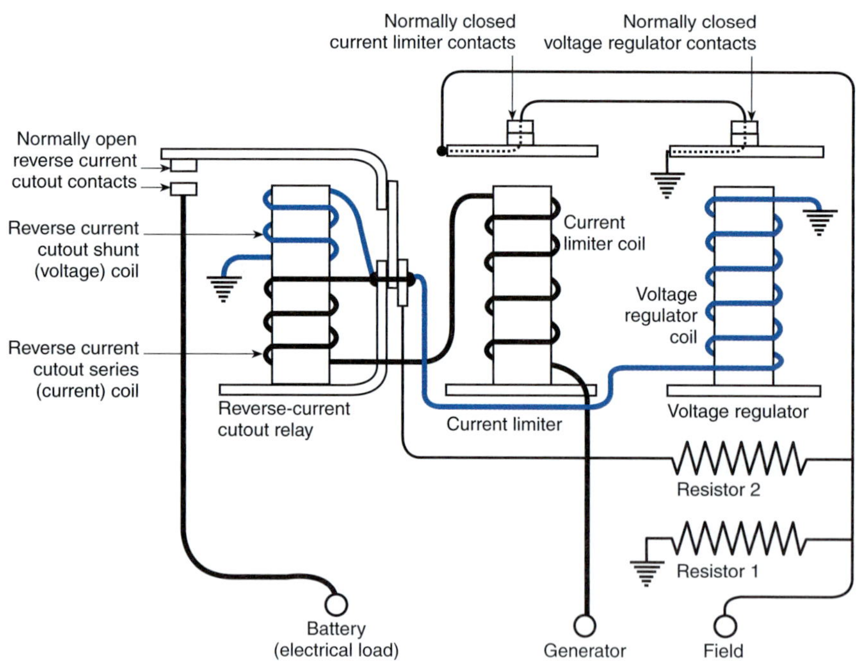

Figure 17-22. *A three-unit electromagnetic generator control for a low-output A-circuit DC generator*

When the engine starts and the generator begins producing voltage, the voltage rises until it is higher than the nominal voltage of the battery. Current flows through the shunt coil until it produces a magnetic field whose pull is strong enough to close the normally open contacts. When they close, current flows to the electrical load through the series coil, and its field adds to the magnetic pull of the shunt coil and holds the contacts tightly closed.

When the generator speed is low and its output voltage is less than that of the battery, current from the battery flows to ground through the generator armature. As this current flows through the series current coil in the reverse direction, its magnetic field cancels the field of the shunt voltage coil and the contacts snap open, disconnecting the generator from the electrical system.

The voltage regulator and current limiter work together to decrease the field current any time the voltage rises above its regulated value or the current exceeds the generator rating.

Figure 17-22 shows how these components work with an A-circuit generator. There are two electromagnetic relays with normally closed contacts, or points. One relay, the current limiter, is wound with a few turns of heavy wire and is in series with the generator armature (GEN) terminal, and the electrical load through the contacts of the reverse-current cutout relay and the battery (BAT) terminal. The coil of the voltage regulator relay has many turns of fine wire, and it is between the reverse-current cutout relay and ground. The two sets of normally closed contacts are in series between the field terminal and ground. Resistor 1 is in parallel with the contacts, and current flows to ground through it when either the voltage regulator or current limiter contacts are open. Resistor 2 is in parallel with the generator field coil, between the field terminal and the cutout relay frame. When either set of contacts opens, the field current immediately stops flowing, and its magnetic field collapses. This collapse induces a high voltage in the field coils that is partially dissipated through the two resistors, thus preventing excessive arcing across the contacts.

When the engine starts, field current flows from the insulated brush, through the field coil (Figure 17-21A) through the two sets of normally closed contacts to ground. The voltage rises to the regulated value, at which time the magnetic pull from the voltage coil becomes strong enough to open the voltage regulator contacts. The field current then flows to ground through resistor 1. This decreases the field current enough to cause the output voltage to drop below the regulated value. The contacts close and the voltage rises. The contacts vibrate open and closed many times a second, alternately grounding the field directly and then through the resistor.

The heavy windings around the current limiter relay are in series with the electrical load. If the load current should exceed the generator current rating, the current limiter contacts will open and place the resistor in the field circuit, decreasing the output voltage until the current decreases to the rated value.

DC Alternators

For many years, the only rectifier that could stand the heat and vibration associated with automotive and aircraft generators was the mechanical commutator and brush rectifier. The basic disadvantages of this mechanical rectifier are that the high current must be produced in the rotating element, and it must be physically interrupted to change the generated AC into DC.

As semiconductor diodes able to withstand the temperature and vibration were developed, the far more practical alternator replaced the generator for producing almost all automotive and aircraft electricity.

DC alternators have three primary advantages over generators:

- They have many more sets of field poles so they can produce current at a lower speed.
- The high current is produced in the stationary component and is rectified by semiconductor diodes.
- An alternator charging system is lighter in weight than a generator charging system.

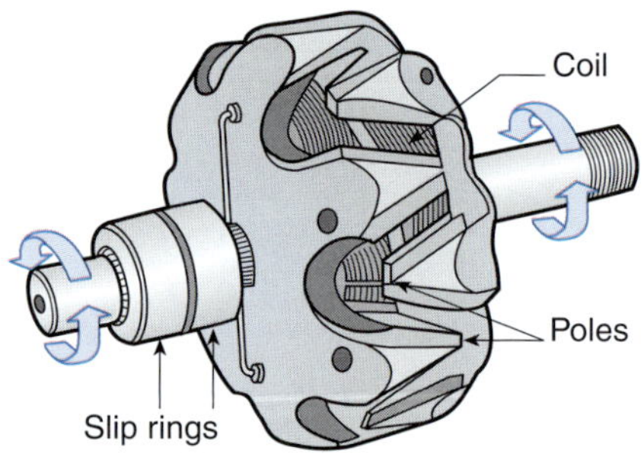

Figure 17-23. *Rotor of a DC alternator*

Figure 17-23 shows a typical DC alternator rotor. This rotor is made of two end pieces of heavy, soft iron with intermeshing fingers that form the magnetic poles. Some rotors have four pairs of poles and others have as many as seven pairs. The voltage regulator supplies current through brushes and smooth slip rings to a coil of insulated copper wire between the two end pieces to control the strength of the magnets.

The stator windings in which the output current is produced consist of a series of coils wound in slots in a laminated soft iron frame. The coils are arranged in a three-phase Y-connected circuit. The AC output of the stator is converted into DC by a six-diode, three-phase rectifier.

distributed pole stator winding. Alternator stator windings wound in a series of slots in the stator frame.

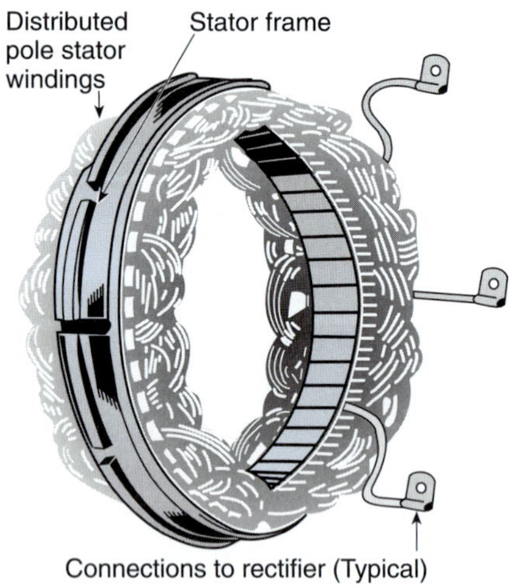

Figure 17-24. *Stator of a DC alternator*

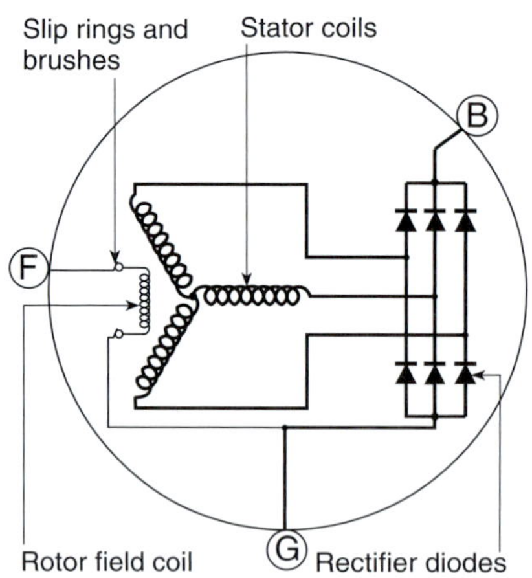

Figure 17-25. *The AC output of the alternator stator is converted into DC by a six-diode, three-phase, solid-state rectifier.*

Current flowing from the main bus through the voltage regulator magnetizes the heavy iron end pieces of the alternator rotor. As the rotor turns, the multiple pairs of poles induce a voltage in the three phases of the stator winding. The three-phase AC thus produced flows into the six-diode rectifier and is converted into DC and sent to the main bus through a high-current circuit breaker.

The 100-amp alternator-output circuit breaker in Figure 17-26 is used as the current limiting device. When current in excess of the alternator rating flows, this circuit breaker opens and stops all current from the alternator to the main bus. This type of current limiting is different from that used in a generator. When excess current flows in a generator system, the current limiter decreases the generator output current by reducing the field current and thus the generator output voltage.

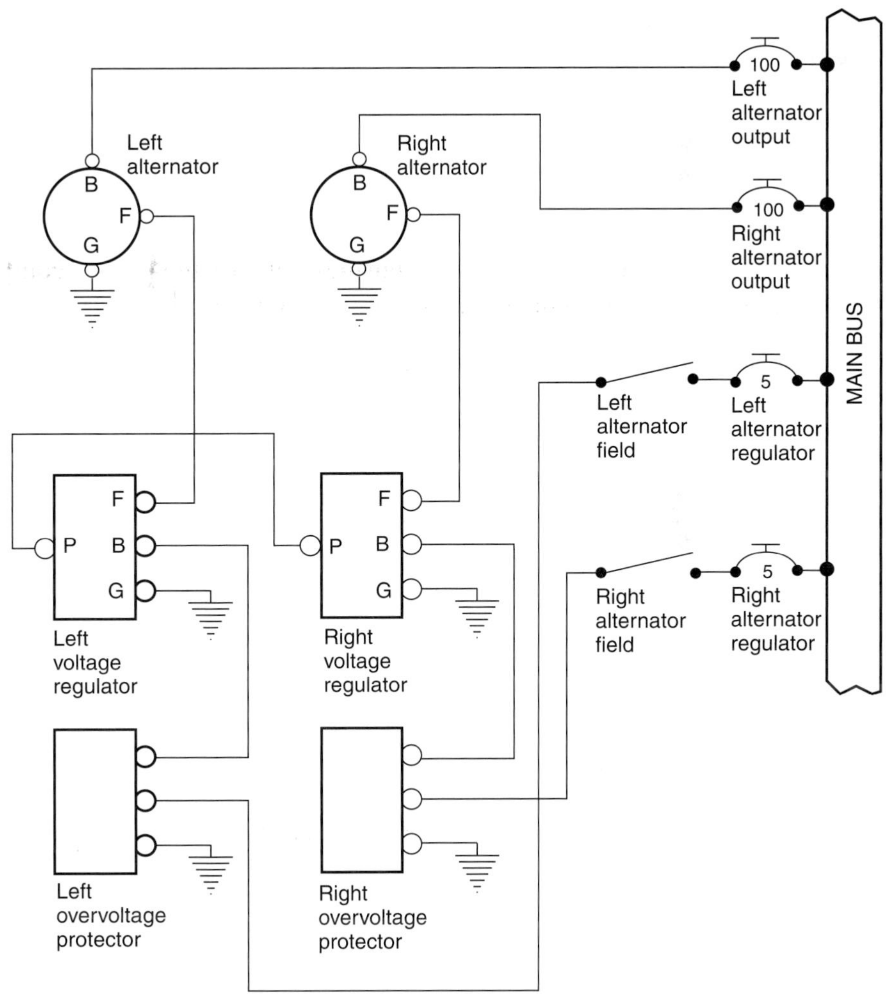

Figure 17-26. *A typical light twin-engine aircraft electrical system, using DC alternators, solid-state voltage regulators, and overvoltage protectors*

DC Alternator Controls

Modern light aircraft that produce their electricity with DC alternators use an electrical system similar to that in Figure 17-26. The output of each alternator goes directly to the main bus through a 100-amp alternator output circuit breaker. Field current is supplied to each alternator through its own 5-amp circuit breaker, alternator field switch, solid-state overvoltage protector, and solid-state voltage regulator.

The two regulators are connected through their P, or paralleling, terminals so circuits inside the regulators can compare the field voltages. If one alternator produces more current than the other, its field voltage is higher. This difference is electronically sensed, and the voltage regulator decreases the field current flowing to the high-output alternator and increases the field current to the other. This adjusts the output voltages so the alternators share the load equally.

The overvoltage protector senses the voltage the alternator produces. If this voltage gets too high, the overvoltage protector opens the field circuit, stopping further output from the alternator. In some installations, a red light on the instrument panel illuminates to show the pilot that the alternator has stopped producing current because of a high-voltage fault.

Starter Generators

Most smaller turbine engines installed in business jet airplanes have a combination starter-generator rather than separate components. These starter generators resemble heavy-duty, compound-wound DC generators, but have an extra set of series windings. These series motor windings are switched into the circuit when the engine is started, but as soon as it is running, they are automatically switched out. In Chapter 13, Figure 13-9 on Page 502 shows a complete starter-generator circuit. Figure 17-27 shows the internal circuitry of the starter-generator itself.

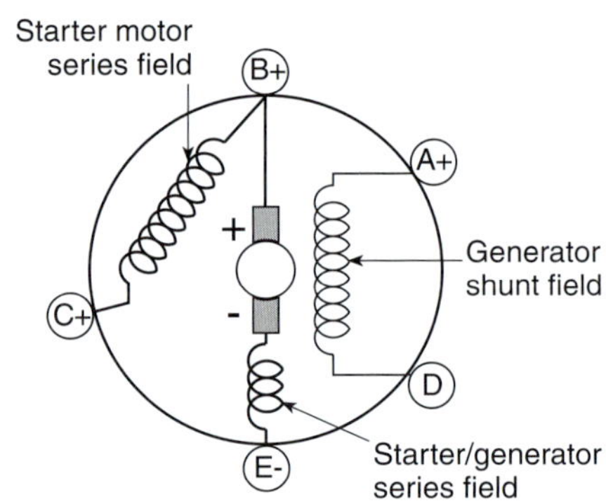

Figure 17-27. *Internal circuit of a starter generator*

AC Power for Large Aircraft

It is important for small aircraft to have the ability to store electrical energy for starting the engine. This is the deciding factor between DC and AC electrical systems. But for large aircraft, the tremendous growth in electrical requirements has made AC systems essential. On-board auxiliary power units and the availability of ground power at all airport terminal gates have allowed these aircraft to exploit the weight-saving advantage of AC. For applications where DC is needed, transformer-rectifier (TR) units reduce the voltage and convert the AC into DC.

Current required to produce power varies inversely with the voltage, and the large amount of current needed to produce power at the low voltage available in DC systems, requires large and heavy wiring. But by increasing

the voltage, as is easily done with AC, much less current will produce the same power. The weight saved by using smaller wires and lighter components more than compensates for the additional complexity of AC systems.

The amount of current needed by an AC component is a function of the frequency. The higher the frequency, the less current is needed, and the components can be smaller for the power they handle. Commercial and residential AC in the United States has a frequency of 60 hertz (cycles per second), but by using a higher frequency, components such as transformers and motors can be made smaller. For this reason, aircraft AC systems use a frequency of 400 hertz.

AC motors are frequency sensitive, and if the frequency changes, speeds will change and motor-driven components will not operate properly. Some components can overheat if the frequency drops and the current increases.

Because frequency is so important, all aircraft engine-driven AC generators have provisions for maintaining a constant frequency as the engine speed varies in normal operation. Under-frequency protection is provided which decreases the generator field excitation if the frequency drops to a certain value; if it drops below that value, it disconnects the generator.

AC Generators

AC generators used in large aircraft electrical systems are often called alternators, and the two terms are used interchangeably.

AC generators used in large aircraft produce three-phase, 400-hertz AC. Their output is normally Y-connected, and the generators are driven by the turbine engines through constant-speed drive (CSD) units. The CSD maintains a constant generator speed as the engine speed varies through its normal operating range.

hertz. A unit of frequency equal to one cycle per second.

CSD (constant-speed drive). A component mounted between an AC generator and an aircraft engine that holds the generator speed and frequency constant as the engine speed varies through its normal operating range.

The relationship between the AC frequency and the speed of the generator is expressed by this formula:

$$\text{Frequency} = \frac{\text{Poles}}{2} \cdot \frac{\text{RPM}}{60}$$

A typical 8-pole generator turning at 6,000 RPM produces 400-Hz AC.

A constant-speed drive unit that mounts between the engine drive pad and the generator is usually a compact unit containing a variable-displacement piston-type hydraulic pump driven by the engine and a fixed-displacement hydraulic motor that drives the generator. The generator speed sensor controls the output of the hydraulic pump. If the generator speed, and thus its frequency, are low, a frequency sensor increases the pump output and the motor speeds up, driving the generator fast enough to produce the required frequency. If the frequency is too high, the sensor causes the pump output to decrease and the motor to slow down enough to produce the correct frequency.

Some modern aircraft use integrated drive generators (IDG). These units contain the generator and the CSD in a single unit. IDGs are smaller and lighter in weight than a conventional generator and CSD.

Because of the high altitude at which these aircraft operate and the low dielectric strength of the low-density air, most modern generators are of the brushless type. The input from the voltage regulator and the output to the electrical system are connected to stationary components. This eliminates arcing that would occur if the high amperage current were taken from rotating components through brushes.

Figure 17-29 is a schematic diagram of a typical brushless alternator. The voltage regulator senses the system voltage and sends the correct amount of current into the fixed exciter field windings to produce the desired AC output.

The rotating portion of this alternator contains a three-phase exciter armature, a six-diode solid-state rectifier, and the main alternator field coil. The strength of the magnetism produced by the rotating main alternator field is proportional to the strength of the fixed exciter field.

The three-phase Y-connected stator windings supply the load current to the aircraft electrical system.

Figure 17-28. *An integrated drive generator combines a brushless alternator with a built-in constant speed drive unit.*

IDG (integrated drive generator). A brushless, three-phase AC generator and a constant-speed drive in a single component.

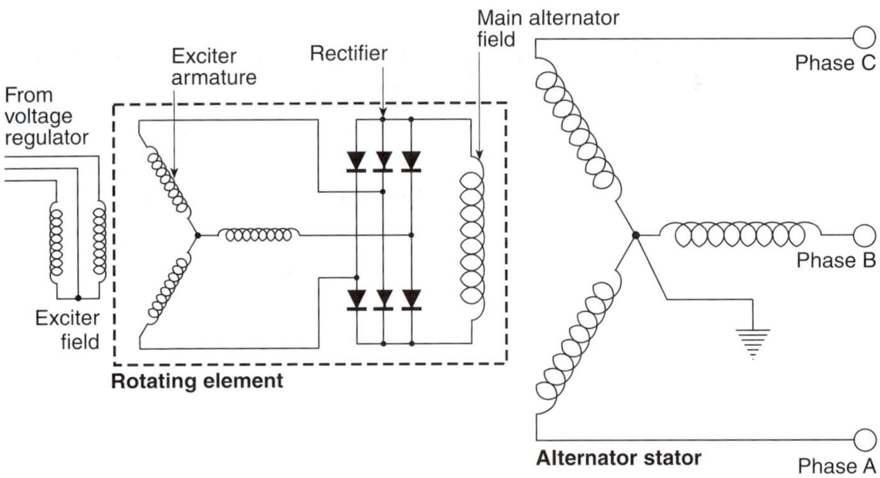

Figure 17-29. *Schematic diagram of a brushless, three-phase alternator*

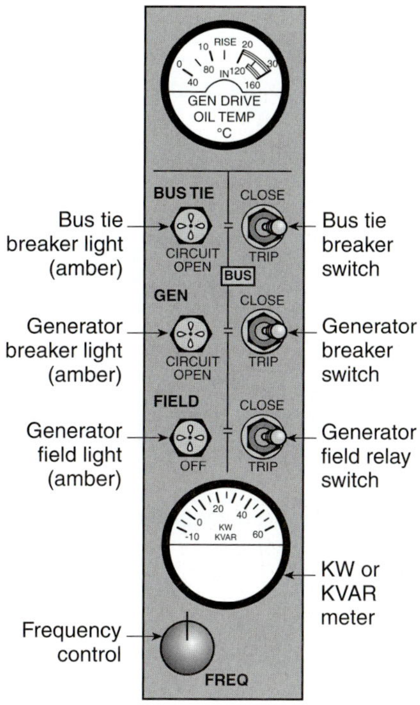

Figure 17-30. *Generator control panel for a Boeing 727*

reactive power. The power in an AC circuit consumed in the inductive and capacitive reactances.

AC Generator Controls

The controls for AC generators are far more complex than those used with DC systems. AC generator controls sense the voltage, current, frequency, and phase sequence, and they control these parameters to keep them within their allowed limits. The generator output is sensed and compared with reference parameters, and the exciter current is adjusted to maintain the output voltage at the proper level. If a fault occurs, the defective generator can be isolated from the circuit and an alternate power source connected.

The generators can be controlled in flight by a generator control panel at the flight engineer's station. Figure 17-30 shows a typical panel for a three-engine jet transport aircraft.

The temperature indicator at the top of the panel shows the oil temperature in the generator drive unit. A switch on the panel above this one allows the flight engineer to select either the temperature rise on the upper scale or the inlet oil temperature on the lower scale.

The bus tie breaker switch allows the flight engineer to connect the generator to the tie, or synchronous, bus for parallel operation. (In the *Airframe* textbook of this *Aviation Maintenance Technician Series*, Figure 7-59 is a block diagram of the Boeing 727 electrical system to which this applies.) The bus tie breaker light illuminates when the generator is not connected to the tie bus.

The flight engineer can close the generator breaker switch to connect the generator to its bus when it comes up to speed. This breaker switch will trip to disconnect the generator from its bus if a fault exists, and a generator breaker light will turn on when it disconnects.

The generator field relay switch is the ON-OFF control for the generator. When it is closed, the generator field is excited, and the generator is ON. When it is in the tripped position, the generator is deactivated, or OFF. The generator field light illuminates when the generator field relay is open and the generator is deactivated.

The KW or KVAR meter normally reads the true power produced by the generator, expressed in kilowatts. When the KVAR selector button located beside this panel is pressed, the meter reads the reactive power the generator is producing.

The frequency control allows the flight engineer to regulate the generator output frequency when the generator is isolated from the tie bus. The frequency, in cycles per second (hertz), is shown on the frequency meter on the AC control panel in Figure 17-31.

A generator cannot be connected to the tie bus with another generator until it is synchronized, with its frequency and voltage the same as those on the bus. When a generator is ready to be connected to the bus, the flight engineer uses the selector switch on the AC control panel to select that generator, and he or she watches the two synchronizing lights. If the lights flash alternately, the phase sequence between the bus and the generator being connected are opposite, and the generator must not be connected. If the lights burn steadily, the generators are not synchronized, and the frequency control on the generator control panel must be adjusted to get the lights to turn off or blink at their slowest rate. When both lights are off, the flight engineer can close the bus tie breaker to connect the generator to the tie bus.

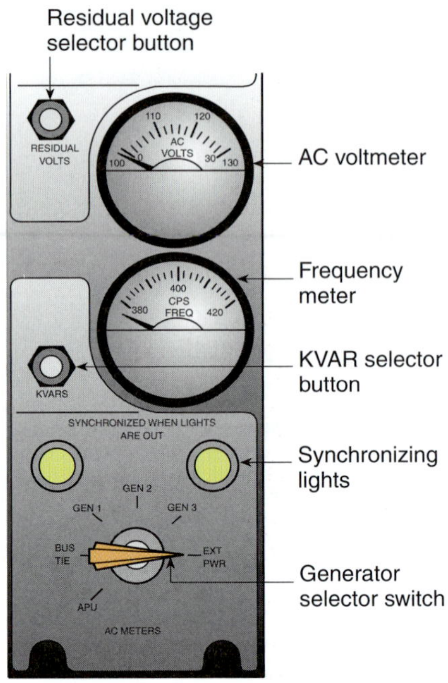

Figure 17-31. *AC control panel for a Boeing 727*

Inverters

Many smaller aircraft have DC systems for their primary electrical power, but require AC for certain instrument and electronic systems. These aircraft use inverters to provide the needed AC.

Some older aircraft use rotary inverters, which are primarily motor-generator units. Rotary inverters have a DC motor with the armature for an AC generator mounted on its shaft. A typical inverter has a 28-volt DC input and a 115-volt single- or three-phase, 400-hertz output.

The modern trend for inverters is to use solid-state, or static, inverters. Static inverters consist of an oscillator that converts DC into 400-hertz AC with the proper waveform. This AC is then passed through a transformer to get the proper voltage.

phase sequence. The sequence with which the output phases of a three-phase generator are connected to the load.

transformer. An electrical component used to change the voltage and current in an AC circuit.

STUDY QUESTIONS: POWERPLANT ELECTRICAL SYSTEMS

Answers are on Page 636. Page numbers refer to chapter text.

1. Two types of current electricity are:
   a. ____________________
   b. ____________________
   Page 590

2. Electricity produced by magnetism is ____________________ (AC or DC). *Page 590*

3. The amount of current that flows in a circuit as a result of magnetic induction is determined by three things. These are:
   a. ____________________
   b. ____________________
   c. ____________________
   Page 591

4. A generator in which slip rings and brushes carry the current from the armature produces ____________________ (AC or DC). *Page 592*

5. When a conductor is moving parallel to lines of magnetic flux, the current in the conductor is ____________________ (maximum or minimum). *Page 592*

6. The electricity produced in the armature coils of a DC generator is ____________________ (AC or DC). *Page 593*

7. A generator in which a commutator and brushes carry the current from the armature produces ____________________ (AC or DC). *Page 593*

8. The current that flows in a DC generator field coil is taken from the ____________________ (generator armature or battery). *Page 594*

9. The magnetism for producing the initial excitation current in a DC generator is ____________________ magnetism in the generator frame. *Page 594*

10. The voltage produced by a generator when no field current is flowing is called ____________________ voltage. *Page 594*

11. Residual voltage in a DC generator is approximately ______________ (how many) volts or less. *Page 594*

12. Residual voltage is lost when a generator is disassembled. It is restored by ______________________ the field. *Page 594*

13. When flashing the field of a DC generator, current is directed through the field coils in the ______________________ (same or opposite) direction as normal flow. *Page 594*

14. In a shunt-connected DC generator, the field coils are in ______________________ (series or parallel) with the armature. *Page 595*

15. In a compound-wound DC generator, the voltage regulator is in the ___________________ (shunt or series) field. *Page 595*

16. The core of the armature used in a DC generator is made of ____________________________________ . *Page 595*

17. The position of the armature in a DC generator in which there is no potential difference between the ends of adjacent coils is called the ______________________ or ___________________________ . *Page 595*

18. The neutral plane of an uncompensated DC generator shifts as the __________________ current changes. *Page 596*

19. The brushes of a DC generator are located in the neutral plane to minimize __________________ . *Page 596*

20. Two components in a high-output DC generator used to minimize armature reaction are:
    a. ____________________________
    b. ____________________________
    Page 597

21. Interpole windings and compensating windings are connected in ______________________ (series or parallel) with the armature. *Page 597*

22. The current limiter for a high-output DC generator is a ______________________ fuse. *Page 598*

23. If battery voltage is applied to the output terminal of a DC generator, the generator will act as a/an ______________________ . *Page 598*

24. The voltage regulator for a DC generator is essentially a ______________________ in the field circuit. *Page 600*

Continued

STUDY QUESTIONS: POWERPLANT ELECTRICAL SYSTEMS *Continued*

25. The variable resistor element in the voltage regulator for a high-output DC generator is normally a/an ______________________ . *Page 600*

26. The generators in a multiengine aircraft are forced to share the electrical load equally by a/an ______________________ circuit in the voltage regulator. *Page 600*

27. Two locations for the voltage regulator with low-output DC generators are between the field and the ___________________ , or the field and the ___________________ . *Page 601*

28. The voltage regulator and current limiter used with low-output DC generators regulate the voltage and limit the current by controlling the generator ____________________________ . *Page 602*

29. The field coil in a DC alternator is in the ______________________ (rotating or stationary) element. *Page 603*

30. The output current produced by a DC alternator is in the ______________________ (rotating or stationary) element. *Page 603*

31. A DC alternator uses a ____________________ (semiconductor or mechanical) rectifier to change its AC into DC. *Page 603*

32. The field of a DC alternator is excited by current from the _____________________ (armature or battery). *Page 605*

33. Field current is supplied to the rotor of a DC alternator through brushes and __________________ (slip rings or commutator). *Page 603*

34. The output of the stator of a DC alternator is _____________________ (single- or three-) phase AC. *Page 603*

35. The rectifier in a DC alternator uses _________________ (how many) solid-state diodes. *Page 603*

36. If a DC alternator produces too high an output voltage, the overvoltage protector opens the _____________________ (field or output) circuit. *Page 605*

37. A starter-generator resembles a heavy-duty DC generator, but it has an additional set of ____________________ (series or shunt) motor windings. *Page 605*

38. The AC power used in aircraft electrical systems has a frequency of _______________ hertz. *Page 606*

39. The frequency of an AC generator is held constant as the engine speed changes through its normal operating range by a/an ____________________________ unit. *Page 606*

40. An AC generator will be automatically disconnected from the electrical system if its frequency becomes too ___________________ (low or high). *Page 606*

41. An integrated drive generator (IDG) has the generator and the _____________________ in a single unit. *Page 607*

42. High-output AC generators used on high-altitude aircraft are normally of the ______________________ type. *Page 607*

43. Four parameters sensed by the control system of an AC generator are:
   a. __________________________
   b. __________________________
   c. __________________________
   d. __________________________
   Page 608

44. An AC generator cannot be connected to the tie, or synchronous, bus unless its ____________________ and ___________________ are the same as that of the bus. *Page 609*

45. Aircraft that have DC systems for the primary electrical power produce the needed AC with a/an _________________________. *Page 609*

46. Two types of inverters used to produce AC are:
   a. __________________________
   b. __________________________
   Page 609

DC Generator and Alternator System Inspection and Maintenance

Battery Maintenance

Since one of the primary purposes of an aircraft generator or alternator is to keep the battery properly charged, any troubleshooting or inspection of the charging system should begin with the battery.

A quick check of the battery electrolyte can be a good indication of the battery's condition and help determine if it is the cause of trouble in the electrical system. The specific gravity of the electrolyte determines the state of charge of the battery; a fully charged cell should be between 1.275 and 1.300 at 80°F. There have been instances of a manufacturer designing a battery to be fully charged with a specific gravity of 1.260, but this is unusual. Final specific gravity value would be determined by the manufacturer's specifications.

Other conditions can affect the state-of-charge of the battery and the operational effectiveness of the charging system.

- *The battery should be clean, dry, and properly secured in the battery compartment.*

  A damp, dirty surface across the top of the battery will cause it to partially discharge in a few days of aircraft inactivity. Inactivity of several weeks can leave the battery with insufficient capacity to close the battery contactor.

- *The battery terminals should be free from any trace of corrosion and the cable connectors tight.*

  If a terminal has corrosion, or added resistance for any other reason, flow to charge the battery will be impeded. With the charging system adjusted for proper system voltage, current flow would heat and increase the resistance of the terminal, dropping the voltage and electron flow to the battery. This resulting voltage drop with current flow will only partially charge the battery and do so at a much slower rate.

- *The ground cable should be properly connected to the aircraft structure for the same reason the battery terminals require a good connection.*

- *The electrolyte in each cell should be at the proper level and all of the cell caps tight.*

  The caps are designed to prevent "flashback" into the battery, in the case of an arc that could occur while working around the terminals when a sufficient amount of hydrogen gas has collected during the battery-charging process.

- *A low state-of-charge could be caused by a lower than normal generator output voltage.*
- *Output voltage being out of specifications on the high side will cause electrolyte to spew out of the battery's vented cell caps and form a damp, acidic residue on the top of the battery.*

  This residue will cause a slow discharge of the battery.

A quick way to check for lower-than-normal generator output voltage is as follows: using a voltmeter, place its positive probe on the bus behind the instrument panel, and place the other probe on a good aircraft ground. With the master switch on, the meter should read battery terminal voltage. After the engine is up and running between 1,800 and 2,000 RPM, the voltmeter should indicate a rise from battery voltage; moreover, the voltage should read within the voltage value range specified by the manufacturer. For example, a normal 24-volt battery, 28-volt electrical system should indicate approximately 27.5 volts with the charging system operating under a light electrical load.

Now would be a good time to check the general capability of the charging system, since its performance is in question. Increase the engine RPM to at least 2000 and begin switching ON as many continuous electrical loads as practicable. As you add these loads, watch the battery charge meter. The charging system should be able to handle most of these loads and maintain system voltage.

Regarding the damage that ensues in a slow, prolonged battery discharge, this process causes a greater portion of the lead sulfate formed during this time to form in a more dense state. The normal charging processes will revert only a very small amount of this denser lead sulfate. The battery then becomes useless when the undissolved sulfate builds up and blocks the chemical reaction of the electrolyte with the lead plates.

This is why many older batteries cannot be recharged back to usefulness after the master switch is left on overnight. Under this condition, the electrolyte concentration is weakened and the internal resistance of the battery is so great that its terminal voltage drops significantly whenever an electrical load, even a minimal one, is applied.

Generator Routine Maintenance

DC generators normally require little routine maintenance as long as they keep the battery charged, but they should be carefully checked on each 100-hour and annual inspection.

- *Check the generator for security of mounting and proper safetying of the hardware involved.*

  If it is belt-driven, check the condition of the belt and the proper belt tension. Normally, the belt should be able to be deflected about a half inch with moderate hand pressure. This pressure should be applied at the center of its span between two belt pulleys, or twistable to only 90 degrees with reasonable single-hand action at the same location. Some installations require this tightness be checked by determining the amount of torque force required to cause the generator pulley to just slip in its drive belt.

- *Inspect all of the electrical connections for security and the condition of the wires.*

  Make sure to check closely for corrosion at the ends of the crimped barrel; corrosion will cause a loss of power for operation of the electrical component it supplies and will eventually burn off the wire at this junction. Gently check the terminal tightness with a gentle twisting action. If the terminal is found to rotate on the electrical stud, verify that the solderless terminal is aircraft quality. Substandard automotive and hardware store terminals expand to a greater extent and are compressed in the process, which creates looseness when the terminal cools down.

- *Remove the band that covers the brush assembly and check the commutator for indication of burned-in pits or roughness from arcing.*

  If just one commutator bar is burned (or possibly two about 180 degrees apart), the armature has an open winding and is usually considered not repairable. Burning and mild pitting may be repaired by turning the commutator on a lathe. If the copper bars are darkened from the normal copper color with darker streaks all over the surface, they may be cleaned by rubbing a commutator cleaning stick across the surface of the bars as the armature is rotated.

- *Blow out any dust accumulated from the normal wearing of the brushes with 35 psi shop pressure.*

  Check the condition of the brushes, making sure that they slide freely in their holders and have sufficient length. Some brushes have a wear-line that must still show; if they do not, most brushes are considered worn-out when they are only 50% of the length of a new one. Check the tension of the brush springs with a small spring scale, pulling perpendicular on the brush spring arm until the arm just pulls free of the brush.

- *Check for any indication of overheating which would normally cause solder to be slung out from the commutator where the coil wires are connected and deposited on the inside of the commutator access cover.*

- *On the post-inspection engine run-up, check the generator output voltage with an accurate voltmeter.*

  If the aircraft has a zero-center ammeter, it should show a charge when the engine is first started, and then the current should decrease as the battery becomes fully recharged.

Generator Overhaul

DC generators are normally overhauled at the time of engine overhaul, or any time they are not operating properly. The generators for light aircraft are simple to overhaul, but all work must be done in accordance with the manufacturer's overhaul manual and any additional service support information such as bulletins, instructions, and letters. This additional service support may be supplied by the airframe engine or generator manufacturer, as well as by the aircraft manufacturer. Any applicable FAA Airworthiness Directives (ADs) *must* be complied with.

Disassembly and Cleaning

After cleaning the exterior with a solvent-dampened rag to minimize internal contamination, place the generator on a clean work bench, disassemble it by removing the band over the brush assembly, and the through bolts that hold the generator together. Lift the brushes from the brush holders and lay out the two end frames with the bearings and/or bushing. Hold the armature in a padded vise and clamp it only tight enough to hold it while you remove the nut that holds the pulley.

Clean all of the parts by wiping them with a cloth dampened with a petroleum solvent, but do not immerse any of the components in a cleaning solvent.

Inspection and Repair

Armature

The armature is the component that is the most likely to need repair. All of the load current is generated as AC in the armature coils, is taken from them, and then rectified into DC by the carbon brushes riding on the copper commutator segments.

growler. A piece of electrical test equipment used to check armatures in DC motors and generators for open or shorted turns in the windings. A growler gets its name from the growling noise made when an armature is being tested.

The armature should be checked for open or shorted coils, and for shorts between the commutator and ground. (*See* Figure 17-32.)

To test for shorts to ground between the commutator and the armature shaft, measure the resistance with an ohmmeter between the commutator segments and the armature frame (Figure 17-33). There should be at least 50,000 ohms resistance between these points.

Place the armature on the growler and turn on the AC power. The growler acts as the primary of a transformer, and the armature windings as the secondary. Place the test prods of a low-voltage AC voltmeter across two adjacent commutator segments at a time and note the voltage of each set (Figure 17-34). Rotate the armature and note the voltage across each pair of adjacent segments. The voltage between any two segments should be essentially the same. A lower voltage between any two segments indicates an open coil.

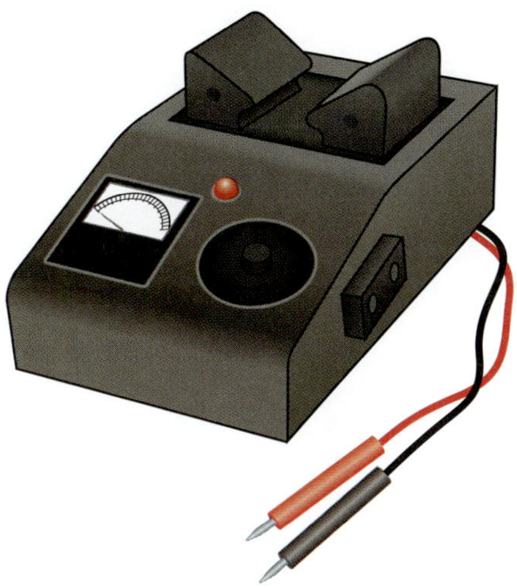

Figure 17-32. *A growler is an essential tool for overhauling DC generators.*

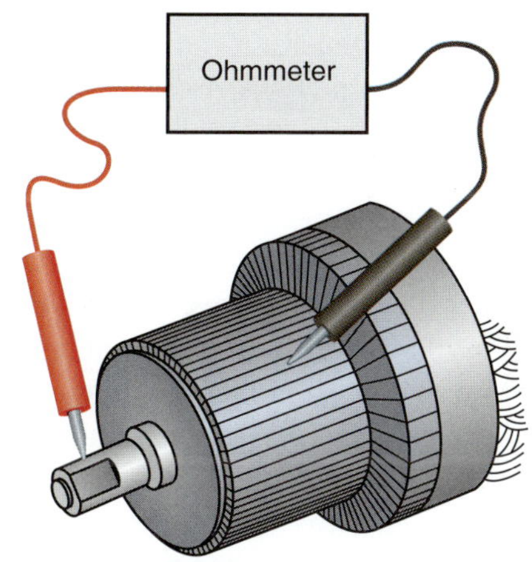

Figure 17-33. *Checking for a short to ground between the commutator and the armature shaft*

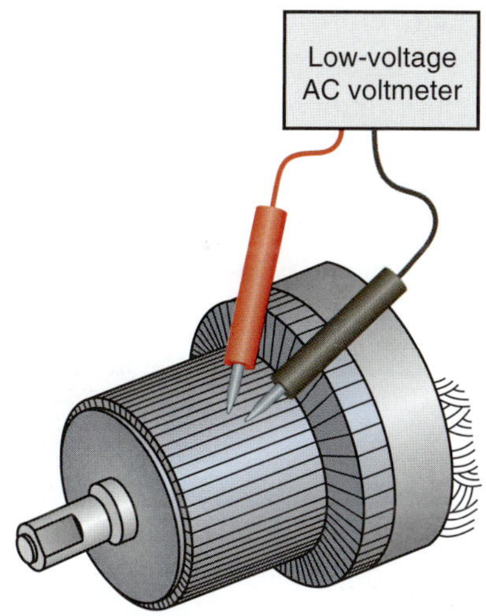

Figure 17-34. *Checking for an open armature coil*

Hold a hacksaw blade over the armature maintaining approximately one eighth inch clearance with the armature frame, and rotate the armature on the growler (Figure 17-35). If there are no shorted coils, a weak magnetic field will cause the blade to vibrate slightly as each coil segment passes under the blade. But if any of the coils are shorted, the field will be much stronger and the blade will vibrate in a noticeably more violent manner.

Pitted or out-of-round commutators can be dressed by turning them in a lathe and taking a very shallow cut, just enough to remove the damage. There are minimum diameter limits that, if exceeded, will cause the commutator bars to overheat and melt the solder that retains the winding ends to the commutator bars.

After the surface is smooth, undercut the mica insulation between the segments to a depth of about $\frac{1}{32}$-inch using a sharp cutting tool, a few thousandths of an inch wider than the thickness of the insulation. When all of the undercutting is done, remove any burrs or sharp edges on the segments with very fine sandpaper. *Do not* use emery paper or any abrasive that is conductive. Some commutators were not undercut at the time of manufacture because the insulation was made softer to mimic the wear rate of the copper. These types of commutators should not be undercut.

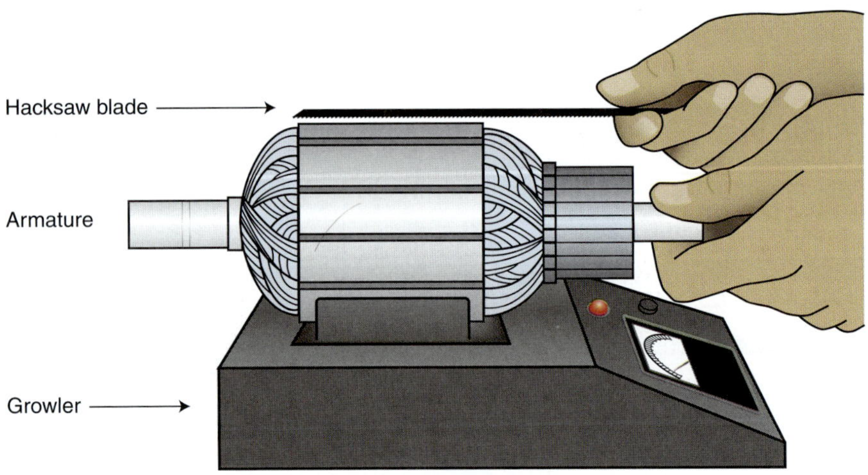

Figure 17-35. *Checking for a shorted armature coil*

Field Coils

Use a 110-volt test lamp, or an ohmmeter, to check for short circuits between the field coils and the generator frame. Place one test prod on the "F" (field) terminal and the other on the generator frame. The lamp should not light up, indicating that there are no shorts between these two components (Figure 17-36A).

For greater safety, it is best to use a 1:1 isolation transformer between the outlet plug and the test lamp. The circuitry would be as follows: the receptacle plug wired to the primary winding and well-shielded from the operator, and the light, probes, and the secondary winding as the test circuit. Caution must still be exercised to keep the operator from becoming part of the secondary circuit, but at least this eliminates the accidental short to ground by the original circuit.

If the older DC generator maintenance manuals were written more recently (the latest existing ones were written in the 1950s), the instructions would suggest using an ohmmeter to find a minimum resistance value of 50,000 ohms. This method is not as exacting as the test lamp which uses

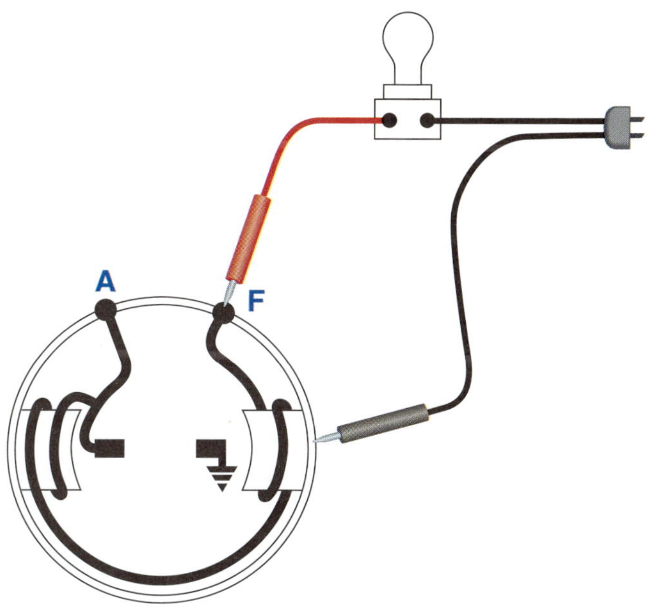

Figure 17-36A. *Checking for a short to ground between the generator frame and the field coils. Note: the commutator and brushes cannot be in place for this test.*

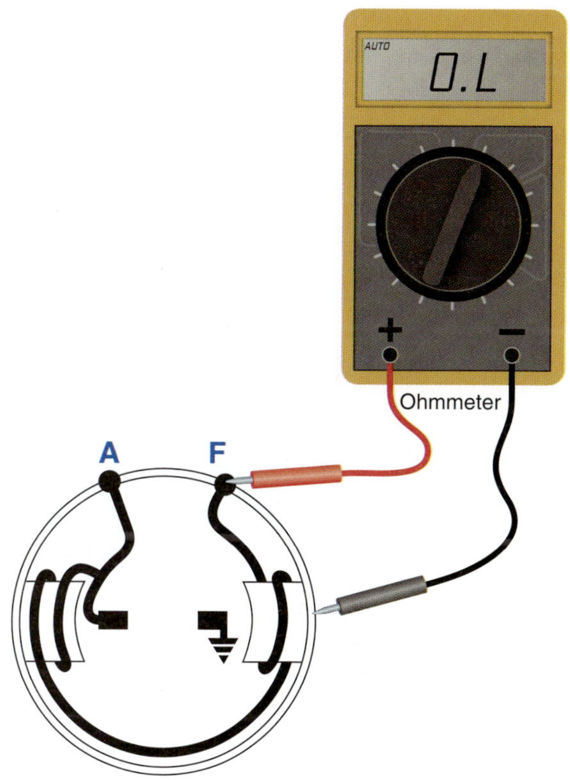

Figure 17-36B. *Alternate method using ohmmeter*

higher voltage, but DC generator field insulation typically has a working voltage of 14 or 28 volts. The editors have yet to see a grounded set of field coils indicated by illumination of the test lamp, that has not *also* indicated significantly less than 50,000 ohms when using the ohmmeter to test for a short-to-ground. It usually takes less than 50,000 ohms in the test lamp circuit for you to be able to distinguish a glow in the test lamp bulb.

By the way of explanation, although DC generators were phased out in the 1960s, many still exist to be serviced in the general aviation market. The methods spelled out in those earlier generator maintenance manuals are the official methods for testing these generators. This is why this text includes these older testing methods as well as the better, more up-to-date methods used throughout this generator maintenance section.

Use the same 110-volt test lamp or ohmmeter to check for open field coils. Place one test prod on the "F" (field) terminal of the generator field frame and the other on the "A" (armature) terminal. The lamp will light up if there is no open circuit in the field coils. (*See* Figures 17-37A and B.)

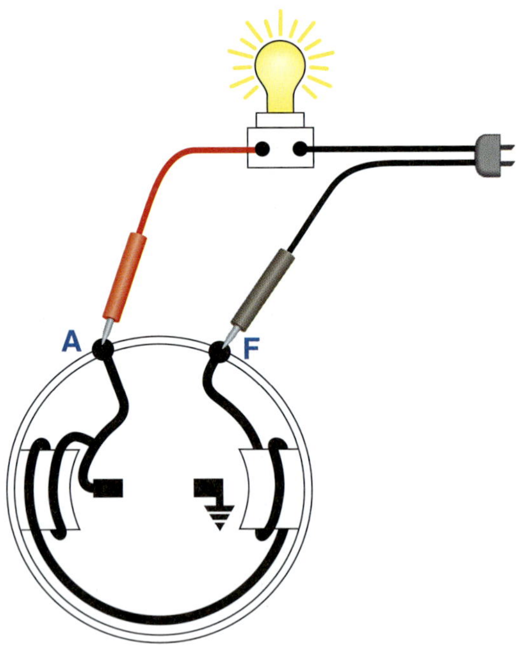

Figure 17-37A. *Checking for open field coils*

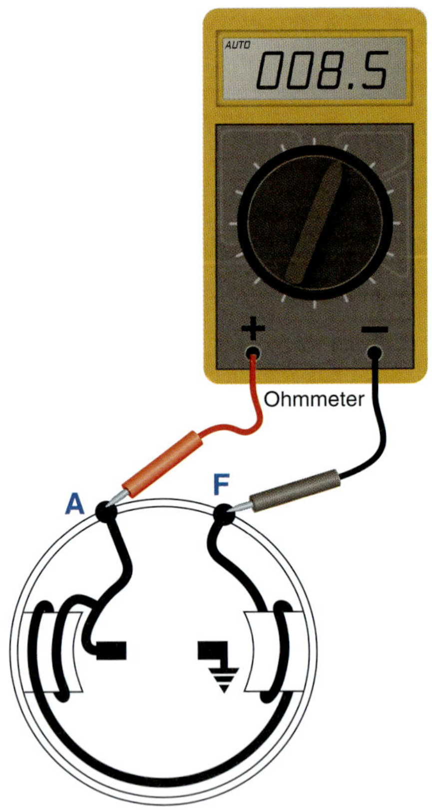

Figure 17-37B. *Alternate method using ohmmeter*

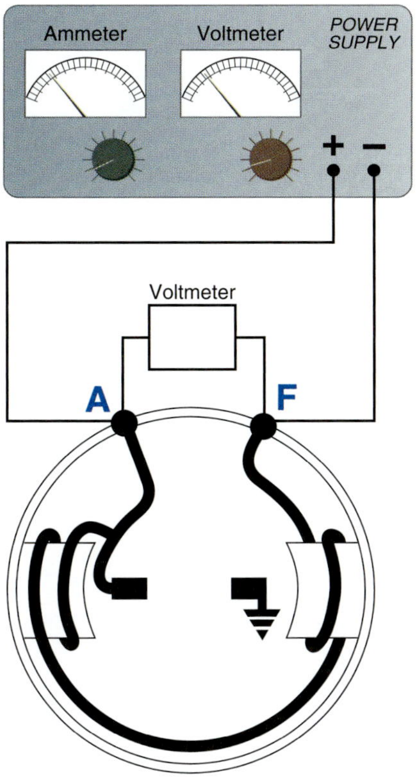

Figure 17-38. *Checking for shorted turns in the field coils*

Shorted turns in the field coils can be detected by using a variable DC power supply capable of approximately 30 volts and about 10 ampers, and a voltmeter connected across the field coils. With the voltage control turned to the minimal setting and the amperage control up, increase the voltage from the power supply until the voltmeter placed directly across the field indicates the value specified in the manufacturer's latest service information for the specific generator. The field current shown on the power supply ammeter should be within the limits stated in the service information. Too high a current through the fields indicates shorted turns; too low a current may be caused by a damaged wire, or a loose or corroded connection. Replacing field coils requires special tools to ensure that the pole shoes are properly secured.

Brushes and Brush Holders

Check the brush holders with the test lamp or an ohmmeter. Hold one test prod on the insulated brush holder and the other on the field frame. The lamp should not light or the resistance should be greater than 50,000 ohms. Then hold one prod on the grounded brush holder, and with the other prod on the field frame, the lamp should light or the ohmmeter should indicate zero ohms resistance.

Brushes are normally replaced when they have worn down to one half of their original length or when the generator is overhauled. When the generator is reassembled, install the new brushes in the brush holders and connect the brush pigtail to the holder. Check the brush arm spring tension with a small spring scale, pulling the brush slightly off the commutator. This should be within the limits specified in the manufacturer's overhaul manual, and is typically about 24 ounces.

Seat the brushes by placing a strip of very fine sandpaper (not emery paper) with the sand side out between the brush and the commutator. Pull the paper in the direction of armature rotation until the brush is properly shaped to fit the arc of the commutator. Lift the brush when the sandpaper is pulled back, so that all of the cutting strokes are in the direction of rotation. When the brushes are seated, blow all of the sand and brush dust from the generator with compressed air.

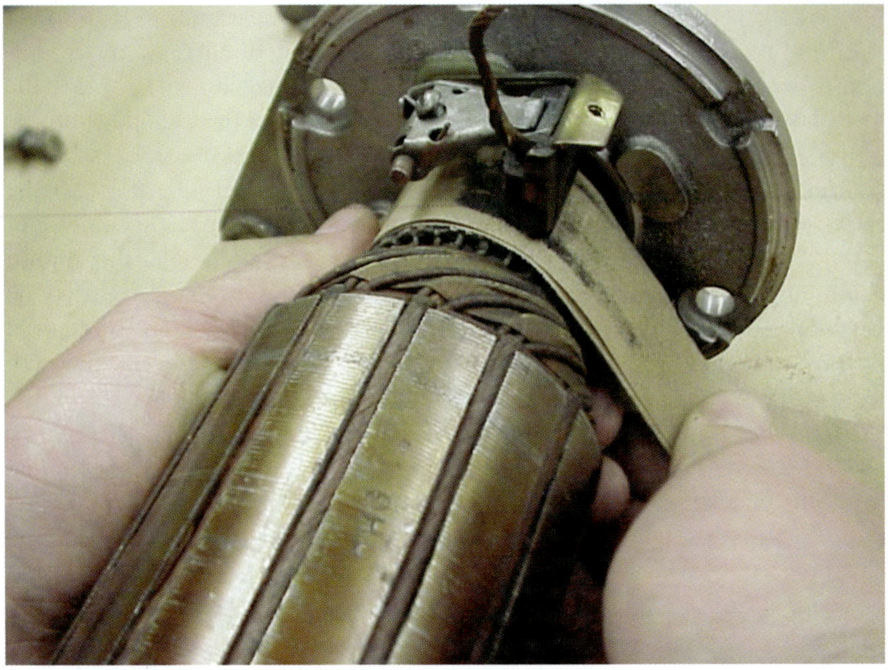

Figure 17-39. *Seating generator brushes*

Bearings

Because there are many kinds of bearings and bushings used in DC generators, the instructions in the manufacturer's manual for the particular generator must be followed in detail. Remember, when a ball bearing is used and the outer race is an interference fit in the end housing, and the inner race is an interference fit with the armature shaft, the order of assembly is critical to avoid applying a pressing force across the ball bearings. If the pressing force is exerted across the balls, this will brinell the bearing and cause it to self-destruct—along with the internal integrity of the generator—soon after it is placed in service.

Polarizing the Generator

When the generator was disassembled, the residual magnetism in the field frame was lost, and it must be restored in the proper polarity by flashing the field. Mount the reassembled generator on the engine or test stand, and if it is an A-type circuit, electrical power of the correct voltage should be placed across the generator field circuit. Place the positive lead on the armature terminal and the negative to the field connection. One way to accomplish this in the aircraft is to ground the generator field terminal, and momentarily short between a positive battery voltage source and the armature (A) terminal of the generator. For this to be successful, the field circuit from the regulator must

not be attached to the generator at the time of the polarization. If the field terminal is connected to the regular field terminal, the current and voltage control contacts will be burned, burned off, or welded shut.

If the field coils are connected to the grounded brush inside the generator (B-circuit), momentarily short between a positive battery potential source and the field (F) terminal of the generator. The generator housing must be grounded to the same ground source as the battery for the circuit to be complete and the polarization to be successful. One way this circuit might not be complete when the generator is mounted in the aircraft is if it is shock-mounted. In this instance, the normal generator ground strap needs to be installed. This flashing the field sends a pulse of current through the field coils in the same direction it normally flows from the armature, and the field frame is given enough residual magnetism with the proper polarity to start the generator producing current when it is rotated on the test stand or on the engine. (*See* Figure 17-40.)

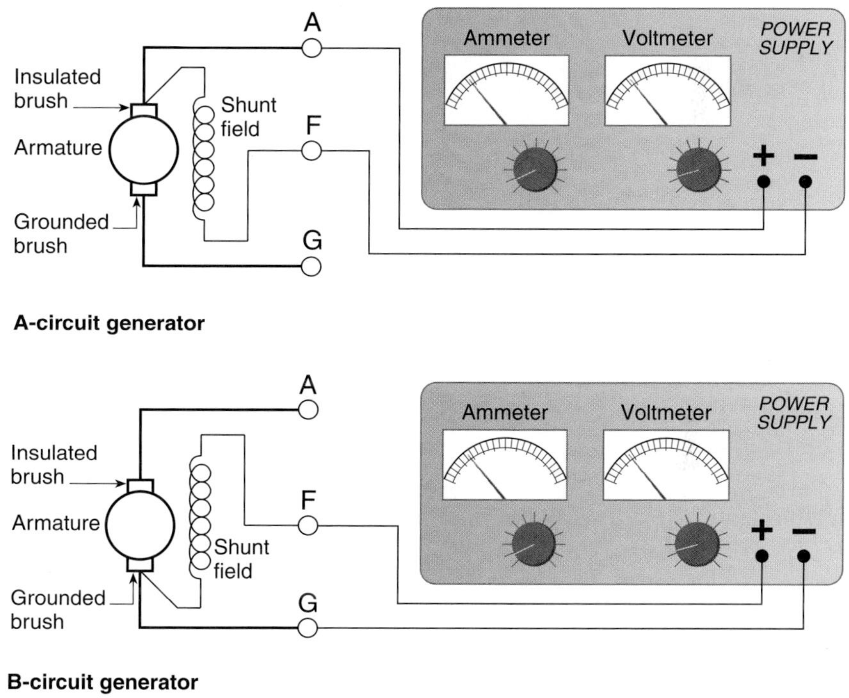

Figure 17-40. *Flashing the field of an A-circuit and B-circuit generators for light aircraft*

Testing

The generator output may be determined by mounting the generator on a test stand. For an A-circuit generator, connect the armature (A) terminal to the positive terminal of the battery through a switch and an ammeter. Connect a high-current carbon-pile variable resistor across the terminals of the battery. Connect the negative terminal of the battery to the field frame, which is the ground point. Ground the field (F) terminal to the field frame and connect a voltmeter between terminal A and ground. (Figure 17-41).

Everything is the same for a B-circuit generator except the field (F) terminal is connected to the armature at the A-terminal. (Figure 17-42).

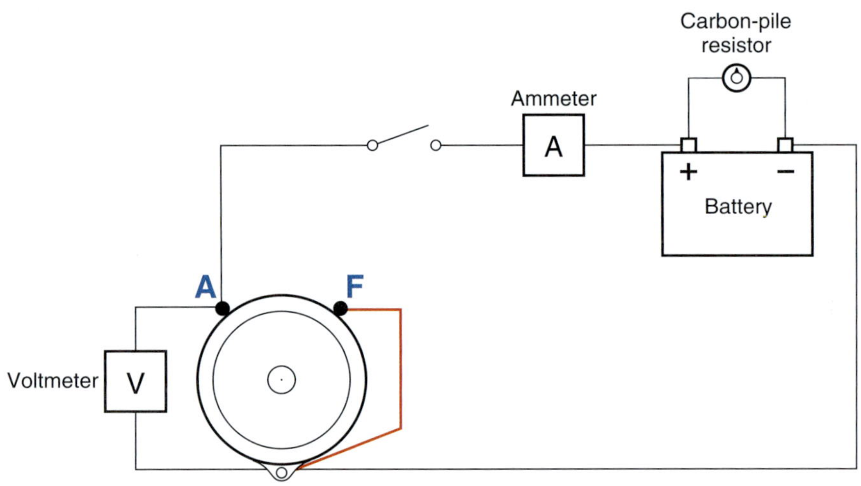

Figure 17-41. *Testing an A-circuit generator using a generator test stand*

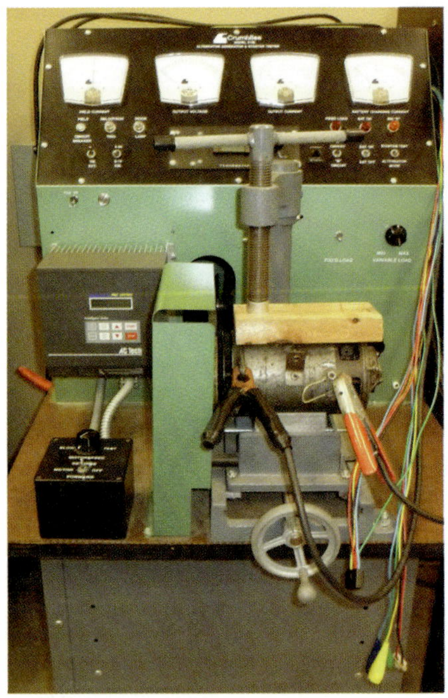

Generator test stand

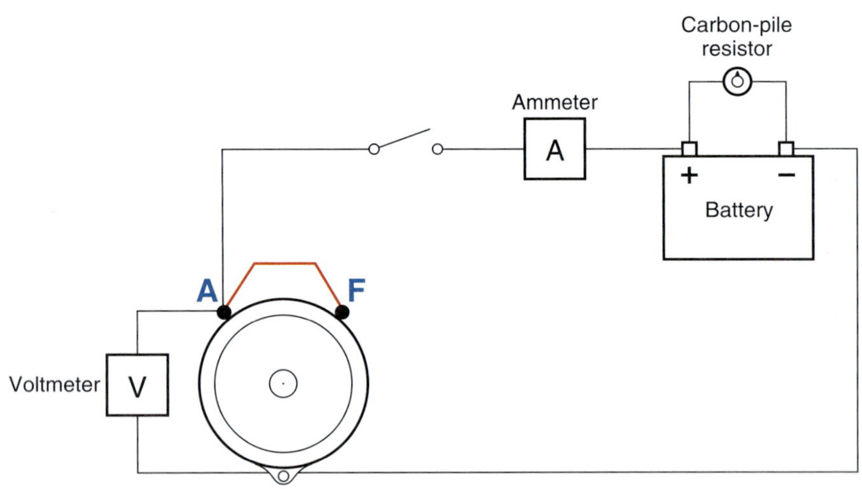

Figure 17-42. *Testing a B-circuit generator using a generator test stand*

Turn the generator at the speed needed to produce the battery voltage and close the switch. Increase the generator speed to that specified in the manufacturer's service information and adjust the carbon pile to get the required voltage. The current should be within the range specified in the service information.

DC Generator Controls

Almost all low-output DC generators installed on small aircraft use a mechanical, three-unit regulator such as the one described on Page 601. These units are not designed to be repaired, but with the proper equipment and instructions they can be adjusted by changing the air gaps and spring tensions.

While most regulators look alike, it is imperative that any adjustments be made in detailed steps, in accordance with the proper manufacturer's service information.

DC Alternators

With their solid-state rectifiers and regulators, DC alternators have just about taken over as the electrical power source of choice for small aircraft use. Some of the more important reasons for this are as follows:

- The multi-poled rotor produces voltage at a lower engine speed than the two or four pole generator.
- The high output current is taken from the stationary stator windings through permanent connection, rather than from the rotating armature through carbon brushes pressing on a commutator.
- Three-phase AC produced by the alternator produces smoother, more efficient DC than the single-phase AC produced by a DC generator.
- The six-semiconductor-diode rectifier is more efficient in converting AC power to DC than the mechanical commutator and brush rectifier used in a DC generator.
- The alternator field is excited by current from the battery rather than being self-excited by residual magnetism, as it is in the generator.
- With the voltage-limiting regulator, the multiple-poled rotor field, the hysteresis limit built into the stator core, and the stator coil density design, the DC alternator is automatically current limiting, eliminating the need for a separate current limiter.
- For the same power output, a DC alternator system is lighter weight and requires less maintenance than a DC generator system.

Because DC alternator systems use semiconductor rectifiers rather than mechanical rectifiers, some special precautions must be observed.

- It is extremely important when installing a battery, or when connecting a booster battery that the proper polarity be observed—negative to negative and positive to positive.
- Never operate a DC alternator without a circuit connected to the output terminal, and all circuit connections must be clean and tight.
- Never short across, or directly ground any of the terminals of an alternator.
- Do not attempt to polarize an alternator.

Routine Maintenance

DC alternators require little maintenance during normal 100-hour/annual inspections, other than closely checking the mounting and condition of all terminals and connecting wiring to be sure that all are tight and clean. This must include all ground straps and the ground wire between the alternator and the alternator regulator. The output voltage should be checked with an accurate voltmeter, and the battery should be examined for any indication of low charge or signs of overcharging, as described in the generator section above.

If the alternator is belt-driven, the belt tension should be checked in accordance with the aircraft service manual.

Alternator Overhaul

At the time the engine is overhauled, or if the alternator is not producing the proper voltage, it should be removed from the engine and disassembled in accordance with the appropriate manufacturer's service information.

Disassembly and Cleaning

Disassemble the alternator into its four parts: rotor and slip-ring assembly, end frame, stator, and drive-end frame. Do this by removing the three or four through bolts and gently prying the sections apart. It is a good idea to make a scribe mark across the end frames and the stator housing so the parts can all be reassembled in the same rotational position. This relationship will be different for varying types of engine mounting.

Place the rotor in a padded vise and tighten it only enough to prevent it turning when remove the nut that holds the drive-belt pulley on the rotor shaft. Remove the shaft nut and washer, the pulley, fan and collar and separate the drive end frame from the rotor shaft. If the rotor shaft must be pressed out of the drive-end bearing, the bearing must be replaced because to accomplish this procedure a pressing force had to be transferred across the balls in the bearing and the races would likely suffer brinelling.

Clean the parts with a soft bristle brush and a little petroleum solvent on a soft, cloth. Do not immerse any of the components in cleaning solutions unless it is specifically indicated by the manufacturer.

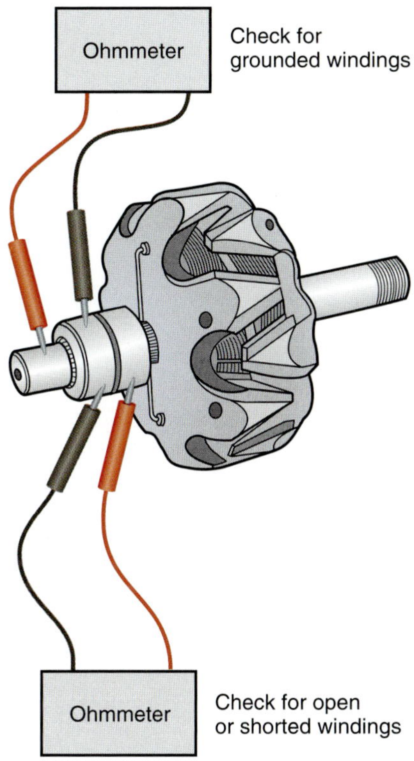

Figure 17-43. *Checking an alternator rotor for an open and field resistance (lower meter), and shorted windings (upper meter).*

Inspection and Repair

Rotor

Check the rotor with an ohmmeter. Place one test prod on each slip ring, being careful not to mar the rings. If the field winding is good, the ohmmeter will indicate the resistance of the winding (*see* Figure 17-43). The proper resistance is found in the manufacturer's service information for the specific alternator. If the actual resistance is not given, it may be found when the proper test voltage and current are given. For example, if the service information specifies 12 volts and 2.2 to 2.6 amps, divide the voltage by the current to get the proper resistance. In this case, the field coil resistance should be between 5.45 and 4.62 ohms (R = E/I, 12/2.2 = 5.45 ohms and 12/2.6 = 4.62 ohms).

This current-voltage method is a more scrutinizing test because it warms the field windings to approximately operating temperature where the insulation is most likely to fail if it is defective. A resistance lower than specified indicates shorted turns, and a higher resistance than specified indicates a bad or corroded connection. Each end of the rotor field winding is soldered to the side of its respective slip ring. If the ohmmeter shows an infinite resistance, the winding is open.

Stator

In an alternator, the load current is taken from the heavy windings of the stator. Disconnect the three stator windings from the diodes and remove the stator assembly from the end frame. Use the ohmmeter to measure the resistance of each pair of windings; they should be the same. (*See* Figure 17-44A.)

Because the resistance of a stator is very low, typically less than 0.3 ohms, it is more satisfactory to calculate the resistance with the currently-used voltage power supply method. The resistance between the windings and the stator frame should be infinite. (*See* Figure 17-44B.)

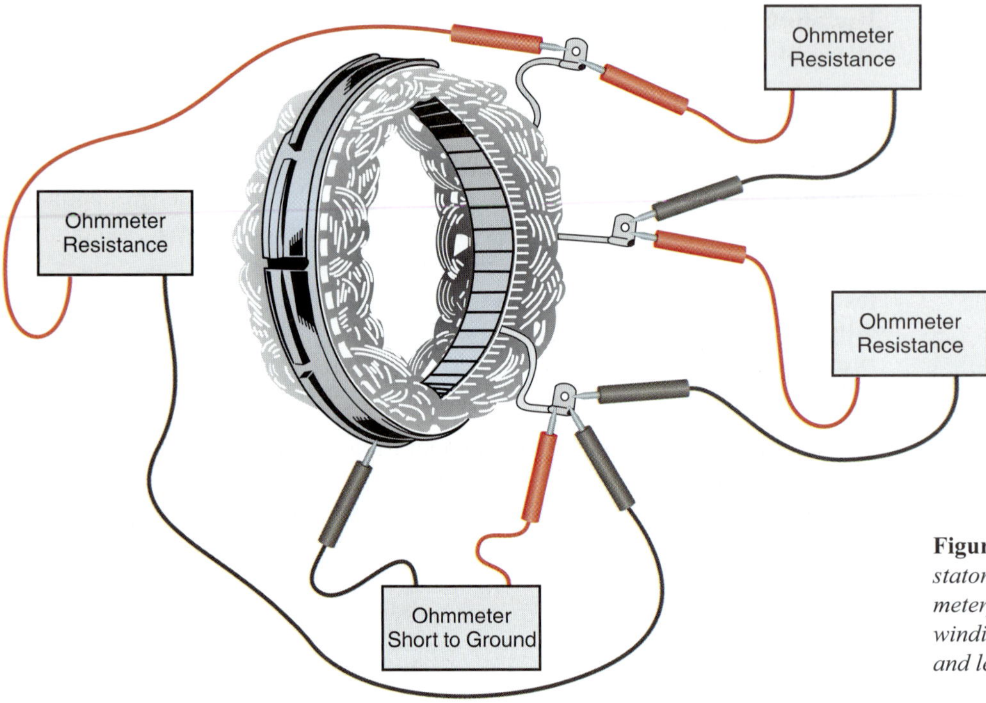

Figure 17- 44A. *Checking an alternator stator for grounded coils (the bottom meter), and open and internally shorted windings (the three meters at top, right, and left).*

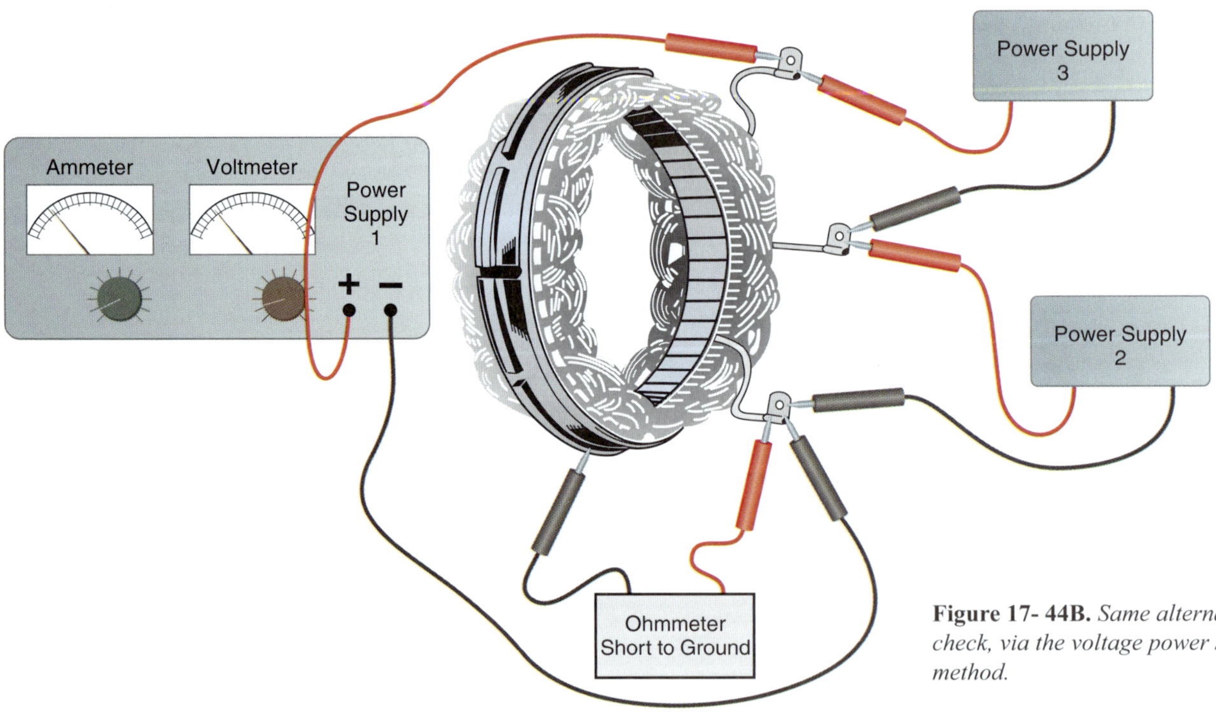

Figure 17- 44B. *Same alternator stator check, via the voltage power supply method.*

Rectifier Diodes

With all of the stator windings disconnected from the diodes, use a low-range ohmmeter, or the diode test position of a multi-meter, to check the front-to-back resistance of each of the six diodes. Place one test probe of the ohmmeter on the diode lead, and the other on the diode case or the heat sink it is mounted in, and note the resistance. Now, reverse the leads and note the resistance. One reading should be low and the other should be high.

If using a diode test option on a multi-meter, one position of the meter test leads will yield an audible signal, and the reverse position of the leads will not. With the ohmmeter probes positioned to produce the audible signal in the diode test mode of the multi-meter, most digital meters will indicate the amount of bias voltage required to forward-bias the diode. All diodes should indicate the same forward-bias voltage in the tenths position. If one does not, it should be considered suspect.

If either type of test is satisfactory, this indicates that the diode is allowing electrons to flow in one direction, but is blocking the flow in the opposite direction. Either a low resistance or a high resistance in both directions, and an audible sound or no sound in both directions identifies the diode as faulty and it must be replaced.

Do not be concerned when the electrical pigtails of three of the diodes indicate forward bias when the negative lead is placed on the pigtail, and the other three are forward bias with the positive lead placed on the pigtail. This is normal, because three of the diodes need for the positive material to be attached to the housing and those housings to be placed in the negative heat sink, and the other three diodes need for the negative material to be attached to the diode housing and that housing attached to the positive heat sink. It may seem unusual, but this is the correct construction of the rectifier unit.

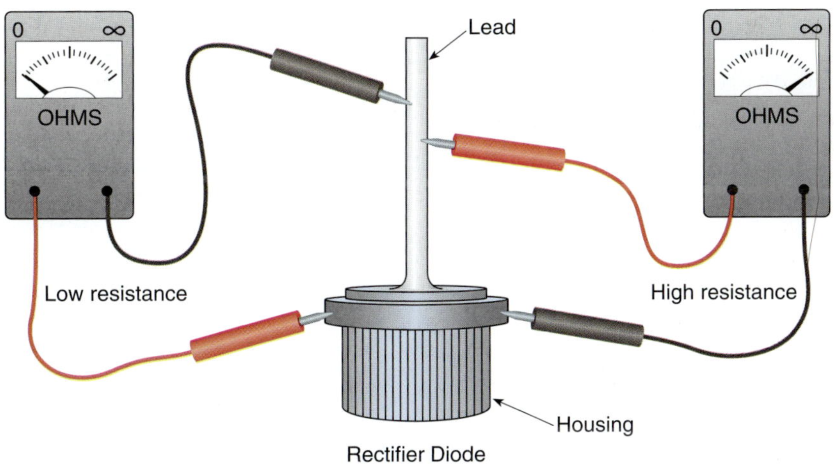

Figure 17-45. *Checking the front-to-back resistance of a rectifier diode. "N"-type diode material is attached to the "lead" of this diode; "P"-type material is attached to the diode housing.*

Slip Rings and Brushes

If the slip rings are dirty, clean them by placing the rotor in a lathe and spinning it while holding a 400-grit or finer polishing cloth to the surface until the slip ring is clean and shiny. Then, blow all of the dust away.

At this point, the slip rings should be dimensioned for minimum diameter and maximum out-of-round. If the diameter is below specification, the copper ring is too thin and will overheat and melt the solder joint where the rotor coil wire is attached. If the rings are excessively out-of-round, the brushes will bounce on and off the rings at higher RPM, causing decreased output and burning of the copper. The manufacturer's service information will provide these limits.

Some manufacturers will allow the slip rings to be turned in a lathe to true them up, as long as the minimum diameter of the ring is not exceeded. This should be accomplished with a narrow cutting tip, a high cutting RPM, a slow feed, and a very light cut per pass. This process should be followed with a nonconductive polishing paper.

The brush and spring assemblies should be replaced. To accomplish this, follow the instructions in the appropriate manufacturer's service information.

Bearings

The bearings must be serviced or replaced by following the manufacturer's specific instructions in the appropriate service information for installation. Most drive-end bearings are of the ball type and they can be easily damaged if they require pressing to install them.

Drive Pulleys

If the pulley is of the single "V" belt style, inspect the "V" sides for a wear step. If one exists, replace the pulley. If the pulley is reusable, deglaze the "V" of the pulley with an abrasive paper.

Reassembly

Assemble the alternator by reversing the disassembly steps. When reassembling the pulley on the rotor shaft, place the rotor in a padded vise, install the drive-end housing and bearing assembly, the appropriate spacer and pulley, and tighten the shaft nut to the specified foot-pounds. The shaft nut will either be a nut intended to be used with a split lock washer, a nut and lock plate, or a Marsden nut without any washers. Be sure to check the service information for the proper assembly.

When installing the slip-ring end frame, hold the brushes in the brush holder with a piece of rigid wire through the hole in the brush holder provided for that purpose. After the end frame is in place, remove the wire. Failure to remove the wire will cause a short between the two brushes when power

is applied to the field terminal. If it is the aircraft's regulator or a test stand regulator when this power is first applied, the regulator may be destroyed with a surge of current.

With the front and aft end frames sandwiching the stator frame, insert the long bolts that hold the units together and evenly snug them to approximately half torque. Tap the stator frame with a light ball-peen hammer as the bolts are brought to full torque. Be sure to safety the bolts.

Testing

Install the alternator on a test stand and connect the positive lead from the battery and carbon-pile load resistor through an ammeter to the output terminal of the alternator, and then the negative battery lead to the "GRD" terminal. Clip a shorting jumper between the output and "FLD" terminals. Operate the test stand at the speed specified in the appropriate service information for the particular alternator. Adjust the load resistor to get the desired output current and voltage.

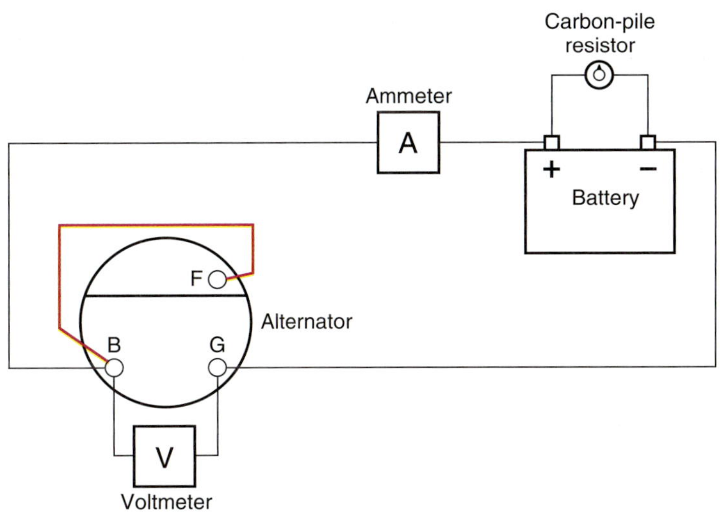

Figure 17-46. *Test circuit for a DC alternator*

Alternator Controls

Currently, aircraft alternators use solid-state voltage regulators that should be checked for proper operation using the equipment and procedure specified in the appropriate manufacturer's service information. The types vary greatly and require a multitude of specific procedures that is beyond the scope of an overall-subject textbook. Lacking specific manufacturer information, the mechanic should at best ensure that the wire connectors are clean, secure, and connected properly. Also be sure that all ground connections are clean and secure; many problems originate from poor or improper connections. The further testing of solid-state regulators usually requires connecting the regulator to special test equipment.

STUDY QUESTIONS: DC GENERATORS/ALTERNATORS INSPECTION & MAINTENANCE

Answers are on Page 636. Page numbers refer to chapter text.

47. The individual generator parts will be cleaned after disassembly, therefore why should the outside of the generator be cleaned prior to opening the unit? *Page 617*

48. Checking the ____________________________ of a battery is a quick method of determining its state of charge. *Page 614*

49. What occurs at a corroded battery terminal that causes a slow charging rate of the battery, and usually prevents the complete charging of the battery? *Page 615*

50. The positive lead of a voltmeter must be connected to a junction with ____________________________ to determine the charging system's output voltage. *Page 615*

51. Substandard electrical terminals will ________________ excessively when heated by current flow, causing the terminal to loosen and usually overheat. *Page 616*

52. What is the prohibition between generator parts and solvent cleaners? *Page 617*

53. Normal belt tension for a belt driven generator or alternator allows the belt to be deflected about a/an _________________ with moderate hand pressure or twisted about _____ degrees. *Page 616*

54. The test equipment that allows you to determine whether or not there are shorted, open, or grounded coils in an armature is called a/an __________________ . *Page 618*

55. Is an indication of 50,000 ohms on an ohmmeter sufficient between a conductor and ground? *Pages 620, 621*

56. When checking an armature on a growler, if a hacksaw blade held slightly above the armature vibrates violently, it indicates that some of the coils are ______________________ (shorted, open, grounded). *Page 619*

57. If a new generator brush is one inch long, it should be replaced when it is worn to _______ inch. *Page 622*

58. Newer generator brushes typically use a/an ___________________ to indicate when a brush is worn to its limit. *Page 617*

59. Generator brushes may be seated by pulling a strip of _____________________ (emery paper or sandpaper) between the commutator and the brush in the ____________ direction of generator rotation. *Page 622*

60. A generator brush is seated-in when its contact surface is ______________ to fit the ______________________________ . *Page 622*

61. If the commutator is lathed below the minimum diameter, the bars will overheat and cause the armature coil wires to come ________________________ and solder to be deposited on the underside of the brush ___________________ . *Page 619*

62. A generator is polarized by sending a pulse of current through the field coils in the _______________________ (same or opposite) direction as it flows in normal operation. *Page 624*

63. Why is it more precise to measure component resistance with the currently-used voltage power supply method rather than with the simple ohmmeter test? *Page 628*

64. Mechanical generator controls are not designed to be repaired, but they may be adjusted by changing the __________________________ and ______________________________. *Page 626*

65. What is the more efficient method of converting AC power to DC between generators and DC alternators? *Page 626*

66. There are _____________ (how many) semiconductor diodes in the rectifier. *Page 626*

67. An alternator "V"-style drive pulley is considered worn out if there is a ____________________ in the "V" surface. *Page 631*

68. Why will the regulator be damaged during alternator testing if the rigid installation wire is not removed? *Pages 631, 632*

69. After a DC alternator has been overhauled, it __________________ (must or must not) be polarized. *Page 627*

70. The field coil in an alternator rotor and stator should be checked for ___________________ , _______________________________ , and ________________________________ . *Pages 628, 629*

71. A DC alternator diode is checked by comparing its ________________________________ resistance with an ohmmeter. *Page 630*

Answers to Chapter 17 Study Questions

1. a. direct current
   b. alternating current
2. AC
3. a. strength of magnetic field
   b. number of conductors passing through field
   c. speed with which field is crossed
4. AC
5. minimum
6. AC
7. DC
8. generator armature
9. residual
10. residual
11. two
12. flashing
13. same
14. parallel
15. shunt
16. laminated soft iron
17. neutral position, neutral plane
18. armature
19. brush arcing
20. a. compensating windings
    b. interpoles
21. series
22. slow-blow
23. motor
24. variable resistor
25. carbon pile
26. paralleling
27. ground, generator armature
28. field current
29. rotating
30. stationary
31. semiconductor
32. battery
33. slip rings
34. three
35. six
36. field
37. series
38. 400
39. constant-speed drive
40. low
41. constant-speed drive
42. brushless
43. a. voltage
    b. current
    c. frequency
    d. phase sequence
44. voltage, frequency
45. inverter
46. a. rotary
    b. static
47. to minimize internal contamination
48. specific gravity
49. resistance, or voltage drop
50. battery positive potential
51. expand
52. do not immerse
53. half inch, 90
54. growler
55. yes
56. shorted
57. ½
58. wear line
59. sandpaper, normal
60. arched, commutator
61. unsoldered, cover
62. same
63. Because it warms the unit to operating temperature where insulation is most apt to fail.
64. air gaps, spring tension
65. using a semiconductor diode rectifier
66. six
67. wear step
68. The brushes are shorted together with the pin, therefore the regulator will be shorted to ground across the brushes and high current will burn the regulator without the field rotor resistance in the circuit.
69. must not
70. an open, internal shorts, a short to ground
71. front-to-back

FIRE PROTECTION SYSTEMS

18

Fire Protection Systems

18

Fire Protection Systems

Fire in flight has always been one of the dangers most feared, and the powerplant is the most likely source. Combustible liquids in the form of fuel and lubricating oil are always present, often under considerable pressure, and the extremely hot exhaust system components have the potential of igniting fluids sprayed on them. The remote location and enclosed installation of most powerplants make systems that quickly detect and efficiently extinguish fires in these vital areas very important to the pilot and aviation maintenance technician.

A complete fire protection system detects a fire or overheat condition as early as possible and allows the flight crew to analyze the situation and discharge efficient agents into the affected area. The discharge systems are not usually fully automatic; they warn the flight crew of an impending problem, then arm the system so the crew can discharge the agent in a minimum of time.

The basic principles of fire protection are discussed in Chapter 13 of the *Airframe* textbook of this *Aviation Maintenance Technician Series*. In this *Powerplant* textbook, the aspects of fire protection as they apply to the powerplant are considered.

Types of Fires

The National Fire Protection Association (NFPA) has categorized fires and identified those extinguishing agents most effective on each class.

Class A fires have as fuels solid combustible materials such as wood, paper, and cloth. These are usually not the source of an aircraft powerplant fire.

Class A fire. A fire with solid combustible materials such as wood, paper, and cloth as its fuel.

Class B fires have combustible liquids such as gasoline, turbine-engine fuel, lubricating oil, and hydraulic fluid as their fuel and usually are associated with powerplants.

Class B fire. A fire that has combustible liquids as its fuel.

Class C fires involve energized electrical equipment, and often occur in powerplants.

Class C fire. A fire which involves energized electrical equipment.

Class D fires are those in which a metal such as magnesium burns. While Class D fires are often associated with brakes and wheels, they do occur in the powerplant when some cast-magnesium housings ignite.

Class D fire. A fire in which a metal such as magnesium burns.

Fire Zones

For the purpose of designing an effective fire protection system, powerplant areas are divided into fire zones based on the volume and smoothness of the airflow. These zones do not relate to the NFPA classes of fires.

Class A fire zones have large volumes of air flowing past regular arrangements of similarly shaped obstructions. An example of a Class A fire zone is the power section of a reciprocating engine installation where the air flows over the cylinders and through the fins.

Class B fire zones have large volumes of air flowing over aerodynamically clean obstructions. In a reciprocating engine installation, Class B fire zones are the heat exchanger ducts and exhaust shrouds. In a turbine engine installation, a Class B fire zone is the inside of the airframe structure that houses an engine covered with a smooth fireproof liner.

Class C fire zones have a relatively small airflow through them, for example, the compartment behind the firewall of a reciprocating engine.

Class D fire zones have little or no airflow through them. An example is the enclosed area inside the wing structure.

Class X fire zones are areas in which large volumes of air flow at an irregular rate. Because of the unpredictability of the conditions in these areas, the extinguishing agent requirements are normally twice those for a similar zone with a predictable air flow.

Fire Detection and Warning Systems

There are several types of fire detection and warning systems used with aircraft powerplant installations. The operating characteristics determine which type of system is installed. Some of the basic requirements for such a system are:

- There must be a separate fire detection system for each engine that actuates an audible alarm and a cockpit light that shows the location of the fire.
- The system must not give false warnings under any flight or ground operating condition. It must continue to indicate the presence of a fire as long as it exists and accurately show when it has been extinguished. It must sound a warning if the fire reignites.
- The detectors must not be damaged by exposure to oil, water, vibration, extremes of temperature, or handling encountered in normal maintenance. They must be light in weight and adaptable to any mounting position.
- The detector circuitry must operate directly from the aircraft electrical system and require a minimum of electrical current when it is not indicating a fire. There must be a means of testing the integrity of the system from the cockpit.

There are two basic classifications of detectors: fire detectors and overheat detectors. A fire detector warns of a fire that raises the temperature of a particular location to a predetermined high value. An overheat detector, on the other hand, initiates a warning when there is a lesser increase in temperature, but over a larger area. Most detection systems installed in an aircraft powerplant turn on a red light and sound a fire-warning bell.

Thermoswitch-Type Fire Detection System

Thermoswitch, or spot-type, fire detectors initiate a warning when the temperature in any protected location in a powerplant installation rises to a predetermined level. There are two types of thermoswitch circuits: single-loop and double-loop.

thermoswitch. An electrical switch that closes a circuit when it is exposed to a specified high temperature.

The single-loop bimetallic thermoswitch detector circuit uses a number of spot detectors like the one in Figure 18-1, installed in a circuit shown in Figure 18-2 on the next page. When the detector is heated, the bimetallic strips on which the contacts are mounted distort and close the contacts, completing the circuit between the loop and ground.

bimetallic strip. A metal strip made of two different types of metal fastened together side-by-side. When heated, the two metals expand different amounts and the strip warps, or bends.

The circuit in Figure 18-2 will signal the presence of a fire even if the loop connecting the detectors is broken. During normal operation the detectors receive their power from both ends of the loop; if the loop should break at any point, the detectors will still have power. If a detector senses a fire, its contacts will close and provide a ground for the warning light and bell.

Closing the fire-warning test switch energizes the test relay. This removes power from one end of the loop and grounds it, turning on the fire-warning light and sounding the bell. If there is an open in the wire between the detectors, there will be no ground for the light and bell, and they will not activate. The test shows the system has a fault, but in normal operation, it will still detect a fire.

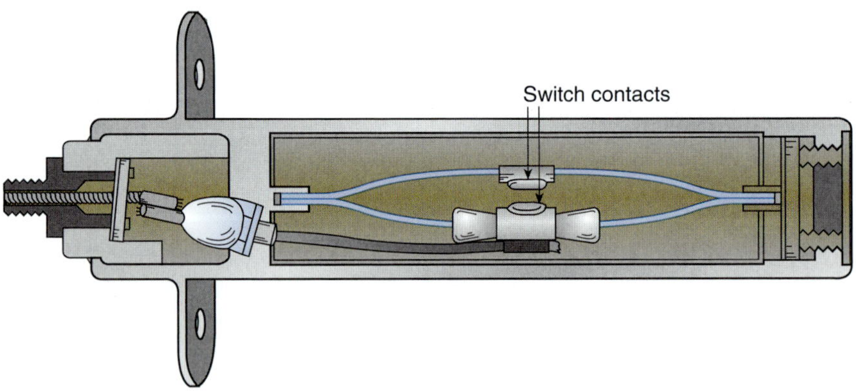

Figure 18-1. *Typical single-loop bimetallic thermoswitch fire detector*

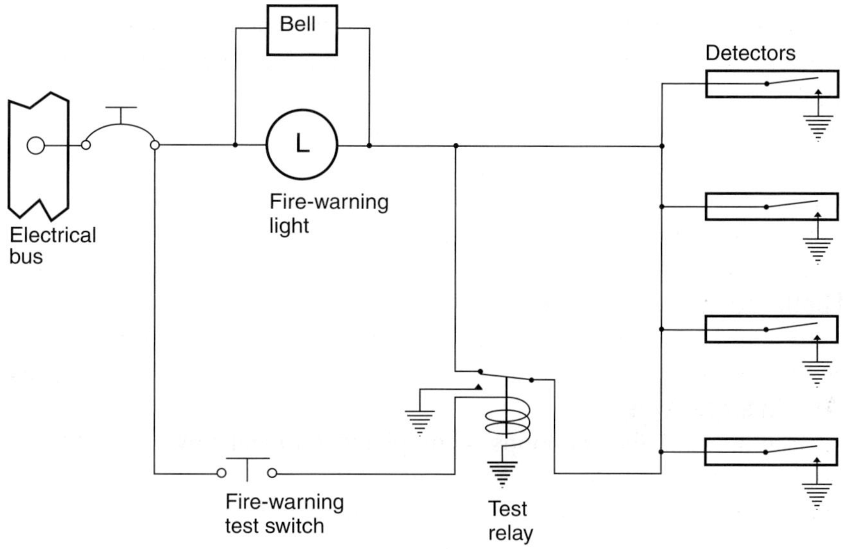

Figure 18-2. *Circuit for a single-loop thermoswitch fire detector*

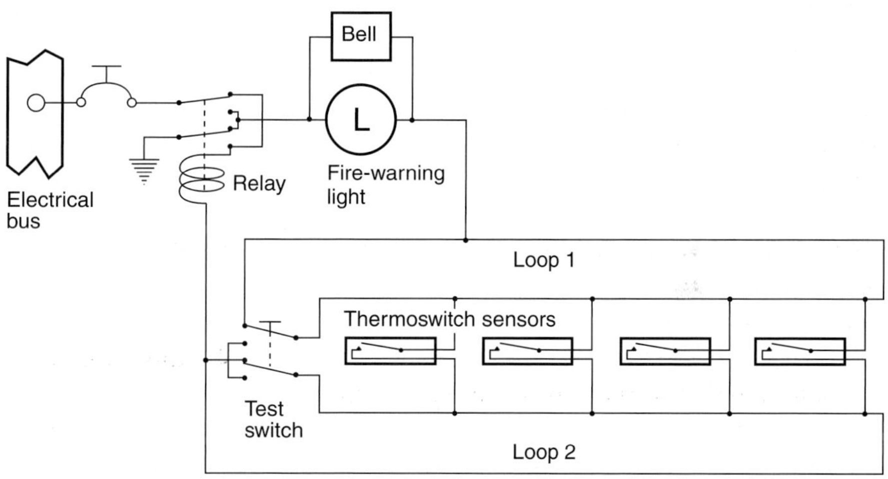

Figure 18-3. *Circuit for a double-loop thermoswitch fire detector*

There is another type of thermoswitch spot detector installed in some aircraft. These detectors have two terminals, and instead of completing the circuit to ground when a fire is detected, the detector completes the circuit between the two conductors connected to their terminals. This type of thermoswitch is connected between two loops, and the system can tolerate either an open circuit or a short to ground without affecting the operation of the system.

Follow the circuit in Figure 18-3 to understand how the double-loop thermoswitch system works. When there is no fire, the relay is not energized, and both ends of loop 2 are connected to the positive voltage at the electrical bus. Both ends of loop 1 are connected to ground through the fire-warning light and bell, and the sensors act as open switches between the two loops.

When a fire occurs and the contacts in any of the thermoswitches close, the circuit between the two loops is completed. Current flows from the bus through part of loop 2, the closed thermoswitch, part of loop 1, and the fire-warning light and bell to ground, initiating the fire warning.

If there is a short to ground in loop 2, the fault current energizes the relay, and its contacts change position. Loop 2 is now at ground potential, and loop 1 is connected to the positive voltage at the bus.

Both ends of the two loops are connected to the test switch so that a single open in either of the loops will not effect normal operation. When the test switch button is depressed, both loops are opened and connected together in series. Current flows through loops 1 and 2 to ground through the fire-warning light and bell. Pressing the test switch button checks the integrity of the entire system, as any open will prevent the actuation of the bell and light.

Rate-of-Temperature-Rise Detection System

A thermoswitch-type detection system initiates a fire warning when a detector reaches a predetermined temperature. But a powerplant fire can have a good start before this temperature is reached. For this reason, thermocouple-type, rate-of-temperature-rise fire-warning systems are used. A thermocouple system initiates a fire warning when the temperature at any specific location within the compartment being monitored rises a great deal faster than the temperature of the entire compartment. These systems are well suited for installation in engine compartments where normal operating temperatures are quite high, but the rise to this temperature is gradual.

thermocouple. A device made of two dissimilar metal wires whose ends are welded together to form a loop. A voltage exists in the loop proportional to the difference in the temperature of the junctions at which the wires are joined.

A thermocouple is made of two different types of wire, typically iron and constantan, welded together to form a junction. When several thermocouples are connected in series to form a loop, a voltage exists that is proportional to the difference in the temperatures of the various junctions. The loop is a complete circuit through which current, proportional to the temperature difference, flows.

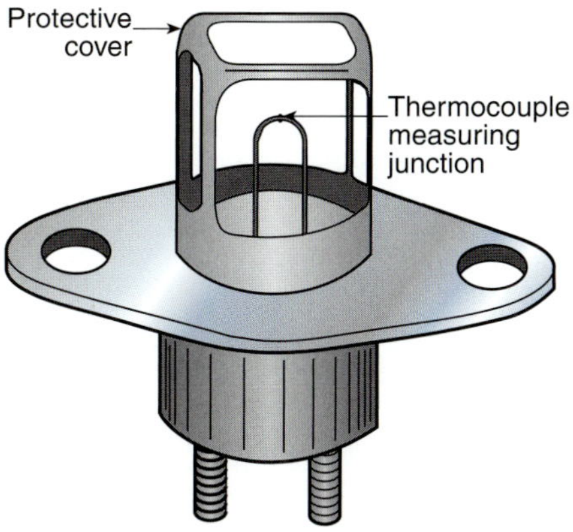

Figure 18-4. *Thermocouple fire sensor*

Sensors, like that in Figure 18-4, consist of a junction mounted inside a metal housing that protects it from physical damage, yet allows free circulation of air.

The sensors are mounted at strategic locations around the powerplant compartment. One sensor is mounted inside a thermal insulating shield that protects it from direct air circulation, yet allows it to reach the temperature of the air in the compartment. This sensor is called the reference junction. The other sensors are called measuring junctions, and all junctions are connected in series with the coil of a sensitive relay and a test thermocouple.

During normal operation all junctions are at the same temperature and no current flows in the thermocouple circuit. When the engine is started and the engine compartment temperature rises, all thermocouple temperatures rise together and still no current flows. But if a fire should occur, the temperature of one or more of the thermocouples will rise immediately, while the temperature of the insulated reference thermocouple rises much slower. As long as there is a temperature difference in any of the junctions, there is a difference in voltage between them, and current flows. The current is small, only a few milliamps, but enough to energize the sensitive relay. The sensitive relay contacts close and carry enough current to the coil of the slave relay to close its contacts and allow current to flow to the fire-warning light and bell.

A thermocouple fire detection system is tested by closing the test switch and holding it closed for a specified number of seconds. Current flows through the heater inside the test thermocouple housing and heats the test junction. Since this junction is in series with all the other junctions, there is a voltage difference, and thus enough current to energize the sensitive relay and initiate a fire warning.

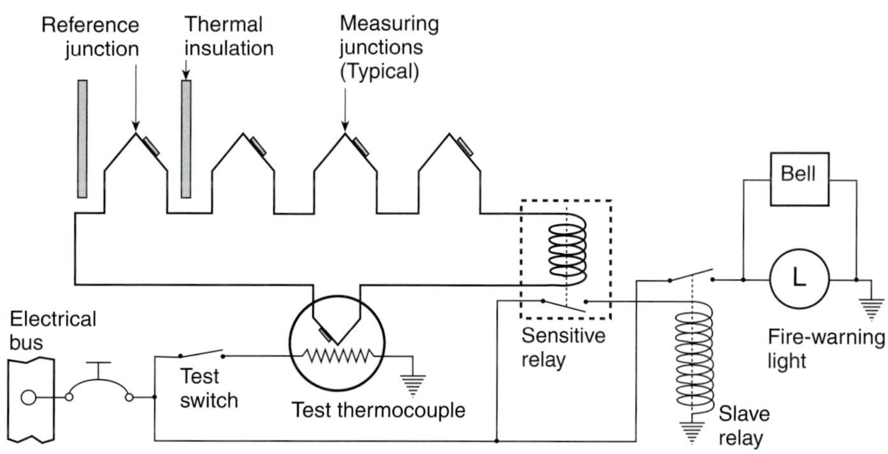

Figure 18-5. *Thermocouple rate-of-temperature-rise fire detection circuit*

Continuous-Loop Fire and Overheat Detection Systems

Because engine compartments are difficult to monitor for fire, continuous-loop type-detectors, with their ability to detect overheat conditions as well as fires, are often used in these areas. There are two basic types of continuous-loop fire and overheat detection systems: thermistor and pneumatic.

thermistor. A semiconductor material whose electrical resistance varies with its temperature.

Inconel. An alloy of chromium, iron, and nickel.

ceramic. A hard, brittle, heat- and corrosion-resistant material made by shaping and then firing a nonmetallic mineral, such as clay, at a high temperature.

Thermistor-Type Continuous-Loop Systems

There are two configurations of thermistor-type continuous loop elements: single-conductor and two-conductor.

The single-conductor element has a center conductor supported in a thin-wall Inconel tube by ceramic beads. An electrical connection is made to the conductor, and the outside tube is grounded to the airframe. The space between the beads is filled with a eutectic (low-melting-point) salt whose resistance drops drastically when it melts. When any portion of the tube gets hot enough to melt the salt, the resistance between the center conductor and the outside tube drops, allowing enough current to flow to ground that a fire warning signal is initiated. When the fire is extinguished, the molten salt solidifies, and its resistance increases enough to stop the fire-warning current.

The two-conductor loop is also mounted in an Inconel tube, but it has two parallel wires embedded in a thermistor material whose resistance decreases as its temperature increases. One of the wires is grounded to the outer tube, and the other wire terminates in a connector that is connected to a control unit that continuously measures the total resistance of the sensing loop. By monitoring the resistance of the sensing loop, this unit can detect a general overheat condition as well as a localized hot spot.

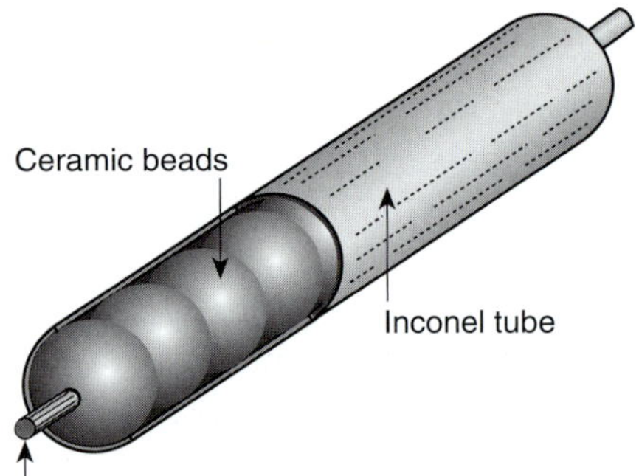

Figure 18-6. *Single-conductor continuous-loop fire detector element*

eutectic. An alloy or solution that has the lowest possible constant melting point.

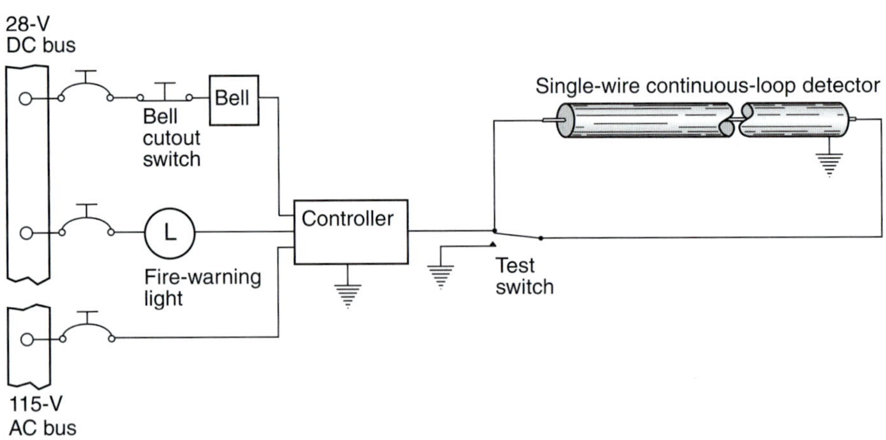

Figure 18-7. *Single conductor continuous-loop fire detector circuit*

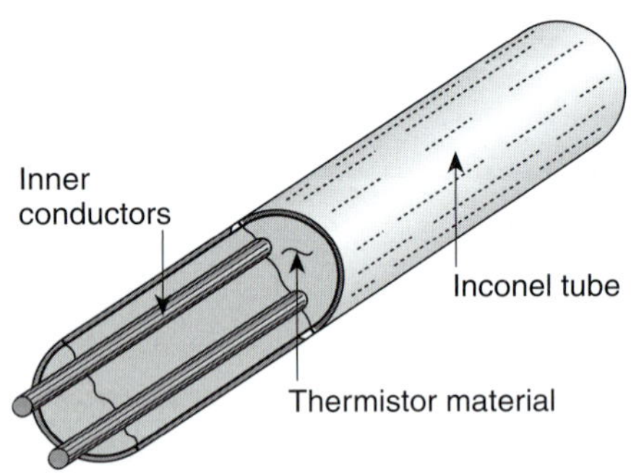

Figure 18-8. *Two-conductor continuous-loop fire detector element*

Pneumatic-Type Continuous-Loop Systems

Pneumatic fire detection systems also use a continuous loop for the detector, but this loop is made of a sealed stainless steel tube that contains an element which absorbs gas when cold, but releases it when heated.

One type of pneumatic fire detection system is the Lindberg system. The stainless steel tube that makes up the loop contains the gas-absorbing element and the gas, and is connected to a pneumatic pressure switch as shown in Figure 18-9. When the loop, which is installed around the area being monitored, is heated in a local area by a fire or over a larger area by an overheat condition, the gas is released which closes the pressure switch. Closing this switch completes a circuit for one of the transformer windings and allows the 115-volt, 400-Hz AC from the aircraft electrical system to illuminate the fire-warning light and sound the bell. The bell can be silenced by depressing the bell-disable switch.

This system is tested by closing the test switch. This allows low-voltage AC to flow through the tubing in the loop and heat it, releasing enough gas to close the pressure switch and initiate a fire warning.

The Systron-Donner pneumatic fire detection system also uses a continuous loop for the detection element, but this loop contains two gases and a titanium center wire that has the capacity to absorb an amount of hydrogen gas proportional to its temperature.

The tube is filled with helium gas under pressure, and at normal temperature, the helium produces a pressure proportional to the average temperature of the entire tube. When the average temperature of the tube reaches the value for which the warning system is set, the pressure of the helium gas becomes great enough to close a set of normally open contacts in the detector housing, which initiates a fire-warning signal.

When a fire increases the temperature of a localized area of the tube, the center wire will release enough hydrogen gas to increase the pressure inside the housing and close the contacts, initiating a fire warning.

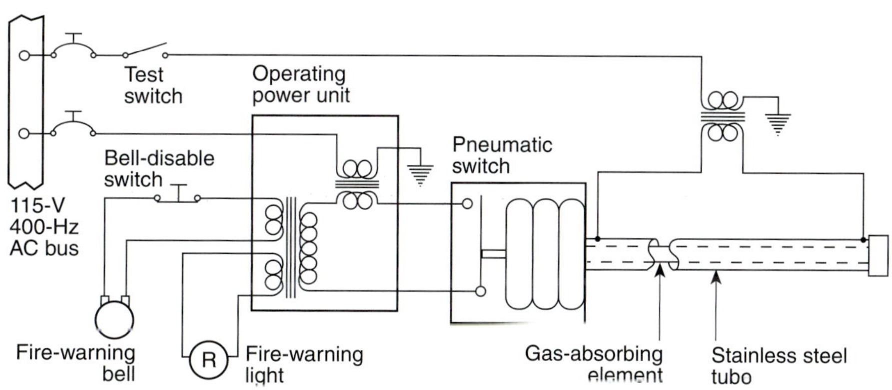

Figure 18-9. *Lindberg pneumatic fire detection system*

When the fire is extinguished, the temperature drops, and the center wire absorbs enough hydrogen gas to lower the pressure in the housing so the contacts can snap open. The system, now restored to its original condition, can detect the fire if it should reignite.

There are two switches in the housing. One is normally open, and it signals the presence of a fire when the pressure of either the hydrogen or helium gas increases enough to close it. The other, called the integrity switch, is held closed by the normal pressure of the helium gas in the tube. If a break should occur in the tube and the helium pressure is lost, the integrity switch will open; then, when the test switch is closed, no current will flow to initiate the fire-warning system. The failure of a warning light indicates a faulty system.

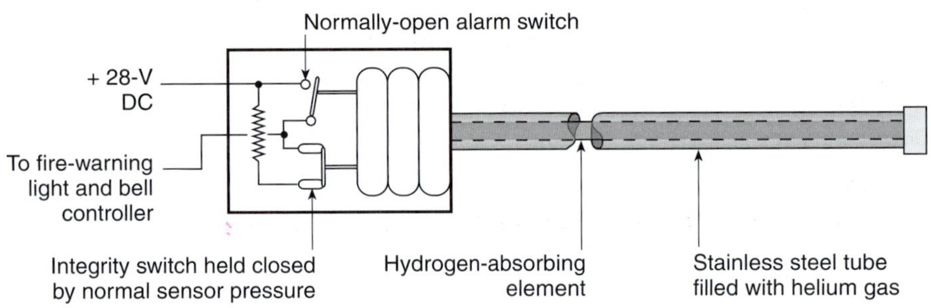

Figure 18-10. *Systron-Donner pneumatic fire detection and overheat detector*

Fire-Extinguishing Systems

Powerplant fire protection systems are divided logically into two divisions: fire detection and fire extinguishing. Here we will consider the various agents applicable to powerplant fire-extinguishing systems.

Fire-Extinguishing Agents

Fire is a chemical reaction between a fuel and oxygen, and it can be controlled by interfering with this reaction. Removing the fuel, smothering the fuel with a substance that excludes oxygen, or lowering the temperature of the fuel are methods of accomplishing this. Another method proven most effective for aircraft fires is the application of a chemical compound that combines with the oxygen to prevent it from reacting with the fuel.

The inert cold gas agents, carbon dioxide (CO_2) and liquid nitrogen (N_2), are both effective fire-extinguishing agents, and both have very low toxicity.

Carbon Dioxide (CO_2)

CO_2 is heavier than air. When sprayed on a fire, it remains on the surface and excludes oxygen from the combustion process. This extinguishes the fire. CO_2 has been used for many years. It is inexpensive, nontoxic, safe to handle, and has a long storage life.

Most older aircraft had CO_2 extinguishing systems installed in the engine nacelles. These extinguishers have been replaced in modern aircraft by more efficient types.

Liquid Nitrogen (N_2)

cryogenic fluid. A liquid which boils at a temperature lower than about 110°K (-163°C) under normal atmospheric pressure.

Dewar bottle. A special container used to store liquid oxygen and liquid nitrogen.

halogenated hydrocarbon. A hydrocarbon compound in which one or more hydrogen atoms have been replaced with atoms of one of the halogen elements such as fluorine, chlorine, or bromine.

N_2 is more effective than CO_2, but because it is a cryogenic liquid, it must be kept in a Dewar bottle. N_2, used for inerting fuel tanks in some military aircraft, is also available for use in powerplant fire-extinguishing systems.

Halogenated Hydrocarbons

This category of fire-extinguishing agents includes those most widely used today, as well as some that were used in the past but are no longer considered suitable. These agents are hydrocarbon compounds in which one or more hydrogen atoms have been replaced with atoms of one of the halogen elements such as fluorine, chlorine, or bromine.

In the combustion process, the molecules of the fuel combine with those of the oxygen in an orderly fashion, but if one of the halogen compounds is mixed with the oxygen, this orderly combination is interrupted or stopped and the fire is extinguished.

The two most widely used halogenated hydrocarbons are bromotrifluoromethane ($CBrF_3$), widely known as Halon 1301, and bromochlorodifluoromethane ($CBrClF_2$), known as Halon 1211. These compounds, often called by the trade name of Freon, have a very low toxicity, with 1301 being the least toxic of all commonly used agents. Both are very effective as fire-extinguishing agents. They are noncorrosive, evaporate rapidly, leave no residue, and require no cleanup or neutralization. Halon 1301 does not require any pressurizing agent, but Halon 1211 may be pressurized with nitrogen.

Powerplant Fire-Extinguishing Systems

There are two types of installed powerplant fire-extinguishing systems: CO_2 systems installed in the engine compartments of many older aircraft, and the high-rate-discharge (HRD) systems used on most modern jet transport aircraft.

Carbon Dioxide Extinguishing Systems

CO_2 fire extinguishers were the most widely used systems installed on twin- and four-engine transport aircraft up through World War II.

CO_2 is carried in steel bottles and often pressurized with compressed nitrogen to aid its expulsion under low temperature conditions. The bottles have a remotely operated valve and are connected to a handle that allows the pilot to select the engine into which the CO_2 will discharge. When it is determined an engine fire actually exists, the pilot selects the affected engine and then pulls the T-shaped handle, or toggle. The bottle discharges CO_2 into the engine through a perforated aluminum tube that surrounds the power section. Some larger systems have two bottles, and the pilot can release the second one if necessary.

CO_2 systems have two indicator disks, one red and one yellow, on the outside of the fuselage near the bottles. If the pilot actuates the T-handle and discharges the bottle, the yellow disk blows out. If the area around the bottles overheats enough to raise the gas pressure to a dangerous level, the red disk blows out, and the system automatically discharges. On the normal walk-around inspection, the flight crewmember can tell, from these disks, the condition of the CO_2 system.

HRD fire extinguisher. A fire extinguisher that carries the extinguishing agent in a sealed sphere or cylinder. The agent is released when an ignited powder charge drives a cutter through a frangible disk. The container is emptied in much less than a second.

frangible. Capable of being broken.

HRD (High-Rate-Discharge) Extinguishing Systems

The powerplant area of most modern turbine-engine-powered aircraft is protected with two or more spherical HRD bottles of Halon 1211 or 1301. The bottles usually include a charge of compressed nitrogen to ensure a quick dispersal. The containers are sealed with a frangible disk. A cutter is fired into the disk by a powder charge, or squib, ignited when the pilot closes the agent discharge switch. The entire contents of the bottle are discharged within about 0.08 second after the switch is closed.

Figure 18-11 shows a cross-sectional view of a typical spherical HRD bottle. The cartridge is electrically ignited and drives the cutter into the disk. The strainer prevents any of the broken disk from getting into the distribution system.

The safety plug is connected to a red indicator disk on the outside of the engine compartment. If the temperature of the compartment in which the bottle is mounted gets high enough to raise the pressure of the gas to a dangerous level, the safety plug melts and releases the gas. As the gas vents to the atmosphere, it blows out the red indicator disk which shows that the

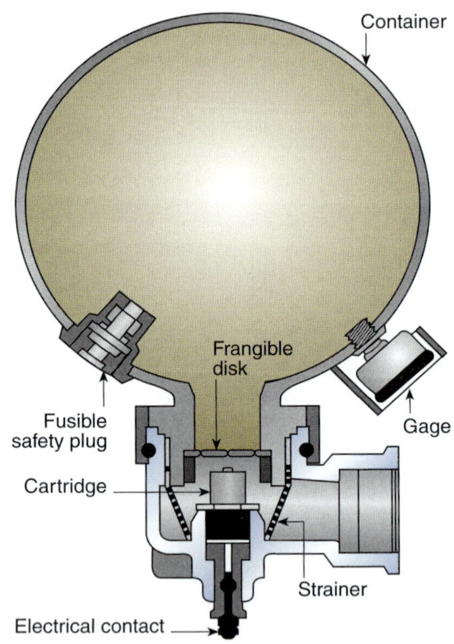

Figure 18-11. *Typical HRD fire extinguisher container for protecting a turbine engine compartment*

bottle has been discharged because of an overheat condition. If the yellow indicator disk blows out, it shows that the bottle has been discharged by normal operation of the system. The gage shows the pressure of the agent and the gas in the container.

Complete Fire Protection System

APU (auxiliary power unit). A self-contained motor-generator and air compressor installed in an aircraft to generate power for ground operation and to start the main engines.

A complete fire protection system incorporates both the detection and the extinguishing systems. Here, we will consider the fire protection system in a typical three-engine jet transport aircraft.

Each of the three engines has two continuous-loop fire detector sensors, one mounted on the firewall, and the other mounted on the engine itself. The wheel wells also have sensors that detect their fire or overheat condition, and the APU has a fire-detector loop and a complete fire-extinguishing system. When any of the engine sensors detect a fire, red fire-warning lights in the fire-pull handles on the captain's and first officer's glare shield illuminate and sound a bell. The bell is silenced when the bell cutout switch is depressed.

If a fire occurs, the light in the fire-pull handle for that engine compartment illuminates, and a bell sounds. The pilot or first officer then pulls the handle. This does six things:

- Closes the engine fuel shutoff valve
- Trips the generator field relay after a delay of 5 to 10 seconds
- Closes the engine bleed-air valve
- Closes the wing or cowl anti-ice valve as appropriate
- Closes the hydraulic supply shutoff valve
- Turns off the hydraulic pump low-pressure warning lights

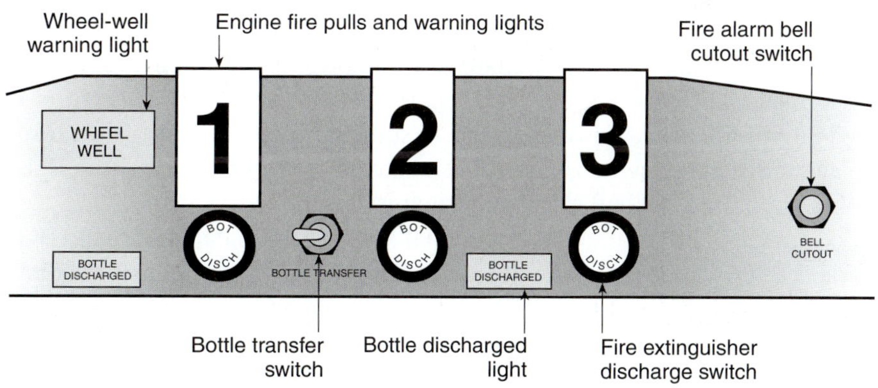

Figure 18-12. *Typical fire warning and fire extinguishing controls on the cockpit glare shield*

If the fire-warning light does not go out, the bottle discharge button is depressed and the selected bottle is discharged. The bottle discharged light then comes on. If the fire-warning light remains illuminated, indicating that the fire has not been extinguished, the bottle transfer switch is changed to select the other bottle and the bottle discharge button again depressed. This will discharge the other bottle, and its bottle discharged light will come on.

Figure 18-13 is a schematic diagram of a typical HRD fire extinguisher system installed in a three-engine jet transport airplane. There are two extinguisher bottles connected through check valves to a manifold to which the three engines are connected. In this diagram the bottle discharge switch for engine 2 has been depressed. Current flows through the switch to ignite the squib to discharge the left bottle and open the solenoid valve for engine 2. When the bottle discharges, the bottle discharged light illuminates and the yellow discharge indicator disk blows out. To discharge the second bottle, the bottle transfer switch is shifted. This not only selects the other bottle, but it takes the current through a different circuit breaker.

The plumbing that carries the fire-extinguishing agent from the bottles to the engines is marked with brown color-coding tape that has a series of diamonds on it to aid technicians who are color-blind and for use in dim light.

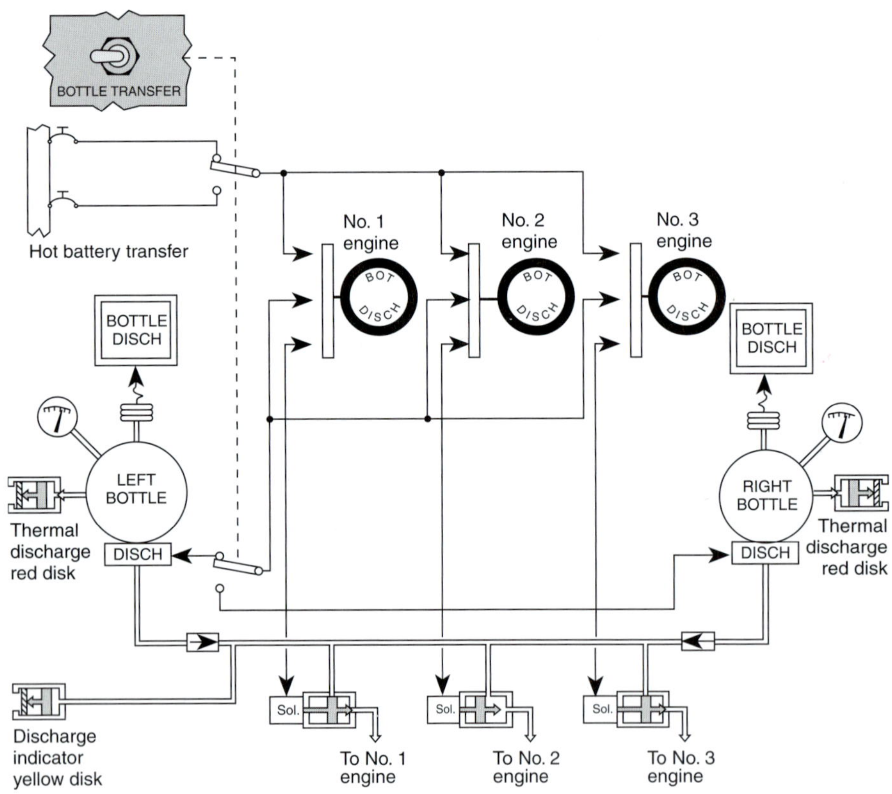

Figure 18-13. *Schematic of a three-engine jet transport fire-extinguishing system. The Bottle Discharge button for engine 2 has been depressed.*

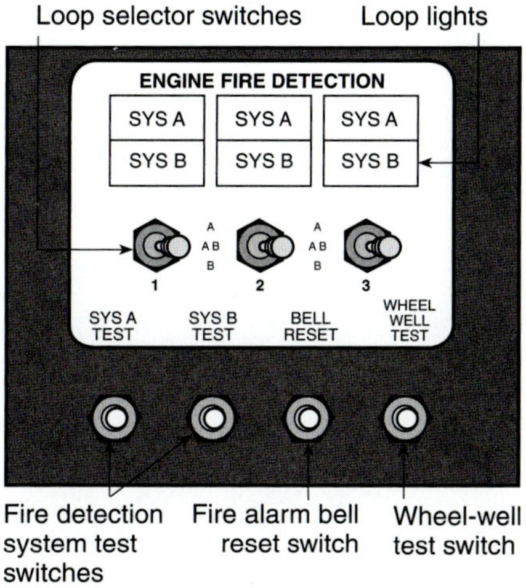

Figure 18-14. *Fire detection test light and switch panel on the flight engineer's panel*

The flight engineer has a panel with engine fire detection test lights and switches. The loop selector switches allow the flight engineer to select the loops separately or together. The loop lights above these switches illuminate to show when any of the loops is energized by a fire signal, a test signal, or a fault. The two fire detection system test switches check the continuity of the selected loop. The fire alarm bell reset switch silences the bell. The wheel-well test switch checks the continuity of the wheel-well detection loop.

Maintenance and Service of Fire-Detection Systems

The detector elements used in fire and overheat protection systems are precision devices that require special care and attention for their installation and servicing. About the only maintenance that can be done to a fire detection system is replacing damaged sensors and ensuring that all wiring is properly supported and in good condition.

The sensors used in the continuous-loop-type systems are particularly subject to damage from careless handling during routine engine maintenance. They should be carefully checked for dented, kinked, or crushed sections, as any damage of this type can cause a false alarm. When replacing continuous-loop sensor elements, be sure to follow the instructions in the aircraft service manual in detail. Support the elements as shown in the service manual and be sure to maintain the required clearance between the elements and the aircraft structure.

The location of all components in a fire detection system has been specially chosen by the engineers of the aircraft manufacturer, and all components must be maintained in the exact location specified. Detectors actuate at different temperatures, and it is especially important that detectors with the correct part number be used when replacing damaged ones.

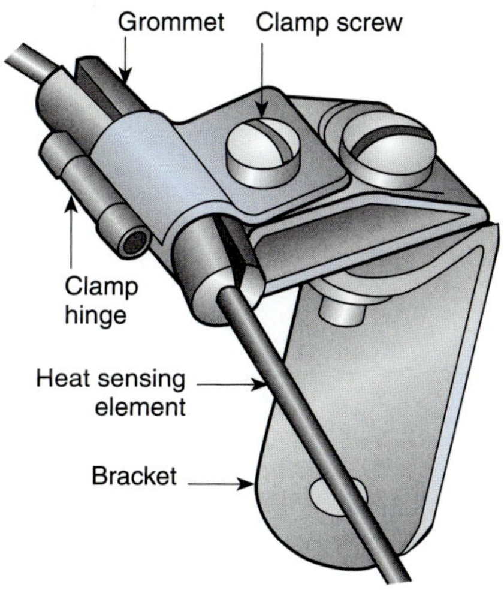

Figure 18-15. *Typical fire-detector loop clamp showing the correct way to install the grommet*

Some specific items to check are:

- Check for abrasion of the loop caused by the elements rubbing on the cowling, accessories, or structural members.
- Be sure there are no pieces of safety wire or metal particles that could short-circuit a spot detector.
- Check all the rubber grommets in the mounting clamps for indication of damage from oil or overheating, and be sure that all grommets are properly installed. If there is a bend in the element, the slit in the grommet should face the outside of the nearest bend to prevent the element chafing on the clamp. *See* Figure 18-15.

- Check for loose nuts or broken safety wire at the ends of the sensing elements. Follow the manufacturer's instructions regarding the torque to be used on the nuts, and the types of washers, if any, to be used. *See* Figure 18-16.
- When replacing a thermocouple sensor, be sure the wires are connected to the proper terminal of the sensor. This is important because the two elements of the thermocouple are made of different metals.

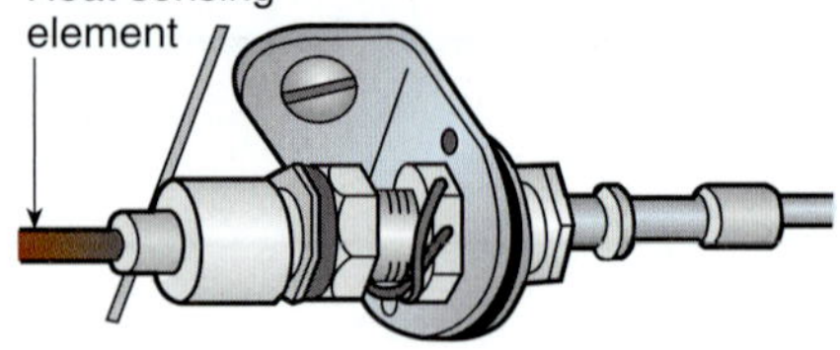

Figure 18-16. *Correct attachment of a fire-detector loop to the aircraft structure*

Maintenance and Service of Fire-Extinguishing Systems

It is extremely important that bottles of fire-extinguishing agent be kept fully charged. Most have gages mounted directly on them. The pressure of the agent varies with its temperature, and for this pressure to be meaningful, a correction must be made for the existing temperature. Figure 18-17 is a typical chart showing the allowable limits for the indicated pressure. If the pressure falls outside of the allowable range, the container must be removed and replaced with one that is properly charged.

The allowable pressure range for an agent temperature other than that in the chart can be found by interpolation. For example, the acceptable pressure range for 75°F is between 337.5 and 424 psig.

75°F is halfway between 70° and 80°.

337.5 psig is halfway between 319 and 356 psig.

424 psig is halfway between 405 and 443 psig.

The discharge cartridges for an HRD container are life-limited components, and the replacement date is measured from the date stamped on the cartridge.

Care must be taken when a cartridge is removed from the discharge valve that the correct one is reinstalled. The cartridges are not normally interchangeable between valves since the distance the contact point protrudes from the cartridge may vary from one cartridge to another. If the wrong cartridge is installed, there is a possibility that there will be no electrical continuity.

It is extremely important when checking the electrical connections to the container to follow the manufacturer's recommendations. This normally means to make sure that the current used to test the wiring is less than that required to detonate the squib.

Container Pressure versus Temperature		
Temperature (°F)	Container Pressure (PSIG)	
	Minimum	Maximum
-40	60	145
-30	83	165
-20	105	188
-10	125	210
0	145	230
10	167	252
20	188	275
30	209	295
40	230	317
50	255	342
60	284	370
70	319	405
80	356	443
90	395	483
100	438	523

Figure 18-17. *Fire extinguisher pressure/temperature chart*

STUDY QUESTIONS: POWERPLANT FIRE PROTECTION SYSTEMS

Answers are on Page 658. Page numbers refer to chapter text.

1. Most powerplant fire protection systems ____________________ (do or do not) automatically discharge the extinguishing agent when the fire is detected. *Page 639*

2. A fire that involves a combustible liquid as its fuel is a Class ______________ fire. *Page 639*

3. Two classes of detectors used in a powerplant fire protection system are:
   a. ____________________________
   b. ____________________________
   Page 641

4. A thermoswitch detector system _____________ (is or is not) an overheat detector. *Page 641*

5. Thermoswitches are connected in _____________________ (series or parallel). *Page 642*

6. If the wire between two single-loop thermoswitch detectors is broken, the system _____________________ (will or will not) detect a fire. *Page 641*

7. A thermocouple-type detector system ____________________ (does or does not) detect a general overheat condition in the engine compartment. *Page 643*

8. A high rate-of-temperature-rise triggers a fire warning by a ___________________ (thermoswitch or thermocouple) detection system. *Page 643*

9. The thermocouple sensors in a thermocouple fire detection system are connected in _____________________ (series or parallel). *Page 643*

10. In a thermocouple fire detection system, current in the thermocouple circuit flows through the coil of the _____________________ (sensitive or slave) relay. *Page 644*

11. The current that flows in a thermocouple circuit is proportional to the ______________________ between the various junctions. *Page 644*

12. Two types of continuous-loop fire and overheat detectors are:
    a. ______________________________
    b. ______________________________
    Page 645

13. The resistance of a thermistor material ____________________ (increases or decreases) as its temperature increases. *Page 645*

14. Pneumatic continuous-loop fire detection systems contain an element that ____________________ (absorbs or releases) gas when it gets hot. *Page 646*

15. The sensor element of a Systron-Donner pneumatic fire detection system is filled with helium gas. The pressure of the helium varies with its temperature, and it is used to close the pressure switch if a/an ____________________ (fire or overheat condition) occurs. *Page 646*

16. The center wire in a Systron-Donner pneumatic fire detector absorbs hydrogen and releases it to close the pressure switch if a/an ____________________ (fire or overheat condition) occurs. *Page 646*

17. If a break should occur in a Systron-Donner pneumatic fire detector, the ____________________ (helium or hydrogen) gas will escape, which opens the integrity switch. *Page 647*

18. The warning light in the Systron-Donner fire detection system will not illuminate if the integrity switch is ____________________ (open or closed). *Page 647*

19. Four fire-extinguishing agents that can be used for powerplant fires are:
    a. ________________________
    b. ________________________
    c. ________________________
    d. ________________________
    Page 648

20. One limitation of liquid nitrogen is that because it is a cryogenic liquid, it must be stored in a ____________________ bottle. *Page 648*

21. If an installed CO_2 fire-extinguishing system is discharged in the normal manner, the ____________________ (yellow or red) disk on the outside of the fuselage is blown out. *Page 649*

22. The agent in an HRD container is discharged when the sealing disk is ruptured by a cutter driven by a ____________________. *Page 649*

23. The Freon, or Halon, fire-extinguishing agent in an HRD bottle is propelled from the bottle by a charge of compressed ____________________. *Page 649*

Continued

STUDY QUESTIONS: POWERPLANT FIRE PROTECTION SYSTEMS *Continued*

24. If an HRD fire extinguisher bottle is discharged because of an overheat condition, the ______________________ (yellow or red) disk on the outside of the fuselage is blown out. *Page 649*

25. Halon 1301 __________________ (is or is not) corrosive to aluminum. *Page 648*

26. Six things that happen when the fire-pull handle in the cockpit is pulled are:
    a. __________________________________
    b. __________________________________
    c. __________________________________
    d. __________________________________
    e. __________________________________
    f. __________________________________
    Page 650

27. The fire-extinguishing agent ____________________ (is or is not) discharged when the fire-pull handle is pulled. *Page 651*

28. The tubing that carries fire-extinguishing agents is color coded with a stripe of ___________________ (what color) tape and a series of ___________________ (what symbol). *Page 651*

29. Maintenance of a fire detection system normally consists of _______________________ (repair or replacement) of damaged components. *Page 652*

30. When installing a grommet around a continuous-loop element, the slit in the grommet should face the _____________________ (inside or outside) of the nearest bend. *Page 652*

31. If the pressure gage on an HRD bottle shows that the container pressure is too low for the existing temperature, the technician must ___________________ (replace or refill) the container. *Page 653*

32. Fire extinguisher discharge cartridges ______________________ (are or are not) normally interchangeable between valves. *Page 653*

Answers to Chapter 18 Study Questions

1. do not
2. B
3. a. fire detectors
   b. overheat detectors
4. is not
5. parallel
6. will
7. does not
8. thermocouple
9. series
10. sensitive
11. temperature difference
12. a. thermistor
    b. pneumatic
13. decreases
14. releases
15. overheat condition
16. fire
17. helium
18. open
19. a. carbon dioxide
    b. liquid nitrogen
    c. Halon 1211
    d. Halon 1301
20. Dewar
21. yellow
22. ignited powder charge
23. nitrogen
24. red
25. is not
26. a. engine fuel shutoff valve closes
    b. generator field relay trips
    c. engine bleed air valve closes
    d. anti-ice valves close
    e. hydraulic supply shutoff valve closes
    f. hydraulic pump low-pressure warning light turns off
27. is not
28. brown, diamonds
29. replacement
30. outside
31. replace
32. are not

Propellers

19

Continued

Propellers for Reciprocating Engines *(Continued)*

PROPELLERS

19

Introduction to Aircraft Propellers

Aircraft propellers are a special type of component. They are a vital part of the aircraft powerplant and must be maintained by aviation maintenance technicians holding a powerplant rating.

The first sections of this chapter discuss propellers for reciprocating engines followed by those for turbine engines. Turbine engine propellers involve more than just the propeller itself, and those additional components are also discussed.

An airplane is propelled by thrust produced by its powerplant accelerating or increasing the momentum of a mass of air.

A turbojet engine takes in a relatively small mass of air and accelerates it by a large amount. An engine driving a propeller accelerates a much larger mass of air but by a smaller amount. Turbofan engines produce thrust in two ways: the core engine imparts a large acceleration to a small mass of air while the fan, acting in much the same way as a propeller, imparts a smaller acceleration to a much larger mass of air. The fan in a high-bypass ratio turbofan engine produces about 75% of the total thrust.

high-bypass ratio engine. A turbofan engine whose bypass ratio is 4:1 or greater.

bypass ratio. The ratio of the mass of air moved by the fan to the mass of air moved by the core engine.

Many of the earliest aeroplanes failed to fly because of the inability of their propellers to convert the power of the engine, as little as it was, into thrust. And one of the reasons the Wright *Flyer* succeeded in 1903 was the careful design of its propellers. The weak 12-horsepower engine drove two propellers that together produced 90 pounds of thrust, enough to fly the 750-pound airplane. The two $8\frac{1}{2}$-foot diameter laminated spruce propellers, turning at 330 RPM, had an efficiency of 66%, which was certainly commendable in view of the fact that the best efficiency of a modern propeller is only about 90%.

thrust. The aerodynamic force produced by a propeller or a turbojet or turbofan engine as it accelerates a mass of air.

The basic principle of propellers has changed very little since 1903, but technology has undergone many evolutionary advances in aerodynamics as well as materials and construction methods.

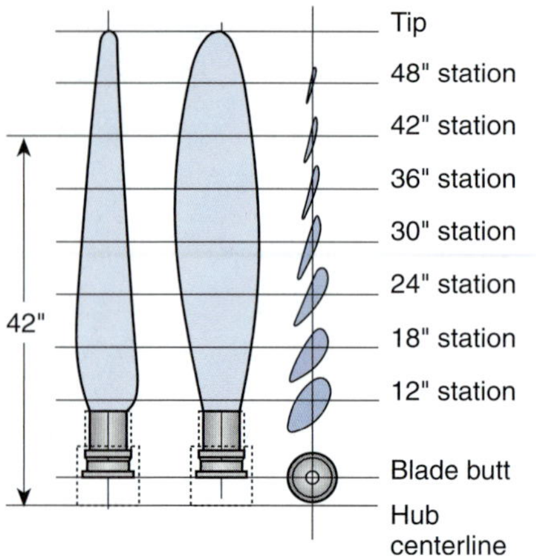

Figure 19-1. *Propeller blade elements*

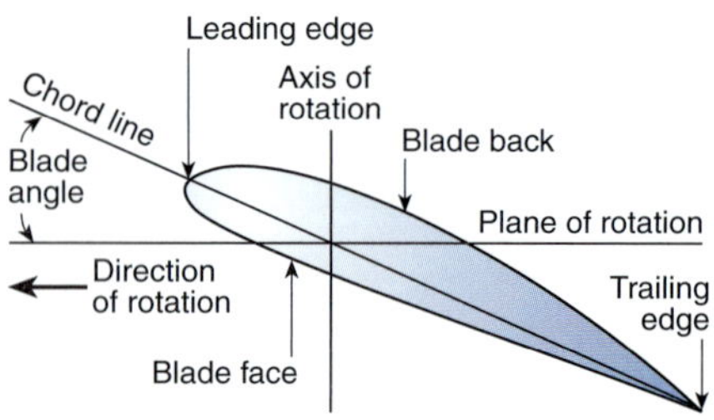

Figure 19-2. *Airfoil section of a propeller blade*

blade. The component of a propeller that converts the rotation of the propeller shaft into thrust.

blade, or pitch angle. The angle between the chord line of a propeller blade and the plane of rotation.

blade shank. The rounded portion of a propeller blade between the root and the airfoil section.

blade station. A reference position on a propeller blade that is a specified number of inches from the center of the propeller hub.

Propeller Theory

The thrust produced by a propeller blade is determined by five things: the shape and area of the airfoil section, the angle of attack, the density of the air, and the speed the airfoil is moving through the air.

There are two aspects of the overall theory that explain the operation of a propeller: the momentum theory and blade-element theory.

The momentum theory considers a propeller blade an airfoil that, when rotated by the engine, produces a pressure differential between its back and face which accelerates and deflects the air. The amount of thrust is determined by the change in the momentum of air passing through the propeller, multiplied by the area of the propeller disc.

The blade element theory considers a propeller blade to be made of an infinite number of airfoil sections, with each section located a specific distance from the axis of rotation of the propeller.

Each blade element travels at a different speed because of its distance from the center of the hub, and to prevent the thrust from increasing along the length of the blade as its speed increases, the cross-sectional shape of the blade and its blade pitch, or blade angle, vary from a thick, high pitch angle near the low-speed shank to a thin, low pitch angle at the high-speed tip.

By using the blade element theory, a propeller designer can select the proper airfoil section and pitch angle to provide the optimum thrust distribution along the blade.

Propeller Pitch

Since the pitch angle of a propeller blade varies along its length, a particular blade station must be chosen to specify the pitch of a blade. This is normally done by specifying the angle and the blade station, as, for example, 14° at the 42-inch station.

Rather than using blade angles at a reference station, some propeller manufacturers express pitch in inches at 75% of the radius. This is the geometric pitch, or the distance this particular element would move forward in one revolution in a solid along a helix, or spiral, equal to its blade angle. The geometric pitch is found by the formula:

$$\text{Geometric pitch} = \text{Tan}^{-1} \text{ pitch angle} \cdot 2\pi \cdot r$$

Tan^{-1} pitch angle = the tangent of the pitch angle

2π = a constant, 6.28

r = radius of the blade element (blade station)

A propeller with a blade angle of 14° at the 42 inch station has a geometric pitch of 65.9 inches.

geometric pitch. The distance a propeller would advance in one revolution if it were turning in a solid.

$$\begin{aligned}\text{Geometric pitch} &= \text{Tan}^{-1}\ \text{pitch angle} \cdot 2\pi \cdot r\\ &= \text{Tan}^{-1}\ 14° \cdot 6.28 \cdot 42\\ &= 0.25 \cdot 6.28 \cdot 42\\ &= 65.9\ \text{inches}\end{aligned}$$

The effective pitch is the actual distance a propeller advances through the air in one revolution. This cannot be determined by the pitch angle alone because it is affected by the forward velocity of the airplane.

The difference between geometric and effective pitch is called slip.

Angle of Attack

Thrust produced by a propeller, in the same way as lift produced by a wing, is determined by the blade's angle of attack.

angle of attack. The acute angle between the chord line of a propeller blade and the relative wind.

Angle of attack relates to the blade pitch angle, but it is not a fixed angle. It varies with the forward speed of the airplane and the RPM of the engine.

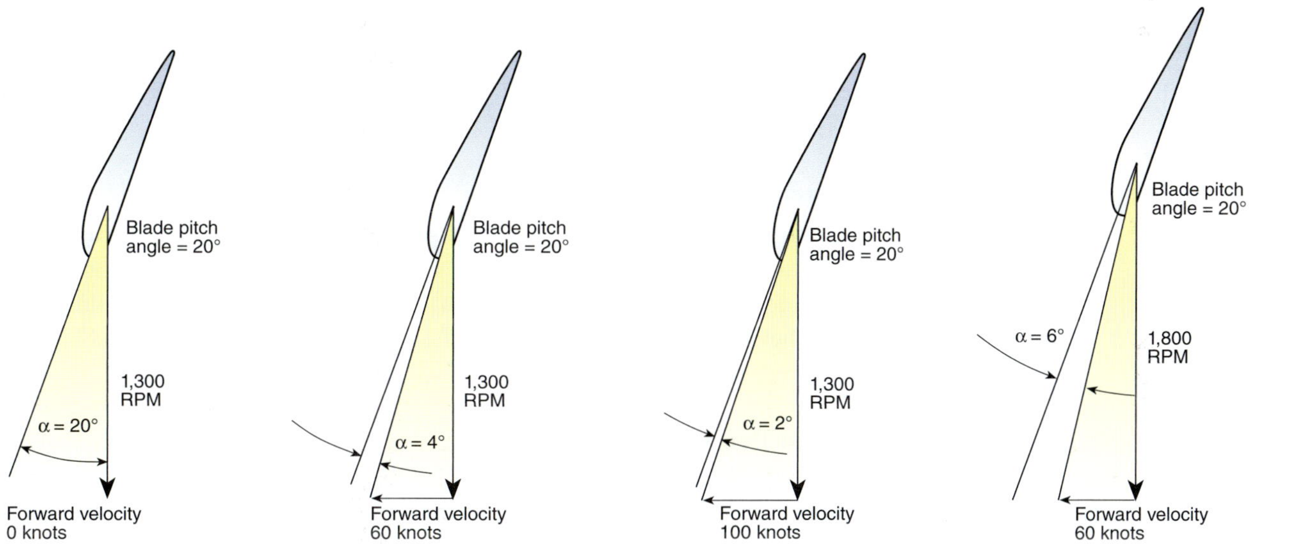

When there is no forward speed, angle of attack (α) and blade pitch angle are the same, 20°.

When the airplane is moving forward at 60 knots, angle of attack becomes much less than blade pitch angle.

Increasing the forward speed to 100 knots with the engine RPM remaining constant at 1,300 RPM decreases angle of attack.

Increasing the engine speed to 1,800 RPM with the forward speed remaining at 60 knots increases angle of attack.

Figure 19-3. *The angle of attack of a propeller blade varies with the forward speed of the airplane and the engine RPM.*

transonic range. Flight at Mach numbers between 0.8 and 1.2. In this range, some air passing over the aircraft is subsonic, and some is supersonic.

thrust horsepower. The actual amount of horsepower an engine-propeller combination transforms into thrust.

stress. A force within an object that tries to prevent an outside force from changing its shape.

Tip Speed

Normally, about half of the noise made by a propeller-driven airplane is made by the propeller itself. When the propeller blade tips approach the speed of sound, vibrations are produced that cause the noise. When the blades operate in the transonic range, they not only produce noise, but the drag becomes excessive and efficiency drops off dramatically.

For the propeller disk to be as large as possible while keeping the tips below the speed of sound, most high-powered reciprocating engines are geared so the propeller turns slower than the crankshaft.

Propeller Efficiency

The efficiency of a propeller is a measure of the ratio of the thrust horsepower to the torque horsepower. Thrust horsepower acts perpendicular to the plane of propeller rotation and is the amount of power the engine-propeller combination converts into thrust. Torque horsepower, on the other hand, acts in the plane of rotation and is the power used by the engine to rotate the propeller.

Forces Acting on Propellers

The propeller is one of the most highly stressed components in an airplane, and five basic forces act on a propeller turning at a high speed. These are centrifugal force, thrust bending force, torque bending force, aerodynamic twisting force, and centrifugal twisting force.

Centrifugal Force

Centrifugal force puts the greatest stress on a propeller as it tries to pull the blades out of the hub. It is not uncommon for the centrifugal force to be several thousand times the weight of the blade. For example, a 25-pound propeller blade turning at 2,700 RPM may exert a force of 50 tons (100,000 pounds) on the blade root.

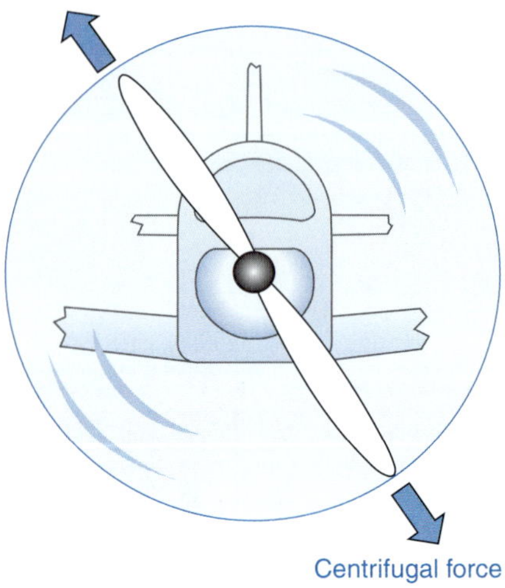

Figure 19-4. *Centrifugal force*

Thrust Bending Force

Thrust bending force is caused by the aerodynamic lift produced by the air-foil shape of the blade as it moves through the air. It tries to bend the blade forward and the force is at its greatest near the tip. Centrifugal force, trying to pull the blade out straight, opposes some of the thrust bending force. The mass of a propeller blade, distributed along its length, causes it to move forward gradually forming a forward coning shape from the blade root to its tip. This design feature reduces the stress and the resulting metal work-hardening effect placed on the blade-to-hub attach area by the thrust bending force. This forward coning aids the centrifugal force in keeping the blades straight as thrust and RPM increase.

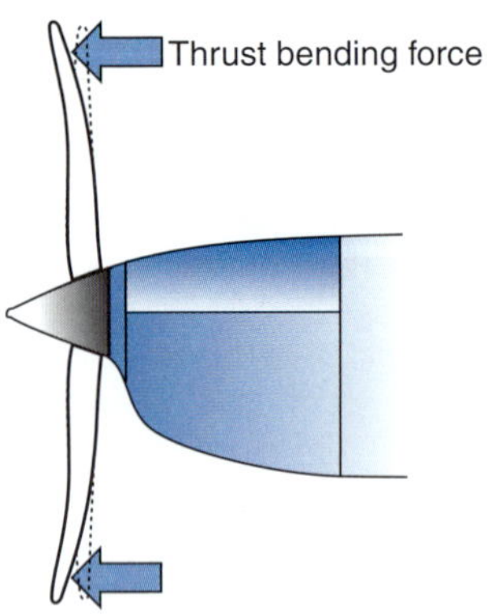

Figure 19-5. *Thrust bending force*

Torque Bending Force

Torque bending force tries to bend a propeller blade in its plane of rotation opposite to the direction of rotation.

torque. A force that produces, or tries to produce, rotation.

Aerodynamic Twisting Force

Centrifugal force, thrust bending force, and torque bending force require a propeller to be strong and heavy, and they serve no useful function. But two twisting forces are useful in the pitch change mechanism of controllable pitch propellers.

Aerodynamic twisting force (ATF) tries to increase the blade angle. The axis of rotation of a blade is near the center of its chord line, and the center of pressure is between the axis and the leading edge. Figure 19-7 shows how the aerodynamic force acting through the center of pressure ahead of the axis of rotation tries to rotate the blade to a higher pitch angle.

ATF (aerodynamic twisting force). The aerodynamic force that acts on a rotating propeller blade to increase its blade angle.

center of pressure. The point on the chord line of an airfoil where all of the aerodynamic forces are concentrated.

Centrifugal Twisting Force

Centrifugal twisting force (CTF) tries to decrease the blade angle. As the propeller turns, centrifugal force acts on all of the blade components and tries to force them to rotate in the same plane as the blade's axis of rotation. This rotates the blade to a lower-pitch angle. CTF opposes ATF, but its effect is greater, and the net result of the twisting forces is a force that tries to move the blades to a lower pitch.

Many controllable-pitch propellers have counterweights on arms clamped around the blade shank to alter the effect of the twisting forces. Practical applications of these forces are discussed in the section on controllable-pitch propellers beginning on Page 675.

CTF (centrifugal twisting force). The force acting about the longitudinal axis of a propeller blade which tries to rotate the blade to a low-pitch angle.

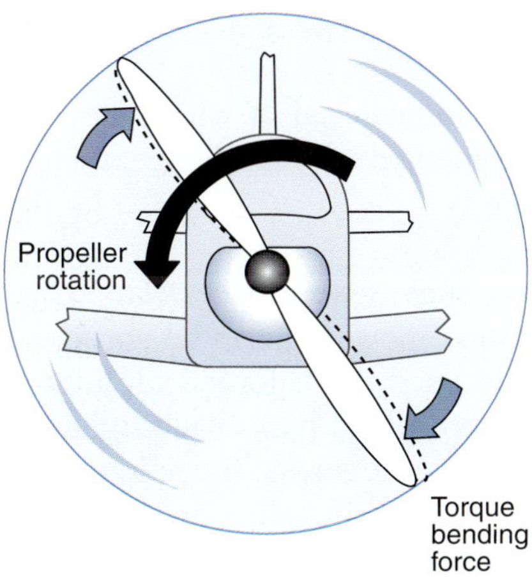

Figure 19-6. *Torque bending force*

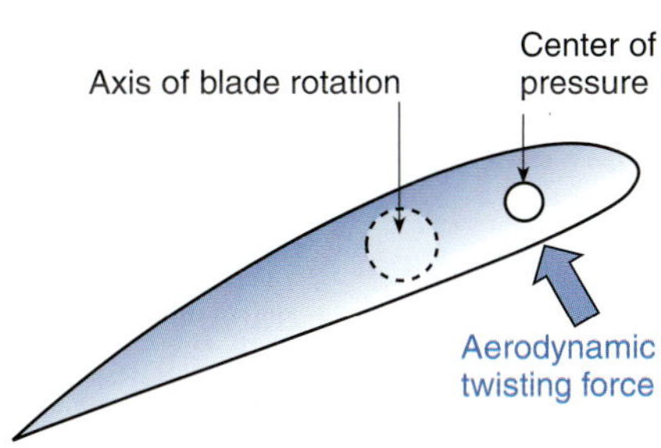

Figure 19-7. *Aerodynamic twisting force tries to increase the blade pitch angle.*

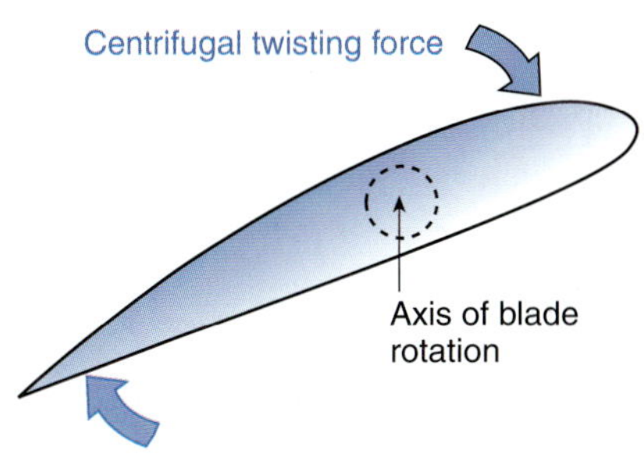

Figure 19-8. *Centrifugal twisting force tries to decrease the blade pitch angle.*

asymmetrical loading. The loading of a propeller disk in such a way that one side produces more thrust than the other side.

yaw. Rotation of an aircraft about its vertical axis.

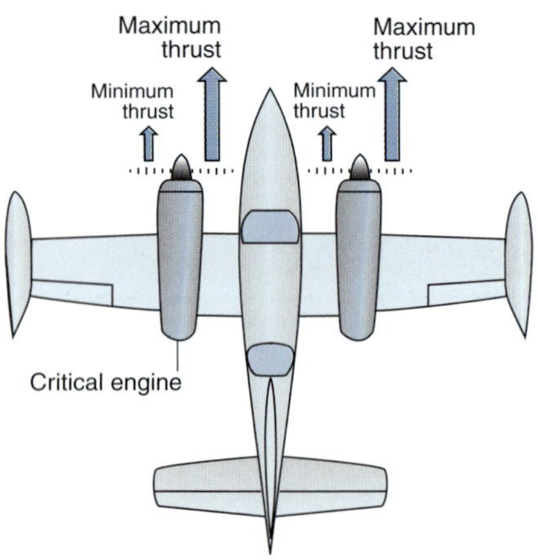

Figure 19-9. *The critical engine of a twin-engine airplane is the one that, if lost, would cause the greatest yawing effect.*

critical engine. The engine of a twin-engine airplane whose loss would cause the greatest yawing effect.

tractor propeller. A propeller mounted on an airplane in such a way that its thrust pulls the aircraft.

pusher propeller. A propeller installed on an aircraft engine in such a way that it faces the rear of the aircraft. Thrust from the propeller pushes the aircraft rather than pulling it.

center-line thrust airplane. A twin-engine airplane with both engines mounted in the fuselage. One, as a tractor, and the other, as a pusher.

Asymmetrical Loading

Most aircraft engines built in the United States rotate in a clockwise direction as viewed from the accessory end. When an airplane is flying in such a way that the relative wind strikes the propeller disk at right angles, all of the blades have the same angle of attack and thrust is uniform throughout the disk. But when the airplane is flying with a nose-high attitude, the downward moving blade on the right side has a higher angle of attack than the upward moving blade on the left side. The propeller is loaded asymmetrically, and the airplane has a tendency to yaw to the left.

On a single-engine airplane, asymmetrical loading, or P-factor, as it is often called, is an interesting phenomenon. But on a twin-engine airplane, it can have serious consequences if the critical engine is lost during takeoff.

Figure 19-9 shows the thrust distribution of the propeller disks of a twin-engine airplane when operating with a nose-high attitude, as it would on takeoff. The maximum thrust is on the right side of both engines. This places the right engine's maximum thrust further from the center line of the airplane than the left engine's maximum thrust. The left engine is therefore the critical one, because if it were to fail with the airplane in a nose-high attitude, the asymmetrical loading of the right-engine propeller would have the greater yawing effect.

To minimize the danger of a critical engine, some twin-engine airplanes have the right engine rotating counterclockwise, as viewed from the accessory end, and the left engine rotating clockwise.

Another expedient for eliminating a critical engine is mounting the engines in tandem. One is installed as a tractor engine in front of the cabin, and the other as a pusher behind the cabin. A center-line thrust airplane does not have a yawing tendency when one engine is inoperative.

Classifications of Propellers

Propellers may be classified according to the material of which they are made, by the number of blades, and by the method of pitch change, if any.

Materials

For decades, propellers used on low-powered engines were made of laminated hardwood and had a fixed pitch. When more power had to be absorbed, propellers made of metal became widely used, with forged aluminum alloy being the most popular metal. Some propellers have blades made of steel with the blade halves stamped of thin sheet steel and brazed together along the leading and trailing edges. The blade shell is then installed over a tubular steel shank.

Some of the most modern blades are made of composite materials. These are discussed in the section on propellers for turbine engines beginning on Page 706.

Number of Blades

The number of blades has been an option for propeller engineers. The logical choice for fixed-pitch wood and forged-metal propellers is two blades, which have the advantage of ease of construction and balancing, low manufacturing cost, and efficient operation. In the mid 1930s a counterbalanced single-blade propeller was approved by the CAA (forerunner of the FAA), but it failed to gain popularity.

When more thrust is needed, the blade area can be increased by lengthening the blades, but only to a point at which tip speeds approach the speed of sound, or if tip clearance from the structure or ground is a factor. To keep the blades short, more blades can be used. Three- and four-blade fixed-pitch propellers have been constructed, but usually, propellers with more than two blades are made so their pitch can be adjusted. Some modern propellers have four, five, or six blades; and Propfan™ and Unducted Fan™ propellers have as many as 12.

Pitch Change Methods

An airplane with a fixed-pitch propeller is no more efficient than an automobile would be if it had only a single transmission gear. It was only when propellers with controllable pitch were introduced that truly efficient operation became possible.

The next section of this text examines adjustable-pitch, controllable-pitch, and constant-speed propellers.

STUDY QUESTIONS: INTRODUCTION TO AIRCRAFT PROPELLERS

Answers begin on Page 733. Page numbers refer to chapter text.

1. The fan in a high-bypass ratio turbofan engine produces about ______________ percent of the total thrust. *Page 663*

2. Two variables in the angle of attack of a propeller blade are:
   a. ______________________________
   b. ______________________________
   Page 665

3. The curved surface of a propeller blade is the __________________ (back or face). *Page 664*

4. The pitch angle and the angle of attack of a propeller blade __________________ (are or are not) the same. *Page 665*

5. Propeller blade stations are measured in inches from the __________________ (hub or tip). *Page 664*

6. The actual distance a propeller advances in one revolution in the air is the ______________________ (effective or geometric) pitch. *Page 665*

7. The difference between the geometric and effective pitch of a propeller is called _________________ . *Page 665*

8. The momentum theory states that the thrust produced by the propeller is determined by the change in the ________________ of the air passing through the propeller multiplied by the ______________ of the propeller disk. *Page 664*

9. Five factors that affect the amount of thrust produced by a propeller blade are:
   a. ______________________________
   b. ______________________________
   c. ______________________________
   d. ______________________________
   e. ______________________________
   Page 664

10. Two things that change on a propeller blade from the root to the tip are:
    a. ____________________________
    b. ____________________________
    Page 664

11. Increasing the engine RPM with the forward speed remaining constant ____________________ (increases or decreases) the angle of attack. *Page 665*

12. The efficiency of a propeller is a measure of the ratio of the _______________ horsepower to the _______________ horsepower. *Page 666*

13. Five forces acting on a propeller turning at a high speed are:
    a. __________________________________
    b. __________________________________
    c. __________________________________
    d. __________________________________
    e. __________________________________
    Page 666

14. The greatest force acting on a propeller is ____________________ force. *Page 666*

15. The force that tries to bend a propeller blade forward is the _______________________ bending force. *Page 666*

16. The force that tries to bend a propeller blade in its plane of rotation is the _______________________ bending force. *Page 667*

17. Aerodynamic twisting force (ATF) tries to ________________________ (increase or decrease) the blade angle. *Page 667*

18. Centrifugal twisting force (CTF) tries to ________________________ (increase or decrease) the blade angle. *Page 667*

19. The twisting force that has the greater effect on propeller operation is ________________________ (ATF or CTF). *Page 667*

20. The critical engine on a twin-engine airplane, when both engines rotate in a clockwise direction (as viewed from the accessory end), is the _______________________ (right or left) engine. *Page 668*

21. Two factors that limit the length of a propeller blade are the __________________ and ______________________ from the ground or structure. *Page 669*

Propellers for Reciprocating Engines

Reciprocating engines have powered so many different types of airplanes that a variety of propellers have had to be designed. Here we will discuss the most common types.

Fixed-Pitch Propellers

Its simplicity, light weight, and low cost have made fixed-pitch propellers the standard for the smallest civilian trainers and personal airplanes.

A fixed-pitch propeller is a compromise. If its pitch angle is low enough to allow the engine to develop maximum RPM for takeoff, it is inefficient in level flight. If the pitch angle is high enough for efficient cruise, the take-off RPM will be too low, and an overly long takeoff run will result.

Aircraft manufacturers normally approve more than one propeller for each aircraft. The standard propeller is designed to give the aircraft the best all-around performance, but if the aircraft is to be operated primarily from short fields, the owner may choose a climb propeller, which has a lower-pitch angle. If the aircraft is always operated from long runways and needs a high cruising speed, a cruise propeller with a higher pitch, may be chosen. Figure 19-10 shows a comparison between the geometric pitch of three propellers approved for a popular four-place private aircraft with a 180-horsepower engine.

Propeller Model	Blade Pitch (inches at 75% radius)
Standard	63
Climb	60
Cruise	65

Figure 19-10. *Blade pitch of three propellers approved for a popular four-place, 180-horsepower airplane*

Wood Propellers

Wood propellers have been used since the Wright *Flyer*'s first flight in 1903 and are still popular on many amateur-built airplanes.

Most wood propellers are made of laminations of carefully selected kiln-dried birch. The laminations are glued together and then machine carved to a rough shape and pitch angle. They are then finished by hand, and fabric is installed over the tip to reinforce it against splitting. A strip of thin brass, monel, or stainless steel is attached by flush-head screws over the leading edge and around the tip to protect the wood from damage caused by sand, small rocks, and erosion from rain. The center hole in the hub and the mounting bolt holes are carefully bored, and the propeller is varnished. When the varnish is fully cured, the propeller is balanced, both horizontally and vertically. Three small holes are drilled with a number 60 drill (0.0400-inch diameter) about $^{3}/_{16}$ inch deep into each blade tip to release moisture and allow the wood to breathe. *See* Figure 19-11.

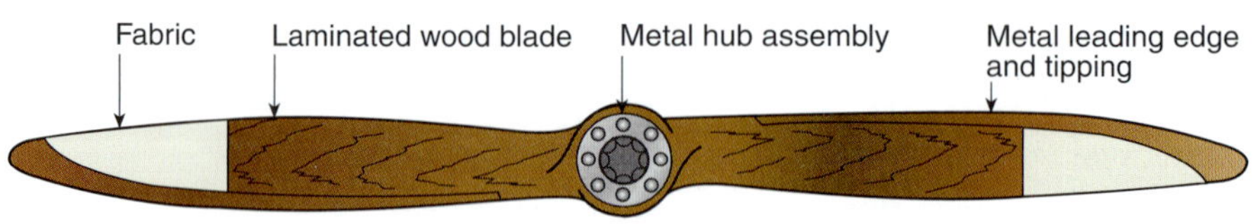

Figure 19-11. *Typical fixed-pitch wood propeller for a low-horsepower engine*

Metal Propellers

Improvements in metallurgy and manufacturing techniques have enabled metal propellers to replace wood propellers for modern commercially manufactured airplanes.

Metal propellers are forged from high-strength aluminum alloy, and after being ground to their finished dimensions and pitch, are anodized to protect them from corrosion. Metal propellers cost more than wood for the same engine and airplane, but their increased durability, resistance to weathering, and ability to be straightened after minor damage have made them more cost effective in the long run.

anodize. A hard, airtight, unbroken oxide film electrolytically deposited on an aluminum alloy surface to protect it from corrosion.

blade root. The ridges or upset portion of a propeller blade that holds it in the hub.

Ground-Adjustable Propellers

To allow a single model of propeller to fit a wide variety of applications, manufacturers made a ground-adjustable propeller that was the most widely used metal propeller throughout the 1930s and during World War II for some of the primary trainers. These propellers have a split, high-strength steel hub held together with two clamps. The blades have upset roots that hold them in the hub against the pull of centrifugal force. *See* Figure 19-12.

The pitch is adjusted by loosening the clamps and rotating the blade in the hub to the correct angle, as measured at the specified blade station with a universal propeller protractor. When both blades are adjusted to the correct angle, the clamps are tightened and safetied. The universal propeller protractor like the one in Figure 19-13 is a precision instrument used to measure the

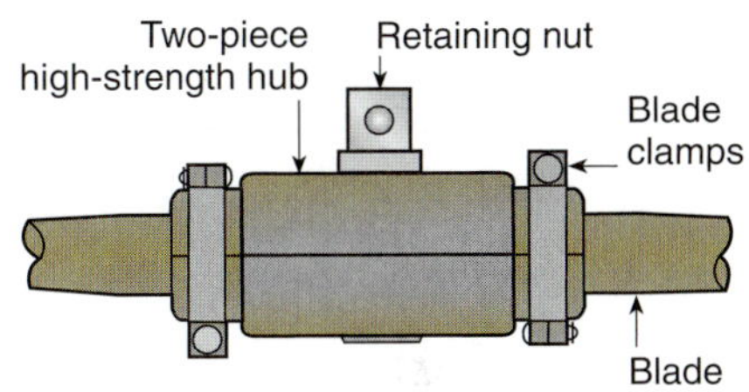

Figure 19-12. *The blades are held tightly in the hub of a ground-adjustable propeller with steel clamps around the ends of the hub.*

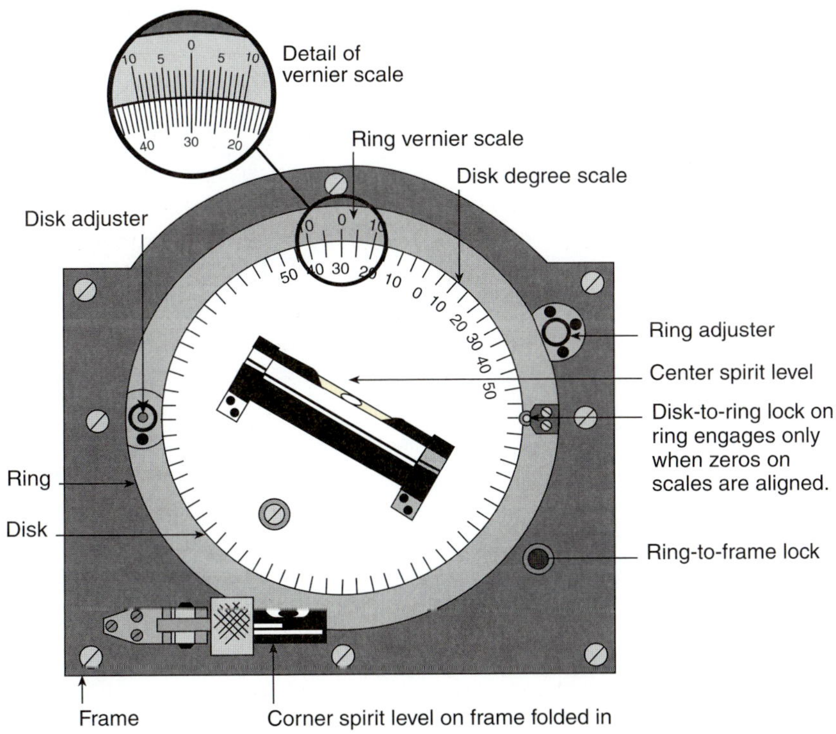

Figure 19-13. *A universal propeller protractor is used to check blade angle.*

propeller pitch, or blade angle, in degrees and tenths of a degree. It is also useful in measuring control surface deflection.

1. Place the propeller blade in a horizontal position and mark the blade station at which the pitch is to be measured. This station is specified in the airplane Type Certificate Data Sheets.
2. Set the disk-to-ring scales so the zeros are lined up and engage the disk-to-ring lock.
3. Hold one of the edges of the protractor frame against a flat area of the propeller hub that is perpendicular to the propeller shaft. Unfold the corner spirit level and hold the protractor so its bubble is centered.
4. Turn the ring adjuster until the bubble in the center spirit level rides in the center of its tube. Lock the outer indexing ring with the ring-to-frame friction lock.
5. Unlock the disk-to-ring lock and hold the edge of the protractor against the face of the blade where the pitch is to be measured and rotate the disk adjuster until the bubble in the center spirit level moves to the center of its tube. *Note:* The protractor must face the same direction as it did in step number four and it must be in the same basic rotational position in order to read the blade angle value directly.
6. Read the degrees of blade pitch angle on the disk degree scale opposite the zero (0) mark on the ring vernier scale.
7. Find the tenths of a degree by noting which line on the vernier scale lines up exactly with a line on the disk scale. Only one line on the vernier scale will line up with a line on the disk scale. *Note:* There are two vernier scales. Use the one that progresses past the whole degree zero index mark.

Controllable-Pitch Propellers

Ground-adjustable propellers were a step in the right direction, but with only minor added weight and complexity, the propeller could be made far more efficient by allowing the pilot to change the pitch of the blades in flight.

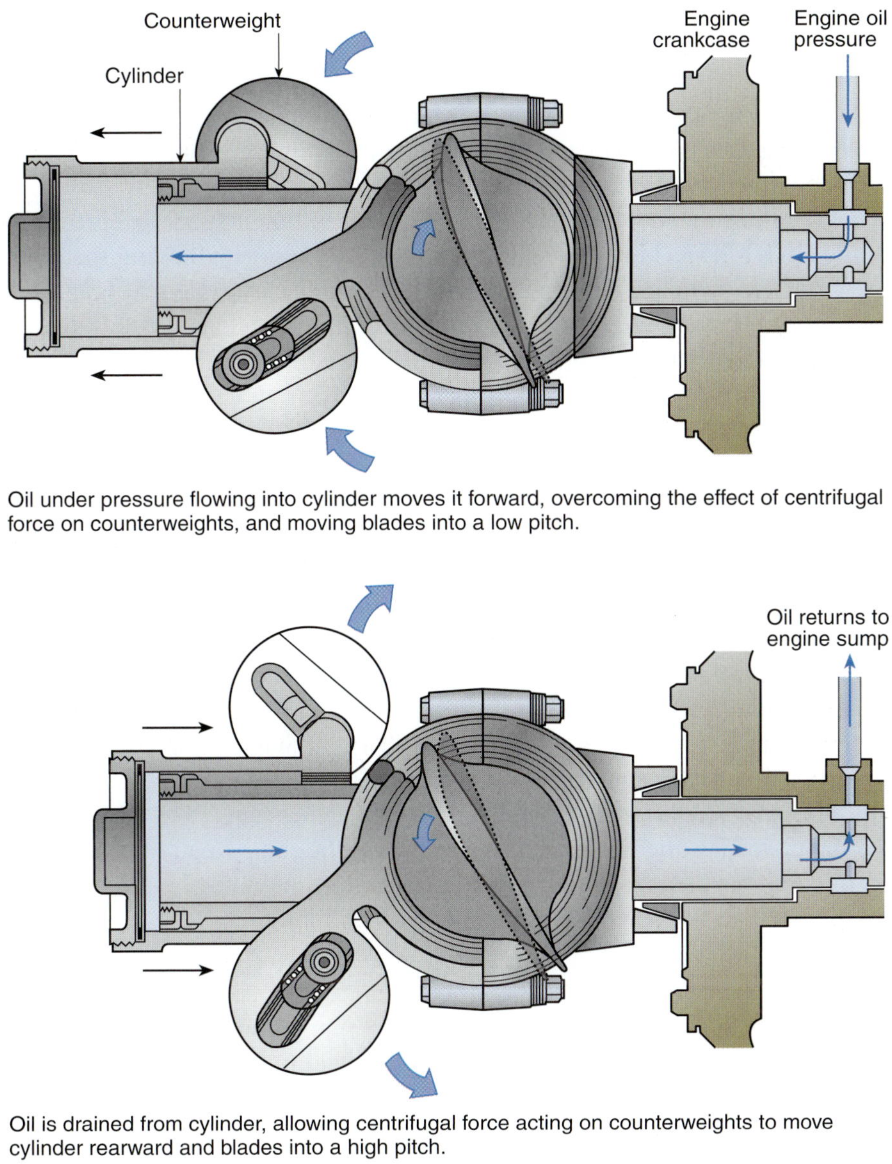

Figure 19-14. *A Hamilton Standard two-position or constant-speed, counterweight propeller*

Two-Position Propellers

The first popular controllable-pitch propellers were hydraulically actuated by engine lubricating oil supplied through a hollow crankshaft. A counterweight on an arm is attached to each blade root so that centrifugal force rotates the blade into a high pitch angle. A fixed piston in the end of the propeller shaft is covered by a movable cylinder attached through bearings to the counterweight arms. *See* Figure 19-14.

For takeoff, the two-position propeller pitch control is moved to the LOW PITCH position that directs engine oil into the cylinder and moves it forward over the piston. This pulls the counterweights in and rotates the blades into their low pitch position.

When the airplane is set up for cruise flight, the pitch control is moved to the HIGH PITCH position. This opens an oil passage, allowing the oil in the propeller cylinder to drain back into the engine sump. Centrifugal force on the counterweights moves them outward into the plane of rotation, and rotates the blades into their high pitch position.

This same configuration of propeller, when equipped with a flyweight governor to control the oil into and out of the cylinder, is a popular constant-speed propeller.

Automatic Propellers

Aeromatic propeller. A patented variable-pitch propeller that automatically maintains a relatively constant RPM for any throttle setting.

At the end of World War II there was a tremendous boom in private airplane, engine, and propeller development and manufacture. One interesting development of that era that became popular but faded away, because its complexities were greater than its advantages, was the Koppers Aeromatic propeller. This propeller was fully automatic and used the balance between the aerodynamic twisting force and the centrifugal twisting force to maintain a relatively constant speed for any given throttle setting.

The two forces were amplified by offsetting the blades from the hub with a pronounced lag angle to increase the effect of the centrifugal twisting force trying to move the blades into a low pitch, and by installing counterweights on the blade roots to help move the blades into high pitch.

Constant-Speed Propellers

governor. A control used to automatically change the pitch of a constant speed propeller to maintain a constant engine RPM as air loads vary in flight.

The tremendous advantage of being able to change pitch in flight opened new possibilities for increased efficiency. Replacing the two-position valve with a flyweight-controlled valve in a governor allows the blade pitch angle to be continuously and automatically adjusted in flight to maintain a constant and efficient engine speed.

Throughout and immediately after World War II an electrically controlled constant-speed propeller was used with some degree of success. A small reversible DC motor mounted in the center of the propeller hub drove a speed reducer with a bevel gear attached. This gear meshed with bevel gears on the root of each blade to change the pitch so the propeller could maintain an air load on the engine that produced the RPM called for by the governor. The large amount of maintenance required for electric propellers caused their demise.

There are only two types of propellers installed on current production airplanes; fixed-pitch propellers for the small and simple airplanes, and hydraulically actuated constant-speed propellers for complex airplanes.

The blades of constant-speed propellers are held in high-strength steel hubs, with roller- or ball-type thrust bearings that enable their pitch angle to be changed with a minimum of force.

Principles of Operation

There are two basic types of constant-speed propellers: counterweight and noncounterweight.

Counterweight propellers have a weight clamped to the blade roots to help move the blades into high pitch. Oil pressure moves the blades, against the force of the counterweights, into low pitch. *See* Figure 19-14 on Page 675.

The pitch of a noncounterweight propeller is controlled by a combination of oil pressure and aerodynamic twisting force to increase the pitch, and centrifugal twisting force and the force of an internal spring to decrease the pitch.

A flyweight-type governor, such as that described beginning on Page 680, senses the engine speed and compares it with the speed selected by the pilot. If an air load on the propeller causes it to slow down, the governor senses this RPM decrease and directs oil into or out of the propeller to decrease the blade pitch. The lowered pitch decreases the load, and the engine returns to the desired speed. If the air load decreases, the RPM increases; the governor senses the increase and directs the oil in the proper direction to increase the pitch and slow the engine.

Counterweight Propellers

One popular type of modern constant-speed propeller is the Hartzell steel hub propeller, which has a counterweight clamped tightly around each blade root, positioned so that as centrifugal force tries to move it into the plane of rotation, it increases the blade pitch angle. These are available as both nonfeathering and full-feathering propellers. *See* Figure 19-15.

When the governor senses that the RPM is lower than that selected, engine oil, boosted in pressure by a pump inside the governor, is sent through the hollow propeller shaft into the propeller cylinder, forcing the piston forward. Pitch-change push rods connecting the piston to a pitch-change block on the counterweight clamp, rotate the blades to a lower-pitch angle, and the engine speeds up to the desired RPM.

Note: In this propeller, the fixed component is called the cylinder, and the movable component which fits around the outside of the cylinder is called the piston.

When the engine is operating at exactly the RPM called for by the pilot, the governor closes the passage between the engine and the propeller. This prevents oil from going to or draining from the propeller.

If the nose of the airplane momentarily drops, the air load decreases and the RPM increases. The governor opens a passage between the propeller shaft and the engine sump, and oil drains from the propeller. Centrifugal force acting on the counterweights moves the blades into a higher pitch and the engine slows down.

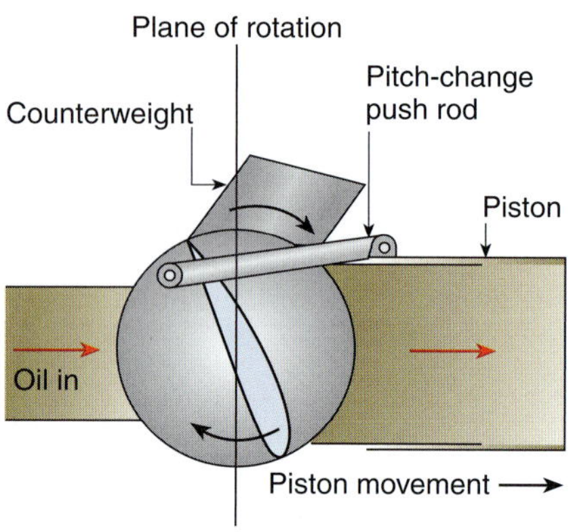

Figure 19-16. *When oil is directed into the propeller, the piston moves forward and the pitch-change push rods rotate the blades into a lower-pitch angle.*

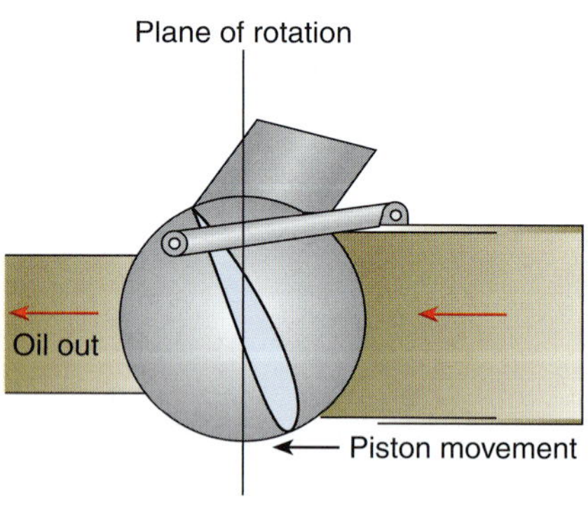

Figure 19-17. *When oil is drained from the propeller, the counterweights move toward the plane of rotation and increase the blade pitch angle.*

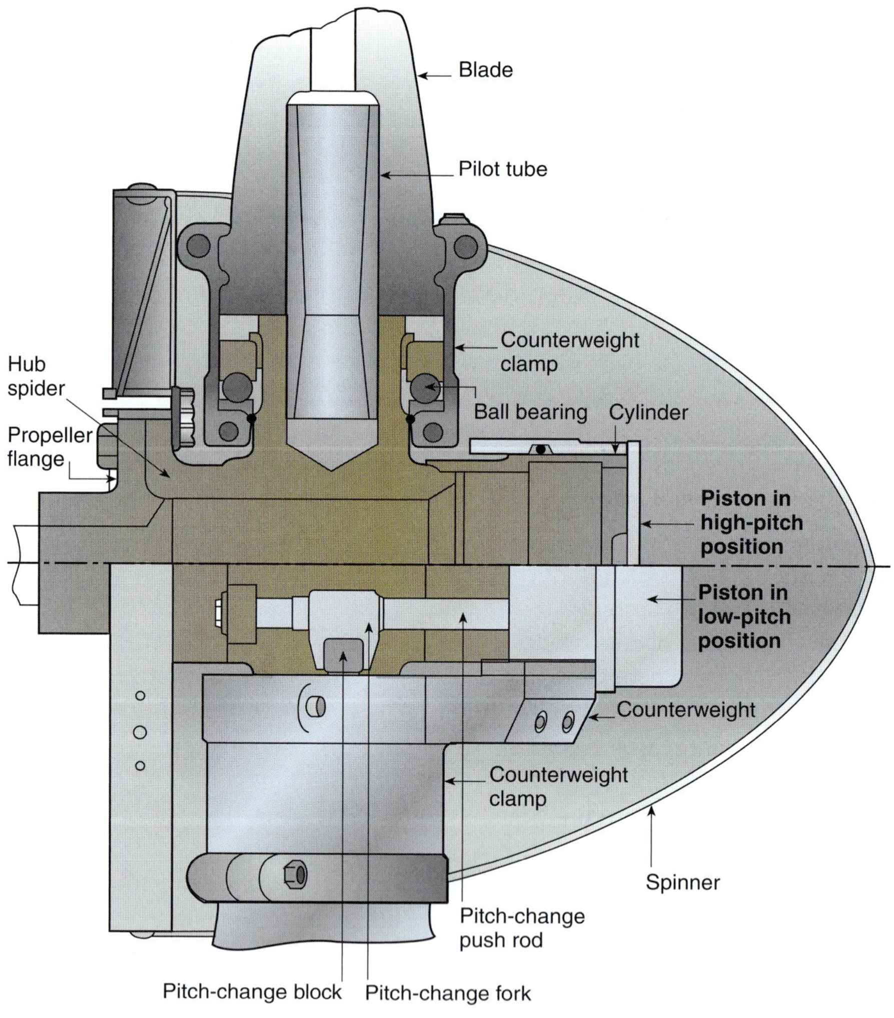

Figure 19-15. *Counterweight-type constant-speed propeller*

Noncounterweight Propellers

Some McCauley and Hartzell constant-speed propellers do not use counterweights, but rather control the pitch with a combination of oil pressure and aerodynamic twisting force to increase the pitch, and centrifugal twisting force, force from an internal spring, and/or gas charge to decrease the pitch.

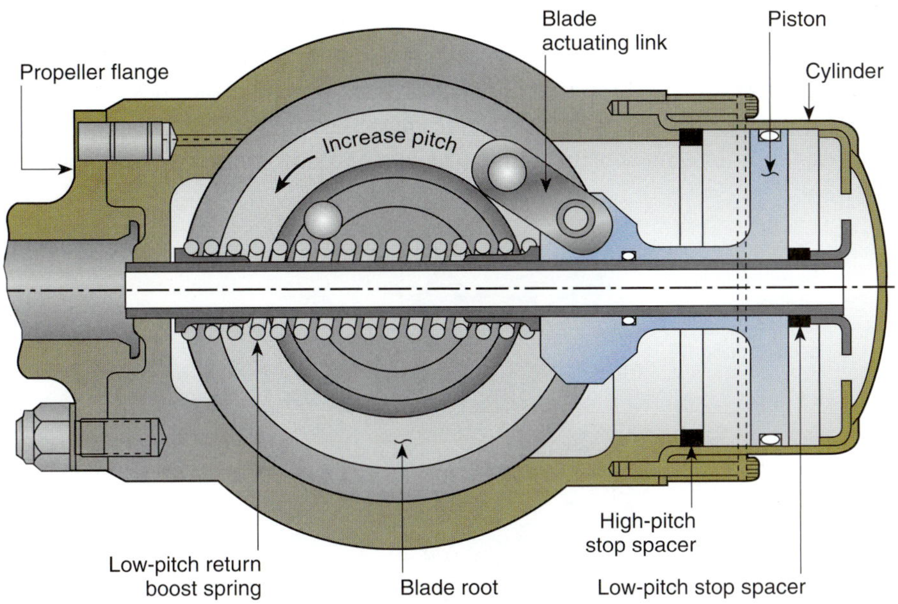

Figure 19-18. *Noncounterweight constant-speed propeller*

When the engine is operating in an underspeed condition, the governor drains the oil from the propeller, and the combination of centrifugal twisting force, the force of a spring, or gas pressure moves the piston forward and the blades to a low pitch angle.

When the air load is low and the engine tries to overspeed, the governor sends oil into the propeller and moves the piston back, overcoming the low angle force and moving the blades into a high pitch angle, which increases the air load and returns the engine to the desired RPM.

When the engine is operating in an on-speed condition, the governor blocks the oil going into the propeller or draining from it.

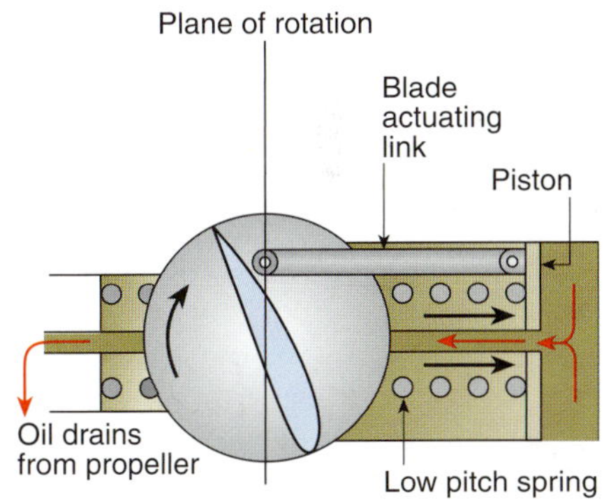

Figure 19-19. *When oil is drained from the propeller, the spring assists the centrifugal twisting force to move the piston forward. The blade actuating links rotate the blade to a lower pitch angle.*

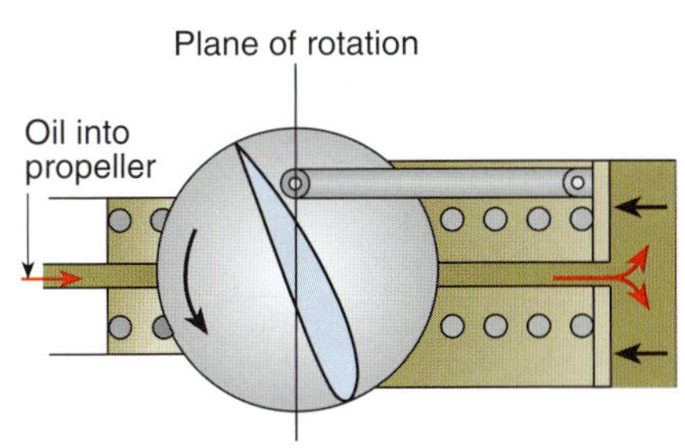

Figure 19-20. *When oil is directed into the propeller, the piston moves aft, and the blade actuating links rotate the blades into a higher pitch angle.*

Propeller Governor

Almost all propeller governors use a pair of L-shaped flyweights, mounted on a flyweight head and driven by the engine, to control the position of the pilot valve in the oil passage between the engine and propeller. A gear-type pump inside the governor boosts engine oil pressure high enough for it to move the propeller piston against the effect of the counterweights or the low pitch spring.

The governor pump and the flyweight head are driven by an accessory gear in the engine. The speeder spring presses down on the toes of the flyweights and, in turn, on the pilot valve plunger. The governor control lever rotates the adjusting worm, which varies the compression of the speeder spring.

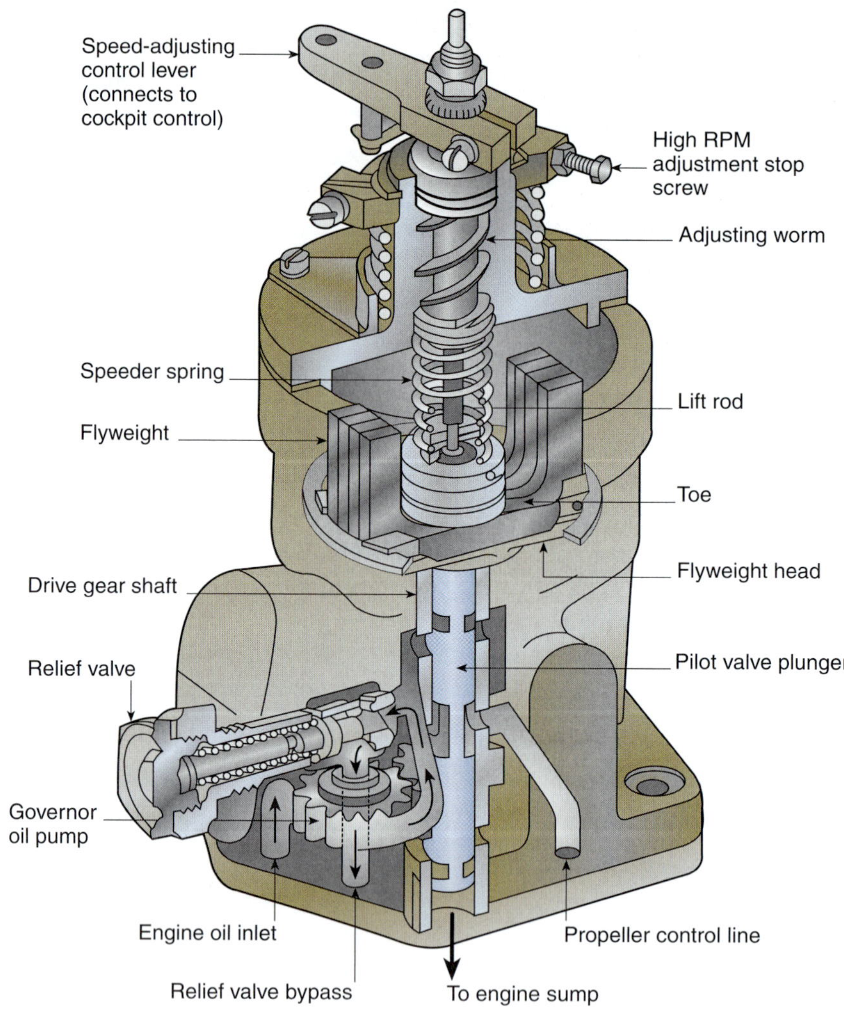

Figure 19-21. *Basic constant-speed propeller governor. The high RPM adjustment stop screw limits the compression that can be put on the speeder spring to limit the maximum engine speed during takeoff.*

First, consider the action of the governor installed with a counterweight propeller like the one in Figure 19-15 on Page 678.

When the engine is running, centrifugal force acting on the flyweights in the governor slings them outward, raising the pilot valve plunger. The centrifugal force is opposed by the speeder spring pressing down on the toes of the flyweights. When the engine is operating at the desired speed (the on-speed condition shown in Figure 19-22A) the centrifugal force slinging the flyweights outward is exactly balanced by the speeder spring pulling the flyweights in. As a result, the flyweights stand upright and all oil flow to and from the propeller is shut off.

If the nose of the airplane drops and the air load on the propeller decreases, the engine will speed up. This is an overspeed condition shown in Figure 19-22B. Centrifugal force acting on the flyweights overcomes the force from the speeder spring and lifts the pilot valve plunger, shutting off the passage to the governor pump and opening a passage between the propeller and the engine sump. Oil drains from the propeller cylinder, and centrifugal force acting on the blade counterweights increases the pitch. The engine slows down to its on-speed condition.

When the air load increases, the engine slows down to an underspeed condition as shown in Figure 19-22C. The force of the speeder spring overcomes the centrifugal force on the flyweights and lowers the pilot valve plunger. Oil from the governor pump flows into the propeller cylinder, moving the piston forward and decreasing the pitch, which returns the engine to its on-speed condition.

overspeed condition. A speed condition in which the engine is turning at an RPM higher than that for which the propeller governor is set.

on-speed condition. The speed condition in which the engine is turning at the RPM for which the propeller governor is set.

underspeed condition. A speed condition in which the engine is turning at an RPM lower than that for which the propeller governor is set.

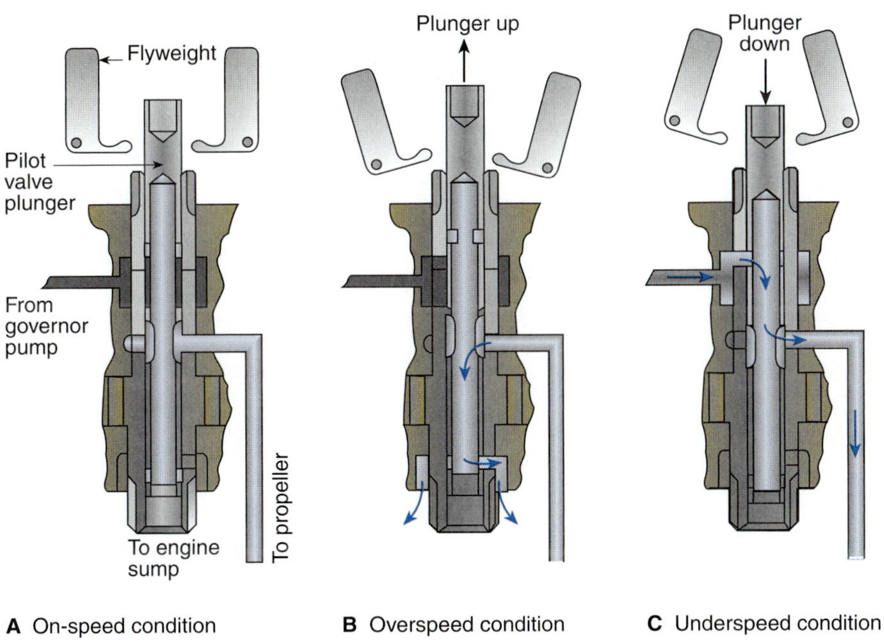

Figure 19-22. *Flyweights and pilot valve in a governor for a counterweight propeller*

When a noncounterweight propeller (*See* Figure 19-18 on Page 679) is operating in its on-speed condition, oil to and from the propeller is blocked by the pilot valve plunger in the governor, as seen in Figure 19-23A.

When the engine is operating in an overspeed condition, the flyweights swing outward and raise the pilot valve plunger. The passage between the propeller and the governor pump opens, allowing governor oil to flow into the propeller to increase its pitch and return the engine to the on-speed condition (Figure 19-23B).

When the engine is in underspeed condition (Figure 19-23C) the speeder spring forces the flyweights in and the pilot valve plunger down. This shuts off the passage to the governor pump and opens the passage between the propeller and the engine sump. Oil drains from the propeller, and the centrifugal twisting force and the force from the low pitch return spring move the piston forward, decreasing the pitch of the blades and allowing the engine to return to its on-speed condition.

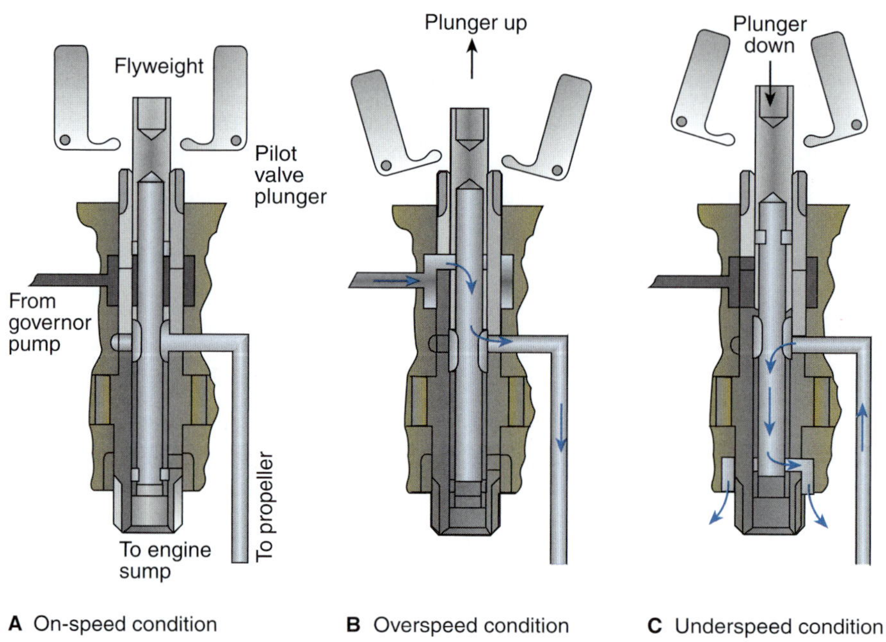

Figure 19-23. *Flyweights and pilot valve in a governor for a noncounterweight propeller*

Feathering Constant-Speed Propellers

Multiengine airplanes have not always been safer than their single-engine counterparts. When an engine failed in flight, before the advent of feathering propellers, aerodynamic forces caused the propeller to continue rotating and producing a tremendous amount of drag. Engine failure on takeoff often resulted in so much yaw that the pilot was unable to control the aircraft and it crashed.

With a feathering propeller, the pilot can press a feather button or move the pitch control to the feather position, and the blades will rotate from low pitch through their high pitch position into an extremely high-pitch angle, normally around 90°. At this angle, there is no aerodynamic force trying to rotate the propeller, and the blades produce the minimum amount of drag.

feathering propeller. A controllable-pitch propeller whose blades can be moved into a high pitch angle of approximately 90°.

Hamilton Standard Hydromatic Feathering Propeller

In the late 1930s, the Hamilton Standard Hydromatic propeller was developed, which gave multiengine airplanes a much needed safety factor. If an engine failed, the pilot could move the blades beyond their normal high pitch position to the feather position, which was normally between 88° and 92°. The blade met the oncoming air at an angle that produced no torque and a minimum of resistance. The propeller stopped turning and the pilot could continue flying on the other engine or engines. *See* Figure 19-24.

The Hydromatic propeller has without a doubt been used more than any other propeller in the history of aviation. It was used on most of the bombers, fighters, and transports during World War II, and it is still seen on large reciprocating engines. Its principle is studied as a classic example of propeller design at its best.

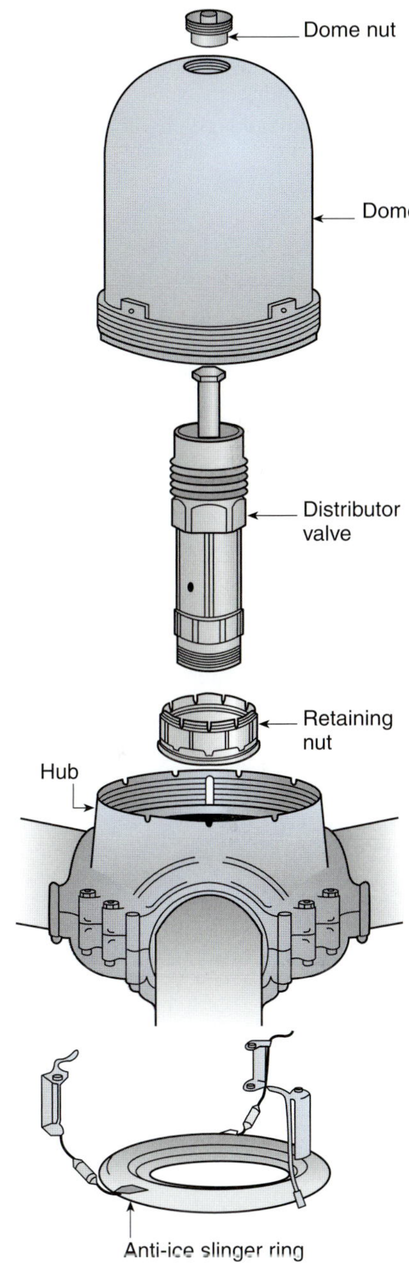

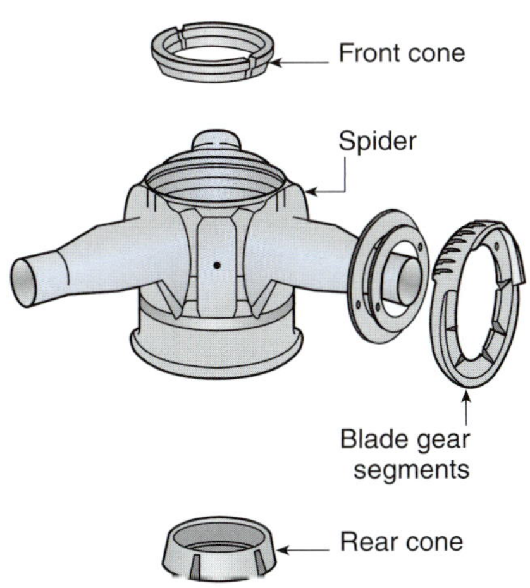

Figure 19-24. *Some basic components of a Hamilton Standard Hydromatic feathering propeller*

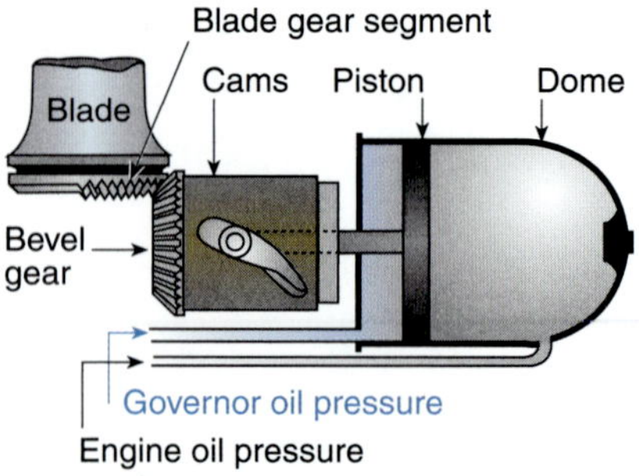

Propeller in unfeathered position

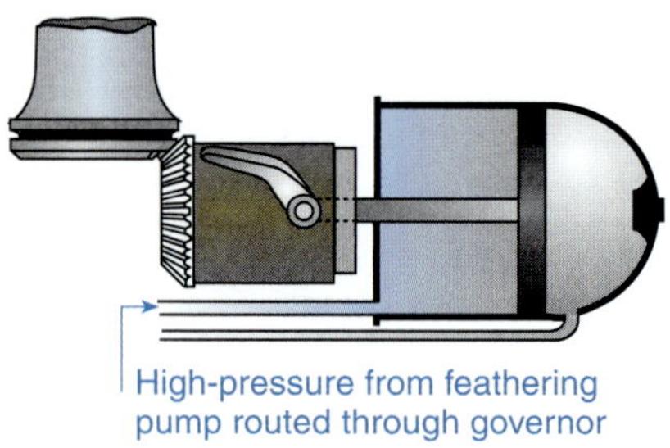

Propeller in feathered position

Figure 19-25. *Basic operating principle of a Hydromatic feathering propeller*

Note: When the cockpit control for a Hydromatic constant-speed propeller is actuated, the compression of the speeder spring is changed. The compression of the speeder spring opposes the centrifugal force on the flyweights in the governor.

To increase the RPM, the compression is increased. This increased spring force requires a higher RPM to overcome it and restore the flyweights to an onspeed condition.

The entire mechanism is enclosed in a sealed hub and dome with no external arms, linkages, or counterweights. Figure 19-25 shows the basic operating principle of this propeller.

The blades of a Hydromatic propeller are secured in a high-strength steel hub with roller-type thrust bearings. Torque from the engine is directed into the blades through arms that extend several inches into the blade butt and are part of a high-strength forged steel spider, splined to the engine propeller shaft. Each blade root is fitted with a segment of a bevel gear.

The dome screws into the propeller hub and houses the piston and two sets of concentric cams. A bevel gear on the inner cam meshes with the gear segments on the blade roots.

A single-acting governor is used with this propeller. In an underspeed condition, oil is sent under engine pump pressure into the dome on the forward side of the piston to aid centrifugal twisting moment (CTM) in moving the piston aft and rotating the cams assembly so that they move the blades into a low pitch angle, so the engine can speed up. Oil from the aft side of the piston drains into the engine sump through the governor.

In an overspeed condition, the passages in the governor are reversed, and engine oil, boosted in pressure by the governor pump, is directed to the aft side of the piston, and the oil from the forward side drains into the engine sump directly. The piston moves forward and rotates the cam to move the blades into a high pitch angle.

To feather a Hydromatic propeller, the pilot presses the feather button on the instrument panel. Once the button is depressed, the magnetic field from a holding coil holds it in that position. An electric-motor-driven feather pump picks up oil from the bottom of the engine oil tank and sends it through the governor, placing the governor into an artificial overspeed condition and thus directing the feathering oil to the back side of the piston. Notice in Figure 19-25 that there are two different slopes to the slot in the cam. The steep portion of the cam slot rotates the blades through their normal pitch range, but it requires the high-pressure oil from the feathering pump to move the piston forward enough for the rollers to move into the shallow angle of the slot. Once the cam is in this portion of the slot, the blades rapidly rotate into an extreme high pitch angle, near 90°. A high-pitch stop ring in the base of the dome is adjusted so the stop lug on the rotating cam contacts the stop on the ring at the proper feather angle. As soon as the lug contacts the stop, the piston can move no further and the oil pressure from the constant-displacement feathering pump builds up until it opens an oil-pressure-cutout switch inside the governor. This de-energizes the holding coil, and a spring pops the feather button out, shutting off the power to the feathering-pump motor.

All of the oil passing into and out of the dome flows through a distributor valve in the end of the propeller shaft. For normal operation and feathering, this valve acts only as an oil passage, but the passages in the valve shift to unfeather the propeller. To unfeather a Hydromatic propeller, the pilot depresses the feathering button. The cutout switch releases the feather button, but the pilot holds it down. The pressure continues to build up until it is high enough to shift the distributor valve and direct oil from the feather pump into the dome on the forward side of the piston to move it back into the normal operating range. When the blade pitch is low enough for the airstream to windmill the propeller, the pilot releases the feathering button, and the propeller resumes its normal operation.

windmilling propeller. A propeller that is rotated by air flowing over the blades rather than powered by the engine.

One of the major safety features of the Hydromatic propeller is the source of its feathering oil. The normal engine oil pump takes its oil from a standpipe or a fitting on the side of the oil tank, as seen in Chapter 3, Figure 3-17 on Page 117. If a major leak should cause all of the oil accessible to the engine oil pump to be pumped overboard, there is still enough oil below the standpipe for the feathering pump to completely feather the propeller.

Because of the large capacity oil cavities in this propeller, any time it has been newly installed on an engine, and as part of the preflight ground run, it must be "exercised" by operating it through its full range of travel a number of times. Exercising the propeller removes any air trapped inside the dome, supplies warm oil to the dome for responsive pitch change in flight, assures the pilot/mechanic of proper propeller operation, and returns excess oil to the engine sump.

Hartzell Steel Hub Feathering Propeller

The Hartzell steel hub feathering propeller in Figure 19-26 operates on the same basic principle as the nonfeathering counterweight propeller illustrated in Figure 19-15, Page 678. The additional features are the feathering spring and the centrifugally actuated high pitch stops.

***Note:** In this propeller, the fixed component is called the cylinder, and the movable component which fits around the outside of the cylinder is called the piston.*

When the engine is operating in its normal constant-speed range, the governor directs oil into the propeller cylinder to move the blades to a lower pitch angle to speed the engine up. To slow the engine down, it drains oil from the cylinder to allow the counterweights and the feathering spring to increase the pitch.

To feather the propeller, the pilot moves the governor control to its full aft position. This raises the lift rod in the governor, which raises the pilot valve plunger and opens the passage which drains the oil in the propeller cylinder back into the engine sump. Centrifugal force acting on the counterweights and the force from the feathering spring move the blades into their full-feathered position, a pitch angle of about 80° to 85°.

If the engine loses oil pressure, the counterweights and feather spring automatically move the propeller into its full-feather position. To prevent the propeller from feathering each time the engine is shut down on the ground, a high pitch stop plate on each blade root contacts a spring-loaded high pitch stop pin in the hub. This pin prevents the blades from moving into the full-feather position when the engine is not rotating. If oil pressure is lost in flight, the propeller will windmill fast enough that centrifugal force will hold the stop pin back against the force of the spring and allow the blades to move into their full-feather position.

To unfeather this propeller, move the governor control into its normal range, and start the engine. As soon as the engine oil pump begins to move oil through the engine and into the propeller cylinder, the piston moves forward and the blades move to a low pitch angle. Unfeathering happens at the same time the main and connecting rod bearings are critically loaded by the feathered propeller and are receiving a diminished oil flow because of the governor's demand on the lubricating system.

Some Hartzell feathering propellers use an accumulator to speed up the unfeathering. A normally open check valve in the governor allows the accumulator to be charged with governor-boosted oil pressure. When the propeller pitch control is moved to its full aft position, the check valve closes to trap the pressure in the accumulator. When the control is returned to its normal range to unfeather the propeller, the check valve opens, and the oil that has been stored under pressure in the accumulator flows into the cylinder to move the piston forward and hasten the unfeathering.

accumulator. A hydraulic component that allows a noncompressible fluid, such as oil, to be stored under pressure.

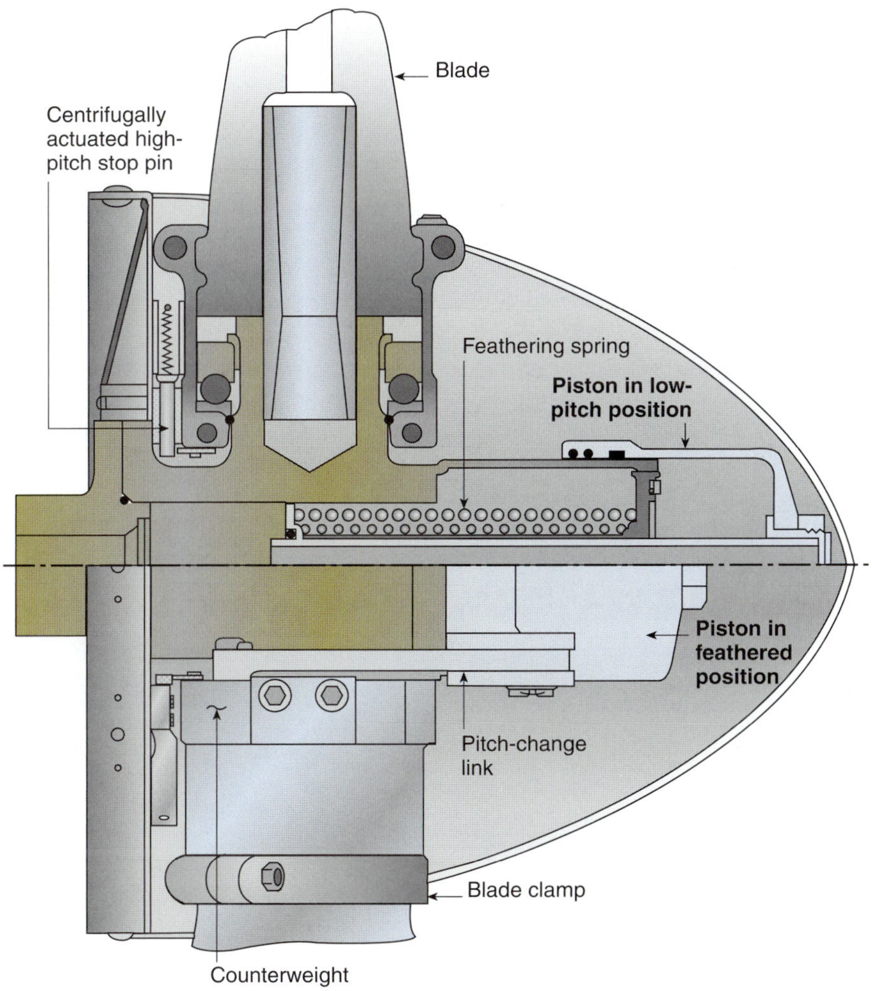

Figure 19-26. *Hartzell steel hub feathering propeller*

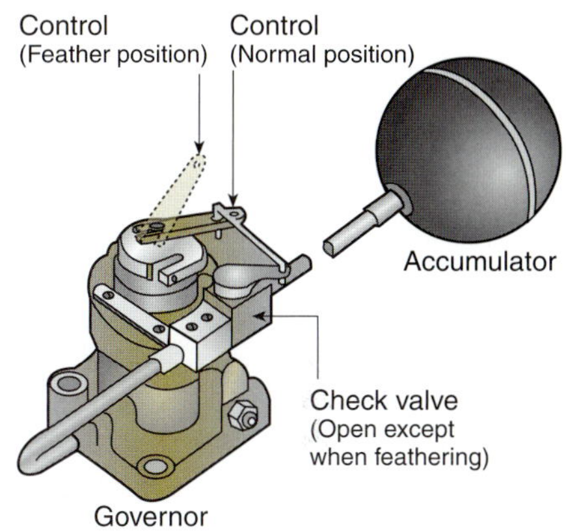

Figure 19-27. *An accumulator is used on some feathering propellers to hasten the unfeathering action.*

McCauley Constant-Speed Feathering Propeller

The McCauley constant-speed feathering propeller operates on a principle similar to that of the noncounterweight propeller described on Page 679. The difference is that these propellers use a counterweight and a spring to assist in feathering.

For normal operation, the governor sends engine oil, boosted in pressure by the internal governor oil pump, into the cylinder to move the piston rearward and the blades into low pitch. As the piston moves rearward, it compresses the feathering spring.

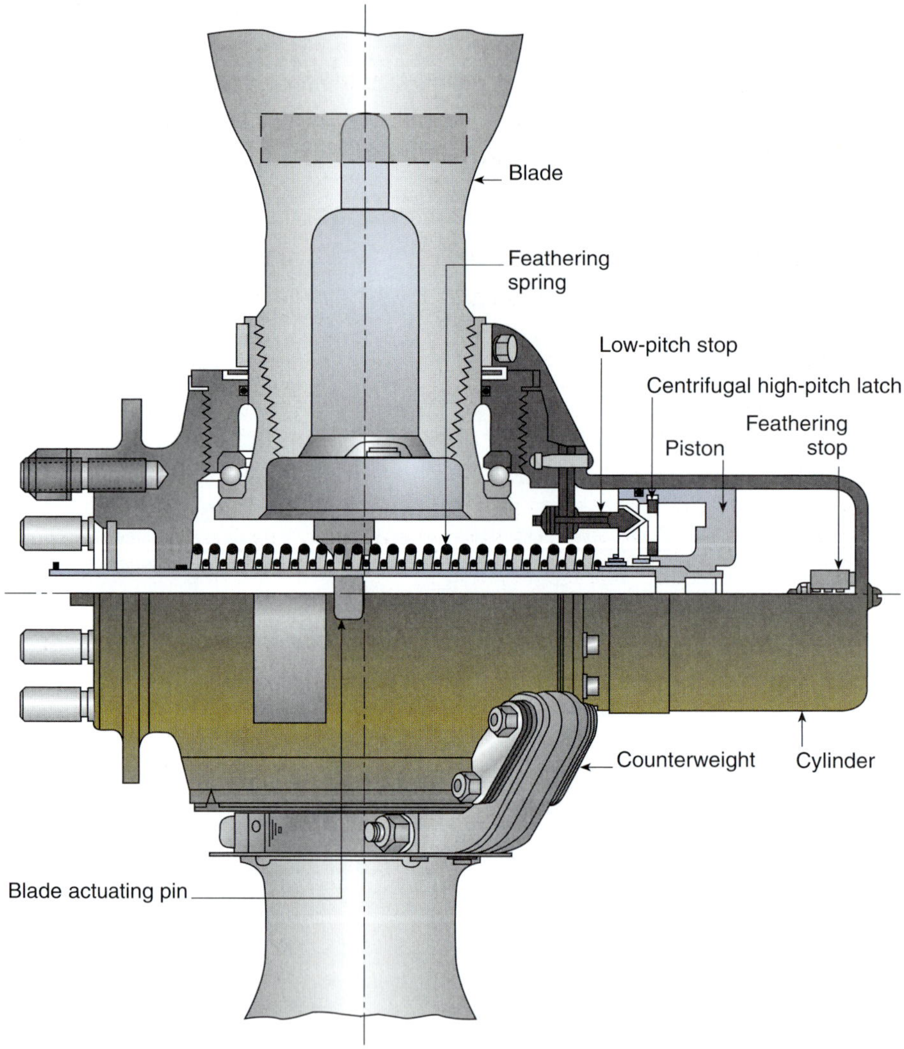

Figure 19-28. *McCauley constant-speed feathering propeller*

To feather the propeller in flight, the governor control is moved to its full aft position, which raises the pilot valve and allows oil to flow from the propeller cylinder back into the engine sump. Centrifugal force holds the high pitch blade latches disengaged, so the combined forces of the counterweights and feathering spring can move the piston forward and rotate the blades into their feathered position.

When the engine is shut down on the ground, it is idling and the centrifugal force is too low to hold the blade latches out, and they stop the blades in their normal high pitch position.

Some McCauley propellers use an accumulator to rapidly move the blades into normal operating range during the unfeathering operation. Governor-boosted oil pressure is trapped in the accumulator by a check valve when the propeller is feathered. When the governor control is returned to its normal operating position for unfeathering, the check valve is opened and the oil stored in the accumulator moves the piston back into its normal operating range.

Some models of McCauley propellers have the hub sealed and partially filled with a light-weight oil that has been colored with a red dye. This dye acts as a leak detector. Any of the dye that leaks out is readily visible and requires action before the next flight.

Reversible Constant-Speed Propellers

As airplanes have increased in size and weight, landing speeds have become so high that wheel brakes alone can no longer be relied upon to slow the airplane in its landing roll. Reversible propellers have been developed for this purpose.

A reversible propeller is a constant-speed propeller whose low pitch stops can be released so the blades can be moved from their normal low pitch range through the zero-thrust angle to a negative blade angle. This allows the propeller to produce thrust in the reverse direction. Almost all turboprop airplanes are equipped with reversible propellers. These propellers and their controls are discussed beginning on Page 693.

STUDY QUESTIONS: PROPELLERS FOR RECIPROCATING ENGINES

Answers begin on Page 733. Page numbers refer to chapter text.

22. A fixed-pitch climb propeller has a ____________________ (higher or lower) pitch angle than a cruise propeller. *Page 672*

23. Centrifugal force acting on the counterweights on a two-position controllable-pitch propeller rotate the blades toward ____________________ (high or low) pitch. *Page 675*

24. The basic difference between a two-position controllable-pitch propeller and a constant-speed propeller is in the control. In a constant-speed propeller, a ____________________ replaces the two-position valve. *Page 676*

25. Multiengine airplanes were made appreciably safer than their single-engine counterpart by the development of the ____________________ propeller. *Page 683*

26. The blade pitch angle when a propeller is feathered is near ____________ degrees. *Page 683*

27. A reversible propeller is a constant-speed propeller whose ____________________ (low or high)-pitch stops can be released so the blades can be moved to an angle that produces thrust in the reverse direction. *Page 689*

28. The wood used for most commercially made wood propellers is ____________________ . *Page 672*

29. Three holes are drilled into the tip of a wood propeller blade to release moisture and allow the wood to breathe. These holes are made with a number _______________ drill. *Page 672*

30. Modern metal propellers are forged from high-strength ________________ alloy, and after being ground to their finished dimensions and pitch, they are _______________ to protect them from corrosion. *Page 673*

31. The two types of propellers that are found on almost all modern airplanes are:
    a. _________________________
    b. _________________________
    Page 677

32. The pitch on all modern constant-speed propellers is changed ____________________ (electrically or hydraulically). *Page 677*

33. Two basic types of constant-speed propellers are:
   a. ______________________________
   b. ______________________________
   Page 677

34. The blade pitch of a noncounterweight constant-speed propeller is increased by two forces. These are:
   a. ______________________________
   b. ______________________________
   Page 679

35. The blade pitch of a noncounterweight constant-speed propeller is decreased by two forces. These are:
   a. ______________________________
   b. ______________________________
   Page 679

36. A flyweight-type propeller governor senses the _______________________, compares it with the speed selected by the pilot, and causes the blade pitch to change to bring these two speeds together. *Page 677*

37. A counterweight nonfeathering propeller uses engine oil boosted by the governor to move the blades into a _____________________ (high or low) pitch angle. *Page 678*

38. A noncounterweight nonfeathering propeller uses engine oil boosted by the governor to move the blades into a _____________________ (high or low) pitch angle. *Page 679*

39. The pilot can increase the RPM of the engine by ________________________ (increasing or decreasing) the compression of the speeder spring in the propeller governor. *Page 681*

40. The governor used with a Hydromatic propeller directs oil to the front side of the piston in the propeller dome to _____________________ (increase or decrease) the blade pitch. *Page 684*

41. A Hydromatic propeller is feathered by oil from a/an _____________________ -driven feathering pump. *Page 684*

42. If the engine-driven oil pump forces all of the oil available to it overboard, a Hydromatic propeller _____________________ (can or cannot) be feathered. *Page 685*

43. A Hartzell steel hub feathering propeller is feathered by ____________________________________ (draining oil from or directing oil into) the propeller. *Page 685*

Continued

STUDY QUESTIONS: PROPELLERS FOR RECIPROCATING ENGINES *Continued*

44. Two forces that move the blades of a Hartzell steel hub feathering propeller into their feather position are:
    a. ______________________________
    b. ______________________________
    Page 686

45. If the engine loses oil pressure in flight, the blades of a Hartzell feathering propeller will move into their ____________________________ (low pitch, normal high pitch, or feather) position. *Page 686*

46. The blades of a Hartzell feathering propeller are prevented from feathering when the engine is shut down on the ground by a/an __________________________ on each blade root contacting a spring-loaded high-pitch stop pin in the hub. *Page 686*

47. Some feathering propellers use an accumulator to speed up the _________________________ (feathering or unfeathering) action. *Page 687*

48. A McCauley feathering propeller is feathered by ___________________________ (draining oil from or directing oil into) the propeller. *Page 689*

49. The blades of a McCauley feathering propeller are prevented from feathering when the engine is shut down on the ground by a centrifugally actuated __________________ (high or low) -pitch blade latches. *Page 689*

50. The accumulator used with some McCauley propellers to speed up the unfeathering action is charged with engine oil whose pressure has been boosted by the pump in the _______________________ . *Page 689*

Propellers for Turbine Engines

The earliest concept of the use of a turbine engine in aircraft was for the turbine to drive a propeller. Turbojet engines showed so much promise that some believed they would make propellers obsolete. Fortunately, this has proven to be untrue. Turboprop powerplants fill an important niche between turbojet or turbofan engines and reciprocating engines. They combine the high propulsive efficiency provided by the propeller with the low weight and high time between overhauls of the turbine engine.

Turboprop powerplant controls are somewhat similar to those used with a reciprocating engine, but there are some important differences. Here we will discuss the turboprop engine, its controls and operation, and the propellers themselves.

turbofan engine. A type of gas turbine engine that has a set of lengthened blades on the low-pressure compressor or low-pressure turbine.

Turboprop Engines

There are two basic types of turboprop engines: single-shaft and free-turbine. The single-shaft engine discussed here is the Garrett TPE331. This engine drives the reduction gears from the same shaft that contains the compressors and the turbines. The free-turbine engine discussed is the Pratt & Whitney of Canada PT6, which drives its propeller reduction gears with a free turbine that is independent of the gas generator turbine. Although similar in operation, there are some important differences between these engines.

free-turbine engine. A gas turbine engine with a turbine stage on a shaft independent of the shaft used to drive the compressor.

single-shaft turbine engine. A turboprop engine in which the propeller reduction gears are driven by the same shaft which drives the compressor for the gas generator.

Garrett TPE331 Engine

The TPE331 engine has an additional turbine stage on the same shaft as the compressor and the gas generator turbines. This shaft, which is coupled to a 26:1 reduction gear system that reduces the low-torque 41,730 RPM turbine speed to a high-torque 1,591 RPM at the propeller shaft, has excess energy beyond that needed to drive the compressor.

gas generator. The basic gas turbine engine, consisting of the compressor, diffuser, combustor, and turbine.

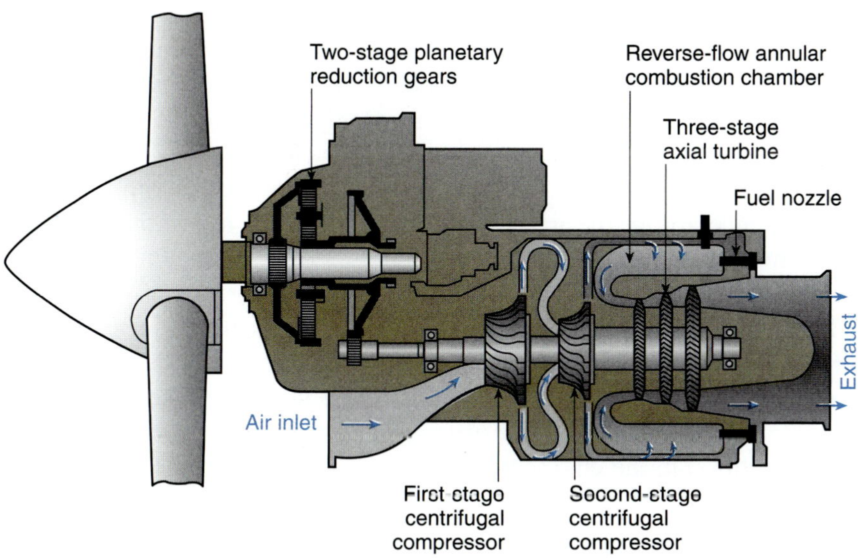

Figure 19-29. *The Garrett TPE331 turboprop engine drives the propeller reduction gears from the same shaft that drives the gas generator compressor.*

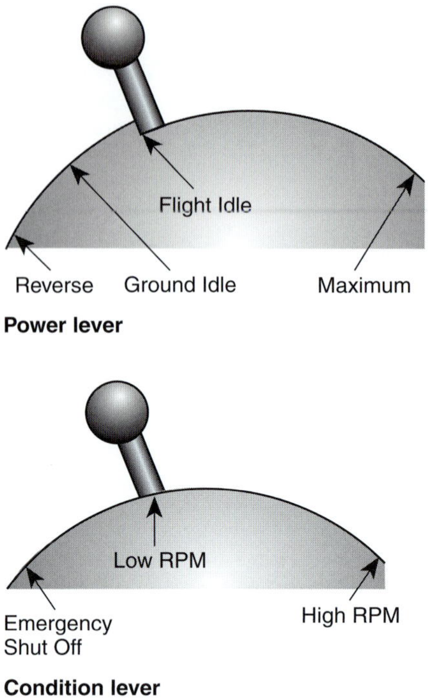

Figure 19-30. *The cockpit quadrant controls for a TPE331 engine*

This is a constant-speed engine, because it operates throughout its operational cycle at near 100% RPM. To hold the RPM constant, the fuel control adjusts the fuel flow in relation to the engine load.

When idling, the RPM remains high, but the propeller pitch is reduced until almost flat, so it produces very little thrust and requires a minimum fuel flow.

The TPE331 uses two engine controls on the cockpit quadrant: the power lever and the speed, or condition, lever. *See* Figure 19-30.

The power lever relates to the throttle of a reciprocating engine, but it also gives the pilot control over the propeller during ground operation. It affects the fuel flow, torque, and EGT, and has four positions: REVERSE, GROUND IDLE, FLIGHT IDLE, and MAXIMUM. The speed, or condition, lever primarily controls the engine RPM, and in some installations it acts as a manual feather and emergency shutoff lever. The condition lever has three positions: EMERGENCY SHUTOFF, LOW RPM, and HIGH RPM.

Four instruments are used to monitor the performance of a TPE331 engine:

Tachometer—Shows the RPM of the compressor in percentage of its rated speed

Torquemeter—Shows the torque or shaft horsepower being developed

Fuel flowmeter—Shows the number of pounds of fuel per hour being delivered to the engine

EGT indicator—Shows the temperature of the exhaust gases as they leave the turbine

When the engine is operating with a given propeller load, and the power lever is moved forward to increase the fuel flow, the RPM will try to increase. To prevent this, the propeller governor increases the blade angle, which causes the RPM to remain constant and the power produced by the engine to increase.

When the power lever is moved back, the fuel flow is reduced, and the RPM begins to decrease. But the propeller governor decreases the blade angle, which causes the RPM to remain constant, and the power to decrease.

The maximum power this engine is allowed to develop may be limited either by the amount of torque the airframe structure can safely accommodate or by the maximum temperature the turbine inlet guide vanes and the first-stage turbine blades can withstand.

flat-rated engine. A turboprop engine whose allowable output power is less than the engine is physically capable of producing.

Turboprop engines are often flat rated. This means the engine is capable of producing more power, or torque, than the airframe can accommodate, and it is limited to a lower torque. Figure 19-32 illustrates flat rating. The limiting turbine inlet temperature of any gas turbine engine decreases as the air flowing through the engine becomes less dense. When the air is dense, the engine is able to move a large mass at a given RPM and fuel flow. As the air becomes less dense, less mass is pumped through the engine at the same RPM, and if the fuel flow were not reduced, the turbine inlet temperature would become excessive. The torque produced by the limiting temperature decreases as the density altitude increases.

In Figure 19-32 the top horizontal line represents the torque the engine is physically capable of producing. The line below that is the maximum torque the engine manufacturer has certificated the engine to produce. The third line is the maximum torque the airframe manufacturer allows for the particular installation. Notice that the engine is flat rated to produce this airframe-limited torque only to the density altitude at which this torque line intersects the turbine temperature limiting line. Above this density altitude, the fuel flow must be decreased to prevent the TIT becoming excessive.

density altitude. The altitude in standard air at which the density is the same as that of the existing air.

At low altitude or cold temperature (low density altitude), the engine is torque limited. At high altitude or hot temperature (high density altitude), it is temperature limited.

Rather than using full throttle for takeoff, the power lever must be advanced only to the point that either the torque or temperature limit is reached.

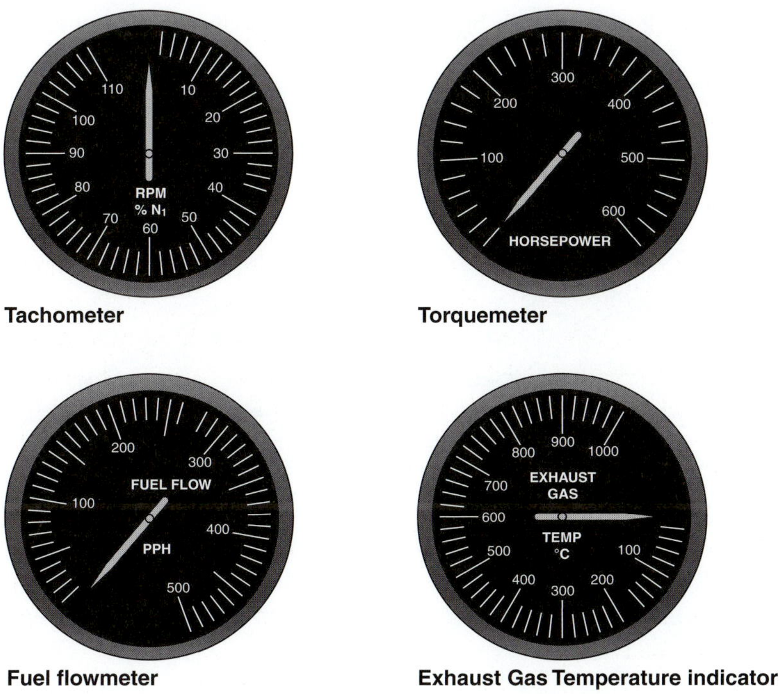

Figure 19-31. *Instruments that monitor turboprop engine operation*

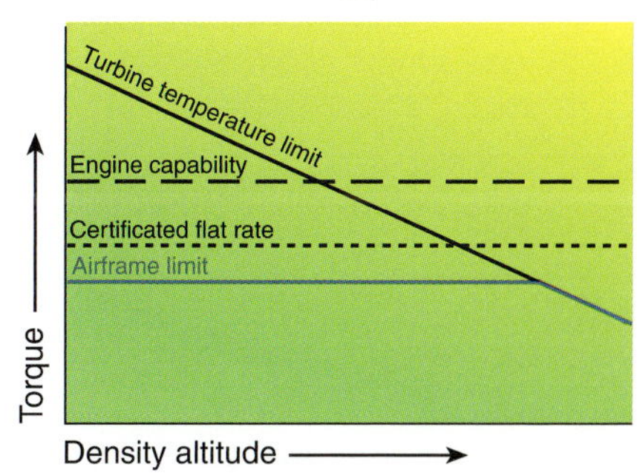

Figure 19-32. *Flat-rate performance of a turboprop engine*

Propeller

A Hartzell three- or four-blade full-feathering reversible steel-hub propeller like the one in Figure 19-33 is approved for installation on this engine. This propeller is similar to the one in Figure 19-15 on Page 678, except that the piston and cylinder are longer so that the blades can be moved into a reverse angle beyond their low pitch stop.

Engine oil, boosted in pressure by the governor pump, flows into the propeller cylinder through the oil transfer, or Beta tube and moves the piston forward, rotating the blades to their low pitch and reverse angles.

Note: *In this propeller, the piston is the movable component that rides over the outside of the cylinder fixed into the propeller hub.*

When the control system drains the oil from the propeller, the combined forces of the feather springs and the counterweights move the blades into their high pitch and feather positions.

A propeller driven by a single-shaft turboprop engine begins turning as soon as the starter begins to rotate the compressor. To minimize the starter load, the propeller blades are held in a flat position (approximately 2° pitch angle) by a start-lock pin on each blade. The engine is shut down, when on

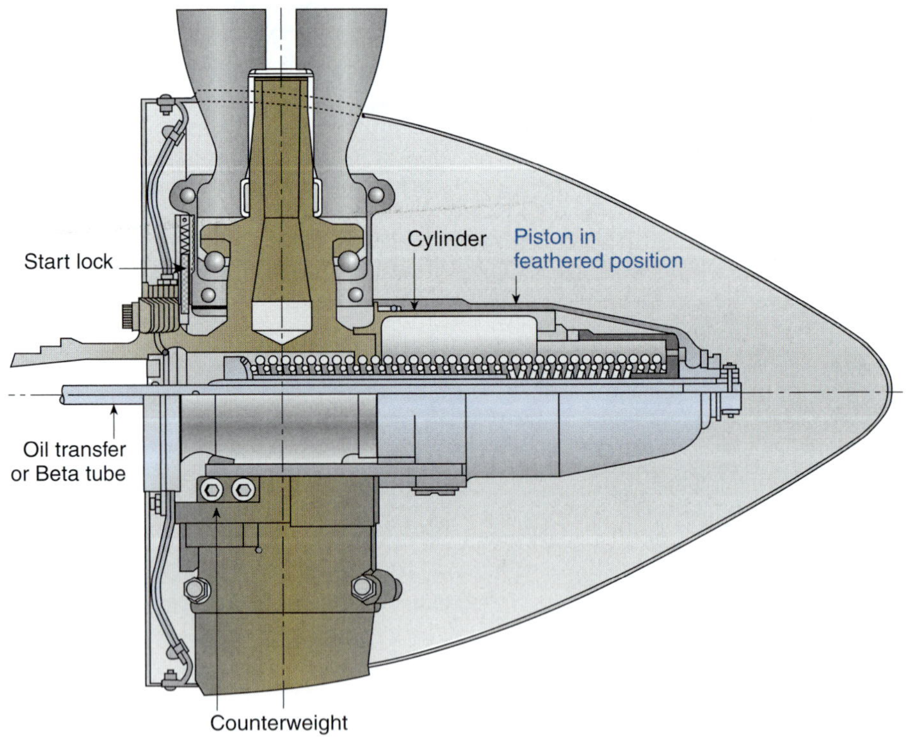

Figure 19-33. *Hartzell steel-hub propeller for a Garrett TPE331 turboprop engine*

the ground, by the pilot moving the power lever toward REVERSE to position a plate on the blade roots behind the start-lock pins. When the engine speed decreases to about 15%, a spring pushes the start-lock pin out, and it holds the blades in their flat, or low pitch, angle.

Power Management

Two basic operating modes of the TPE331 engine are the Beta mode and Alpha mode. Beta is the ground operations mode and includes start, taxi, and reverse operations. Alpha is the flight mode, and it includes all operations from takeoff through landing. Typically, the Beta mode includes operations from 65% to 95% and Alpha from 95% to 100% of the engine's rated RPM.

Figure 19-35 is a simplified diagram of the power management controls for a TPE331 engine.

Beta mode. The range of operation of a turboprop powerplant that is normally used for in-flight approach and ground handling of the engine and aircraft.

Alpha mode. The flight operating mode for a turboprop engine from takeoff through landing.

Cockpit Control	Function	
	Alpha Mode	Beta Mode
Power lever	Manual fuel valve	Propeller pitch control
Condition lever	Propeller governor	Underspeed governor

Figure 19-34. *TPE331 cockpit controls and their functions*

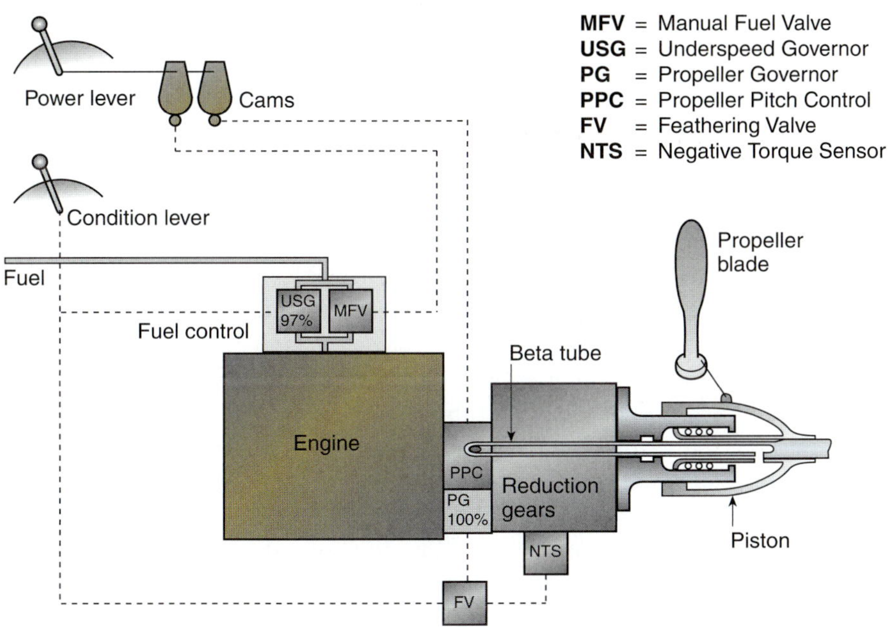

Figure 19-35. *The power management system for a TPE331 turboprop engine*

The propeller pitch control (PPC) is mounted on the reduction gear assembly in line with the center of the propeller shaft. It directs oil into and out of the propeller to change the blade angles during ground operations. One end of the oil transfer, or Beta, tube rides back and forth inside the PPC to act as a feedback device. The PPC is cam operated from the power lever, and during flight operations it serves no function other than acting as an oil passage between the propeller governor (PG) and the propeller.

The propeller governor is mounted on the reduction gearbox, and it operates in the same way as other flyweight governors, controlling the engine speed from 97% to 100% RPM.

Below 97% RPM, the underspeed governor (USG), which is inside the fuel control and operated by the condition lever, regulates the amount of fuel allowed to flow to the engine, to maintain the selected RPM when the engine is below the speed controlled by the PG.

The manual fuel valve (MFV), also inside the fuel control, meters fuel in response to the power lever demands for high power.

A feathering valve (FV) is operated manually by moving the condition lever to its extreme aft position or automatically by the negative torque sensor (NTS). It shuts off the oil from the propeller governor and the oil drains from the propeller, allowing the feathering spring to move the blades into their feathered position.

When the engine is started, the power lever is in its GROUND IDLE position and the condition lever in the LOW RPM position. *See* Figure 19-30 on Page 694. When the engine starts, the start-lock pins in the propeller retract, and the power lever positions the PPC over the Beta tube, causing the propeller to move to a 0° blade angle. The Beta tube is attached to the propeller piston and it moves forward with the piston as the propeller blades move to their low pitch angle. The blade angle stops changing when the Beta tube moves into the PPC's neutral position.

The condition lever is used to set the desired RPM through the USG during ground operations, and the power lever varies the blade angle to move the aircraft forward or rearward.

Beta tube. A tube in a Garrett TPE331 turboprop powerplant that extends into the propeller pitch control to act as a follow-up device.

When the power lever is moved forward, a cam in the PPC uncovers an oil port on the end of the Beta tube, which allows oil in the propeller to drain into the reduction gearbox. The feathering spring and the force from the counterweights move the piston and the Beta tube rearward, increasing the propeller pitch until the Beta tube, with its oil port covered, reaches a new neutral position inside the PPC. The PPC and the Beta tube cause the propeller pitch to respond proportionally to movement of the power lever.

When the blade angle increases, the engine begins to slow down, but the USG, controlled by the position of the condition lever, increases the fuel flow to the engine to maintain the selected RPM.

When the power lever is moved rearward, a cam in the PPC opens a passage in the Beta tube that directs governor-boosted oil pressure into the propeller piston. The piston and the Beta tube move forward, and the blade angle decreases until the Beta tube finds a new neutral position. This lower blade angle causes the engine RPM to increase, but the USG reduces the fuel flow to maintain the selected RPM.

When the power lever is moved to the FLIGHT IDLE position, and the condition lever is moved to a higher RPM setting (97% to 100%), the USG is fully open and no longer affects the system operation. RPM control is now accomplished by the PG. At this time, the PPC covers the opening in the Beta tube so it no longer moves oil into or out of the propeller, and the pitch remains fixed. The power lever then controls the fuel flow through the MFV.

When the power lever is moved forward from its FLIGHT IDLE position, it opens the MFV, and at the same time, the PPC cam holds the propeller in a fixed-pitch position, allowing the RPM to increase.

The condition lever controls both the PG and the USG. When it is moved to the TAKEOFF, or HIGH-RPM, position, its mechanical linkage adjusts the USG to 97% RPM and the PG to 100%. The USG meters additional fuel into the engine to increase its speed to 97%. This sets the engine up for takeoff.

With the engine running at 97%, the power lever is moved toward its MAXIMUM position. As the RPM increases above 97%, it approaches the setting of the PG. When the PG senses 100% RPM, it takes control of the propeller and increases the blade angle to absorb the increased engine power and to maintain the set RPM.

When the power lever is moved aft, fuel flow decreases, and the PG decreases the blade angle to maintain the selected RPM.

On landing, the pilot retards the power lever to reduce the fuel flow, and when there is no longer enough fuel to maintain the speed set by the PG, the RPM drops to the range of the USG. At this point, the power lever controls the propeller through the PPC, and the RPM is controlled by the USG metering enough fuel to prevent the engine speed from dropping below that called for by the condition lever position.

When the power lever is moved all the way to its aft stop, the propeller blades move into a preset negative angle to produce reverse thrust.

When the engine is shut down in flight, the propeller moves into its feather position. The condition lever is moved to its full aft position. This shifts the feathering valve and allows oil from the propeller to drain into the gearbox.

The combined forces from the feather springs and counterweights on the blade shanks move the piston rearward, forcing the oil out of the propeller and moving the blades into their feather angle.

The feathering valve may also be operated automatically by the negative torque sensing (NTS) system. When the negative torque sensor in the reduction gearing senses a loss of positive torque, oil pressure is directed into the feathering valve, shifting it to the feather position.

An electric motor-driven oil pump is used to unfeather this propeller. Oil from the engine lubricating oil tank is boosted in pressure by the unfeathering pump and directed through the Beta tube into the propeller. This moves the piston forward and the blades into their low pitch position.

Pratt & Whitney of Canada PT6 Engine

The Pratt & Whitney of Canada PT6 is a free-turbine turboprop engine in the 750 to 1,000 horsepower range and is popular for commuter airliners and business aircraft. For the gas generator, 100% RPM is in the neighborhood of 38,000 RPM and at this speed, the propeller turns at about 2,000 RPM.

Figure 19-36 illustrates this popular engine. Air enters near the accessory end and flows forward through three stages of axial compression and one stage of centrifugal compression. It then flows through an annular reverse-flow combustor where fuel is added and burned. The hot gases reverse direction again and flow forward through a single stage of compressor turbine and a single stage of free, or power, turbine, and exit through pipes at the forward end of the engine.

One of the operational differences between the PT6 free-turbine engine and the TPE331 single-shaft engine is that the TPE331 is shut down with the propeller blades held against low pitch stops to minimize the load on the starter when the engine is being started. The propeller on the PT6 is allowed to go to its feather position when the engine is shut down because the starter rotates only the gas generator turbine and is not loaded by the propeller and power turbine during an engine start. The turbine that drives the propeller is turned by the hot exhaust from the gas generator.

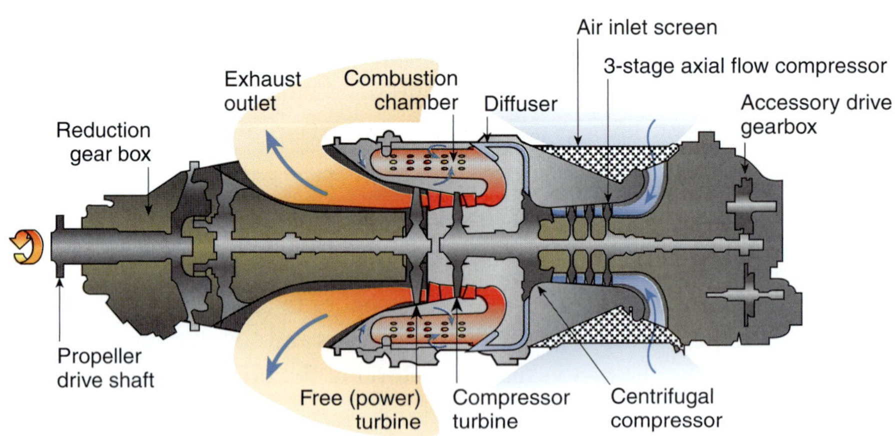

Figure 19-36. *The Pratt & Whitney of Canada PT6 free-turbine turboprop engine*

Propeller

A propeller commonly used with the PT6 engine is a flange-mounted, three-, four-, or five-blade Hartzell steel-hub reversing propeller with composite blades. Counterweights on the blade shanks move the blades into their high pitch position and move the piston inward. Heavy coil feathering springs inside the cylinder move the piston fully inward and the blades into their full-feather position. Engine oil boosted in pressure by a gear-type pump inside the governor moves the piston outward to decrease the blade angle and take the blades into their reverse-pitch angle. A feedback ring on the rear of the propeller functions as a follow-up mechanism to give the propeller proportional response to control inputs while operating in the Beta mode.

Note: *In this propeller, like the other Hartzell propellers, the piston is the movable component that rides over the outside of the cylinder that is fixed into the hub.*

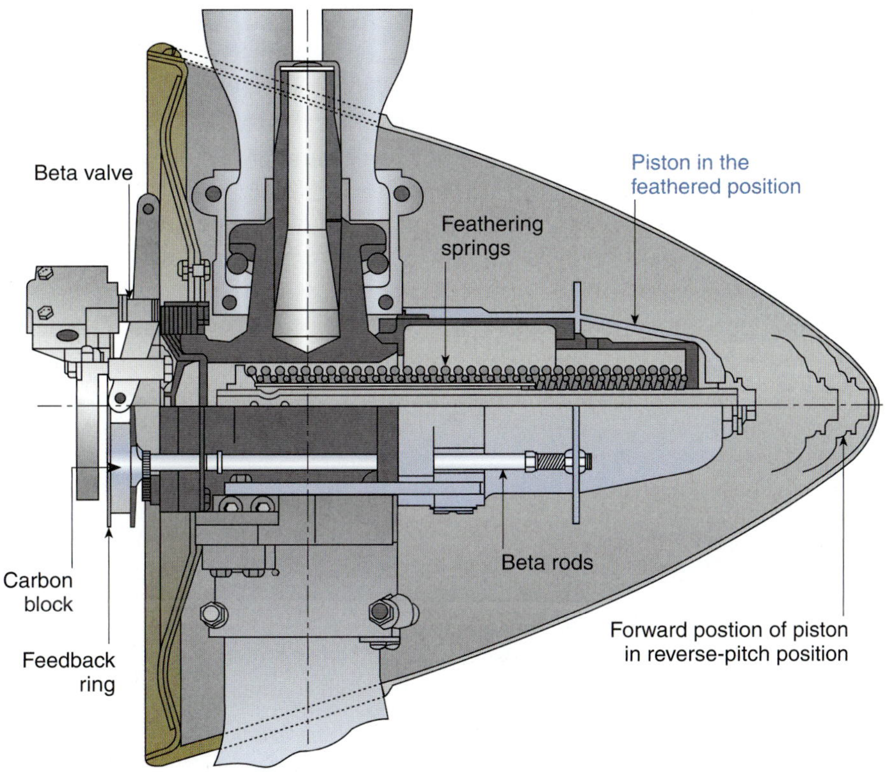

Figure 19-37. *Hartzell steel-hub propeller for a PT6 turboprop engine*

Engine Controls

The PT6 engine is controlled by engine and propeller control systems that are operated by three levers: a power control lever, a propeller control lever, and a condition lever.

The power control lever is connected to the fuel control and is used to control the engine power from full reverse thrust, through idle, to takeoff.

The propeller control lever is connected to the propeller governor and sets the desired propeller RPM. When moved to the extreme aft position, it causes the propeller to feather.

The condition lever attaches to the fuel control and it has three positions: Cutoff, Idle, and Run.

Governor

Power management of the PT6 engine centers around the constant-speed, single-acting hydraulic governor shown in Figure 19-38. This governor senses propeller speed and controls the pitch of the blades to maintain the RPM desired by the pilot.

The governor mounts on the reduction gear housing, and the flyweights are pivoted on the flyweight head which is driven by the propeller reduction gears. Rotation of the head causes centrifugal force to act on the flyweights and swing them outward, lifting the pilot valve plunger. The speeder spring,

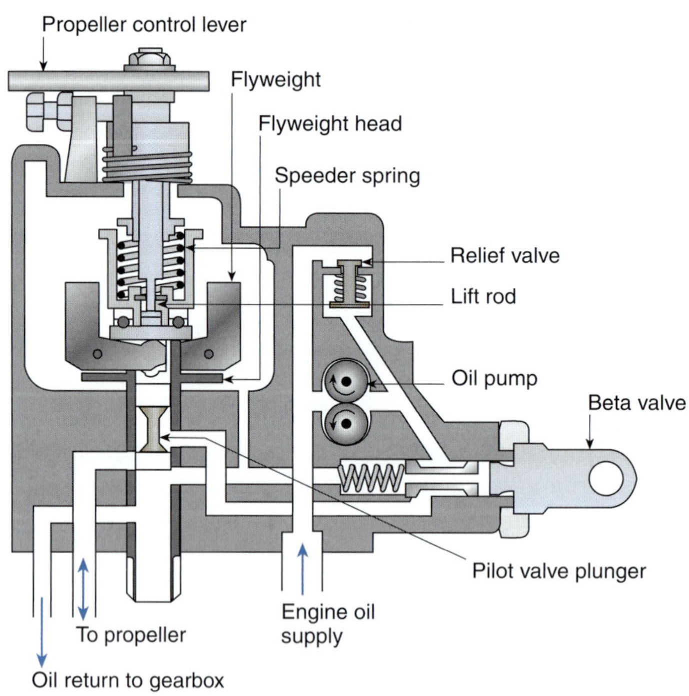

Figure 19-38. *Flyweight governor used to control PT6 propeller speed*

whose compression is controlled by the propeller control lever, balances the centrifugal force on the flyweights and enables the pilot to set the governor for the desired RPM. A lift rod connected to the propeller lever enables the pilot to override the governor and dump the oil from the propeller so the feathering springs can take the blades to their feathered position.

Oil from the engine lubrication system is supplied to the oil pump in the governor where its pressure is increased to the point that it forces the relief valve off its seat. Oil under this regulated governor pump pressure flows through the Beta valve and finally into the area above the control land on the pilot valve plunger. The passage below the control land returns oil to the reduction gearbox. The port in the governor that is covered by the land on the plunger when the flyweights are in their vertical, or on-speed, condition is connected to the propeller cylinder.

If the propeller speed is less than that for which the governor is set, the flyweight force is less than the speeder spring force, and the pilot valve plunger moves down. This allows governor oil to flow into the propeller cylinder and move the piston outward, decreasing the blade pitch. As the pitch decreases, the propeller speeds up until the flyweights again reach their on-speed condition, and oil is shut off to the propeller.

If the propeller speed becomes greater than desired, the flyweights sling outward. This is an overspeed condition, and it raises the pilot valve plunger, opening the passage between the propeller cylinder and the reduction gearbox. Oil drains from the propeller cylinder, and the combined counterweight action and force of the feathering spring move the piston inward, increasing the blade angle. The increased pitch slows the propeller and allows the governor to return to its on-speed condition.

Overspeed or underspeed conditions can be caused either by a change in the air load or a change in the propeller control setting.

For operation in the Beta mode, the governor contains a Beta control valve operated by the power lever linkage that directs oil into or out of the propeller to change the blade angle.

Beta Operation

Beta operation is a ground mode of operation between 50% and 85% RPM. During Beta operation, the propeller governor is in an underspeed condition with the pilot valve lowered. This allows the power lever to control both fuel flow and propeller blade angle. *See* Figure 19-39.

Moving the power control lever forward increases the fuel flow to the engine, and at the same time, the linkage to the propeller governor pulls the Beta valve outward and releases oil from the propeller into the engine gearbox. The combination of centrifugal force acting on the counterweights and the force from the feathering springs moves the piston inward and increases the blade angle.

As the propeller piston moves inward, the steel Beta rods move the feedback ring, on the back of the propeller, rearward. This movement causes the carbon block that rides in this ring to return the Beta valve to its neutral position and stop the flow of oil from the cylinder. This action makes the propeller pitch proportional to the movement of the power lever.

When the power lever is moved rearward, fuel flow is reduced and the mechanical linkage moves the Beta valve inward, directing oil from the propeller governor into the propeller to decrease blade angle. As the cylinder moves outward, the feedback ring moves forward, and the Beta control valve is returned to its neutral position.

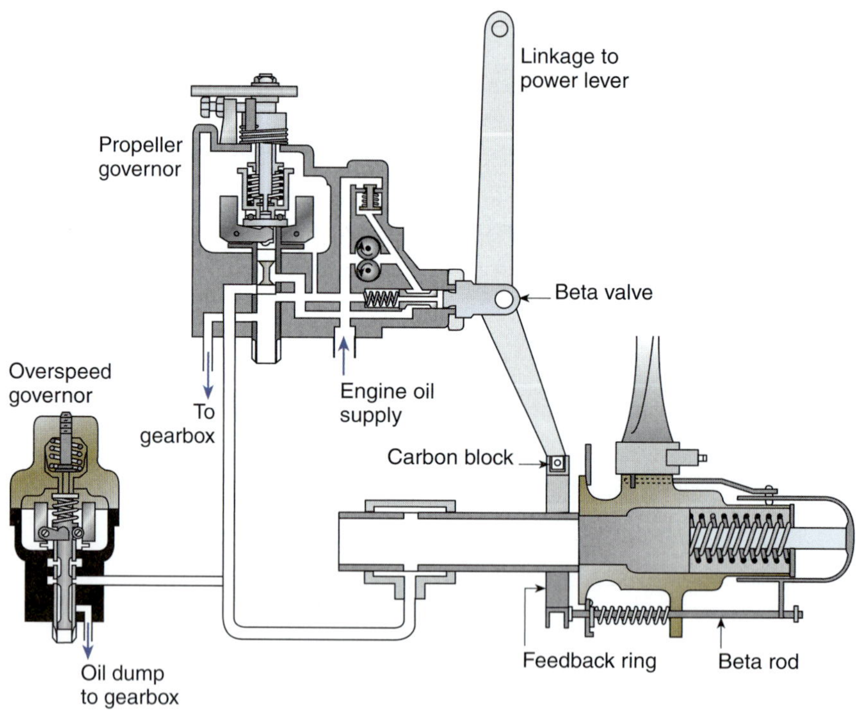

Figure 19-39. *Beta operation of the PT6 propeller*

Continued movement of the power lever into the REVERSE position causes a mechanical linkage to position the Beta valve so that oil from the governor moves the propeller piston fully outward, placing the blades in a reverse-pitch angle. At the same time, the cam-operated fuel control increases the fuel flow, producing a variable amount of reverse thrust.

When operating in the flight, or Alpha mode, the propeller RPM is high enough for the propeller governor to operate in its constant-speed mode. As the power lever is moved forward, more fuel flows to the engine to increase horsepower, and the propeller governor increases the blade angle to absorb the added power and maintain the selected RPM. When the power lever is moved aft, the blade angle decreases to maintain the selected RPM.

To feather the propeller, the propeller control lever is moved fully aft, causing a lift rod to raise the pilot valve in the governor and open the oil passage between the propeller and the reduction gearbox. The feather springs move the piston inward, forcing the oil from the propeller and the blades into their full-feather position.

The propeller is unfeathered by moving the propeller control into the idle position and starting the engine. Gases produced by the gas generator rotate the power turbine, and engine oil boosted in pressure by the governor pump moves the propeller piston forward and the blades into their low pitch-range.

Overspeed Governor

An overspeed governor is a backup for the propeller governor and is mounted on the reduction gearbox. It has its own flyweights and pilot valve, and it releases oil from the propeller whenever the propeller RPM exceeds a preset limit above 100%. Releasing the oil allows the blades to move to a higher pitch angle, which reduces the RPM. The overspeed governor is adjusted when installed and cannot be adjusted in flight—there are no cockpit controls for it.

Autofeather System

Some PT6 installations have autofeathering capability. This system reduces the risk of inadvertently feathering the propeller on an operating engine. Feathering the wrong propeller could cause a serious accident if an engine should fail during takeoff or climb-out.

The autofeather system uses torquemeter oil pressure as its sensing element. It operates when the system is armed and the torque pressure drops below a predetermined value. On one particular installation, the following events occur when the sensor detects a loss of torque oil pressure:

- When the system is armed, green annunciator lights illuminate to advise the pilot that the autofeather system is operational.
- If the torque on one of the engines drops below about 400 foot-pounds, the autoignition pressure switch closes and energizes the igniters.
- The autofeather system is de-energized and the green annunciator goes out, allowing the engine a chance to restart. If, instead of restarting, the torque drops below about 200 foot-pounds, the system re-energizes and the autofeather valve opens, dumping the oil from the propeller cylinder into the reduction gearbox and allowing the feathering spring to rapidly move the blades into their full-feathered position.

Composite Propeller Blades

composite propeller blade. A propeller blade made from several materials such as metal, graphite, glass or aramid fibers, and foam.

graphite fibers. An advanced composite fiber made by drawing filaments of carbon at a high temperature and in a controlled atmosphere.

aramid fibers. Fibers made from an organic compound of carbon, hydrogen, oxygen, and nitrogen. It has high strength and low density and is flexible under load.

Kevlar™. The registered trade name by DuPont for a patented aramid fiber.

matrix. The material that bonds the fibers together in an advanced composite structure.

thermosetting resin. A plastic resin that, once it has been hardened by heat, cannot be softened by heating again.

Laminated wood, forged aluminum alloy, and brazed sheet steel propellers have been the standard for decades. But the powerful turbopropeller engines and the demands for higher-speed flight and quieter operation have caused propeller manufacturers to exploit the advantages of modern advanced composite materials.

Composite materials used in propeller manufacturing consist of two constituents: the fibers and the matrix. The fibers most generally used are glass, graphite, and aramid (Kevlar™), and the matrix is a thermosetting resin such as epoxy.

The strength and stiffness of the blades are determined by the material, diameter, and orientation of the fibers. The matrix material supports the fibers, holds them in place, and completely encapsulates them to protect them from the environment.

Because the fibers have strength only parallel to their length, they are laid up in such a way that they are placed under tensile loads.

Three popular composite propeller blades are discussed here: Hartzell, Hamilton-Standard, and Dowty Rotol.

The typical Hartzell composite propeller, like that in Figure 19-40, has a machined aluminum alloy shank, and molded into this shank is a low-density foam core. Slots are cut into the foam core and unidirectional Kevlar shear webs are inserted. The leading and trailing edges are made of solid sections of unidirectional Kevlar, and laminations of prepreg material are cut and laid up over the core foundation to provide the correct blade thickness, airfoil shape, pitch distribution, planform, and ply orientation. The outer shell is held in place on the aluminum alloy shank by Kevlar filaments impregnated with epoxy resin wound around the portion of the shell that grips the shank.

Some Hartzell blades have a stainless steel mesh under the final layer of Kevlar to protect against abrasion, and a nickel leading edge erosion shield is bonded in place.

The entire blade is put into a blade press and cured under computer-controlled heat and pressure.

unidirectional fibers. Fibers in a piece of composite material that are arranged in such a way that they can sustain loads in only one direction.

prepreg. Preimpregnated fabric.

pitch distribution. The gradual change in pitch angle of a propeller blade from the root to the tip.

leading edge. The thick edge at the front of a propeller blade.

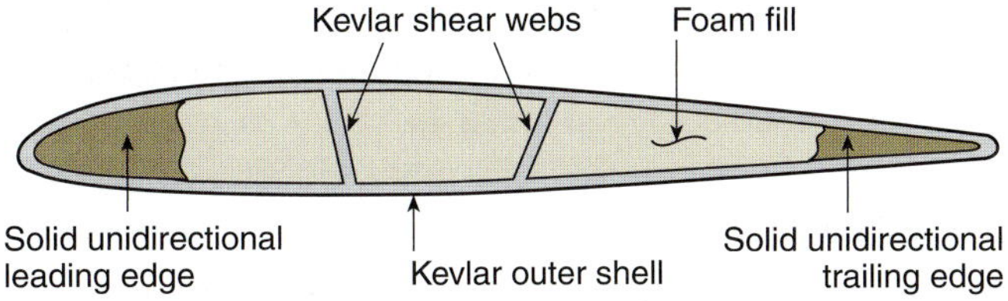

Figure 19-40. *Cross section of a Hartzell composite propeller blade*

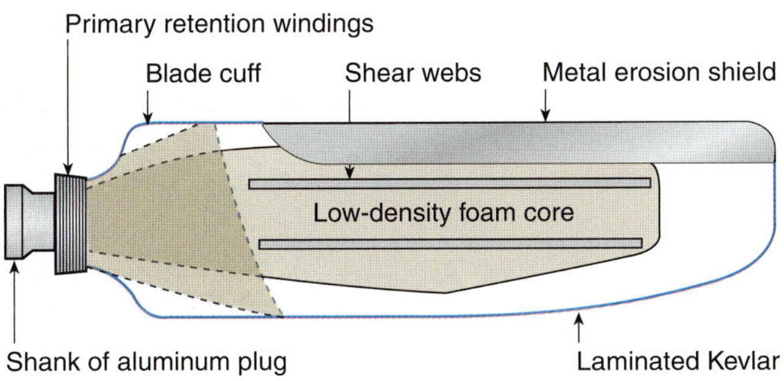

Figure 19-41. *Plan view of a Hartzell composite propeller blade*

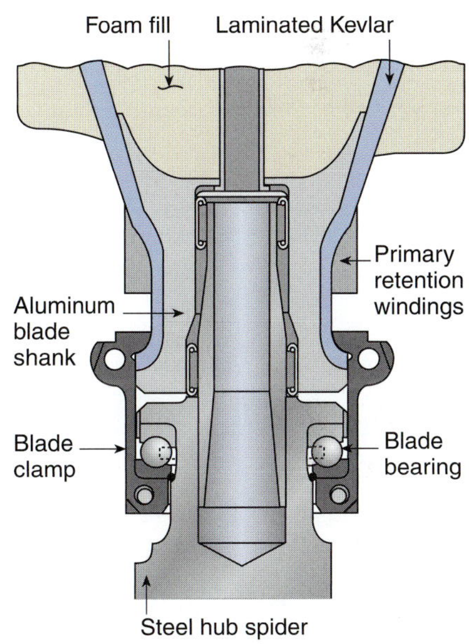

Figure 19-42. *Method of blade retention of a Hartzell composite propeller blade*

The Hamilton-Standard blade has tremendous strength and fatigue resistance because of its solid aluminum alloy spar enclosed in a glass fiber shell.

The spar is machined to its correct configuration and placed in a mold cavity, and the core foam is injected around it. The foam is cured and is removed from the mold. Glass fiber cloth, with the correct number of plies and the proper ply orientation, is then laid up over the cured core. The complete lay-up is then placed in a second mold that has the shape of the finished blade. The resin matrix is injected to impregnate all the fibers, and is cured with heat and pressure. *See* Figure 19-43.

The Dowty Rotol composite propeller blade has two carbon fiber spars that run the length of the blade on both the face and back and come smoothly together at the blade root. The carbon fibers and preimpregnated glass fiber cloth are laid up with the proper number of plies and in the correct ply orientation and are placed in a mold. Polyurethane foam is injected into the inside of the blade, and the entire unit is cured under heat and pressure.

The Dowty Rotol blade is secured in the hub by expanding the carbon fiber spars with tapered glass fiber wedges and locking them between the inner and outer sleeves. *See* Figures 19-44, and 19-45.

notch sensitivity. A measure of the loss of strength of a material caused by the presence of a notch, or a V-shaped cut.

Composite propeller blades are much lighter than metal blades capable of absorbing the same amount of power. The lighter blades impose less centrifugal loading on the hub, allowing it to be made lighter. Composite blades have very low notch sensitivity, and their foam cores absorb much of the vibration that would damage metal propellers. While composite blades currently cost more than metal blades, their greater efficiency and longer life make them much more cost effective.

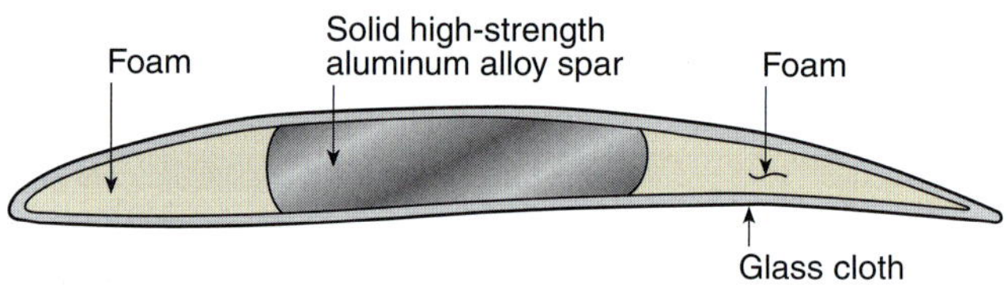

Figure 19-43. *Cross section of a Hamilton-Standard composite propeller blade*

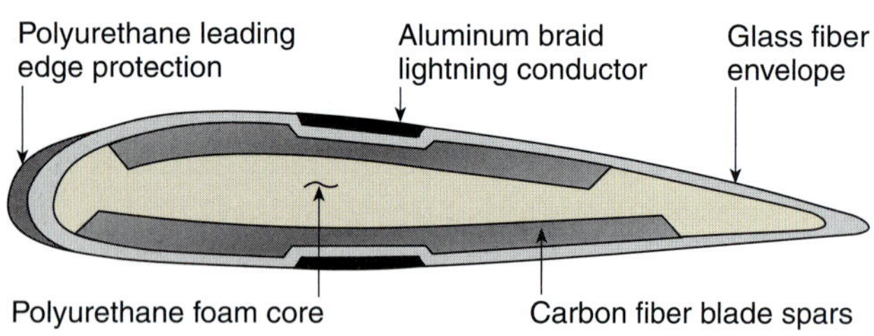

Figure 19-44. *Cross section of a Dowty Rotol composite propeller blade*

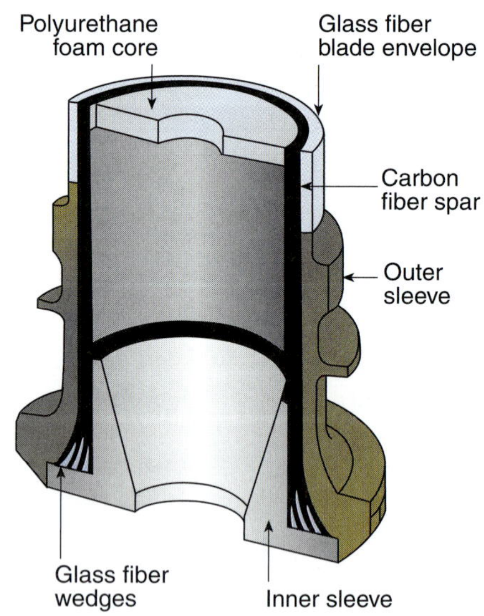

Figure 19-45. *Method of blade retention of a Dowty Rotol composite propeller blade*

UHB (Ultrahigh Bypass Ratio) Engines

Until World War II, only Frank Whittle and a handful of other engineers envisioned the turbine engine doing anything other than driving a propeller. But in the following decades, the tremendous versatility of the gas turbine as a prime mover in the form of turbojet, turbofan, turboprop, and turboshaft engines revolutionized transportation.

Another application of the gas turbine engine for transportation has appeared. This is the ultrahigh bypass (UHB) turbine engine which is built under such names as Prop-Fan™ (a trademark of Hamilton-Standard), and Unducted Fan™ (a trademark of The General Electric Company). This type of engine has evolved from the turbojet through the turbofan. It has a bypass ratio of more than 30, and it incorporates some of the characteristics and advantages of a turboprop engine while allowing higher flight speeds.

The propulsive efficiency of a turbojet engine is so low that the bypass engine, which evolved into the turbofan, took over. As the bypass ratios begin exceeding 1:1, it became evident that the fan was capable of producing the major part of the thrust.

Modern turbofan engines have bypass ratios between 6:1 and 8:1, with some of the newer ones reaching 9:1. The fans on these high bypass ratio turbofan engines perform like a many-bladed, ducted propeller. Lower-speed airplanes that routinely fly short stage lengths find the high propulsive efficiency of the propeller to be such an advantage that turboprops are their powerplant of choice.

UHB engine. A gas turbine engine in which a fan, either ducted or unducted, has a bypass ratio greater than 30:1.

Figure 19-46. *The GE Unducted Fan™ engine uses multiblade contrarotating propellers with blades having scimitar shape for better high-speed aerodynamic efficiency.*

Turbojet: Bypass ratio = 0
Very low propulsive efficiency, high specific fuel consumption and undesirable noise level.
Low bypass engine: Bypass ratio = less than 2:1
Slight improvement in specific fuel consumption and slightly decreased noise level.
Medium bypass engine: Bypass ratio between 2:1 and 4:1
Increased propulsive efficiency, improved specific fuel consumption, decreased noise level.
High bypass engine: Bypass ratio greater than 4:1, some as high as 9:1
Further increased propulsive efficiency, improved specific fuel consumption, decreased noise level. This is the type of engine used in many current generation large transport aircraft.
Turboprop
High propulsive efficiency but limited in speed to approximately Mach 0.6.
UHB engines: Bypass ratio greater than 30:1
Further increased propulsive efficiency, and ability to fly in the range of Mach 0.75 to 0.8; Exceptionally low thrust specific fuel consumption (TSFC).

Figure 19-47. *Evolution of the UHB engine*

Propfan™ engine. The trade name registered by Hamilton Standard for an ultrahigh-bypass turbine engine.

UDF engine (Unducted Fan™). The trade name registered by General Electric for a type of ultra-high-bypass turbofan engine.

stage length. The distance between landing points in airline operation.

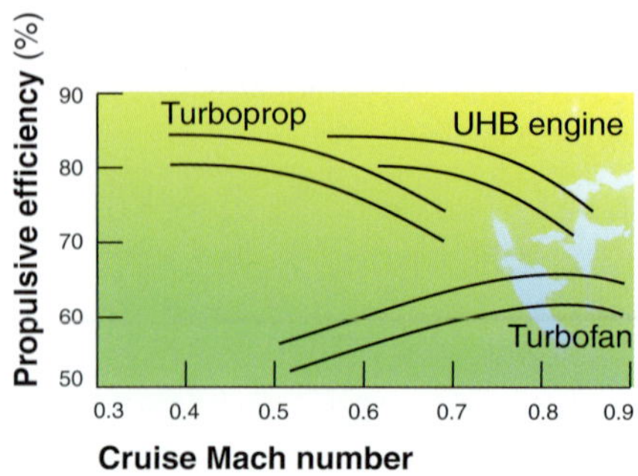

Figure 19-48. *Propulsive efficiency trends for turbofan, turboprop, and UHB engines*

Figure 19-48 compares the propulsive efficiency trends for three types of propulsion systems. Turboprop efficiency drops off considerably beyond Mach 0.5, and turbofan engines do not gain peak efficiency until about Mach 0.8, and even at this speed it is low. UHB engines have high propulsive efficiency in the range between Mach 0.6 and 0.8, an important speed range for transport aircraft operation.

When an airplane is flying at about Mach 0.6, the tips of a traditional propeller are traveling supersonic and their efficiency drops off dramatically. The noise made by the propeller increases to a level intolerable for airline operation.

UHB engines, which are a cross between a turbofan and a turboprop, use about 95% of the power available from the turbine to drive the propeller, which bypasses a tremendous amount of air around the core engine.

The concept of the UHB engine has existed for some time, but as with many developments in aviation, it had to wait for the availability of proper composite materials and high-strength, lightweight metal alloys to make the propellers.

The propellers for these engines have between eight and twelve very short blades to keep the tip speed and noise down, while providing the same amount of thrust as a propeller with fewer but longer blades. The blades of these propellers are very thin, but have wide chords and a scimitar shape to increase their critical Mach number. *See* Figure 19-49.

There are two basic configurations of UHB engines: pushers and tractors. Both configurations use contrarotating multiblade variable-pitch propellers.

Contrarotating propellers are used to recover some of the energy lost in the swirl of air leaving a propeller. Much of the swirl energy leaving the first set of blades is captured by the second set and converted into thrust.

The tractor propeller in Figure 19-50 is driven by a long shaft from the free turbine, and the speed is reduced by a planetary gear system which produces torque for the two sets of contrarotating blades. The propeller is enclosed in a shroud that reduces the noise level and vibration and provides some degree of safety in the event of a lost blade.

The advantage of tractor propellers is that they supercharge the core engine, but the disadvantage is that they drive the propeller through a long shaft from the power turbine.

There are two types of pusher engines: those that drive the propellers directly from extensions on the free-turbine blades, and those that drive them through a set of reduction gears.

The pusher configuration in Figure 19-51 drives its two unshrouded contrarotating propellers from a multistage free-power turbine through a reduction gear and pitch-change mechanism.

Pusher propellers have a higher propulsive efficiency than tractor propellers, but they must operate in the wing turbulence and the hot exhaust gases from the engine.

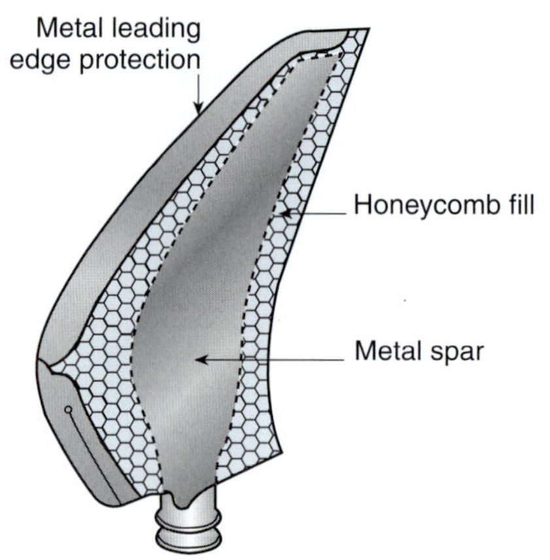

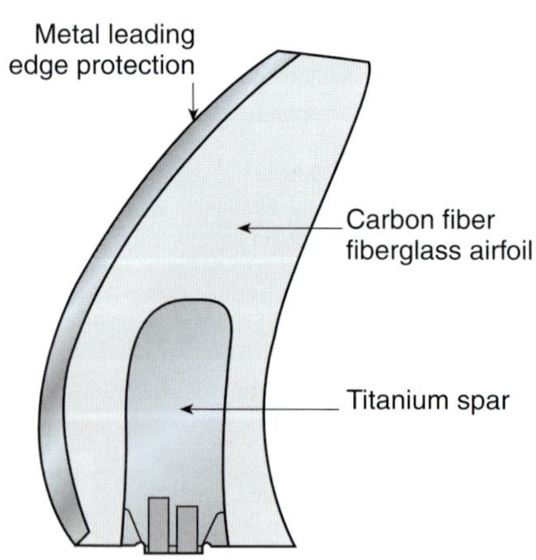

Figure 19-49. *Blade shapes of propellers used on UHB engines*

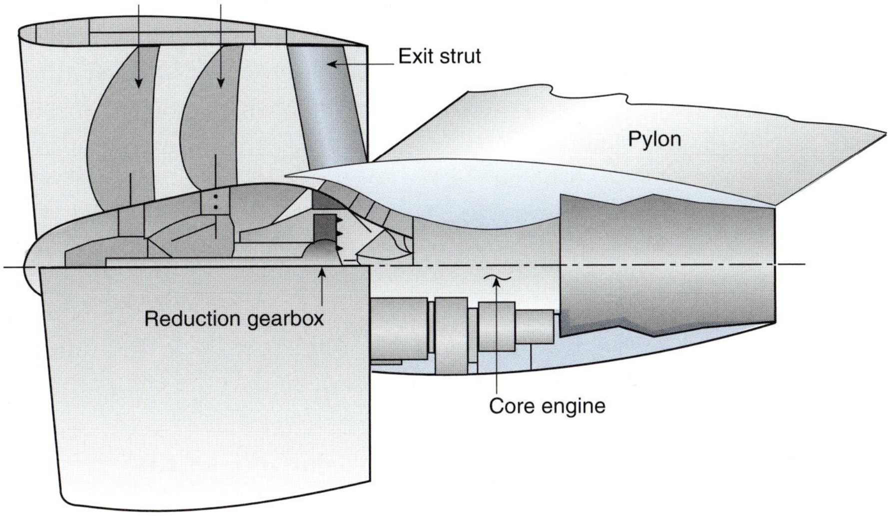

Figure 19-50. *Tractor configuration UHB engine with the contrarotating blades enclosed in a shroud*

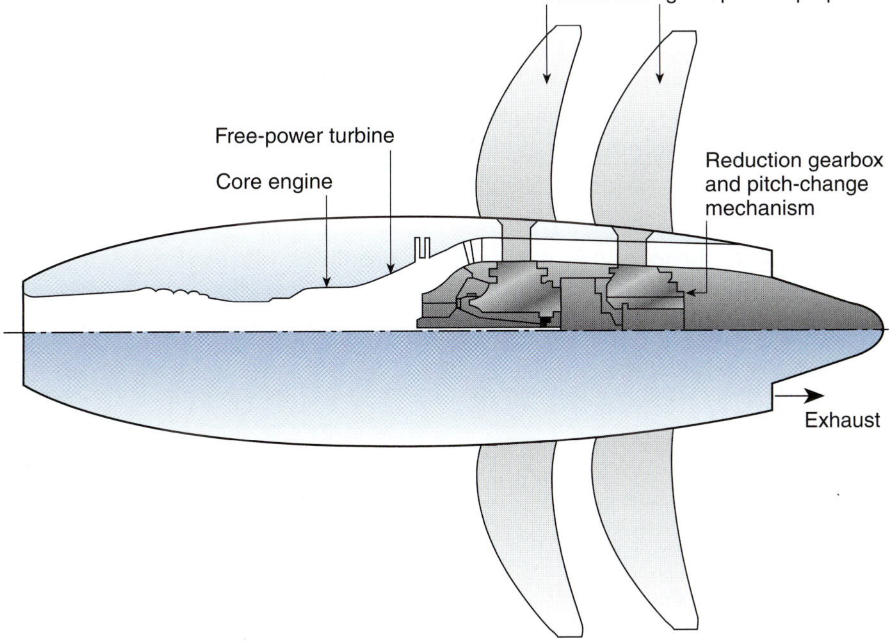

Figure 19-51. *Pusher configuration UHB engine with unshrouded contrarotating blades*

propulsive efficiency. A measure of the effectiveness with which an aircraft engine converts the fuel it burns into useful thrust.

core engine. The gas generator portion of a gas turbine engine.

scimitar shape. The shape of the propeller blades mounted on an unducted-fan engine.

critical Mach number. The flight Mach number at which there is the first indication of air flowing over any part of the structure at the local speed of sound.

Mach number. The ratio of the speed of an object through the air to the speed of sound under the same atmospheric conditions.

STUDY QUESTIONS: PROPELLERS FOR TURBINE ENGINES

Answers begin on Page 733. *Page numbers refer to chapter text.*

51. Two basic types of turboprop engines are:
    a. ____________________________
    b. ____________________________
    Page 693

52. The two engine controls for a TPE331 engine are:
    a. ____________________________
    b. ____________________________
    Page 694

53. The RPM of a turboprop engine is controlled by the ________________________ (power or speed) lever. *Page 694*

54. Four instruments used to monitor the performance of a turboprop engine are:
    a. ____________________________
    b. ____________________________
    c. ____________________________
    d. ____________________________
    Page 694

55. The maximum power a turboprop engine is allowed to produce at low altitude is normally limited by the ____________________________ (airframe structure or turbine inlet temperature). *Page 695*

56. The amount of torque a temperature-limited turboprop engine is allowed to produce ____________________ (increases or decreases) as the altitude increases. *Page 695*

57. The two basic operating modes of a turboprop engine are:
    a. ____________________________
    b. ____________________________
    Page 697

58. The propeller on a TPE331 engine is feathered by directing oil ________________________ (into or out of) the propeller cylinder. *Page 698*

59. Two forces that move the TPE331 propeller blades toward high pitch are:
    a. ____________________________
    b. ____________________________
    Page 698

60. When a TPE331 engine is operating at 100% RPM, the propeller pitch is controlled by the ______________________ . *Page 698*

61. When a TPE331 engine is operating at a speed below that which is controlled by the propeller governor, the propeller pitch is controlled by the ______________________ . *Page 698*

62. When a TPE331 engine is operating at a speed below that which is controlled by the propeller governor, the engine speed is controlled by the ______________________ . *Page 698*

63. The propeller on a TPE331 engine is unfeathered with oil pressure produced by a/an ______________________ -driven unfeathering pump. *Page 700*

64. When a TPE331 engine is shut down on the ground, the propeller blades remain in ______________________ (low or high) pitch. *Page 699*

65. When a TPE331 engine is shut down in flight, the propeller blades move into their ______________________ (reverse-pitch, low-pitch, high-pitch, or feather) position. *Page 699*

66. When a TPE331 engine is shut down on the ground, the propeller blades are prevented from moving into their feather position by spring-loaded ______________________ . *Page 696, 697*

67. When a Pratt & Whitney of Canada PT6 engine is shut down on the ground, the propeller blades move into ______________________ their (reverse-pitch, low-pitch, high-pitch, or feather) position. *Page 700*

68. The three engine controls used on a PT6 engine are:
    a. ______________________
    b. ______________________
    c. ______________________
    Page 702

69. The propeller governor used on a PT6 engine is a ______________________ (single-acting or double-acting) governor. *Page 702*

70. Movement of the ______________________ valve causes the pitch of the propeller installed on a PT6 engine to be proportional to the movement of the power control lever. *Page 704*

71. A PT6 engine is normally started with the propeller in the ______________________ (reverse-pitch, low-pitch, or feather) position. *Page 700*

72. The sensor used for the autofeather system on a PT6 engine is the ______________________ . *Page 706*

Continued

STUDY QUESTIONS: PROPELLERS FOR TURBINE ENGINES *Continued*

73. Three advantages of composite blades over metal blades for turboprop engine propellers are:
    a. ______________________
    b. ______________________
    c. ______________________
    Page 708

74. The powerplant with the greater propulsive efficiency is the ________________ (turbofan or turboprop). *Page 709*

75. The upper practical speed limit for an airplane driven by a conventional propeller is approximately Mach ________ . *Page 710*

76. The propeller blades installed on a UHB engine have a scimitar shape to increase their ____________________ . *Page 710*

77. Contrarotating propellers are effective because the rear set of blades converts some of the ________________ energy from the front blades into thrust. *Page 710*

Propeller Installation, Inspection, and Maintenance

spline. Parallel slots cut in the periphery of a propeller shaft, parallel to its length.

Installation and removal procedures for a propeller depend upon the type of propeller shaft, and there are three types found on reciprocating engines: flanged, splined, and tapered. The larger turboprop engines have splined shafts, and the smaller ones have flanged shafts.

Installation on a Flanged Shaft

Most modern engines, both reciprocating and turbine, have flanged crankshafts or propeller shafts. Some of these flanges have integral internally threaded bushings that fit into counterbores in the rear of the propeller hub around each bolt hole. Propellers with these bushings are attached to the shaft with long bolts that pass through the propeller. A stub shaft, or pilot, fits into the center bore of the propeller to center it on the shaft. Figure 19-52 shows a cross section of the flanged crankshaft of a direct-drive reciprocating engine.

Notice that this shaft is fitted for a fixed-pitch propeller. If a constant-speed propeller were installed, the expansion plug in the end of the shaft would be removed so propeller control oil from the governor could flow into and out of the propeller.

Some engines have index pins in the propeller flange so the propeller can be installed in only one position relative to the crankshaft. Some aircraft manufacturers specify the position related to the top dead center mark on the crankshaft flange or the starter ring gear housing. Nowadays, this position more than likely will not place the propeller blades in a convenient position for hand propping. But if there are no such pins or instructions, install the propeller so the blades will be in the 10 o'clock and 4 o'clock position when the engine stops. Stopping the propeller in this position makes hand propping easier and safer.

Slide the propeller on the end of the shaft, making sure it fits properly over the bushings and flat against the face of the flange. Install the correct bolts and washers as specified in the illustrated parts list for the particular aircraft. Torque the bolts to the specified value in the recommended sequence, and safety them as prescribed in the service manual. The standard sequence for torquing a six-bolt propeller to a flanged shaft (Figure 19-54) is shown in Figure 19-53.

Hartzell flange-mounted compact hub propellers are mounted with bolts constructed from a drilled stud with a castle nut pinned to it by a roll-pin. A standard bolt will not fit into the propeller flange bolt recess as a single piece unit. The fabricated bolt is safety wired through the center of the roll-pin.

McCauley flange-mounted constant-speed propellers have studs anchored in the hub that protrude through threadless crankshaft flange holes. Washers and self-locking nuts are threaded onto the studs to retain the propeller.

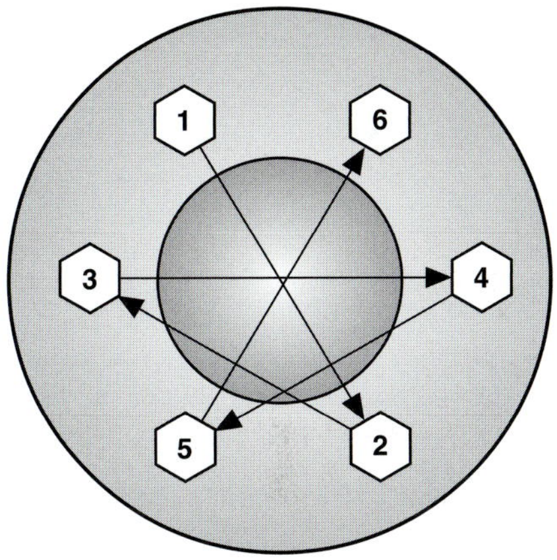

Figure 19-53. *Typical sequence for torquing a six-bolt propeller if no sequence is specified in the aircraft maintenance manual*

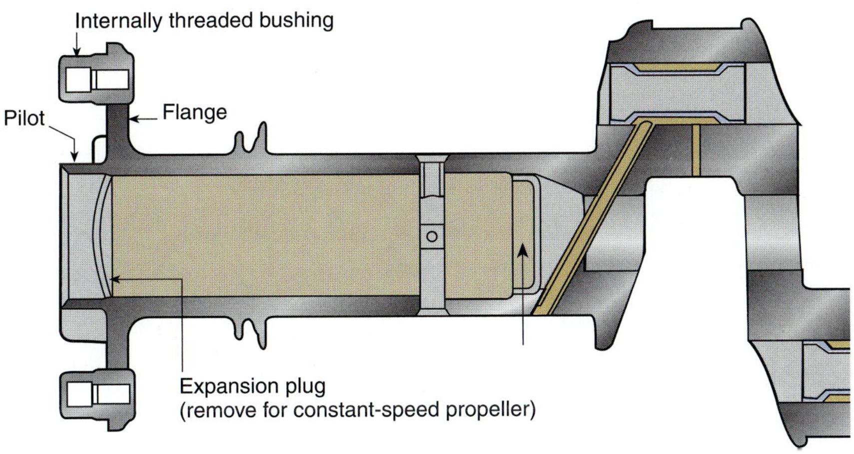

Figure 19-52. *Propeller flange on the crankshaft of a direct-drive reciprocating engine*

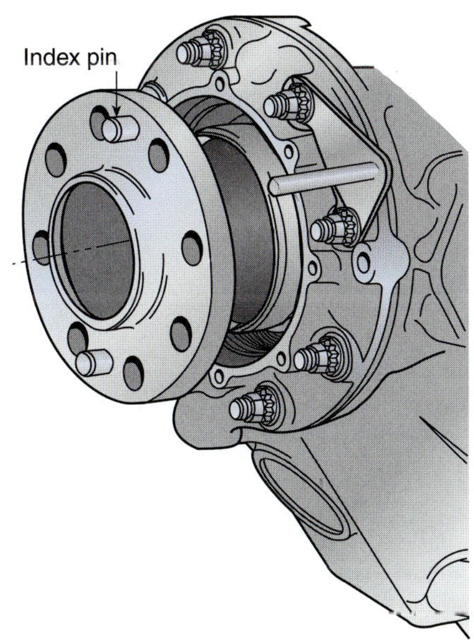

Figure 19-54. *Flanged propeller shaft for a turboprop engine*

Propeller Spinners

All modern propeller-driven airplanes have spinners over their propeller hubs. These spinners have the dual aerodynamic function of streamlining the engine installation and directing cool air into the openings in the cowling. Figure 19-55 shows a typical spinner installation over a constant-speed propeller. The spinner bulkhead is installed on the propeller shaft flange and held in place by six spinner attaching bolts. The propeller is then installed so that the dowel pins in the propeller hub align with the holes in the flange. The propeller attaching nuts are installed and tightened to the torque value specified in the airframe maintenance manual.

If a spinner support is required, it is installed and the spinner is secured to the bulkhead with the proper machine screws.

Be sure to consult the maintenance manual for the proper spinner preload. Failure to install the proper thickness of preload shims will likely result in the spinner being violently thrown from the propeller at high RPM.

If a spinner support is not used, a forward spinner bulkhead is used. The forward bulkhead is attached to the spinner with screws into self-locking nut plates attached to the front bulkhead.

The propeller spinner and bulkhead are critical components, and cracks in either one can be repaired only if they do not exceed the allowable limits. Repair them using the procedures specified in the airframe maintenance manual and take special care to not add weight where it could cause vibration.

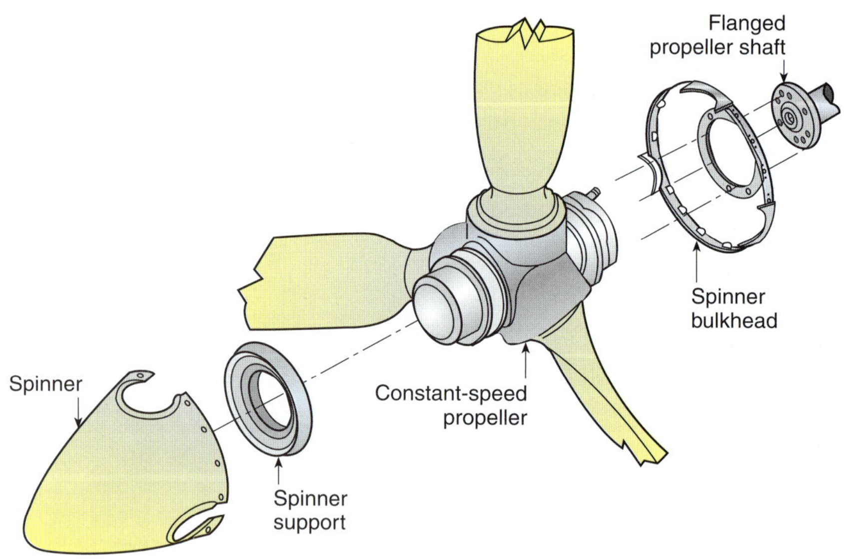

Figure 19-55. *Constant-speed propeller and spinner installation on a flanged propeller shaft*

Installation on a Splined Shaft

The most popular type of propeller shaft on reciprocating engines built through World War II and on the larger turboprop engines is the splined shaft. The sizes of splined shafts are identified by an SAE (Society of Automotive Engineers) number. SAE 20 splines are used on engines in the 200-horsepower range; SAE 30 splines are used in the 300- and 400-horsepower range, and SAE 40 in the 500- and 600-horsepower range. SAE 50 in the 1,000-horsepower range and SAE 60 and 70 are used for larger engines.

Splines are longitudinal grooves cut in the periphery of the shaft. The grooves and lands (the space between the grooves) are the same size, and one groove is either missing or has a screw in it to form a master spline. The purpose of the master spline is to ensure the propeller's correct installation in relation to the throws of the crankshaft. This orientation is critical to minimize vibration. *See* Figure 19-56.

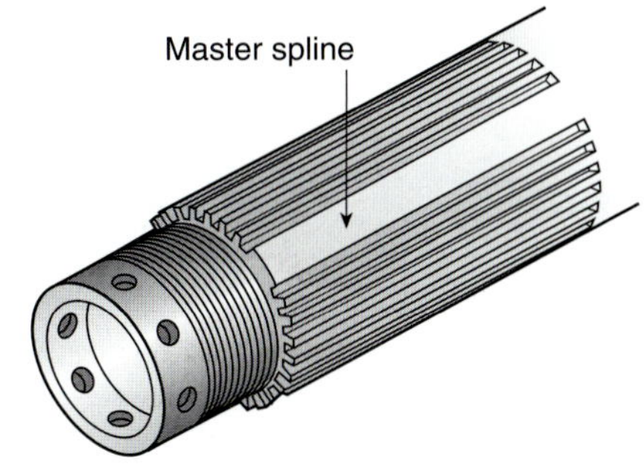

Figure 19-56. *Splined propeller shaft showing the master spline*

The inside of the propeller hub is splined to match the shaft, and the hub is centered on the shaft with two cones. The rear cone is a single-piece split bronze cone, and is considered to be part of the engine. The front cone is a two-piece hardened steel cone and is considered to be part of the propeller. The two halves are marked with the same serial number to ensure that only a matched set is used.

To install a propeller, lightly coat the shaft with clean engine oil or anti-seize compound, and slide the propeller in place. Assemble the two-piece front cone over the flange of the propeller retaining nut, and screw the nut onto the end of the shaft. Tighten the nut to the specified torque, using a steel bar through the holes in the nut. Install the snap ring in the hub, and safety the hub to the shaft with a clevis pin, washer, and cotter pin. The head of the clevis pin should be inside the hollow crankshaft as shown in Figure 19-57.

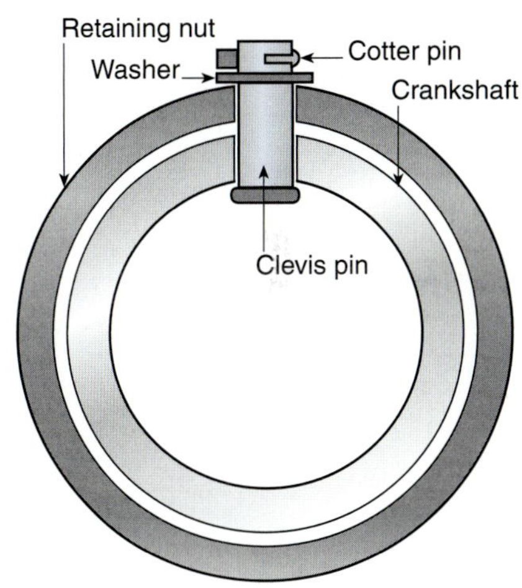

Figure 19-57. *Propellers on splined and tapered shafts are safetied by passing a clevis pin through matching holes in the end of the crankshaft and the propeller retaining nut. The clevis pin is safetied with a washer and cotter pin.*

There is a possible problem when installing a propeller on a splined shaft that prevents the propeller seating on the cones. Either the front or rear cone can bottom.

A rear cone bottoms if its point contacts the step at the junction of the smaller inside diameter of the cone taper seat and the internal propeller splines. Correct this condition is by machining off the tip of the cone. *See* Figure 19-58A on the next page.

A front cone bottoms if its tip contacts the ends of the splines on the propeller shaft, as shown in Figure 19-58B. To correct front cone bottoming, place a bronze spacer on the propeller shaft behind the rear cone. This moves the propeller forward and allows the cone to seat in the propeller rather than contacting the shaft splines. Front cone bottoming will cause the required torque to be reached abruptly when tightening the propeller retaining nut. If you suspect this condition, perform a blueing ink transfer (Prussian Blue) inspection. If bottoming is not corrected, propeller and crankshaft splines will be damaged in flight due to loose attachment. A resulting imbalance could cause additional severe damage.

bottom. *(v)* A condition in the installation of a propeller on a splined shaft when either the front or rear cone contacts an obstruction that prevents the cone from properly seating inside the propeller hub.

Propellers installed on either splined or tapered shafts can sometimes be extremely difficult to loosen from the shaft. To correct this, the manufacturer has made provisions for removing the propeller by installing a snap ring inside the hub. To remove the propeller, first remove the safetying clevis pin and loosen the retaining nut with a steel bar through the holes. The nut will back off for a turn or two and pull the front cone with it. Then the front cone contacts the snap ring and further turning pulls the propeller away from the rear cone and off of the shaft.

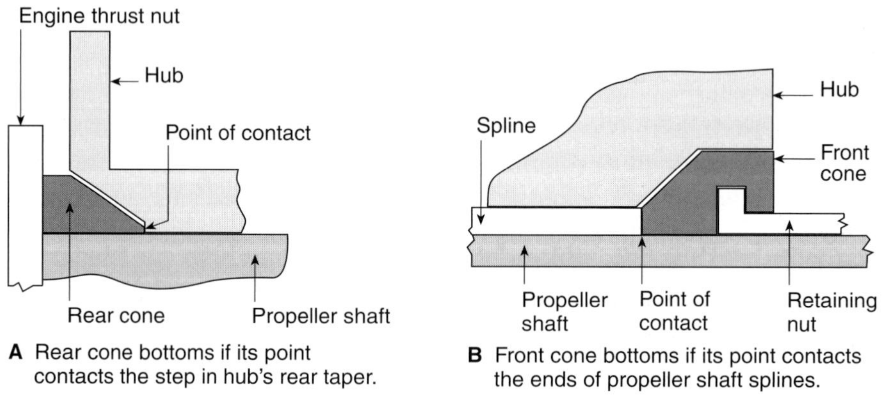

Figure 19-58. *Propeller cone bottoming*

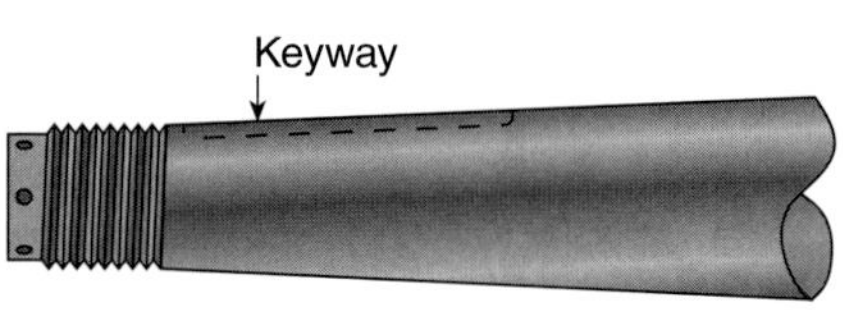

Figure 19-59. *A tapered propeller shaft*

Installation on a Tapered Shaft

Some small engines have a tapered crankshaft onto which the propeller is locked with a steel key fitting in a keyway slot cut into the tapered surface.

Most propellers installed on tapered shafts are made of wood, which requires a steel hub with a tapered hole and a keyway slot cut into it.

The hub is installed in the propeller and fitted to the tapered shaft. With the key installed in the slot in the shaft, apply Prussian blue transfer dye to the shaft and install the hub, tightening the retaining nut to the required torque. Remove the nut and hub, and note the amount of dye that has transferred to the taper inside the hub. If less than 70% of the surface is covered, remove the key and lap the hub to the shaft, using a fine-grit lapping compound. Be sure to follow the engine manufacturer's instructions in detail when lapping the hub to the shaft. When the lapping is completed, clean the hub and shaft to remove all traces of lapping compound and install the propeller.

Apply a very light coat of clean engine oil or antiseize compound to the shaft and slide the propeller and hub into place. Be sure that the threads on the end of the shaft are clean and dry, and install the retaining nut and torque it to the correct value, using a steel rod through the holes in the nut as the wrench. Install the snap ring in the groove in the hub. Safety the propeller using a clevis pin through holes in the propeller shaft and nut, positioning the head of the pin on the inside of the shaft. Install a washer on the pin and safety it with a cotter pin. On some propellers, the retaining nut is placed in the hub and loosely retained with a snap ring before the propeller is installed on the crankshaft.

Propeller Vibration

Vibration has always been a major problem in aircraft operation. The lightweight structure has so little mass that it cannot dampen or absorb vibrations that disturb the occupants, fatigue the structure, wear components, and cause cracks.

There are two sources of propeller-induced vibration: those caused by an out-of-track condition and those caused by an out-of-balance condition.

Propeller Track

Propeller track is the path followed by a blade segment in one rotation. If one blade does not follow in the same track as the others, its angle of attack and thus the thrust it produces, is different, and vibration will result.

track. The path followed by a blade segment of a propeller or helicopter rotor in one rotation.

Propeller track should be checked on every annual and 100-hour inspection and any time vibration is a problem. To make this check, chock the wheels so the airplane cannot move and place a board under the propeller so the blade tip nearly touches it. Put a mark on the board at the tip of the propeller, and rotate the propeller until the next blade is near the mark. The amount that blades can be out of track is specified by the propeller manufacturer, but for a light airplane, $\frac{1}{16}$ inch is normally the maximum allowed for a metal propeller and $\frac{1}{8}$ inch for a wood propeller.

Checking the track on some constant-speed propellers is a little more involved. For example, some McCauley propellers require that each blade be positioned vertically downward and the blade loaded lightly aft and then forward with a reference mark made on the board for each position. Then a primary mark is to be scribed at the midpoint between these two. After each blade has been represented with a midpoint reference, the difference between these reference marks is the amount of out-of-track. This extra effort is required because of the play in the blade-retaining bearings when they are not loaded by centrifugal force. The difference between the loaded forward and aft reference marks for each blade is caused by the bearing play.

A slightly out-of-track condition on a wood or metal fixed-pitch propeller can be corrected by placing thin metal shims between the propeller and the crankshaft flange. Any out-of-track condition on a constant-speed propeller should be referred to a certificated propeller repair shop.

Propeller Balance

There are two types of balance of importance when working with propellers: static and dynamic.

mandrel. A precision steel bar on which a propeller is mounted for balancing.

Static balance is checked and corrected at a propeller repair shop. The propeller is mounted on a mandrel and placed across perfectly level knife edges. For a two-bladed propeller, the balance is checked in two planes, one with the blades horizontal and one with them vertical.

A propeller with three or more blades is only checked for horizontal balance. Horizontal balance is attained when one or two blades will rest horizontally with no tendency for the propeller to rotate.

Vertical balance is the balance between the two sections of the propeller when it is divided along a straight line running longitudinally through each blade and the hub. A two-bladed propeller is in vertical balance when it rests with the blades vertical on the balance stand and there is no tendency for it to rotate.

When a two-bladed propeller is placed on a balance stand and it rotates so one blade always comes to rest downward, the propeller is out of horizontal balance. If it always comes to rest horizontally, it is out of balance vertically.

Horizontal imbalance of a wood propeller is corrected by adding solder to the metal tipping of the light blade. Vertical imbalance is corrected by attaching a brass weight with countersunk screws to the lightweight side of the hub.

Fixed-pitch metal propellers are balanced in a propeller repair station by removing some of the metal from the heavy side and then refinishing the propeller.

Some constant-speed propellers are balanced by placing a lead washer on a balancing stud inside the hollow blade shank. Small amounts of unbalance are corrected by packing lead wool in the hollow shanks of the bolts that fasten the halves of the propeller barrels together. This type of balancing can only be done by a certificated propeller repair station. Some of the other propellers use plates attached to the exterior of the hub with Fillister-head machine screws for both vertical and horizontal balance.

Dynamic balance is the most effective type of balancing as it takes all of the factors into consideration. It is done with the propeller installed on the engine in the airplane.

There are several aircraft balancers/analyzers on the market that are essential for helicopter maintenance and extremely valuable for propeller balancing. The ACES ProBalancer by TEC Aviation Division is a microprocessor-controlled instrument that measures the amount of vibration and shows the position and amount of weight needed on the propeller spinner bulkhead to correct the out-of-balance condition. *See* Figure 19-60.

The procedure for correcting a dynamic out-of-balance condition is simple and effective. Face the airplane into the wind, remove the cowling as necessary, and attach the vibration sensing unit to a crankcase bolt as near the front bearing or propeller as possible. For best performance, the sensor should be attached vertically, pointing either up or down. Mount the photo-tach with duct tape to the cowling about 12 to 18 inches behind the propeller. Secure the leads from both the vibration sensor and the photo-tach every eight to twelve inches with wire ties or tape to prevent movement and subsequent damage from the propeller slipstream.

Figure 19-60. *The ACES ProBalancer is a self-contained microprocessor-controlled system for determining the amount of balance weight and the location of the weight to attach to the propeller spinner bulkhead.*

After mounting the photo-tach on the cowling, attach a two-inch strip of reflective tape to the target propeller blade directly in front of the photo-tach lens. The retroreflective beam from the photo-tach shines on the tape and is reflected back to the photo-tach. This induces an electrical pulse that is sent to the balancer for speed and phase-angle reference. At the same time, the vibration sensor sends its electrical signal to the balancer indicating the severity of the out-of-balance condition in IPS, or inches per second, of movement.

Connect the cables from the vibration sensor and photo-tach to the ProBalancer. Turn the power switch to ON, press the ABORT key, and select NEW JOB. Follow the instructions on the screen for entering the balancing RPM, horsepower, and number of vibration sensors being used. Start the engine and allow it to stabilize at its idle RPM. This allows the unit to check itself for proper operation. Now advance the throttles to the balancing RPM you entered into the unit, and press the START button. The instrument displays the angular location of the out-of-balance condition and the current RPM. Press the STOP key, and the instrument will give the option to accept the current data or retake it. If you accept the data, the instrument will display a suggested test weight, in grams, and a mounting location, in degrees. Shut the engine down and install the suggested weight as near as possible to the suggested location. The test weights are normally attached under a spinner attaching screw. Enter the exact number of grams and angular location of the test weight into the balancer and make a second run at the same RPM. Repeat these steps (usually three runs total) until the vibration is reduced to an acceptable level. A vibration level of less than 0.07 IPS is usually considered acceptable; however, you can reduce the out-of-balance even further if you desire. When the vibration is reduced to the desired level, install the permanent weights using only the type of weights and instructions specified by the airframe or propeller manufacturer.

The ProBalancer offers an option to split weights. If the location for mounting the permanent weights or the weight allowed per location is limited, enter the angles that are available and the unit will give a weight correction for each new location.

Install the permanent weights, and make a final engine run to ensure that they have the same balancing effect as the test weights.

Propeller Inspections

Routine propeller inspections are made as part of the overall powerplant inspections, and while the propeller is only a small part of the powerplant, its importance cannot be overemphasized.

Technically, the propeller is a major component of the aircraft, and information concerning it is to be entered into the airframe log. Often, the mistake is made of entering the maintenance record in the powerplant log only.

Wood Propellers

At one time wood propellers were installed on a large percentage of personal airplanes. But metal propellers have proven so much more cost effective that practically the only wood propellers an AMT is likely to encounter on an FAA-certificated airplane will be on an antique, homebuilt, or ultralight. As primary category aircraft become more popular, however, there will probably be a resurgence in wood propellers.

An AMT with a powerplant rating is responsible for determining the airworthiness of a wood propeller, and FAA-certificated propeller repair stations approved for wood propellers can repair certain damages. If there is any doubt about the airworthiness of a propeller, contact one of these shops.

Advisory Circular 43.13-1A *Acceptable Methods, Techniques and Practices, Aircraft Inspection and Repair* lists a number of faults that could render a propeller irreparable. If any of the conditions shown in Figure 19-61 are found, the propeller should be referred to an approved repair station.

- A crack or deep cut across the grain of the wood
- Dry rot
- Split blades
- Separated laminations, except the outside lamination of fixed-pitch propellers
- More screw or rivet holes, including holes filled with dowels, than used to attach the metal leading edge strip and tip
- An appreciable warp
- An appreciable portion of wood missing
- A crack, cut, or damage to the metal shank or sleeve of blades
- Broken lag screws which attach the metal sleeve to the blade
- Oversize shaft hole in fixed-pitch propellers
- Cracks between the shaft hole and bolt holes
- Cracked internal laminations
- Excessively elongated bolt holes

Figure 19-61. *Defects that could render a wood propeller irreparable*

Metal Propellers

Most metal propellers have forged aluminum alloy blades protected from corrosion by anodizing and often further protected by polyurethane enamel. A few propellers with hollow steel blades are still flying, but these are usually found only on special-purpose airplanes.

One of the most important things to inspect on a propeller is the condition of the leading edge. When an engine is run up in an area with dirt or loose gravel, some small pebbles will likely be drawn by the propeller vortex into the whirling blades and make small impressions, or nicks, in the leading edge. It takes only a very small nick to concentrate the stresses enough to cause blade failure. The area most likely to fail because of a leading-edge nick is between six and twelve inches from the tip. In this area, the mass beyond the nick is appreciable, the centrifugal force is extremely high, the bending forces are great, and the blade is relatively thin.

The FAA has not specified limits to the depth of a nick a technician can dress out of the leading or trailing edge of a propeller blade, as so many variables are involved. AC 43.13-1B, Figures 8-27 and 8-28 specify the amount of metal that may be removed from an aluminum propeller blade. This is expressed in percent of thickness and width at specific blade stations. This data may be used only if the propeller manufacturer does not provide instructions for the same repair.

It would be illegal to use the AC 43.13-1B limits and methods if the manufacturer prescribes a repair method and/or repair limits. As a precautionary note, with the rising probability of lawsuits from an aircraft accident for anyone having performed maintenance on the aircraft, regardless of the accident's cause, it would behoove the mechanic to call the manufacturer and the FAA when the legality of the repair is in question.

If the propeller manufacturer does not indicate the method of making the repair, the 43.13-1B method follows. Small nicks in the leading or trailing edge can be removed by carefully filing them parallel to the blade length. Form a smooth dish-shaped depression whose length is about 10 times the depth of the nick. After you have filed all the damage out, carefully remove the file marks with progressively finer abrasive paper with a straight sanding motion along the blade length. Finish the polishing with 600-grit paper. It is important to not use paper with a grit that if imbedded in the blade surface, will set up anodic/cathodic corrosion cells. Treat the repaired area with a conversion coating material such as Alodine, as described in Chapter 8 of the *General* textbook of this *Aviation Maintenance Technician Series*. Small nicks in the face of a blade can be filed out with a spoon-shaped file. The depression should have a diameter of 20 times the depth of the damage. All file marks must be removed with fine sandpaper or crocus cloth and the repaired area Alodine treated and restored to its original finish.

After any repair of this type has been made, check the dynamic balance of the propeller.

Propeller Storage

Propellers should be stored with their blades in a horizontal position. If a wood propeller is left with its blades in a vertical position, moisture can collect in the lower blade and cause an out-of-balance condition.

When an aircraft with three-blade constant-speed propellers is left outside for an extended period of time, position the propeller with one of the blades pointing down. This prevents water from collecting around a blade seal and entering the hub.

Composite Propeller Blades

Blades made of composite materials rather than metal enhance the efficiency of modern turbopropellers. Composite blades have a long life, and you may be able to repair some types of minor damage in the field, but because of the high stresses involved in these blades, it is advisable to consult a person experienced with the particular propeller before attempting any repair.

Propeller Repairs and Alterations

The highly stressed nature of propellers has caused the FAA to have special requirements for their repairs and alterations. Minor repairs, alterations, and inspections can be performed by an AMT holding a powerplant rating, but major repairs and alterations can be done only by the propeller manufacturer or by an FAA-certificated repair station approved for the particular repair or alteration. A list of major propeller repairs and alterations taken from 14 CFR Part 43 *Maintenance, Rebuilding, and Alteration*, *Appendix A* are included in Figure 19-62.

minor repair. Any repair that does not fit the definition of a major repair.

minor alteration. Any alteration that does not fit the definition of a major alteration.

major repair. A repair to a component that if improperly done might appreciably affect weight, balance, structural strength, performance, powerplant operation, flight characteristics or other qualities affecting airworthiness; a repair not done according to accepted practices, or one that cannot be done by elementary operations.

major alteration. An alteration not listed in the aircraft, aircraft engine, or propeller specifications that might appreciably affect weight, balance, structural strength, powerplant operation, flight characteristics, or other qualities affecting airworthiness; an alteration not done according to accepted practices or one that cannot be done by elementary operations.

Propeller Major Repairs
- Any repairs to, or straightening of steel blades
- Repairing or machining steel hubs
- Shortening of blades
- Retipping wood propellers
- Replacement of outer laminations of fixed-pitch wood propellers
- Inlay work on wood blades
- Repairs to composition blades
- Replacement of tip fabric
- Replacement of plastic covering
- Repair of propeller governors
- Overhaul of controllable-pitch propellers
- Repairs to deep dents, cuts, scars, nicks, etc., and straightening of aluminum blades
- Repair or replacement of internal elements of blades

Propeller Major Alterations
- Changes in blade design
- Changes in hub design
- Changes in governor or control design
- Installation of a propeller governor or feathering system
- Installation of a propeller deicing system
- Installation of parts not approved for propeller

Figure 19-62. *Major propeller repairs and alterations*

Determining Propeller Repairability

The most frequent major damage to a propeller is bent blades. No straightening is allowed by anyone other than the propeller manufacturer or an FAA-certificated repair station approved for the particular operation. It is, how-ever, the responsibility of an AMT to know the repairable limits of a propeller so a decision can be made to either scrap the propeller or to send it to a repair station. When in doubt, consult technicians at the repair station, because they may have additional approved techniques and procedures that give them latitude in making repairs.

Advisory Circular 43.13-1B *Acceptable Methods, Techniques, and Practices — Aircraft Inspection and Repair* lists the limits a repair station must observe when repairing a propeller.

STUDY QUESTIONS: PROPELLER INSTALLATION, INSPECTION, AND MAINTENANCE

Answers begin on Page 733. *Page numbers refer to chapter text.*

78. Three types of propeller shafts used on aircraft engines are:
    a. ______________________________
    b. ______________________________
    c. ______________________________
    Page 714

79. The most popular type of propeller shaft for both reciprocating and turbine engines is the _____________________ shaft. *Page 714*

80. Two functions of a propeller spinner are:
    a. ______________________________
    b. ______________________________
    Page 716

81. A propeller must always be installed in correct relation to the throws of the crankshaft. This is ensured on a propeller installed on a splined shaft by the ___________________________ . *Page 717*

82. A propeller is centered on a splined shaft by two _____________________ . *Page 717*

83. The snap ring inside a propeller hub installed on a splined or tapered shaft is there to assist in _______________________ (installing or removing) the propeller. *Page 718*

84. A spacer may be placed behind the rear cone on a splined-shaft propeller installation to prevent _______________________ (front or rear) cone bottoming. *Page 717*

85. When fitting a propeller hub to a tapered shaft, the amount of contact is determined by dye transferred from the shaft to the inside of the hub. For a fit to be satisfactory, at least __________% of the surface inside the hub must show contact with the dye. *Page 718*

86. Two sources of propeller-induced vibration are:
    a. ______________________________
    b. ______________________________
    Page 719

87. If there are no other specifications, the maximum amount the blades of a fixed pitch metal propeller are allowed to be out of track are ___________________ inch. A wood propeller is allowed to be out of track by ____________________ inch. *Page 719*

Continued

STUDY QUESTIONS: PROPELLER INSTALLATION, INSPECTION, AND MAINTENANCE *Continued*

88. There are two types of balance that are important to a propeller. These are:
    a. ____________________________
    b. ____________________________
    Page 720

89. The portion of a propeller blade that must be most carefully inspected is the _____________________ edge. *Page 723*

90. A small nick in the leading edge of a propeller will cause a __________________________ that can break the blade. *Page 723*

91. If a small nick is removed from the leading edge of a propeller, the cleaned-out depression should have a length of ________________ times the depth of the nick. *Page 724*

92. A cleaned-out depression in the face of a propeller should have a diameter ___________________ times the depth of the damage. *Page 724*

93. Propellers must be maintained by an aviation maintenance technician who holds a/an _____________________ (airframe or powerplant) rating. *Page 723*

94. An AMT holding a powerplant rating _____________________ (can or cannot) perform a major repair to a propeller. *Page 723*

95. Shortening an aluminum propeller blade by one inch is a _____________________ (major or minor) repair. *Page 724*

Propeller Auxiliary Systems

Two systems that enable the propeller to operate more efficiently are the synchronizer system and its elaboration, the synchrophaser, and the ice control system which includes anti-icing and deicing.

Synchronizer Systems

As previously mentioned, vibration has always been a problem with aircraft because the lightweight structure does not have sufficient mass to absorb it. Some annoying and harmful vibration in multiengine airplanes is caused by the propellers being slightly out of synchronization, that is, not turning at exactly the same speed. This type of vibration has a low fundamental frequency that is the approximate difference between the RPM of the engines. To prevent this vibration, the propellers can be synchronized with an electronic synchronizer system shown in Figure 19-63.

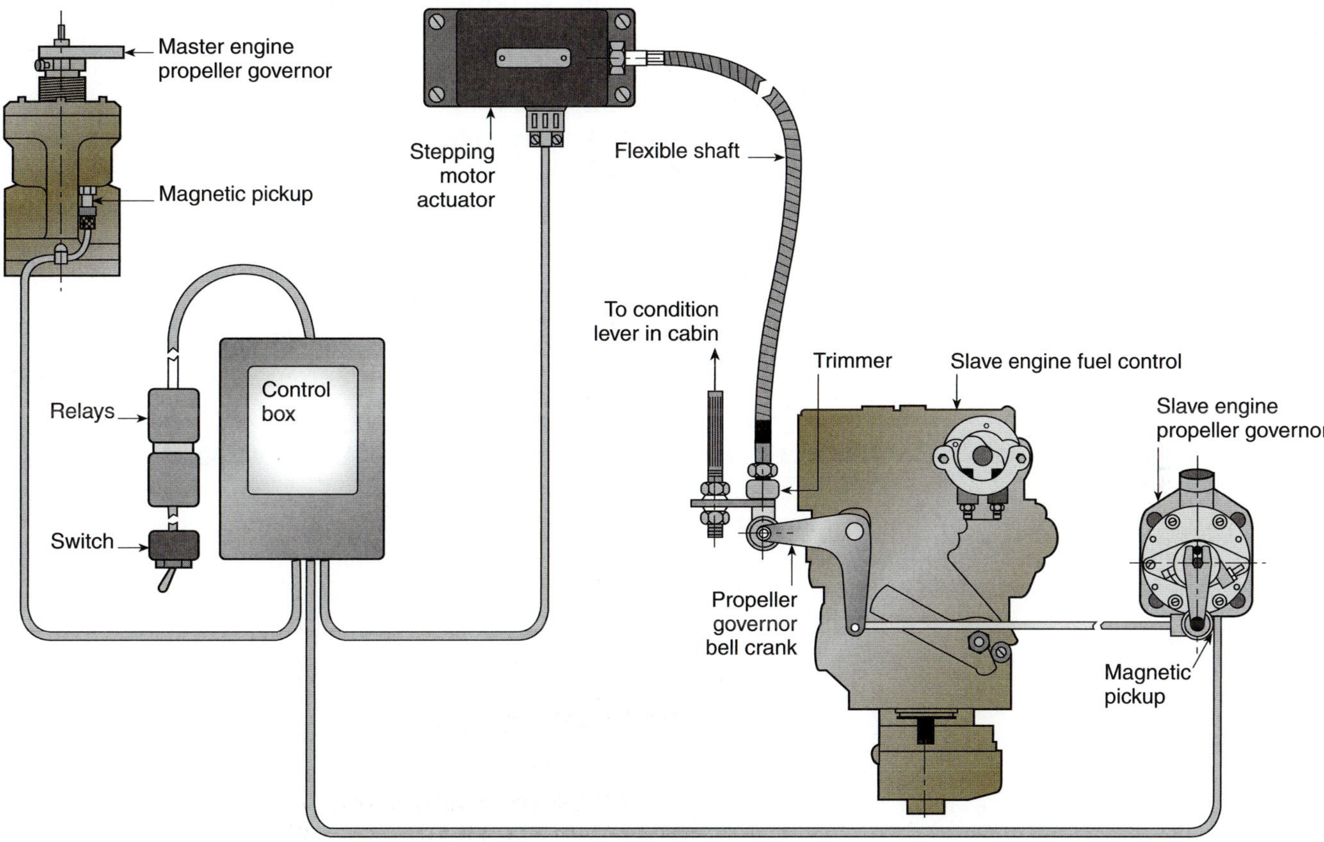

Figure 19-63. *Schematic diagram of a propeller synchronizer system for a twin-engine turboprop airplane*

One engine is designated as the master engine. When the RPM of this engine is adjusted by the pilot and the synchronizer system is ON, the RPM of the slave engine will automatically adjust to the same RPM.

stepping motor. A precision electric motor whose output shaft position is changed in steps by pulses from the control device.

Each propeller governor contains a rotating magnet and a magnetic pickup which produces alternating current as the governor rotates. The frequency of this AC is proportional to the speed of the governor. The outputs from the two governors are compared in the synchronizer control box, and an output signal is sent to the DC stepping motor actuator. A flexible steel shaft connects the actuator to the propeller governor bell crank on the fuel control of the slave engine. If the slave engine is slower than the master engine, the control box will drive the actuator motor in a direction that will move the bell crank and connection arm on the slave motor fuel control and the propeller governor, in the correct direction to increase its RPM.

The operation of the synchronizer system is simple. It is left OFF during takeoff and landing. When the aircraft is trimmed for cruise flight, the condition levers of the engines are manually adjusted to bring their RPM close enough to the same speed that the engines will be within the synchronizing range. Then the synchronizer is turned ON. Any difference in RPM is sensed, and the slave engine fuel control and propeller governor are adjusted so the slave engine RPM matches that of the master engine.

When making power changes in flight, adjust both condition levers together to keep the RPM within synchronizing range. If the engines get out of synchronization beyond the limits of the system, the actuator will be driven to the limit of its travel. Turn the system OFF and the actuator will return to its center position. Manually synchronize the engines and turn the system ON. It will fine tune the synchronization and hold the engines together.

Synchrophasing System

A refinement to the synchronizer is the synchrophasing system. In this system, a pulse generator is keyed to a master blade in each propeller. This generator produces a pulse each time the master blade is in its reference position. The pulses from the two pickups are directed into the control box and compared. With a knob on the synchrophaser panel the pilot can command the actuator to fine tune the slave engine propeller governor just enough to keep the master blades in the angular relationship that produces the minimum vibration and noise.

Ice Control Systems

anti-icing. Prevention of the formation of ice on a surface.

Ice on a propeller blade changes its airfoil shape and decreases the thrust it produces. Though rare to minimal, it can also create an unbalanced condition which produces vibration that can damage the engine as well as the airframe. The earliest propeller ice control, a system that is still in use, is chemical anti-icing. Isopropyl alcohol is carried in a tank in the aircraft, and when icing conditions are anticipated, some of it is pumped into a slinger ring around the propeller hub. The ring-supply tubes deliver the alcohol to the propeller leading edge by centrifugal force and grooved deicer boots help carry it out the length of the blades. Centrifugal force slings it out along the leading edges of the blades. Some propellers have molded rubber feed shoes bonded to the blade roots to concentrate the flow of fluid along the portions of the blade most susceptible to ice formation. Keeping the blade surfaces perfectly smooth and waxed helps prevent ice from sticking when it forms. *See* Figure 19-64.

Propellers are deiced with an electrothermal system that has rubber boots bonded to the leading edges of the blades. These boots contain embedded electrical heating elements supplied with current from a propeller deicing timer.

deicing. The removal of ice that has formed on a surface.

Figure 19-65 on the next page is the schematic diagram of a typical system used on a twin turboprop airplane. Current flows from the bus through the 20-amp Auto Prop Deice circuit breaker/switch into the deicer timer unit. When the manual-override relays are not energized, this current flows through brushes riding on slip rings mounted on the propeller spinner bulkhead and into the heating elements bonded to the propeller blades. The slip rings are connected to the heater elements through flexible conductors that allow the blades to change their pitch angle.

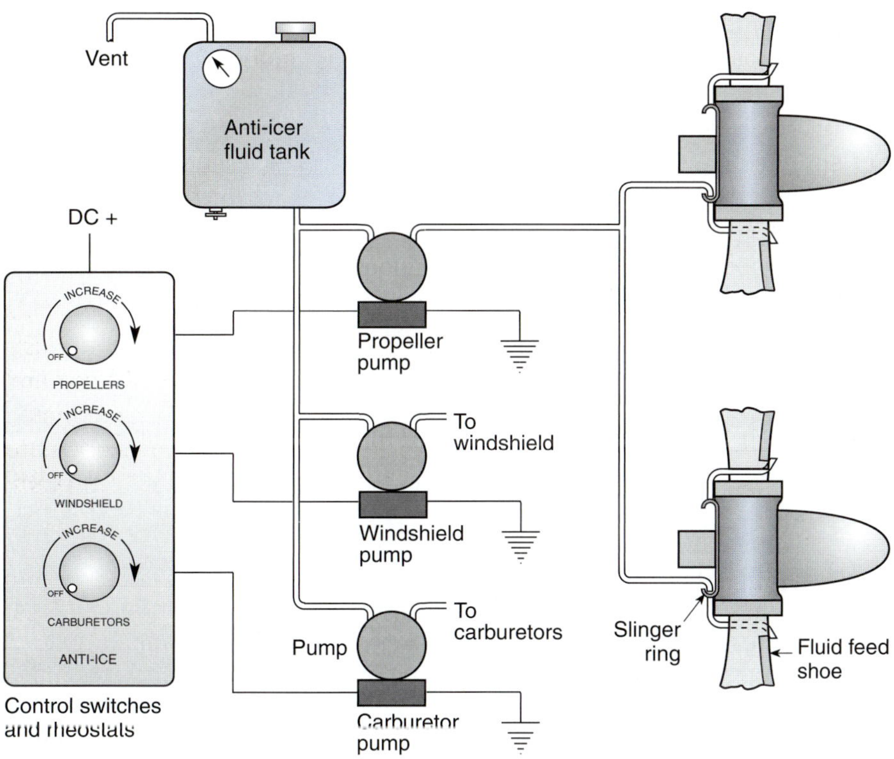

Figure 19-64. *Typical chemical anti-icing system for propellers*

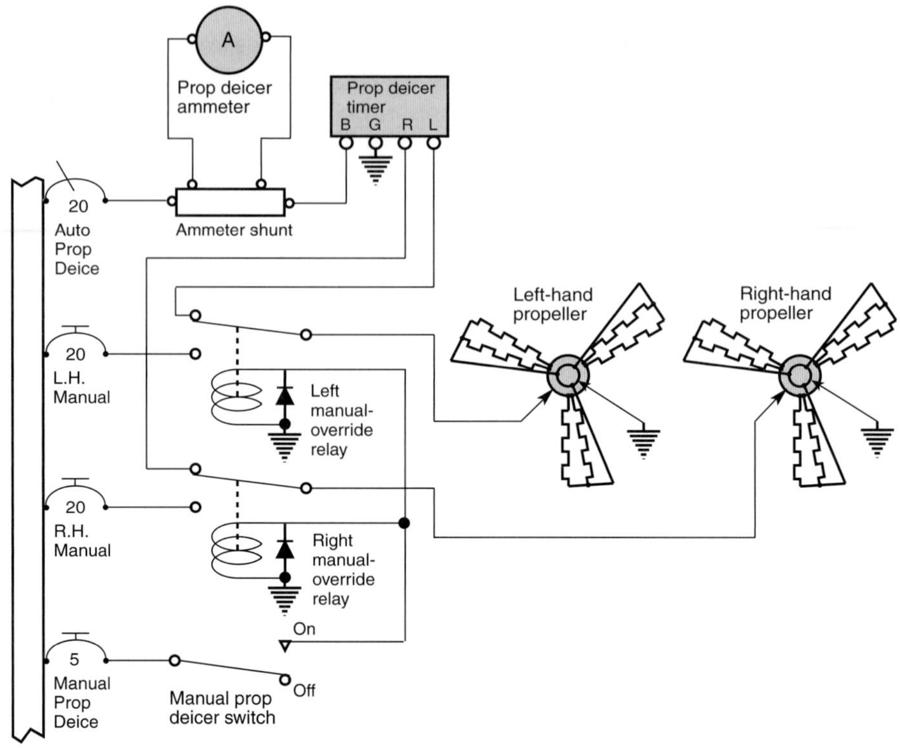

Figure 19-65. *Electrothermal propeller deicing system*

The timer sends current through the right propeller for about 90 seconds, then switches over and sends current through the left propeller for 90 seconds.

Some propeller deicing systems have two separate heating elements on each blade. Current flows through the right propeller outboard element for about 30 seconds, then through the right propeller inboard element for the same length of time. After the right propeller is deiced, the timer shifts over and sends current through the left propeller outboard elements and then the left propeller inboard elements.

Current cycles of the two propellers are controlled by the timer as long as the propeller Auto Prop Deice switch is ON. When the Manual Prop Deicer switch is held in its momentary ON position, the two manual-override relays are energized, and current flows directly from the bus to the blades without going through the timer.

A ground check can easily tell whether or not the deicing system is operating correctly in the AUTOMATIC mode by observing the ammeter for the proper amount of current flow, and by feeling the boots in their proper sequence to see if they are all heating. All of the boots should have a similar heat rise in the same length of time.

Manufacturers limit the operation of the propeller deicing system on the ground because the inflight air cooling is not present and prolonged use will cause damage.

STUDY QUESTIONS: PROPELLER AUXILIARY SYSTEMS

Answers begin below. Page numbers refer to chapter text.

96. The vibration that is caused by the propellers being out of synchronization has a ____________________ (high or low) fundamental frequency. *Page 729*

97. The pickups that determine whether or not the two engines are in synchronization are in the ____________________ . *Page 730*

98. Two types of ice control systems for propellers are:
    a. ____________________
    b. ____________________
    Page 730, 731

Answers to Chapter 19 Study Questions

1. 75
2. a. engine RPM
   b. forward speed of aircraft
3. back
4. are not
5. hub
6. effective
7. slip
8. momentum, area
9. a. shape of the airfoil section
   b. area of the blade
   c. angle of attack
   d. density of the air
   e. speed of the blade through the air
10. a. cross section of the blade
    b. blade pitch angle
11. increases
12. thrust, torque
13. a. centrifugal force
    b. thrust bending force
    c. torque bending force
    d. aerodynamic twisting force
    e. centrifugal twisting force
14. centrifugal
15. thrust
16. torque
17. increase
18. decrease
19. CTF
20. left
21. tip speed, tip clearance
22. lower
23. high
24. governor
25. feathering
26. 90
27. low
28. birch
29. 60
30. aluminum, anodized
31. a. fixed-pitch
    b. constant-speed
32. hydraulically
33. a. counterweight
    b. noncounterweight
34. a. aerodynamic twisting force
    b. oil pressure
35. a. centrifugal twisting force
    b. internal spring force
36. engine speed
37. low
38. high
39. increasing
40. decrease

Continued

Answers to Chapter 19 Study Questions *(Continued)*

41. electric motor
42. can
43. draining oil from
44. a. centrifugal force on the counterweights
    b. force from a feathering spring
45. feather
46. high-pitch stop plate
47. unfeathering
48. draining oil from
49. high
50. governor
51. a. single-shaft
    b. free-turbine
52. a. power lever
    b. speed, or condition, lever
53. speed
54. a. tachometer
    b. torquemeter
    c. fuel flowmeter
    d. EGT indicator
55. airframe structure
56. decreases
57. a. Alpha, or flight
    b. Beta, or ground
58. out of
59. a. centrifugal force on the blade counterweights
    b. feather spring
60. propeller governor
61. propeller pitch control
62. underspeed governor
63. electric motor
64. low
65. feather
66. start-lock pins
67. feather
68. a. power control lever
    b. propeller control lever
    c. condition lever
69. single-acting
70. Beta
71. feather
72. torquemeter oil pressure
73. a. lighter weight
    b. lower notch sensitivity
    c. more cost effective
74. turboprop
75. 0.6
76. critical Mach number
77. swirl
78. a. flanged
    b. splined
    c. tapered
79. flanged
80. a. streamline engine installation
    b. direct cooling air into cowling
81. master spline
82. cones
83. removing
84. front
85. 70
86. a. out-of-track condition
    b. out-of-balance condition
87. $\frac{1}{16}$, $\frac{1}{8}$
88. a. static
    b. dynamic
89. leading
90. stress concentration
91. 10
92. 20
93. powerplant
94. cannot
95. major
96. low
97. propeller governors
98. a. anti-ice systems
    b. deice systems

GLOSSARY

abradable strip. A strip of material in the compressor housing of some axial-flow gas turbine engines. The tip of the compressor blade touches the abradable strip, and wears, or abrades, a groove in it. This groove ensures the minimum tip clearance.

abradable tip (compressor blade tip). The tip of some axial-flow compressor blades constructed so that it will abrade, or wear away, upon contact with the compressor housing, which ensures the minimum tip clearance between the blade and the housing.

absolute pressure. Pressure referenced from zero pressure or a vacuum.

AC (alternating current). Electrical current in which the electrons continually change their rate of flow and periodically reverse their direction.

ACC (active clearance control). A system for controlling the clearance between tips of the compressor and turbine blades and the case of high-performance turbofan engines. When the engine is operating at maximum power, the blade tip clearance should be minimum, and the ACC system sprays cool fan discharge air over the outside of the engine case. This causes the case to shrink enough to decrease the tip clearance. For flight conditions that do not require such close clearance, the cooling air is turned off, and the case expands to its normal dimensions. The control of the ACC system is done by the FADEC, or full-authority digital electronic control.

acceleration. The amount the velocity of an object is increased by a force during each second it is acted upon by that force. Acceleration is usually measured and expressed in terms of feet per second, per second (fps^2).

accessory end. The end of a reciprocating engine on which many of the accessories are mounted. Also called the antipropeller end.

accumulator. A hydraulic component that stores a noncompressible fluid, such as oil, under pressure. An accumulator has two compartments separated by a flexible or movable partition with one compartment containing compressed air. When oil is pumped into the other compartment, the partition moves over, further compressing the air which holds pressure on the oil.

AD (ashless dispersant) oil. A mineral-base lubricating oil used in reciprocating engines. This oil does not contain any metallic ash-forming additives, but has additives that disperse the contaminants and hold them in suspension until they can be removed by filters.

ADC (air data computer). An electronic computer in an aircraft that senses pitot pressure, static pressure, and total air temperature. It produces an indication of altitude, indicated airspeed, true airspeed, and Mach number.

The output of the ADC is usable by any of the engine or flight control computers.

ADI (antidetonation injection) fluid. A mixture of water and methanol with a small amount of water-soluble oil added for corrosion prevention. Methanol is used primarily to prevent water from freezing at high altitude.

ADI (antidetonation injection) system. A system used with some large reciprocating engines in which a mixture of water and alcohol is sprayed into the engine with the fuel when operating at extremely high power. The fuel-air mixture is leaned to allow the engine to develop its maximum power, and the ADI fluid absorbs excessive heat when it vaporizes.

adiabatic change. A physical change that takes place within a material in which heat energy is neither added to the material, nor taken away. If a container of gas is compressed, with no heat energy added to or taken from it, the gas will become hotter; its temperature will rise.

Aerodrome. The name given by Dr. Samuel Langley to the flying machines built under his supervision between the years of 1891 and 1903.

Aeromatic propeller. A patented variable-pitch propeller that has counterweights around the blade shanks and the blades angled back from the hub to increase the effects of aerodynamic and centrifugal twisting forces. This propeller automatically maintains a relatively constant RPM for any throttle setting.

aft-fan engine. A turbofan engine with the fan mounted behind the compressor section. The blades of an aft-fan are normally extensions of the free turbine blades.

afterburner. A component in the exhaust system of a turbojet or turbofan engine used to increase the thrust for takeoff and for special flight conditions.

Since much of the air passing through a gas turbine engine is used only for cooling, it still contains a great deal of oxygen. Fuel is sprayed into the hot, oxygen-rich exhaust in the afterburner, where it is ignited and burned to produce additional thrust.

air bleed (carburetor component). A small hole in the fuel passage between the float bowl and the discharge nozzle of a float carburetor. Air drawn into the liquid fuel through the air bleed breaks the fuel up into an emulsion, making it easy to atomize and vaporize.

air cooling. The removal of unwanted heat from an aircraft engine by transferring the heat directly into the air flowing over the engine components.

air-fuel mixture ratio. The ratio of the weight of the air to that of the fuel in the mixture fed into the cylinders of an engine.

air impingement starter. A turbine engine starter that basically consists of a nozzle that blows a stream of compressed air against the turbine blades to rotate the compressor for starting the engine.

air-oil separator. A component in a turbine engine lubrication system that removes the air from the scavenged oil before it is returned the oil tank.

Airworthiness Directive. A regulatory notice sent out by the FAA to the registered owner of an aircraft informing him or her of the discovery of a condition that keeps the aircraft from continuing to meet its conditions for airworthiness. Also called AD notes, they are covered by Federal Aviation Regulations Part 39 *Airworthiness Directives*. They must be complied with within the specified time, and the fact, date, and method of compliance must be recorded in the aircraft maintenance records.

all-weather spark plug. A shielded spark plug designed for high altitude operation. The ceramic insulator is recessed into the shell to allow a resilient grommet on the ignition harness to provide a watertight seal.

All-weather spark plugs, also called high-altitude spark plugs, are identified by their ¾-20 shielding threads.

Alpha control range (Alpha mode). The flight operating mode from takeoff through landing for a turboprop engine. Alpha mode includes operations from 95% to 100% of the engine's rated RPM.

altitude engine. An aircraft reciprocating engine equipped with a supercharger that allows it to maintain its rated sea-level horsepower to an established higher altitude.

amateur-built aircraft. Aircraft built by individuals as a hobby rather than by factories as commercial products. Amateur-built or homebuilt aircraft do not fall under the stringent requirements imposed by the FAA on commercially built aircraft.

ambient air pressure. The pressure of the air that surrounds an object.

analog indicator. An indicator that shows the value of the parameter being measured by a number marked on a graduated dial aligned with a movable pointer.

angle of attack. The acute angle between the chord line of a propeller blade and the relative wind. The angle of attack is affected by both the engine RPM and the forward speed of the aircraft.

annual inspection. A complete inspection of the airframe and powerplant required for FAA-certificated aircraft operating under 14 CFR Part 91 *General Operating and Flight Rules*, and not on one of the authorized special inspection programs. An annual inspection must be conducted every 12 calendar months, and it must be conducted by an aviation maintenance technician who holds an Airframe and Powerplant rating and an Inspection Authorization.

The scope of an annual inspection is the same as that of a 100-hour inspection.

annular duct. A duct, or passage, that surrounds an object. The annular fan-discharge duct surrounds the core engine.

annular orifice. A ring-shaped orifice, normally one that surrounds another orifice.

annulus. A ring or groove around the outside of a circular body or shaft, or around the inside of a cylindrical hole.

annunciator panel. A panel of warning lights visible to the flight crew. The lights are identified by the name of the system they represent and are often covered with colored lenses. Red lights indicate a dangerous condition, amber indicates a system is armed, and green indicate a safe condition.

anodizing. A hard, airtight, unbroken oxide film electrolytically deposited on an aluminum alloy surface to protect it from corrosion.

anti-icing. Prevention of the formation of ice on a surface.

antipropeller end. The end of a reciprocating engine that does not attach to the propeller. Also called the accessory end.

APC. Absolute pressure controller.

APU (auxiliary power unit). A small turbine- or reciprocating-engine-powered generator, hydraulic pump, and air pump. APUs are installed in the aircraft and are used to supply electrical power, air, and hydraulic pressure when the main engines are not running.

aramid fiber. Fiber made from an organic compound of carbon, hydrogen, oxygen, and nitrogen. It has high strength and low density. It is flexible under load and is able to withstand impact, shock, and vibration. Kevlar® is a well-known aramid fiber.

aromatic compound. A chemical compound such as toluene, xylene, and benzene that is blended with gasoline to improve its antidetonation characteristics.

articulating rod. *See* link rod.

aspect ratio. The ratio of the length of an airfoil, such as a compressor blade, to its width.

asymmetrical loading. The loading of a propeller disk that causes one side to produce more thrust than the other side.

ATF (aerodynamic twisting force). The aerodynamic force that acts on a rotating propeller blade to increase its blade angle. The axis of rotation of a blade is near the center of its chord line, and the center of pressure is between the axis and the leading edge. Aerodynamic lift acting through the center of pressure tries to rotate the blade to a higher pitch angle.

atomize. The process of breaking a liquid down into tiny droplets or a fine spray. Atomized liquids vaporize easily.

augmentor tube. A long, specially shaped stainless steel tube mounted around the exhaust tail pipe of a reciprocating engine. As exhaust gases flow through the augmentor tube, they produce a low pressure in the engine compartment that draws in cooling air through the cylinder fins.

automatic intake valve. An intake valve opened by low pressure created inside the cylinder as the piston moves down. There is no mechanical means of opening it.

automatic mixture control (AMC). The device in a fuel metering system such as a carburetor or fuel injection system that keeps the fuel-air mixture ratio constant as the density of air changes with altitude.

Autosyn system. The registered trade name of a remote indicating instrument system. An Autosyn system uses an electromagnet rotor, excited with 400-hertz AC, and a three-phase distributed-pole stator.

axial bearing load. The load on a bearing parallel to the shaft on which the bearing is mounted. Thrust produces an axial load on a bearing.

axial turbine. A turbine that is turned by a fluid flowing through it in a direction that is approximately parallel to the shaft on which the turbine wheel is mounted.

axial-flow compressor. A type of compressor used in gas turbine engines. Air passes through the compressor in essentially a straight line, parallel to the axis of the compressor. The compressor is made of a number of stages of rotating compressor blades between stages of stationary stator vanes.

axis of rotation. The center line about which a propeller rotates.

babbitt. A soft, silvery metal used for main bearing inserts in aircraft reciprocating engines. Babbitt is made of tin with small amounts of copper and antimony.

back (propeller nomenclature). The curved surface of a propeller blade. The back of a propeller blade corresponds to the upper surface of an airplane wing.

back-suction mixture control. A type of mixture control used in some float carburetors that regulates the fuel-air mixture by varying the air pressure above the fuel in the float bowl.

baffle. A thin sheet metal shroud or bulkhead used to direct the flow of cooling air between and around the cylinder fins of an air-cooled reciprocating engine.

bayonet stack. An exhaust stack with an elongated and flattened end. The gases leave the stack through a slot perpendicular to its length. Bayonet stacks decrease both exhaust back pressure and noise.

BDC (bottom dead center). The position of a piston in a reciprocating engine when the piston is at the bottom of its stroke, and the wrist pin, crankpin, and center of the crankshaft are all in line.

bell mouth. The shape of the inlet of an augmentor tube that forms a smooth converging duct. The bell mouth shape allows the maximum amount of air to be drawn into the tube.

bell mouth inlet duct. A form of convergent inlet-air duct used to direct air into the compressor of a gas turbine engine. It is extremely efficient, and is used where there is little ram pressure available to force air into the engine. Bell mouth ducts are used in engine test cells and on engines installed in helicopters.

benzene. A colorless, volatile, flammable, aromatic hydrocarbon liquid which has the chemical formula C_6H_6.

Benzene, which is sometimes called benzoil, is used as a solvent, a cleaning fluid, and a fuel for some special types of reciprocating engines.

Bernoulli's principle. A physical principle that explains the relationship between kinetic and potential energy in a stream of moving fluid. When energy is neither added to nor taken from the fluid, any increase in its velocity (kinetic energy) will result in a corresponding decrease in its pressure (potential energy).

Beta control range (Beta mode). The range of operation of a turboprop powerplant used for in-flight approach and ground handling of the engine and aircraft. Typically, the Beta mode includes operations from 65% to 95% of the engine's rated RPM.

Beta tube. A tube in a Garrett TPE331 turboprop powerplant that extends into the propeller pitch control to act as a follow-up device. It provides movement of the propeller blades in proportion to movement of the power lever.

bezel. The rim which holds the glass cover in the case of an aircraft instrument.

BHP (brake horsepower). The actual horsepower delivered to the propeller shaft of a reciprocating or turboprop engine.

bidirectional fibers. Fibers in a piece of composite material arranged to sustain loads in two directions.

bimetallic hairspring. A flat, spiral-wound spring made of two strips of metal laid side-by-side and welded together. The two metals have different coefficients of expansion, and as the temperature changes, the spiral either tightens or loosens. A bimetallic hairspring is used in a thermocouple temperature indicator to compensate for temperature changes at the reference junction.

bimetallic strip. A metal strip made of two different types of metal fastened together side by side. When heated, the two metals expand different amounts and the strip warps or bends.

BITE (built-in test equipment). A troubleshooting system installed in many modern jet aircraft. BITE equipment monitors engine and airframe systems, and when a fault is found, isolates it and provides maintenance personnel with a code that identifies the LRU (line replaceable unit) that contains the fault.

blade. The component of a propeller that converts the rotation of the propeller shaft into thrust. The blade of a propeller corresponds to the wing of an airplane.

blade angle. The acute angle between the chord line of a propeller blade and the plane of rotation.

blade butt. The end of a propeller blade that fits into the hub.

blade root. The ridges or upset portion of a propeller blade that holds it in the hub.

blade shank. The rounded portion of a propeller blade between the root and the airfoil section.

blade station. A reference position on a propeller blade that is a specified number of inches from the center of the propeller hub.

blade tip. The opposite end from the root of a propeller blade.

blending. A method of repairing damaged compressor and turbine blades. The damage is removed and the area is cleaned out with a fine file to form a shallow depression with generous radii. The file marks are then removed with a fine abrasive stone so the surface of the repaired area will match the surface of the rest of the blade.

blisk. A turbine wheel machined from a single slab of steel. The disk and blades are an integral unit.

blow-in doors. Spring-loaded doors in the inlet duct of some turbojet or turbofan engine installations that are opened by differential air pressure when inlet air pressure drops below that of the ambient air. Air flowing through the doors adds to the normal inlet air passing through the engine and helps prevent compressor stall.

bluckets. The portions of aft-fan blades that are in the exhaust of the core engine. Bluckets drive the fan from energy received from hot gases leaving the core engine.

BMEP (brake mean effective pressure). The average pressure inside the cylinder of a reciprocating engine during the power stroke. BMEP, measured in pounds per square inch, relates to the torque produced by the engine and can be calculated when you know the brake horsepower.

boost. A term for manifold pressure that has been increased above the ambient atmospheric pressure by a supercharger.

bootstrapping. An action that is self-initiating or self-sustaining. In a turbocharger system, bootstrapping describes a transient increase in engine power that causes the turbocharger to speed up, which in turn causes the engine to produce more power.

bore. The diameter of a reciprocating engine cylinder.

borescope. An inspection tool for viewing the inside of a turbine engine without disassembling it. The instrument consists of a light, mirror, and magnifying lens mounted inside a small-diameter tube that is inserted into a turbine engine through borescope inspection ports.

boss. An enlarged area in a casting or machined part. A boss provides additional strength to the part where holes for mounting or attaching parts are drilled.

bottom. *(verb)* A condition in the installation of a propeller on a splined shaft when either the front or rear cone contacts an obstruction that prevents the cone from properly seating inside the propeller hub.

Bourdon tube. The major component in a gage-pressure measuring instrument. It is a thin-wall metal tube that has an elliptical cross section and is formed into a curve. One end of the tube is sealed and connected to an arm that moves the pointer across the instrument dial, and the open end is anchored to the instrument case. The pressure to be measured is directed into the open end, which causes the elliptical cross section to become more circular. As the cross section changes,

Continued

Bourdon tube *(continued)*
the curve straightens and moves the pointer over the dial by an amount proportional to the amount of pressure.

Brayton cycle. The constant-pressure cycle of energy transformation used by gas turbine engines.

Fuel is sprayed into the air passing through the engine and burned. Heat from the burning fuel-air mixture expands the air and accelerates it as it moves through the engine.

The Brayton cycle is an open cycle in that the intake, compression, combustion, expansion, and exhaust events all take place at the same time but in different locations within the engine.

British thermal unit (Btu). The basic unit of heat energy in the English system. One Btu is the amount of heat energy needed to raise the temperature of one pound of pure water from 60° to 61°F.

BSFC (brake specific fuel consumption). A measure of the amount of fuel used for a given amount of power produced by a heat engine. BSFC is expressed in pounds of fuel burned per hour for each brake horsepower the engine is producing.

bungee cord. An elastic cord made of small strips of rubber encased in a loosely braided cloth tube that holds and protects the rubber, yet allows it to stretch. The energy in a stretched bungee cord may be used to crank a large aircraft engine.

burner. *See* combustor.

burnish. To smooth the surface of a metal part that has been damaged by a deep scratch or gouge. Metal piled at the edge of the damage is pushed back into the damage with a smooth, hard steel burnishing tool.

butterfly valve. A flat, disk-shaped valve used to control the flow of fluid in a round pipe or tube. When the butterfly valve is across the tube, the flow is shut off, and when it is parallel with the tube, the obstruction caused by the valve is minimum, and the flow is at its greatest.

Butterfly-type throttle valves are used to control the airflow through the fuel metering system.

bypass engine. Another name for a turbofan engine. *See* turbofan engine.

bypass ratio. The ratio of the mass of air moved by the fan to the mass of air moved by the core engine.

calendar month. The measurement of time used by the FAA for inspection and certification purposes. One calendar month from a given date extends from that date until midnight of the last day of that month.

cam. An eccentric, or lobe, on a rotating shaft that changes rotary motion into linear motion. A cam is mounted on the magnet shaft in a magneto to push upward on the insulated breaker point to separate, or open, the points when the magnet is in a particular location.

cam engine. A reciprocating engine with axial cylinders arranged around a central shaft. Rollers on the pistons in the cylinders press against a sinusoidal cam mounted on the shaft to produce rotation of the shaft.

cam-ground piston. A reciprocating engine piston that is not round, but is ground so that its diameter parallel to the wrist pin is slightly smaller than its diameter perpendicular to the pin.

The mass of metal used in the wrist pin boss, the enlarged area around the wrist pin hole, expands when heated, and when the piston is at its operating temperature, it is perfectly round.

can-annular combustor. A type of combustor used in some large turbojet and turbofan engines. It consists of individual cans into which fuel is sprayed and ignited. These cans mount on an annular duct which collects the hot gases and directs them uniformly into the turbine.

capacitance afterfiring. The continuation of the spark across the gap in a shielded spark plug after the fuel-air mixture in the cylinder is ignited. Afterfiring is caused by the return of electrical energy stored in the capacitance of the shielded ignition leads. Capacitance afterfiring is eliminated by the use of a resistor in the spark plug.

capacitor. An electrical component, formerly called a condenser, that consists of two large-area conductors, called plates, separated by an insulator.

Electrons stored on one of the plates produces an electrostatic pressure difference between the plates.

capillary tube. A glass or metal tube with a tiny inside diameter. Capillary action causes the fluid to move within the tube.

carbon pile voltage regulator. A voltage regulator for a high-output DC generator that uses a stack of pure carbon disks for the variable resistance element. A spring holds pressure on the stack to reduce its resistance when the generator output voltage is low. This allows maximum field current to flow.

The field from an electromagnet whose strength varies directly with the generator voltage opposes the spring to loosen the stack and increase its resistance when the generator voltage needs to be decreased. The increased resistance decreases the field current and reduces the output voltage.

carbon track. A trail of carbon deposited by an arc across a high-voltage component such as a distributor block. Carbon tracks have a relatively low resistance to the high voltage and can cause misfiring and loss of engine power.

cartridge starter. A self-contained starter used on some military aircraft. A cartridge similar in size to a shotgun shell is ignited in the starter breech. The expanding gases drive a piston attached to a helical spline that converts the linear movement of the piston into rotary motion to rotate the crankshaft.

cascade effect. The cumulative effect that occurs when the output of one series of components serves as the input to the next series.

catalyst. A substance used to change the speed, or rate, of a chemical action without being chemically changed itself.

cavitating. The creation of low pressure in an oil pump when the inlet system is not able to supply all of the oil the pump requires. Prolonged cavitation can damage pump components.

center of pressure. The point on the chord line of an airfoil where all aerodynamic forces are concentrated.

center-line thrust airplane. A twin-engine airplane with both engines mounted in the fuselage. One is installed as a tractor in the front of the cabin, and the other as a pusher behind the cabin. The empennage is mounted on booms.

centrifugal compressor. A type of compressor that uses a vaned plate-like impeller. Air is taken into the center, or eye, of the impeller and slung outward by centrifugal force into a diffuser where its velocity is decreased and its pressure increased.

ceramic. Any of several hard, brittle, heat-resistant, noncorrosive materials made by shaping and then firing a mineral, such as clay, at a high temperature.

channel-chromed cylinders. Reciprocating engine cylinders with hard chromium-plated walls. The surface of this chrome plating forms a spider web of tiny stress cracks. Deplating current enlarges the cracks and forms channels that hold lubricating oil on the cylinder wall.

cheek (crankshaft). The offset portion of a crankshaft that connects the crankpin to the main bearing journals.

chip detector. A component in a lubrication system that attracts and holds ferrous metal chips circulating with the engine oil.

Some chip detectors are part of an electrical circuit. When metal particles short across the two contacts in the detector, the circuit is completed, and an annunciator light is turned on to inform the flight crew that metal particles are loose in the lubrication system.

choke of a cylinder. The difference in the bore diameter of a reciprocating engine cylinder in the area of the head and in the center of the barrel.

choke-ground cylinder. A cylinder of a reciprocating engine that is ground so that its diameter at the top of the barrel is slightly smaller than the diameter in the center of the stroke.

The large mass of metal in the cylinder head absorbs enough heat to cause the top end of the barrel to expand more than the rest of the barrel. At normal operating temperature, the diameter of a choke-ground cylinder is uniform throughout.

choked nozzle. A nozzle in a gas turbine engine that limits the speed of gases flowing through it. The gases accelerate until they reach the speed of sound, and a normal shock wave forms that prevents further acceleration.

chord line. An imaginary line, passing through a propeller blade, joining the leading and trailing edges.

cigarette. A commonly used name for a spark plug terminal connector used with a shielded spark plug.

circular magnetism. A method of magnetizing a part for magnetic particle inspection. Current is passed through the part, and the lines of magnetic flux surround it. Circular magnetism makes it possible to detect faults that extend lengthwise through the part.

circumferential coil spring (garter spring). A coil spring formed into a ring. This type of spring is used to hold segmented ring-type carbon seals tightly against a rotating shaft.

claret red. A dark purplish pink to a dark gray purplish red color.

Class A fire. A fire with solid combustible materials such as wood, paper, and cloth as its fuel.

Class B fire. A fire that has combustible liquids as its fuel.

Class C fire. A fire which involves energized electrical equipment.

Class D fire. A fire in which a metal such as magnesium burns.

closed-loop control. A type of control in which part of the output is fed back to the input. This allows the input to continually compare the command signals with the output to determine the extent to which the commands have been complied with.

coke. The solid carbon residue left when all volatile parts of a mineral oil have been evaporated by heat.

cold-cranking simulation. A method used for specifying the characteristics of a lubricating oil at low temperature. Oils rated by this test have the letter W (standing for Winter) in their designation. For example, SAE 15W50.

cold section. The portion of a gas turbine engine ahead of the combustion section. The cold section includes the inlet duct or ducts, compressor, and diffuser.

cold-tank lubrication system. A turbine engine lubricating system in which the oil cooler is in the scavenge subsystem.

collector ring. A ring made of thin corrosion-resistant steel tubing that encircles a radial engine and collects exhaust gases from each cylinder. The ring ends with a connection to the exhaust tail pipe.

combustor (combustion chamber). The section of a gas turbine engine in which fuel is injected. This fuel mixes with air from the compressor and burns. The intense heat from the combustion expands the air flowing through the combustor and directs it out through the turbine. Combustors are also called burners.

commutation. The process by which DC voltage is taken from an armature in which AC voltage has been induced.

commutator. A mechanical rectifier mounted on the armature shaft of a DC generator or motor. It consists of a cylindrical arrangement of insulated copper bars connected to the armature coils. Carbon brushes ride on the copper bars to carry current into or out of the commutator, providing a unidirectional current from a generator or a reversal of current in the motor coils.

compensating winding. A series winding in a compound-wound DC generator. The compensating windings are embedded in the faces of the field poles, and their varying magnetic field works with the fields from the interpoles to effectively cancel the field distortion caused by armature current.

composite propeller blade. A propeller blade made from several materials such as metal, graphite, glass or aramid fibers, and foam.

compression ratio (reciprocating engine). The ratio of the volume of a cylinder of a reciprocating engine with the piston at the bottom of its stroke to the volume of the cylinder with the piston at the top of its stroke.

compression ratio (turbine engine). The ratio of the pressure of the air at the discharge of a turbine engine compressor to the pressure of the air at its inlet.

compressor bleed air. Air that is tapped off from a turbine engine compressor. Compressor bleed air is used for anti-icing the inlet ducts and for cooling the turbine inlet guide vanes and first stage turbine blades. Bleed air is also used for certain airframe functions. *See* customer bleed air.

compressor pressure ratio. *See* compression ratio (turbine engine).

compressor stall. A condition in a turbine engine axial-flow compressor in which the angle of attack of one or more blades is excessive and the smooth airflow through the compressor is disrupted.

compressor surge. A stall that affects the entire compressor and seriously restricts the airflow through the engine.

con-di ducts. The British name for a convergent-divergent duct. *See* convergent-divergent duct.

condenser. *See* capacitor.

"Contact." The term used between a person hand-propping an aircraft engine and the person in the cockpit. When the person is ready to spin the propeller, he calls "contact." The person in the cockpit turns on the fuel, slightly opens the throttle, applies the brakes, and replies "contact," and then turns the ignition switch to BOTH. The propeller is then pulled through to start the engine.

constant-displacement pump. A fluid pump that moves a specific volume of fluid each time it rotates.

constant-pressure cycle of energy release. The cycle of energy transformation of a gas turbine engine. *See* Brayton cycle.

constant-volume cycle of energy release. The cycle of energy transformation of a reciprocating engine. *See* Otto cycle.

continuous magnetic particle inspection. A method of magnetic particle inspection in which the part is inspected by flowing a fluid containing particles of iron oxide over the part while the magnetizing current is flowing.

contrarotating. Rotating in opposite directions. Turbine rotors are contrarotating when the different stages have a common center, but turn in opposite directions.

convergent-divergent duct. A duct that has a decreasing cross section in the direction of flow (convergent) until a minimum area is reached. After this point, the cross section increases (divergent).

Convergent-divergent ducts are called CD ducts or con-di ducts.

convergent duct. A duct that has a decreasing cross section in the direction of flow.

core engine. The gas generator portion of a turboshaft, turboprop, or turbofan engine. The core engine consists of the portion of the compressor used to supply air for the engine operation, diffuser, combustors, and turbine or turbines used to drive the compressor.

The core engine provides the high-velocity gas to drive the fan and/or any free turbines that provide power for propellers, rotors, pumps, or generators.

cowling. The removable cover that encloses an aircraft engine.

crankcase. The housing that encloses the crankshaft, camshaft, and many of the accessory drive gears of a reciprocating engine. The cylinders are mounted on the crankcase, and the engine attaches to the airframe by the crankcase.

crankshaft. The central component of a reciprocating engine. This high-strength alloy steel shaft has hardened and polished bearing surfaces that ride in bearings in the crankcase. Offset throws, formed on the crankshaft, have ground and polished surfaces on which the connecting rods ride. The connecting rods change the in-and-out motion of the pistons into rotation of the crankshaft.

creep. The deformation of a metal part that is continually exposed to high centrifugal loads and temperatures.

critical altitude. The altitude above which a reciprocating engine will no longer produce its rated horsepower with its throttle wide open.

critical engine. The engine of a twin-engine airplane whose loss would cause the greatest yawing effect.

critical Mach number. The flight Mach number at which there is the first indication of air flowing over any part of the structure at a speed of Mach one, the local speed of sound.

CRT (cathode ray tube). An electronic display tube in which a stream of electrons is attracted to the charged inner surface of the tube face. Acceleration grids and focusing grids speed the movement of the electrons and shape the beam to a pin-point size. Electrostatic or electromagnetic forces caused by deflection plates or coils move the beam over the face of the tube.

The inside of the tube face is treated with a phosphor material that emits light when the electrons strike it.

cryogenic fluid. A liquid which boils at a temperature lower than about 110°K (-163°C) under normal atmospheric pressure.

CSD (constant-speed drive). A component used with either aircraft gas turbine or reciprocating engines to drive AC generators.

The speed of the output shaft of the CSD is held constant while the speed of its input shaft varies. The CSD holds the speed of the generator, and the frequency of the AC constant as the engine speed varies through its normal operating range.

CTF (centrifugal twisting force). The force acting about the longitudinal axis of a propeller blade, and which tries to rotate the blade to a low-pitch angle. As the propeller rotates, centrifugal force tries to flatten the blade so all of its mass rotates in the same plane.

Curtiss Jenny (Curtiss JN4-D). A World War I training airplane powered by a Curtiss OX-5 engine. It was widely available after the war and helped introduce aviation to the general public.

customer bleed air. Air that is tapped off a turbine engine compressor and used for such airframe functions as the operation of air conditioning and pressurization systems.

cylinder. The component of a reciprocating engine which houses the piston, valves, and spark plugs and forms the combustion chamber.

data. The input for computer processing in the form of numerical information that represents characters or analog quantities.

dataplate specifications. Specifications of each gas turbine engine determined in the manufacturer's test cell when the engine was calibrated. This data includes the engine serial number with the EPR that produced a specific RPM. The technician refers to this information when trimming the engine.

dataplate performance. The performance specifications of a turbine engine observed and recorded by the engine manufacturer or overhauler and recorded on the engine dataplate. This data includes the engine speed at which a specified EPR is attained. When trimming the engine, the technician uses this data as the goal.

DC (direct current). Electrical current in which the electrons always flow in the same direction.

deaerator. A component in a turbine engine lubrication system that removes air from the scavenged oil before it is returned to the tank.

deceleration. The amount the velocity of an object, measured in feet per second, is decreased by a force during each second it is acted upon by that force. Deceleration is usually expressed in terms of feet per second, per second (fps^2).

DeHaviland DH-4. An English designed observation airplane built in large quantities in America during World War I. After the war, surplus DH-4s were used for carrying the U.S. Mail.

deicing. The removal of ice that has formed on a surface.

density. A measure of the amount of mass in a unit volume, normally measured in such terms as pounds per cubic foot or grams per cubic centimeter.

density altitude. The altitude in standard air at which the density is the same as that of the existing air.

detergent oil. A type of mineral oil with metallic-ash-forming additives that protects the inside of an engine from sludge and varnish buildup. Used in automotive engines, it has proven unsuitable for use in aircraft engines.

detonation. An uncontrolled explosion inside the cylinder of a reciprocating engine. Detonation occurs when the pressure and temperature of the fuel inside the cylinder exceeds the critical pressure and temperature of the fuel. Detonation may be caused by using fuel that has a lower octane rating or performance number than is specified for the engine.

Dewar bottle. A special container used to store liquid oxygen and liquid nitrogen. A Dewar bottle has an inner and an outer container, and the space between them forms a vacuum. The two surfaces within the vacuum are silvered to reflect heat away from the container walls.

differential pressure. A single pressure that is the difference between two opposing pressures.

diffuser. A component in a gas turbine engine that decreases the velocity of air flowing through it and increases its pressure.

digitized image. A modified image picked up by the miniature TV camera in the end of a fiber-optic probe. This image is converted into a digital electronic signal that eliminates unwanted portions of the viewed area and allows the desired image to be enhanced for a clearer view of the inside of a turbine engine.

dipstick. A gage, in the form of a thin metal rod, used to measure the level of liquid in a reservoir.

The dipstick is pushed into the reservoir until it contacts a built-in stop; then it is removed and visually inspected. The level of liquid in the reservoir is indicated by the amount of the dipstick wet by the liquid.

dirigible. A large, cigar-shaped, lighter-than-air flying machine. Dirigibles differ from balloons in that they are powered and can be steered.

distributed pole stator winding. Alternator stator windings wound in a series of slots in the stator frame. A distributed pole stator is distinguished from a salient pole stator whose coils are wound around separate pole shoes that project inward from the field frame toward the rotor.

distributor. A high-voltage selector switch that is gear driven from the shaft of the rotating magnet in a magneto. The distributor rotor picks up the high voltage from the secondary winding of the coil and directs it to high-voltage terminals. From here it is carried by high-tension ignition leads to the spark plugs.

divergent duct. A duct that has an increased cross-sectional area in the direction of flow.

downdraft carburetor. A carburetor that mounts on the top of a reciprocating engine. Air entering the engine flows downward through the carburetor.

ΔP (delta P). Differential pressure.

droop. A progressive decrease in RPM with load in a gas turbine engine whose speed is governed with a flyweight-type governor in the fuel control.

As the load increases, the pilot valve drops down to meter more fuel. The lower position of the valve decreases the compression of the speeder spring and allows the flyweights to assume an on-speed position at a lower RPM.

dry-sump engine. An engine that carries its lubricating oil supply in a tank external to the engine.

dual ignition. An ignition system of an aircraft reciprocating engine that has two of every critical unit, including two spark plugs in each cylinder. Dual ignition provides safety in the event of one system malfunctioning, but more important, igniting the fuel-air mixture inside the cylinder at two locations provides more efficient combustion of the fuel-air mixture in the cylinder.

dual-spool gas-turbine engine. An axial-flow turbine engine that has two compressors, each driven by its own stage or stages of turbines.

duct heater. A thrust augmentation system similar to an afterburner where fuel is added to the fan-discharge air and burned.

duct losses. A decrease in pressure of the air flowing into a gas turbine engine caused by friction.

durability. A measure of engine life. Durability is usually measured in TBO hours.

duty cycle. A schedule that allows a device to operate for a given period of time, followed by a cooling down period before the device can be operated again.

dwell chamber. A chamber in a turbine engine oil tank into which the scavenged oil is returned. Entrained air separates from the oil in the dwell chamber before it is picked up by the pressure pump.

dynamometer. A device used to measure the amount of torque being produced by an engine. The drive shaft of the engine is loaded with either an electric generator or a fluid pump, and the output of the generator or pump is measured and converted into units of torque. Torque at a specific RPM can be converted into brake horsepower.

dyne. The unit of force that imparts an acceleration of one centimeter per second per second to a mass of one gram. One dyne is equal to $2.248 \cdot 10^{-6}$ pounds.

eddy current. Current induced into a conductor due to a moving or nonuniform magnetic field.

EEC (electronic engine control). An electronic fuel control for a gas turbine engine. The EEC senses the power-lever angle (PLA), engine RPM, bleed valve, and variable stator vane position, and the various engine pressures and temperatures. It meters the correct amount of fuel to the nozzles for all flight conditions, to prevent turbine overspeed and overtemperature.

effective pitch. The actual distance a propeller advances in one revolution through the air.

E-gap angle. The position of the rotating magnet in a magneto when the breaker points are timed to open. The E-gap (efficiency gap) angle is several degrees of magnet rotation beyond the magnet's neutral position. At this point the magnetic field stress is the greatest, and the change in flux is the greatest, inducing the maximum voltage in the secondary winding.

EGT (exhaust gas temperature). The temperature of the gases as they leave the cylinder of a reciprocating engine or the turbine of a gas turbine engine.

EICAS (engine indicating and crew alerting system). An electronic instrumentation system that monitors airframe and engine parameters and displays the essential information on a video display on the instrument panel. Only vital information is continually displayed, but when any sensed parameters fall outside of their allowable range of operation, they are automatically displayed.

elastic limit. The maximum amount of tensile load, in pounds per square inch, that a material is able to withstand without permanent deformation.

electrical potential. The electrical force caused by a deficiency of electrons in one location and an excess of electrons in another. Electrical potential is measured in volts.

electrical steel. A low-carbon iron alloy that contains some silicon. It is used as the core for transformers, field frames for generators and alternators, and the magnetic circuit of magnetos.

electromagnet. A magnet produced by an electrical current flowing through a coil of wire. The coil is normally wound around a core of soft iron which has an extremely low retentivity, allowing it to lose its magnetism as soon as the current stops flowing.

electromagnetic radiation. A method of transmitting energy from one location to another.

Current caused by high voltage in the secondary winding of a magneto produces electric and magnetic fields which oscillate back and forth at a high frequency and extend out into space in the form of waves. These waves of electromagnetic radiation are received as interference by the radio receivers in the aircraft.

electromotive force. A force that causes electrons to move from one atom to another within an electrical circuit. An electromotive force, or EMF, is the difference in the electrical pressure, or potential, that exists between two points.

An EMF may be produced by converting mechanical movement, pressure, chemical, light, or heat energy into electrical energy. The basic unit of EMF is the volt.

emulsion. A suspension of small globules of one material in another, when the two materials will not mix. Oil and water will not mix, but they can be formed into an emulsion. An emulsion will separate into its components when it is allowed to sit.

engine trimming. A maintenance procedure in which the fuel control on a gas turbine engine is adjusted to cause the engine to produce the required EGT or EPR at a specified RPM.

entrained water. Water suspended in jet fuel. The amount of entrained water that can be held in the fuel is determined by the temperature of the fuel. When the fuel becomes cold, the water precipitates out and forms ice crystals on the fuel filter element.

epicyclic reduction gears. A gear train in which a series of small planetary gears rotate around a central gear. More commonly called a planetary gear train.

EPR (engine pressure ratio). The ratio of the turbine discharge total pressure to the compressor inlet total pressure. EPR is normally used as the parameter to determine the amount of thrust an axial-flow turbojet or turbofan engine is producing.

ESHP (equivalent shaft horsepower). A measure of the power produced by a turboprop engine. ESHP takes into consideration both the shaft horsepower delivered to the propeller and the thrust developed at the engine exhaust. Under static conditions, one shaft horsepower is approximately equal to 2.5 pounds of thrust.

ethanol. Alcohol made from cereal grains such as corn.

ether. A volatile, highly flammable liquid that may be used to prime the cylinders of an aircraft engine when starting under extremely cold conditions.

ethylene dibromide. A colorless poisonous liquid $BrCH_2CH_2Br$ that is blended with leaded gasoline to help scavenge lead oxides.

ethylene glycol. A form of alcohol used as a coolant for liquid-cooled aircraft engines. It is also used in automobile engines as a permanent antifreeze.

eutectic. An alloy or solution that has the lowest possible constant melting point.

evaporative cooling. *See* steam cooling.

exceedance condition. A condition in which a parameter sensed by the EICAS exceeds the limits for which it is programmed.

exhaust back pressure. The pressure in the exhaust system of a reciprocating engine that opposes the flow of exhaust gases as they are forced out of the cylinders when the exhaust valve is open.

exhaust cone. The fixed conical fairing centered in the exhaust stream immediately aft of the last-stage turbine wheel. The exhaust cone straightens the flow and prevents the hot gases from circulating over the rear face of the turbine wheel.

exhaust nozzle. The opening at the rear of the exhaust pipe.

expansion wave. The change in pressure and velocity of supersonic air as it passes over a surface that drops away from the flow. As the surface drops away, the air tries to follow it, and in changing its direction, the air speeds up to a higher supersonic speed, and its static pressure decreases. There is no change in the total amount of energy as air passes through an expansion wave.

external-combustion engine. A form of heat engine in which the fuel releases its energy outside of the engine. This released heat expands air which is used to perform useful work. Steam engines are a popular type of external combustion engine.

extreme pressure (EP) lubricant. A lubricant that reacts with iron to form iron chlorides, sulfides, or phosphides on the surface of a steel part. These compounds reduce wear and damage to surfaces in heavy rubbing contact. EP lubricants are specially suited for lubricating gear trains.

FAA Form 337. The *Major Repair and Alteration* form that must be completed when an FAA-certificated aircraft or engine has been given a major repair or major alteration.

face (propeller nomenclature). The flat surface of a propeller that strikes the air as the propeller rotates. The face of a propeller corresponds to the bottom of an airplane wing.

FADEC (full-authority digital electronic control). A digital electronic fuel control for a gas turbine engine that is functioning during all engine operations, hence full authority. It includes the EEC (*see* EEC) and functions with the flight management computer. FADEC schedules the fuel to the nozzles in such a way that prevents overshooting power changes and over-temperature conditions. FADEC furnishes information to the EICAS (engine indication and crew alerting system).

fan pressure ratio. The ratio of the fan-discharge pressure to the fan inlet pressure.

14 CFR Part 33. Federal Aviation Regulation Part 33 *Airworthiness Standards: Aircraft Engines.*

feathering propeller. A controllable-pitch propeller whose blades can be moved into a high pitch angle of approximately 90°. Feathering the propeller of an inoperative engine prevents it from windmilling and greatly decreases drag.

feeler gages. A type of measuring tool consisting of strips of precision-ground steel of accurately measured thickness.

Feeler gages are used to measure the distance between close-fitting parts, such as the clearances of a mechanical system or the distance by which moving contacts are separated.

FHP (friction horsepower). The amount of horsepower used to turn the crankshaft, pistons, gears, and accessories in a reciprocating engine and to compress the air inside the cylinders.

fiber optics. The technique of transmitting light or images through long, thin, flexible fibers of plastic or glass. Bundles of fibers are used to transmit complete images.

fire sleeve. A covering of fire-resistant fabric used to protect flexible fluid lines that are routed through areas subject to high temperature.

flame tubes. Small-diameter metal tubes that connect can-type combustors in a turbine engine to carry the ignition flame to all of the combustion chambers.

The British call combustion liners flame tubes.

flameout. A condition of turbine engine operation when the fire unintentionally goes out. Improper fuel-air mixture or interruption of the air flow through the engine can cause a flameout.

flash point. The temperature to which a liquid must be raised for it to ignite, but not continue to burn when a flame is passed above it.

flashing the field. A maintenance procedure for a DC generator that restores residual magnetism to the field frame. A pulse of current from a battery is sent through the field coils in the direction in which current normally flows. The magnetic field produced by this current magnetizes the steel frame of the generator.

flashover. An ignition system malfunction in which the high voltage in the magneto distributor jumps to the wrong terminal. Flashover causes the wrong spark plug to fire. This reduces the engine power and produces vibration and excessive heat.

flat-rated engine. A turboprop engine whose allowable output power is less than the engine is physically capable of producing.

float carburetor. A fuel metering device that uses a float-actuated needle valve to maintain fuel level slightly below the edge of the discharge nozzle.

flock. Pulverized wood or cotton fibers mixed with an adhesive. Flock, attached to a wire screen, acts as an effective induction air filter for small reciprocating engines.

flow divider (reciprocating engine). The valve in an RSA fuel injection system that divides the fuel from the fuel control unit and distributes it to all of the cylinders. It compares with the manifold valve in a Teledyne-Continental fuel injection system.

flow divider (turbine engine). A component in a turbine engine fuel system that routes all of the fuel to the primary nozzles or primary orifices when starting the engine or when the RPM is low. When the engine speed builds up, the flow divider shifts and opens a passage to send the majority of the fuel to the secondary nozzles or orifices.

FMC (flight management computer). An electronic flight instrumentation system that enables the flight crew to initiate and implement a given flight plan and monitor its execution.

FOD (foreign object damage). Damage to components in the gas path of a turbine engine, caused by ingested objects. Debris from the runway or ramp cause FOD on the ground. Ice and birds cause most in-flight FOD.

four-stroke cycle. A constant-volume cycle of energy transformation that has separate strokes for intake, compression, power, and exhaust.

fractional distillation. Procedure used for separating various components from a physical mixture of liquids.

Crude oil is a mixture of many different types of hydrocarbon fuels which can be separated by carefully raising its temperature. The first products to be released, those having the lowest boiling points, are some of the gaseous fuels; next are gasoline, kerosine, diesel fuel, heavy fuel oils, lubricating oils, and finally, tar and asphalt.

frangible. Capable of being broken.

free-turbine engine. A gas turbine engine with a turbine stage on a shaft independent of the shaft used to drive the compressor. Free turbines are used to drive the propeller reduction gear in a turboprop engine and the rotor transmission in a helicopter.

freezing point. The temperature at which solids, such as wax crystals, separate from a hydrocarbon fuel as it is cooled.

fuel-air mixture ratio. Ratio of the number of pounds of fuel to the number of pounds of air in the mixture burned in cylinders of a reciprocating engine.

full-register position. The position of a magnet in a magneto when its poles are aligned with the pole shoes and the maximum amount of magnetic flux is flowing through the magnetic circuit.

gage pressure. Pressure referenced from existing atmospheric pressure.

gas generator. The basic gas turbine engine. It consists of the compressor, diffuser, combustor, and turbine. The gas generator is also called the core engine.

gas turbine engine. An internal combustion engine that burns its fuel in a constant-pressure cycle and uses the expansion of the air to drive a turbine which, in turn, rotates a compressor. Energy beyond that needed to rotate the compressor is used to produce torque or thrust.

General Aviation Airworthiness Alerts. Documents published by the FAA as Advisory Circulars 43.16. These are used to alert technicians of problems that have been found in specific models of aircraft, and reported on Malfunction and Defect Reports. Airworthiness Alerts suggest corrective action, but compliance with the suggestion is not mandatory.

general aviation. A term used to describe the total field of aviation operation except the military and airlines.

geometric pitch. The distance a propeller would advance in one revolution if it were rotating in a solid.

geopotential of the tropopause. The point in the standard atmosphere where the temperature stops dropping and becomes constant. This is the tropopause, or the dividing line between the troposphere and the stratosphere.

gerotor pump. A form of constant-displacement pump that uses an external-tooth drive gear that meshes with and drives an internal-tooth gear that has one more space for a tooth than there are teeth on the drive gear. Both gears turn inside a close-tolerance housing.

As the gears rotate, fluid flows between the teeth that are beginning to unmesh, and is carried around the pump as the space continues to open up. On the discharge side of the pump, the teeth begin to mesh, and as the space between the teeth becomes smaller, fluid is forced out of the pump.

glass cockpit. An aircraft instrument system that uses a few color cathode-ray-tube displays to replace a large number of mechanically actuated instruments.

governor. A control used to automatically change the pitch of a constant speed propeller to maintain a constant engine RPM as air loads vary in flight.

GPU (ground power unit). A service component used to supply electrical power and compressed air to an aircraft when it is operating on the ground.

graphite fibers. An advanced composite fiber made by drawing filaments of carbon at a high temperature and in a controlled atmosphere. Graphite fibers are strong and stiff.

gross thrust. The thrust produced by a turbojet or turbofan engine when the engine is static or not moving. The air is considered to have no inlet velocity, and the velocity of the gas leaving the engine is considered to be the acceleration factor.

ground-boosted engine. An aircraft reciprocating engine with a built-in supercharger that boosts the sea-level rated horsepower of the engine.

gudgeon pin. The British name for a wrist pin, or piston pin. *See* wrist pin.

half-wave rectifier. An electrical rectifier circuit that converts AC into pulsating DC. Only one alternation of each cycle is present in the output.

halogenated hydrocarbon. A hydrocarbon compound in which one or more hydrogen atoms have been replaced with atoms of one of the halogen elements such as fluorine, chlorine, or bromine.

head of pressure. Pressure exerted by a column of fluid and created by the height of the column.

heat engine. A mechanical device that converts the chemical energy in a fuel into heat energy. The heat energy is then converted into mechanical energy and useful work.

Heli-Coil insert. The registered trade name of a special helical insert used to restore threads stripped from a bolt hole, or to reinforce the threads in an aluminum casting.

The damaged threads are drilled out and new threads are cut with a special oversize tap. A coil of stainless steel wire, with a cross section in the shape of a diamond, is screwed into the hole and serves as the new threads.

Heli-Coil inserts are also used to provide durable threads in soft metal castings.

Some spark plug holes in aluminum alloy cylinder heads are fitted with Heli-Coil inserts to minimize the wear caused by repeated removal and installation of the spark plugs.

helical spline. A spline that twists, or winds, around the periphery of a shaft. Helical splines are used to change linear motion into rotary motion of the shaft on which the splines are cut.

helical spring. A spring wound in the form of a helix, or coil.

helix. A spiral.

heptane. An organic compound, $CH_3(CH_2)_5CH_3$, that is used as the low reference fuel for rating the antidetonation characteristics of aviation gasoline.

hermetically sealed. A complete seal, especially against the escape or entry of air.

hertz. A unit of frequency equal to one cycle per second.

high-bypass ratio engine. A turbofan engine whose bypass ratio is 4:1 or greater.

high-pressure compressor. The second-stage compressor in a dual-spool gas turbine engine. The high-pressure compressor is called the N_2 compressor and is the one that is rotated by the starter for starting, and the one whose RPM is controlled by the fuel control.

high unmetered fuel pressure. Pressure in a Teledyne-Continental fuel injector pump that is adjusted by the variable orifice.

homebuilt aircraft. *See* amateur-built aircraft.

honing (cylinder wall treatment). Scratching the surface of the cylinder wall with an abrasive to produce a series of grooves of microscopic depth and uniform pattern. The honed pattern holds oil to lubricate the cylinder walls.

horsepower. The most commonly used unit of mechanical power. One horsepower is equal to 33,000 foot-pounds of work done in one minute, or 550 foot-pounds of work done in one second.

hot section. The portion of a gas turbine engine that operates at a high temperature. The hot section includes the combustion, turbine, and exhaust sections.

hot-tank lubricating system. A turbine engine lubricating system in which the oil cooler is located in the pressure subsystem. The oil is returned to the tank without being cooled.

HRD fire extinguisher. A fire extinguisher that carries the extinguishing agent in a sealed sphere or cylinder. When the agent-discharged switch is closed, an ignited powder charge drives a cutter through a frangible disk which releases the agent. The entire contents of the container is emptied in much less than a second.

hub (propeller component). The high-strength component in a propeller that attaches the blades to the engine propeller shaft.

hybrid compressor engine. A gas turbine engine that has both centrifugal and axial-flow compressors.

hybrid spark plug. A fine-wire spark plug that has a platinum center electrode and iridium ground electrodes.

hydraulic lock. A condition in which oil drains into the lower cylinders of a reciprocating engine and leaks past the piston rings to fill the combustion chamber. If the oil is not removed before the engine is started, it can cause serious damage.

hydromechanical. Any device that combines fluid pressures with mechanical actions to achieve a desired result. In a hydromechanical fuel control used for a turbine engine, hydraulic servos are used in conjunction with mechanical linkages.

ice bridging. A spark plug failure that occurs when starting a reciprocating engine in extremely cold weather. When a cylinder fires, the fuel-air mixture is converted into carbon dioxide and water vapor. The water vapor condenses on the spark plug electrodes and forms ice that bridges the electrode gap and prevents the plug firing until the ice is melted. This normally requires removing the spark plugs from the engine.

IDG (integrated drive generator). An AC generator installed on turbine engines. An IDG incorporates a brushless, three-phase AC generator and a constant-speed drive in a single component.

igniter. The component in a turbine-engine ignition system that provides a high-energy spark for igniting the fuel-air mixture in the combustion chamber for starting.

IHP (indicated horsepower). The theoretical horsepower a reciprocating engine develops.

IMEP (indicated mean effective pressure). The average pressure existing inside the cylinder of a reciprocating engine during its power stroke.

impulse coupling. A spring-loaded coupling between a magneto shaft and the drive gear inside the engine. When the engine is rotated for starting, the impulse coupling locks the magnet so it cannot turn. The spring in the coupling winds up as the crankshaft continues to turn, and when the piston is near top center, the coupling releases and spins the magnet, producing a hot and retarded spark.

in-line engine. A reciprocating engine with all of the cylinders arranged in a straight line.

incandescent. Glowing because of intense heat.

Inconel. The registered trade name for an alloy of chromium, iron and nickel. Inconel is similar to stainless steel, but cannot be hardened by heat treatment.

inductive reactance. An opposition to the flow of alternating current or changing direct current caused by inductance in the circuit.

Inductive reactance, whose symbol is X_L, is measured in ohms and it varies directly as the frequency and the amount of inductance. X_L causes a voltage drop, but it does not use power nor produce heat.

inertia. The tendency of a body to resist acceleration. A body at rest will remain at rest or a body in motion will stay in motion in a straight line unless acted on by an outside force.

inertia starter. A starter for a large reciprocating engine that uses energy stored in a rapidly spinning flywheel to turn the crankshaft.

inlet guide vanes. A set of stator vanes in front of the first stage of compression in a gas turbine engine. The inlet guide vanes deflect the air entering the compressor in the correct direction for optimum operation.

Inlet guide vanes may be fixed, or their angle may be controlled hydraulically by fuel from the fuel control.

integral fuel tank. An aircraft fuel tank made by sealing off part of the structure so fuel can be carried in the structure itself.

intercooler. An air-to-air heat exchanger installed between a turbosupercharger and the carburetor. Intercoolers decrease the temperature of compressed air to prevent detonation.

interference angle (poppet valve dimension). The difference between the valve seat and the valve face angles. Normally the valve seats are ground with between 0.5° and 1° greater angle than the valve face. This allows the face to touch the seat with a line contact that provides the best sealing.

interference fit. A type of fit used when assembling certain mechanical devices. The hole is made smaller than the part that fits into it. The material containing the hole is heated to expand the hole, and the part that fits into the hole is chilled to shrink it. The parts are assembled, and when they reach the same temperature, their fit is so tight they will not loosen in service.

internal-combustion engine. A form of heat engine in which the fuel and air mixture is burned inside the engine to heat and expand the air so it can perform useful work.

internal timing. The adjustment of the breaker points of a magneto so they will begin to open at the time the magnet is in its E-gap position.

interpole. A field pole in a compound-wound DC generator used to minimize armature reaction. Interpoles are located between each of the regular field poles, and their coils are in series with the armature winding so all of the armature current flows through them. The magnetic field produced by the interpole coils cancels the distortion caused by the armature field and allows the brushes to remain in the neutral plane where there is no potential difference between the commutator segments. Keeping the brushes in the neutral plane minimizes sparking.

inverted engine. An in-line or V-engine in which the cylinders are mounted below the crankshaft.

iridium. A very hard, brittle, highly corrosion-resistant, whitish-yellow, metallic chemical element. Iridium is used for the fine-wire electrodes in spark plugs that must operate in engines using fuel with an exceptionally high lead content.

iso-octane. An organic compound used as the high reference fuel for rating the antidetonation characteristics of aviation gasoline. $(CH_3)_2CHCH_2C(CH_3)_3$

isothermal change. A physical change that takes place within a material in which heat energy is added to or taken from the material as needed to keep its temperature constant.

jet fuel. Fuel designed and produced to be used in aircraft gas turbine engines.

jet propulsion. A method of propulsion by accelerating a relatively small mass of air through a large change in velocity.

jeweler's file. A small, fine-cut metalworking file used by jewelry manufacturers.

joule. A measure of energy. In terms of electrical energy, one joule is equal to one watt-second.

journal (bearing). A hardened and polished surface on a rotating shaft that rides in a plain bearing.

kerosine. A light, almost colorless hydrocarbon liquid obtained from crude oil through the fractional distillation process. Kerosine is the base for turbine engine fuel.

Kevlar®. The registered trade name by DuPont for a patented aramid fiber.

kinematic viscosity. The ratio of the absolute viscosity of a fluid to its density. Kinematic viscosity is measured in centistokes.

labyrinth seal. A type of air and/or oil seal used around the main-shaft bearings in a gas turbine engine. The seal consists of a series of rotating blades that almost contact the seal land. A small amount of air flows between the seal and the land to prevent oil flowing past the seal.

land (piston). The portion of a piston between the ring grooves.

land (splined shaft). The portion of a splined shaft between the grooves.

laser tachometer. A highly accurate tachometer that shines a laser beam on a rotating element that has reflective tape or a contrasting mark. The reflected laser beam is converted into electrical pulses which are counted and displayed on a monitoring instrument.

last-chance oil filter. A small filter installed in the oil line to the bearing jet in a gas turbine engine. This filter traps any contaminants that have passed the main filter and holds them until the engine is disassembled for overhaul.

LCD (liquid crystal display). A digital display that consists of two sheets of glass separated by a sealed-in, normally transparent liquid crystal material. The outer surface of each glass sheet has a transparent conductive coating with the viewing side etched into character-forming segments with leads going to the edges of the display. A voltage applied between the front and back coatings disrupts the orderly arrangement of molecules and causes the liquid to darken so that light cannot pass through it. The segment to which the voltage is applied appears as black against a reflected background.

leading edge. The thick edge at the front of a propeller blade.

lean die-out. A condition in which the fire in a gas turbine engine goes out because the fuel-air mixture ratio is too lean to sustain combustion.

lean mixture. A fuel-air mixture that contains more than 15 parts of air to 1 part of fuel, by weight.

liaison aircraft. A type of light military aircraft made popular during World War II because of its ability to land and takeoff from unimproved terrain. Liaison aircraft have been replaced with helicopters.

line boring. A method of assuring concentricity of bored holes. A boring bar extends through all of the holes and cuts the inside diameters so they all have the same center.

link rod. The rod in a radial engine that connects one of the piston wrist pins to a knuckle pin on the master rod. Also called articulating rods.

liquid cooling. The removal of unwanted heat from an aircraft engine by transferring the heat into a liquid and then passing the heated liquid through a liquid-to-air heat exchanger (radiator) to transfer the heat into the ambient air.

longitudinal magnetism. A method of magnetizing a part for magnetic particle inspection. Current flows through a solenoid, or coil, that encircles the part so the lines of magnetic flux pass lengthwise through the part. Longitudinal magnetism makes it possible to detect faults that extend across the part.

low-bypass ratio engine. A turbofan engine whose bypass ratio is less than 2:1.

low-pressure compressor. The first-stage compressor in a dual-spool gas turbine engine. The low-pressure compressor is called the N_1 compressor and its speed is not governed. It seeks its own best speed as the atmospheric conditions change so it can furnish a relatively constant mass of air to the inlet of the second-stage compressor.

low unmetered fuel pressure. Pressure in a Teledyne-Continental fuel injector pump that is adjusted by the relief valve.

LRU (line replaceable unit). Aircraft components designed to be replaced as a unit while the aircraft is on the flight line.

M & D (Malfunction and Defect) report. A small postcard-like form (FAA Form 8330) used by repair stations, maintenance shops, and technicians to report an unacceptable condition to the FAA. Information on these forms provides the basis for the General Airworthiness Alerts and subsequent Airworthiness Directives.

Mach number. The ratio of the speed of an object through the air to the speed of sound under the same atmospheric conditions. An object traveling at the speed of sound is traveling at Mach one (M1.0).

Magnesyn system. The registered trade name of a remote indicating instrument system. A Magnesyn system uses a permanent magnet as its rotor and a toroidal coil excited by 400-hertz AC as its stator.

A small magnet in the center of the indicator coil follows the movement of a larger magnet in the transmitter coil.

magnetic field. The invisible, but measurable, force surrounding a permanent magnet or current-carrying conductor. This field is produced when the orbital axes of the electrons of the atoms in the material are all in alignment.

magnetic flux. Lines of magnetic force that are assumed to leave a magnet at its north end and return to its south end. Lines of flux tend to be as short as possible and cannot cross each other.

magnetic particle inspection. A method of nondestructive inspection for ferrous metal components.

The part being inspected is magnetized and then flooded with a solution of iron oxide suspended in a light oil much like kerosine. Any flaw, either on the surface or just below the surface, forms a north and south pole, and the iron oxide attracted to these poles helps locate the flaw.

The iron oxide is normally treated with a fluorescent dye, and the inspection is conducted in a darkened booth. When an ultraviolet light (black light) is shone on the part, the treated iron oxide shows up as a brilliant line.

major alteration. An alteration not listed in the aircraft, aircraft engine, or propeller specifications that might appreciably affect weight, balance, structural strength, powerplant operation, flight characteristics, or other qualities affecting airworthiness; an alteration not done according to accepted practices, or one that cannot be done by elementary operations.

major overhaul. The disassembly, cleaning, and inspection of an engine and the repair and replacement of all parts that do not meet the manufacturer's specifications.

major repair. A repair to a component that if improperly done might appreciably affect weight, balance, structural strength, performance, powerplant operation, flight characteristics, or other qualities affecting airworthiness; a repair not done according to accepted practices, or one that cannot be done by elementary operations.

mandrel. A precision steel bar on which a propeller is mounted for balancing. The mandrel is placed across two perfectly level knife-edge plates, and the propeller is allowed to rotate until it stops with its heavy point at the bottom.

manifold pressure. The absolute pressure of the air inside the induction system of a reciprocating engine.

manifold valve. *See* flow divider (reciprocating engine).

MAP (manifold absolute pressure). The absolute pressure that exists within the induction system of a reciprocating engine. It is the MAP that forces air into the cylinders of the engine. MAP is commonly called manifold pressure.

mass. A measure of the amount of matter in an object. For the purpose of measuring the mass of air flowing through a turbine engine, the weight of the air, in pounds per second, is divided by the acceleration due to gravity (32.2 feet per second^2).

matrix (advanced composites). The material that bonds the fibers together in an advanced composite structure. The matrix carries the stresses into the fibers.

matter. Something that has mass, takes up space, and exists as a solid, liquid, or gas.

medium-bypass ratio engine. A turbofan engine whose bypass ratio is between 2:1 and 4:1.

MEK (methyl ethyl ketone). A volatile, water soluble, organic chemical compound that is used as a solvent to remove oily contaminants from ignition system components.

methanol. Alcohol made from wood.

MFD (multifunction display). A liquid crystal or CRT display that shows a number of parameters and replaces several analog-type indicators.

microinches rms. A measure used for cylinder wall surface roughness. Twenty microinches rms means that the highest and lowest deviation from the average surface is 20 millionths of an inch.

micron. A measurement used to identify the size of particles trapped by filters. One micron is a micro meter, or one millionth of a meter. It is 0.000 039 inch.

microprocessor. A single silicon chip that contains the arithmetic and logic functions of a computer.

milliammeter. An instrument that measures electrical current in units of thousandths of an ampere.

millibar. A unit of pressure in the metric system. One bar is a pressure of 14.5 psi, or 29.52 in. Hg. One millibar is one thousandth of a bar, or 0.014 69 psi, or 0.029 52 in. Hg.

minor alteration. Any alteration that does not fit the definition of a major alteration. *See* major alteration.

minor repair. Any repair that does not fit the definition of a major repair. *See* major repair.

module (modular engine construction). The method of construction for most modern gas turbine engines. The engine is made of several modules, or units, that can be removed and replaced or serviced independent of the rest of the engine.

momentum. A force caused by the inertia of a moving body as it tries to keep the object moving in the same direction, at the same speed.

motor. *(verb)* The act of rotating a turbine engine using the starter, with the ignition system deactivated. An engine is motored to force air through it to purge fuel fumes.

multiple-can combustor. A combustor used in a gas turbine engine that consists of a series of individual burner cans, each made of an inner liner and an outer case. The individual cans are arranged around the periphery of a centrifugal compressor. Hot gases flow directly from the cans into the turbine.

N_1. A symbol representing the rotational speed of the low-pressure compressor in a dual-spool gas turbine engine.

N_2. A symbol representing the rotational speed of the high-pressure compressor in a dual-spool gas turbine engine.

NACA. National Advisory Committee for Aeronautics. This organization, dedicated to the technical development of aviation, has been superseded by NASA, the National Aeronautics and Space Administration.

NACA cowling. A long-chord cowling used over a radial engine. The forward portion of this cowling has an aerodynamic shape that produces a forward pull, and the rear portion extends back to fair in with the fuselage. There is a narrow peripheral gap between the rear of the cowling and the fuselage for the cooling air to escape. Some NACA cowlings have controllable flaps over this opening to control the amount of cooling air that flows through the engine.

nacelle. An enclosed compartment, normally in the leading edge of the wing, in which an aircraft engine is mounted.

naphtha. A volatile, flammable liquid distilled from petroleum. It is used as a cleaning agent and solvent, and is present in some blended turbine-engine fuels.

NASA. National Aeronautics and Space Administration.

naturally aspirated engine. A reciprocating engine that depends upon atmospheric pressure to force the fuel-air mixture into the cylinders. Naturally aspirated engines are neither supercharged nor turbocharged.

net thrust. The thrust produced by a turbojet or turbofan engine in which the acceleration factor is the difference between the velocity of the incoming air and the velocity of the exhaust gases leaving the engine.

neutral position. The position of the magnet in a magneto when its poles are between the pole shoes and no lines of flux are flowing through the magnetic circuit.

newton. The unit of force needed to accelerate a mass of one kilogram one meter per second per second. One newton is equal to 100,000 dynes, or $2.248 \cdot 10^{-1}$ pound.

Nichrome. The registered trade name for an alloy of nickel and chromium. Nichrome wire is used for making electrical heater elements and precision wire-wound resistors. Nichrome's resistance is approximately 65 times that of copper.

nitriding. A method of case hardening steel. Steel is placed in a retort (a sealed, high-temperature furnace), and heated to a specified temperature while surrounded by ammonia gas (NH_3). The ammonia breaks down into nitrogen and hydrogen, and the nitrogen unites with some of the alloying elements in the steel to form an extremely hard surface.

Nitriding hardens crankshaft bearing surfaces and cylinder walls in reciprocating engines. It takes place at a lower temperature than other forms of case hardening, and does not cause warping.

normal category airplane. An aircraft that is certificated under 14 CFR Part 23 that is not certificated under the acrobatic, utility, or commuter category.

normal shock wave. A type of pressure wave that forms at right angles to a surface when air moves at the speed of sound.

notch sensitivity. A measure of the loss of strength of a material caused by the presence of a notch, or a V-shaped cut.

nozzle guide vanes. *See* turbine inlet guide vanes.

oblique shock wave. A pressure wave that forms on a sharp-pointed object when air flows past it at a supersonic speed.

octane rating. A system used to rate the antidetonation characteristics of a reciprocating engine fuel.

Fuel with an octane rating of 80 performs in a laboratory test engine the same as fuel made of a mixture of 80% iso-octane and 20% heptane.

odometer. The portion of an automobile speedometer that indicates the distance traveled.

offset throw (crankshaft design). Crank arms on a reciprocating engine crankshaft. The arms, or throws, to which the connecting rods and pistons are attached are offset from the center of the crankshaft to move the pistons in and out of the cylinder. The amount of the offset determines the stroke of the engine.

oil analysis. A method of measuring the contents in parts per million of various chemical elements in oil.

A sample of the oil is burned in an electric arc, and the resulting light is analyzed with a spectroscope which identifies the chemical elements in the oil and gives an indication of the amount of each element. This type of oil analysis is called a spectrometric oil analysis program, or SOAP.

oil dilution. A method of temporarily decreasing the viscosity of the lubricating oil to make it possible to start a reciprocating engine when the temperature is very low.

Before shutting the engine down, enough gasoline from the fuel system is mixed with the lubricating oil in the engine to dilute it so the starter can turn the engine over when the oil is cold and viscous.

When the engine starts and the oil warms up, the gasoline evaporates.

oil-damped bearing. A type of roller bearing installation in a gas turbine engine in which the outer race is installed in an oil damper compartment whose inside diameter is a few thousandths of an inch larger than the outside diameter of the outer race. Oil under pressure fills the oil damper compartment and allows the bearing to compensate for sight misalignment and to absorb vibrations of the shaft.

on-condition maintenance. A maintenance program that closely monitors the operating condition of an engine and allows major repairs or replacements to be made when engine performance deteriorates to a specific level.

on-speed condition. The speed condition in which the engine is turning at the RPM for which the propeller governor is set.

one-hundred-hour inspection. An inspection required by 14 CFR §91.409 for FAA-certificated aircraft operated for hire or used for flight instruction for hire. A 100-hour inspection is identical in content to an annual inspection, but can be conducted by an aviation maintenance technician who holds an

Airframe and Powerplant rating, but does not have an Inspection Authorization.

See 14 CFR Part 43, Appendix D for list of the items that must be included in an annual or 100-hour inspection.

operating cycle. One complete series of events in the operation of a turbine engine that consists of starting the engine, taking off, landing, and shutting the engine down.

optoelectronic device. An electronic device that produces, modulates, or senses electromagnetic radiation in the ultraviolet, visible light, or infrared portions of the energy spectrum.

Otto cycle. The constant-volume cycle of energy transformation used by reciprocating engines. A mixture of fuel and air is drawn into the cylinder as the piston moves to the bottom of its stroke. The mixture is compressed as the piston moves upward in the cylinder, and when the piston is near the top of its stroke, the mixture is electrically ignited and burns. The burning mixture heats and expands the air inside the cylinder and forces the piston down, performing useful work. The piston then moves back up, forcing the burned gases out of the cylinder.

overboost. A condition of excessive manifold pressure in a reciprocating engine. Overboosting occurs when the supercharger is operated at too high a speed.

overrunning clutch. A type of clutch that couples an input shaft with an output shaft. When the input shaft is driven, the output shaft rotates with it. But when the output shaft is driven, the input shaft does not turn.

overspeed condition. A speed condition in which the engine is turning at an RPM higher than that for which the propeller governor is set.

P-lead. Primary lead. The wire that connects the primary winding of a magneto to the ignition switch. The magneto is turned OFF by grounding its P-lead.

pascal. The unit of pressure produced when one newton of force acts uniformly over an area of one square meter. One pascal is equal to $14.503 \cdot 10^{-5}$ (0.000 145 03) psi. The kilopascal (kPa) is easier to manipulate. 1 kPa = 1,000 Pa = 0.14503 psi.

PCB (plenum chamber burning). A method of thrust augmentation used on engines with vectored nozzles. Fuel injected into the fan-discharge air is burned to increase thrust.

peak voltage. The voltage of AC electricity that is measured from zero voltage to the peak of either alternation.

penetrant dwell time. The length of time a part is left in the penetrant when preparing it for inspection by the fluorescent or dye penetrant method. The hotter the part and the longer the penetrant dwell time, the smaller the fault that will be detected.

performance number. The rating of antidetonation characteristics of a reciprocating engine fuel that is better than the high rating reference fuel, iso-octane. Performance numbers are greater than 100.

permanent magnet. A piece of hardened steel that has been exposed to a strong magnetizing force which has aligned the spin axes of the electrons surrounding its atoms. The high retentivity of the material causes the electrons to retain their magnetic orientation.

permanent-mold casting. A casting made in a reusable metal mold. The walls of permanent-mold castings can be made thinner than similar walls made by sand casting.

permeability. A measure of the ease with which lines of magnetic flux can pass through a material.

phase sequence, or phase rotation. The sequence with which the output phases of a three-phase generator are connected to the load. Reversing the phase sequence of a generator from A-B-C to A-C-B prevents the generator from being synchronized with the others on the bus.

pi (π) filter. An electronic filter used to prevent radio frequency energy produced in the ignition exciter from feeding back into the aircraft electrical system. The filter is made of an inductor with a capacitor on its input and output. The name is derived from the resemblance of the three components on a schematic diagram to the Greek letter pi (π).

pinion. A small gear that meshes with and drives a larger gear.

piston (reciprocating engine component). The movable plug inside the cylinder of a reciprocating engine. The piston moves in and out to compress the fuel-air mixture and to transmit the force from the expanding gas in the cylinder to the crankshaft.

piston displacement. The total volume, in cubic inches, cubic centimeters, or liters, swept by all of the pistons of a reciprocating engine as they move in and out in one revolution of the crankshaft.

piston pin. *See* wrist pin.

pitch angle. The angle between the chord line of a propeller blade and the plane of rotation. *See* blade angle.

pitch distribution. The gradual change in pitch angle of a propeller blade from the root to the tip.

plane of rotation. The plane in which a propeller blade rotates. The plane of rotation is perpendicular to the propeller shaft.

planetary gears. A type of large-ratio reduction gearing. A series of small planetary gears are mounted on a spider attached to the output shaft. The planetary gears rotate between a fixed sun gear and a driven ring gear.

plenum chamber. An enclosed chamber in which air can be held at a pressure slightly higher than that of the surrounding air. Plenum chambers are used to stabilize the pressure of the air before it enters a double-entry centrifugal compressor.

POH (Pilot's Operating Handbook). A document published by the airframe manufacturer and approved by the FAA that lists the operating conditions for a particular model of aircraft. Engine operating parameters are included in the POH.

pole shoe. Inward extensions from the field frame of a generator around which the field coils are wound.

poppet valve. A T-shaped valve with a circular head. Poppet valves are used to cover the intake and exhaust openings in the cylinder head of a reciprocating engine. The valves are held closed by one or more coil springs and are opened by a cam lobe or a rocker arm pushing on the end of the valve stem.

porcelain. A hard, white, translucent ceramic material that was used as the insulator in some of the early aircraft spark plugs.

positive-displacement pump. A fluid pump that moves a specific volume of fluid each time it rotates. Spur-gear pumps, gerotor pumps, and vane pumps are all positive-displacement pumps.

power. The time rate of doing work. Power is found by dividing the amount of work done, measured in foot-pounds, by the time in seconds or minutes used to do the work.

Power may be expressed in foot-pounds of work per minute or in horsepower. One horsepower is 33,000 foot-pounds of work done in one minute, or 550 foot-pounds of work done in one second.

power-assurance check. A test run made of a gas turbine engine to determine how its performance compares with its previous performance as new or freshly overhauled.

powerplant. The complete installation of an aircraft engine, propeller, and all accessories needed for its proper function.

preignition. Ignition of the fuel-air mixture inside the cylinder of an engine before the time for normal ignition. Preignition is often caused by incandescent objects inside the cylinder.

prepreg. Preimpregnated fabric. A type of composite material in which the reinforcing fibers are encapsulated in an uncured resin. Prepreg materials are cut to size and shape and laid up with the correct ply orientation, and the entire component is cured with heat and pressure.

pressure. A measure of force applied uniformly over a given unit of surface area.

pressure altitude. The altitude in standard atmosphere at which the pressure is the same as the existing pressure.

pressure carburetor. A carburetor installed on some aircraft reciprocating engines that uses the pressure difference between air inside the venturi and ram air entering the carburetor to produce a fuel-metering force. Pressure carburetors have been generally replaced with continuous-flow fuel injection systems.

pressure cooling. A method of air cooling a reciprocating engine in which the cylinders are enclosed in tight-fitting shrouds. The cowling is divided into two compartments by baffles and seals, with half of each cylinder in each compartment. Ram air is directed into one compartment, and the pressure in the other is decreased by air flowing over a flared exit or adjustable cowl flaps. The pressure difference across the cylinders causes cooling air to be drawn through the fins to remove the unwanted heat.

pressure-injection carburetor. A multibarrel pressure carburetor used on large radial and V-engines. Fuel is metered on the basis of air mass flowing into the engine and is sprayed under pressure into the eye, or center, of the internal supercharger impeller.

prevailing torque. The torque required to turn a threaded fastener before it contacts the surface it is intended to hold.

primary winding. The winding in a magneto or ignition coil that is between the source of voltage and the breaker points. The primary winding is normally made of comparatively large diameter wire, and has a small number of turns, typically about 200.

profile tip (compressor blade tip). The tip of an axial-flow compressor blade whose thickness is reduced to give it a higher resonant frequency so it will not be subject to the vibrations that would affect a blade with a squared tip. The profile tip also provides a more aerodynamically efficient shape for the high-velocity air that is moved by the blade.

Profile tips often touch the housing and make a squealing noise as the engine is shut down. For this reason, profile tips are often called squealer tips.

profilometer. A precision measuring instrument used to measure the depth of the hone marks in the surface of a cylinder wall.

prony brake. An instrument used to measure the amount of horsepower an engine is delivering to its output shaft. The engine is operated at a specific RPM, and a brake is applied to its output shaft. The amount of torque applied to the brake is measured, and this, with the RPM, is converted into brake horsepower.

propeller end. The end of a reciprocating engine to which the propeller is attached.

PropFan™ engine. The registered trade name by Hamilton Standard of an ultra-high-bypass turbine engine. *See* UHB engine.

propulsive efficiency. A measure of the effectiveness with which an aircraft engine converts the fuel it burns into useful thrust. It is the ratio of the thrust horsepower produced by a propeller to the torque horsepower of the shaft turning the propeller. The nearer the speed of the aircraft is to the speed of the exhaust jet or propeller wake, the less kinetic energy is lost in the jet or wake, and the higher the propulsive efficiency.

PRT (power recovery turbine). A turbine driven by exhaust gases from several cylinders of a reciprocating engine. Energy extracted from exhaust gases by the turbine is coupled, through a fluid clutch, to the engine crankshaft.

pulsating DC. Direct current whose voltage periodically changes, but whose electrons flow in the same direction all of the time.

pulse-jet engine. A type of air-breathing reaction engine used during World War II to power jet-propelled missiles. Fuel is sprayed into the combustion chamber and ignited. As the heated air expands, it closes the one-way shutter valve in the front of the engine and exits the engine through the nozzle at the rear. As soon as the pressure inside the combustion chamber decreases, air enters through the shutter valve and more fuel is ignited. The thrust is produced in a series of pulses.

push fit. A fit between pieces in a mechanical assembly that is close enough to require the parts to be pushed together.

A push fit is looser than a press fit, but closer than a free fit.

pusher engine. An engine installed with the propeller facing the rear of the aircraft. Thrust produced by the propeller mounted on a pusher engine pushes rather than pulls the aircraft.

pusher propeller. A propeller installed on an aircraft engine so that it faces the rear of the aircraft. Thrust from the propeller pushes rather than pulls the aircraft.

PV diagram. A diagram showing the relationship between the volume of a cylinder and the pressure during a cycle of engine operation.

quill shaft. A type of shaft used to couple parts of an engine that are subject to torsional loads.

A quill shaft is a long, hardened steel shaft with splines on each end. One end splines into the drive shaft and the other end splines into the device being driven. Torsional vibrations are absorbed by the quill shaft twisting.

radial bearing load. The load on a bearing perpendicular to the shaft on which the bearing is mounted. Centrifugal loads are radial loads.

radial engine (static radial). A form of reciprocating engine in which the cylinders radiate out from a small central crankcase. The pistons in the cylinders drive a central crankshaft which in turn drives the propeller.

radial-inflow turbine. A turbine, similar in appearance to a centrifugal compressor rotor. Radial-inflow turbines are used to drive the compressor in reciprocating engine turbochargers and some of the smaller APU turbine engines. Hot gases flow into the turbine from its outside rim, then radially inward through the vanes and out of the turbine at its center.

radiation. *See* electromagnetic radiation.

ram air. Air whose pressure has been increased by the forward motion of the aircraft. Ram air pressure is the same as pitot pressure.

ram drag. The loss of thrust produced by a turbojet or turbofan engine caused by the increase of velocity of air entering the engine. Ram drag is the difference between gross thrust and net thrust.

ram pressure. Pressure produced when a moving fluid is stopped.

ram-recovery speed. The speed of an aircraft at which the ram effect caused by the forward movement increases the air pressure at the compressor inlet so that it is the same as that of the ambient air.

ramjet engine. The simplest type of air-breathing reaction engine. Air entering the front of the engine at a high velocity has fuel sprayed into it and ignited. A barrier formed by the incoming air forces the expanding gases to leave through the nozzle at the rear. The energy added by the burning fuel accelerates the air and produces a forward thrust.

Ramjet engines are used in some military unmanned aircraft that are initially boosted to a speed high enough for the engine to function.

ratiometer indicator. An analog temperature measuring instrument in which the pointer deflection is proportional to the ratio between the current flowing in an internal reference circuit and that flowing through the temperature-sensing probe.

reach (spark plug specification). The length of the threads on the shell of a spark plug.

reaction engine. A form of heat engine that produces thrust by heating a mass of air inside the engine and discharging it at a high velocity through a specially shaped nozzle. The amount of thrust is determined by the mass of the air and the amount it is accelerated.

reactive power. Wattless power in an AC circuit. It is the power consumed in the inductive and capacitive reactances. Reactive power is expressed in volt-amps reactive (VAR) or in kilovolt-amps reactive (KVAR).

reamed fit. The fit of a shaft in a hole in which the hole is drilled undersize and cut with a reamer to the correct diameter. Reamed holes have smooth walls and a consistent diameter.

rebuilt engine. A used engine that has been completely disassembled, inspected, repaired as necessary, and reassembled, tested, and approved in the same manner and to the same tolerances and limits as a new engine, using either new or used parts. However, all parts used must conform to all production drawings, tolerances, and limits for new parts, or be of approved oversize or undersize dimensions for a new engine.

According to 14 CFR §91.421, a rebuilt engine is considered to have no previous operating history and may be issued a zero-time logbook. Only the engine manufacturer or an agent approved by the manufacturer can rebuild an engine and issue a zero-time record.

reciprocating engine. A type of heat engine that changes the reciprocating (back-and-forth) motion of pistons inside the cylinders into rotary motion of a crankshaft.

rectifier. A device that allows electrons to flow in one direction while preventing their flow in the opposite direction. Rectifiers are used to change alternating current into direct current.

reheat system. The British name for an afterburner. *See* afterburner.

Reid vapor pressure. The amount of pressure that must be exerted on a liquid to keep it from vaporizing. Reid vapor pressure is measured at 100°F.

reliability. The ability of an aircraft engine to perform its designed functions under widely varying operating conditions.

residual magnetic particle inspection. A form of magnetic particle inspection for small steel parts that have a high degree of retentivity. The part is magnetized, removed from, and inspected away from the magnetizing machine.

residual magnetism. The magnetism that remains in the field frame of a generator when no current is flowing in the field coils.

residual voltage. The voltage produced in a generator armature when the armature is rotated in the residual magnetism.

resistor spark plug. A shielded spark plug with a resistor between the ignition lead terminal and the center electrode. The resistor stops the flow of secondary current when its voltage drops to a specified value. The resistor prevents capacitive afterfiring.

retarded sparks. The timing of the firing of the spark plugs used to start a reciprocating engine. The sparks for starting occur later in terms of crankshaft rotation than those used for normal operation. Retarding the sparks prevent the engine from kicking back when it is being started.

retentivity. The ability of a magnetizable material to retain the alignment of the magnetic domains after the magnetizing force has been removed. Hard steel normally has a high retentivity, while soft iron and electrical steel both have very low retentivity.

reverse-flow combustor. A type of combustor in which the air from the compressor enters the combustor outer case and reverses its direction as it flows into the inner liner. It again reverses its direction before it flows through the turbine. Reverse-flow combustors are used where engine length is critical.

RF energy (radio frequency) energy. Electromagnetic energy with a frequency high enough to radiate from any conductor through which it is flowing.

rich blowout. A condition in which the fire in a gas turbine engine goes out because the fuel-air mixture ratio is too rich to sustain combustion.

rich mixture. A fuel-air mixture that contains less than 15 parts of air to 1 part of fuel, by weight.

riffle file. A hand file with its teeth formed on a curved surface that resembles a spoon.

rms. Root mean square. A dimension that is the square root of the average of an infinite number of varying values. An rms dimension is used to indicate the allowable surface roughness of a reciprocating engine cylinder wall.

rocker arm. A pivoted arm on the cylinder head of a reciprocating engine. The pushrod forces one end of the rocker arm up, and as the other end moves down, it forces the poppet valve off of its seat.

rocker box. The enclosed part of a reciprocating engine cylinder that houses the rocker arm and valve mechanism.

rocket engine. A form of reaction engine whose fuel and oxidizer contain all of the oxygen needed for the release of heat energy. The released heat expands the gases which are ejected at a high velocity from a nozzle at the rear of the rocket. Because rocket engines carry their own oxygen, they can operate in outer space where there is no atmosphere.

rotary radial engine. A form of reciprocating engine used in some early aircraft. The crankshaft is rigidly attached to the airframe, and the propeller, crankcase, and cylinders all revolve as a unit.

rotating combustion (RC) engine. A form of internal combustion engine in which a rounded, triangular-shaped rotor with sliding seals at the apexes forms the combustion space inside an hourglass-shaped chamber. Expanding gases from the burning fuel-air mixture push the rotor around and turn a geared drive shaft in its center. The RC engine was conceived in Germany by Felix Wankel in 1955.

RPM (revolutions per minute). A measure of rotational speed. One RPM is one revolution made in one minute.

run in. A time of controlled operation of a new or freshly overhauled engine that allows the moving parts to wear together.

run up. A procedure in which an aircraft engine is operated on the ground to determine its condition and performance.

runout. A measure of the amount a shaft, flange, or disk is bent or fails to run true. Runout is normally measured with a dial indicator.

SAE (Society of Automotive Engineers). A professional organization that has formulated standards for the automotive and aviation industries.

safety gap. A location in a magneto that allows a spark to jump to ground from the secondary circuit before the voltage rises high enough to damage the secondary insulation.

sand casting. A method of molding metal parts in a mold made of sand. A pattern that duplicates the part to be molded is made of wood and is covered with a special casting sand that contains a resin to bind it. The mold is separated along a special parting line, and the pattern is removed. The mold is put back together, and molten metal is poured into the cavity. When the metal cools, the sand is broken away from the molded part.

Sand casting is less expensive than permanent-mold casting.

Saybolt Seconds Universal viscosity. A measurement of viscosity (resistance to flow) of a lubricating oil.

The number of seconds needed for 60 milliliters of oil at a specified temperature to flow through a calibrated orifice is measured and is called the Saybolt Seconds Universal (SSU) viscosity of an oil.

The viscosity number used for commercial aviation engine lubricating oil relates closely to the SSU viscosity of the oil at 210°F.

scavenge subsystem. The subsystem in the lubrication system of a gas turbine engine that collects oil after it has lubricated the bearings and gears and returns it to the oil tank.

scimitar shape. The shape of the blades of the propellers mounted on UHB engines. The name is derived from the shape of a curved Asian sword that has its edge on the convex side. *See* UHB engine.

scramjet (supersonic combustion ramjet). A special type of ramjet engine whose fuel can be ignited while the vehicle is moving at a supersonic speed.

scuffing. Severe damage to moving parts caused when one metal part moves across another without sufficient lubricant between them. Enough heat is generated by friction to cause the high points of the surfaces to weld together; continued movement tears, or scuffs, the metal.

sea-level boosted engine. A reciprocating engine that has had its sea-level rated horsepower increased by supercharging. This is the same as a ground-boosted engine.

secondary winding. The winding in a magneto or ignition coil that connects to the distributor rotor.

The secondary winding is normally made of very small diameter wire and has a large number of turns, typically about 20,000.

self-accelerating speed. The speed attained by a gas turbine engine during start-up that allows it to accelerate to its normal idling speed without assistance from the starter.

semiconductor transducer. A piezoelectric crystal that converts input energy of one form, such as pressure, into output energy of another, such as an electrical signal.

series-wound motor. An electric motor with field coils connected in series with the armature.

serviceable limits. Limits included in a reciprocating engine overhaul manual. If a part measures outside of the new-parts limits, but within the serviceable limits, it will not likely wear to the point of causing engine failure within the next TBO interval.

servo system. A type of automatic control system in which part of the output is fed back into the input.

shaft horsepower. The horsepower actually available at a rotating shaft.

shielding. A metal braid that encloses wires carrying high-frequency alternating current or high-voltage DC that has radio-frequency energy superimposed on it. The shielding intercepts any electromagnetic radiation and conducts it to the engine structure so it will not interfere with any installed electronic equipment.

shielding. The electrically conductive covering placed around an electrical component to intercept and conduct to ground any electromagnetic energy radiated from the device.

short circuit. A low-resistance connection between two points in an electric circuit.

Shower of Sparks ignition system. A patented ignition system for reciprocating engines. An induction vibrator sends pulsating direct current into a set of retard breaker points in one of the magnetos. This provides a hot and retarded spark for starting the engine.

single-shaft turbine engine. A turboprop engine in which the propeller reduction gears are driven by the same shaft that drives the compressor for the gas generator.

single-spool gas-turbine engine. A type of axial-flow-compressor gas turbine engine that has only one rotating element.

skin radiator. A type of radiator used on some early liquid-cooled racing airplanes. The radiator was made of two thin sheets of brass, slightly separated so the heated coolant could flow between them. Skin radiators were mounted on the surface of the wing, on the sides of the fuselage, or on the floats of seaplanes. Air flowing over the smooth surface of the radiator removed heat from the coolant.

slip (propeller specification). The difference between the geometric and effective pitch of a propeller.

slip ring. A smooth, continuous ring of brass or copper mounted on the rotor shaft of an electrical generator or alternator. Brushes riding on the smooth surface of the slip ring carry current into and out of the rotor coil.

slow-blow fuse. A special type of electrical circuit protection device that allows a momentary flow of excess current, but opens the circuit if the excessive flow is sustained.

sludge. A heavy contaminant that forms in an aircraft engine lubricating oil because of oxidation and chemical decomposition of the oil.

sludge plugs. Spool-shaped sheet metal plugs installed in the hollow throws of some engine crankshafts.

slug. The unit of mass equal to that which experiences an acceleration of one foot per second, per second when a force of one pound acts on it. It is equal to 32.174 pounds, or 14.5939 kilograms, of mass. Also called a G-pound.

SOAP (spectrometric oil analysis program). An oil analysis program in which a sample of oil is burned in an electric arc and an analysis is made of the wavelength composition of the resulting light. Each chemical element in the oil, when burned, produces light containing a unique band of frequencies. A computer analyzes the amount of each band of frequencies and prints out the number of parts of the element per million parts of the entire sample.

Continued

SOAP *(continued)*

SOAP can predict engine problems by warning the engine operator of an uncharacteristic increase of any elements in the oil.

sound suppressor. The airframe component that replaces the turbine engine tail pipe. It reduces the distance the sounds made by the exhaust gases propagate by converting low-frequency vibrations into high-frequency vibrations.

specific gravity. The ratio of the density of a material to the density of pure water.

specific weight. The ratio of the weight of an aircraft engine to the brake horsepower it develops.

spline. Parallel slots cut in the periphery of a shaft, parallel to its length. Matching slots, cut into the hub or wheel that fits on the shaft, lock the shaft into the device to transmit torque.

sprag clutch. A freewheeling, nonreversable clutch that allows torque to be applied to a driven unit in one direction only.

springback. A condition in the rigging of an aircraft engine control in which the stop at the engine is reached before the stop in the cockpit. The cockpit control moves slightly after the stop in the engine is reached, and when it is released, it springs back slightly.

spur-gear pump. A form of constant-displacement fluid pump that uses two meshing spur-gears mounted in a close-fitting housing.

Fluid is taken into the housing where it fills the space between the teeth of the gears and is carried around the housing as the gears rotate. On the discharge side of the pump, the teeth of the two gears mesh, and the fluid is forced out of the pump.

squat switch. An electrical switch actuated by the landing gear scissors on the oleo strut. When no weight is on the landing gear, the oleo piston is extended and the switch is in one position; but when weight is on the gear, the oleo strut compresses and the switch changes its position.

squealer tip (compressor blade tip). *See* profile tip.

squeeze film bearings. Another name for oil-damped bearings. *See* oil-damped bearings.

stage length. The distance between landing points in airline operation.

stage of a compressor. One disk of rotor blades and the following set of stator vanes in an axial-flow compressor.

staggered timing. Ignition timing that causes the spark plug nearest the exhaust valve to fire a few degrees of crankshaft rotation before the spark plug nearest the intake valve.

standard day conditions. Conditions that have been decided upon by the ICAO (International Civil Aeronautics Organization) for comparing all aircraft and engine performance. The most basic standard day conditions are: temperature, 15°C or 52°F; altitude, mean sea level; pressure, 29.92 inches of mercury.

Standard J-1. A World War I training airplane powered by a Curtiss OX-5 engine.

standpipe. A pipe which protrudes upward from the base of an oil tank and through which oil used for normal engine lubrication is drawn. In the event of a catastrophic leak when all oil available to the engine-driven pump is lost overboard, enough oil is available from an outlet below the standpipe to feather the propeller.

starter-generator. A single-component starter and generator used on many smaller gas-turbine engines. It is used to start the engine, and when the engine is running, its circuitry is shifted so that it acts as a generator.

static pressure. The pressure of an unmoving fluid.

static RPM. The number of revolutions per minute an aircraft engine can produce when the aircraft is not moving.

steam cooling. A method of liquid cooling in which the coolant, normally water, is allowed to absorb enough heat that it boils. The steam gives up its heat when it condenses back into a liquid.

stellite. A nonferrous alloy of cobalt, chromium, and tungsten. Stellite is hard, wear resistant, and corrosion resistant, and it does not soften until its temperature is extremely high.

Stellite is welded to the faces of many reciprocating engine exhaust valves that operate at very high temperatures.

stepping motor. A precision electric motor whose output shaft position is changed in steps by pulses from the control device. Stepping motors can make high-torque changes in small angular increments to their output shaft.

stoichiometric mixture. The fuel-air mixture ratio that, when burned, leaves no uncombined oxygen nor any free carbon. It releases the maximum amount of heat, and therefore produces the highest exhaust gas temperature. A stoichiometric mixture of gasoline and air contains 15 pounds of air for 1 pound of gasoline.

straight-through combustor. A combustor in a gas turbine engine through which the air from the compressor to the turbine flows in an essentially straight line.

stratosphere. The upper part of the earth's atmosphere. The stratosphere extends upward from the tropopause, which is about 36,000 feet above the surface of the earth, to about 85,000 feet. The temperature of the air in the stratosphere remains constant at -56.5°C (-69.7°F).

stress. A force within an object that tries to prevent an outside force from changing its shape.

stroboscopic tachometer. A tachometer used to measure the speed of any rotating device without physical contact. A highly accurate variable-frequency oscillator triggers a high-intensity strobe light. When the lamp is flashing at the same frequency the device is rotating, the device appears to stand still.

stroke. The distance the piston moves inside the cylinder.

sump (aircraft engine component). A low point in an aircraft engine in which lubricating oil collects and is stored or transferred to an external oil tank.

A removable sump attached to the bottom of the crankcase of a reciprocating engine is often called an oil pan.

sump (fuel tank component). A low point in an aircraft fuel tank in which water and other contaminants collect and are held until they can be drained out.

supercharged engine. A reciprocating engine that uses a mechanically driven compressor to increase the air pressure before it enters the engine cylinders.

supercharger. An air compressor used to increase the pressure of the air being taken into the cylinders of a reciprocating engine.

surface roughness. The condition of the surface of a reciprocating engine cylinder wall that has been honed to make it hold lubricating oil. Surface roughness is measured in microinches RMS.

surge. A condition of unstable airflow, through the compressor of a gas turbine engine, in which the compressor blades have an excessive angle of attack. Surge usually affects an entire stage of compression.

synthetic oil. Oil made by chemical synthesis of a mineral, animal, or vegetable base. Synthetic oils have appropriate additives that give them such characteristics as low volatility, low pour point, high viscosity index, good lubricating qualities, low coke and lacquer formation, and low foaming.

tachometer. An instrument that measures the rotational speed of an object.

TAI (thermal anti-ice). A system used to prevent the formation of ice on an aircraft by flowing heated air inside the structure.

tail pipe. The portion of the exhaust system of a gas turbine engine through which the gases leave. The tail pipe is often called the exhaust duct, or exhaust pipe.

TBO (time between overhauls). A time period specified by the manufacturer of an aircraft engine as the maximum length of time an engine should be operated between overhauls without normal wear causing parts of the engine to be worn beyond safe limits.

Continued

TBO *(continued)*

TBO depends upon proper operation and maintenance in accordance with the engine manufacturer's recommendations.

The overhaul of an engine when it reaches its TBO hours is not mandatory except for certain commercial operators that have this requirement written into their operations manual.

TDC (top dead center). The position of a piston in a reciprocating engine when the piston is at the top of its stroke and the wrist pin, crankpin, and center of the crankshaft are all in line.

TEL. Tetraethyl lead.

test club. A wide-blade, short-diameter propeller used on a reciprocating engine when it is run in a test cell. A test club applies a specific load to the engine and forces the maximum amount of air through the engine cooling fins.

thermal efficiency. The ratio of the amount of useful work produced by a heat engine, to the amount of work that could be done by all of the heat energy available in the fuel burned.

thermal expansion coefficient. A number that relates to the change in the physical dimensions of a material as the temperature of the material changes. The thermal expansion coefficient of aluminum is approximately twice that of steel.

thermal shock. The sudden change in engine operating temperature that occurs when engine power is suddenly reduced at the same time the airspeed, thus the cooling, is increased. Thermal shock occurs when an aircraft is required to rapidly descend to a lower altitude.

thermistor. A semiconductor material whose electrical resistance varies with its temperature.

thermocouple. A device used to generate an electrical current. A thermocouple is made of two dissimilar metal wires whose ends are welded together to form a loop. A voltage exists in the loop proportional to the difference in temperature of the junctions at which the wires are joined.

The amount of current flowing in the loop is determined by the types of metals used for the wires, the temperature difference between the junctions, and the resistance of the wires.

thermosetting resin. A plastic resin that, once it has been hardened by heat, cannot be softened by heating again.

thermostatic valve. A temperature-sensitive valve that controls the temperature of oil in an aircraft engine. When the oil is cold, the valve directs it through the engine, but when it is hot, the valve shifts and directs the oil through the oil cooler.

thermoswitch. An electrical switch that closes a circuit when it is exposed to a specified high temperature.

three-dimensional cam. A drum-shaped cam in a hydromechanical fuel control whose outer surface is ground so that followers riding on the surface, as the cam is moved up and down and rotated, can move mechanical linkages to control the fuel according to a preprogrammed schedule.

throttle. The control in an aircraft that regulates the power or thrust the pilot wants the engine to produce.

throw (crankshaft design). *See* offset throw.

thrust horsepower. The horsepower equivalent of the thrust produced by a turbojet engine. Thrust horsepower is found by multiplying the net thrust of the engine, measured in pounds, by the speed of the aircraft, measured in miles per hour, and then dividing this by 375.

thrust. The aerodynamic force produced by a propeller or turbojet engine as it forces a mass of air to the rear, behind the aircraft.

A propeller produces its thrust by accelerating a large mass of air by a relatively small amount. A turbojet engine produces its thrust by accelerating a smaller mass of air by a much larger amount.

Time-Rite indicator. A patented piston-position indicator used to find the position of the piston in the cylinder of a reciprocating engine.

The body of the Time-Rite indicator screws into a spark plug hole, and as the piston moves outward in the cylinder, it contacts the arm of the indicator.

A pointer contacted by the arm moves across a calibrated scale to show the location of the piston in degrees of crankshaft rotation before top center.

timing light. An indicator light used when timing magnetos to an engine to indicate when the breaker points open. Some timing lights incorporate an oscillator or buzzer that changes its pitch when the points open.

TIT (turbine inlet temperature). The temperature of the gases from the combustion section of a gas turbine engine as they enter the turbine inlet guide vanes or the first stage of the turbine.

toggle. A T-shaped handle fitted onto the end of a cable used to engage a simple starter with an overrunning clutch.

top overhaul. An overhaul of the cylinders of an aircraft engine. The valves, pistons, and cylinders are overhauled, but the crankcase is not opened.

torque. A force that produces or tries to produce rotation.

total pressure. The pressure a column of moving fluid would have if it were stopped from its motion. Total pressure is the sum of dynamic pressure and static pressure.

total temperature. The temperature of moving fluid that has been stopped from its motion. Total temperature is the sum of the static temperature and the temperature rise caused by the ram effect as the fluid was stopped.

Townend ring. A type of ring cowling used over a single-row radial engine. The cross section of the ring is in the form of an airfoil that produces enough forward thrust to compensate for the cooling drag of the engine. In the United States, Townend rings are often called speed rings.

track. The path followed by a blade segment of a propeller or helicopter rotor in one rotation.

tractor engine. An engine installed with the propeller facing the front of the aircraft. Thrust produced by the propeller mounted on a tractor engine pulls the aircraft through the air.

tractor propeller. A propeller mounted on an airplane in such a way that its thrust pulls the aircraft.

trailing edge. The thin edge at the rear of a propeller blade.

transducer. A device that changes energy from one form to another. Commonly used transducers change mechanical movement or pressures into electrical signals.

transformer. An electrical component used to change the voltage and current in an AC circuit.

transonic range. Flight at Mach numbers between 0.8 and 1.2. In this range, some air passing over the aircraft is subsonic, and some is supersonic.

trend monitoring. A system for comparing engine performance parameters with a baseline of these same parameters established when the engine was new or newly overhauled.

Parameters such as EGT, RPM, fuel flow, and oil consumption are monitored on every flight, and the difference between the current indication and the baseline is plotted. Any deviation from a normal increase or decrease warns the technician of an impending problem.

tricresyl phosphate (TCP). A colorless, combustible compound, $(CH_3C_6H_4O)_3PO$, that is used as a plasticizer in aircraft dope and an additive in gasoline and lubricating oil. TCP aids in scavenging lead deposits left in the cylinders when leaded fuel is burned.

TSFC (thrust specific fuel consumption). A measure of efficiency of a turbojet or turbofan engine. It is a measure of the number of pounds of fuel burned per hour for each pound of thrust produced.

turbine. A wheel fitted with vanes, or buckets, radiating outward from its circumference. The reactive or aerodynamic force caused by the fluid flowing through the vanes is converted into mechanical power that spins the shaft on which the wheel is mounted.

turbine engine. *See* gas turbine engine.

turbine inlet guide vanes. A series of stator vanes immediately ahead of the first-stage turbine. The function of the inlet guide vanes is to divert the hot gases in the proper direction to enter the turbine, and to provide a series of convergent ducts which increase the velocity of the gases.

turbine nozzle. Another name for turbine inlet guide vanes.

turbocharger. An exhaust-driven air compressor used to increase the power of a reciprocating engine. A turbocharger uses a small radial inflow turbine in the exhaust system to drive a centrifugal-type air compressor on the turbine shaft. The compressed air is directed into the engine cylinders to increase power.

turbocompound engine. A reciprocating engine that has power recovery turbines in its exhaust system. The power extracted from the exhaust by these turbines is directed into the engine crankshaft through a fluid coupling.

turbofan engine. A type of gas turbine engine that has a set of lengthened blades on the low-pressure compressor or low-pressure turbine. Air moved by these special blades bypasses the core engine and produces between 30% and 75% of the total thrust.

turbojet engine. A gas turbine engine that produces thrust by accelerating the air flowing through it. A minimum of energy is extracted by the turbine, with the majority used to produce an exhaust velocity much greater than the inlet velocity. The amount of thrust produced by the engine is determined by the amount the air is accelerated as it flows through the engine.

turboprop engine. A turbine engine in which several stages of turbines are used to extract as much energy as possible. The turbines drive reduction gears which in turn drive a propeller.

turboshaft engine. A turbine engine in which several stages of turbines are used to extract as much energy as possible. The turbines drive shafts which are used to drive helicopter rotors, generators, or pumps.

turbosupercharger. A centrifugal air compressor driven by exhaust gases flowing through a turbine. The compressed air is used to increase the power produced by a reciprocating engine at altitude.

two-spool engine. *See* dual-spool gas-turbine engine.

two-stroke cycle. A constant-volume cycle of energy transformation that completes its operating cycle in two strokes of the piston, one up and one down.

When the piston moves up, fuel is pulled into the crankcase, and at the same time the fuel-air mixture inside the cylinder is compressed. When the piston is near the top of its stroke, a spark plug ignites the compressed fuel-air mixture, and the burning and expanding gases force the piston down. Near the bottom of the stroke, the piston uncovers an exhaust port and the burned gases leave the cylinder. When the piston moves further down, it uncovers the intake port, and a fresh charge of fuel and air are forced from the crankcase into the cylinder.

UDF engine (Unducted Fan™). The trade name registered by General Electric for a type of ultra-high-bypass turbofan engine that drives one or more wide-blade propellers that have between eight and twelve blades. These blades, which are not enclosed in a duct or shroud, are very thin, have wide chords, and are highly swept back in a scimitar shape that enables them to power airplanes flying in the speed range near Mach 0.8.

UHB (ultra-high-bypass) engine. A turbine engine that drives a pair of ducted or unducted contrarotating propellers which have eight to 12 variable-pitch blades. These blades are very thin, have wide chords, and are swept back with a scimitar shape that allows them to power airplanes flying in the speed range of Mach 0.8. The blades are made of advanced composites for high strength and light weight. UHB engines may be of either the tractor or pusher type, and have a bypass ratio in excess of 30:1.

underspeed condition. A speed condition in which the engine is turning at an RPM lower than that for which the propeller governor is set.

unidirectional fibers. Fibers in a piece of composite material arranged so that they sustain loads in only one direction.

updraft carburetor. A carburetor that mounts on the bottom of a reciprocating engine. Air entering the engine flows upward through the carburetor.

upper-deck pressure. The absolute pressure of air at the inlet to the fuel metering system of a turbocharged engine. Upper-deck pressure is the same as the turbocharger discharge pressure.

V-blocks. A fixture that allows a shaft to be centered and rotated to measure any out-of-round condition.

V-engine. A form of reciprocating engine in which the cylinders are arranged in two banks. The banks are separated by an angle of between 45° and 90°.

Pistons in two cylinders, one in each bank, are connected to each throw of the crankshaft.

valence electrons. Electrons in the outer shell, or ring, around the nucleus of an atom. It is the valence electrons that give an atom its electrical characteristics and are the electrons that may be pulled loose from an atom to cause electrical current.

valve overlap. The portion of the operating cycle of a four-stroke-cycle reciprocating engine during which both the intake and exhaust valves are off of their seats at the same time.

vapor lock. A condition of fuel starvation that can occur in a reciprocating engine fuel system. If the fuel in the line between the tank and carburetor is heated enough for the fuel to vaporize, a bubble will form in the line. If the vapor pressure of the bubble is high enough, it will block the fuel and keep it from flowing to the engine.

vapor pressure. The amount of pressure needed above a liquid to prevent it from evaporating.

vaporize. The changing of a liquid into a vapor.

vectored-thrust engine. A turbojet or turbofan engine with the fan and/or exhaust nozzles mounted in such a way that they may be rotated in flight to produce forward, vertically upward, or rearward thrust.

velocity. A vector quantity that expresses both the speed an object is moving and the direction in which it is moving.

velocity turbine. A turbine driven by forces produced by the velocity, rather than the pressure, of gases flowing through the vanes.

venturi. A specially shaped restrictor in a tube designed to speed up the flow of fluid passing through it. According to Bernoulli's principle, any time the flow of fluid speeds up without losing or gaining any energy from the outside, the pressure of the fluid decreases.

vernier coupling. A timing coupling used with base-mounted magnetos. The vernier coupling allows the timing to be adjusted in increments of considerably less than one degree.

vertical tape instrument. A tall rectangular instrument that displays the quantity of the parameter being measured by a movable strip of colored tape. The presentation resembles a vertical bar graph.

vibration loop. A loop in a rigid fluid line used to prevent vibration from concentrating stresses that could cause the line to break.

VIFF (vectoring in forward flight). A method of enhancing the maneuverability of an airplane by vectoring the exhaust gases and/or fan-discharge air to produce thrust components not parallel to the longitudinal axis of the aircraft.

viscosimeter. An instrument used to measure the viscosity of a liquid. The time required for a given volume of liquid at a specified temperature to flow through a calibrated orifice is used to indicate the viscosity of the liquid.

viscosity. The resistance of a fluid to flow. Viscosity is the stiffness of the fluid, or its internal friction.

viscosity index (VI). A measure of change in viscosity of an oil as it changes temperature. The higher the viscosity index, the less the viscosity changes.

viscosity index improver. An additive used to produce a multiviscosity lubricating oil. The polymer additive expands as temperature increases and contracts as temperature decreases. VI improvers cause viscosity to increase as oil heats and decrease as it cools.

volatile memory. Computer memory that is lost when the power to the computer is turned off.

volatility. The characteristic of a liquid that relates to its ability to vaporize or change into a gas.

volumetric efficiency. The ratio of the volume of the charge of the fuel and air inside the cylinder of a reciprocating engine to the total physical volume of the cylinder.

von Ohain, Dr. Hans Pabst. The designer and developer of the first turbojet engine to power an airplane. His HeS3b engine was built in Germany by the Heinkel Company, and it flew in a Heinkel He178 airplane on August 27, 1939.

vortex. A whirling mass of air that sucks everything near it toward its center.

vortex dissipater. A high-velocity stream of compressor bleed air blown from a nozzle into an area where vortices are likely to form. Vortex dissipaters destroy the vortices that would otherwise suck debris from the ground into engines mounted in pods that are low to the ground.

wake. The high-velocity stream of turbulent air behind an operating aircraft engine.

Wankel engine. *See* rotating combustion (RC) engine.

waste gate. A controllable butterfly valve in the exhaust pipe of a reciprocating engine equipped with an exhaust-driven turbocharger.

When the waste gate is open, exhaust gases leave the engine through the exhaust pipe, and when it is closed, they leave through the turbine.

watt. The basic unit of power in the metric system. One watt is the amount of power needed to do one joule (0.7376 foot-pound of work) in one second. One watt is 1/746 horsepower.

wet-sump engine. An engine that carries its lubricating oil supply in a reservoir that is part of the engine itself.

wet-sump lubrication system. A lubrication system in which the oil supply is carried within the engine itself. Return oil drains into the oil reservoir by gravity.

Whittle, Sir Frank. The British Royal Air Force flying officer who in 1929 filed a patent application for a turbojet engine. Whittle's engine first flew in a Gloster E.28 on May 15, 1941. The first jet flight in America was made on October 2, 1942, in a Bell XP-59A that was powered by two Whittle-type General Electric I-A engines.

windmilling propeller. A propeller that is rotated by air flowing over the blades rather than powered by the engine.

work. The product of a force times the distance the force is moved.

worm gear. A helical gear mounted on a shaft. The worm meshes with a spur gear whose teeth are cut at an angle to its face.

A worm gear is an irreversible mechanism. The rotation of the shaft, on which the worm gear is mounted, rotates the spur gear, but the worm gear locks the spur gear so its shaft cannot be rotated.

wrist pin. The hardened steel pin that attaches a piston to the small end of a connecting rod.

yaw. Rotation of an aircraft about its vertical axis.

zero-lash valve lifter. A hydraulic valve lifter that maintains zero clearance in the valve actuating mechanism.

INDEX

B

C

D

E

F

G

H

I

J

K

L

M

N

O

P

Q

R

S

Aviation Maintenance Technician Series
Powerplant

Editing	Jennie Trerise
Copy Editing	Kristina Anderson, Kitty Crane
Design & Art Direction	Dora McClurkin Muir
Illustration & Production	Jennifer Shontz
Typesetting & Production	Cynthia Wyckoff, Sandi Harner
Color Illustration, 2nd Edition	Tyra Menzel, Kelly Burch

Produced at
ASA's Graphic Design & Publications Department
in Times New Roman and Helvetica Black using Microsoft Word,
Macromedia FreeHand, Adobe Illustrator, Adobe PhotoShop,
and Adobe InDesign

Reader Response

Dear Aviation Maintenance Technician:

You have made an investment in your future by purchasing this textbook from ASA's *Aviation Maintenance Technician Series*. We hope you were pleased with your selection. Your input is invaluable to us. Please take a moment to provide us with your comments and suggestions. Include your name and address so we can thank you.

—Aviation Supplies & Academics, Inc.

Please print clearly.

Name ______________________ Date ____________

Title *(Student, Instructor, other)* ______________________

Business/School ______________________

Address ______________________

City ______________ State ________ Zip code/Postal code ____________

Country ______________ Email *(optional)* ______________

Where did you purchase this **Powerplant** Aviation Maintenance text? ______________

Was the text recommended to you? ________ By whom? ______________

Do you intend to purchase ASA's two additional texts in the series? General ____________ Airframe ____________

Comments and Suggestions

Please tell us what you liked or disliked about the text: content, subject matter, ease-of-use, illustrations and figures, etc.

____

____

____

____

____

____

____

Mail this form to:

Aviation Supplies & Academics, Inc.
7005 132nd Place SE
Newcastle, Washington 98059-3153

Please photocopy or remove this page.

ASA-AMT-P3 06/13